MATHEMATICS

Applications and Connections

Course 3

GLENCOE

McGraw-Hill

New York, New York Columbus, Ohio Mission Hills, California Peoria, Illinois

Send all inquiries to:
Glencoe/McGraw-Hill
936 Eastwind Drive
Westerville, Ohio 43081

ISBN: 0-02-824625-X (Student Edition)
ISBN: 0-02-824628-4 (Teacher's Wraparound Edition)

5 6 7 8 9 10 VH/LP 03 02 01 00 99 98 97 96

Letter from the Authors

Dear Students, Teachers, and Parents,

Mathematics students are special! That's why we've written a mathematics program designed especially for students your age. The layout will hold your interest and the exciting content will show you why you need to study mathematics every day.

As you page through *Mathematics: Applications and Connections*, you'll notice that the mathematics content is presented in a variety of relevant and interesting ways. You'll see the many connections made among mathematical topics and note how mathematics naturally fits into other subject areas and with technology.

You will note that content for each lesson is clearly labeled up front. And you'll appreciate the lesson format that introduces each new concept with an interesting application followed by clear examples.

As you read the text and complete the activities, you will also become aware of how frequently mathematics is used in real-world situations that relate directly to your life. If you don't already realize the importance of mathematics in your life, you soon will!

Sincerely, The Authors

Linda Dritsas

Kay McClain

David DeMalma

Patricia Frey-Mason

Ron Pelfrey

Beatrice Moore-Harris

Jack M. Ott

Patricia S. Wilson

Barbara D. Smith

Authors

William Collins teaches mathematics at James Lick High School in San Jose, California. He has served as the Mathematics Department Chairperson at James Lick and Andrew Hill High Schools. He received his B.A. from Herbert H. Lehman College and is a Masters candidate at California State University, Hayward. Mr. Collins has been a consultant for the National Assessment Governing Board. He is a member of the National Council of Teachers of Mathematics and is active in several professional mathematics organizations at the state level. Mr. Collins is currently a mentor teacher for the College Board's EQUITY 2000 Consortium in San Jose, California.

Linda Dritsas is the Mathematics Coordinator for the Fresno Unified School District in Fresno, California. She also taught at California State University at Fresno for two years. Ms. Dritsas received her B.A. and M.A. (Education) from California State University at Fresno. Ms. Dritsas has published numerous mathematics workbooks and other supplementary materials. She has been the Central Section President of the California Mathematics Council and is a member of the National Council of Teachers of Mathematics and the Association for Supervision and Curriculum Development.

Patricia Frey-Mason is the Mathematics Department Chairperson at the Buffalo Academy for Visual and Performing Arts in Buffalo, New York. She received her B.A. from D'Youville College in Buffalo, New York, and her M.Ed. from the State University of New York at Buffalo. Ms. Frey-Mason has published several articles in mathematics journals. She is a member of the National Council of Teachers of Mathematics and is active in other professional mathematics organizations at the state, national, and international levels. Ms. Frey-Mason was named a 1991 Woodrow Wilson Middle School Mathematics Master Teacher.

Arthur C. Howard is Consultant for Secondary Mathematics at the Aldine School District in Houston, Texas. He received his B.S. and M.Ed. from the University of Houston. Mr. Howard has taught in grades 7–12 and in college. He is Master Teacher in the Rice University School Mathematics Project in Houston. Mr. Howard is also active in numerous professional organizations at the national and state levels, including the National Council of Teachers of Mathematics. His publications include curriculum materials and articles for newspapers, books, and *The Mathematics Teacher.*

Kay McClain received her B.A. from Auburn University and her Educational Specialist degree from the University of Montevallo. She is currently working on a Ph.D. at Vanderbilt University. While a teacher at Mountain Brook Middle School in Birmingham, Ms. McClain received a Presidential Award for Excellence in the Teaching of Mathematics. She is a Woodrow Wilson fellow and an active member of the National Council of Teachers of Mathematics.

David D. Molina is a professor at Trinity University in San Antonio, Texas. He received his M.A. and Ph.D. in Mathematics Education from the University of Texas at Austin. Dr. Molina has been a speaker both at national and international mathematics conferences. He has been a presenter for the National Council of Teachers of Mathematics, as well as a conductor of workshops and in services for other professional mathematics organizations and school systems.

Authors

Beatrice Moore-Harris is the EQUITY 2000 Project Administrator and former Mathematics Curriculum Specialist for K-8 in the Fort Worth Independent School District in Fort Worth, Texas. She is also a consultant for the National Council of Teachers of Mathematics. Ms. Moore-Harris received her B.A. from Prairie View A & M University in Prairie View, Texas. She has also done graduate work there and at Texas Southern University in Houston, Texas, and Tarleton State University in Stephenville, Texas. Ms. Moore-Harris is active in many state and national mathematics organizations. She also serves on the Editorial Board of NCTM's *Mathematics and the Middle Grades* journal.

Ronald S. Pelfrey is the Mathematics Coordinator for the Fayette County Public Schools in Lexington, Kentucky. He has taught mathematics in Fayette County Public Schools, with the Peace Corps in Ethiopia, and at the University of Kentucky in Lexington, Kentucky. Dr. Pelfrey received his B.S., M.A., and Ed.D. from the University of Kentucky. He is also the author of several publications about mathematics curriculum. He is an active speaker with the National Council of Teachers of Mathematics and is involved with other local, state, and national mathematics organizations.

Barbara Smith is the Mathematics Supervisor for Grades K-12 at the Unionville-Chadds Ford School District in Unionville, Pennsylvania. Prior to being a supervisor, she taught mathematics for thirteen years at the middle school level and three years at the high school level. Ms. Smith received her B.S. from Grove City College in Grove City, Pennsylvania and her M.Ed. from the University of Pittsburgh in Pittsburgh, Pennsylvania. Ms. Smith has held offices in several state and local organizations, has been a speaker at national and state conferences, and is a member of the National Council of Teachers of Mathematics.

Jack Ott is a Professor of Mathematics Education at the University of South Carolina in Columbia, South Carolina. He has also been a consultant for numerous schools in South Carolina as well as the South Carolina State Department of Education and the National Science Foundation. Dr. Ott received his A.B. from Indiana Wesleyan University, his M.A. from Ball State University, and his Ph.D. from The Ohio State University. Dr. Ott has written articles for *The Mathematics Teacher* and *The Arithmetic Teacher* and has been a speaker at national and state mathematics conferences.

Jack Price has been active in mathematics education for over 40 years, 38 of those in grades K-12. He is currently the Co-Director of the Center for Science and Mathematics Education at California State Polytechnic University at Pomona, California, where he teaches mathematics and methods courses for preservice teachers and consults with school districts on curriculum change. Dr. Price received his B.A. from Eastern Michigan University, and has a Doctorate in Mathematics Education from Wayne State University. He is president of the National Council of Teachers of Mathematics and is an author of numerous mathematics instructional materials.

Patricia S. Wilson is an Associate Professor of Mathematics Education at the University of Georgia in Athens, Georgia. Dr. Wilson received her B.S. from Ohio University and her M.A. and Ph.D. from The Ohio State University. She has received the Excellence in Teaching Award from the College of Education at the University of Georgia and is a published author in several mathematics education journals. Dr. Wilson has taught middle school mathematics and is currently teaching middle school mathematics methods courses. She is on the Editorial Board of the *Journal for Research in Mathematics Education,* published by the National Council of Teachers of Mathematics.

Editorial Advisor

Dr. Piyush C. Agrawal
Supervisor for Mathematics Programs
Dade County Public Schools
Miami, Florida

Consultants

Winifred G. Deavens
Mathematics Supervisor
St. Louis Public Schools
St. Louis, Missouri

Leroy Dupee
Mathematics Supervisor
Bridgeport Public Schools
Bridgeport, Connecticut

Marieta W. Harris
Curriculum Coordinator
Memphis City Schools
Memphis, Tennessee

Deborah Haver
Mathematics Supervisor
Chesapeake Public Schools
Chesapeake, Virginia

Dr. Alice Morgan-Brown
Assistant Superintendent for Curriculum Development
Baltimore City Public Schools
Baltimore, Maryland

Dr. Nicholas J. Rubino, Jr.
Program Director/Citywide Mathematics
Boston Public Schools
Boston, Massachusetts

Telkia Rutherford
Mathematics Coordinator
Chicago Public Schools
Chicago, Illinois

Phyllis Simon
Supervisor, Computer Education
Conway Public Schools
Conway, Arkansas

Beverly A. Thompson
Mathematics Supervisor
Baltimore City Public Schools
Baltimore, Maryland

Jo Helen Williams
Director, Curriculum and Instruction
Dayton Public Schools
Dayton, Ohio

Reviewers

David Bradley
Mathematics Teacher
Thomas Jefferson Junior High School
Kearns, Utah

Joyce Broadwell
Mathematics Teacher
Dunedin Middle School
Dunedin, Florida

Charlotte Brummer
K-12 Mathematics Supervisor
School District Eleven
Colorado Springs, Colorado

Richard Buckler
Mathematics Coordinator K-12
Decatur School District
Decatur, Illinois

Josie A. Bullard
MSEN PCP Lead Teacher/Math Specialist
Lowe's Grove Middle School
Durham, North Carolina

Joseph A. Crupie, Jr.
Mathematics Department Chairperson
North Hills High School
Pittsburgh, Pennsylvania

Kathryn L. Dillard
Mathematics Teacher
Apollo Middle School
Antioch, Tennessee

Rebecca Efurd
Mathematics Teacher
Elmwood Junior High School
Rogers, Arkansas

Nevin Engle
Supervisor of Mathematics
Cumberland Valley School District
Mechanicsburg, Pennsylvania

Rita Fielder
Mathematics Teacher
Oak Grove High School
North Little Rock, Arkansas

Henry Hull
Adjunct Assistant Professor
Suffolk County Community College
Selden, New York

Elaine Ivey
Mathematics Teacher
Adams Junior High School
Tampa, Florida

Donna Jamell
Mathematics Teacher
Ramsey Junior High School
Fort Smith, Arkansas

Augustus M. Jones
Mathematics Teacher
Tuckahoe Middle School
Richmond, Virginia

Marie Kasperson
Mathematics Teacher
Grafton Middle School
Grafton, Massachusetts

Larry Kennedy
Mathematics Teacher
Kimmons Junior High School
Fort Smith, Arkansas

Patricia Killingsworth
Math Specialist
Carver Math/Science Magnet School
Little Rock, Arkansas

Al Lachat
Mathematics Department Chairperson
Neshaminy School District
Feasterville, Pennsylvania

Kent Luetke-Stahlman
Resource Scholar Mathematics
J. A. Rogers Academy of Liberal Arts & Sciences
Kansas City, Missouri

Dr. Gerald E. Martau
Deputy Superintendent
Lakewood City Schools
Lakewood, Ohio

Nelson J. Maylone
Assistant Principal
Maltby Middle School
Brighton, Michigan

Irma A. Mayo
Mathematics Department Chairperson
Mosby Middle School
Richmond, Virginia

Daniel Meadows
Mathematics Consultant
Stark County Local School System
Canton, Ohio

Dianne E. Meier
Mathematics Supervisor
Bradford Area School District
Bradford, Pennsylvania

Rosemary Mosier
Mathematics Teacher
Brick Church Middle School
Nashville, Tennessee

Judith Narvesen
Mathematics Resource Teacher
Irving A. Robbins Middle School
Farmington, Connecticut

Raymond A. Nichols
Mathematics Teacher
Ormond Beach Middle School
Ormond Beach, Florida

William J. Padamonsky
Director of Education
Hollidaysburg Area School District
Hollidaysburg, Pennsylvania

Delores Pickett
Instructional Supervisor
Vera Kilpatrick Elementary School
Texarkana, Arkansas

Thomas W. Ridings
Team Leader
Gilbert Junior High School
Gilbert, Arizona

Sally W. Roth
Mathematics Teacher
Francis Scott Key Intermediate School
Springfield, Virginia

Dr. Alice W. Ryan
Assistant Professor of Education
Dowling College
Oakdale, New York

Fred R. Stewart
Supervisor of Mathematics/Science
Neshaminy School District
Langhorne, Pennsylvania

Terri J. Stillman
Mathematics Department Chairperson
Boca Raton Middle School
Boca Raton, Florida

Marty Terzieff
Secondary Math Curriculum Chairperson
Mead Junior High School
Mead, Washington

Tom Vogel
Mathematics Teacher
Capital High School
Charleston, West Virginia

Joanne Wilkie
Mathematics Teacher
Hosford Middle School
Portland, Oregon

Larry Williams
Mathematics Teacher
Eastwood 8th Grade School
Tuscaloosa, Alabama

Deborah Wilson
Mathematics Teacher
Rawlinson Road Middle School
Rock Hill, South Carolina

Francine Yallof
Mathematics Teacher
East Middle School
Brentwood, New York

Table of Contents

Chapter 1

Tools for Problem Solving

Chapter 2

An Introduction to Algebra

High Interest Features

Chapter 3

Integers

Chapter 4

Statistics and Data Analysis

High Interest Features

Applications and Connections

Have you ever asked yourself this question?

"When am I ever going to use this stuff?"

It may be sooner than you think! Here's two of the many ways this textbook will help you answer that question.

Applications

You'll find mathematics in all of the subjects you study in school and in your life outside of school. In Lesson 3-5 on page 98, subtracting integers is related to geography. In Example 2 on page 345, solving proportions is applied to fitness.

These and other applications provide you with fascinating information that connects math to the real world and other school subjects and gives you a reason to learn math.

On pages 665–668, you will find a **Data Bank.** You'll have the opportunity to use the up-to-date information in it to answer questions throughout the book.

The **Extended Projects Handbook** consists of interesting long-term projects that involve issues in the world around you.

Five **Decision Making** features further enable you to connect math to your real-life experience as a consumer.

Connections

You'll discover that various areas of mathematics are very much interrelated. For example, Lesson 8-7 on page 325 connects measurement with distance on the coordinate plane. Connections to algebra, geometry, statistics, measurement, probability, and number theory help show the power of mathematics.

Connections to algebra, geometry, statistics, measurement, probability, and number theory help show the power of mathematics.

The **Mathematics Labs** and **Mini-Labs** also help you connect what you've learned before to new concepts. You'll use counters, measuring tapes, algebra tiles, and many other objects to help you discover these concepts.

Chapter 5

Investigations in Geometry

Chapter 6

Patterns and Number Sense

High Interest Features

Chapter 7 Rational Numbers

Chapter 8 Real Numbers

High Interest Features

Chapter 9

Applications with Proportion

Chapter 10

Applications with Percent

High Interest Features

Technology

Labs, examples, computer-connection problems, and other features help you become an expert in using computers and calculators as problem-solving tools. You'll also learn how to read data bases, use spreadsheets, and use BASIC and LOGO programs. On many pages, **Calculator Hints** and printed keystrokes illustrate how to use a calculator.

Here are some highlights.

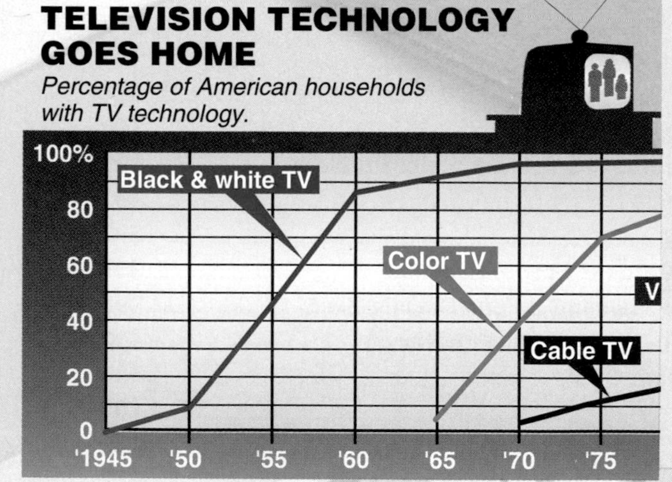

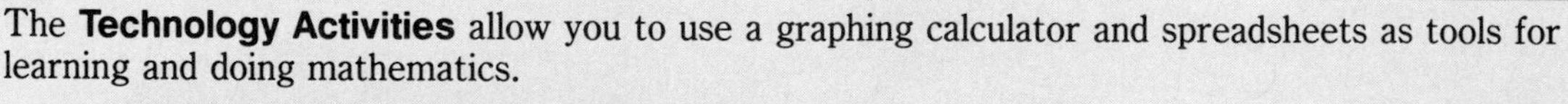

The **Technology Activities** allow you to use a graphing calculator and spreadsheets as tools for learning and doing mathematics.

Chapter 11

Algebra: Functions and Graphs

Chapter 12

Area and Volume

High Interest Features

Chapter 13 Discrete Math and Probability

Chapter 14 Algebra: Investigations with Polynomials

High Interest Features

Setting The Scene

To help chart their journeys, wise travelers consult a map before they begin. Just as maps lead travelers to their destinations, the script on the next five pages points out the ways that you use the mathematics in this text in your daily lives.

Narrator:
Dia and Theresa have just received the newest issue of Fashion Teen magazine and are flipping through it.

Dia:
Wow! Look at this outfit!

Theresa:
Yeah, I love those colors. Turn the page; maybe that designer has more outfits on the next page.

Dia:
No, there's nothing but this gross cigarette ad.

Theresa:
I know. I'm sick of that ad—it's all over the place.

Dia:
I thought it was illegal to advertise cigarettes in magazines.

Theresa:
I think it's only illegal to advertise cigarettes on TV.

Dia:
Well, it should be illegal, especially for magazines that teens read. I think it's a nasty habit.

Narrator:
The girls are joined by their friend Dylan.

Dylan:
Hey!
What's up?

Theresa:
Oh, we were just flipping through this magazine. Can you believe it's got cigarette ads in it?!

Dia:
Smoking is so uncool.

Dylan:
Hey, in science class today, we saw a video about the effects of smoking. I'll tell you one thing—I'll never smoke.

Dia:
Somebody should do something about all of these cigarette advertisements geared to teens.

Theresa:
Maybe that somebody should be us.

Dylan:
What do you mean? What can we do?

Dia:
We should at least write a letter to the editor and complain.

Dylan:
Who's gonna listen to us?

Theresa:
I've got an idea. We could get facts from the American Lung Association and use them in our letter.

Dia:
That's a great idea!

Dylan:
Okay. I'll go down to the American Lung Association and see what kind of information they have.

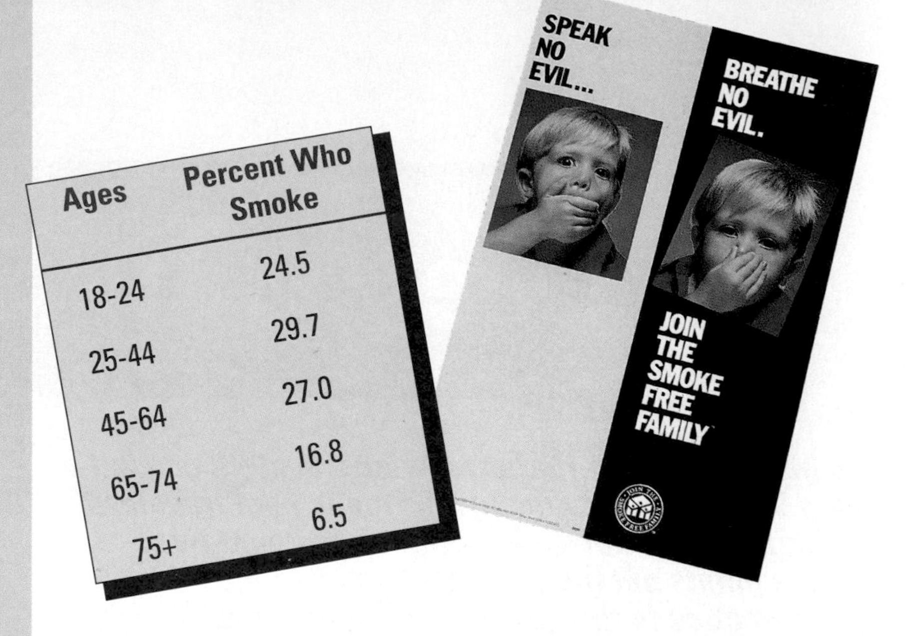

Ages	Percent Who Smoke
18-24	24.5
25-44	29.7
45-64	27.0
65-74	16.8
75+	6.5

Narrator:
Dylan searched the library at the American Lung Association and found lots of brochures that contained useful information. He wrote down the following data.

1.3 million out of 45.8 million people quit smoking each year, while 13 million people try to quit, but fail.

90% of all adults who smoke started by the age of 21, and half of them had become regular smokers by their 18th birthday.

For people ages 12-18, 15.7% have smoked in the last month, and 11.5% have smoked in the last week. For people over age 18, 25.5% are current smokers.

Over 1 billion packs of cigarettes are sold to teenagers each year. A million teenagers take up smoking each year.

434,000 Americans die prematurely each year from diseases caused by smoking. That's 20% of all deaths annually.

Cigarette companies spend over $3 billion on advertising each year. It is illegal to advertise tobacco products on TV or radio.

Stop the Script...

What does this information tell you? How can you use it to make a case against advertising cigarette smoking in magazines?

Narrator:
The students get together to discuss Dylan's findings.

Dia:
Look at all this stuff! I don't know where to start.

Dylan:
Well, we should definitely show that most people start smoking as teenagers.

Theresa:
Okay, it says that 90% of smokers start before they're 21 and half of those start before they're 18. Since $\frac{1}{2}$ of 90% is 45%, that means that 45% of smokers started when they were 17 or younger. Another 45% started when they were 18 to 20. Only 10% started when they were 21 or older.

Dia:
Right. It also says that 45.8 million people now smoke. 45% of 45.8 million is—wait, let me get my calculator—20.6 million. That's how many people started when they were in middle or high school. Only 4.6 million started as adults.

Dylan:
Why don't we make a bar graph of the data? Graphs make a big impact.

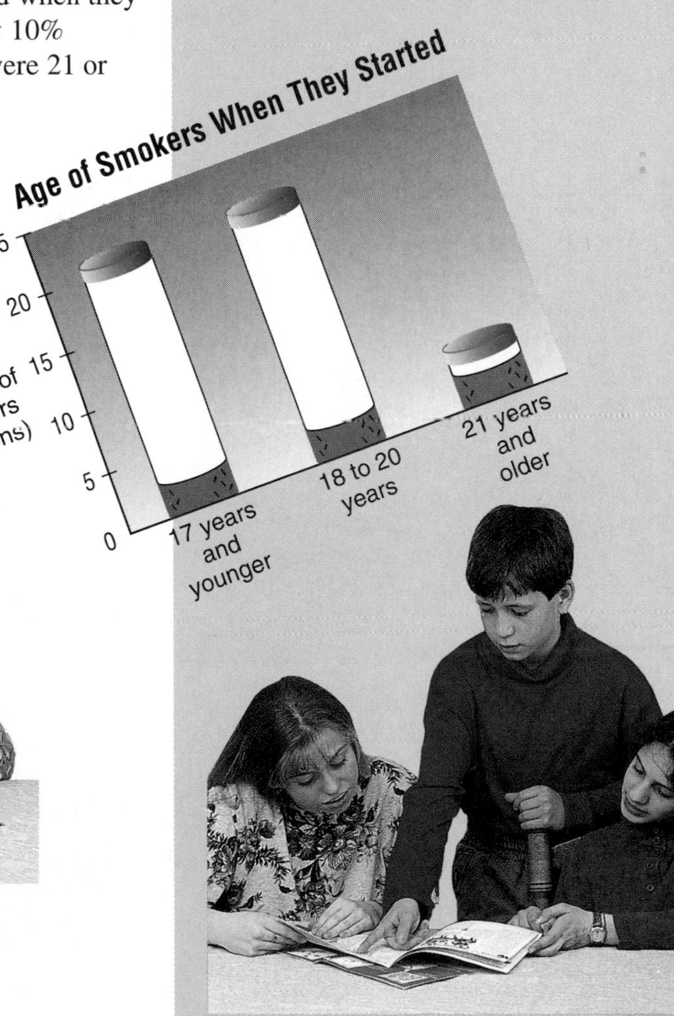

Dia:
We should also show how hard it is to stop smoking.

Dylan:
The information says that 14.3 (13 + 1.3) million people try to quit each year, but only 1.3 million actually do.

Theresa:
So what percent of those who try actually quit?

Dia:
1,300,000 divided by 14,300,000 is 0.091. That means only 9.1% of those who try to quit, succeed. And only 31% of smokers even try to quit.

Dylan:
Wait a minute. Only 9.1% of 31% of the people who smoke actually quit. That means only 9.1% times 31%, or 2.8%, of all smokers quit each year. That's awful!

Theresa:
I'll make a circle graph by multiplying the percentages by 360°: 2.8% times 360° is about 10° and 31% of 360° is about 112°.

Narrator:
The friends agree that Dia should write the letter. After they all read it and make several suggestions, Dia types the letter, and they all sign it. (The final version is shown on the following page.) The students send the letter to the magazine, and it appears in the next issue in the Letters to the Editor. Underneath it is the editor's note shown below.

Editorial Note: These thoughtful readers make an excellent point. We at **Fashion Teen** are concerned with the health of our readership. Therefore, starting with our next issue, we will no longer advertise tobacco products in our magazine.

This concludes Setting the Scene. Throughout this text, you will encounter new ways to make mathematics real to you. From time to time, read through this script to remind yourself how relevant mathematics can be in your life.

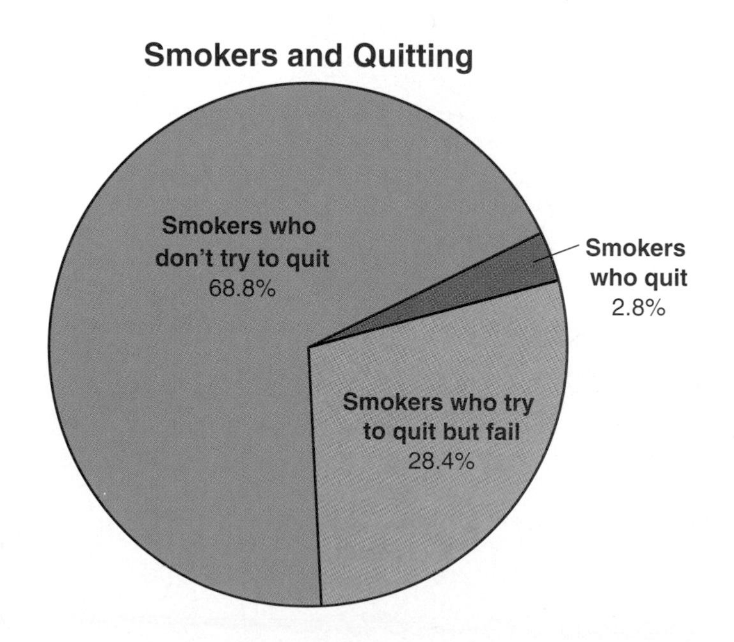

Dear Fashion Teen Editor:

We really enjoy reading your magazine and seeing the great fashions. However, in your last issue, we were troubled by your ads. As we see it, you allow tobacco advertisers to promote smoking to your readers.

As you know, the majority of your readers are teenagers. We feel that this is very irresponsible on your part. Hundreds of studies show that smoking causes premature death. In fact, 434,000 Americans die each year from diseases related to smoking. That is 20% of all deaths annually.

More people will continue to die premature deaths if cigarette companies are allowed to advertise to teenagers. In fact, most smokers start as teenagers. The graph below illustrates this fact.

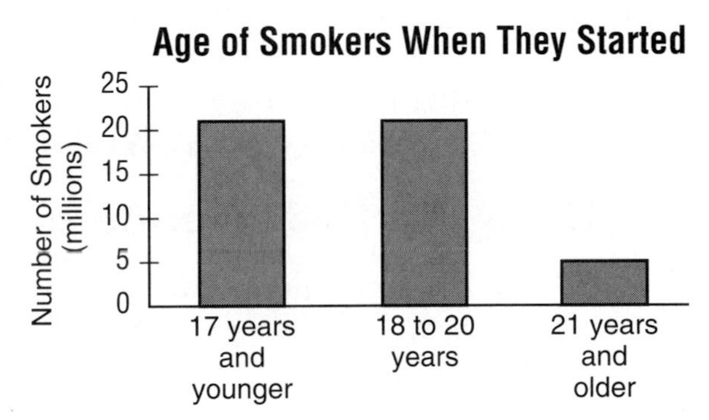

Each year, over 1 billion packs of cigarettes are sold to teenagers. One million teenagers take up smoking each year. Once a person has started to smoke, it is very difficult for that person to quit.

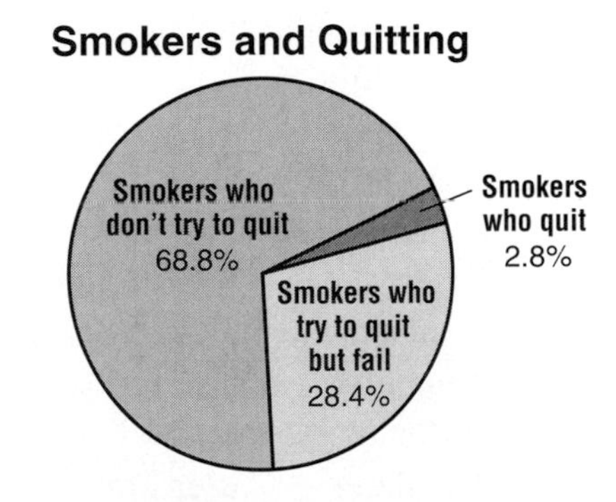

Cigarette companies spend over $3 billion on advertising each year. It is already illegal to advertise on TV or radio. We believe that Fashion Teen should take a similar position and ban all tobacco advertisements from your magazine.

Sincerely,

Dia Arroyo *Theresa Jones* *Dylan Rubini*

Dia Arroyo, Theresa Jones, and Dylan Rubini

Chapter 1

Tools for Problem Solving

Spotlight on Athletes

Have You Ever Wondered . . .

- About how much money Kirby Puckett earns per game since he signed a three-year, $9 million deal with the Minnesota Twins?
- How many different ways can a baseball manager arrange the batting order of the nine starting players on the team?

World Swimming Records
(as of July 1991)

Event	Record Holders	Country	Time
50-meter freestyle	Tom Jager	USA	0:21.81
	Yang Wenyi	China	0:24.98
100-meter breaststroke	Adrian Moorhouse	Great Britain	1:01.49
	Silke Hoerner	Germany	1:07.91
200-meter butterfly	Michael Gross	Germany	1:56.24
	Mary T. Meagher	USA	2:05.96
200-meter backstroke	Igor Polianskiy	USSR	1:58.14
	Betsy Mitchell	USA	2:08.60
400-meter medley	Tamas Darnyi	Hungary	4:15.42
	Petra Schneider	Germany	4:36.10

1930 1940 1950 1960

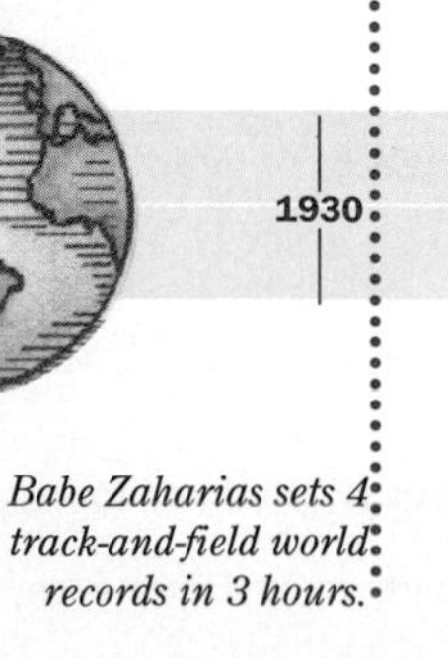

1932 *Babe Zaharias sets 4 track-and-field world records in 3 hours.*

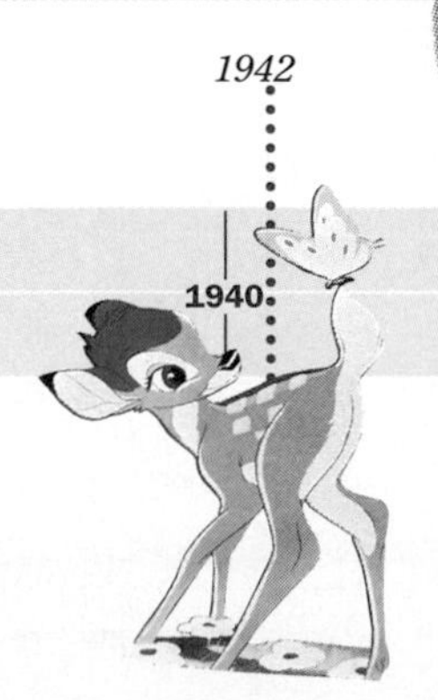

1942

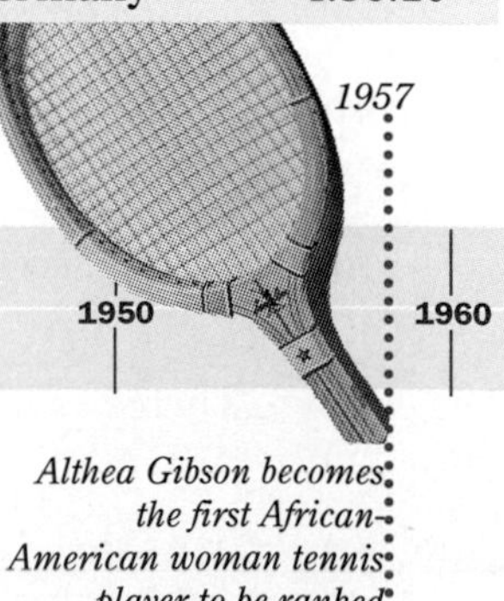

1957 *Althea Gibson becomes the first African-American woman tennis player to be ranked number one in the world.*

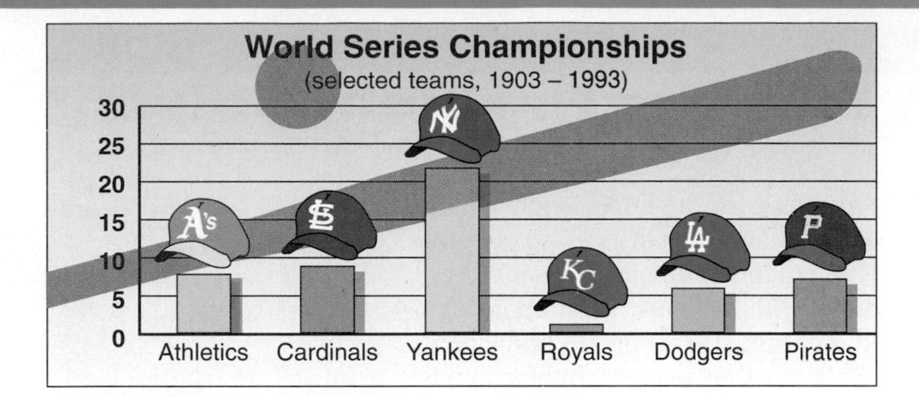

Looking Ahead

In this chapter, you will see how mathematics can be used to answer the questions about estimating player salaries and solving problems. The major objectives of this chapter are to:

- solve problems using the four-step plan
- compute mentally using properties
- estimate sums and differences of whole numbers and decimals
- solve problems by using guess and check

Chapter Project

Athletes

Work with a partner.

1. Choose a prominent athlete that plays a professional team sport near where you live. Write a short biography about his or her life.
2. Collect statistics about your athlete's career over several seasons. In how many games did he or she play? How much money did he or she earn? How many points did he or she score?
3. Suppose you and your partner will represent this player in contract negotiations. Display your findings graphically in such a way that will highlight your client's contribution to the team.

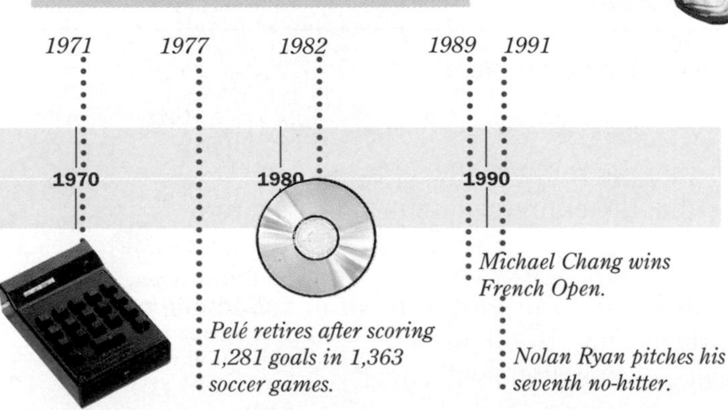

Pelé retires after scoring 1,281 goals in 1,363 soccer games.

Michael Chang wins French Open.

Nolan Ryan pitches his seventh no-hitter.

1-1 A Plan for Problem Solving

Objective

Solve problems using the four-step plan.

The committee for the Fort Couch Middle School's fall dance has decided to decorate the ceiling of the multi-purpose room by covering it with balloons. Their sponsor agreed, but said she needed to know how many balloons they needed. If the room is 40 feet by 60 feet, what is the least number of balloons they will need?

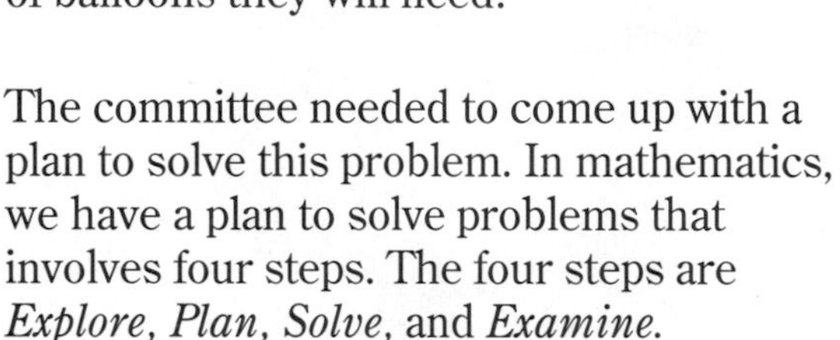

The committee needed to come up with a plan to solve this problem. In mathematics, we have a plan to solve problems that involves four steps. The four steps are *Explore, Plan, Solve,* and *Examine.*

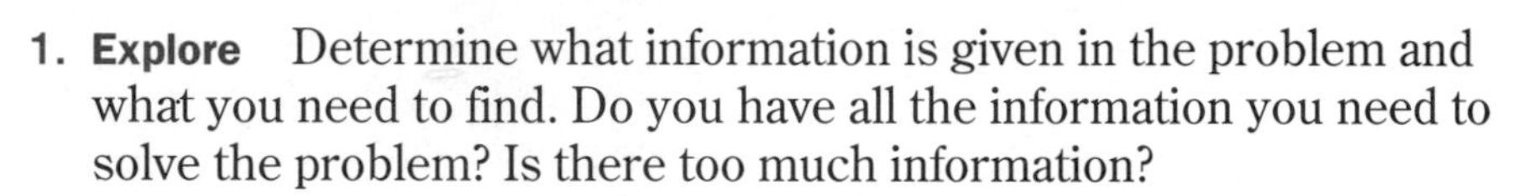

1. **Explore** Determine what information is given in the problem and what you need to find. Do you have all the information you need to solve the problem? Is there too much information?

2. **Plan** Make an estimate of what you think the answer should be. Then select a strategy for solving the problem. There may be any number of ideas or strategies that you can use.

3. **Solve** Solve the problem by carrying out your plan. If your plan doesn't work, try another, and maybe even another.

4. **Examine** Examine your answer carefully. See if it fits the facts given in the problem. Compare it to your estimate. If the answer is not reasonable, make a new plan. You may also want to check your answer by solving the problem again in a different way.

Example 1 ***Problem Solving***

Let's try our plan on the opening problem.

Explore *What do we know?*
We know that the room is 40 feet by 60 feet.
We know that the entire ceiling is to be covered.

What are we trying to find out?
We need to find out the least number of balloons it will take to cover the ceiling. But to do that, we need to know how much space each balloon will cover.

Plan The committee guessed they needed about 2,000 balloons. Let's make a drawing. If you knew how many balloons fit in one row across the room and how many rows you had, you could find the total number of balloons needed.

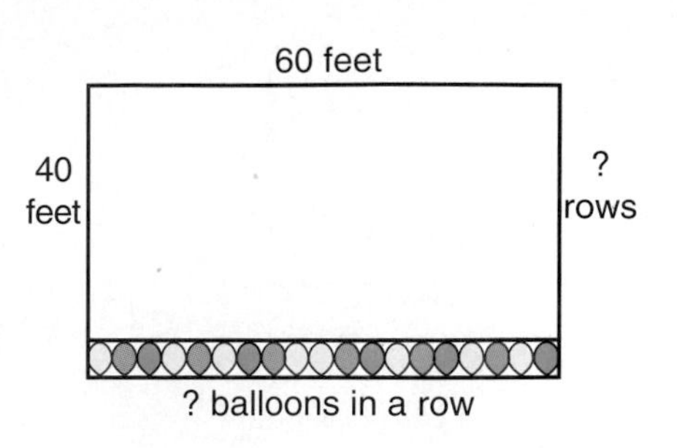

Solve Suppose an inflated balloon is 1 foot across. Then 60 balloons would fit in a row across the length of the room. Since the room is 40 feet wide, there would be 40 rows. Multiply to find how many balloons there would be.

$40 \times 60 = 2{,}400$

The committee will need at least 2,400 balloons to cover the ceiling of the room for the fall dance.

Examine The committee's estimate of 2,000 was close. The answer 2,400 assumes that the balloons are in rows and columns. It would be difficult to keep that many balloons in that pattern. The committee may want to get slightly more than 2,400 to account for this and other unforeseen problems. The committee decides to tell their sponsor that they will need 2,500 balloons.

Now let's try another problem using a different strategy but the same four steps of problem solving.

DID YOU KNOW

Despite headlines hailing 1989 as the best box-office year ever with hits like *Batman,* the motion picture industry was healthier in both admissions and earnings in the 1930s–1950s.

Example 2 *Problem Solving*

Entertainment A feature film lasts two hours ten minutes. There are 15 minutes between showings of the film and 10 minutes of previews before each showing. The previews for the first showing start at 5:00 P.M. If your curfew on the weekend is 11:00 P.M. and it takes 15 minutes to go home after the film, can you attend the second showing of the film?

Explore We know that the previews for the first show start at 5:00 and last 10 minutes. The film is two hours ten minutes long. There is a 15-minute intermission after the first show, and there are more previews before the second showing.

We need to know what time the second showing ends so we will know if that leaves enough time to get home by 11:00.

Plan Estimate how long it will take. The two showings of the film last a little more than 4 hours. Intermissions and previews are less than 1 hour. The total should be about 5 hours. From 5:00 to 11:00 is 6 hours, so you should be able to attend the second showing. Let's make a chart to solve this problem.

Solve The chart should list each event, how long it takes, and the time it is completed.

Event	How long does it take?	When is it over?
Start time	—	5:00
Previews	10 minutes	5:00 + 0:10 = 5:10
1st showing	2 hours 10 minutes	5:10 + 2:10 = 7:20
Intermission	15 minutes	7:20 + 0:15 = 7:35
Previews	10 minutes	7:35 + 0:10 = 7:45
2nd showing	2 hours 10 minutes	7:45 + 2:10 = 9:55

The show is over at 9:55. You can attend the second showing and still be home before your curfew of 11:00.

Examine *Does your answer seem reasonable?*
The first showing including previews is about $2\frac{1}{2}$ hours. The intermission and the second showing plus previews is about $2\frac{1}{2}$ hours. At most, the total is about 5 hours. Five hours after 5:00 would be 10:00. That gives you an hour to get home by 11:00.

Checking for Understanding

Communicating Mathematics **Read and study the lesson to answer each question.**

1. **Write** what each step in the four steps of problem solving means.
2. **Tell** what to do when your plan doesn't work.
3. **Show** how you can use the information in your local newspaper to determine whether you can see the second evening showing of a movie and still make it home before your curfew.

Guided Practice

4. Look at Example 2. What time does the third showing end?
5. **Chemistry** A chemist pours sodium chloride (salt) into a beaker. If the beaker plus the sodium chloride have a mass of 84.8 grams and the beaker itself has a mass of 63.3 grams, what is the mass of the sodium chloride that was poured into the beaker?
   a. Write the *Explore* step. What do you know?
   b. Write the *Plan* step. What strategy will you use?
   c. *Solve* the problem using your plan. What is the answer?
   d. *Examine* your solution. Is it reasonable?

Exercises

Problem Solving and Applications

Use the four-step plan to solve each problem.

6. **Photography** The photography class needs to enlarge a 15-centimeter by 25-centimeter picture so that the shorter side is 30 centimeters long. How long will the longer side be after the enlargement?
7. **Engineering** Geothermal energy is heat from inside the earth. Underground temperatures generally increase 9° C for every 300 feet of depth. For the ground temperature to rise 90° C, how deep would you have to dig?
8. **Transportation** A DC-11 jumbo jet carries 342 passengers with 36 in first class seating and the rest in coach class seating. A first class ticket to fly from Los Angeles to Chicago costs $750, and a coach class ticket costs $450. What will be the ticket sales for the airline if the flight is full?
9. **Money** Maya has only nickels in her pocket. Kim has only dimes in hers. Kareem has only quarters in his. Marta approached them for a donation for the Heart Association drive. What is the least each could donate so that each one gives the same amount?
10. **Smart Shopping** At the school bookstore, a ball point pen costs 28¢ and a small tablet costs 23¢. What could you buy and spend exactly 74¢?
11. **Fitness** Jonna walks along a path around the lake at Triangle Park. One trip around the lake is 4,700 feet. The lengths to two parts of the path are shown. What is the length of the third part?

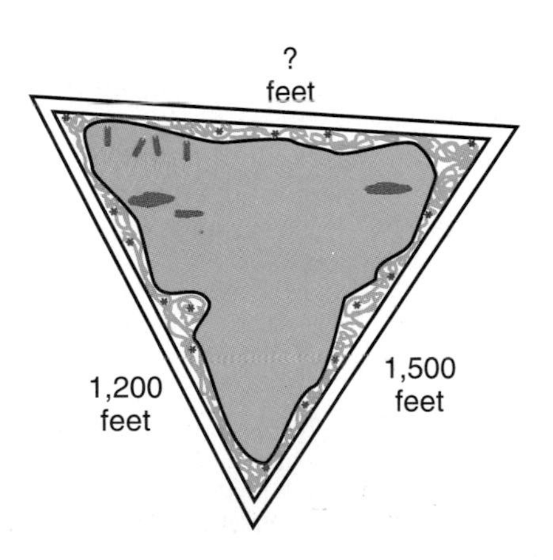

12. **Critical Thinking** Refer to the middle school dance problem at the beginning of the lesson. Joel and Sara decided to estimate the number of balloons needed by finding how many were needed to fill a smaller space. They used cardboard and the corner of a room to make a space 6 feet by 4 feet. They found that 28 balloons cover this space. How could they use this information to solve the problem?
13. **Make a Model** Suppose you had 100 sugar cubes. What would be the largest cube you could build with the sugar cubes? *Remember that in a cube, all of the sides are the same length.*
14. **Journal Entry** In this course, you will be required to keep a journal. Write two or three sentences in your journal that describe what you expect to learn in this course.

1-2 Mental Math Strategies

Objective

Compute mentally using compensation and properties of numbers.

Words to Learn

compensation
commutative
associative
distributive

Carl Friedrich Gauss (1777–1855), a famous mathematician, entered his first class of arithmetic at age 10. One day, the teacher asked the students to find the sum of the numbers 1 through 100. Young Gauss amazed his teacher by writing the correct answer immediately. Gauss used a mental math strategy to find the sum.

Mental math strategies help you compute an answer quickly and easily when a calculator or pencil and paper are not handy. They can also help to cut calculation time when you are working with pencil and paper. Unlike estimation, mental math allows you to find an exact answer.

One mental math strategy is **compensation.** In compensation, you change a problem so it is easy to solve mentally. Then you make an adjustment to compensate for the change you made.

There are many ways to compensate in finding an answer. A number line and basic facts can show you how compensation works.

Addition:

$$\begin{array}{ccccc} 6 & + & 7 & = & 13 \\ +4 & & -4 & & \updownarrow \\ \hline 10 & + & 3 & = & 13 \end{array}$$

6 units
7 units
0 1 2 3 4 5 6 7 8 9 10 11 12 13 14
6 + 4 = 10 units
7 − 4 = 3 units

The total distance, 13, must stay the same. So if you add units to one number, you must subtract units from the other number.

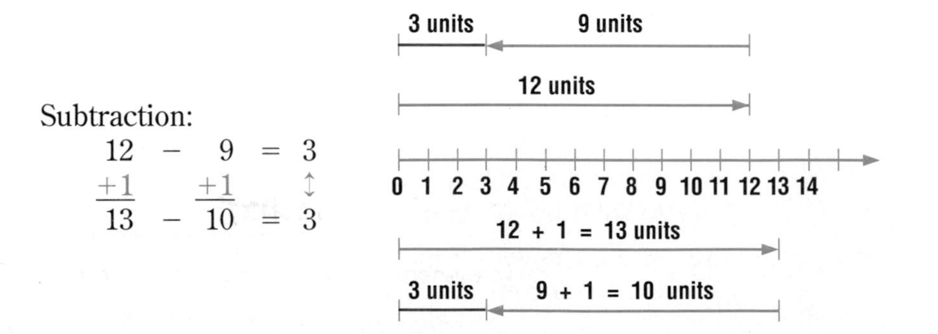

Subtraction:

$$\begin{array}{ccccc} 12 & - & 9 & = & 3 \\ +1 & & +1 & & \updownarrow \\ \hline 13 & - & 10 & = & 3 \end{array}$$

The distance 3, representing the difference of the numbers, must stay the same. If you add units to one number, you must add units to the other also.

Examples

1 Find 298 + 147.

$$\begin{array}{rlcr} 298 & +2 & \rightarrow & 300 \\ \underline{+147} & -2 & \rightarrow & \underline{+145} \\ & & & 445 \end{array}$$

2 Find \$5.00 − \$1.43.

$$\begin{array}{rlcr} \$5.00 & -0.01 & \rightarrow & \$4.99 \\ \underline{-1.43} & -0.01 & \rightarrow & \underline{-1.42} \\ & & & \$3.57 \end{array}$$

You may remember some properties of numbers from your previous mathematics courses that you can also use to do mental math. These are the **commutative, associative,** and **distributive** properties for addition and multiplication.

The symbol · means to multiply. 6 · 3 means 6 × 3.

Property	Arithmetic	Algebra
Commutative	$1 + 2 = 2 + 1$ $3 \cdot 2 = 2 \cdot 3$	$a + b = b + a$ $a \cdot b = b \cdot a$
Associative	$3 + (4 + 10) = (3 + 4) + 10$ $5 \cdot (2 \cdot 12) = (5 \cdot 2) \cdot 12$	$a + (b + c) = (a + b) + c$ $a \cdot (b \cdot c) = (a \cdot b) \cdot c$
Distributive	$6 \cdot (5 + 7) = (6 \cdot 5) + (6 \cdot 7)$	$a \cdot (b + c) = a \cdot b + a \cdot c$

Examples

3 Find 5 · 27 · 2.

It is easier to multiply by 10. Use the commutative property.

$$\begin{aligned} 5 \cdot 27 \cdot 2 &= 5 \cdot 2 \cdot 27 \\ &= 10 \cdot 27 \\ &= 270 \end{aligned}$$

4 Find (7 · 25) · 4.

It is easier to multiply by 100. Regroup the numbers by using the associative property.

$$\begin{aligned} (7 \cdot 25) \cdot 4 &= 7 \cdot (25 \cdot 4) \\ &= 7 \cdot 100 \\ &= 700 \end{aligned}$$

5 Find 7 · 15.

Use the distributive property to rewrite 15 as the sum 10 + 5.

$$\begin{aligned} 7 \cdot 15 &= 7 \cdot (10 + 5) \\ &= (7 \cdot 10) + (7 \cdot 5) \\ &= 70 + 35 \text{ or } 105 \end{aligned}$$

Checking for Understanding

Communicating Mathematics

Read and study the lesson to answer each question.

1. **Tell** how you could compensate to add \$1.98 and \$3.98.
2. **Write** the expression 25 + 80 + 75 so it is easier to add mentally.
3. **Show** several examples of how compensation and mental math involve changes that result in numbers ending in 0.

Guided Practice Tell how you would use mental math to find each answer. Then find the answer.

4. $15 + 37 + 15$
5. $47 + 98$
6. $700 - 23$
7. $(23 + 17) \cdot 8$
8. $83 + 48 + 17$
9. $20 \cdot 93 \cdot 50$
10. $7 \cdot 35$
11. $43 \cdot 50 \cdot 2$
12. $57 + 69 + 43$

Exercises

Independent Practice Use mental math to find each answer.

13. $65 + 73 + 15$
14. $(13 \cdot 25) \cdot 4$
15. $9 \cdot 15$
16. $600 - 243$
17. $53(90 + 10)$
18. $37 + 29 + 3$
19. $498 + 329$
20. $740 + 987 + 60$
21. $5(15 + 5)$
22. $(21 \times 2) \times 50$
23. $14 \cdot 4 \cdot 25$
24. $69 + 93 + 31$
25. $43 + 29 + 7$
26. $702 - 598$
27. $(623 + 420) + 80$

28. Find $(423 \times 50) \times 2$. Tell how you found the answer.
29. Suppose ■ $= 12 \cdot 25$. Use mental math to find the missing number.
30. Suppose $a = 14 \times (7 + 10 + 13)$. If a represents the answer to the problem, find a.

Mixed Review

31. **Communication** The cost of a long distance phone call is 15¢ for the first 5 minutes and 10¢ for each additional minute. If Laura makes a 30-minute long distance phone call, how much will it cost? *(Lesson 1-1)*
32. **Production** A furniture company produces 15 rocking chairs in one hour. How long will it take to produce 45 chairs? *(Lesson 1-1)*
33. **Hobbies** Romo likes to collect baseball cards. Every week he buys 20 baseball cards and sells 16 of them. How many baseball cards will he have after 7 weeks? *(Lesson 1-1)*

Problem Solving and Applications

34. **Geometry** Three sides of a four-sided figure are 51 inches, 49 inches, and 38 inches long. How long is the fourth side if the sum of the lengths of all the sides is 200 inches?

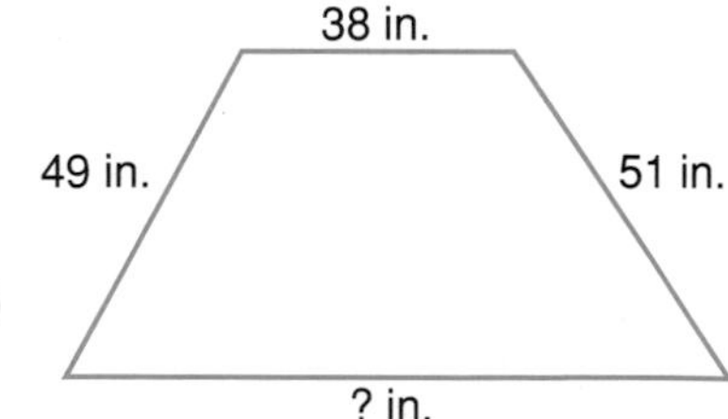

35. **Critical Thinking** José was given 30 capsules by the doctor to help him recover from strep throat. He was to take two capsules with every meal for the first two days and then one capsule with each meal until they are all taken. If he starts the medication with breakfast on Monday and eats three meals a day, on what day and with which meal did he take the last capsule?

36. **Journal Entry** Find the definition of *compensate* in a dictionary. Write one or two sentences to explain how the definition describes the compensation strategy.

1-3 Estimation Strategies

Objective

Estimate answers by using different strategies.

Words to Learn

rounding
front-end
compatible numbers
clustering

Whatever happened to the gum wrapper you threw in the wastebasket the other day? Chances are it ended up in a landfill somewhere. Experts estimate that Americans throw away about 160 million tons of garbage each year. How do they get this estimate? It's not by weighing each family's garbage each week. They estimate the total by using amounts from sample days in different parts of the country.

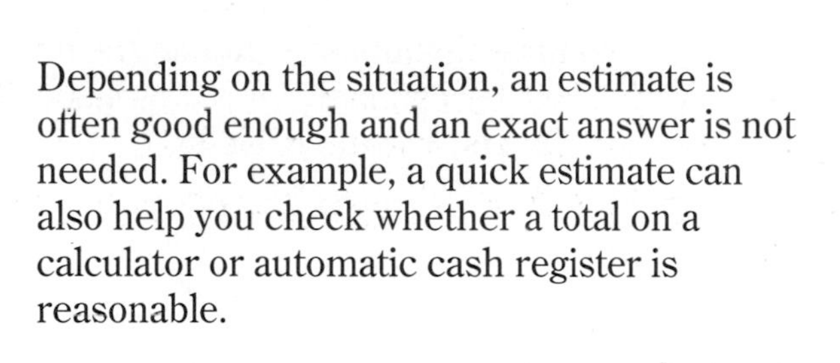

Depending on the situation, an estimate is often good enough and an exact answer is not needed. For example, a quick estimate can also help you check whether a total on a calculator or automatic cash register is reasonable.

In this lesson, we will show four different ways you can estimate. They are by **rounding,** by **front-end estimation,** by using **compatible numbers,** and by **clustering.** The following examples will show how each type can be used.

Example 1

Estimate 6,337 + 2,875.

Use rounding.

$$\begin{array}{r} 6{,}337 \\ +\ 2{,}875 \\ \hline \end{array} \rightarrow \begin{array}{r} 6{,}000 \\ +\ 3{,}000 \\ \hline 9{,}000 \end{array}$$

One estimate for 6,337 + 2,875 is 9,000.

Use front-end estimation.

Add the left column of digits. Then add the next column of digits.

$$\begin{array}{r} 6{,}337 \\ +\ 2{,}875 \\ \hline 8\ 000 \end{array} \rightarrow \begin{array}{r} 6{,}3\ 37 \\ +\ 2{,}8\ 75 \\ \hline 1100 \end{array}$$

Add.

9,100

Another estimate for 6,337 + 2,875 is 9,100.

Examples

2 **Estimate 118.1 − 57.5.**

Use rounding.

$$\begin{array}{r} 118.1 \\ -\ 57.5 \\ \hline \end{array} \begin{array}{c} \rightarrow \\ \rightarrow \\ \end{array} \begin{array}{r} 120 \\ -\ 60 \\ \hline 60 \end{array}$$

One estimate for 118.1 − 57.5 is 60.

Use front-end estimation.

$$\begin{array}{r} 118.1 \\ -\ 57.5 \\ \hline 60.0 \end{array} \begin{array}{c} \\ \rightarrow \\ \end{array} \begin{array}{r} 118.1 \\ -\ 57.5 \\ \hline 1.0 \end{array}$$

Add.

61

Another estimate for 118.1 − 57.5 is 61.

3 **Estimate 4,321 ÷ 73.**

Use compatible numbers. Round the divisor to 70. Replace 4,321 with 4,200. By estimating 4,321 as 4,200, the division becomes easier. That is, 42 is divisible by 7.

$4{,}321 \div 70 \rightarrow 4{,}200 \div 70 = 60$

An estimate for 4,321 ÷ 73 is 60.

When using a calculator, it is easy to hit a wrong key. This is especially true if there are a lot of numbers to enter or if the numbers are large. You should use estimation to check your work.

Example 4 *Problem Solving*

Technology Jesse uses his calculator to find the sum of 7,129 + 6,859 + 7,523 + 6,792 + 6,928. How can he tell if the answer 35231 showing on the calculator is reasonable?

Use clustering. Each of the numbers in Jesse's problem is very close to 7,000. You can estimate the sum of these five numbers by multiplying 5 and 7,000.

$5 \times 7{,}000 = 35{,}000$

The estimate of 35,000 is very close to his answer of 35,231. The calculator answer is reasonable and is probably correct.

Estimation Hint

Clustering works best with numbers that all round to approximately the same number.

Some estimation methods work better in certain situations than in others. Occasionally you may want to combine different estimation methods. The purpose of estimation is to give you a guide for easily determining a number that is close to the actual number.

Checking for Understanding

Communicating Mathematics

Read and study the lesson to answer each question.

1. **Write** two reasons for using estimation.
2. **Tell** what to do if you have several numbers to add and clustering does not work.
3. **Show** how front-end estimation differs from rounding.

Guided Practice

Round each number to its greatest place value.

4. 5,432
5. 723.9
6. 0.08547

Tell which method you would use to estimate each answer. Explain why you chose that method. Then estimate.

7. $628 + 547 + 432$
8. $4,423 - 2,983$
9. $5,231 - 4,108$
10. $\$12.99 - \$4.33$
11. $5.593 \div 0.74$
12. $2.3 + 2.5 + 2.8 + 2.4$

Exercises

Independent Practice

Estimate first by rounding and then by front-end estimation.

13. $5,293 + 3,733$
14. $6.59 + 4.65 + 2.28$
15. $0.7829 - 0.5392$
16. $7,623 - 5,450$

Estimate by using compatible numbers.

17. $3,593 \div 62$
18. $7.347 \div 0.79$

Estimate each sum by clustering.

19. $5,473 + 4,987 + 5,129 + 4,873$
20. $\$2.10 + \$1.89 + \$2.15 + \$1.98 + \$2.09$

Estimate. Use an appropriate strategy.

21. $576 - 395$
22. $5.247 - 3.258$
23. $3,500 \div 62$
24. $527 + 915 + 467$
25. $82.43 + 79.28 + 37.41$
26. $\$5.99 + \$6.94 + \$7.15$

Mixed Review

27. **Inventory** A restaurant serves 240 cups of coffee each day. If 1 pound of coffee is used to make 40 cups, how many pounds will the restaurant need in a week? *(Lesson 1-1)*
28. Ted weighs 119 pounds. If Ted's weight plus Tisha's weight is 215 pounds, how much does Tisha weigh? *(Lesson 1-1)*
29. Use mental math to find $21(9 + 6)$. *(Lesson 1-2)*
30. Use mental math to find $534 + 88$. *(Lesson 1-2)*

Problem Solving and Applications

31. **Inventory** Mark works in the bookstore during his free period. He learned that a ream of paper has 500 sheets. At the end of the month, he counted 423 reams of paper left. Approximately how many sheets of paper is this?

32. **Critical Thinking** Sam wanted to use front-end estimation to estimate 8,325 − 4,986. He says the process presented in Example 2 doesn't work.
    a. How could Sam change the process to use the front-end method to estimate this difference?
    b. Write a rule for your method of front-end estimation.
    c. Try your rule in estimating 5,300 − 2,654.

33. **Geometry** Mrs. Hawks owns a farm next to the school. She donated the use of a large five-sided field to the school's agriculture program.
    a. If the field has sides that are 524 feet, 498 feet, 519 feet, 502 feet, and 486 feet long, about how much fence must they order to enclose the field?
    b. What factors must you consider when buying fence to enclose a given area?

34. **Arts and Crafts** At an arts festival, the Art Club set up a booth to sell their ceramic pots. Mr. Wilburn selected pots that were priced at $5.98, $7.25, $3.25, $8.75, $9.85, $2.50, and $7.25. He has $50 in his wallet. How could he use estimation to see if he can pay cash or if he has to write a check?

Save Planet Earth

Recycling Newspapers Americans use 50 million tons of paper each year. This means we consume more than 850 million trees. It is estimated that the average American uses about 420 pounds of paper each year.

How You Can Help

- Don't throw away newspapers with the trash anymore. Select a spot in your home, stack them, tie them together, and weigh them for one week. If one tree produces 120 pounds of paper, how many trees would you save by recycling?
- Find a recycling program in your neighborhood.

If everyone in this country recycled even one-tenth of their newspapers, we would save about 25 million trees every year.

Problem-Solving Strategy

1-4 Determine Reasonable Answers

Objective

Determine whether answers to problems are reasonable.

You are going to paint one wall and a door of the school rec room with two coats of paint. The wall is 19 feet by 9 feet, and the door is 3 feet by 7 feet. At the store, you find that a gallon will cover about 400 ft^2 and a quart will cover about 100 ft^2, when using two coats. Which size should you buy?

$1\ ft^2 = 1$ sq ft = 1 ft □ 1 ft (a square with sides labeled 1 ft)

Explore You know that the areas to be painted are 19 feet by 9 feet and 3 feet by 7 feet. You know that a gallon will cover about 400 ft^2 and a quart will cover about 100 ft^2.

You want to find out which size is more reasonable to buy.

Plan Estimate to find the total area. Then see which size has enough paint for that area.

Solve The area of the door is 7×3 or 21 ft^2, which rounds to 20 ft^2. The area of the wall is about 10×20 or 200 ft^2. The total area is about $20 + 200$ or 220 ft^2.

A quart will not be enough to cover this area.

Examine The total area to be painted is more than a quart will cover. It makes sense to buy the gallon.

Wearing eye shadow began in ancient times. The Egyptians painted their eyelids to shield their eyes from the sun.

Example

Smart Shopping You spend \$5.55 plus \$0.44 tax at the store for makeup and pay with a \$10 bill. Would it be more reasonable to expect about \$3.00 or \$4.00 in change?

You can estimate how much change you should receive.
\$5.55 rounds to \$5.60 and \$0.44 rounds to \$0.40, so you spent about \$5.60 + \$0.40, or \$6.00. Since \$10.00 − \$6.00 = \$4.00, it is more reasonable to expect about \$4.00 in change.

Checking for Understanding

Communicating Mathematics

Read and study the lesson to answer each question.

1. **Write** another way to estimate the amount of change in the Example. How does this estimate affect the answer to the problem?
2. **Tell** why it is good to round up when estimating the area of a wall that needs to be painted.

Guided Practice

Solve by determining reasonable answers.

3. When Vic divided 45,109.5 by 1,479 the calculator showed 305. Is this answer reasonable?
4. An orange grower harvested 1,260 pounds of oranges from one grove, 874 pounds from another, and 602 pounds from another. What is a reasonable number of crates to have on hand if each crate holds 14 pounds of oranges?
5. You need to buy 3 cans of cat food at 39¢ each, 5 pounds of tomatoes at 89¢ a pound, and a quart of milk at 75¢. Do you need to take $5.00 or $10.00 with you?

Problem Solving

Practice

Solve. Use any strategy.

6. The space shuttle can carry about 65,000 pounds of cargo. A compact car weighs about 2,450 pounds. What is a reasonable number of compact cars that could be carried on the space shuttle?
7. On your 14th birthday, your aunt says you can have twice your age in dollars each year for 5 years or you can have $32 each year for 5 years. Which way do you get more money overall?
8. If 610,184 people attended 8 Denver Bronco home games during the 1990 football season, which is a reasonable estimate for the number of people that attended each game: 7,500 or 75,000?
9. As part of your exercise program, you decide to do 5 more sit-ups each day. If you start with 5 sit-ups, how many sit-ups will you do on the 10th day?
10. Yesterday, you noted that the mileage on the family car read 60,094.8 miles. Today it read 60,099.1 miles. Was the car driven about 4 or 40 miles?
11. **Data Search** Refer to page 665. What is the pattern in the rates for first class postage? Would a 7-ounce letter cost more or less than $1.60?

Strategies

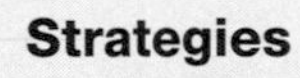

Look for a pattern.
Solve a simpler problem.
Act it out.
Guess and check.
Draw a diagram.
Make a chart.
Work backward.

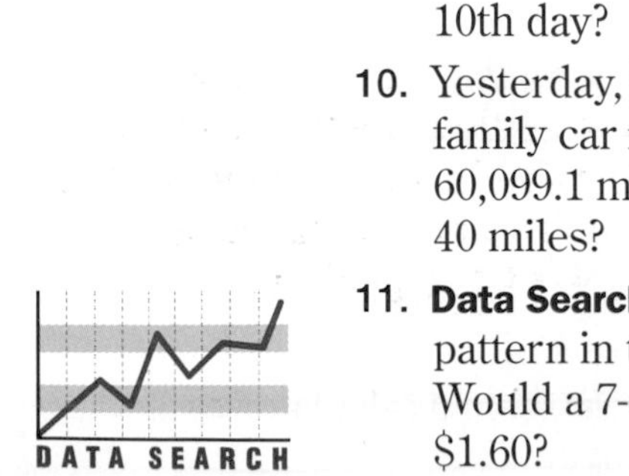

Problem-Solving Strategy

1-5 Eliminate Possibilities

Objective

Solve problems by using estimation to eliminate possibilities.

Three after-school jobs are posted on the bulletin board. The first job pays \$4.25 per hour for 15 hours of work each week. The second job pays \$10.95 a day for 3 hours, 5 days each week. The third job pays \$59.75 for 15 hours total each week. If you want to apply only for the job that pays the most, which should you choose?

Explore You know that each job is for 15 hours a week. One job pays \$4.25 per hour; one pays \$10.95 per day; and one pays \$59.75 per week.

You want to compare the amount each pays, so you will know which job to apply for.

Plan First, estimate how much each job pays a week by multiplying the hours or days by the amount of pay. Then you can eliminate the lowest paying job or jobs.

Solve \$4.25 can be rounded to \$4.00. Since this is the amount for an hour, multiply by 15 to find the amount for the week.

$$\$4.00 \times 15 = \$60.00$$

\$10.95 can be rounded to \$11.00. Since this is the amount for a day, multiply by 5 to find the amount for the week.

$$\$11.00 \times 5 = \$55.00$$

\$59.75 can be rounded to \$60.00. This is the amount for a week.

The job that pays \$10.95 a day can be eliminated, since it pays the lowest. Since \$4.25 is a little more than \$4.00, that job will actually pay more than \$60.00 a week. Since \$59.75 is less than \$60.00, that job can also be eliminated. So, you should apply for the job that pays \$4.25 per hour.

Examine You can check your answer by using a calculator to find the actual pay.

Example

Suppose you have the following division problem on a multiple-choice test. How can you eliminate possibilities to choose the correct answer?

$16{,}340 \div 19 =$

a. 80 b. 860 c. 8,600

You can estimate the quotient by rounding to compatible numbers. 16,340 can be rounded to 16,000 and 19 can be rounded to 20. $16{,}000 \div 20 = 800$.

Choice **a** can be eliminated because 80 is too small to be the correct answer.
Choice **c** can be eliminated because 8,600 is too great to be the correct answer.
The correct answer must be choice **b.**
Check by multiplying. $860 \times 19 = 16{,}340$.

Checking for Understanding

Communicating Mathematics

Read and study the lesson to answer each question.

1. **Tell** how to use the eliminate possibilities strategy to solve a multiple-choice problem like the one in the example.
2. **Explain** why, in the introductory problem, you might want to choose one of the other jobs.

Guided Practice

Solve by eliminating possibilities.

3. Last month's gas meter reading was 16,800 ft^3. This month the reading is 17,900 ft^3. The number of cubic feet (ft^3) of gas used was about:
   a. 100 b. 1,000 c. 10,000
4. Sheila uses her calculator to divide 685,300 by 86.3. She should expect the result to be about:
   a. 80 b. 800 c. 8,000
5. The Meadowlands Arena seats 20,039 people. If all tickets sell for $12.75 each, the gross receipts for a sell-out would be:
   a. $2,554,972 b. $255,497.25 c. $25,549.72

Problem Solving

Practice

Solve using any strategy.

6. A speeding bullet travels at about 886.4 mi/h, and a rocket in orbit travels about 17,500 mi/h. The rocket travels about how many times faster than a speeding bullet?
   a. 20 b. 25 c. 200 d. 2,000

7. Maria's house number has four digits and is divisible by 3. The second digit is the square of the first digit. The last two digits are the square of the second digit. What is her house number?

Strategies

Look for a pattern.
Solve a simpler problem.
Act it out.
Guess and check.
Draw a diagram.
Make a chart.
Work backward.

8. Basil is painting a border of stars along three walls of his room. Each star measures 0.25 meters wide and the walls are 3.6 meters, 2.7 meters, and 3.3 meters long. About how many stars will he have to paint?
   a. 4 b. 40 c. 400
9. Robin has five different colors of ribbons. She uses three colors for each hairbow she is making. How many different color combinations for the bows are there?
10. Is the average of 342.7, 54.3, 119.45, 909.8, 17.6, and 81.75 about 25, 100, or 250?
11. The human heart beats an average of 72 times in one minute. In one year, the number of times a human heart beats is about:
   a. 37,800,000 b. 378,000 c. 37,800

1

Assessment: Mid-Chapter Review

1. **Trucking** A service station along I-64 in Lynchburg, Virginia, charges $1.39 a gallon for diesel fuel and $1.10 a quart for oil. If a truck driver bought 38 gallons of diesel fuel and 2 quarts of oil, how much did she pay? *(Lesson 1-1)*

Use mental math to find each answer. *(Lesson 1-2)*

2. $12 + 35 + 18$
3. $2.99 + $4.98
4. $6(25 + 5)$

Estimate. Use an appropriate strategy. *(Lesson 1-3)*

5. $6.99 + 7.12 + 6.88 + 7.05$
6. $6{,}245 \div 81$
7. $8{,}101 - 7{,}002$
8. 8.987×11
9. **Geography** The population of North America is 278,000,000 and its area is 7,466,890 mi^2. Which is a reasonable estimate for the population density (number of people per unit) of this continent, 370 people or 37 people per square mile? *(Lesson 1-4)*
10. **Recycling** The recycling center pays 5¢ for every two aluminum cans turned in. If a group of students brought in 1,200 cans, how much money would they get back? *(Lesson 1-5)*
   a. $3 b. $6 c. $30 d. $300

Buying a Personal Computer

Situation

Your computer club just completed its fund-raising drive and now has $2,082 in its treasury. You have agreed to buy a new PC (personal computer). The question is, can you afford a 16MHz system with a hard drive and a color monitor? Heather brought in a catalog from a national supplier of discounted computers to help the group make a decision.

Hidden Data

Cost of extended warranty: Do you want more than one year of protections?
Cost of shipping: Do you know the cost of shipping, which will be included in your final bill?
Cost of partial payment: If you consider this option, you will want to figure the additional charges.
State tax: Will you have to pay state tax for your state or the state from which you are ordering?

Analyzing the Data

1. **Is there** more than one option for a system with 16MHz, a hard drive, and a color monitor?
2. **Does the cost** of shipping and tax, if any, increase the total cost over your spending limit?
3. **How much** is the cost for an extended warranty?

Outstanding $avings

ON OUR NEW LINE OF COMPUTER SYSTEMS

12 MHz Computer with 40 mb Hard Drive, 3¼" Floppy, and Mono Amber or Mono Rite Monitor.
$1049.88
w/80 mb Drive.....1259.88
w/124 mb Drive...1399.88

12 MHz with 800 X 600 Super Rite Color Monitor.
$1279.88
w/80 mb Drive.....1489.88
w/124 mb Drive...1629.88
For 1024 X 786 SuperRite Color Monitor, add $150

16 MHz Computer with 40 mb Hard Drive, 3¼" Floppy, and Mono Amber or Mono Rite Monitor.
$1099.88
w/80 mb Drive.....1309.88
w/124 mb Drive...1449.88

16 MHz with 800 X 600 Super Rite Color Monitor.
$1329.88
w/80 mb Drive.....1539.88
w/124 mb Drive...1679.88
For 1024 X 786 SuperRite Color Monitor, add $150

All items shipped in two cartons. Shipping weight 74 lbs. Shipping extra.

ZZ Computer Sales **1-800-555-5656**

Making Decisions in the Real World

4. **What if** this system is not compatible with your software? Is there another system that is?
5. **Is this** system completely compatible with your present hardware? Is there any additional hardware needed to install this system?
6. **Which system** should your club buy if you are interested in using the equipment to make money?

Making a Decision

7. **Investigate** the cost of a similar system purchased in separate components from a local source. Get catalogs from other companies that make computers.
8. **Contact** the supplier and see if you can have an additional discount because you are a student group.

Merchandise Covered	1 yr.	2 yr.
Peripherals priced under $250	$24.98	$39.98
Peripherals priced $250 to $499.99	49.98	69.98
Peripherals priced $500 to $999.99	109.98	199.98
Peripherals priced $1000 to 1999.99	199.98	269.98
Computers priced under $1000	$79.98	$199.98
Computers priced $1000 to $1999.99	119.98	179.98
Computers priced $2000 to $3499.99	199.98	269.98

Cooperative Learning

1-6A Using Nonstandard Units

A Preview of Lesson 1-6

Objective

Measure items using a nonstandard unit of measure.

Materials

string
marker/pen
scissors

"What's a cubit?" someone asked while reading a story about Cleopatra. Cubits are no longer used today, but were commonly used in the countries located in what is now southwestern Asia over 2,000 years ago. The cubit was the measure along the forearm from the elbow to the end of the longest finger. How precise is this measure?

Try this!

Work with a partner.

- Wrap a piece of string around your wrist. Cut it so that it is as long as the distance around your wrist. This will be your unit of measure.
- Take another long piece of string. This will be your measuring tape.
- Measure around your partner's head using your measuring tape. How many "wrists" would it take to go around your partner's head? Estimate to the nearest $\frac{1}{4}$ wrist. Record your measure.
- Have your partner measure your head and find how many "wrists" around it is. Record your measure.
- Take turns measuring and recording the wrist measures for the distance around your knee, the length of your arm, and the distance around your neck.

What do you think?

1. Accumulate the data for the class and find an "average" size for the head, knee, and arm to the nearest $\frac{1}{4}$ wrist. Does this measure tell you how long something really is?
2. Measure your partner's wrist. Describe how your data would be different if you had used that measure.
3. Suppose you had to crawl 1,000 wrists down the hallway. Whose wrist would you choose? Suppose you received a roll of \$1 bills 100 wrists long. Whose wrist would you choose? Are these different wrists? Explain your answer.

Measurement Connection

1-6 The Metric System

Objective

Use metric units of measurement.

Words to Learn

metric system
meter
liter
gram

DID YOU KNOW

Scientists use only SI (International System) units. SI includes the metric system. In this way, scientists can easily share information with scientists in other countries. Look in your science book. What system do you find?

A baseball bat is a little less than a meter long.
A nickel has a mass of about 5 grams.
A quart of milk is a little less than a liter.

At Laddie Creek and Dead Indian Creek, sites in southern Montana, anthropologists have found arrowheads that date from 4000 B.C. to A.D. 500. The average arrowhead has a length of 35 millimeters, a width of 18 millimeters, and a neck width of 13 millimeters. The *millimeter* (mm) is a unit of length in the **metric system.**

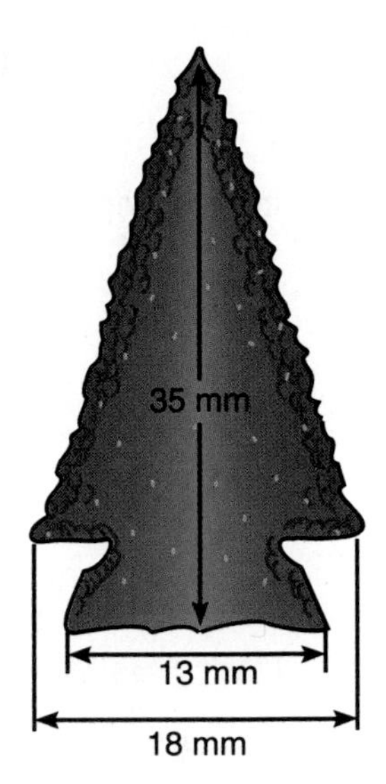

The metric system was created by French scientists in the late 18th century as a standard of measurement. The United States is the only large nation of the world that does not commonly use the metric system. The metric system is a decimal system. That means it is based on 10.

The standard unit of length in the metric system is the **meter** (m). The standard unit of capacity is the **liter** (L), and the standard unit of mass is the **gram** (g).

There are other measurements of length, capacity, and mass in the metric system, but they are all defined using the basic units (meter, liter, gram) and a prefix.

Prefix	Symbol	Meaning	Example
kilo-	k	1,000	1 km = 1,000 m
hecto-	h	100	1 hL = 100 L
deka-	da	10	1 dag = 10 g
deci-	d	0.1	1 dg = 0.1 g
centi-	c	0.01	1 cm = 0.01 m
milli-	m	0.001	1 mL = 0.001 L

The diagram below shows how you change from one unit to another by multiplying or dividing by 10.

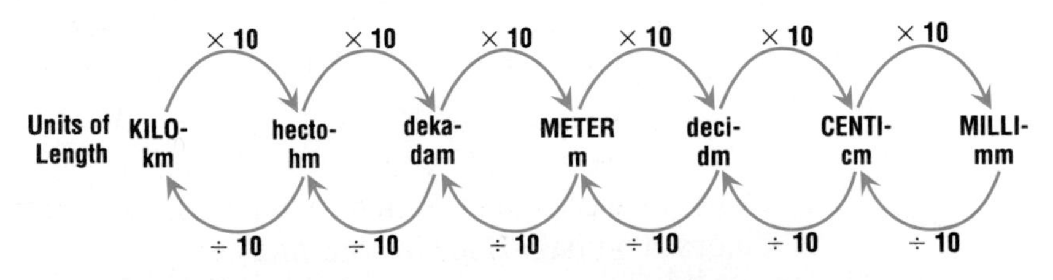

Study the patterns shown below.

- When you multiply by 10, 100, and 1,000 you can determine the answer by moving the decimal point right as many places as there are zeros in the multiplier.

  $1 \cdot 6.54 = 6.54$

  $10 \cdot 6.54 = 65.4$

  $100 \cdot 6.54 = 654.$

  $1{,}000 \cdot 6.54 = 6{,}540.$

- When you divide by 10, 100, and 1,000 you move the decimal point left as many places as you have zeros in the divisor.

  $78 \div 1 = 78$

  $78 \div 10 = 7.8$

  $78 \div 100 = 0.78$

  $78 \div 1{,}000 = 0.078$

This is very helpful in changing from one metric unit to another.

Mental Math Hint

If you are changing from one type of unit to a heavier, longer, or larger one, divide. If you are changing from one type of unit to a lighter, shorter, or smaller one, multiply.

Examples

1 5 g = ? mg

Since a gram is heavier than a milligram, you multiply. There are 1,000 mg in a gram, so multiply mentally by 1,000.

5.000 → 5,000

5 g = 5,000 mg

2 59 cm = ? m

A centimeter is shorter than a meter, so divide. There are 100 cm in a meter. Divide mentally.

59. → 0.59

59 cm = 0.59 m

3 1.5 L = ? mL

A liter is more than a milliliter, so multiply by 1,000.

1.500 → 1,500

1.5 L = 1,500 mL

4 1,035 cm = ? km

You can use two steps.

1,035 cm = ? m = ? km

1,035 cm = 10.35 m   *1,035*

10.35 m = 0.01035 km   *010.35*

1,035 cm = 0.01035 km

Checking for Understanding

Communicating Mathematics

Read and study the lesson to answer each question.

1. **Tell** why we divide when changing from grams to kilograms.
2. **Write** how to move the decimal point when multiplying by 100.
3. **Show** how many times longer a kilometer is than a centimeter by using the diagram on page 23.

Guided Practice Tell which metric unit you would probably use to measure each item.

4. length of a book
5. contents of a soda can
6. a load of bricks
7. distance between cities
8. water in a swimming pool
9. width of the letter A

Tell whether you multiply or divide and what number you use. Then complete each sentence.

10. 10 g = __?__ mg
11. 1,000 km = __?__ m
12. 4.39 mL = __?__ L
13. 1.5 mL = __?__ L
14. 5.93 g = __?__ kg
15. 7.89 m = __?__ km

Exercises

Independent Practice Complete each sentence.

16. 3.54 m = __?__ mm
17. 525 g = __?__ kg
18. 4.23 L = __?__ mL
19. __?__ m = 1.37 km
20. __?__ mg = 5.23 g
21. 9.24 kL = __?__ L
22. __?__ L = 2,354 mL
23. 0.924 m = __?__ cm
24. 427 m = __?__ km
25. The width of the tail of a microorganism is 0.00723 cm wide. How many millimeters is this?
26. How many grams are in 0.875 kilograms?
27. A mayonnaise jar holds 0.947 liters. How many milliliters does it hold?

Mixed Review

28. **Smart Shopping** Lora wants to buy some goldfish that cost 59¢ each. How many fish can she afford to buy with her $2.50 allowance? *(Lesson 1-1)*
29. Use mental math to find 9 · 42. Tell what method you used. *(Lesson 1-2)*
30. Estimate 7,103 − 2,980 by rounding. *(Lesson 1-3)*
31. Estimate 2.381 ÷ 0.39 by using compatible numbers. *(Lesson 1-3)*

Problem Solving and Applications

32. **Archeology** Look at the drawing of the arrowhead on page 23. Would it fit in a box 3.8 cm by 0.20 m? Why or why not?
33. **Critical Thinking** Some basic metric units also are closely related. One milliliter of water at 4° C has a mass of 1 gram and a volume of 1 cubic centimeter (cm^3). A 520-gram metal bar is placed in a beaker full of water. When immersed, 260 grams of water spills out.
    a. How many milliliters of water spilled out?
    b. The volume of water that spilled out equals the volume of the bar. What is the volume of the bar in cubic centimeters?
    c. *Density* is defined as mass divided by volume. What would be the density of the bar in grams per cubic centimeter?
    d. **Research** Use the encyclopedia to find information about Archimedes. How does his work relate to this problem?

Measurement Connection

1-7 The Customary System

Objective

Use customary units of measurement.

Words to Learn

customary system

In 1992, the EPA estimated there were about 1,200 hazardous waste sites in the United States. Scientists are often called to assist in the detoxification of these sites.

A college degree in engineering or science is usually required to be a specialist in this field.

To obtain more information, contact: EPA, 401 M. St. SW, Washington, DC 20460.

In 1990, the Environmental Protection Agency (EPA) announced that the four most polluting states released 1,730 million pounds of toxic chemicals into the environment. Pounds are units of weight in the **customary system.**

The customary system developed from a series of nonstandard measurements. The unit of length, the foot, evolved from the length of a king's foot. The inch was the length of 3 barleycorns laid end to end. The basic unit of weight is a pound, and the basic unit of capacity is the quart.

Changing from one unit to another in the customary system is not as easy as in the metric system. The customary system is *not* a decimal system. The following chart lists some common units and their equivalents.

Length:	12 inches (in.) = 1 foot (ft)
	3 ft = 1 yard (yd)
	5,280 ft = 1 mile (mi)
Capacity:	8 fluid ounces (oz) = 1 cup (c)
	2 c = 1 pint (pt)
	2 pt = 1 quart (qt)
	4 qt = 1 gallon (gal)
Weight:	16 dry ounces (oz) = 1 pound (lb)
	2,000 lb = 1 ton (T)

Changing from one unit to another in the customary system may involve more than one step. The rules about when to multiply and when to divide are like those you used with the metric system. The actual computation may not be as easy since the metric system is based on multiples of ten and the customary system is not.

Example 1

18 ft = __?__ yd

A foot is shorter than a yard, so divide. There are 3 feet in every yard, so divide by 3.

$18 \div 3 = 6 \quad \rightarrow \quad 18 \text{ ft} = 6 \text{ yd}$

Example 2

How many ounces are in 105 pounds?

A pound is heavier than an ounce, so multiply.
There are 16 ounces in each pound.

105 [×] 16 [=] 1680

There are 1680 ounces in 105 pounds.

Sometimes you may need to use more than one measurement to complete a problem.

Example 3 ***Problem Solving***

> **Problem-Solving Hint**
>
> Drawing a diagram may help you solve the problem.

Construction **The highway department is putting a railing along a highway that edges a ravine. The railing is attached to posts placed a yard apart. If the railing is 4 miles long, how many posts do they need?**

One post is needed to start with and then we need to know how many yards there are in 4 miles. We know that there are 5,280 feet in a mile. To find the number of yards, divide by 3. There are 1,760 yards in each mile.

Now, draw a diagram.

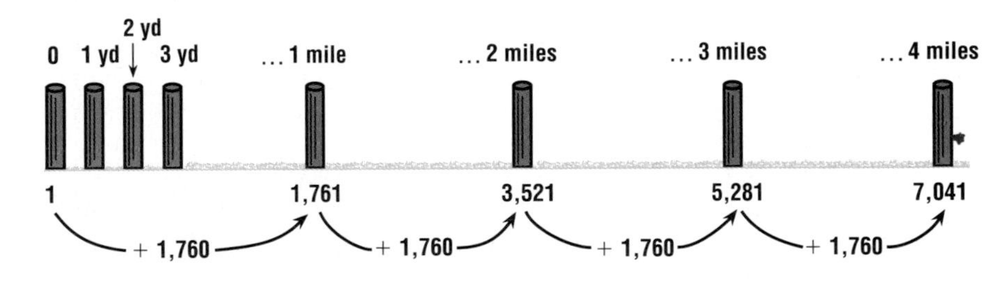

The highway department needs 7,041 posts.

Checking for Understanding

Communicating Mathematics

Read and study the lesson to answer each question.

1. **Tell** one advantage the metric system has over the customary system.
2. **Write** how you would change 6 cups to gallons.
3. **Show** another way to solve the problem in Example 3.

Guided Practice

Tell whether you multiply or divide and by what number to complete each sentence. Then complete.

4. 3 lb = __?__ oz
5. 6,000 lb = __?__ T
6. 24 c = __?__ qt
7. $2\frac{1}{2}$ yd = __?__ in.
8. 8,000 ft = __?__ mi
9. 10 gal = __?__ pt

Exercises

Independent Practice

Complete each sentence.

10. 12 ft = __?__ yd
11. 5 mi = __?__ ft
12. 2.5 gal = __?__ qt
13. 3.5 lb = __?__ oz
14. 2 T = __?__ lb
15. 7 c = __?__ pt
16. __?__ gal = 20 qt
17. $3\frac{1}{2}$ ft = __?__ in.
18. 2 gal = __?__ oz

19. How many gallons are there in 32 ounces of orange juice?
20. How many feet are in 15 inches?
21. An inchworm travels 4 yards. How many inches is that?
22. How many feet are in $7\frac{1}{2}$ yards?

Mixed Review

23. Tell how you could use compensation to find 404 − 153. *(Lesson 1-2)*
24. Estimate 4,521 − 3,158 using front-end estimation. *(Lesson 1-3)*
25. How many centimeters are in 2.79 meters? *(Lesson 1-6)*
26. A coffee mug safely holds 375 milliliters. How many coffee mugs could you safely fill with 12 liters of coffee? *(Lesson 1-6)*

Problem Solving and Applications

27. **Critical Thinking** Marva filled a one-cup measuring cup with sand. She said it weighed eight ounces. Is she right? Explain your answer.
28. **Ecology** Every ton of recycled office paper saves about 17 trees. Glencoe Publishing's office in Westerville, Ohio, recycled 8,000 pounds of paper in 1991. How many trees did this save?
29. **Design** Martina designed a new cheerleader uniform that requires $3\frac{1}{2}$ yards of material. All 10 cheerleaders wear the same design. How long is the piece of cloth needed to make all their uniforms if it is measured in feet?

30. **Portfolio Suggestion** A portfolio contains representative samples of your work collected over a period of time. Begin your portfolio by selecting an item that shows something you learned in this chapter.

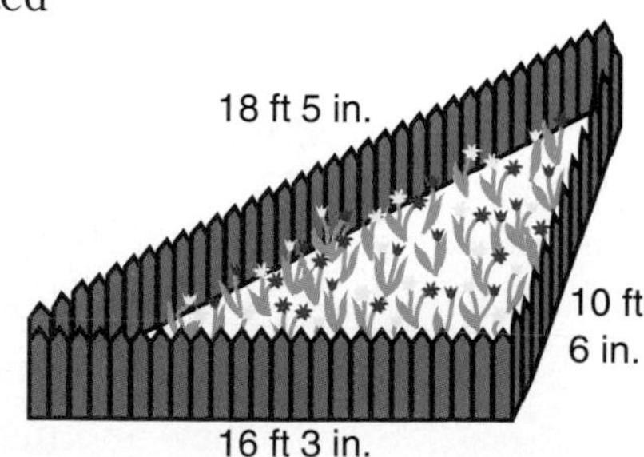

31. **Geometry** The fence around a triangular flower garden in front of the school has sides of 16 ft 3 in., 18 ft 5 in., and 10 ft 6 in. Approximately how many feet long is the fence?
32. **Save Planet Earth** In a lifetime, the average person will throw away 600 times his or her adult weight in trash. Suppose a person weighs 150 pounds. How many tons of trash will that person leave behind?

DATA SEARCH

33. **Data Search** Refer to pages 2 and 3. How much faster is the record in the men's 200-meter butterfly than the women's record?

Cooperative Learning

1-7B Customary Measures

A Follow-Up of Lesson 1-7

Objective

Explore the origin of the mile unit of measure.

Materials

yardstick

A legion was a division of the Roman army that varied from 4,000 to 6,000 soldiers. Legions were divided into units of 100 men, each led by a veteran officer called a centurion.

The word "mile" comes from the Latin *mille passus,* which means 1,000 paces. The distance the Roman legions could travel in 1,000 paces became the standard length now known as a mile.

Try this!

Work in groups of two.

- A pace is the distance walked from the heel of one foot to the heel of the same foot when it hits the ground.

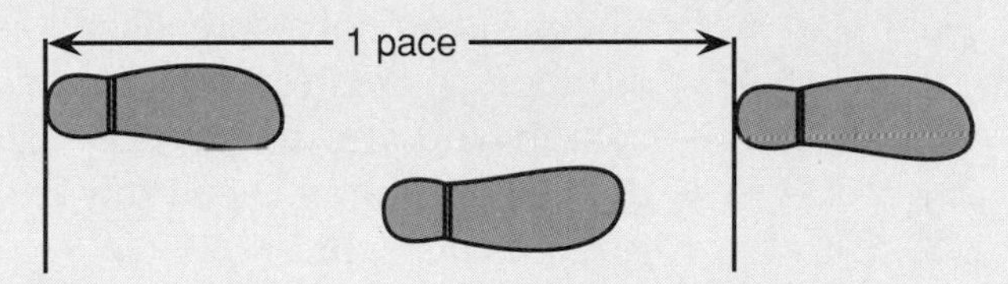

  Walk one pace. Have your partner measure the length of your pace in feet. Record the measurement. Have your partner walk one pace. Then measure the length of your partner's pace. Record this measurement.
- Repeat walking paces and measuring two more times. Record your measurements.
- Find the average pace for you and your partner.

	Trial 1	Trial 2	Trial 3	Average
Your Pace				
Your Partner's Pace				
Class Average				

What do you think?

1. Compare your group's average to the class average. Are they similar?
2. If you multiplied your average pace by 1,000, how close does it come to 5,280 feet?
3. Why do you think that the paces of the Roman legions would be nearly the same every time?

Extension

4. **Research** Investigate the origins of other customary units of measure.

Cooperative Learning

1-8A Spreadsheets

A Preview of Lesson 1-8

Objective

Explore the use of a computer spreadsheet.

Materials

calculators

Computer **spreadsheets** are important tools for organizing and analyzing data for problem solving. A spreadsheet is made of rows and columns. The columns are identified by letters and the rows are identified by numbers. Most spreadsheets have up to 2,048 rows available, numbered 1 through 2,048. There are 256 columns available. After column Z, the columns are named AA, AB, AC, and so on, through IV. However, only eight columns can be seen at a time on the computer screen.

The basic unit of a spreadsheet is called a **cell.** A cell may contain numbers, words (called labels), or a formula. Each cell is named by the column and row that describes its location, or **address.** The cell D2 is the cell in column D and row 2. *Cell D2 in the spreadsheet below contains the number 1,650.*

	A	B	C	D	E	F	G	H
1		JAN.	FEB.	MARCH	APRIL	MAY	JUNE	JULY
2	REVENUE	2,500	5,000	1,650	10,000	3,100	8,200	12,500
3	EXPENSES	700	1,200	225	3,550	1,800	2,300	5,600
4	PROFIT	1,800	3,800	1,425	6,450	1,300	5,900	6,900
5								
6								
7								

Activity One

Work with a partner. Use the spreadsheet shown above.

- Locate cell E3. What is printed in this cell?
- Give the address of all cells that contain labels.
- Locate cell B2, B3, and B4. What is the relationship among the numbers in these cells?

What do you think?

1. Write a sentence to fully describe the information in cell E3.
2. What type of information is stored in cells A2 through A4?
3. Does the relationship that you found for B2, B3, and B4 hold true for C2, C3, and C4? Is it true for all columns?

When cells are related, a formula can be used to generate this relationship. In the spreadsheet on the previous page, in every column the same pattern can be observed.

$$\text{row } 4 = \text{row } 2 - \text{row } 3$$

Instead of subtracting each time, you can use a formula to calculate the difference. The formula uses the cell names and symbols for the operations.

+ (add) − (subtract) * (multiply)
/ (divide) ^ (exponent)

When you type the formula C2 − C3 into cell C4, the cell will not show the formula, but the result of the subtraction. When you move the cursor to a cell, the formula in that cell appears at the bottom of the screen along with the cell address.

Activity Two

	A	B	C	D	E	F	G	H
1	CONCERT TICKETS	NUMBER SOLD	TICKET PRICE	TOTAL SALES				
2	FLOOR SEATS	150	$25	$3,750				
3	BALCONY	210	$20					
4	MEZZANINE	300	$15					
5								
6	TOTAL SEATS		TOTAL SALES					

- Locate B2, C2, and D2. How are these cells related?
- Suppose the value in B2 changed to 175. How would the value in D2 change?

What do you think?

4. Write spreadsheet formulas for cells D2, D3, and D4.
5. Write a formula for cell B6 to find the total number of seats sold.
6. Write a formula for finding the total sales for all seats.
7. Find the values that would appear in the cells of the spreadsheet.
8. Suppose the information *Standing Room Only, 100, $10* was inserted in cells A5, B5, and C5, respectively. How would the formulas in B6 and D6 change to include this information?

Application

9. Describe possible uses of spreadsheets at home and at school.
10. Create your own spreadsheet. Explain what each row and column represent and how you use formulas to save calculation time.

Problem-Solving Strategy

1-8 Guess and Check

Objective

Solve problems by using guess and check.

Judith noticed the due dates for the library books she returned. All but four books were due yesterday. All but four books are due today. All but four books are due tomorrow. How many books did Judith return?

Explore

What do you know?
All but 4 books were due yesterday.
All but 4 books are due today.
All but 4 books are due tomorrow.

You want to find out how many books Judith returned.

Plan

Guess the number of books Judith returned. Then check this number against the due date statements. If the guess doesn't check out, you can use what you learn to make a better guess.

Solve

First guess: Judith returned 10 books.

Check:

- *All but 4 were due yesterday.*

  Make a sketch and label it. Use Ys for books due yesterday.

  All but 4 due yesterday means 6 were due yesterday.

- *All but 4 are due today.*

  The 6 books labeled for yesterday contradicts this statement. Revise your guess downward to 8 books.

Second guess: 8 books

Check:

- *All but 4 were due yesterday.*

- *All but 4 are due today.*

  Use Ds for books due today.

- *All but 4 are due tomorrow.*

  No more books to label contradicts this statement. Revise your guess downward to 6 books.

Third guess: 6 books
Check:

- *All but 4 were due yesterday.*

- *All but 4 are due today.*

- *All but 4 are due tomorrow.*
  Use Ms for books due tomorrow.

Judith returned 6 books.

Examine Out of 6 books, if exactly 2 of them are due on separate days, then all but 4 are due on each day. You have not tested all numbers. But with fewer books, you run out of books to label. With more books, you have too many books to label.

Example

Inventory Terry's job at the cafeteria includes ordering supplies. Twelve-ounce cups come in cases of 1,600, and 16-ounce cups come in cases of 1,200. Terry ordered 7,200 cups in full cases. How many of each case were ordered?

Use a table to show what you know and what you are trying to find out. You should leave a place for guesses.

About how many 12-oz cases? (Guess.)	How many 12-oz cups is that? (Multiply by 1,600.)	How many 16-oz cups are still needed? (Subtract.)	How many 16-oz cases is that? (Divide by 1,200).
1	1,600	7,200 − 1,600 = 5,600	4.7 cases
2	3,200	7,200 − 3,200 = 4,000	3.3 cases
3	4,800	7,200 − 4,800 = 2,400	2 full cases

Terry ordered 3 cases of the 12-ounce cups and 2 cases of the 16-ounce cups.

You can check by referring to the problem. Two 16-ounce cases at 1,200 cups per case is 2,400 cups. Three 12-ounce cases at 1,600 cups per case is 4,800 cups. Together, 2,400 plus 4,800 is 7,200, a check.

Checking for Understanding

Communicating Mathematics

Read and study the lesson to answer each question.

1. **Tell** why it is important to use estimation with the guess-and-check strategy.
2. How do you know if your guess doesn't work?
3. Make a table that can be used to solve the problem in the lesson about Judith and the returned library books.

Guided Practice

Use the guess-and-check strategy to find the number.

4. The sum of the number and its double is 21.
5. The product of the number and itself is 196.
6. One hundred more than the number is twice the number.
7. The quotient of the number and itself is always 1.
8. The product of the number and its next two consecutive whole numbers is 60.

Problem Solving

Practice

Solve using any strategy.

Strategies

Look for a pattern.
Solve a simpler problem.
Act it out.
Guess and check.
Draw a diagram.
Make a chart.
Work backward.

9. José's mother is five times as old as José. Five years from now she will be just three times as old. How old is José now?
10. In 1991, stamps for postcards cost \$0.19. Stamps for first class letters cost \$0.29. Louise used both to write to 11 friends. If she spent \$2.59 for stamps, how many postcards and how many letters did she send?
11. Forty-two inches of molding are needed to frame a picture. If molding sells for \$5.80 a foot, how much will the molding cost to frame the picture?
12. The graph below shows the busiest airports in the United States, categorized by number of passengers in 1990.
    a. Which airport has 12.5 million more flyers than Phoenix?
    b. Which airport has nearly half as many flyers as Atlanta?
    c. If this list was divided into east and west by using the Mississippi River as the dividing line, which group would have the most passengers?

Algebra Connection

1-9 Powers and Exponents

Objective

Use powers and exponents in expressions.

Words to Learn

factor
exponent
power
base
evaluate

Every person has 2 biological parents. Each of their parents has 2 parents, so every person has $2 \cdot 2$ or 4 grandparents. Each grandparent had 2 parents, so every person has $2 \cdot (2 \cdot 2)$ or 8 great grandparents. How many great-great grandparents does every person have?

Generation	Family Tree	Number
Person		1
Parents		$1 \cdot 2$ or 2
Grandparents		$2 \cdot 2$ or 4
Great Grandparents		$2 \cdot 2 \cdot 2$ or 8
Great-Great Grandparents	?	?

Study the pattern in the number column of the table. The next value would be $2 \cdot 2 \cdot 2 \cdot 2$ or 16. Each person has 16 great-great grandparents.

When two or more numbers are multiplied, these numbers are called **factors** of the product. When the same factor is repeated, you may use an **exponent** to simplify the notation.

$$16 = 2 \cdot 2 \cdot 2 \cdot 2 \rightarrow 2^4$$ *4 is the exponent.*

An expression like 2^4 is called a **power** and is read *2 to the fourth power.* The 2 in this expression is called the **base.** The base names the factor being repeated.

Powers are often used to write a product in a shorter form. For example, $3 \cdot 4 \cdot 4 \cdot 3 \cdot 4 \cdot 3 \cdot 3 \cdot 3$ can be written as $4^3 \cdot 3^5$.

Examples

1 **Write $5 \cdot 5 \cdot 5 \cdot 5 \cdot 5 \cdot 5$ using exponents.**

There are six factors of 5.

$5 \cdot 5 \cdot 5 \cdot 5 \cdot 5 \cdot 5 = 5^6$

2 **Write $3 \cdot 3 \cdot 5 \cdot 5 \cdot 6 \cdot 6 \cdot 6$ using exponents.**

There are two factors of 3, two factors of 5, and three factors of 6.

$3 \cdot 3 \cdot 5 \cdot 5 \cdot 6 \cdot 6 \cdot 6 = 3^2 \cdot 5^2 \cdot 6^3$

> **Calculator Hint**
>
> Some calculators will repeat the last operation you did. If your calculator doesn't have a [yˣ] key, try finding 4^3 by pressing
>
>  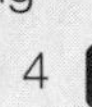 

You can use your calculator to find the values of powers. Use the [yˣ] key to **evaluate,** or find the value of, a power. Here's how to evaluate the expression $4^3 \cdot 3^5$.

4 [yˣ] 3 [×] 3 [yˣ] 5 [=] 15552

$4^3 \cdot 3^5 = 15{,}552$ *You can find the product of $4 \cdot 4 \cdot 4 \cdot 3 \cdot 3 \cdot 3 \cdot 3 \cdot 3$ to check this answer.*

Examples

3 Find 8^3.

Use paper and pencil.

$$8^3 = 8 \cdot 8 \cdot 8 = 64 \cdot 8 = 512$$

Use a calculator.

4 Find $2^3 \cdot 3^4$.

Use paper and pencil.

$$2^3 \cdot 3^4 = \underbrace{2 \cdot 2 \cdot 2}_{8} \cdot \underbrace{3 \cdot 3 \cdot 3 \cdot 3}_{81} = 8 \cdot 81 = 648$$

Use a calculator.

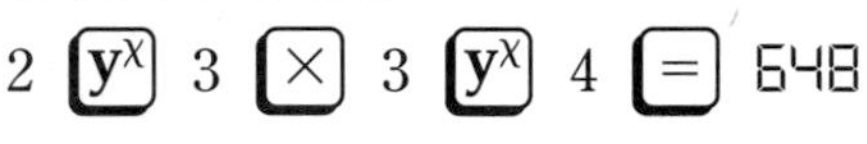

Checking for Understanding

Communicating Mathematics

Read and study the lesson to answer each question.

1. **Tell** which is the base and which is the exponent in 10^5.
2. **Write** the expression *5 to the fourth power* using exponents.
3. **Show** how you can use paper and pencil to verify the calculator example at the top of this page.
4. **Show** that 3^2 and 2^3 are not equal.

Guided Practice

Write each product using exponents.

5. $5 \cdot 5 \cdot 5$
6. $10 \cdot 10 \cdot 10 \cdot 10$
7. $8 \cdot 8 \cdot 8 \cdot 8 \cdot 8$

Evaluate each expression.

8. 5^3
9. 7^2
10. $3^2 \cdot 4^2$
11. $5 \cdot 6^3 \cdot 10^3$

Exercises

Independent Practice

Write each product using exponents.

12. $3 \cdot 3 \cdot 5 \cdot 5 \cdot 5$
13. $6 \cdot 6 \cdot 7 \cdot 6 \cdot 7$
14. $2 \cdot 3 \cdot 5 \cdot 3 \cdot 2$

Evaluate each expression.

15. 5^4
16. 4^3
17. 2^5
18. 1^{40}
19. $2^2 \cdot 7^2$
20. $5^2 \cdot 8^2 \cdot 3^3$
21. 100^3
22. 56^4

Use your calculator to evaluate each expression.

23. $6^2 - 2^2$
24. $2 \cdot 2^3 \cdot 3^2$
25. $2 \cdot 3^2 + 9 \cdot 6^2$

Mixed Review

26. Estimate \$10.19 + \$9.89 + \$10.09 +\$9.99 by clustering. *(Lesson 1-3)*
27. How many kilograms are in 5,734 grams? *(Lesson 1-6)*
28. How many pints are in 6 cups? *(Lesson 1-7)*
29. **Smart Shopping** Which is the better purchase price, 39 ounces of flour for \$2.70 or 2.5 pounds of flour for \$2.70? *(Lesson 1-7)*

Problem Solving and Applications

30. **Geometry** To find the volume of a cube, you multiply the length times thc width times the depth of the box. Suppose a cube is 30 cm long, 30 cm wide, and 30 cm tall.

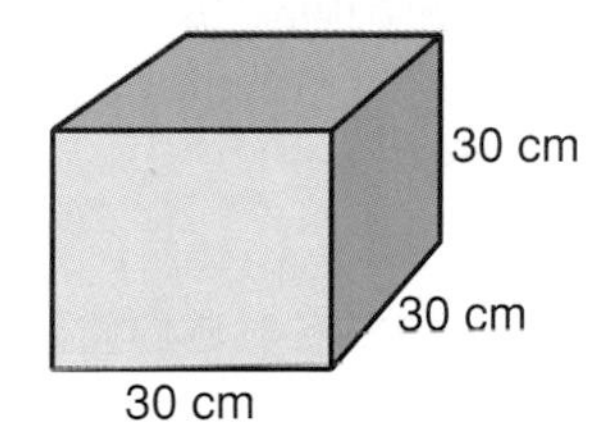

a. Write an expression for the volume using exponents.

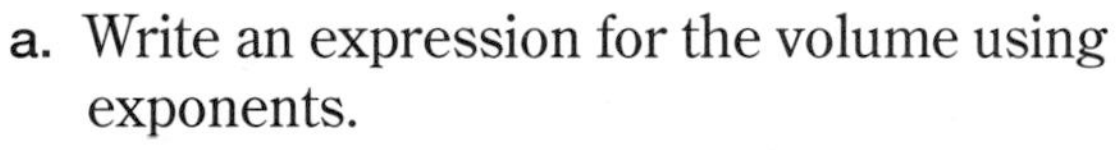

b. Find the volume of the cube.

31. **Critical Thinking** Complete each sentence with the correct base or exponent.

a. $27 = 3^{\blacksquare}$ b. $36 = \blacksquare^2$ c. $16 = \blacksquare^2$ or $\blacksquare^4 =$

32. **Automobiles** The formula to determine the horsepower for a 4-cylinder car can be found by evaluating the expression $6^2 \cdot 4 \div 2.5$. Use your calculator to find the horsepower rounded to the nearest 10.
33. **Number Sense** If $2^{10} = 1{,}024$, find 2^{11} mentally.
34. **Journal Entry** Write how you could state the equivalent values in the metric system by using exponents, such as 1 km $=10^{\blacksquare}$ m.
35. **Mathematics and Astronomy** Read the following paragraph.

> Some scientists believe that our solar system was formed when a cloud of gas condensed to form the sun. Parts of the cloud formed small bodies called planetoids. The planetoids crashed into each other, breaking up into smaller pieces and attaching themselves to larger ones.

a. The average distance from the sun to Uranus is 2.87×10^9 km. Write this distance in standard form.

b. The average distance from Venus to the sun is 67,200,000 mi. Between which powers of 10 is the distance?

Chapter 1

Study Guide and Review

Communicating Mathematics

Choose the correct term to complete each sentence.

1. The (quart, liter) is a unit of capacity in the metric system.
2. The (ounce, gram) is a unit of weight in the customary system.
3. $2 + 3 = 3 + 2$ is an example of the (commutative, associative) property of addition.
4. In the expression 5^3, the (base, exponent) is 3.
5. Estimating $219.8 - 78.2$ by computing $220 - 80$ is an example of (clustering, rounding).
6. Give one advantage that the metric system has over the customary system.

Self Assessment

Objectives and Examples

Upon completing this chapter, you should be able to:

Review Exercises

Use these exercises to review and prepare for the chapter test.

- solve problems using the four-step plan *(Lesson 1-1)*

The four steps are:

- **Explore** What do I know? What do I need to find out?
- **Plan** How will I go about solving this?
- **Solve** Carry out your plan. Does it work? Do you need another plan?
- **Examine** Does your answer seem reasonable? If not, find another.

Use the four-step plan to solve each problem.

7. **Time** The distance between John's house and Bill's house is 90 feet. If it takes John 3 seconds to walk 10 feet, how long will it take him to walk to Bill's house?
8. **Construction** The concrete slab for a floor is 47 feet long, 50 feet wide, and 2 feet deep. If the volume of the concrete is measured in cubic feet and is found by multiplying the length, width, and height, find the volume of the concrete used in the slab.

- compute mentally using compensation *(Lesson 1-2)*

Find $300 - 128$.

$$\begin{array}{lcr} 300 - 1 & \rightarrow & 299 \\ 128 - 1 & \rightarrow & -127 \\ \hline & & 172 \end{array}$$

Use compensation to find each answer.

9. $57 + 88$
10. $600 - 461$
11. $398 + 119$
12. $402 - 237$
13. $\$2.97 + \$4.59$
14. $903 - 584$

Objectives and Examples	Review Exercises

Objectives and Examples

- compute mentally using the commutative, associative, and distributive properties *(Lesson 1-2)*

Find $25 + 47 + 25$.

$$\begin{aligned} 25 + 47 + 25 &= 25 + 25 + 47 \\ &= 50 + 47 \\ &= 97 \end{aligned}$$

Review Exercises

Use the commutative, associative, and distributive properties to find each answer.

15. $7 \cdot 12$
16. $8(10 + 40)$
17. $34 + 49 + 66$
18. $(19 \cdot 4) \cdot 25$
19. $5 \cdot 34 \cdot 20$
20. $18(47 + 53)$

- estimate answers by using different strategies *(Lesson 1-3)*

Four ways to estimate are by rounding, by front-end estimation, by using compatible numbers, and by clustering.

Estimate. Use an appropriate strategy.

21. $2{,}794 \div 68$
22. $628 + 482 + 817$
23. $5{,}128 + 4{,}984 + 5{,}021 + 4{,}898$
24. $9.872 - 5.344$

- use metric units of measurement *(Lesson 1-6)*

2.81 m = __?__ mm

A meter is longer than a millimeter, so multiply.

1 m = 1,000 mm, so multiply by 1,000.

2.810 = 2,810

2.81 m = 2,810 mm

Complete each sentence.

25. 3.4 m = __?__ km
26. 2.71 L = __?__ mL
27. 620 g = __?__ kg
28. 0.748 m = __?__ cm
29. __?__ L = 8.62 mL
30. __?__ mg = 4.25 g

- use customary units of measurement *(Lesson 1-7)*

12 qt = __?__ gal

A quart is less than a gallon, so divide.

4 qt = 1 gal, so divide by 4.

$12 \div 4 = 3$

12 qt = 3 gal

Complete each sentence.

31. __?__ ft = 54 in.
32. 2.5 T = __?__ lb
33. 15 ft = __?__ yd
34. 6 pt = __?__ c
35. __?__ oz = 4 lb
36. 22 gal = __?__ qt

- use powers and exponents in expressions *(Lesson 1-9)*

Find $5^2 \cdot 2^3$.

$$\begin{aligned} 5^2 \cdot 2^3 &= \underbrace{5 \cdot 5}_{} \cdot \underbrace{2 \cdot 2 \cdot 2}_{} \\ &= \ \ 25 \ \ \cdot \ \ \ 8 \\ &= 200 \end{aligned}$$

Write each product as an expression using exponents.

37. $3 \cdot 3 \cdot 8 \cdot 8 \cdot 8$
38. $4 \cdot 6 \cdot 4 \cdot 4 \cdot 6 \cdot 7$

Find the value of each expression.

39. 2^4
40. $3^2 \cdot 4^3$
41. $5^3 \cdot 100^2$

Study Guide and Review

Applications and Problem Solving

42. **Consumer Math** At a restaurant, entrees cost from \$8.95 to \$15.95; appetizers cost from \$2.25 to \$5.95; a salad costs \$2.75; and soup costs \$1.95. A full meal includes soup, salad, appetizer, and an entree. Without tax and tip, would a couple expect to pay more or less than \$30 for a full meal at this restaurant? *(Lesson 1-4)*

43. **Astronomy** When Venus is closest to Earth, it is about 25 million miles away. Radio waves, like light waves, travel through space at about 186,272 miles per second. Would it take about 0.225, 2.25, or 22.25 minutes to send a radio signal from Earth to Venus? *(Lesson 1-5)*

44. Melba has 15 coins, consisting of quarters and dimes. Their total value is \$3. How many of each coin does she have? *(Lesson 1-8)*

45. **Sports** During a football game, Steve had four punt returns that were 11 yards, 23 yards, 19 yards, and 32 yards long. Approximately how many total yards did Steve gain on punt returns during that game? *(Lesson 1-3)*

46. **Mail** Lisa mailed three packages. The first one weighed 186 grams, the second one weighed 198,000 milligrams, and the third one weighed 0.221 kilograms. What was the total weight of all three packages in grams? *(Lesson 1-6)*

Curriculum Connection Projects

- **Science** Find the steps of the scientific process from a science book. Write a short paper comparing it to the four-step plan for problem solving in math.
- **Consumer Math** From a trip to the mall or from a catalog, select five complete outfits. Record the price of each item. Estimate the total cost of each outfit, the whole wardrobe, and the amount left over from \$500.
- **Physical Education** From research on the 1992 Summer Olympics, make a five-column table with the following headings: Event, Kilometers, Meters, Centimeters, and Millimeters. Record distances from five running events and five field events. Convert each to the missing metric measurements on your paper.

Read More About It

Mamioka. *Who's Hu.*
Raskin, Ellen. *The Westing Game.*
Smullyan, Raymond. *The Lady or the Tiger? And Other Logic Puzzles.*

Chapter

1 Test

Estimate. Use an appropriate strategy.

1. $8.25 + 3.59 + 5.76$
2. $2,948 \div 49$
3. $7,548 - 4,139$
4. $5,842 + 3,112$
5. $2.015 + 1.893 + 2.121 + 1.902$
6. Estimate the product of 489 and 19.
7. Use mental math to find $(650 + 793) + 50$. Tell how you found the answer.
8. Explain how you would use mental math to find $500 - 232$.

Complete each sentence.

9. ___?___ ft = 5 yd
10. ___?___ kg = 735 g
11. 8.2 L = ___?___ mL
12. 3.5 lb = ___?___ oz
13. ___?___ cm = 0.52 m
14. 4.21 km = ___?___ m
15. Josh's recipe makes 9 cups of lemonade. How many pints is this?
16. How many quarts are in 6 gallons of oil?

Write each product using exponents.

17. $6 \cdot 6 \cdot 6 \cdot 7 \cdot 7$
18. $3 \cdot 4 \cdot 8 \cdot 4 \cdot 3 \cdot 4$

Evaluate each expression.

19. $2^3 \cdot 4^2$
20. $2^2 \cdot 5^2 \cdot 4^3$

Use the four-step plan to solve each problem.

21. **Music** Michelle wants to listen to 5 compact discs. Two compact discs are 51 minutes long each and the other three compact discs are 46 minutes long each. How long will it take Michelle to listen to all 5 compact discs?
22. **Earning Money** Maria and Emilio shovel driveways in the winter for $5.00 each. If Emilio can shovel 5 driveways per hour and Maria can shovel 3 driveways per hour, how long will it take them to earn a total of $80?
23. Mrs. Martinez bought a house for $44,800 in 1974. She sold it in 1991 for 3.25 times more. Did the value of the house increase to $1,456,000 or $145,600?
24. Mr. Higgins' credit card purchases for this month are $129.47 and $169.72. His previous balance was $2,147.91 on which he had to pay $33.83 in finance charges. He also made a payment of $150. His new balance should be about:

    a. $2,500 b. $2,300 c. $230 d. $200
25. Four pears cost $2.18. Three cans of apple juice cost $3.29. About how much money do you need to buy one of each?

Bonus Suppose x represents a number. Find the value of x so that $x^3 = 64$.

Chapter 2

An Introduction to Algebra

Spotlight on the Human Body

Have You Ever Wondered....

- How many high school students actually see themselves as thinner or heavier than they are?
- What chemical elements make up the weight of the human body?

The Elements in a 150-pound Individual

Element	Weight (lb)	Element	Weight (lb)
Oxygen	97.5	Sodium	0.165
Carbon	27.0	Magnesium	0.06
Hydrogen	15.0	Iron	0.006
Nitrogen	4.5	Cobalt	0.00024
Calcium	3.0	Copper	0.00023
Phosphorus	1.8	Manganese	0.00020
Potassium	0.3		
Sulfur	0.3	+ other elements in minute quantities	
Chlorine	0.3		

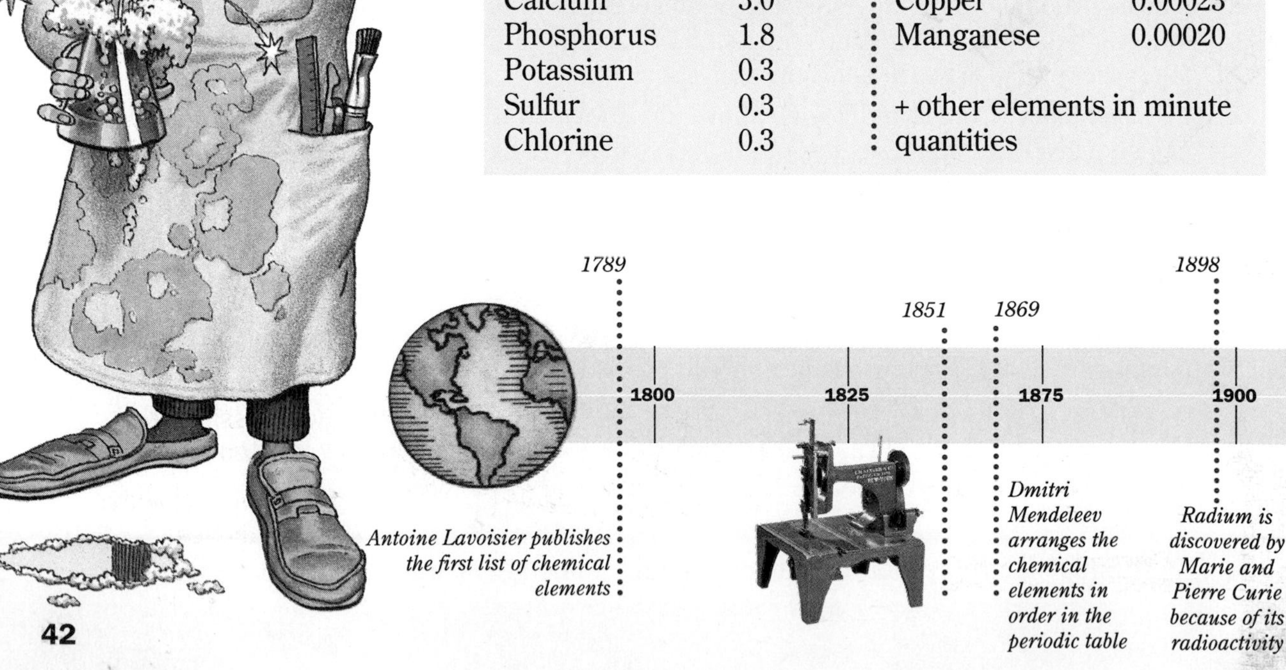

Chapter Project

The Human Body

Work in a group.

1. Fill a jar with different candies or coins.
2. Weigh the full jar.
3. What are the individual weights of each of the groups of candies or coins? What is the weight of the jar?
4. Calculate what percent of the human body each element represents by dividing each element's weight by the total body weight. Then change the decimal to a percent.
5. Create at least two more examples like the jar. Make a chart showing your data.

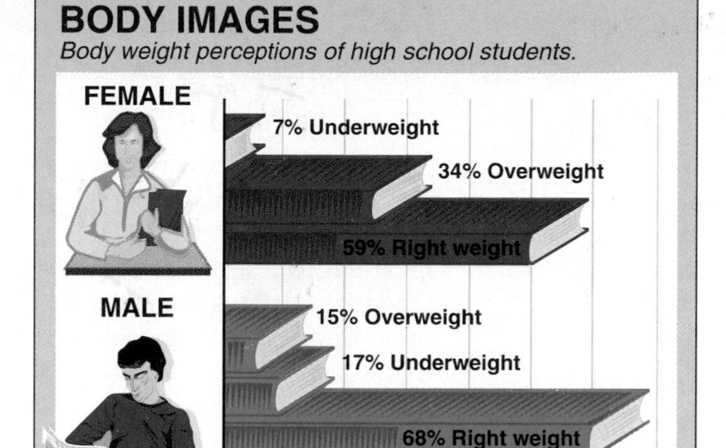

Looking Ahead

In this chapter, you will see how mathematics can be used to answer the questions about weight and the elements. The major objectives of the chapter are to:

- evaluate numerical and algebraic expressions
- identify and solve equations
- evaluate expressions using the order of operations
- write and solve two-step expressions and equations
- identify and solve inequalities

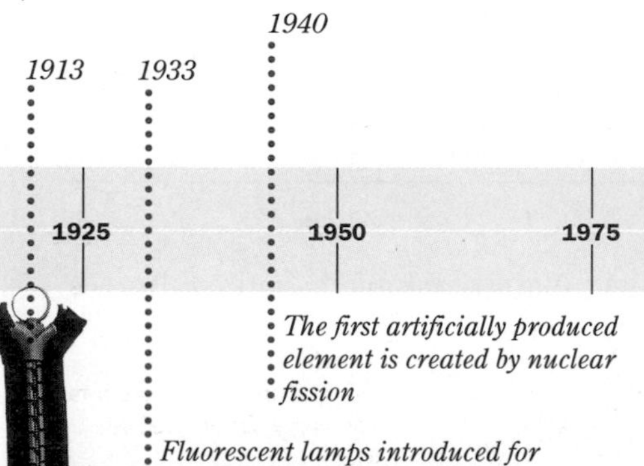

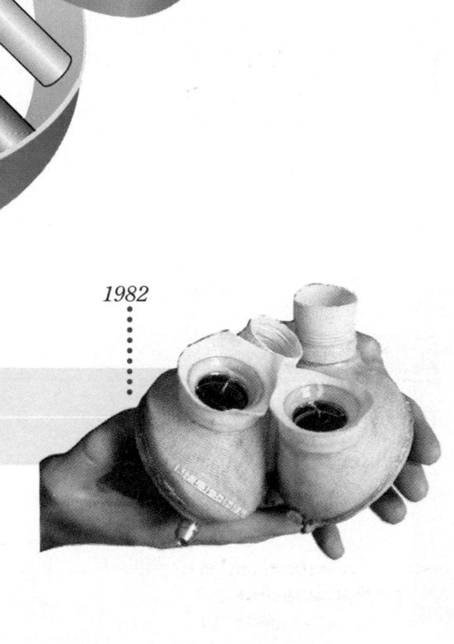

1913 1933 1940 1982

1925 1950 1975

1940 The first artificially produced element is created by nuclear fission

1933 Fluorescent lamps introduced for floodlighting and advertising

2-1 Variables and Expressions

Objective

Evaluate numerical and algebraic expressions by using the order of operations.

Words to Learn

numerical expression
substitute
evaluate
order of operations
variable
algebraic expression

Did you know that when you see an infant on television or in a movie, the infant is often played by not one baby but by a set of identical twins? One twin can be substituted for the other during a filming sequence. In this way, production can go on, and it appears to be the same baby in each scene.

In mathematics, we also use substitution. Consider the **numerical expression** $5 + 6$. It has a value of 11. However, the expression $x + 4$ does not have a value until a value for x is given. Suppose you let $x = 12$. You can **substitute,** or put 12 in place of x in the expression. It becomes the numerical expression $12 + 4$, which has a value of 16. Therefore, if $x = 12$, then $x + 4 = 16$.

In order to **evaluate,** or find the value of, a numerical expression, we need to follow an **order of operations.** That is, you need to know which operation to do first when there is more than one operation in the expression. The following rules are used when evaluating numerical expressions.

Grouping symbols include:

- *parentheses ()*
- *brackets [].*
- *fractions bars, as in* $\frac{6 + 4}{2}$*, which means* $(6 + 4) \div 2$.

Order of Operations	
	1. Do all operations within grouping symbols first; start with the innermost grouping symbols. 2. Do all powers before other operations. 3. Next, do all multiplication and division in order from left to right. 4. Then, do all addition and subtraction in order from left to right.

Example 1

Evaluate $(4 + 8) \div 3 \times 5 + (2^2 + 9)$.

Do operations in the parentheses first. You may have to follow the order of operations to do this step.

$$(4 + 8) \div 3 \times 5 + (2^2 + 9) = 12 \div 3 \times 5 + 13 \quad \textit{2^2 + 9 = 4 + 9 or 13}$$
$$= 4 \times 5 + 13 \quad \textit{12 \div 3 = 4}$$
$$= 20 + 13 \text{ or } 33 \quad \textit{4 \times 5 = 20}$$

Mini-Lab

Work with a partner.

Materials: calculator

Does your calculator follow the order of operations automatically?

- Enter this expression in your calculator.

  2 [+] 3 [×] 4 [−] 6 [÷] 2 [=]

- Evaluate this expression using pencil and paper and the order of operations.

Talk About It

a. Are the answers the same?

b. If the answers are not the same, why are they different?

Technology Activity

You can learn how to evaluate expressions with a graphing calculator in Technology Activity 1 on page 658.

Most scientific calculators follow the order of operations. Arithmetic calculators do not.

Calculator Hint

If you have an arithmetic calculator, you can still use it to evaluate expressions. Just enter the numbers as you would evaluate the expression with pencil and paper. Use [=] after completing each operation and continue without clearing.

Algebra is a language of symbols. In algebra, we use letters, called **variables,** to represent unknown quantities. In the expression $x + 4$, x is a variable. Expressions that contain variables are called **algebraic expressions.** In order to evaluate algebraic expressions, you must know how to read algebraic expressions.

$3a$	means	$3 \times a$
ab	means	$a \times b$
$5 \cdot 6d$	means	$5 \times 6 \times d$
$3xy^2$	means	$3 \times x \times y \times y$
$a[b(cd)]$	means	$a \times (b \times c \times d)$
$\frac{t}{3b}$	means	$t \div (3 \times b)$

$3a$, $3 \times a$, $3 \cdot a$, and $3(a)$ all are ways to write the product of 3 and a.

Example 2

Evaluate $x + y - 4$ if $x = 7$ and $y = 3$.

First, use substitution. Replace each variable in the expression with its value. Then use the order of operations.

$x + y - 4 = 7 + 3 - 4$ *Replace x with 7 and y with 3.*

$= 10 - 4$

$= 6$ *Check your answer mentally.*

Examples

3 Evaluate $2a + 3b$ if $a = 4$ and $b = 12$.

$$\begin{aligned} 2a + 3b &= 2(4) + 3(12) && \textit{Replace a with 4 and b with 12.} \\ &= 8 + 36 && \textit{Do multiplication before addition.} \\ &= 44 && \textit{Add.} \end{aligned}$$

4 Evaluate $\frac{y^2}{3x}$ if $y = 6$ and $x = 3$.

The bar, which means division, is also a grouping symbol. Evaluate the expressions in the numerator and denominator separately before dividing.

$$\begin{aligned} \frac{y^2}{3x} &= \frac{6^2}{3 \cdot 3} && \textit{Replace y with 6 and x with 3.} \\ &= \frac{36}{9} && \textit{Evaluate the numerator and the denominator separately.} \\ &= 36 \div 9 \text{ or } 4 && \textit{Then divide.} \end{aligned}$$

Checking for Understanding

Communicating Mathematics

Read and study the lesson to answer each question.

1. **Write** three expressions that all mean *the product of 6 and the variable x.*
2. **Tell** what the difference is between a numerical expression and an algebraic expression.
3. **Show** how to evaluate the expression $(4 + 3)^2 - 5 \cdot 6$.
4. **Tell** why the expressions $2 \cdot 5 + 3$ and $2 \cdot (5 + 3)$ have different values, even though they have the same numbers and operations.

Guided Practice

Tell which operation should be done first. Then evaluate each expression.

5. $21 + 2 \cdot 9$
6. $3(5) + 4 \div 2$
7. $18 \div 3 - 5$
8. $(3 + 7) \div 2$
9. $5 \cdot (6 + 3^2)$
10. $(9 - 7)^3 - \frac{12}{3 \cdot 2}$

Evaluate each expression if $a = 7$, $b = 6$, $c = 4$, and $d = 3$.

11. $3a + 4b - 2d$
12. $abc \div 21$
13. $(3b + 2c) \cdot d$
14. $3b + (2c \cdot d)$
15. cd^2
16. $(cd)^2$

Exercises

Independent Practice

Evaluate each expression.

17. $3 \cdot (19 - 4) + \frac{23 - 7}{2}$
18. $4^2 - 3 \cdot 4 + (6 - 6)$
19. $(7^2 + 1) \div 5$
20. $\frac{21}{3^2 - 2}$
21. $[4 + (7 - 3)^2] + 2$
22. $[7 - (8 - 6)^2] - 1$
23. $[(4 + 3)\ 2] \div 7$
24. $17 - 2[8 - (17 - 14)]$

25. Evaluate $7a - 3 + 2c$ if $a = 12$ and $c = 10$.

26. Find the value of the expression $a^2 + b^2 - c^3$ if $a = 10$, $b = 5$, and $c = 4$.

27. If $y = 36$, $x = 25$, and $w = 20$, find the value of $\frac{xy}{w} - (x + w)$.

Copy each arithmetic sentence. Use your calculator to find out where to place the grouping symbols so that the sentence is true. You may need to use more than one set of grouping symbols.

28. $5^2 + 4 \cdot 3 = 37$

29. $72 \div 6 + 3 = 8$

30. $46 + 4 - 36 \div 4 + 5 = 46$

31. $7 + 4^2 \div 2 + 6 = 9$

32. $3 \cdot 8 - 5 + 1 = 6$

33. $4 + 8 - 7 - 5 = 10$

Mixed Review

34. **Number Sense** Carlos and Juanita want to share in the cost of a 45¢ candy bar. He has dimes. She has nickels. If each put in the same number of coins, what is that number? *(Lesson 1-1)*

35. Estimate 328 + 671 + 459. *(Lesson 1-3)*

36. How many meters are in 1.725 kilometers? *(Lesson 1-6)*

37. How many pounds of corn are in a bin that holds 2 tons? *(Lesson 1-7)*

38. Find the value of $2^3 \cdot 6^2$. *(Lesson 1-9)*

Problem Solving and Applications

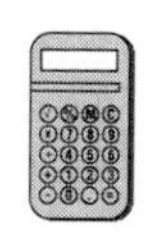

39. **Nutrition** You can use the expression $\frac{C}{8p}$ to determine how many grams of protein you should eat each day if your total calorie intake is C and there are p calories in each gram of protein. Find how many grams of protein you should eat if your calorie intake in a day is 2,176 and a gram of protein has 4 calories.

40. **Critical Thinking** The variables M, A, T, and H represent the numbers 2, 4, 6, and 48, but not necessarily in that order. Use the sentence $M \div (A \cdot T) = H$ to determine the value of each variable.

41. **Number Patterns** Use each of the numbers 1, 2, 3, 4, 5, 6, and 7 exactly once and any of the four operations to find an expression that equals 100.

COMPUTER CONNECTION

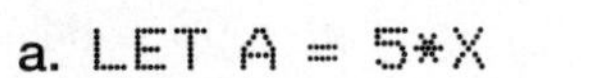

42. **Computer Connection** In the BASIC computer language, LET statements assign values to variables. BASIC uses the symbol * for multiplication and the symbol / for division. Find the value of A for each statement if $X = 4$ and $Y = 12$.

a. `LET A = 5*X` b. `LET A = 2*Y` c. `LET A = X/2`

d. `LET A = Y/3` e. `LET A = Y/X` f. `LET A = X*Y`

2-2 Equations

Objective

Identify and solve equations.

Words to Learn

equation
open sentence
solution
replacement set

In 1917, the United States had just 7,000 miles of concrete roads. The growing popularity of the automobile called for more concrete roads to be built. By 1927, a network of 50,000 miles of roads had been built. Today there are over 4 million miles of paved highways in the United States.

Suppose you wanted to know how many miles of concrete roads were constructed between 1917 and 1927. This situation can be represented by a mathematical sentence called an **equation.** If r stands for the miles of roads built between 1917 and 1927, the equation that solves this problem would be $50{,}000 - 7{,}000 = r$.

An equation that contains a variable is an **open sentence.** When numbers are used to replace the variable, the sentence may be either true or false.

Problem Solving Hint

Use the guess-and-check strategy.

Equation:		$50{,}000 - 7{,}000 = r$
Replace r with 42,000.	→	$50{,}000 - 7{,}000 = 42{,}000$ This sentence is false.
Replace r with 43,000.	→	$50{,}000 - 7{,}000 = 43{,}000$ This sentence is true.

The value for the variable that makes the sentence true is called the **solution** of the equation. In the equation $50{,}000 - 7{,}000 = r$, 43,000 is the solution of the equation. The process of finding the solution is called *solving the equation.*

Some equations do not have the variable alone on one side of the equals sign. For example, let's look at the equation $86 + a = 94$.

Example 1

Suppose the value of a can be selected from the set {12, 10, 8}. Find the solution of $86 + a = 94$.

Try each value from the set to see which is a solution.

The $\stackrel{?}{=}$ symbol means that you question whether the equation is true or false.

Try 12.

$86 + a = 94$

$86 + 12 \stackrel{?}{=} 94$ *Replace a with 12.*

$98 = 94$ *False*

12 is *not* a solution.

Try 10.

$86 + a = 94$

$86 + 10 \stackrel{?}{=} 94$ *Replace a with 10.*

$96 = 94$ *False*

10 is *not* a solution.

Try 8.

$86 + a = 94$

$86 + 8 \stackrel{?}{=} 94$ *Replace a with 8.*

$94 = 94$ *True*

8 is a solution of $86 + a = 94$.

Whenever you are given a set of numbers from which to choose the value of the variable, the set of numbers is called the **replacement set** for the equation.

Example 2 ***Problem Solving***

Smart Shopping Lamar had \$4.93 to buy stamps. What is the greatest number of 29¢ stamps he can buy? Use the replacement set {15, 16, 17, 18}.

Let *s* represent the number of stamps.
Then the equation $0.29s = 4.93$ represents this situation.

Let's estimate first. Round 29¢ to 30¢.

30¢ × 16 = \$4.80

30¢ × 17 = \$5.10

Both are close to \$4.93. Try 16 first.

$0.29s = 4.93$

$0.29 \cdot 16 \stackrel{?}{=} 4.93$ *Replace s with 16.*

$4.64 = 4.93$ *False* 16 is *not* a solution.

Now try 17.

$0.29\ s = 4.93$

$0.29 \cdot 17 \stackrel{?}{=} 4.93$ *Replace s with 17.*

$4.93 = 4.93$ *True*

17 is a solution. Lamar can buy 17 stamps for \$4.93.

Checking for Understanding

Communicating Mathematics

Read and study the lesson to answer each question.

1. **Tell** which symbol is always in an equation.
2. **Tell** how you would solve the equation $15 \cdot 6 = w$.
3. **Write** the definition of a replacement set. How is the solution related to this set?

Guided Practice

Find the solution for each equation from the given replacement set.

4. $x = 14 + 38$, {52, 42, 24}
5. \$1.30 − c = \$0.50, {\$0.80, \$1.80, \$5.30}
6. $3m = 48$, {14, 16, 19}
7. $\frac{792}{w} = 8$, {44, 54, 99}

Solve each equation.

8. $5y = 100$
9. $34 + 16 = x$
10. \$2.99 − \$1.25 = p

Exercises

Independent Practice

11. Solve $y + 45 = 60$ if the replacement set is {10, 15, 20, 25}.
12. Solve $4x = 124$ if the replacement set is {29, 31, 33, 35}.
13. Solve $p = \$5 - \$2.33$ if the replacement set is {\$3.33, \$2.77, \$2.67}.
14. Solve $\frac{456}{y} = 76$ if the replacement set is {4, 6, 7, 9}.

Solve each equation.

15. $\$4.50 + \$1.56 = y$
16. $6 \cdot 34 = z$
17. $\frac{98}{14} = q$
18. $t + 8 = 18$
19. $d - 7 = 24$
20. $2y = 24$
21. $42 = 6m$
22. $\frac{42}{g} = 4 + 3$

Mixed Review

23. Use mental math to find $2 \cdot (50 \cdot 78)$. *(Lesson 1-2)*
24. Estimate 9,728 − 6,284. Use an appropriate strategy. *(Lesson 1-3)*
25. Write $5 \cdot 5 \cdot 8 \cdot 8 \cdot 8$ using exponents. *(Lesson 1-9)*
26. Evaluate $\frac{45 - 9}{3^2 + 3}$. *(Lesson 2-1)*

Problem Solving and Applications

27. **Lawn Care** The Green Grow company determines how much fertilizer to use on a lawn by finding its area. They use the formula $A = lw$ where A is the area, l is the length of the lawn, and w is the width of the lawn. What is the area of the front lawn if its length is 120 feet and its width is 90 feet?

28. **Critical Thinking** Suppose the replacement set for an equation contains the whole numbers less than 10.
    a. Write the replacement set.
    b. Use this set to find the solution of $3x + 5 = 14$.
29. **Advertising** The *Lancaster Gazette* charges \$5.50 per column inch (ci) for its employment ads. The People Line agency for temporary employment wants to put four ads in the Sunday paper. The ads are 6.5 ci, 5 ci, 3.5 ci, and 3 ci long. What would be the total cost for running the four ads?
30. **Weather** The formula $C = \frac{5(F - 32)}{9}$ relates degrees Celsius *(C)* and degrees Fahrenheit *(F)*. Tell how you would use the order of operations to find a solution for C if $F = 32$.

31. **Data Search** Refer to pages 42 and 43. Suppose a person's body contained the exact amounts of all the chemical elements listed in the chart except for potassium and calcium. The person's body has only half as much calcium as listed, but twice as much potassium. Will the person weigh more or less than 150 pounds?

2-3 Solving Subtraction and Addition Equations

Objective

Solve equations using the subtraction and addition property of equality.

Words to Learn

addition property
subtraction property
inverse operation

In 1988, there were 280,000 foster children living in foster homes. By the early 1990s, the number of foster children increased to 370,000. How many more foster children were there in the early 1990s than there were in 1988?

Suppose we let f represent the increase in foster children. The equation $280{,}000 + f = 370{,}000$ can be used to solve this problem. *You will solve this problem in Exercise 3.*

In Lesson 2-2, you learned to find a solution for an equation by using a replacement set. Sometimes the replacement set is not given or it contains too many numbers to try all of them. Two properties of algebra can help us solve equations when no replacement set is given.

DID YOU KNOW

ACTION, an independent organization created in 1971, develops volunteer service opportunities, including the Foster Grandparents Program (FGP) where adults can spend time with children without grandparents.

Addition Property of Equality	**In words:** If you add the same number to each side of an equation, then the two sides remain equal. **Arithmetic**: $5 = 5$, $5 + 2 = 5 + 2$, $7 = 7$ **Algebra**: $a = b$, $a + c = b + c$
Subtraction Property of Equality	**In words:** If you subtract the same number from each side of an equation, then the two sides remain equal. **Arithmetic**: $5 = 5$, $5 - 2 = 5 - 2$, $3 = 3$ **Algebra**: $a = b$, $a - c = b - c$

Let's see how these properties can be used to solve an equation like $x + 5 = 8$.

Mini-Lab

Work with a partner to solve $x + 5 = 8$.

Materials: counters, cups, mats

- Let a cup represent x. Put a cup and 5 counters on one side of the mat and 8 counters on the other side. These two quantities are equal.
  Our goal is to get the cup by itself on one side of the mat.

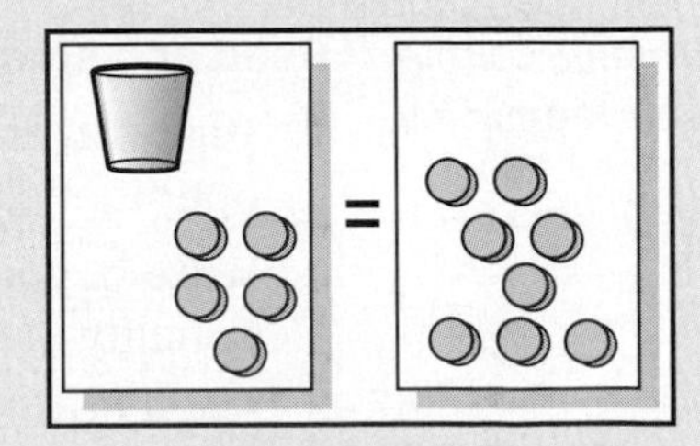

- Take 5 counters away from each side. What you have left is the value of the cup, which is also the value of x.

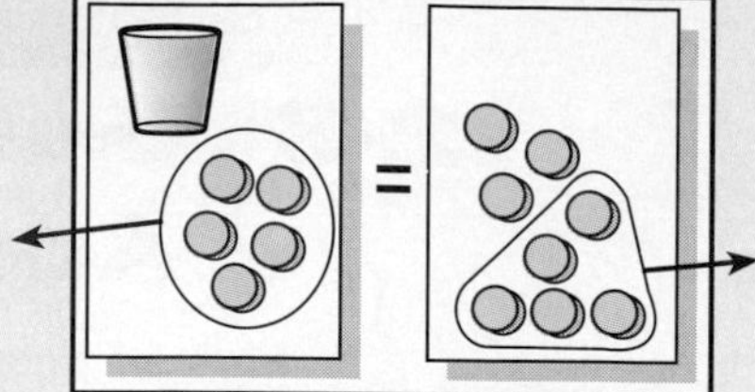

Talk About It

a. What is the value of x (the cup)?

b. Which property lets you take counters away from each side?

c. In the equation $x + 5 = 8$, 5 is added to x. To solve it, you subtract 5. Suppose you were solving the equation $y - 3 = 8$. What operation would you use with 3 to find y?

Mental Math Hint

You can also solve an equation using related facts. You may recall that $5 - 2 = 3$ because $2 + 3 = 5$. So, to solve $x + 35 = 46$, the related fact is $46 - 35 = x$. So, $x = 11$.

To solve an equation in which a number is added or subtracted to the variable, you must use the opposite, or **inverse,** operation. Remember, it is always wise to check your solution.

Examples

1 **Solve $y - 34 = 15$.**

34 is subtracted from y. To solve, add 34 to each side.

Solve the equation.

$$y - 34 = 15$$
$$y - 34 + 34 = 15 + 34$$
$$y = 49$$

Check the solution.

Replace y with 49.

$$y - 34 = 15$$
$$49 - 34 \stackrel{?}{=} 15 \quad \textit{Is the sentence true?}$$
$$15 = 15 \checkmark$$

The solution is 49.

2 **Solve $w + \$4.35 = \$7.23$.**

\$4.35 is added to w. To solve, subtract \$4.35 from each side.

$$w + 4.35 = 7.23$$
$$w + 4.35 - 4.35 = 7.23 - 4.35$$
$$w = 2.88$$

Check:

$$w + \$4.35 = \$7.23$$
$$\$2.88 + \$4.35 \stackrel{?}{=} \$7.23$$
$$\$7.23 = \$7.23 \checkmark$$

The solution is \$2.88.

Checking for Understanding

Communicating Mathematics

Read and study the lesson to answer each question.

1. **Tell** what is meant when we say addition and subtraction are inverse operations.
2. **Write** in your own words how the addition and subtraction properties of equality are used in solving equations.
3. **Show** how to solve the equation presented at the beginning of the lesson. What was the increase in foster children?

4. **Model** the equation $x + 1 = 6$ by using cups and counters. Then solve it.

Guided Practice

Solve each equation. Check your solution.

5. $15 + x = 21$
6. $40 + n = 70$
7. $p - 82 = 142$
8. $y = 34 + 89$
9. $r - 14 = 19$
10. $81 - 56 = z$
11. $24 = x - 5$
12. $m + 1.2 = 1.5$
13. $19 = 13 + s$

Exercises

Independent Practice

Solve each equation. Check your solution.

14. $m + 30 = 110$
15. $173 = x + 83$
16. $2.34 + 1.22 = p$
17. $s - 5.8 = 14.3$
18. $14 + r = 23$
19. $p - 72 = 182$
20. $125 - 52 = q$
21. $0.5 + x = 2.72$
22. $24 = r - 18$
23. $x + 1.4 = 11.2$
24. $w = 312 + 120$
25. $6.01 - 3.12 = t$

Mixed Review

26. Use mental math to find $800 - 356$. *(Lesson 1-2)*
27. How many milligrams are in 4.28 grams of aspirin? *(Lesson 1-6)*
28. How many gallons of juice are in 10 quarts? *(Lesson 1-7)*
29. Solve $x - 32 = 20$ if the replacement set is $\{47, 52, 57, 63\}$. *(Lesson 2-2)*

Problem Solving and Applications

30. **Critical Thinking** Find the solution to the equation $x + 3 = x + 7 - 4$. Explain how you arrived at your solution.
31. **Geometry** Let the measure of angle $A = a$ and the measure of angle $B = b$. Angle A and angle B are complementary if $a + b = 90°$. Suppose angle A has a measure of 30°. Use the equation to find the measure of angle B.

CULTURAL KALEIDOSCOPE

Blandina Cardenas Ramirez

Blandina Cardenas Ramirez could speak and read Spanish and English before she entered school. Education continued to be of primary importance. Although she graduated from college with a degree in journalism, she continued her education after going to work for the government and received her Doctorate in Education.

She was involved with programs in Texas where she developed and directed a center concerned with multicultural education and equal educational opportunities for all children. In 1977, back in Washington, President Jimmy Carter appointed her Commissioner of the Administration for Children, Youth and Families in the Department of Health, Education and Welfare.

2-4 Solving Division and Multiplication Equations

Objective

Solve equations using the division and multiplication property of equality.

Words to Learn

division property
multiplication property

In $d = rt$, d = total distance, r = rate, and t = time.

To work off the number of calories you consume when you eat a Burger King Whopper® with cheese, you would have to ride your bicycle at a rate of 13 miles per hour for a total of 31 miles. How much time would it take you to do this?

The formula $d = rt$ can be used to solve this problem. Substitute 31 for d and 13 for r. The formula becomes the equation $31 = 13t$. *You will solve this equation in Exercise 2.*

In the last lesson, you learned that equations could be solved by using the inverse operation. This holds true for multiplication and division equations too. The Mini-Lab below shows how this works.

Mini-Lab

Work with a partner. Use models to solve $3y = 12$.

Materials: cups, counters, mats

- Let each cup represent $1y$. So $3y$ means 3 cups. Put 3 cups on one mat. On the other mat, put 12 counters.
- Arrange the counters into 3 equal groups to correspond to the 3 cups.

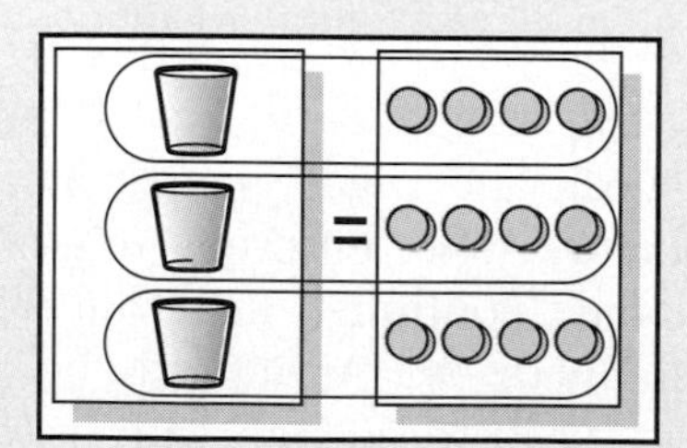

Talk About It

a. What number of counters matches with each cup?

b. If $3y = 12$, what is the value of y?

c. What operation does your model suggest? Explain.

In the Mini-Lab, you solved the multiplication equation $3y = 12$ by separating the counters into three equal groups. You would get the same result if you divided each side of the equation by 3.

This division property is another property of algebra that you can use to solve equations. There is also a multiplication property.

Multiplication Property of Equality	**In words:** If you multiply each side of an equation by the same number, then the two sides remain equal. **Arithmetic:** $6 = 6$, $6 \cdot 2 = 6 \cdot 2$, $12 = 12$ **Algebra:** $a = b$, $ac = bc$
Division Property of Equality	**In words:** If you divide each side of an equation by the same number (except 0), then the two sides remain equal. **Arithmetic:** $6 = 6$, $\frac{6}{2} = \frac{6}{2}$, $3 = 3$ **Algebra:** $a = b$, $\frac{a}{c} = \frac{a}{c}$, $c \neq 0$

Estimation Hint

Any number, except 0, divided by itself is 1. Since the equation equals a number greater than 1, the value of *p* must be greater than 4.

Example 1

Solve $\frac{p}{4} = 2$.

$\frac{p}{4} = 2$

$\frac{p}{4} \cdot 4 = 2 \cdot 4$ *Multiply to undo division by 4.*

$p = 8$

The solution is 8.

Check: $\frac{p}{4} = 2$

$\frac{8}{4} \stackrel{?}{=} 2$

$2 = 2$ ✓

Example 2 *Connection*

Geometry In an equilateral triangle, all sides have the same length. If the sum of the lengths of three sides of an equilateral triangle is 36.435 meters, what is the length of one side of that triangle?

This problem can be solved with the equation $3s = 36.435$.

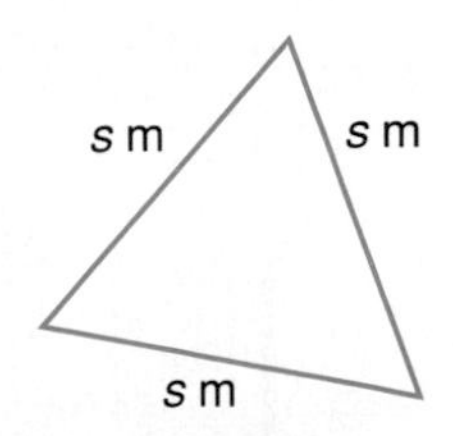

$3s = 36.435$ *Divide to undo multiplication by 3.*

36.435 [÷] 3 [=] 12.145

Each side of the equilateral triangle is 12.145 meters long.

Checking for Understanding

Communicating Mathematics

Read and study the lesson to answer each question.

1. **Tell** how solving equations that involve multiplication or division is similar to solving equations that involve addition or subtraction.
2. **Show** how to use a calculator to solve the equation for the opening problem. Round to the nearest tenth.
3. **Write** the equation shown by the model at the right. Then find the solution.

Guided Practice

Solve each equation. Check your solution.

4. $16x = 48$
5. $24 = 6p$
6. $q = \frac{240}{30}$
7. $\frac{s}{7} = 18$
8. $23 \cdot 4 = w$
9. $9 = \frac{c}{60}$
10. $18.4 = 0.2q$
11. $\frac{p}{0.6} = 3.6$
12. $243 \div 6 = t$

Exercises

Independent Practice

Solve each equation. Check your solution.

13. $3m = 183$
14. $\frac{f}{7} = 56$
15. $14 \cdot 25 = w$
16. $102 = 17p$
17. $1.4t = 3.22$
18. $2.45 \div 5 = q$
19. $8 = \frac{s}{25}$
20. $\frac{d}{0.11} = 5$
21. $\$5.44 = \$0.34b$
22. The product of 8 and a number r is 480. Find the number.

Mixed Review

23. Estimate $5,412 \div 58$ by using compatible numbers. *(Lesson 1-3)*
24. Evaluate $(17 - 8) \div 3 + 5^2$. *(Lesson 2-1)*
25. Solve $y - 28 = 65$. Check your solution. *(Lesson 2-3)*

Problem Solving and Applications

26. **Critical Thinking** If n is greater than 0, which is greater, $\frac{1}{n}$ or $\frac{n}{1}$? Give examples to support your answer.
27. **Space Science** Outer space is becoming a giant junkyard. There are more than 7,000 items trackable by radar floating around the earth. Only about 5% of them are working spacecraft. Let j represent the number of items. Use the equation $0.05j = W$ to find W, the number of working spacecraft, if $j = 7,000$.
28. **Auto Mechanics** Larry thinks his car needs a tune-up. He uses the formula $m = rg$, where m = total miles driven, r = miles per gallon, and g = total gallons of gas used. If he drove 2,450 miles on 112 gallons, how many miles per gallon did his car get?

29. **Journal Entry** How is solving a division or multiplication equation similar to solving an addition or subtration equation? Do you have a method for remembering the steps to solve each type?

Problem-Solving Strategy

2-5 Work Backward

Objective

Solve problems by working backward.

Toshi, Joe, and Al are talking. Along comes Jorge, who places a sticker on each of their foreheads. "What are you doing?" Toshi asks. "I'm playing the spot game," Jorge replies. "At most, two of you have a red spot on your foreheads. If you know that you have a red spot, you can't tell. But if you know that you don't have a red spot, you win." Toshi sees that Al has a red spot, but Joe has a blue spot. Does Toshi have a red spot on his forehead?

Explore *What do you know?*

One or two of the boys has a red spot on his forehead. If Toshi knows that he has a red spot, he can't tell. If Toshi knows that he doesn't have a red spot, he wins.

What are you trying to find?

You are trying to find out if Toshi has a red spot on his forehead.

Plan Work backward by thinking about what the boys are seeing and thinking.

Solve Joe sees at least one red spot, Al's. He's hesitating because there can be two red spots, and he's not sure what color the spot is that's on his forehead. That means that he is not seeing two red spots. Aha! Toshi knows that he doesn't have a red spot. Joe's hesitation is his clue.

Examine Al is quiet. If he doesn't see any red spots, he knows that he has one, but can't tell. If he sees one red spot, he's wondering if there are two.

Joe sees Al's red spot. If he sees a red spot on Toshi's forehead, then he knows that he doesn't have a red spot, since there are at most two.

Example

Sylvia is thinking of a number. She divides her number by 4 and subtracts 6. The result is 3. What number is Sylvia thinking of?

Explore You know that Sylvia divided her number by 4 and then subtracted 6. You know that the result was 3. You want to find Sylvia's number.

Plan Work backward by reversing the operations.

Sylvia's number	Find the number.
↓	↓
Divide by 4.	Start with 3.
↓	↓
Subtract 6.	Add 6.
↓	↓
The result is 3.	Multiply by 4.

Solve Start with 3. Add 6. Multiply by 4.
Sylvia's number is 36.

Examine Work forward in order to check.

$(36 \div 4) - 6 = 3$

This verifies that Sylvia was thinking of 36.

Checking for Understanding

Communicating Mathematics

Read and study the lesson to answer each question.

1. **Tell** where you begin when you are solving a problem by working backward.
2. **Explain** how you solve a problem by working backward.
3. **Show** the order that you would need to use to solve this problem by working backward. Start with 25. Multiply by 4, divide by 10, and add 14. Then double.

Guided Practice

Solve by working backward.

4. Jan says, "I am thinking of a number. If I increase it by 11, and then subtract 10, I get 2." What is Jan's number?
5. You are thinking of a number. You divide it by 20. Then you add 34 and divide the result by 13. Finally you triple the number. The result is 9. What was your number?
6. I have a number. I halve it. I multiply the result by itself. I now have 9,801. What was my original number?

Problem Solving

Practice **Solve using any strategy.**

Strategies

Look for a pattern.
Solve a simpler problem.
Act it out.
Guess and check.
Draw a diagram.
Make a chart.
Work backward.

7. Robert picked out a pair of trousers. When he saw the price, he said, "These are a third as much as I paid for my new suit last month!" The trousers cost \$45.50. How much did his suit cost?

8. Paul is thinking of a number. He divides by 11, takes a third of the quotient, divides the result by 10, and arrives at 13. What is Paul's number?

9. Forty-four students take the bus to Grand Lapere. Each student pays \$1.20. How much did the driver collect?

10. Mrs. Lachinsky's class bought food for the Adopt-a-Family program. They spent \$127.68. The 19 students divided the cost equally. How much did each student contribute?

11. Mr. Jacobs signed an installment contract to pay \$400 a month for 5 years for a new car. The car's actual price is \$15,900. How much extra is Mr. Jacobs paying for spreading the payments over 5 years?

12. **Portfolio Suggestion** Select one of the assignments from this chapter that you found especially challenging. Place it in your portfolio.

2 Assessment: Mid-Chapter Review

Evaluate each expression. *(Lesson 2-1)*

1. $2 + 3 \cdot 5 + 3^3 - 4$

2. $6 \cdot (6 + 3 - 2) \div (12 - \frac{24}{3})$

3. Evaluate $2a + b^2 - 5c + abd$ if $a = 2$, $b = 3$, $c = 0$, and $d = 1$. *(Lesson 2-1)*

4. Solve $x - 51 = 91$ if the replacement set is {40, 101, 141, 142, 151}. *(Lesson 2-2)*

5. **Geometry** The sum of the lengths of the sides of the pentagon at the right is 38 meters. Find the value of x, if x is the missing length of one side. *(Lesson 2-2)*

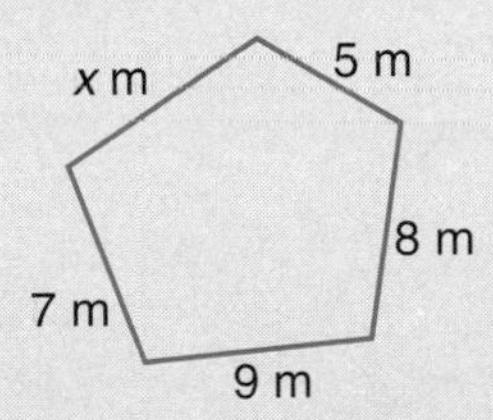

Solve each equation. Check your solution. *(Lessons 2-3 and 2-4)*

6. $45 + x = 140$

7. $y - 8.9 = 10.2$

8. $3w = 45.6$

9. $50 = \frac{p}{3}$

10. Rosa is tall. Add 30 centimeters to her height and take a third of the sum. Now you have half of Alice's height. Alice is 140 centimeters tall. How tall is Rosa? *(Lesson 2-5)*

Cooperative Learning

2-6A Writing Expressions and Equations

A Preview of Lesson 2-6

Objective

Write algebraic expressions and equations from verbal problems.

Materials

red, green, and yellow counters
sheets of paper

The History Club at school is assembling bags of canned goods to give to needy families in the neighborhood. They have cans of soup, vegetables, and fruit to distribute. Joey asked, "How do we know which cans to put in each bag?" Mr. Boyarski, their sponsor, told them he wrote clues about the contents of each bag on the outside of the bag.

Joey picked up the first bag and was puzzled. The clues on the bag read as follows.

> There are cans of soup, vegetables, and fruit in this bag. There is one more can of soup than vegetables. There are 9 cans in all. There are 2 cans of vegetables.

Try this!

Problem Solving Hint

Use the work backward strategy. Begin with the total and work backward to find the answer.

Let's help Joey out by using a model.

- The bag says there are 9 cans in all. Let's draw 9 circles on a piece of paper.
- Let each color of counter represent a different type of can. Suppose red is soup, green is vegetable, and yellow is fruit.
- What clues do we have?
  The bag says there are 2 vegetables. Put green counters on 2 of the circles.

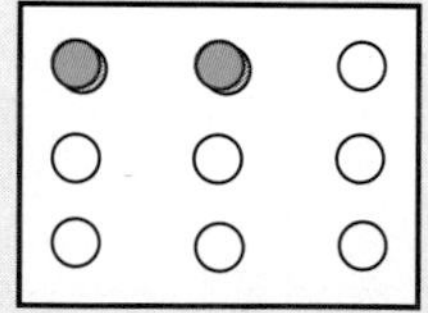

- What other clues do we have?
  It says that there is one more soup than vegetable. So, if there are 2 vegetables, there must be 3 soups. Put red counters on 3 of the circles.

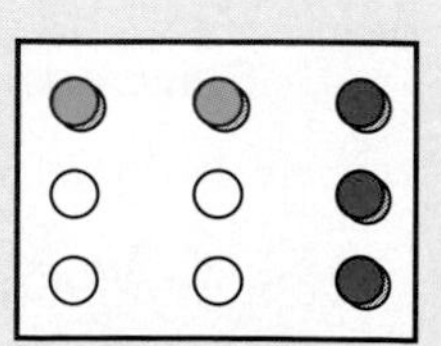

- The rest of the circles represent cans of fruit since there are 9 cans in all. *How many yellow counters will you use?*

Your Turn

Here are some of the clues on bags to be filled by other students. Use a model to figure out how many of each type of can is to be placed in each bag. Each bag contains at least one of each type of can.

Bag A There are 6 vegetables. There are 14 cans in this bag. There is one less can of fruit than there are of vegetables.

Bag B There are 8 cans in this bag. Two are fruits. There are 2 more soups than fruits.

Bag C There are 2 vegetables. There are 3 times as many fruits as vegetables. There are 10 cans in all.

Bag D There are 19 cans in this bag, 3 of which are fruit. There are 4 times as many soups as fruits.

What do you think?

1. How many fruit cans went in the bag Joey was preparing?
2. When making a model, how do you know how many circles to draw for each bag?
3. Which clues do you look for first when figuring out each bag's contents?

Extension

4. Look at Joey's bag. We can write an expression to name each clue. Study the following equations.

   There are cans of soup, vegetables, and fruit in this bag.

   Let S = soup, V = vegetables, and F = fruit.

There is one more soup than vegetable.	$S = V + 1$
There are 9 cans in all.	$S + V + F = 9$
There are 2 vegetables.	$V = 2$

   Do your values for S, V, and F make these sentences true? Explain your answer.

5. Write equations for each step for bags A, B, C, and D listed in the *Your Turn* section. Check your solutions with your equations.

2-6 Writing Expressions and Equations

Objective

Write algebraic expressions and equations from verbal phrases and sentences.

The United Nations was established on October 24, 1945, shortly after World War II. Today the UN is composed of 159 nations, representing dozens of languages. When in official session, only six languages—Arabic, Chinese, English, French, Russian, and Spanish—are offered by translators. The delegates wear earphones to listen to the translation they choose while skilled interpreters translate the words as they are spoken.

In mathematics, we often act as interpreters, translating words and ideas into mathematical expressions and equations. There are many words and phrases that suggest arithmetic operations. Here are some examples of translated phrases. Any variable can be used to represent a number.

Verbal Phrase	Algebraic Expression
8 more than a number	$n + 8$
a number decreased by 10	$x - 10$
the sum of twice a number and 4	$2y + 4$
a number separated into 5 groups	$\frac{n}{5}$

"When am I ever going to use this?"

Today, the total number of animal collections in the world exceeds 1,000.

Zoo design and architecture must meet two conflicting needs, those of the animals and those of the visitors.

A college degree and mathematical skills in measurement, design, drafting, and estimation are essential for this nonstandard form of architecture.

For more information, contact the zoological society in your area.

Examples

Write each sentence or phrase as an algebraic expression.

1 One year, the number of zoos in the United States increased by 3. Let z represent the number of zoos before the increase.
The word *increased* means the U.S. added to the number of zoos.
The algebraic expression is $z + 3$.

2 three less than four times the number of eggs required for a chocolate cake

Let c represent the number of eggs required for a chocolate cake.
Four times means multiply by 4. → $4 \cdot c$ or $4c$
Three less means subtract 3. → $4c - 3$

The algebraic expression is $4c - 3$.

Verbal sentences may be translated into equations. The equation can often be used to solve a problem.

Verbal sentence	Algebraic Equation
Three is five more than a number.	$3 = n + 5$
Four times a number is one hundred.	$4n = 100$

Example 3

Jesse has $5 more than twice the amount Rosa has. Jesse has $15. Write an equation to represent this problem.

Let r = the amount Rosa has.

twice the amount Rosa has	$\rightarrow$	$2r$
$5 more than that	$\rightarrow$	$2r + 5$
Jesse's amount, $15, equals this.	$\rightarrow$	$15 = 2r + 5$

Checking for Understanding

Communicating Mathematics

Read and study the lesson to answer each question.

1. **Write** two different verbal phrases that could be represented by the algebraic expression $y + 7$.
2. **Tell** what the expression $t - 5$ represents if t is the scheduled blastoff time of a space shuttle flight.
3. **Tell** what word usually occurs in verbal sentences that can be written as equations.

Guided Practice

Write each phrase or sentence as an algebraic expression or equation.

4. 17 more than p
5. the quotient of x and 3
6. the product of 6 and r
7. 4 less than m
8. three more than twice the total number of turtles t
9. Six less m is 25.
10. The difference between 8 and the quotient of a and 4 is 19.
11. How many apples are there in each bag if you separate 37 apples into each of five bags and there are 2 apples left?

Exercises

Independent Practice

Write each phrase or sentence as an algebraic expression or equation.

12. the sum of p and 4
13. 18 less n
14. the sum of 9 and five times y
15. the difference between 24 and twice a number
16. five dollars more than Leroy made
17. her salary plus a $200 bonus

Write each phrase or sentence as an algebraic expression or equation.

18. three times as many hits as the Pirates
19. Two less than the quotient of 24 and x is 2.
20. the product of 5 and x decreased by their sum
21. six more than two times the number of pizzas ordered yesterday
22. The product of a number and 5 is 45.
23. Steve inventoried the paper left in the bookstore. There were 14 less than 3 cases of packages of paper. Steve reported there were 46 packages of paper.
24. Six less than the number of students divided into three groups is 47.

Mixed Review

25. Use mental math to find $64 + 28 + 6$. *(Lesson 1-2)*
26. How many liters are in 34.7 milliliters of water? *(Lesson 1-6)*
27. How many ounces are in $3\frac{1}{2}$ pounds? *(Lesson 1-7)*
28. Solve $t - 9 = 23$. Check your solution. *(Lesson 2-2)*
29. Solve $35 = 5m$. Check your solution. *(Lesson 2-4)*

Problem Solving and Applications

30. **Statistics** Use the information below to answer each question.

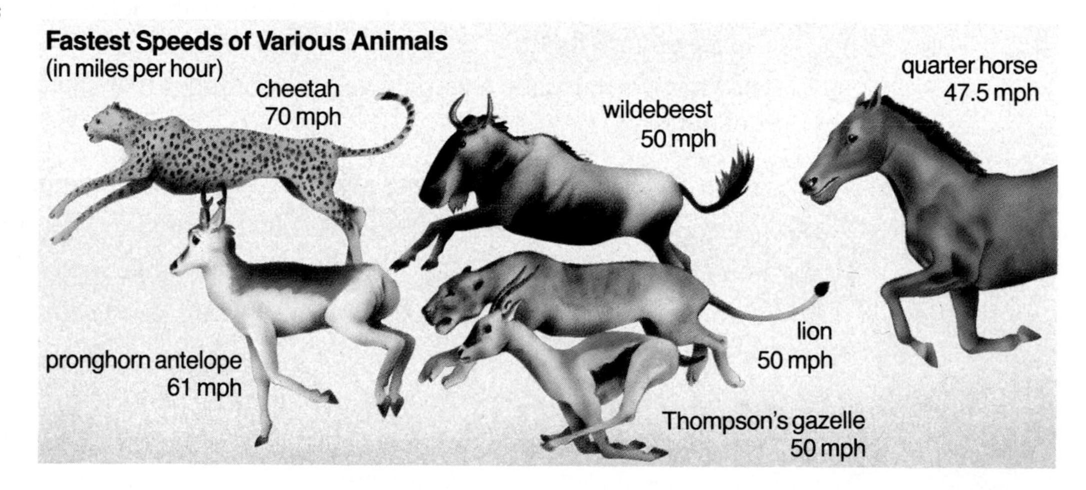

a. If s represents the fastest speed of a cheetah, which animal's speed can be represented by $s - 9$?
b. If x represents the speed of the lion, which other animals have speeds that can be represented by x?
c. If w represents the speed of the wildebeest, write an expression that represents the speed of the pronghorn antelope.

31. **Critical Thinking** How would you write an algebraic equation to represent *5 times the sum of twice x and 4 less 5 is 2 more than 15?*

32. **Journal Entry** Write what the sentence $B = P + 12$ means if B represents Bob's test score and P represents Paulo's test score.

Cooperative Learning

2-7A Function Input and Output

A Preview of Lesson 2-7

Objective

Use function machines to find output from a given input and then to find input from a given output.

Today you can put flour, yeast, eggs, and other ingredients into a machine and 4 hours later, you get a loaf of bread. The machine performs all the functions of mixing, kneading, rising, and baking to complete the process of bread making.

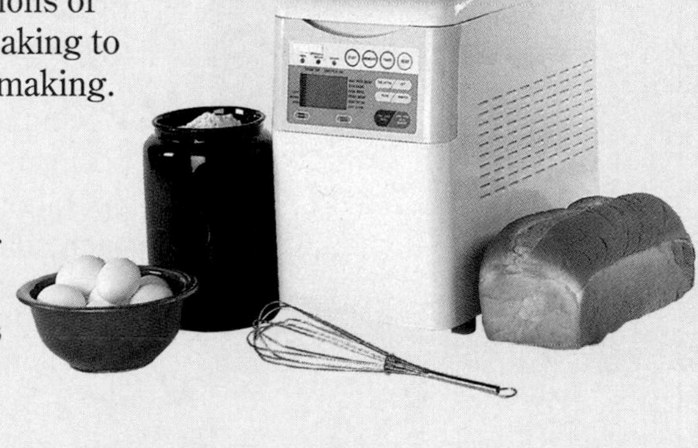

In mathematics, we have **functions** that work like machines. You **input** a number and the **output** is the function of the number. If the number is represented by x, the function of the number is represented by $f(x)$. *This is read "f of x."*

Activity One

Look at the function machine below. You put a number into the top of the machine and at each stage an operation is performed. Then the result moves onto the next stage.

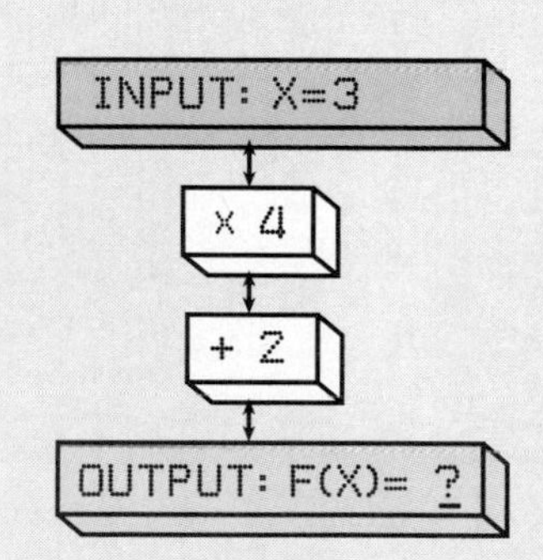

- What number is being used as input?
- What is the first operation?
  What is the result after the first operation is performed?
- Move this result to the next stage.
- What is the second operation?
  What is the result after this operation?
- What is your final output?

What do you think?

1. Write the complete process performed by the function machine from input to output.
2. Each stage of this function machine represents a mathematical expression. If the input is represented by x, write an algebraic equation that summarizes what the function machine does.
3. Suppose you were given the output and had to determine the input. How do you think you would go about finding the input?

Activity Two

One day a comet passed very close to Earth and all the function machines went berserk. Everything started working backwards. The conveyor belt reversed. The machines started sucking up the outputs and spewing out inputs from the tops of the machines.

- Look at the function machine at the right. What was the output when the machine worked normally?
- When the machine went berserk, it performed the inverse of each operation. Describe what happens in each chamber of the berserk machine.
- What number was the input for the machine to get an output of 24?

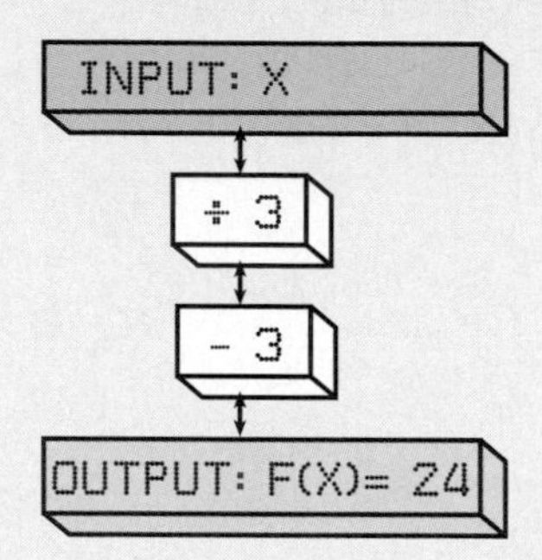

What do you think?

4. Write about the process the berserk machine followed after the comet passed Earth. Compare this to the process the machine followed when it worked properly.
5. This machine could be the model of an equation.
   a. Write the equation represented if the input is x and the output is 24.
   b. What is the solution to the equation?

Find the missing input or output for each function machine.

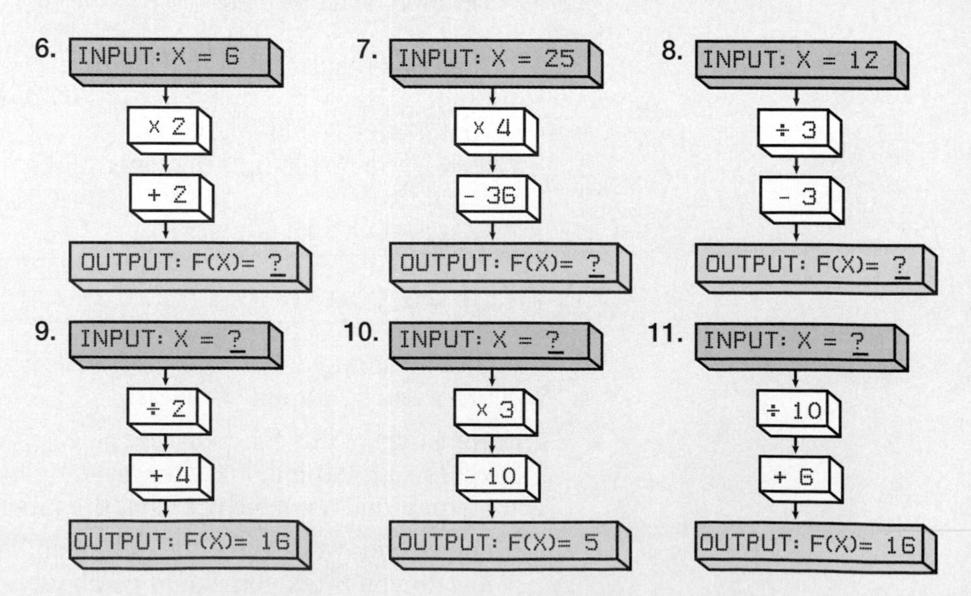

2-7 Solving Two-Step Equations

Objective

Solve two-step equations.

TEEN SCENE

The custom of sending valentine cards and gifts began in the 16th century. Many believe the valentine was the first of all greeting cards.

When you wrap a present, you put the paper on first and then you add the ribbon. To open a present you usually do the opposite—take the ribbon off first and then the paper.

You can solve an equation in a similar way. Look at the equation $4 + 2x = 8$. In this equation, the x is first multiplied by 2 and then 4 is added.

To solve the equation, you work backwards using the reverse of the order of operations. That is, you subtract 4 from each side of the equation and then divide each side by 2.

Mini-Lab

Work with a partner. Solve $4 + 2x = 8$ using models.
Materials: cups, counters, mats.

- Put 2 cups and 4 counters on one side of the mat. Put 8 counters on the other side of the mat. The two sides of the mat represent equal quantities.

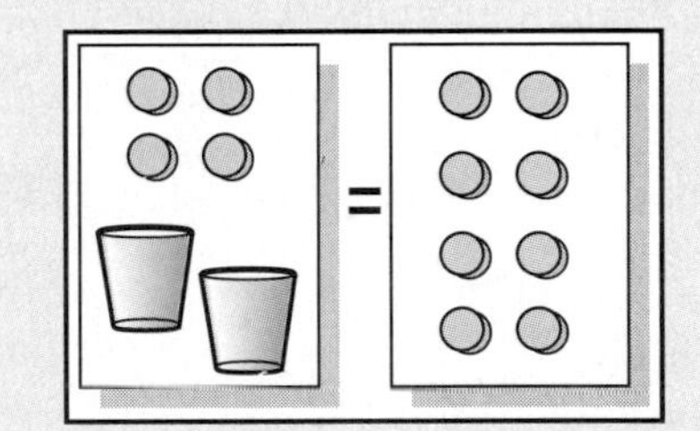

- Take 4 counters from each side.

- Separate the remaining counters into 2 groups to correspond to the 2 cups.

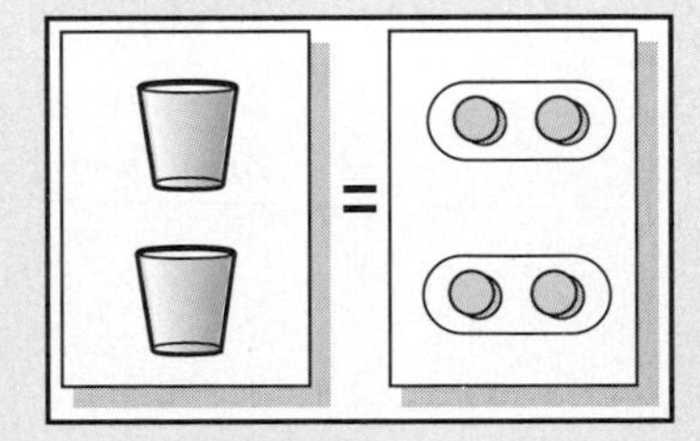

Talk About It

a. What operation is represented by removing 4 counters from each mat?

b. What operation is represented by separating the remaining counters into groups?

c. How many counters correspond to each cup?

d. What is the solution to $4 + 2x = 8$?

Many equations can be solved algebraically in a similar manner.

Examples

1 **Solve $3k - 4 = 17$. Check your solution.**

$$\begin{aligned} 3k - 4 &= 17 \\ 3k - 4 + 4 &= 17 + 4 && \textit{Add to undo subtraction.} \\ 3k &= 21 \\ \frac{3k}{3} &= \frac{21}{3} && \textit{Divide to undo multiplication.} \\ k &= 7 \end{aligned}$$

Check:

$$\begin{aligned} 3k - 4 &= 17 \\ 3(7) - 4 &\stackrel{?}{=} 17 && \textit{Replace k with 7.} \\ 21 - 4 &\stackrel{?}{=} 17 && \textit{Multiply before subtracting.} \\ 17 &= 17 \checkmark && \textit{It checks. The solution is 7.} \end{aligned}$$

2 **Solve $\frac{d}{3.5} + 4.9 = 12.65$.**

Problem Solving Hint

Use the work backward strategy to solve an equation.

You can use your calculator to solve this equation.

12.65 [−] 4.9 [=] [×] 3.5 [=] 27.125

Subtract to undo addition. *Multiply to undo division.*

So, $d = 27.125$.

Check: $\frac{d}{3.5} + 4.9 \stackrel{?}{=} 12.65$ if $d = 27.125$

27.125 [÷] 3.5 [+] 4.9 [=] 12.65 *It checks.*

Checking for Understanding

Communicating Mathematics

Read and study the lesson to answer each question.

1. **Tell** what step you would do first in solving $3e - 5 = 25$.
2. **Write** a sentence to explain how the work-backward strategy is used in solving two-step equations.
3. **Show** how to use the model to solve the equation $3x + 5 = 14$. How can you check your solution?

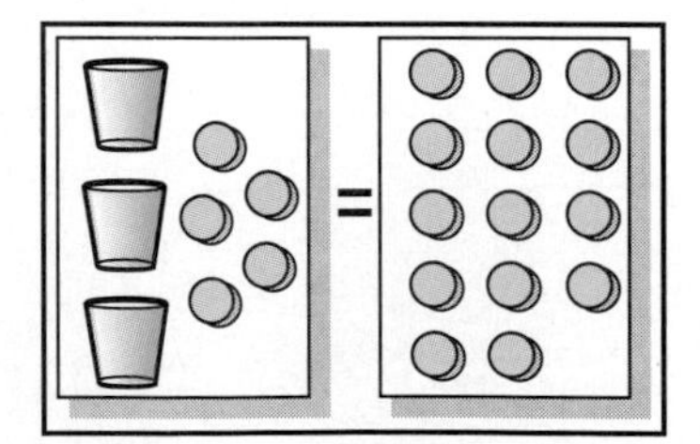

Guided Practice

Tell the first step you would do to solve each equation. Then solve the equation. Check your solution.

4. $7c + 3 = 17$
5. $8 + 3d = 17$
6. $11e - 5.3 = 5.7$
7. $\frac{b}{9} + 0.8 = 1.3$
8. $\frac{f}{4} - 3 = 4$
9. $7 + \frac{g}{2} = 8$

Exercises

Independent Practice

Solve each equation. Check your solution.

10. $3x + 4 = 7$
11. $4r - 9 = 7$
12. $1.2 + 6t = 3.6$
13. $5 + 14s = 33$
14. $\frac{p}{5} - 1 = 13$
15. $\frac{f}{8} + 3 = 27$

Solve each equation. Check your solution.

16. $\frac{d}{12} - 4 = 8$
17. $2e + 5.0 = 6.0$
18. $3f - 15 = 6$
19. $4g + 1.7 = 2.3$
20. $35 = 18h - 1$
21. $0.42 = 0.17 + 0.5k$

Write each sentence as an equation. Then solve the equation.

22. Twice a number less five is 46.
23. The quotient of a number and six, decreased by seven is twelve.
24. Suki gets \$3 more than half of Tommie's allowance. Suki gets \$5 a week. What is Tommie's allowance?

Mixed Review

25. Estimate $3.21 + 2.95 + 3.12 + 2.89$ by clustering. *(Lesson 1-3)*
26. Evaluate $3^3 \cdot 4^2 \cdot 1^5$. *(Lesson 1-9)*
27. Solve $p + 84 = 196$. Check your solution. *(Lesson 2-3)*
28. Write an algebraic expression to represent *eight less than four times the number of people in Africa.* *(Lesson 2-6)*

Problem Solving and Applications

29. **Sales** Chet Morris is a salesperson for the Beechwood Bottling Company. On Monday, he sold 625 cases of soft drinks. One case contains 24 cans. How many cans of soft drinks did he sell?
    a. Write an equation involving division that represents this situation.
    b. Solve the equation to find the total number of cans sold.
30. **Critical Thinking** Use what you have learned about two-step equations to solve the equation $\frac{(t + 3)}{6} + 5 = 7$. How could you show that your solution is correct?
31. **Keyboarding** Lawanda began a class to learn to type on her computer. She began at 10 words per minute. Each week she increased her speed at a steady rate. At the end of six weeks, she was typing 40 words per minute. Write an equation to find the number of words she increased per week. Then solve it.

32. **Medicine** Dr. Poloma recommended that Anna take six tablets on the first day and then three tablets each day until the prescription ran out. The prescription contained 21 tablets. Use the equation $3d + 6 = 21$ to find how many days she will be taking pills after the first day.
33. **Make Up a Problem** that can be solved by using the equation $2x + 3 = 15$.

Problem-Solving Strategy

2-8 Use an Equation

Objective

Solve problems by using an equation.

Les receives \$2 from his grandfather for each test he passes. If he fails a test, he pays his grandfather \$3. Les has made \$5 on his tests so far this year. He has passed four times as many tests as he has failed. How many tests has Les passed? How many has he failed?

Explore What do you know?
Les has \$5.
He gets \$2 for passed tests.
He pays back \$3 for failed tests.
He passed four times as many as he failed.

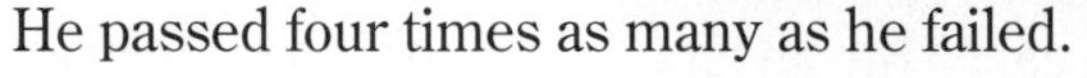

What are you trying to find out?
We want to know how many tests Les has passed and failed.

Plan Let's use an equation.
Let f = the number of tests failed.
Then $4f$ = the number of tests passed.

Subtracting 8f − 3f is like subtracting 3 fruits from 8 fruits. The answer is 5 fruits. So, 8f − 3f is 5f.

$$\underbrace{\text{money earned}}_{\$2 \cdot 4f} - \underbrace{\text{money paid back}}_{\$3 \cdot f} = \underbrace{\text{money he has}}_{\$5}$$

That is, $8f - 3f = 5$ or $5f = 5$.

Solve

$5f = 5$

$\frac{5f}{5} = \frac{5}{5}$ *Divide each side by 5.*

$f = 1$

So, Les failed one test and passed four tests.

Examine Passing four tests at \$2 per test is \$8. Failing one test at \$3 is the money that Les had to pay back. Since \$8 − \$3 = \$5, the answer checks.

Example

Barbara is driving to her sister's house 425 miles away. She drove 253 miles the first day. The next morning, she drove 109 miles. How far must she drive that afternoon to reach her sister's house later that day?

Use an equation. The sum of the distances Barbara drives will equal the total distance to her sister's house.

$$253 + 109 + d = 425 \qquad \textit{Let } d = \textit{the afternoon miles.}$$
$$362 + d = 425$$
$$d = 63$$

Barbara must drive 63 more miles to reach her sister's house.

Checking for Understanding

Communicating Mathematics

Read and study the lesson to answer each question.

1. **Explain** how you represent an unknown quantity in a problem in order to use it in an equation.
2. **Show** how you would change the problem about the tests if Les paid back $2 for failing a test.
3. **Explain** how the equation would change if Les failed more tests than he passed. Use the $2 rate from Exercise 2 for passing and failing.

Guided Practice

Solve by using an equation.

4. Paul bought 2 pairs of skates for $52.22 each and 2 pairs of goggles at $9.94 each. He handed the cashier $125. How much change did he get?
5. Three times a number minus twice the number plus one is 6. What is the number?
6. Adult tickets for a play cost $5.50. Student tickets cost $5. Twice as many students as adults attended. How many tickets of each kind were sold if the total sales were $1,953?

Problem Solving

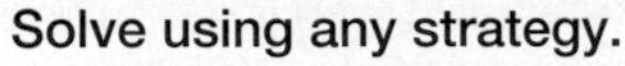

Practice

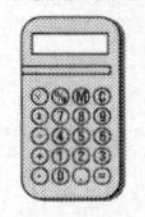

Solve using any strategy.

7. The Zoological Gardens in Philadelphia is home to 15,000 species, many of them rare, in natural settings. If you could walk through the zoo for 8 hours a day and see a different species every minute, how many days would it take you to see all of the species?

Strategies

Look for a pattern.
Solve a simpler problem.
Act it out.
Guess and check.
Draw a diagram.
Make a chart.
Work backward.

8. Morita's car gets 29.5 miles per gallon. How far can the car travel on 17 gallons of fuel?

9. Seventeen times a number plus 55 is 4 more than the product of 51 and 5. What is the number?

10. Melissa earns $6.25 per hour for regular hours worked and double time for overtime hours. Last week, she worked 3 hours more than her regular 35-hour week. How much did she earn?

11. The New Jersey shore is a 127-mile stretch of beaches, wildlife preserves, amusement parks, and boardwalks. There are more than 60 resorts available for fun-loving families. About how many resorts is that per mile of beach?

12. Roy worked 610 hours at Grant Beach last summer for $3.90 per hour. Lyla earned $5.50 per hour in the business across the street. How much more would Roy have earned if his rate had been $5.50 per hour?

13. **Data Search** Refer to page 666.
In 1993, how did the median home price in each region compare to the median U.S. home price?

14. Annette is thinking of a number. She triples the number and divides the result by 3. Then she subtracts 2 from the quotient. "Now I have 10," she says. What is Annette's number?

15. **Mathematics and Science** Read the following paragraph.

> Gold has been used to make jewelry for thousands of years. However, the Incas used gold for everyday objects, like nails, combs, dishes, and cups. Gold was also prized by the ancient Egyptians, who recognized its beauty. Gold does not rust, tarnish, or stain. Pure gold is a soft metal that can be molded in your hands. Because of its softness, jewelers mix copper with it, to give it added strength. The mixture is measured in *karats*. Twenty-four karat gold is considered pure. Twelve-karat gold is $\frac{12}{24}$ or $\frac{1}{2}$ pure gold.

a. If there are 45 grams of gold on one pan of a balance scale, and 5.2 grams on the other, how much gold is needed to balance the scale?

b. If a piece of jewelry is $\frac{2}{3}$ gold, how would you describe it in karats?

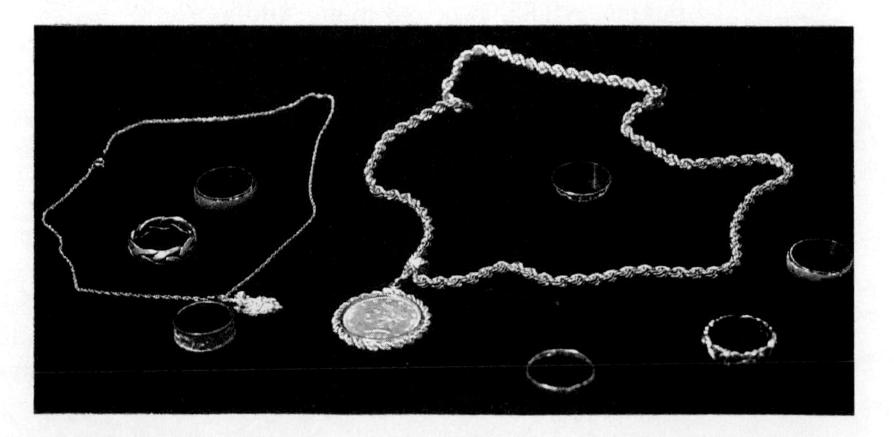

Geometry Connection

2-9 Perimeter and Area

Objective

Find the perimeters and areas of rectangles, squares, and parallelograms.

Words to Learn

rectangle
perimeter
square
area
parallelogram
base
height
altitude

Less than a month after the John Hancock Tower was completed in 1972, dozens of its windows started popping out. The windows were replaced with 400-pound sections of half-inch tempered glass to solve the problem. However, the windows still pop out occasionally.

The exterior of the John Hancock Tower in Boston, Massachusetts contains 10,334 huge 4 foot by 11 foot **rectangles** of glass. In order to calculate the sealant needed to go around each pane of glass, you need to find the **perimeter** of each rectangle.

The perimeter of a geometric figure is the sum of the measures of all of its sides. To find the perimeter of the rectangular piece of glass you can add up the lengths of all the sides or you can use an equation. The equation for the perimeter of any rectangle is $P = 2\ell + 2w$, where ℓ represents the length and w represents the width.

Example 1 *Problem Solving*

Construction **Find the perimeter of each pane of glass in the John Hancock Tower.**

4 ft

11 ft

The length of each pane is 11 feet and the width is 4 feet.

$P = 2\ell + 2w$
$P = 2(11) + 2(4)$ *Replace ℓ with 11 and w with 4.*
$P = 22 + 8$ *Multiply before adding.*
$P = 30$

The perimeter of each pane is 30 feet.

A **square** is a special rectangle in which the lengths of all the sides are equal. The values for ℓ and w in the perimeter equation are the same number. For this reason, the perimeter equation for a square is often written as $P = 4s$, where s is the length of a side. In the square shown at the right, the perimeter is 4(5) or 20 centimeters.

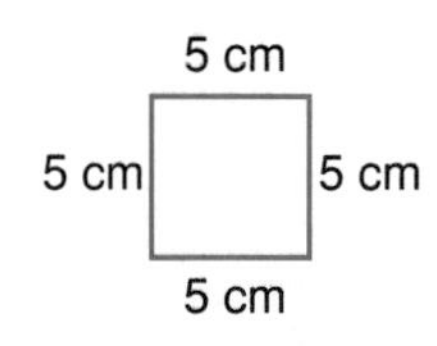

Squares and rectangles are special types of **parallelograms.** Each pair of opposite sides of a parallelogram are parallel and have the same length. To find the perimeter of a parallelogram, you add the lengths of the sides. The parallelogram at the right has a perimeter of $2(7) + 2(6)$ or 26 feet.

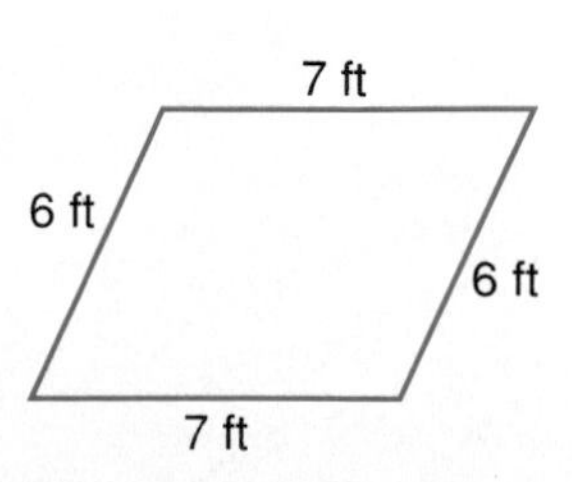

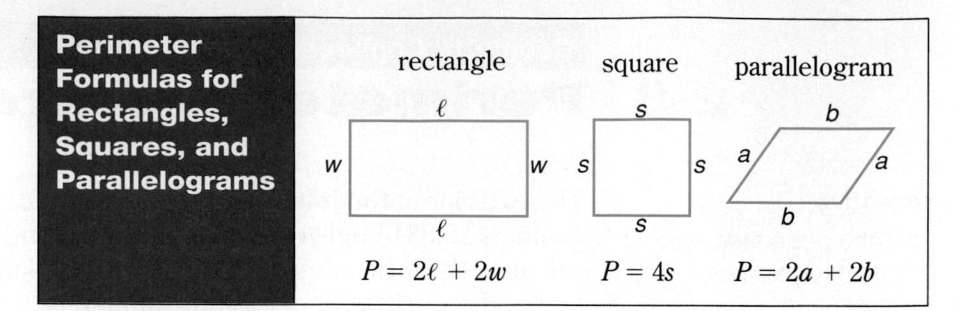

In addition to the perimeter, we often solve problems by using the **area** of a geometric figure. The area is the measure of the surface enclosed by the figure. The area of any rectangle can be found by multiplying the width and length.

Examples

Find the area of each rectangle.

2

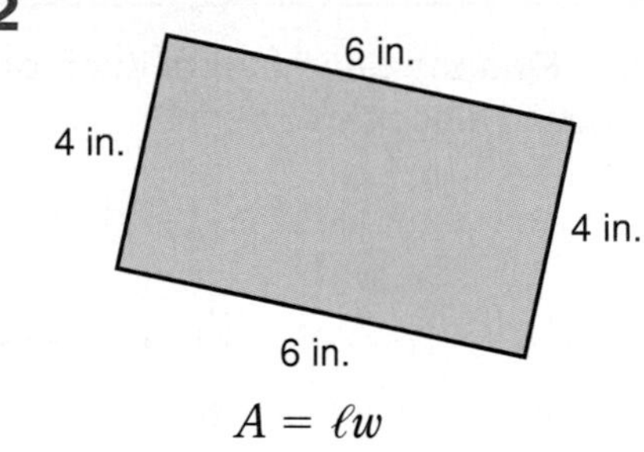

$A = \ell w$
$A = 6 \cdot 4$
$A = 24$

The area is 24 square inches.

3

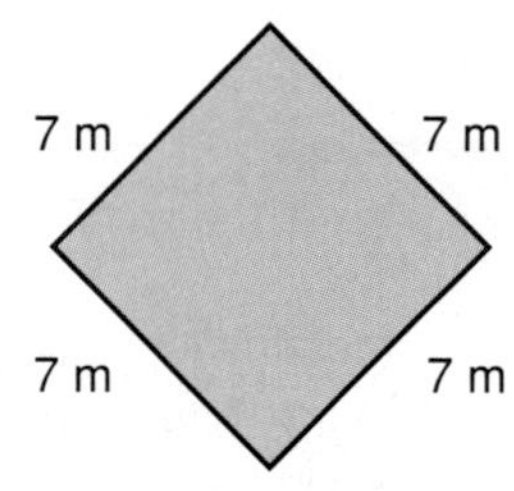

$A = s \cdot s$ or s^2
$A = 7 \cdot 7$
$A = 49$

The area is 49 square meters.

The formula for the area of a parallelogram is *not* the product of the sides. However, it is related to the formula for the area of a rectangle.

Mini-Lab

Work with a partner.

Materials: graph paper, scissors

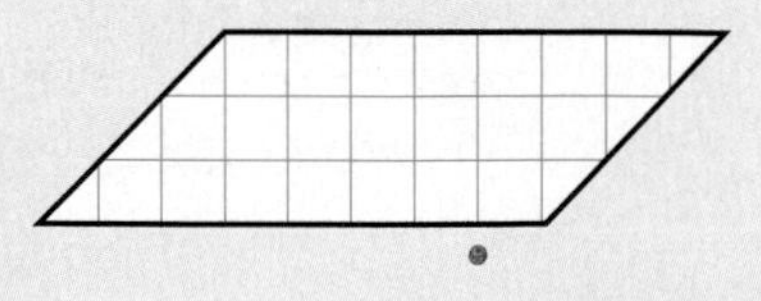

- Copy the parallelogram at the right on a piece of graph paper. Cut out the parallelogram.

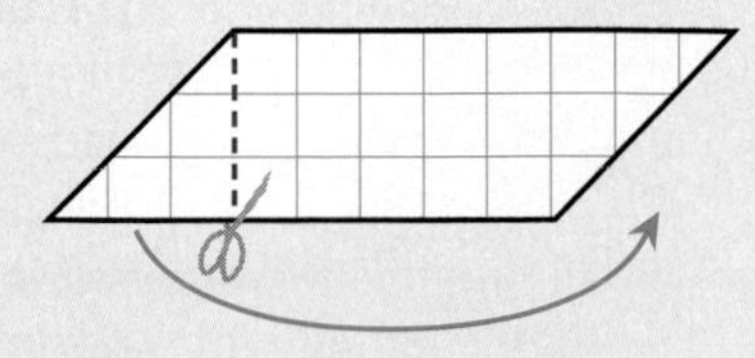

- Make a cut along the dashed line. Move the parts so that they form a rectangle.

Talk About It

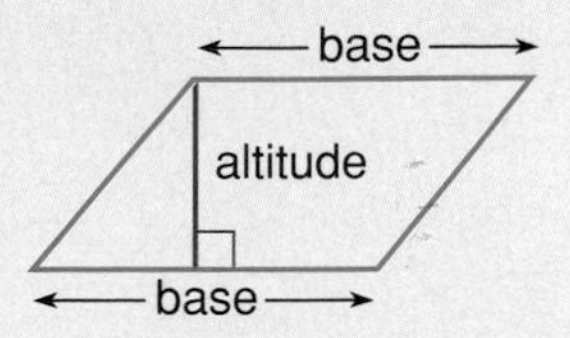

a. What is the area of the rectangle you formed?

b. Study the parts labeled on the figure at the right. Compare the length and width of the rectangle with the **base** and **altitude** of the parallelogram.

c. The length of the altitude is the **height** of a parallelogram. If b represents the length of the base and h represents the height, write a formula for the area of a parallelogram.

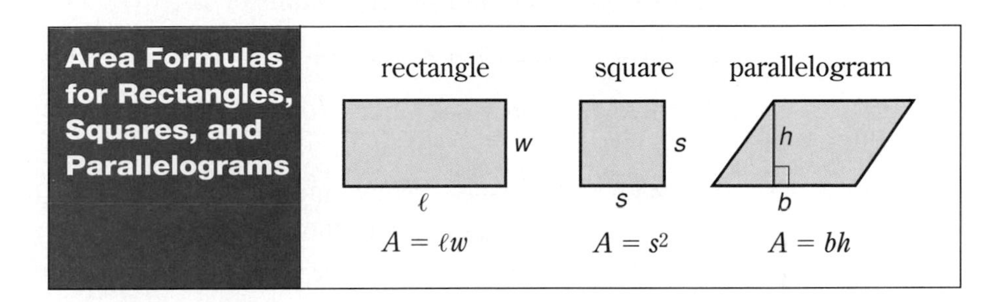

Checking for Understanding

Communicating Mathematics

Read and study the lesson to answer each question.

1. **Draw** and label a rectangle that has a length of $2\frac{1}{2}$ inches and a width of 1 inch.
2. **Tell** how to find the perimeter and area of the rectangle you drew in Exercise 1.
3. **Write** in your own words the difference between perimeter and area.

Guided Practice

Find the perimeter and area of each figure.

4. 6.5 m, 6.5 m, 6.5 m, 6.5 m

5. 6 yd, 5 yd, 5 yd, 6 yd

6.

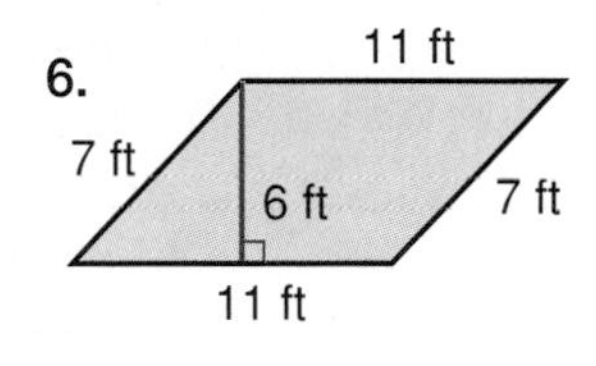

Exercises

Independent Practice

Find the perimeter and area of each figure.

7.

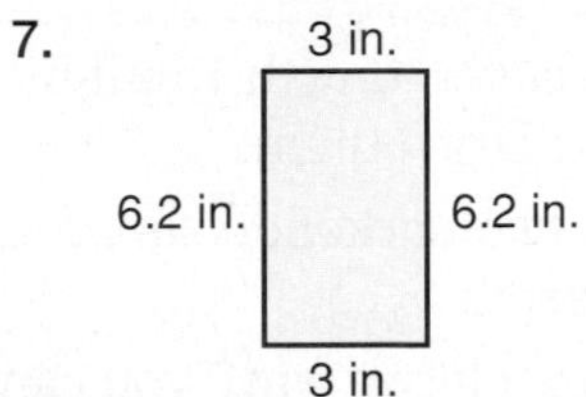

8.

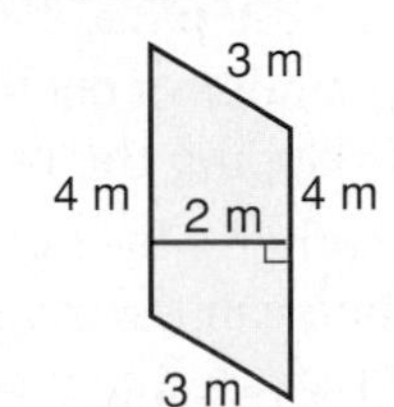

9.

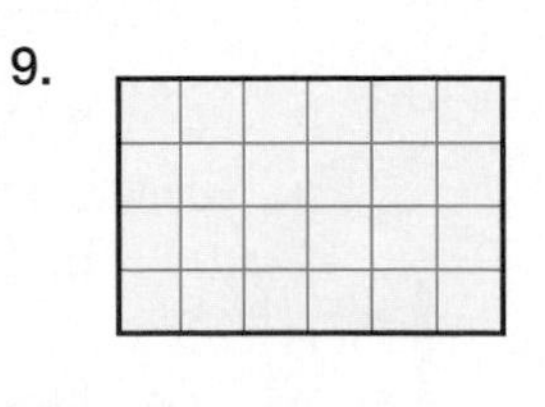

10. 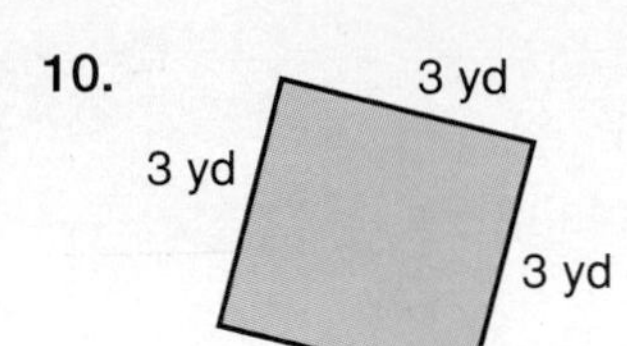

11. 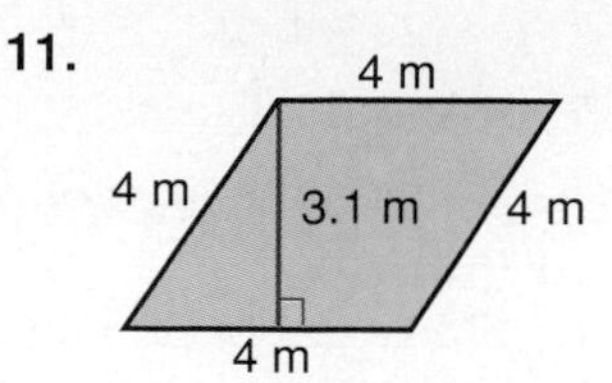

12. 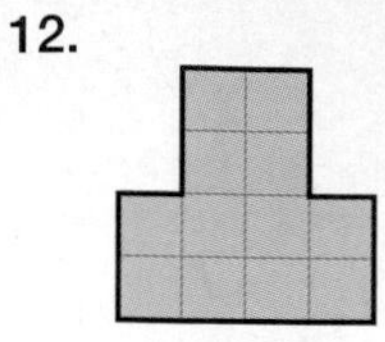

13. Find the perimeter of a rectangle whose length is twice its width. Its width is 8 centimeters.

14. Use an equation to find the base of a parallelogram whose height is 7 yards and whose area is 91 square yards.

Mixed Review

15. How many pounds are in 6.5 tons of salt? *(Lesson 1-7)*

16. Solve $\frac{r}{5} = 20$. Check your solution. *(Lesson 2-4)*

17. Solve $12 + 5d = 72$. Check your solution. *(Lesson 2-7)*

Problem Solving and Applications

18. **Home Maintenance** How much will it cost to tile the floor of a room if the tiles cost $0.79 per square foot and the room is a rectangle 20 feet by 15 feet?

19. **Critical Thinking** Find the area of the shaded part in each rectangle.

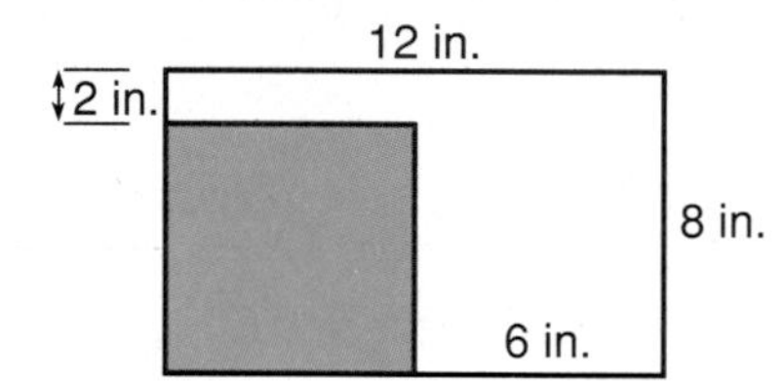

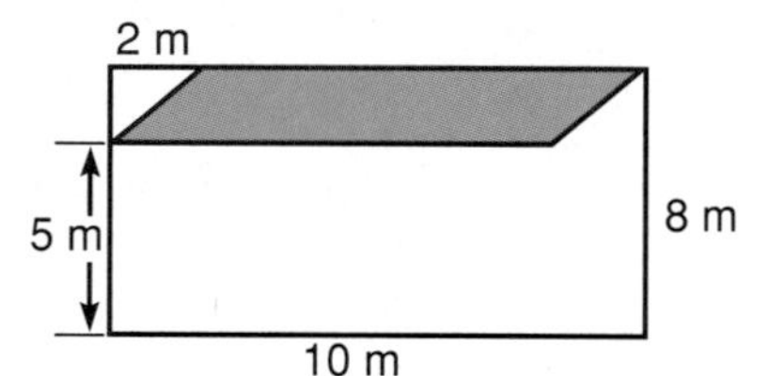

20. **Shipping** The Minto Company ships candy to mail order customers. To ensure each package's safety, they place a strip of tape around the box and seal the edges. If the average box is 1 foot by 8 inches, approximately how much tape is used on each box?

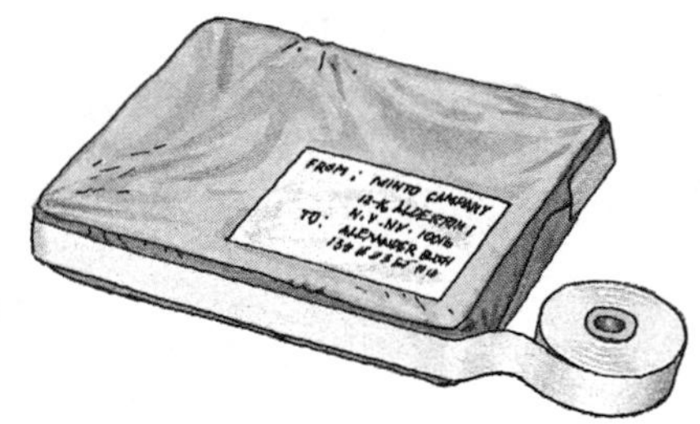

21. **Make a Drawing** Copy each of the 4-by-4 square grids below.

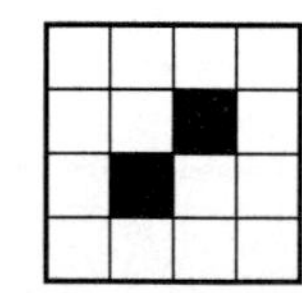 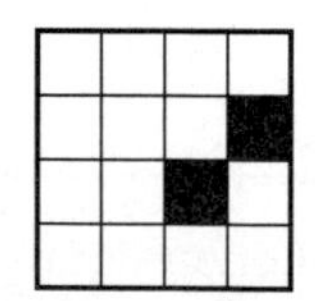 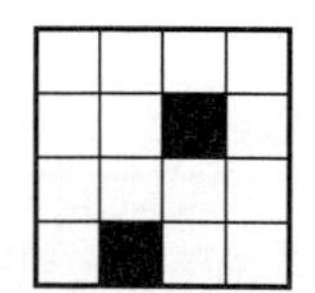 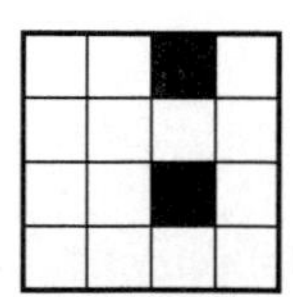

a. Can the remainder of each grid be covered with 1 unit-by-2 unit rectangles without overlapping the rectangles? Draw them.

b. Draw some other 4-by-4 grids with two darkened squares. Look for a pattern. When can the grid be covered?

22. **Journal Entry** Write a few sentences to tell two ways you can use perimeter and area at home.

2-10 Solving Inequalities

Objective

Identify and solve inequalities.

Words to Learn

inequality

ZIP stands for the Zoning Improvement Plan, which the Postal Service implemented in 1963.

Why is it so important to use the correct ZIP code when you mail a letter? The numbers indicate where each letter is to be sent. Mail is sorted according to the information the ZIP code provides.

The first automatic sorter takes only those letters with ZIPs that begin with 0. All others go to the next sorter. That sorter takes only the 1s. For numbers greater than 1, the letters go to the next sorter, and so on.

If ℓ represents the first number of the ZIP, then $\ell > 1$ represents all letters that go to the next sorter. The sentence $\ell > 1$ is called an **inequality.** Inequalities are sentences that contain symbols like > or <. You may remember that > is read *is greater than* and < is read *is less than.*

Words:	6 is greater than 4. 5 is less than 10.	**Arithmetic:**	$6 > 4$ $5 < 10$
Words:	$3x + 7$ is greater than 10. $\frac{y}{4} - 2$ is less than 4.	**Algebra:**	$3x + 7 > 10$ $\frac{y}{4} - 2 < 4$

Equations often have one solution. Unlike equations, inequalities may have many solutions. The solution can be written as a set of numbers.

Mini-Lab

Work with a partner. Solve $\frac{x}{2} - 1 > 2$.

- Draw a number line like the one shown below.

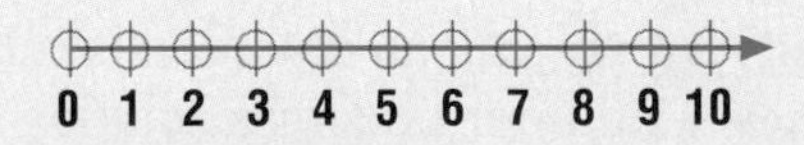

- Use the whole numbers 0 through 10 as your replacement set. Color in each circle that represents a solution for $\frac{x}{2} - 1 > 2$.

LOOKBACK

You can review solutions and replacement sets on page 48.

Talk About It

a. What does the number line suggest about the solution set?

b. Are there other solutions not shown on the graph? Explain.

c. Solve the equation $\frac{x}{2} - 1 = 2$. How does its solution relate to the solution of $\frac{x}{2} - 1 > 2$?

As we learned in solving equations, the guess-and-check method is not always the quickest way to solve an equation. Likewise, it is usually not the best way to solve an inequality. You can use your knowledge of solving equations to solve an inequality.

Example

Solve $4x + 17 < 50$. Show the solution on a number line.

Solve the related equation, $4x + 17 = 50$. Use your calculator. Remember to subtract to undo addition, then divide to undo multiplication.

$4x + 17 = 50$ ➡ 50 [−] 17 [=] [÷] 4 [=] 8.25

The solution will either be numbers greater than 8.25 or numbers less than 8.25. Let's test a number to see which is correct.

numbers greater than 8.25

Try 9. $4(9) + 17 < 50$
$36 + 17 < 50$
$53 < 50$ *false*

numbers less than 8.25

Try 6. $4(6) + 17 < 50$
$24 + 17 < 50$
$41 < 50$ *true*

The numbers less than 8.25 make up the solution set. So, $x < 8.25$.

To show the solution, draw an empty circle at 8.25. Then draw a large arrow to indicate the numbers that are solutions.

The arrow shows that the numbers continue. To show numbers greater than 8.25, the arrow would go in the opposite direction.

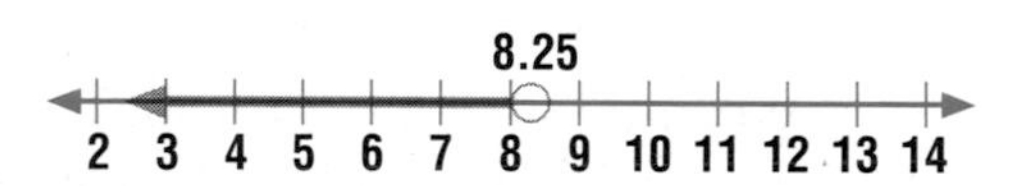

Try other numbers in the set to check your solution.

Checking for Understanding

Communicating Mathematics

Read and study the lesson to anwer each question.

1. **Write** some of the numbers in the solution set if the solution set is all numbers greater than 5.
2. **Tell** some other numbers in the solution of the inequality in Example 1.
3. **Tell** how solving an equation is related to solving an inequality.
4. **Show** a number line for a solution set that is all numbers less than 8.

Guided Practice

Solve each inequality. Show the solution on a number line.

5. $6 + t > 11$
6. $a - 4 > 3$
7. $12 + d < 21$
8. $3y < 15$
9. $2m - 3 > 7$
10. $\frac{p}{3} + 8 < 11$

Exercises

Independent Practice

Solve each inequality. Show the solution on a number line.

11. $x - 3 < 14$
12. $y + 4 > 9$
13. $b - 9 > 6$

14. $2e < 16$

15. $3g < 27$

16. $\frac{s}{8} > 5$

17. $2a - 5 > 9$

18. $5 + 3b < 11$

19. $5c - 8 > 7$

20. $9d + 4 < 22$

21. $\frac{a}{4} + 5 > 6$

22. $\frac{c}{5} - 8 < 2$

Write an inequality for each sentence. Then solve the inequality.

23. Five times a number is greater than sixty.
24. A number less three is less than fourteen.
25. The sum of four times a number and five is greater than thirteen.

Mixed Review

26. Use mental math to find 698 + 471. *(Lesson 1-2)*
27. Write an equation to represent *20 more than the number of pencils that is divided into three groups is 25.* *(Lesson 2-6)*
28. Solve $\frac{n}{2} + 31 = 45$. Check your solution. *(Lesson 2-7)*
29. **Geometry** Find the perimeter and area of a rectangle that is 2 centimeters wide and 5 centimeters long. *(Lesson 2-9)*

Problem Solving and Applications

30. **Critical Thinking** Find the least whole number that is in the solution set of $3x - 5 > 12$.

31. **Portfolio Suggestion** Select an item from this chapter that you feel shows your best work and place it in your portfolio. Explain why you selected it.

32. **Football** The graph shows which universities have had the most Heisman Trophy winners as of 1991.

   a. If H represents the number of Heisman trophies, name all universities that are in the solution set for $H > 3$.
   b. How many winners could USC add and still have fewer than Notre Dame?

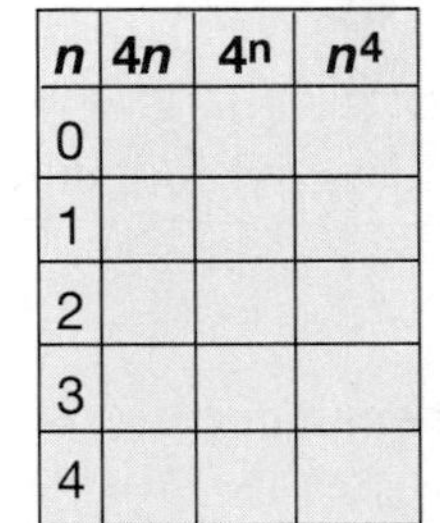

n	$4n$	4^n	n^4
0			
1			
2			
3			
4			

33. **Calculator** Copy the table at the left.
   a. Use your calculator to complete the table.
   b. For which values of n is each statement true?
   $4^n = n^4$ $\quad$ $4^n < n^4$ $\quad$ $4^n > n^4$

34. **Measurement** Measure your height and the height of a classmate.
   a. Write an inequality that compares these heights.
   b. Suppose each of you grew four inches in the next year. Write an inequality that would compare your new heights.

Chapter

2 Study Guide and Review

Communicating Mathematics

State whether each sentence is true or false. If false, replace the underlined word or number to make a true sentence.

1. $8 + 12$ is an <u>algebraic</u> expression.
2. The sentence $x > 2$ is called an <u>equation</u>.
3. In the expression $t - 9$, <u>t</u> is a variable.
4. The <u>first</u> step in solving $4b + 3 = 27$ is to divide each side of the equation by 4.
5. 6 is the <u>solution</u> of $x + 4 = 10$.
6. The <u>area</u> of a rectangle is found by multiplying the width and length.
7. Write a sentence that explains the difference between an algebraic expression and an equation.

Self Assessment

Objectives and Examples	***Review Exercises***
Upon completing this chapter, you should be able to:	*Use these exercises to review and prepare for the chapter test.*
• evaluate expressions by using the order of operations *(Lesson 2-1)* Evaluate $3x + y$ if $x = 6$ and $y = 4$. $3x + y = 3(6) + 4$ $= 18 + 4$ $= 22$	**Evaluate each expression if $a = 3$, $b = 8$, $c = 5$, and $d = 2$.** 8. $cd^2 + (4b - 3)$ 9. $abd - 6a$ 10. $(2c + b) \div (ad)$ 11. $4a + 2c - d$
• identify and solve equations *(Lesson 2-2)* Find the solution of $8 + m = 14$. Let $m = 6$. $8 + 6 \stackrel{?}{=} 14$ 6 is the solution. $14 = 14$ ✓	**Find the solution from the replacement set for each equation.** 12. $4z = 520$, {120, 130, 90} 13. $x + 12 = 68$, {56, 46, 62} 14. $84 - b = 47$, {47, 60, 37}
• solve equations using the subtraction and addition properties of equality *(Lesson 2-3)* Solve $d - 24 = 18$. $d - 24 + 24 = 18 + 24$ $d = 42$	**Solve each equation. Check your solution.** 15. $n + 50 = 80$ 16. $s - 12 = 61$ 17. $145 = a + 32$ 18. $7.2 = 3.6 + t$ 19. $r - 13 = 29$ 20. $1.2 = p - 2.7$

Objectives and Examples

- solve equations using the division and multiplication properties of equality *(Lesson 2-4)*

Solve $3f = 42$.

$\frac{3f}{3} = \frac{42}{3}$ *Divide to undo multipliclation.*

$f = 14$

Review Exercises

Solve each equation. Check your solution.

21. $2.3m = 11.5$ **22.** $\frac{s}{6} = 45$

23. $\frac{r}{0.4} = 6.2$ **24.** $4 = \frac{d}{26}$

25. $72 = 12i$ **26.** $8.68 = 0.62j$

- write algebraic expressions and equations *(Lesson 2-6)*

Write an algebraic expression to represent *the sum of twice a number and 5.*

The algebraic expression is $2n + 5$.

Write each phrase as an algebraic expression or equation.

27. eight less a number is 31

28. the product of 7 and x

29. the sum of 8 and six times u

- solve two-step equations *(Lesson 2-7)*

Solve $4d + 6 = 34$.

$4d = 28$ *Subtract 6 from each side.*

$d = 7$ *Divide each side by 4.*

Solve each equation. Check your solution.

30. $6f - 17 = 37$ **31.** $\frac{h}{5} + 20 = 31$

32. $18 = \frac{t}{2} - 6$ **33.** $\frac{m}{8} - 12 = 14$

34. $2k + 15 = 83$ **35.** $30 = 9 + 3c$

- find the area and perimeter of rectangles, squares, and parallelograms *(Lesson 2-9)*

	Perimeter formula	Area formula
rectangle	$P = 2\ell + 2w$	$A = \ell w$
square	$P = 4s$	$A = s^2$
parallelogram	$P = 2a + 2b$	$A = bh$

Find the perimeter and area of each figure.

36. **37.**

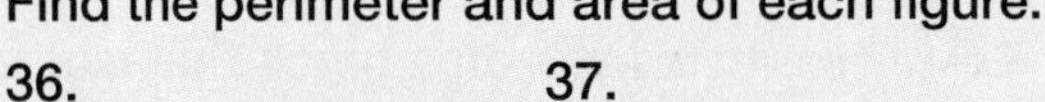

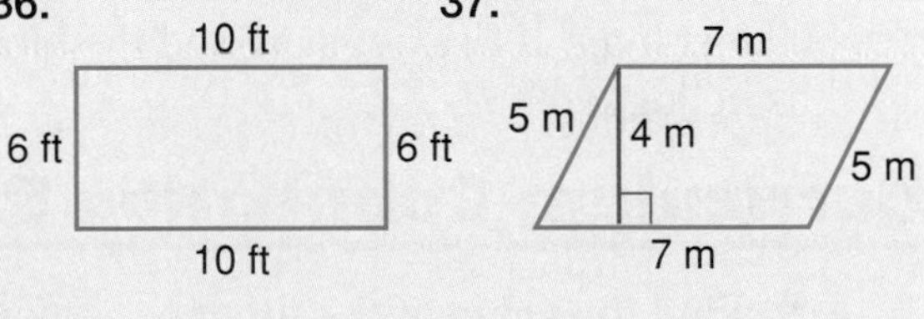

- identify and solve inequalities *(Lesson 2-10)*

Solve $3x - 5 > 10$.

Solve $3x - 5 = 10$. $\rightarrow$ $x = 5$

Test a number greater than 5.

Try 6. $3(6) - 5 > 10$

$18 - 5 > 10$

$13 > 10$ *true*

The solution to $3x - 5 > 10$ is all numbers greater than 5.

Solve each inequality. Show the solution on a number line.

38. $2r - 4 < 10$

39. $4g > 24$

40. $6w + 5 < 29$

41. $\frac{a}{3} + 7 > 8$

Applications and Problem Solving

42. A cup, two saucers, and three bowls cost $38. Two bowls cost as much as one saucer and three cups. If a saucer costs $3, how much is each bowl and cup? *(Lesson 2-5)*

43. A substitute taxi driver earns time and a half for overtime driving. What is her regular rate and her overtime rate if she earns $88 for a 10-hour shift, 2 hours of which were overtime? *(Lesson 2-8)*

44. **Shopping** The Spreindale Produce Market charges $1.99 a pound for red grapes. Mrs. Miller paid $5.97 for a bag of red grapes. How many pounds of grapes did she buy? *(Lesson 2-4)*

45. Kimiko has four less than twice the number of necklaces that Vicki has. Kimiko has six necklaces. How many necklaces does Vicki have? *(Lesson 2-7)*

46. **Exercise** Eric lives on a block that is a square. Each side of the block is 500 feet long. If Eric jogs around the block four times, how many feet has he jogged? *(Lesson 2-9)*

Curriculum Connection Projects

- **Art** Find the value of the ten most expensive paintings in the world. Find the current exchange rates of the U.S. dollar. Write an equation that converts the value of each painting into the currency of the artist's country.
- **Consumer Awareness** Find your state's tax on a gallon of gasoline. Write an equation to find the cost to fill a car's tank as the cost per gallon changes, plus the gas tax.
- **Meteorology** A *degree day* is the difference between 64°F and the day's average temperature. Write an expression to find the degree days, if you are given the average temperatures. Find yesterday's degree day.

Read More About It

D'Ignazio, Fred. *Invent Your Own Computer Games.*
Kaufmann, John. *Fly It.*
Mitsumasa, Anno. *Socrates and the Three Little Pigs.*

Chapter

2 Test

1. Solve $8x = 112$ if the replacement set is $\{20, 16, 14, 11\}$.

Evaluate each expression.

2. $3(6 - 4)^2 + 13 \cdot 5$
3. $[6^2 - (24 \div 8)] - 20$
4. $\frac{47 + 3}{5} + 48 \div 6$

Solve each equation. Check your solution.

5. $k - 20 = 55$
6. $0.4m = 9.6$
7. $\frac{x}{2} + 39 = 49$
8. $\$3.50 = \$1.90 + b$
9. $\frac{n}{1.5} = 6$
10. $11 = \frac{a}{4} - 9$
11. Find the value of d if $d = 6.34 - 2.96$.
12. Find the value of g if $30 + g = 61$.
13. A rectangle is 14 meters wide. Find its perimeter if its length is half its width.

Write each phrase or sentence as an algebraic expression or equation.

14. the quotient of 6 and y, decreased by 12
15. Eight more than five times the number of cats is 48.
16. Find the value of the expression $\frac{xy}{z} + x^2z$ if $x = 3$, $y = 6$, and $z = 9$.

Find the perimeter and area of each figure.

17.

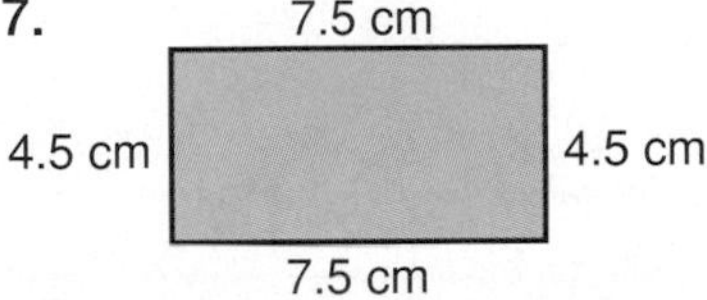

18.

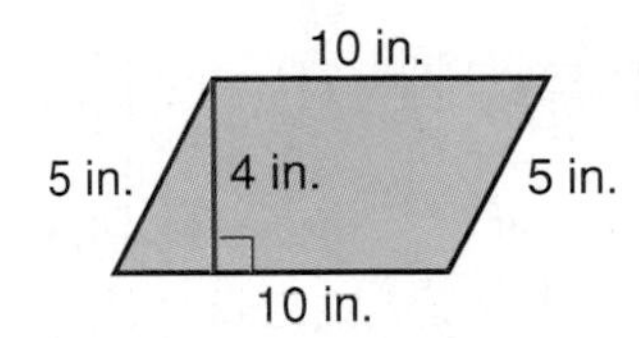

19. Five less than three times the number of hamburgers is 31. How many hamburgers are there?

Solve each inequality. Show the solution on a number line.

20. $x + 4 > 16$
21. $4c - 5 < 7$
22. $\frac{z}{3} + 8 < 10$
23. $3f > 45$
24. Rita has \$3,193 in her savings account. She deposits \$75. How much will she have in her account after the deposit?
25. Phil bought a CD at a one-third-off sale. The cost to Phil was \$9.90. What was the original price of the CD?

Bonus Solve $4[(7 - 3) + 3x] = \frac{156}{3}$

Chapter Test

Chapter 3

Integers

Spotlight on the Telephone

Have You Ever Wondered ...

- How many telephone calls are made by all the people in the United States each day?
- How much time people in the United States spend on the telephone with people in other countries?

Annual Telephone Calls (in millions of minutes)

	Calls to the U.S.	Calls from the U.S.
Canada	1,420	1,460
Great Britain	483	627
Germany	201	494
France	137	201
Italy	70	169
Dominican Republic	42	165
Mexico	374	780
Philippines	51	163
South Korea	57	164
Japan	268	314

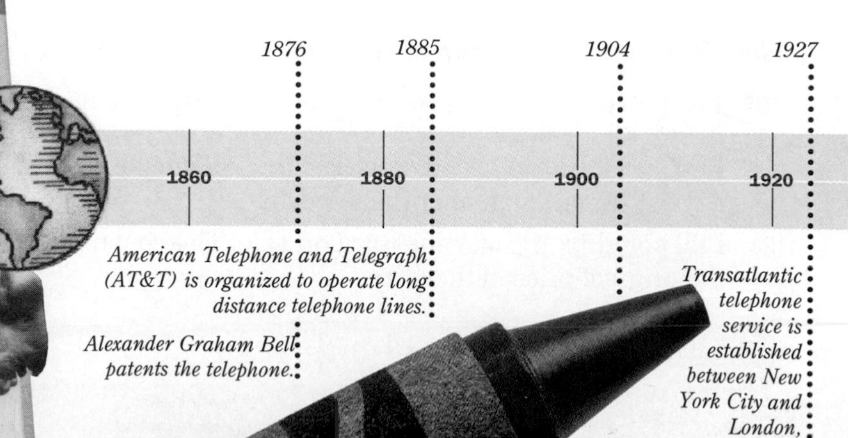

Looking Ahead

In this chapter, you will see how mathematics can be used to answer questions about telephone usage in the United States. The major objectives of the chapter are to:

- compare and order integers.
- add, subtract, multiply, and divide integers.
- solve problems by classifying information.
- solve equations involving integers.

Chapter Project

The Telephone

Work in a group.

1. Keep a log of how many telephone calls you make and receive each day for one week. Also keep track of the number of minutes you spend talking.
2. Make a creative chart or graph showing your results.
3. Investigate the cost of a 3-minute call to each of the countries in the international table.

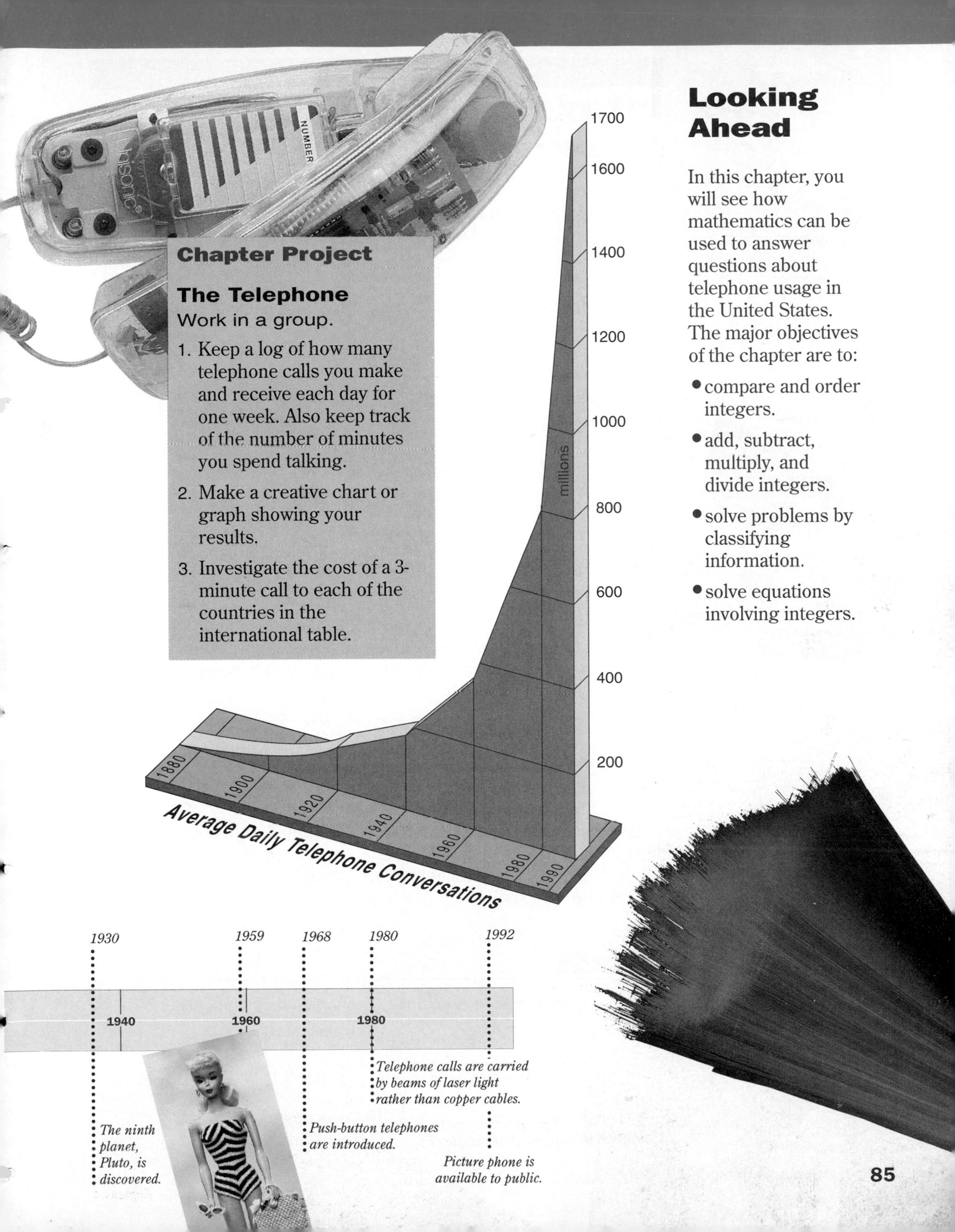

3-1 Integers and Absolute Value

Objective

Graph integers on a number line and find absolute value.

Words to Learn

integer
graph
coordinate
absolute value

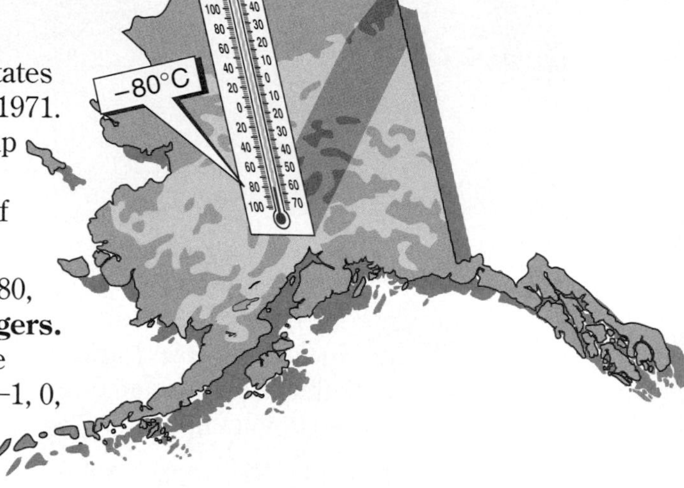

The coldest temperature recorded in the United States occurred on January 23, 1971. The Prospect Creek Camp weather station in Alaska reported a temperature of −80° F.

Negative numbers, like −80, are part of the set of **integers.** The set of integers can be written as {..., −4, −3, −2, −1, 0, +1, +2, +3, +4, ...}.

The ..., called ellipses, mean that the set continues without end, following the same pattern.

The positive integers are often written without the + sign. So, +2 and 2 are the same.

Integers can be graphed on a number line. On the number line, 0 is considered the starting point with the positive numbers to the right and the negative numbers to the left. Zero is neither negative nor positive.

−4 is read "negative 4." +2 or 2 is read "positive 2."

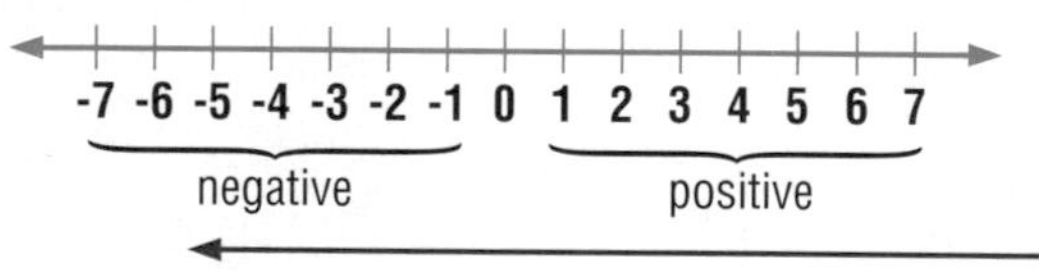

The arrows, like ..., show that the numbers continue without end.

To **graph** an integer, you locate the number and draw a dot at that point on the line. Letters are sometimes used to name points on a number line. The integer that corresponds to the letter is called the **coordinate** of the point.

Example 1

Name the coordinates of each point graphed on the number line.

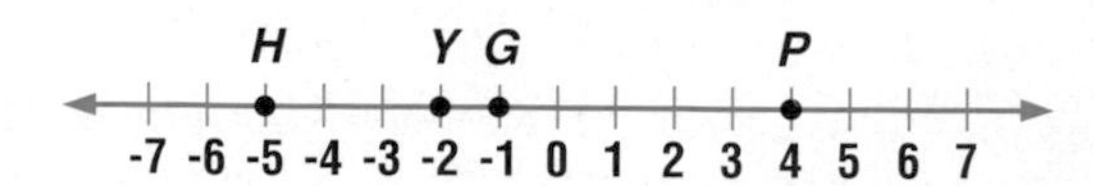

The coordinate of H is −5.
The coordinate of Y is −2.
The coordinate of P is 4.
The coordinate of G is −1.

Example 2

Graph points *M, A, T,* and *H* on a number line if *M* has coordinate 8, *A* has coordinate –3, *T* has coordinate –7, and *H* has coordinate 3.

Find each number. Draw a dot there. Write the letter above the dot.

The absolute value of 0 is 0.

The **absolute value** of a number is the distance it is from 0 on the number line. We write *the absolute value of –3* as $|-3|$. Let's find $|-3|$.

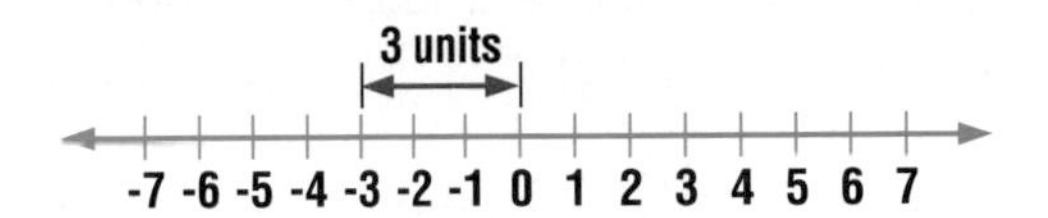

–3 is 3 units from 0. So, $|-3| = 3$.

Example 3

Find $|6|$ and $|-6|$.

First locate each number on a number line.
Then count how many units each number is from 0.

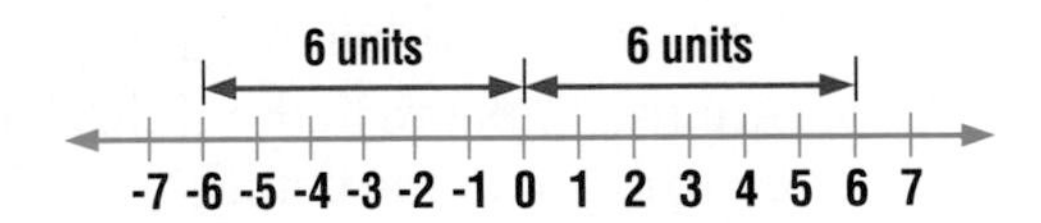

6 is 6 units from 0. So, $|6| = 6$.
–6 is 6 units from 0. So, $|-6| = 6$.

Checking for Understanding

Communicating Mathematics

Read and study the lesson to answer each question.

1. **Tell** some situations in the real world where negative integers are used.
2. **Write** how you would graph –12.
3. **Write** an equation using absolute value that describes the value shown on the number line.

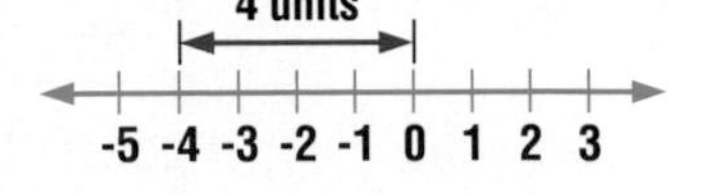

4. **Draw** a number line. Graph two points on it so that the coordinates of both points have an absolute value of 4.

Guided Practice 5. Name the coordinate of each point graphed on the number line.

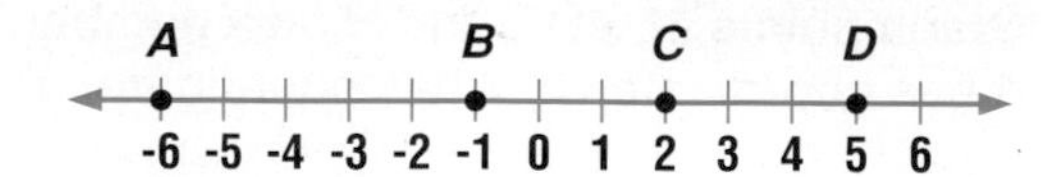

Find each absolute value.

6. $|4|$ 7. $|-3|$ 8. $|-23|$ 9. $|124 + 5|$ 10. $|0|$

Graph each set of numbers on a number line.

11. $\{-1, -3, -5\}$ 12. $\{4, 5, 7\}$ 13. $\{-3, -1, 0, 2\}$

Exercises

Independent Practice

Graph each set of numbers on a number line.

14. $\{-5, -7, 9, 12\}$ 15. $\{-6, -8, -10\}$ 16. $\{-3, -2, 0, 2, 3\}$

Find each absolute value.

17. $|34|$ 18. $|-93|$ 19. $|-87|$ 20. $|132 - 20|$

21. Evaluate $b + |a - c|$ if $a = 1$, $b = 2$, and $c = 0$.
22. Evaluate $xy - |-30|$ if $x = 10$ and $y = 4$.

Mixed Review

23. Use mental math to find 297 + 478. *(Lesson 1-2)*
24. Solve $q + 6.8 = 15.2$. Check your solution. *(Lesson 2-3)*
25. Solve $\frac{b}{2} - 1 > 2$. Show the solution on a number line. *(Lesson 2-10)*

Problem Solving and Applications

26. **Number Sense** Graph any point on a number line.
    a. Are the integers to the right of that point greater than or less than the coordinate of the point?
    b. What can you conclude about the integers to the left of that point?
27. **Critical Thinking** A small snail in the forest is climbing a tree 10 feet tall. Each day it climbs 3 feet. Because of the dampness of the forest, it slips down 2 feet each evening. Use a number line to determine how many days it will take the snail to reach the top of the tree.
28. Copy the chart below. Complete the chart with a phrase that fits each situation.

Situation	Negative	Positive	Neither Positive nor Negative
altitude of a city		above sea level	
football	5 yard loss		
time			today
money		profit	

3-2 Comparing and Ordering

Objective

Compare and order integers.

In 1988, advertisers spent over $1,504 million advertising in women's magazines.

What magazines do you have at your house? The list below shows five top selling magazines for 1990. The number after each magazine tells the increase (+) or decrease (−) in the circulation from the previous year.

Sports Illustrated	+94,978
People	−78,566
Seventeen	+13,853
Rolling Stone	−56,900
'Teen	+7,186

If you ordered these magazines from the greatest increase to the greatest decrease, would the order be the same as above? Which magazine would be last on your list?

Think about the number line. Remember that the values increase as you go right and decrease as you go left. Let's sketch a number line for the circulation values.

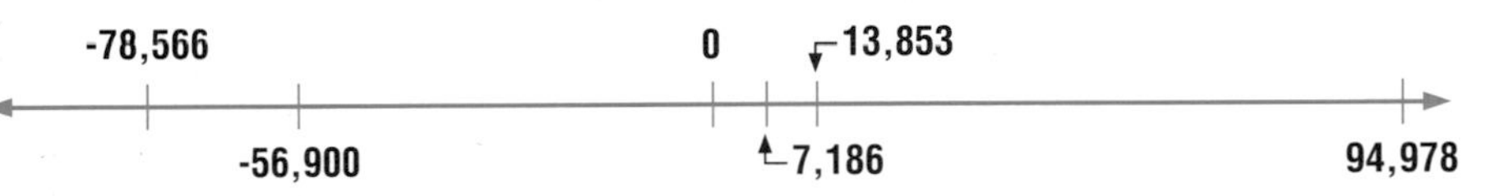

Which integer is graphed farthest to the right? This would be the greatest integer. What is the least integer?

Let's list the numbers from greatest to least. Use the number line to confirm the order.

{94,978, 13,853, 7,186, −56,900, −78,566}

Any two numbers can be compared using one of three symbols.

$=$ *is equal to* $>$ *is greater than* $<$ *is less than*

Examples

Use the number line to compare integers. Replace each ● with $>$, $<$, or $=$.

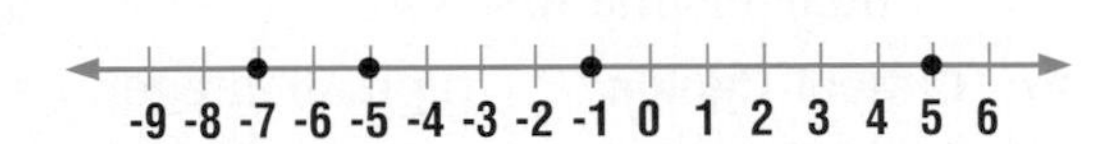

1 -7 ● -1

-7 is to the left of -1 on the number line.
So, $-7 < -1$.

2 $|-5|$ ● 5

$|-5|$ has a value of 5.
So, $|-5| = 5$.

Checking for Understanding

Communicating Mathematics

Read and study the lesson to answer each question.

1. **Write** two inequalities that relate -245 and 612.
2. **Tell** how you can easily remember what $>$ and $<$ mean.
3. **Show** how it is possible that $-3 > -6$, but $|-3| < |-6|$. Use a number line.

Guided Practice

Replace each ● with $>$, $<$, or $=$.

4. -19 ● -22	5. 0 ● -7	6. 4 ● 87
7. -56 ● 0	8. $\|-9\|$ ● $\|-3\|$	9. -459 ● -23

10. Order the integers in the set $\{34, 0, -7, -1, 8, -99, 123\}$ from least to greatest.
11. Order the integers in the set $\{78, -665, 1, 34, -99, 129, 65, -6\}$ from greatest to least.

Exercises

Independent Practice

Replace each ● with $>$, $<$, or $=$.

12. -14 ● 0	13. -23 ● 9	14. -99 ● -789
15. 0 ● $\|-7\|$	16. 90 ● 21	17. $\|-34\|$ ● $\|-9\|$
18. -632 ● -347	19. -56 ● 56	20. $\|214\|$ ● $\|-214\|$

21. Order the integers in the set $\{34, 0, -7, 99, -56, -9, -33\}$ from least to greatest.
22. Order the integers in the set $\{8, -999, 12, 0, 40, -50, 93, -66\}$ from greatest to least.

Mixed Review

23. Evaluate $(13 - 9)^2 + \frac{15}{21 \div 7}$. *(Lesson 2-1)*
24. Solve $4m - 28 = 68$. *(Lesson 2-7)*
25. Find $|4^2|$. *(Lesson 3-1)*

Problem Solving and Applications

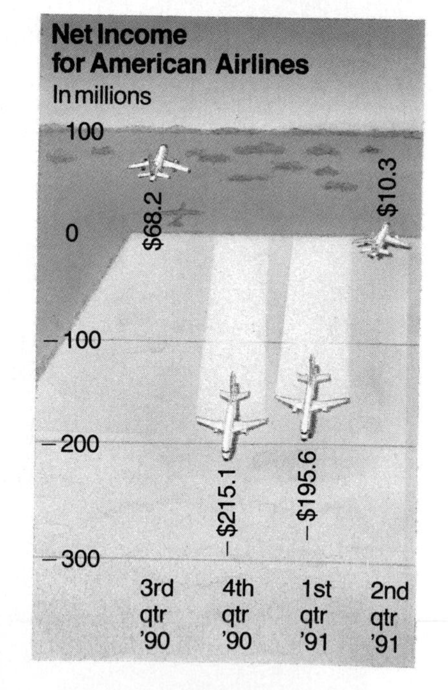

26. **Transportation** Use the graph at the right to answer each question.
    a. Describe the net income for each quarter of American Airlines as a gain or loss.
    b. Write a mathematical sentence to compare each quarter with the quarter that follows it.
27. **Critical Thinking** Graph all integer solutions for $|x| < 5$ on a number line.

28. **Computer Connection** In BASIC, the INT(X) function finds the greatest integer that is not greater than X. INT(2.5) = 2, because 2 is the greatest integer and is not greater than 2.5. Find INT(−5.35).

3-3 Adding Integers

Objective
Add integers.

The first bicycle was built in the late 1700s by a man named DeSivrac. However, the title of "father of the bicycle" is given to Baron Karl von Drais, who invented a two-wheeler in 1816.

Steve wanted to buy a new bicycle, but he did not have enough money for the 21-speed model he wanted. His parents offered him a deal. They said, "We'll lend you the money, but you must work the debt off by doing extra jobs around the house." Steve agreed. He borrowed \$65 from his parents. This debt can be written as -65.

After shopping around, he realized that he needed to borrow more money from his parents. He borrowed an additional \$50. This debt can be written as -50.

What is the total amount that Steve owes his parents? Let a = the total amount he owes.

$$-65 + (-50) = a$$

Addition can be represented on a number line. First graph -65. Since -50 is negative, you will move 50 units to the left.

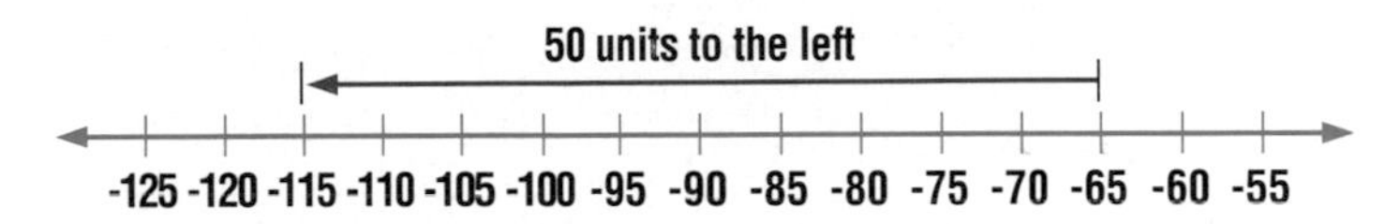

$$-65 + (-50) = a$$
$$-115 = a$$
Steve owes his parents \$115.

You already know how to add two positive integers. For example, to solve $w = 45 + 30$, you simply add 45 and 30. The solution is 75. The solution for $-65 + (-50) = a$ is the same as the solution for $-(65 + 50) = a$. In each of these equations, you add the absolute value of the addends. The sum has the same sign as the integers.

Adding Integers with Same Sign	The sum of two positive integers is positive. The sum of two negative integers is negative.

Examples

1 **Solve $p = 4 + 56$.**

Use a number line.

56

4 60

$p = 4 + 56$
$p = 60$

The solution is 60.

2 **Solve $-8 + (-3) = x$.**

Use counters. Put in 8 negative counters. Add 3 more negatives.

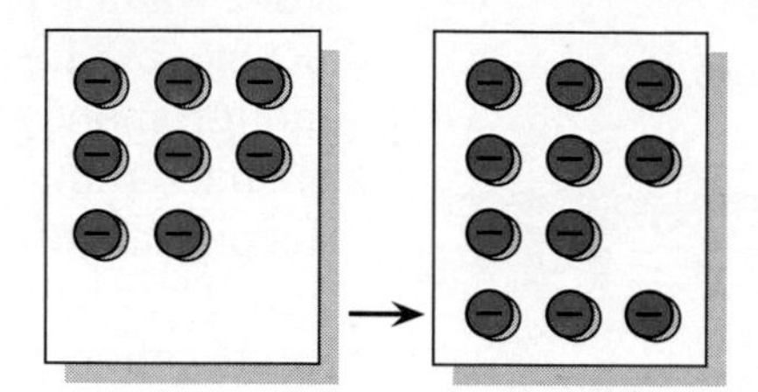

There are 11 negative counters. Therefore, $x = -11$. The solution is -11.

What do you suppose happens when you add a negative integer and a positive integer? Let's use counters to find a rule.

Mini-Lab

Work with a partner to solve $x = 7 + (-4)$.
Materials: two colors of counters, mats

- Let one color of counter represent positive and another color represent negative. We need 7 positive counters and 4 negative counters.

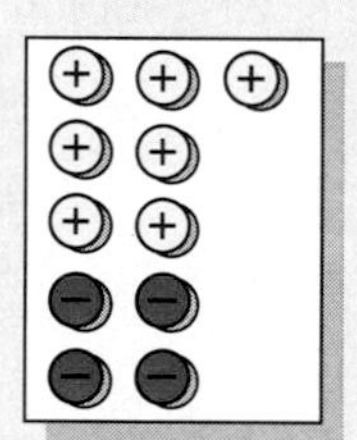

- When a positive counter is paired with a negative counter, the result is called a **zero pair.** You can add or remove zero pairs without changing the value of the set. Remove all the zero pairs from the mat.

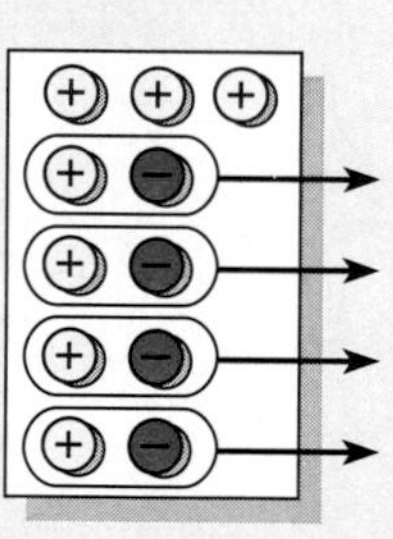

- The counters you have left represent the solution.

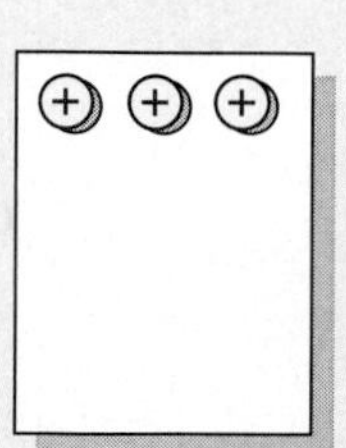

Talk About It

a. How many zero pairs did you find?
b. What kind of counters did you have left?
c. What is the solution of $x = 7 + (-4)$?

The results of the Mini-Lab suggest the following rule for adding two integers with different signs.

Adding Integers with Different Signs	To add integers with different signs, subtract their absolute values. The sum is: • positive if the positive integer has the greater absolute value. • negative if the negative integer has the greater absolute value.

Examples

3 **Solve $76 + (-9) = c$.**

$|76| > |-9|$, so the sum is positive.

The difference of 76 and 9 is 67, so $c = 67$.

4 **Solve $a = -34 + 12$.**

$|-34| > |12|$, so the sum is negative.

The difference of 34 and 12 is 22, so $a = -22$.

Checking for Understanding

Communicating Mathematics

Read and study the lesson to answer each question.

1. **Show** two methods to solve $x = -15 + 23$.
2. **Tell** how you know the sign of the sum of two integers with different signs.
3. **Tell** how you know whether to add or subtract the absolute values to find the sum of two integers. Give examples.
4. **Write** the addition sentence shown by each model.

a. 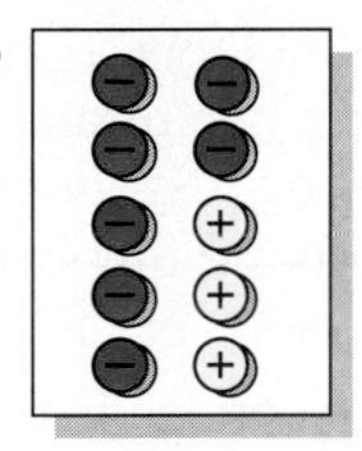

b. 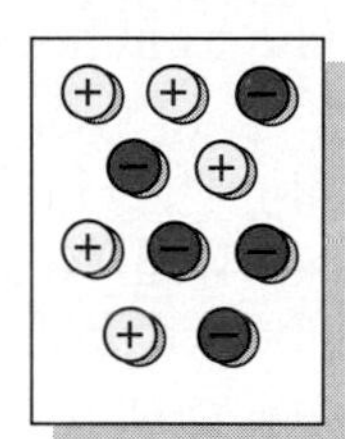

c. 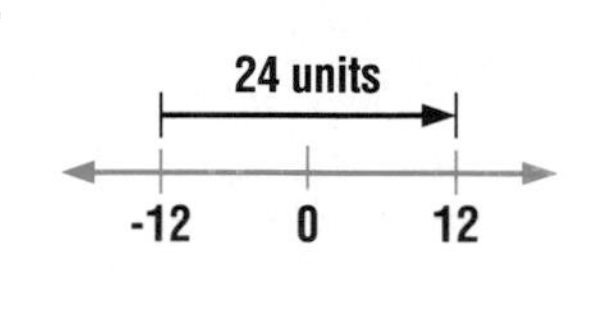

Guided Practice

Tell the sign of each sum.

5. $-45 + (-5)$
6. $-9 + 3$
7. $456 + 12$
8. $-32 + 40$
9. $12 + (-12)$
10. $34 + (-60)$

Solve each equation.

11. $29 + (-9) = e$
12. $5 + (-12) = p$
13. $z = -34 + 75$
14. $-41 + (-18) = w$
15. $-42 + 42 = q$
16. $f = 63 + 45$

Exercises

Independent Practice

Solve each equation.

17. $-54 + 21 = y$
18. $-456 + (-23) = j$
19. $z = 60 + 12$
20. $35 + (-32) = m$
21. $n = -98 + (-32)$
22. $s = -34 + 56$
23. $r = -19 + (-37)$
24. $r = -319 + (-100)$
25. $56 + (-2) = b$
26. $-60 + 30 = v$
27. $409 + 309 = a$
28. $c = 76 + (-45)$

Evaluate each expression if $r = 5$, $t = -5$, and $w = -3$.

29. $r + 45$
30. $w + (-7)$
31. $t + w$
32. $-9 + t$
33. $-2 + w$
34. $(r + t) + w$

Mixed Review

35. Jamal put 4 pounds of sunflower seeds in his bird feeder on Sunday. On Friday, the bird feeder was empty, so Jamal put 4 more pounds of seed in it. The following Sunday, the seeds were half gone. How many pounds of sunflower seeds were consumed by the birds that week? *(Lesson 1-1)*
36. Write $4 \cdot 4 \cdot 6 \cdot 6 \cdot 6$ using exponents. *(Lesson 1-9)*
37. Solve $6x = 42$. Check your solution. *(Lesson 2-4)*
38. José has four times as many coins as Rebecca. Write an algebraic expression to represent José's coins. *(Lesson 2-6)*
39. Replace ● with >, <, or = in -114 ● -97. *(Lesson 3-2)*

Problem Solving and Applications

40. **Sales** Madeline is selling candy for the band. Her strategy is to work one side of the street and then work the other side of the street. She begins at Avenue H and Meadow Street. She travels east on Meadow Street for 4 blocks. Then she crosses the street and goes 6 blocks west.
    a. Draw a diagram to find Madeline's location now.
    b. Write an equation to show Madeline's path.
    c. How far is she from her original starting point?
41. **Marketing** The Foot Locker® bought a certain brand of tennis shoe for \$38.98 wholesale. They increased the price by \$25 to make a profit. The shoes did not sell well. So to reduce inventory, the store is having a half-off sale. What will be the profit on each pair of shoes? Explain your answer.
42. **Critical Thinking** The teacher gave the equation $-34 + 16 + (-23) + 7 + (-50) + 121 + (-300) = x$ to her students as a challenge. Marc and Teresa discussed how each of them solved the equation. Marc said he couldn't believe how many times he had to subtract to solve the equation. Teresa said she only subtracted once. How did she do this?

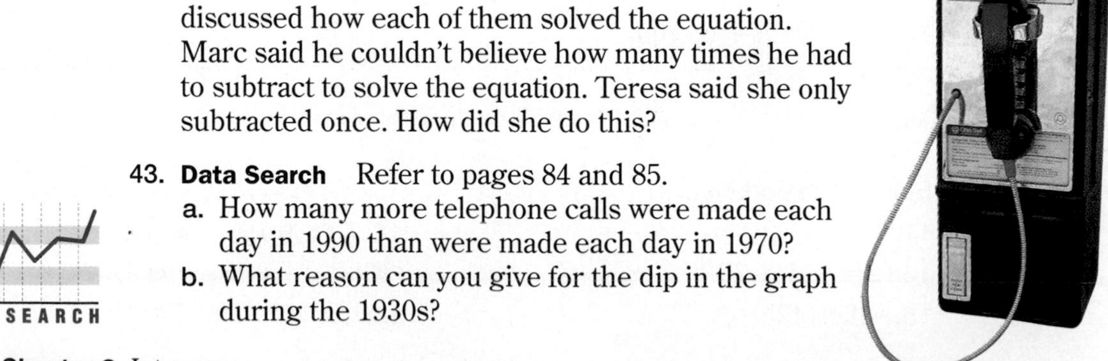

43. **Data Search** Refer to pages 84 and 85.
    a. How many more telephone calls were made each day in 1990 than were made each day in 1970?
    b. What reason can you give for the dip in the graph during the 1930s?

3-4 More About Adding Integers

Objective

Add more than two integers.

What is your favorite television show? *The Cosby Show* has been a favorite show of TV viewers since its creation in 1984. In November of 1985, about 32 of every 100 viewers watched *The Cosby Show.* By 1986, the number rose by 2 viewers per hundred. By 1990, the number dropped by 11 viewers per hundred. What was the number of viewers (per hundred) that watched *The Cosby Show* in 1990?

The number of viewers can be found using a sum of integers. Let v = the number of viewers per hundred in 1990.

$v = 32 + 2 + (-11)$

$v = (32 + 2) + (-11)$ *Use the associative property to group the first two addends.*

$v = 34 + (-11)$

$v = 23$

About 23 viewers per hundred watched *The Cosby Show* in 1990.

LOOKBACK

You can review the commutative property and associative property on page 9.

The equation above was solved by using the associative property to group the first two addends. The commutative property can also be used when working with integers.

Each of the following examples is solved one way and checked by using these properties in another way.

Example 1

Solve $r = 12 + (-3) + 7$.

Use the associative property to group the first two addends.

$r = [12 + (-3)] + 7$

$r = (9) + 7$

$r = 16$

Use the associative property to group the last two addends.

Check: $r = 12 + [(-3) + 7]$

$r = 12 + (4)$

$r = 16$ ✓

Mental Math Hint

Look for groupings that make the sum easier to find.
3 + (−8) + (−3)
Since adding 0 is easy, group 3 and (−3) first.

Calculator Hint

To enter −92, press 92 and then press the [+/−] key. A negative sign will appear.

Examples

2 **Solve $y = -4 + 2 + (-8)$.**

Use the commutative property to change order.

$y = -4 + (-8) + 2$
$y = -12 + 2$
$y = -10$

Use the associative property to group the first two addends.

Check: $y = (-4 + 2) + (-8)$
$y = -2 + (-8)$
$y = -10$ ✓

3 **Solve $t = -92 + 73 + (-51) + 100$.**

$t = -92 + 73 + (-51) + 100$
$t = -92 + (-51) + 73 + 100$ *Use the commutative property.*
$t = [-92 + (-51)] + [73 + 100]$ *Use the associative property.*
$t = -143 + 173$
$t = 30$

Check: Use your calculator.

92 [+/−]  73 [+] 51 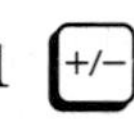 [+] 100 [=] 30 ✓

Checking for Understanding

Communicating Mathematics

Read and study the lesson to answer each question.

1. **Write** a sentence to tell how you can use the associative and commutative properties to solve a problem with more than two addends.
2. **Tell** how mental math can be used to find the sum of two or more integers.
3. **Show** three ways to solve $x = -4 + 8 + 12 + (-6) + (-3) + 13$.

Guided Practice

Solve each equation.

4. $y = 21 + 3 + (-6)$
5. $(-3) + 8 + 9 = c$
6. $(-2) + 3 + (-10) + 6 = f$
7. $d = 7 + 20 + (-5)$
8. $w = (-4) + (-3) + 4 + 3$
9. $(-8) + 4 + 12 + (-11) = r$
10. $s = 9 + 10 + (-6) + 6$
11. $21 + 3 + (-9) = g$
12. $(-7) + 12 + 9 = q$
13. $p = (-6) + 12 + (-11) + 1$

Exercises

Independent Practice

Solve each equation. Check by solving another way.

14. $a = 6 + 9 + (-11)$
15. $c = (-8) + 4 + 21$
16. $19 + 23 + (-8) + 12 = f$
17. $(-4) + 5 + 7 + 12 = g$
18. $w = -32 + 32 + 70$
19. $x = 5 + (-12) + 7 + 3$
20. $j = -8 + 6 + (-20)$
21. $m = (-50) + 9 + 3 + 50$
22. $p = -13 + (-5) + 7 + (-20)$
23. $8 + 30 + 21 + (-5) = r$

24. $14 + 7 + (-23) + 10 = t$

25. $v = (-30) + 5 + 12 + (-23)$

26. $r = 4 + (-10) + 6$

27. $-21 + 17 + 10 + (-17) = z$

Evaluate each expression if $c = 5$, $x = -4$, and $h = 6$.

28. $c + 3 + 4 + (-8)$

29. $(-6) + x + 2 + 10$

30. $(-5) + h + 1$

31. $(-5) + 16 + c + h$

Mixed Review

32. How many ounces are in 2.25 pounds of gelatin mix? *(Lesson 1-7)*

33. The replacement set for $\frac{448}{y} = 14$ is {22, 29, 41, 32}. Find the solution. *(Lesson 2-2)*

34. Solve $4s < 36$. Show the solution on a number line. *(Lesson 2-10)*

35. Solve $p = 85 + (-47)$. *(Lesson 3-3)*

Problem Solving and Applications

36. **Science** By studying the records of earthquakes, scientists have learned that the inside of the earth is divided into three parts: the mantle, the outer core, and the inner core in that order. The outer core begins at about −1,800 miles; that is, 1,800 miles below the earth's surface. It is about 1,400 miles thick. Name an integer to tell where the inner core begins.

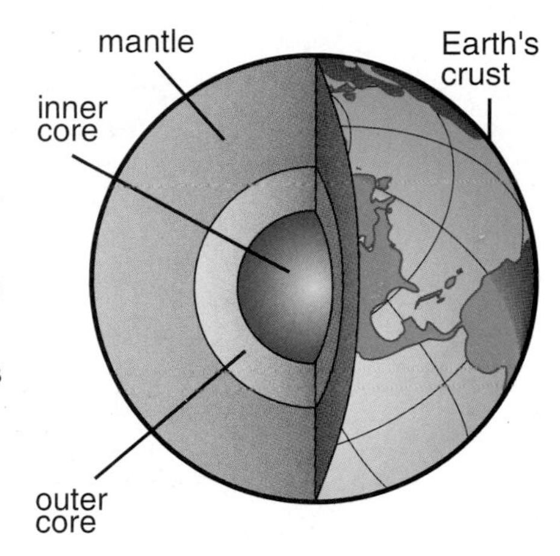

37. **Critical Thinking** Explain how you might estimate the sum $326 + (-76) + 210 + (-330) + (-215)$. Give more than one method.

38. **Geography** The lowest elevation on Earth is the shore of the Dead Sea. It is about 1,310 feet below sea level. If you traveled so that you gained 400 feet altitude each hour, how many hours would it take you to reach sea level?

39. **Mathematics and Sports** Read the following paragraph.

There are at least six kinds of football. Soccer is probably the most popular. There are 11 players on each team, and the game is played with a round ball that cannot be handled, except by the goalie. Rugby has either 13 or 15 players and is played with an oval ball that can be carried. American football has 11 players per side and is played with an oval ball that can be carried. Canadian football is similar to American football but is played with 12 players. Australian Rules is played on an oval field, and each side has 18 players. Gaelic football is like a combination of soccer and rugby and has 15 players per team.

a. Vance lost 5 yards on the first play and then gained 8 yards on the next play. Find the total number of yards gained or lost.

b. The Bruins' offense advanced the football 15 yards. On the next play, the quarterback was sacked and lost 23 yards. Find the total number of yards gained or lost.

3-5 Subtracting Integers

Objective
Subtract integers.

Words to Learn
opposite
additive inverse

The highest point on Earth is Mount Everest in the Himalaya Mountains on the border of India and China. It has an altitude of 29,028 feet above sea level. The lowest verified point on Earth is in the Mariana Trench in the western Pacific Ocean. It is 35,840 below sea level. You could write these altitudes as 29,028 feet and −35,840 feet.

Suppose we want to find the difference between the altitudes of Mount Everest and the Mariana Trench. If d represents the difference, then $d = 29{,}028 - (-35{,}840)$. *You will solve this in Exercise 2.*

Remember that we used counters to show how to add integers. You can use these counters to show subtraction, too.

Mini-Lab

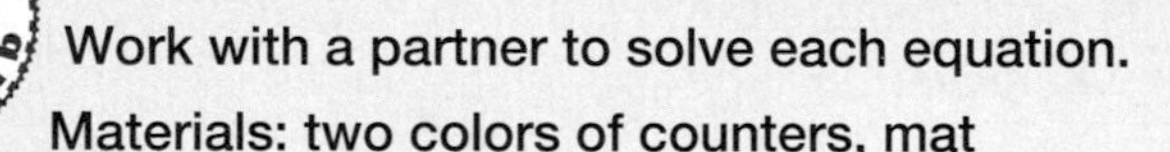

Work with a partner to solve each equation.

Materials: two colors of counters, mat

a. Solve $x = -8 - (-2)$.

- Start with 8 negative counters.
- Remove 2 negative counters.

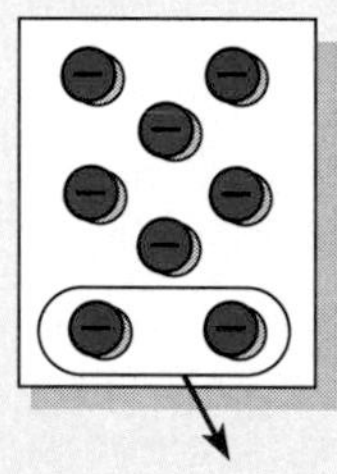

b. Solve $y = 7 - 3$.

- Start with 7 positive counters.
- Remove 3 positive counters.

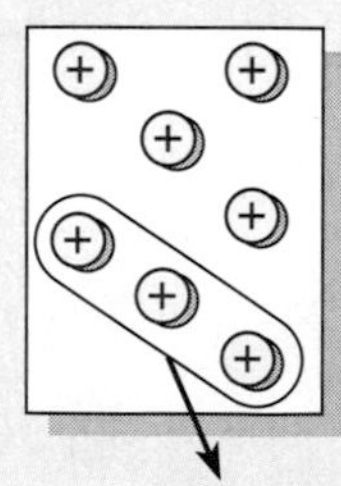

c. Solve $6 - (-3) = z$.

- Start with 6 positive counters. There are no negative counters, so we can't remove 3 negatives.
- Add 3 zero pairs to the mat.
- Now remove the 3 negatives.

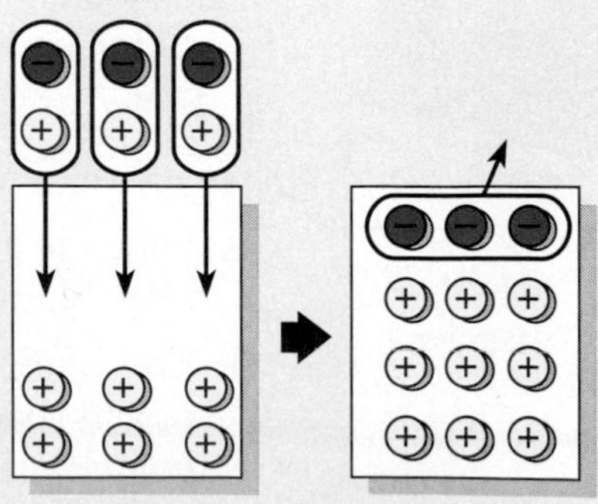

Talk About It

a. Compare the solution of $x = -8 - (-2)$ with the solution of $x = -8 + 2$.

b. Compare the solution of $y = 7 - 3$ with the solution of $y = 7 + (-3)$.

c. Compare the solution of $6 - (-3) = z$ with the solution of $6 + 3 = z$.

In Chapter 2, you learned that adding and subtracting were opposite, or inverse, operations. Each integer also has an **opposite.** The opposite of an integer is called its **additive inverse.**

Additive Inverse	**In words:** The sum of an integer and its additive inverse is 0.	
	Arithmetic	**Algebra**
	$6 + (-6) = 0$	$a + (-a) = 0$

In the Mini-Lab, you were asked to compare the result of subtracting an integer with the result of adding its inverse.

Subtraction	*Addition of the Additive Inverse*
$-8 - (-2) = -6$	$-8 + 2 = -6$
$7 - 3 = 4$	$7 + (-3) = 4$
$6 - (-3) = 9$	$6 + 3 = 9$

Adding the additive inverse of an integer produces the same result as subtracting the integer.

Subtracting Integers	To subtract an integer, add its additive inverse.

Examples

1 Solve $p = -5 - 3$.

$p = -5 - 3$

$p = -5 + (-3)$ *To subtract 3, add −3.*

$p = -8$

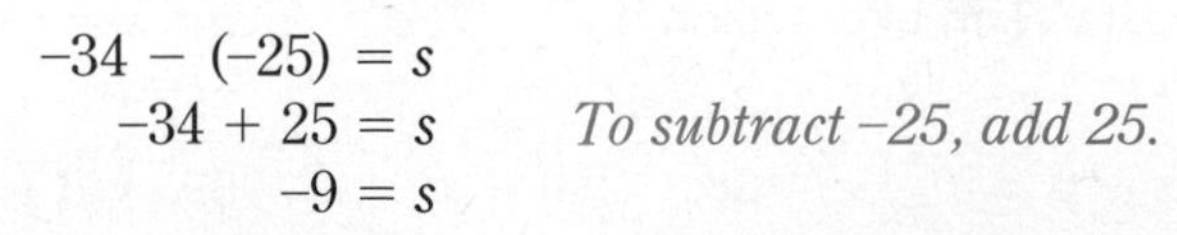

2 Solve $-34 - (-25) = s$.

$-34 - (-25) = s$

$-34 + 25 = s$ *To subtract −25, add 25.*

$-9 = s$

Example 3 *Problem Solving*

Accounting The balance of credit card accounts is figured at the end of each month. The balance can be negative or positive. Additional charges are subtracted from the balance, and payments are added to the balance. Find the final balance on the spreadsheet below.

DATE	CHARGES	PAYMENTS	BALANCE
10/1			–$12.30
10/6	$56.78		
10/15	$21.20		
10/30		$75.00	

Use pencil and paper.

$-12.30 - 56.78 = -69.08$

$-69.08 - 21.20 = -90.28$

$-90.28 + 75.00 = -15.28$

Use a calculator. 12.30 [+/–] [–] 56.78 [–] 21.20 [+] 75 [=] –15.28

The final balance is –$15.28. This means the customer owes $15.28.

Checking for Understanding

Communicating Mathematics

Read and study the lesson to answer each question.

1. **Write** three examples of integers and their additive inverses.
2. **Show** how you would solve the equation to find the difference between the altitudes of Mount Everest and the Mariana Trench.
3. **Tell** whether every integer has an additive inverse. Which integer is its own inverse?
4. **Draw** a model that shows $-3 - (-5) = q$.

Guided Practice

Find the additive inverse of each integer.

5. 10
6. –9
7. 30
8. –29

Rewrite each equation using the additive inverse. Then solve.

9. $4 - (-7) = y$
10. $n = -43 - 99$
11. $p = -23 - (-2)$
12. $53 - 78 = z$
13. $y = 14 - 14$
14. $11 - (-19) = p$
15. $x = 17 - (-26)$
16. $123 - (-33) = n$
17. $b = -345 - 67$

Exercises

Independent Practice

Solve each equation.

18. $j = 44 - (-11)$
19. $4 - (-89) = u$
20. $56 - (-78) = p$
21. $435 - (-878) = u$
22. $k = -99 - 4$
23. $w = -43 - 88$
24. $x = -78 - (-98)$
25. $-5 - 3 = k$
26. $63 - 92 = q$
27. $m = 56 - (-22)$
28. $r = -9 - (-4)$
29. $-789 - (-54) = s$
30. $x = -351 - 245$
31. $v = 89 - (-54)$
32. $-109 - (-34) = g$

Evaluate each expression if $y = -7$, $p = 9$, and $x = -10$.

33. $45 - y$
34. $67 - p$
35. $x - (-23)$
36. $y - x$
37. $x - y$
38. $-240 - x$
39. $y - p - x$
40. $x - y - p$

Mixed Review

41. Find the value of 5^3. *(Lesson 1-9)*
42. Evaluate $8x - 2y$ if $x = 6$ and $y = 14$. *(Lesson 1-3)*
43. **Geometry** Find the area of a square if its side is 8 cm long. *(Lesson 2-9)*
44. Order the set {52, –3, 128, 4, –22, 15, 0, –78} from greatest to least. *(Lesson 3-2)*
45. Solve $44 + 8 + (-20) + 15 = s$. Check your solution. *(Lesson 3-4)*

Problem Solving and Applications

Use the graph below for Exercises 46-47.

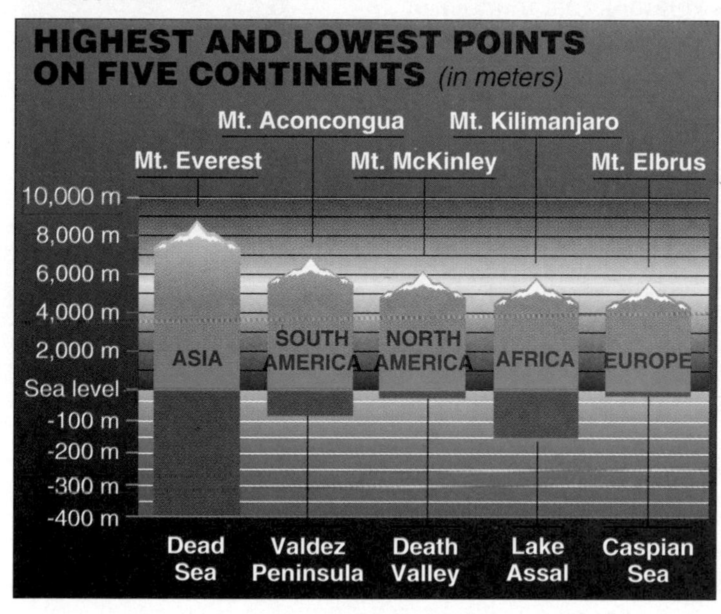

46. **Geography** Estimate the difference between the highest and lowest points on each continent.
47. **Research** Use an encyclopedia or almanac to find the exact elevations of the highest and lowest points given in the graph.
    a. What is the actual difference between the altitudes of the highest and lowest points on each continent?
    b. What is the difference between the altitudes of the lowest point in Africa and the lowest point in South America?
48. **Critical Thinking** Is the subtraction of integers associative? That is, does $-53 - (23 - 37) = (-53 - 23) - 37$? Explain.
49. **Business** Accountants use the formula $P = I - E$ to find the profit (P) when income (I) and expenses (E) are known.
    a. Find P if $I = \$18,345$ and $E = \$25,000$.
    b. What does this answer mean?

3 Assessment: Mid-Chapter Review

Find each absolute value. *(Lesson 3-1)*

1. $|64|$ 2. $|-31|$ 3. $|-4|$

4. Graph the set {4, 8, –3, 2, 0, –8, –1} on a number line. *(Lesson 3-1)*
5. Write the set {6, 5, –2, 0, –3, 8, –7} in order from least to greatest. *(Lesson 3-2)*

Solve each equation. *(Lessons 3-3, 3-4, and 3-5)*

6. $x = 3 + (-5)$ 7. $y = 2 + (-4) + (-6) + 8$ 8. $514 - 600 = r$
9. $90 + (-90) = p$ 10. $w = 67 - (-32)$ 11. $m = -89 - 25$

3-6 Multiplying Integers

Objective

Multiply integers.

Have you ever played golf or watched a golf tournament? In golf, the score for each hole is often stated in relation to the number of strokes the course says is standard for that hole. *Par* means you took that number of strokes. Other scores for one hole are shown below.

eagle	−2	two strokes under par
birdie	−1	one stroke under par
par	0	even par
bogey	+1	one stroke over par
double bogey	+2	two strokes over par

If you took 2 strokes on a par-3 hole, you would make a birdie. Add −1 to your score in relation to par.

DID YOU KNOW

The winner of a golf tournament is the person who took the fewest strokes. TV statisticians use the number under or over par to indicate who is leading during any given round of the tournament.

At the 1991 Western Open, Greg Norman entered the last day of the tournament with a score of 11 under par, or −11. On the first nine holes that day, he scored 5 birdies and was even par for the other four holes. How did this affect his score?

Even par scores did not affect his score. The 5 birdies can be expressed as 5(−1).

$$\begin{aligned} 5(-1) &= (-1) + (-1) + (-1) + (-1) + (-1) \\ &= -5 \end{aligned}$$

His score on the last day at the end of the nine holes was −11 + (−5) or −16. That is, 16 under par.

Mini-Lab

Work with a partner to solve each equation.

Materials: two colors of counters, mats

a. Solve $x = 5 \cdot 2$.

- Begin with an empty mat.
- The 5 means to put 5 sets of counters on the mat. Each set will contain 2 positive counters.

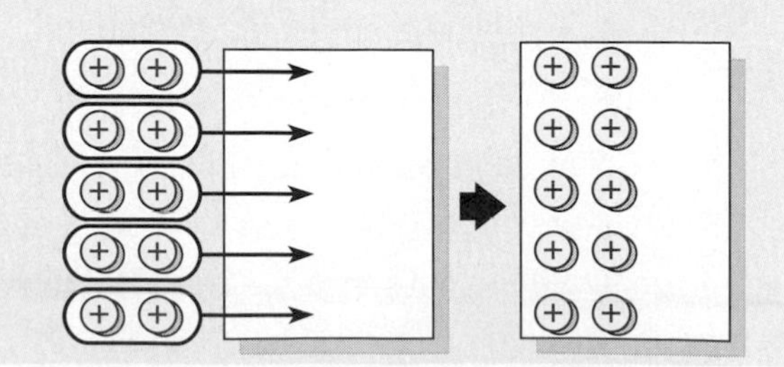

b. Solve $y = 3 \cdot (-4)$.

- Begin with an empty mat.
- The 3 means to put 3 sets on the mat. Each set will contain 4 negative counters.

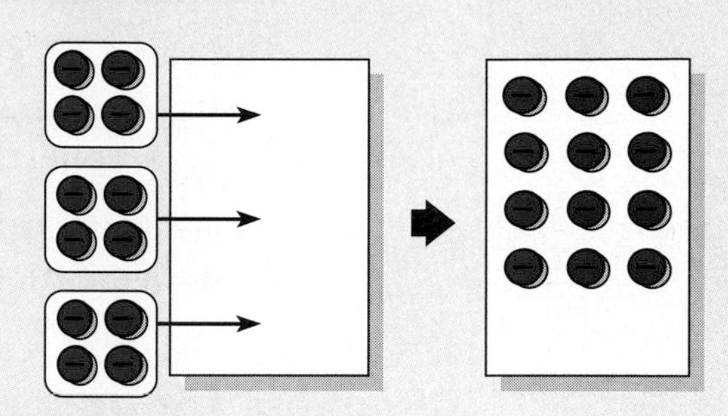

c. Solve $z = -2 \cdot (-4)$.

- Start with an empty mat.
- The −2 means to take 2 sets from the mat. Each set will contain 4 negative counters.
- Since the mat contains no counters, you must first place enough zero pairs on the mat so that it will be possible to remove 2 sets of 4 negative counters.

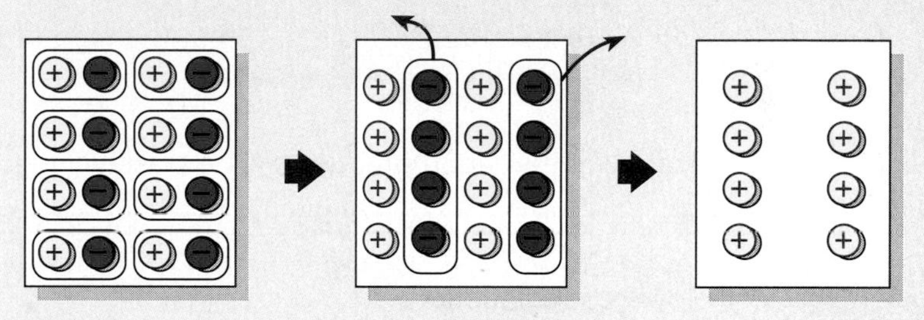

Talk About It

a. What are the values of x, y, and z?
b. When were the products positive?
c. When were the products negative?

The Mini-Lab suggests the following rule for multiplying integers.

Multiplying Integers	The product of two integers with the same sign is positive. The product of two integers with different signs is negative.

Examples

1 Solve $p = 5(-6)$.

The two integers have different signs. The product will be negative.

$p = 5(-6)$
$p = -30$

2 Solve $x = -7(-7)$.

The two integers have the same sign. The product will be positive.

$x = -7(-7)$
$x = 49$

Examples

3 Solve $y = (7)(-9)(6)$.

Use the associative property to group the factors.

$y = [7(-9)](6)$
$y = (-63)(6)$ or -378

4 Solve $r = (-6)^2$.

The exponent says there are two factors of -6.

$r = (-6)^2$
$r = (-6)(-6)$ or 36

Checking for Understanding

Communicating Mathematics

Read and study the lesson to answer each question.

1. **Tell** why $3(-8)$ and $-8(3)$ have the same product.
2. **Show** another way to solve the equation in Example 3.
3. **Show** how you would use your calculator to solve the equation in Example 4.
4. **Draw** a model to show the solution for $w = (-6)(2)$.
5. **Write** a sentence that tells what the sign of y is in $y = x^2$, no matter what value of x, besides 0, you choose.

Guided Practice

State whether each product will be positive, negative, or zero.

6. $7(-8)$
7. $98(-2)$
8. $(23)(-3)(-7)$
9. $(5)(6)(0)$

Solve each equation.

10. $t = 9(-3)$
11. $g = 5(-30)$
12. $q = 9(-11)$
13. $-3(-7) = k$
14. $g = -8(-3)$
15. $-12(-8) = p$
16. $b = -9(12)$
17. $w = 4(30)$
18. $d = 6(-3)$
19. $(-8)^2 = y$
20. $-5(-14) = f$
21. $y = 7(5)$

Exercises

Independent Practice

Solve each equation.

22. $p = 6(8)$
23. $w = -9(7)$
24. $7(-14) = y$
25. $9(-9) = k$
26. $r = 55(-11)$
27. $w = -6(-13)$
28. $-5(80)(-2) = m$
29. $(-6)(7)(-12) = p$
30. $u = 9(-10)(3)$
31. $q = 7(23)(5)$
32. $-4(-50)(-1) = j$
33. $(-21)^2 = t$
34. $(-8)(9)(6)^2 = k$
35. $(3)(12)(-2) = k$
36. $m = (6)^2 \cdot (-3)^2$

Evaluate each expression if $a = -3$, $b = -6$, and $c = 10$.

37. $3ab$
38. $-10ac$
39. ab^2
40. $12abc$

Mixed Review

41. **Car Rental** Jackson Auto Rental charges \$20 per day and \$0.15 per mile to rent a car. Find the cost of renting a car for two days and driving 100 miles. *(Lesson 1-1)*
42. How many kilometers did Katra walk if she walked 39.4 meters? *(Lesson 1-6)*

43. Solve $8.2 \div 0.2 = t$. Check your solution. *(Lesson 2-4)*

44. Graph $\{-3, -1, 0, 2\}$ on a number line. *(Lesson 3-1)*

45. Solve $57 - (-26) = d$. *(Lesson 3-5)*

Problem Solving and Applications

46. **Health** Mr. Tu has lost an average of three pounds a week on his diet. He has dieted for 11 weeks. He weighed 268 pounds at the beginning of his diet.
    a. How much does he weigh now?
    b. Mr. Tu's goal is to weigh 168 pounds. If he continues at this rate, how much longer will it take him to reach his goal?
    c. Draw a graph to show Mr. Tu's weight loss for the first eight weeks.

47. **Data Search** Refer to page 667.
    Which state has the greatest range in record high and low temperatures? Which state has the least range?

48. **Critical Thinking** Find the value of each expression.
    a. $(-1)^2$ b. $(-1)^3$ c. $(-1)^4$ d. $(-1)^5$ e. $(-1)^6$
    f. What do you think is the value of $(-1)^{5,280}$?

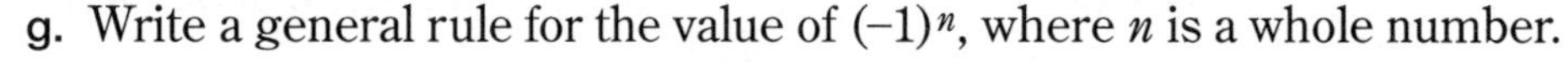

    g. Write a general rule for the value of $(-1)^n$, where n is a whole number.

49. **Journal Entry** How can you determine the sign of the product of two integers? Can you write a rule for determining the sign of a product based on the signs of its factors?

Save Planet Earth

Aluminum Recycling The first aluminum beverage can appeared in 1963, and today it accounts for the largest single use of aluminum. In 1992, more than 92 billion beverage cans were used, and 97% of them were aluminum.

According to the Aluminum Association, Americans recycled 63 billion aluminum cans in 1992. This cuts related air pollution by 95% and uses 90% less energy than making aluminum from scratch. It is estimated that the energy saved from recycling just one aluminum can will operate a television for 3 hours.

How You Can Help

- Check the yellow pages to find out which kinds of recycling programs exist in your area. The most popular are reverse vending machines, curbside pickup, and drop-off centers with bins for recycling.
- You can also recycle aluminum foil, pie plates, and frozen food trays.

3-7 Dividing Integers

Objective
Divide integers.

When you watch the news, do you ever wonder what the stock market reports have to do with the economy? The greatest event in the stock market's history is probably the crash of 1929. After the stock market crash, the economy of the United States fell to an all-time low. The percent of employed people fell from 91 percent in 1930 to 75 percent in 1932. What was the average fall per year in employed people?

The change in the percent of employed people can be expressed by 75 − 91, or −16. This occurred over 2 years. The average can be found by dividing −16 by 2.

Division of integers is related to multiplication of integers. That is, for $b \neq 0$, $a \div b = c$ if $b \cdot c = a$. Let's use this rule to solve $y = -16 \div 2$.

Write a related multiplication sentence.

$y = -16 \div 2$ if $2y = -16$

Think: 2 times what number equals −16?

$2 \cdot 8 = -16$ *no* $\quad 2 \cdot (-8) = -16$ ✓

So, $y = -8$. The average fall in employment was 8 percent per year.

Division is related to multiplication, so it uses the same rules of signs.

DID YOU KNOW

The greatest fall in the Dow-Jones Averages occurred on October 19, 1987. The average fell over 500 points.

Dividing Integers	The quotient of two integers with the same sign is positive. The quotient of two integers with different signs is negative.

Examples

1 Solve $-40 \div (-5) = z$.

$-40 \div (-5) = z$ *The signs are the same.*

$8 = z$ *The quotient is positive.*

2 Solve $v = -36 \div 9$.

$v = -36 \div 9$ *The signs are different.*

$v = -4$ *The quotient is negative.*

Remember that fractions are also a way of showing division. Another way to show $c = a \div b$ is $c = \frac{a}{b}$.

Examples

3 Solve $p = \frac{45}{9}$.

$p = \frac{45}{9}$

$p = 5$

4 Solve $\frac{81}{-9} = t$.

$\frac{81}{-9} = t$

$-9 = t$

Checking for Understanding

Communicating Mathematics

Read and study the lesson to answer each question.

1. **Write** the verbal sentence *forty-five divided by negative nine equals t* as an equation in two different ways.
2. **Tell** a related multiplication sentence for $320 \div (-8) = p$.
3. **Show** how you can check a solution found by dividing.

Guided Practice

State whether each quotient is *positive* or *negative*.

4. $\frac{-24}{3}$
5. $\frac{66}{-11}$
6. $\frac{-145}{-11}$
7. $\frac{200}{-25}$

Solve each equation.

8. $240 \div (-60) = y$
9. $365 \div (-5) = r$
10. $-12 \div (-4) = h$
11. $f = \frac{245}{-5}$
12. $d = \frac{224}{32}$
13. $\frac{-88}{44} = t$
14. $\frac{90}{10} = z$
15. $-56 \div (-2) = p$
16. $w = 49 \div 7$

Exercises

Independent Practice

Solve each equation.

17. $\frac{62}{-2} = n$
18. $\frac{700}{-100} = k$
19. $m = 564 \div (-3)$
20. $-26 \div (-13) = b$
21. $295 \div 5 = y$
22. $t = -930 \div (-30)$
23. $\frac{588}{-6} = g$
24. $k = \frac{-195}{65}$

Evaluate each expression if $c = -9$, $r = 3$, and $t = -10$.

25. $\frac{99}{r}$
26. $-\frac{99}{c}$
27. $\frac{800}{t}$
28. $\frac{c}{-3}$
29. $t \div (-2)$
30. $50 \div t$
31. $342 \div c$
32. $-342 \div r$
33. $cr - 4$
34. $rt \div (-5)$
35. $5ct \div r$
36. $(crt)^2 \div t$

Mixed Review

37. Use mental math to find $42 + 86 + 58$. *(Lesson 1-2)*
38. Solve $c + 9 = 27$. *(Lesson 2-2)*
39. **Geometry** Find the perimeter of a rectangle whose length is three times its width. Its width is 4 inches. *(Lesson 2-9)*
40. Solve $z = -5 + 24 + (-8)$. Check your solution. *(Lesson 3-4)*
41. Solve $u = -8(10)(-12)$. *(Lesson 3-6)*

Problem Solving and Applications

42. **Critical Thinking** Find values for a, b, and c, so that all of the following statements are true.
    (1) $b > a$, $c < b$, and $a < 0$.
    (2) a has three digits.
    (3) b and c each have two digits.
    (4) c is divisible by 2 and 3.
    (5) b can only be divided by 1 and itself.
    (6) $a \div b = c$

43. **Banking** The Westminster Savings and Loan invested poorly and lost \$36,048 in seven weeks. About how much did they lose on average each week?
44. **Business** A check written by a company is recorded as a negative value on its monthly spreadsheet. The Neat Lawn Company hired several teenagers for one day to pick up the trash in a park. Each teen received the same pay. Each teen's entry on the spreadsheet was –\$38.35. The total of all entries was –\$345.15. Use your calculator to find out how many teens they hired to clean the park.
45. **Make a Model** You have learned how to model addition, subtraction, and multiplication of integers using colored counters. Use colored counters to develop a model for dividing integers.
    a. $y = -12 \div 3$ b. $x = 8 \div (-2)$ c. $-6 \div (-3) = z$
46. **Weather** The chart at the right is a record of falling temperatures every 3 hours on a November day in Roanoke, Virginia.

Time	Temperature
Midnight	60°F
3:00 A.M.	50°F
6:00 A.M.	42°F
9:00 A.M.	38°F
Noon	35°F

    a. How much did the temperature fall from each time period to the next? Write each fall as a negative temperature.
    b. The average temperature fall can be calculated by finding the sum of the temperature falls and dividing by the number of time periods. What is the average temperature fall?

47. **Journal Entry** Write a sentence telling why the rules for signs when dividing two integers are the same as those for multiplying two integers.

Problem-Solving Strategy

3-8 Classify Information

Objective

Solve problems by identifying important information.

Temperate climates usually do not have extreme hot or cold weather. Only 7% of Earth's land surface has a temperate climate, yet nearly half the world's population lives in this area.

The winter temperatures in Yakutsk, Yakut Republic (formerly part of the U.S.S.R.), have fallen as low as −64° C, and in the summer they have reached a high of 39° C. What is the possible range of temperatures in a year?

Explore What do you know?

- The lowest temperature is −64° C.
- The highest temperature is 39° C.
- Yakutsk is in the Yakut Republic, which used to be part of the U.S.S.R.

You need to find the range of temperatures in a year.

Plan Classify the information into what you need to know to solve the problem and what you do not need to know.

What you need to know:

- To find the range, you need to know the highest temperature and the lowest temperature.

lowest temperature	−64° C
highest temperature	39° C

What you do not need to know:

- Yakutsk is in the Yakut Republic.

There is too much information given in the problem.

Solve To find the range, subtract the lowest temperature from the highest temperature.

Let r represent the range.

$r = 39 - (-64)$

$r = 39 + 64$ *Rewrite using the additive inverse.*

$r = 103$

The possible range of temperatures in a year is 103° C.

Examine If the range is correct, you should be able to subtract the range from the highest temperature to get the lowest temperature.

$$39 - r = -64$$
$$39 - 103 \stackrel{?}{=} -64 \qquad \textit{Replace r with 103.}$$
$$39 + (-103) \stackrel{?}{=} -64 \qquad \textit{Rewrite using the additive inverse.}$$
$$-64 = -64 \checkmark \qquad \textit{The answer checks.}$$

Checking for Understanding

Communicating Mathematics

Read and study the lesson to answer each question.

1. **Tell** how you know if a problem does not contain enough information.
2. **Write** a paragraph about a real-life situation where you did not have all the facts to solve a problem.

Guided Practice

Solve, if possible. Classify information in each problem by writing *not enough information* or *too much information.*

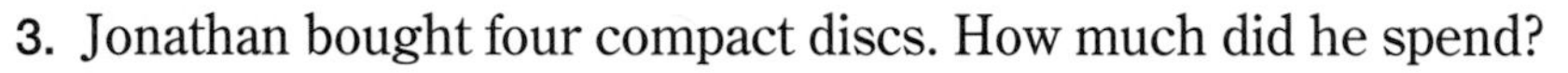

3. Jonathan bought four compact discs. How much did he spend?

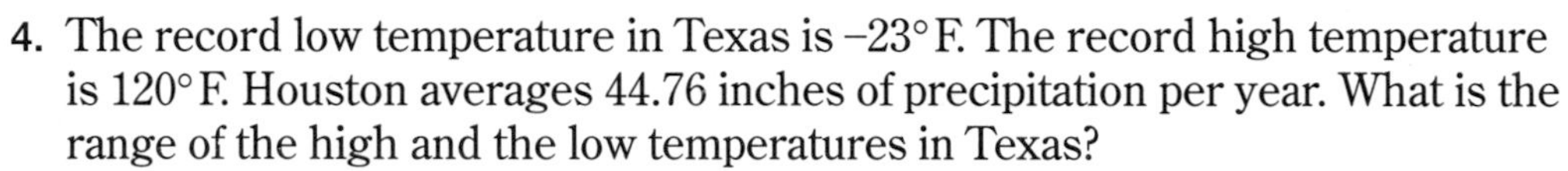

4. The record low temperature in Texas is –23° F. The record high temperature is 120° F. Houston averages 44.76 inches of precipitation per year. What is the range of the high and the low temperatures in Texas?
5. Denver is known as the Mile High City. Its elevation is 5,283 feet. The elevation of Boise, Idaho, is 2,838 feet. About how much higher is Denver?

Problem Solving

Practice

Solve using any strategy.

Strategies

Look for a pattern.
Solve a simpler problem.
Act it out.
Guess and check.
Draw a diagram.
Make a chart.
Work backward.

6. The Nepalese jawa, produced in 1740, is the smallest known coin ever minted. This sliver of silver weighs in at a mere 0.014 gram. How much would a stack of 50 jawas weigh?
7. Papua, New Guinea, an independent state in the southwest Pacific, has a population of 3,221,000. How many people are there per square mile?
8. The product of 6 and a number is –36. What is the number?
9. Kimiko's second bowling score was 82 pins less than twice her first score. If her second score was 168, what was her first score?
10. Find the area of a rectangular-shaped living room if its length is 15 feet.
11. Last year, Mrs. Penny spent $400 on season tickets to the Los Angeles Raiders home games. This year she bought two tickets per game at $35 each for a total of eight home games. How much did she spend this year?

Cooperative Learning

3-9A Solving Equations

A Preview of Lesson 3-9

Objective

Solve equations by using models.

Materials

two colors of counters
cups
mats

You used colored counters to model integers in Lessons 3-3, 3-5, and 3-6. These can also be used to solve equations that involve integers.

Activity One

Work with a partner. Solve $x + (-5) = 8$.

- Start with an empty mat.
- Let a cup represent the unknown x value. Put a cup and 5 negative counters on one side of the mat. Place 8 positive counters on the other side.

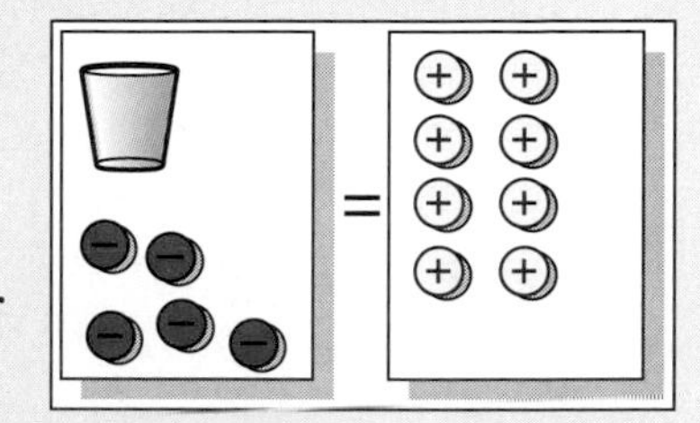

- Our goal is to get the cup by itself on one side of the mat. Then the counters on the other side will be the value of the cup, or x.
- Add 5 positive counters to each side.

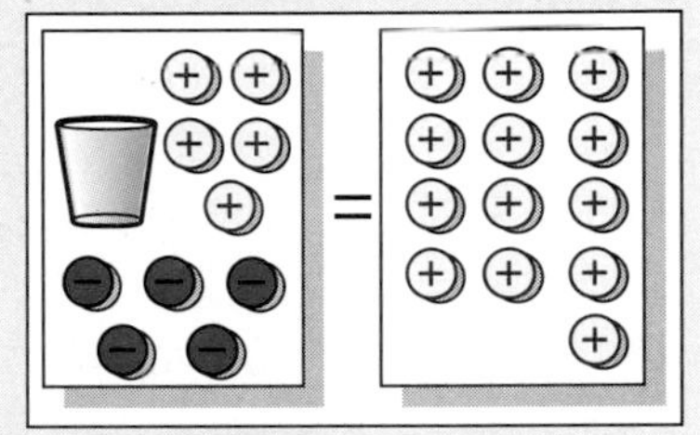

- Group the counters to form zero pairs. Then remove all zero pairs.

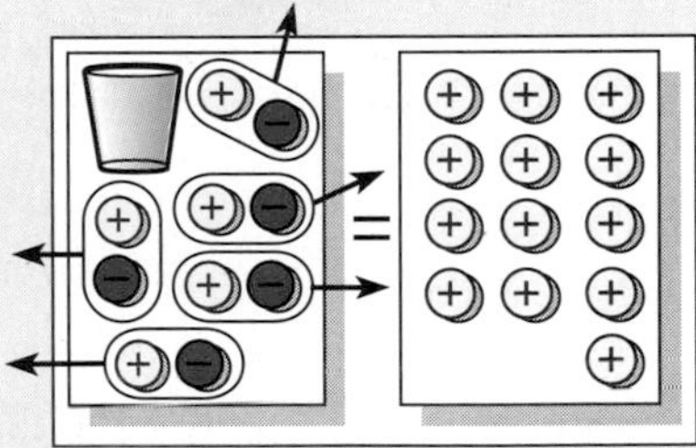

- The cup is now by itself on one side of the mat. The counters are on the other side.

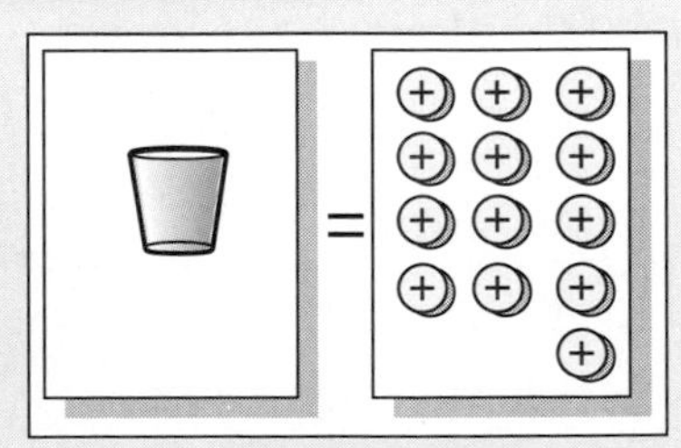

What do you think?

1. What is the solution of $x + (-5) = 8$?
2. How do you know what type of counter to add to each side of the mat?

Activity Two

Solve $3x - (-2) = 11$.

First rewrite the expression using the additive inverse. Then $3x - (-2) = 11$ becomes $3x + 2 = 11$.

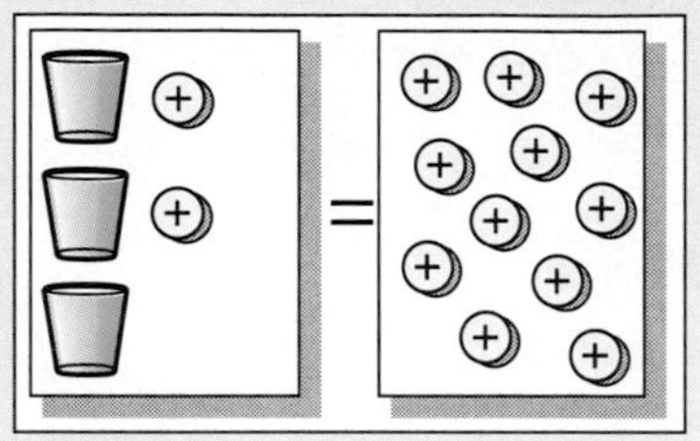

- Begin with an empty mat.
- To represent $3x$, put 3 cups on one side of the mat. Also add 2 positive counters on that side. Put 11 positive counters on the right mat.

Our goal is still to get the cups alone on one side and all counters on the other side.

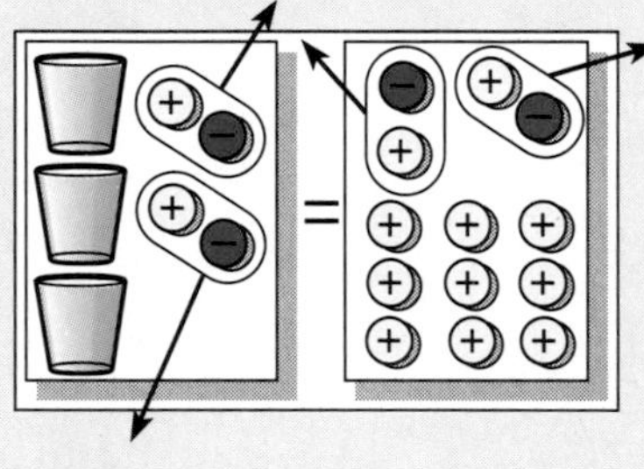

- Add 2 negative counters to each side.
- Remove any zero pairs that can be formed on each side of the mat.

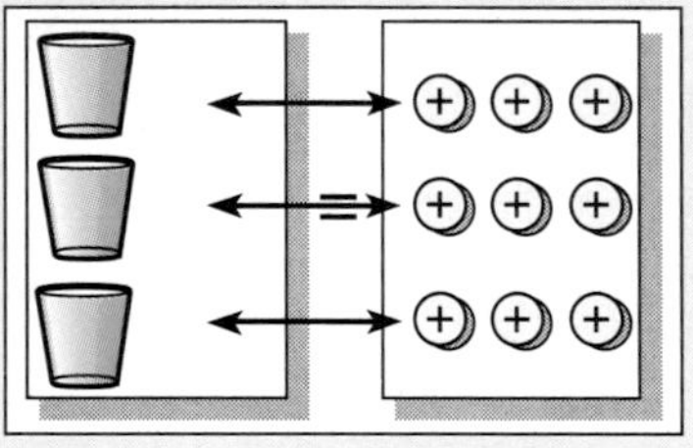

- Arrange the remaining counters on the right side of the mat into 3 equal groups so that they correspond to the 3 cups.

What do you think?

3. How many counters correspond to each cup?
4. What is the solution of $3x - (-2) = 11$?

Write an equation for each model.

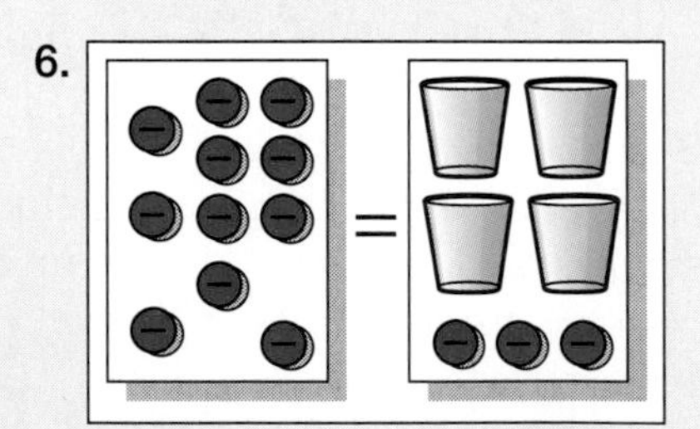

5. 6.

Solve these equations by using models.

7. $x + 3 = 9$
8. $x - 6 = 2$
9. $x + (-4) = 6$
10. $x + 6 = -2$
11. $x - (-2) = 4$
12. $x - (-4) = -3$
13. $3x = -15$
14. $2x = 14$
15. $4x + 6 = 18$
16. $2x - 1 = 11$
17. $2x + 3 = -5$
18. $3x - 5 = -5$

Algebra Connection

3-9 Solving Equations

Objective

Solve equations with integer solutions.

When chemical reactions occur, the resulting molecules may not resemble the elements being combined. For example, oxygen and hydrogen are gases. But when they react, they can form water, a liquid.

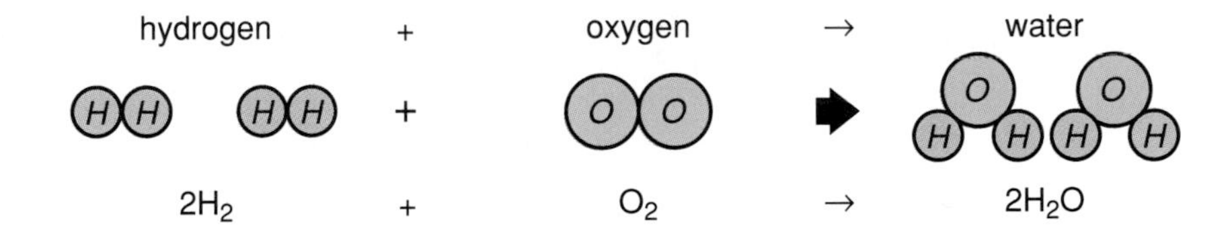

Chemists describe these reactions by using chemical equations. The equations are balanced. That is, they always have the same number of atoms on each side.

Equations in mathematics are also balanced. Whatever you do to one side of the equation, you must also do to the other.

You solve equations involving integers in the same way you solve equations involving whole numbers or decimals.

Examples

1 **Solve $m - (-45) = 35$.**

$m - (-45) = 35$

$m + 45 = 35$ *Rewrite the equation using the additive inverse.*

$m + 45 - 45 = 35 - 45$ *Subtract 45 from each side.*

$m = -10$

Check: $m - (-45) = 35$

$-10 - (-45) \stackrel{?}{=} 35$ *Replace m with −10.*

$-10 + 45 \stackrel{?}{=} 35$

$35 = 35$ ✓

2 **Solve $-15 = -5w$.**

$-15 = -5w$

$\frac{-15}{-5} = \frac{-5w}{-5}$ *Divide to undo multiplication.*

$3 = w$

Check: $-15 = -5w$

$-15 \stackrel{?}{=} -5(3)$ *Replace w with 3.*

$-15 = -15$ ✓

"When am I ever going to use this?"

Suppose you are planting flowers in a garden. Read the package of seeds to see how far apart the seeds need to be planted. If you want 10 seeds in a row, use algebra to find out how long the row would be if you wanted 4 in. of extra space on each end of the row.

Integers are also used in equations that require two steps to solve. You work backward to solve for the variable.

Example 3

LOOKBACK

You can review two-step equations on page 67.

Solve $4x - (-7) = -17$.

First rewrite the equation using the additive inverse.
Then $4x - (-7) = -17$ becomes $4x + 7 = -17$.

$$
\begin{aligned}
4x - (-7) &= -17 \\
4x + 7 &= -17 && \textit{Rewrite using the additive inverse.} \\
4x + 7 - 7 &= -17 - 7 && \textit{Subtract 7 from each side.} \\
4x &= -24 \\
\frac{4x}{4} &= \frac{-24}{4} && \textit{Divide each side by 4.} \\
x &= -6
\end{aligned}
$$

Check: Use your calculator. Replace x with –6.

4 [×] 6 [+/–] [–] 7 [+/–] [=] -17

Equations with integers can be used to represent real-life problems.

Example 4 ***Problem Solving***

Personal Finance Morty borrowed $63 from his parents to buy a video game. He promised to pay it back. Then he borrowed money two times to go to the movies. He now owes his parents $87. If he borrowed equal amounts each time he went to the movies, how much did he borrow each time?

Estimation Hint

The total is about $90 and he borrowed about $60 the first time. Think: $\frac{90 - 60}{2} = 15$. He borrowed about $15 each time to go to the movies.

Explore We know his original debt is $63 and he now owes $87. He borrowed two equal amounts. We need to find out what those amounts were.

Plan Let m represent the money he borrowed to go to the movies. Since he went twice, the amount borrowed is $2m$. Write an equation to solve this problem.

The total owed equals the sum of what he owed before and the new amount.

$$-63 + 2m = -87$$

Solve *Use pencil and paper.*

$$\begin{aligned} -63 + 2m &= -87 \\ -63 + 63 + 2m &= -87 + 63 \quad \text{Add 63 to each side.} \\ 2m &= -24 \\ \frac{2m}{2} &= \frac{-24}{2} \quad \text{Divide each side by 2.} \\ m &= -12 \end{aligned}$$

Use a calculator.

87 [+/−] [+] 63 [=] [÷] 2 [=] −12

The solution, −12, means Morty borrowed \$12 each time.

Examine We can check our solution by adding.

amount he owed	\$63
amount for first movie	\$12
amount for second movie	+ \$12
total borrowed	\$87

This total matches the total given in the problem.

Checking for Understanding

Communicating Mathematics

Read and study the lesson to answer each question.

1. **Tell** the first step in solving $4y + 5 = -15$.
2. **Show** how you could solve the equation in Example 2 by using counters, cups, and mats.
3. **Write** a sentence to explain why it is always a good idea to check your solution to an equation.

Guided Practice

Solve each equation. Check your solution.

4. $2y = -90$
5. $c + 16 = -64$
6. $\frac{m}{-14} = 32$
7. $x - (-35) = -240$
8. $36 = -12 + 8m$
9. $-3y + 15 = 75$
10. $\frac{t}{6} - 5 = -13$
11. $2x - (-34) = 16$

Exercises

Independent Practice

Solve each equation. Check your solution.

12. $2t = -98$
13. $s - (-350) = 32$
14. $\frac{y}{15} = 22$
15. $2w + 35 = 105$
16. $45 = y - 13$
17. $-200 = \frac{z}{3}$
18. $4p - 15 = -75$
19. $-69t = -4{,}968$
20. $q + (-367) = 250$
21. $-30 = 42 + c$
22. $5d + 120 = 300$
23. $4m - 15 = 45$

Write an equation for each problem and solve.

24. The sum of two integers is –24. One of the integers is –13. What is the other integer?

25. The product of two integers is 35. One of the integers is –7. What is the other integer?

26. Twice a number plus 7 is –21. What is the number?

Mixed Review

27. Estimate 4,286 + 3,716 by rounding. *(Lesson 1-3)*

28. Solve $30 = k - 141$. Check your solution. *(Lesson 2-3)*

29. Solve $9 = \frac{h}{4} - 6$. Check your solution. *(Lesson 2-7)*

30. Solve $g = -56 - 77$. *(Lesson 3-5)*

31. Solve $f = 320 \div (-40)$. *(Lesson 3-7)*

Problem Solving and Applications

32. **Critical Thinking** Solve each equation below.

$$2x = -2$$
$$3x = -6$$
$$4x = -16$$
$$5x = -40$$
$$6x = -96$$

a. What pattern do you notice about the multiplier of x?

b. What pattern do you notice in the solutions?

c. Use the patterns you found to write the next five equations that follow in this pattern.

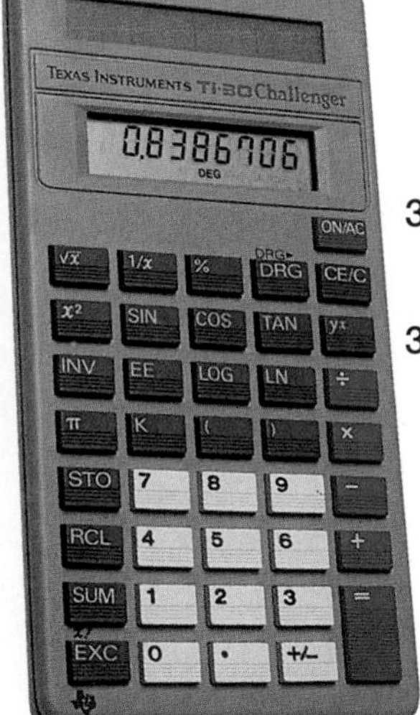

33. **Football** A football team lost 73 yards in 14 plays. Estimate the average yards lost on each play.

34. **Critical Thinking** Use your calculator to solve these equations.

a. $3.1 + (-2.1) = x$

b. $4.5 - (4.5) = z$

c. $5(-2.50) = w$

d. $\frac{-8.758}{2} = m$

e. $(-3.470)(-2.11) = r$

f. $\frac{-5.5555}{-1.1111} = p$

g. Write a sentence to compare solving equations with positive and negative decimals and solving equations with integers.

35. **Marketing** The profits for one day at Safety First Car Rental can be found by solving the equation $35n - 150 = P$, where n is the number of cars rented and P is the profit. On Tuesday, there were 4 cars rented.

a. Find what the profit was for that day.

b. What does your answer mean?

36. **Journal Entry** Make up a problem that can be solved by using the equation $2w + 24 = 75$.

3-10 Coordinate System

Objective

Graph points on a coordinate plane.

Words to Learn

coordinate system
origin
x-axis
y-axis
quadrant
x-coordinate
y-coordinate
ordered pair

You have probably used a map at some time in your life. Did you know that the first person to use a coordinate system on a map was a second century Chinese scientist named Chang Heng? He did this so that positions, distances, and pathways could be studied in a more scientific way.

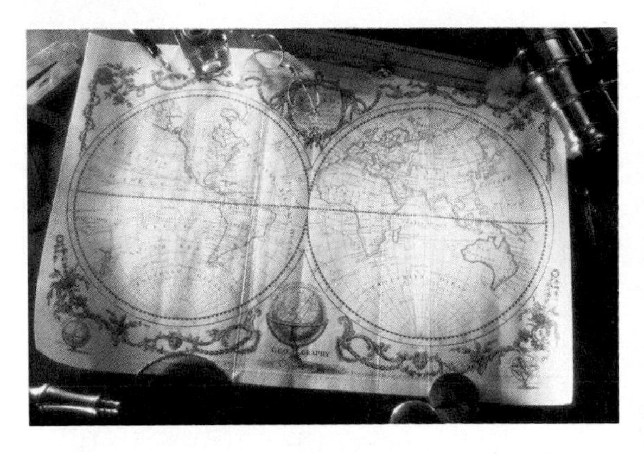

When reading maps, cities are often located by codes, such as B-3, that designate in which section of the map you can find the city. Sometimes you still have to hunt for that city because the section contains many cities.

The coordinate system is also called the coordinate plane.

In mathematics, we can locate a point more exactly by using a **coordinate system.** The coordinate system is formed by the intersection of two number lines that meet at their zero points. This point is called the **origin.** The horizontal number line is called the ***x*-axis,** and the vertical number line is called the ***y*-axis.** The two axes separate the coordinate plane into four sections called **quadrants.**

y-axis
Quadrant II
Quadrant I
origin
x-axis
Quadrant III
Quadrant IV

You can graph any point on the coordinate plane by using an **ordered pair** of numbers. The first number in the pair is called the ***x*-coordinate.** The second number is called the ***y*-coordinate.** The coordinates are your directions to find the point.

Example 1

Graph the point whose coordinates are (–6, 3).

Begin at the origin. The *x*-coordinate is –6. This tells you to go 6 units left of the origin.

The *y*-coordinate is 3. This tells you to go up 3 units.

Draw a dot. You have now graphed the point whose coordinates are (–6, 3).

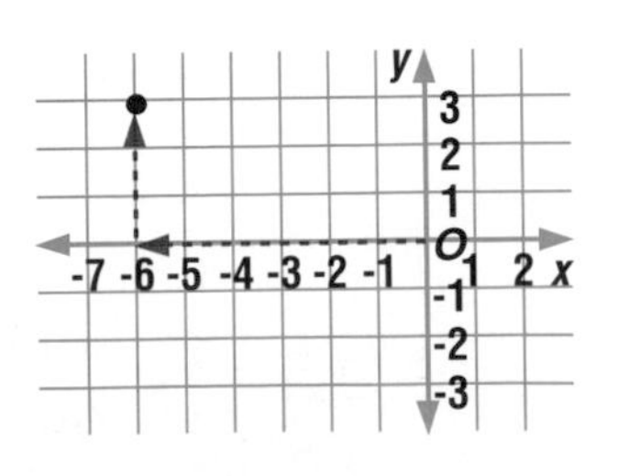

Sometimes points are named by using letters. The symbol $B(3, 4)$ means point B has an x-coordinate of 3 and a y-coordinate of 4.

Example 2

Name the ordered pair for point C.

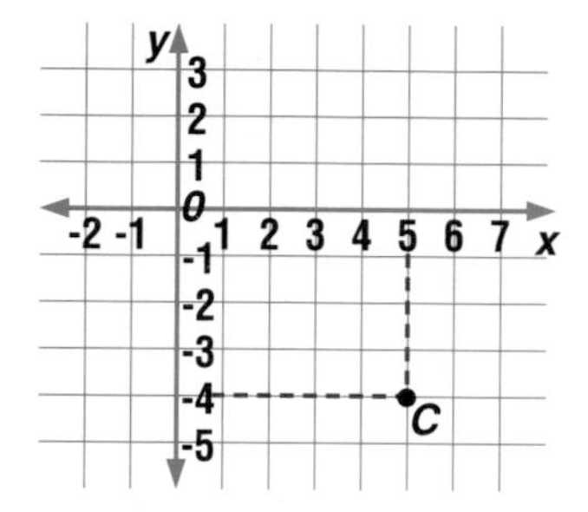

Go right on the x-axis to find the x-coordinate of point C. The x-coordinate is 5.

Go down along the y-axis to find the y-coordinate. The y-coordinate is -4.

The ordered pair for point C is $(5, -4)$.

Checking for Understanding

Communicating Mathematics

Read and study the lesson to answer each question.

1. **Draw** a coordinate system and label the origin, x-axis, y-axis, and each of the quadrants.
2. **Tell** the coordinates of the origin.
3. **Tell** in your own words how to graph the point whose coordinates are $(5, -7)$.
4. **Write** a sentence to tell what the symbol $E(6, 10)$ means.

Guided Practice

Name the ordered pair for the coordinates of each point graphed on the coordinate plane at the right.

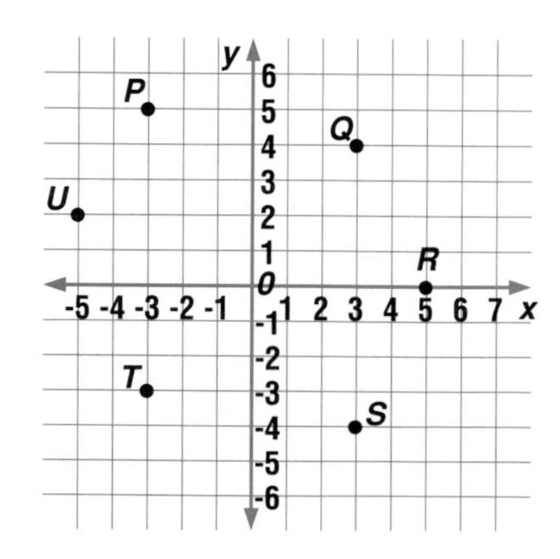

5. P
6. Q
7. R
8. S
9. T
10. U

Graph each point on the same coordinate plane.

11. $A(-3, -9)$	12. $B(-4, 3)$	13. $C(5, 0)$
14. $D(9, -7)$	15. $E(7, 7)$	16. $F(0, 0)$

Exercises

Independent Practice

Name the ordered pair for the coordinates of each point graphed on the coordinate plane at the right.

17. R 18. L 19. K
20. M 21. C 22. X
23. J 24. B 25. T

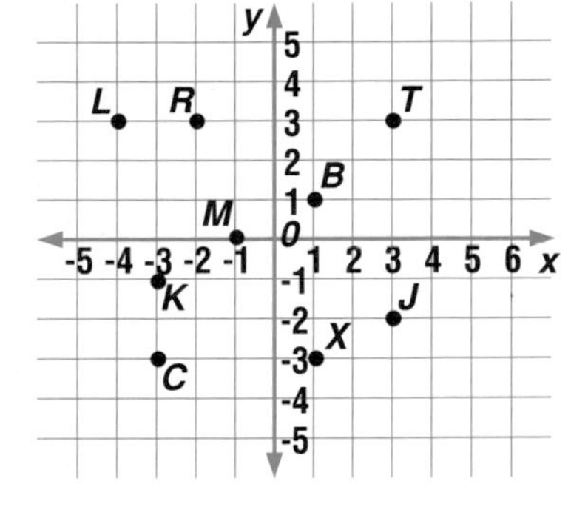

Graph each point on the same coordinate plane.

26. $J(1, 0)$ 27. $I(4, -7)$
28. $H(0, 7)$ 29. $G(-10, -3)$
30. $F(-4, -7)$ 31. $E(7, -8)$
32. $D(-6, 0)$ 33. $C(9, 9)$

Mixed Review

34. **Number Sense** The product of 9 and a number r is 54. Find the number. *(Lesson 2-4)*
35. **Number Sense** Write an inequality for *Four times a number is less than 20.* Then solve the inequality. *(Lesson 2-10)*
36. Solve $-7(-31) = w$. *(Lesson 3-6)*
37. Solve $\frac{-108}{12} = a$. *(Lesson 3-7)*
38. Solve $-2y + 15 = 55$. *(Lesson 3-9)*

Problem Solving and Applications

39. **Critical Thinking** Without graphing, tell in which quadrant each point lies.
    a. $C(4, 5)$ b. $F(-1, -4)$ c. $R(-3, 2)$
    d. $W(3, -10)$ e. $S(-4, -12)$ f. $Z(800, 400)$
40. **Geometry** Graph the points $R(-2, 2)$, $S(4, 2)$, $T(2, -1)$, and $U(-4, -1)$ on the same coordinate plane. Draw line segments from R to S, S to T, T to U, and U to R. What shape is formed?

41. **Video Entertainment** In Nintendo's *The Legend of Zelda*™, Link must travel through many rooms arranged in a coordinate maze to find the silver arrow. From the entry room, he goes right 3 rooms, up 4 rooms, left 5 rooms, down 2 rooms, left 1 room, up 3 rooms, and right 4 rooms to find the chamber that has the silver arrow. If the entry room has coordinates (0, 0) and all directions are as you view the maze, find the coordinates of the room with the arrow.
42. **Critical Thinking** Graph the points $A(6, 1)$, $B(6, -3)$, $C(-1, -3)$, and $D(-1, 1)$. Connect the points with line segments and find the area of the rectangle formed.
43. **Portfolio Suggestion** Review the items in your portfolio. Make a table of contents of the items, noting why each item was chosen. Replace any items that are no longer appropriate.

Chapter

3 Study Guide and Review

Communicating Mathematics

Choose the letter of the correct word or words to complete each sentence.

1. The ___?___ of a number is the distance it is from zero on a number line.
2. To subtract an integer, add its ___?___ .
3. The product of two integers with the same sign is ___?___ .
4. The quotient of two integers with different signs is ___?___ .
5. The two axes separate the coordinate plane into four ___?___ .
6. The ___?___ is the first number in an ordered pair.

a. positive
b. negative
c. x-coordinate
d. y-coordinate
e. absolute value
f. additive inverse
g. quadrants

7. Write a sentence that explains why the additive inverse of zero is zero.

Self Assessment

Objectives and Examples	***Review Exercises***
Upon completing this chapter, you should be able to:	*Use these exercises to review and prepare for the chapter test.*
• graph integers on a number line and find absolute value *(Lesson 3-1)* Graph −2 and 1. Find each number on a number line. Draw a dot there. 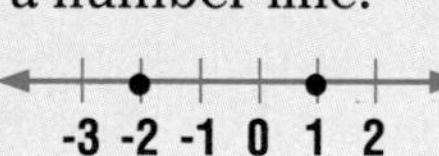 	**Graph each set of numbers on a number line.** 8. {2, 4, 7} 9. {−3, −1, 0} 10. {−6, −2, 1, 4} 11. {−5, −3, −1, 3}
• compare and order integers *(Lesson 3-2)* Replace ● with >, <, or = in −6 ● −11. −6 is to the right of −11 on a number line, so −6 > −11.	**Replace each ● with >, <, or =.** 12. −29 ● −345 13. $\lvert -481 \rvert$ ● 481 14. −15 ● 1 15. −8 ● $\lvert -8 \rvert$
• add integers *(Lesson 3-3)* Solve $-15 + (-25) = d$. $\lvert -15 \rvert + \lvert -25 \rvert = 15 + 25$ or 40 So, $d = -40$.	**Solve each equation.** 16. $128 + (-75) = z$ 17. $-64 + (-218) = j$ 18. $m = -47 + 29$

Objectives and Examples	*Review Exercises*
• add more than two integers *(Lesson 3-4)* Solve $s = 32 + (-18) + 6$. $s = [32 + (-18)] + 6$ $s = 14 + 6$ $s = 20$	**Solve. Check by solving another way.** **19.** $21 + 15 + (-7) + 3 = k$ **20.** $-16 + 38 + (-25) + 1 = x$ **21.** $v = -54 + 81 + 54$ **22.** $29 + (-60) + 11 + (-5) = p$
• subtract integers *(Lesson 3-5)* Solve $c = -20 - 12$. $c = -20 - 12$ $c = -20 + (-12)$ $c = -32$	**Solve each equation.** **23.** $46 - (-62) = b$ **24.** $y = -59 - 33$ **25.** $j = -86 - (-96)$ **26.** $-17 - 28 = n$
• multiply integers *(Lesson 3-6)* Solve $t = (-8)(4)(6)$. $t = [(-8)(4)](6)$ $t = (-32)(6)$ $t = -192$	**Solve each equation.** **27.** $u = -8(12)$ **28.** $w = -5(-20)$ **29.** $(-10)(-2)(4) = q$ **30.** $(25)(-3)(1) = g$
• divide integers *(Lesson 3-7)* Solve $a = -42 \div (-7)$. $a = -42 \div (-7)$ $a = 6$	**Solve each equation.** **31.** $\frac{-66}{6} = c$ **32.** $\frac{-280}{-7} = h$ **33.** $z = \frac{160}{5}$ **34.** $b = \frac{360}{-24}$
• solve equations with integer solutions *(Lesson 3-9)* Solve $y - (-40) = 275$. $y - (-40) = 275$ $y + 40 = 275$ $y + 40 - 40 = 275 - 40$ $y = 235$	**Solve each equation.** **35.** $-3s = -54$ **36.** $3p - (-18) = 63$ **37.** $28 + m = -94$ **38.** $\frac{r}{4} - 13 = -15$
• graph points on a coordinate plane *(Lesson 3-10)* Graph the point whose coordinates are $(4, -2)$. 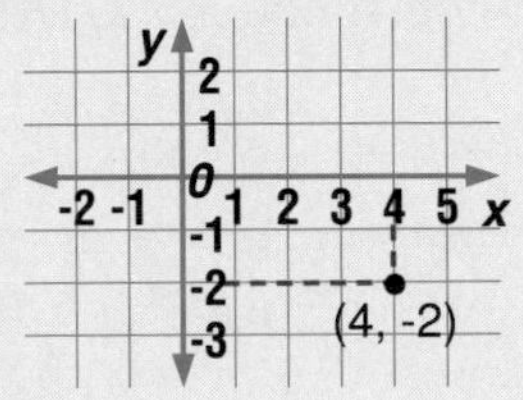 	**Graph each point on the same coordinate plane.** **39.** $A(-2, 8)$ **40.** $B(-4, -1)$ **41.** $C(5, 7)$ **42.** $D(3, -6)$

Study Guide and Review

Study Guide and Review

Applications and Problem Solving

43. **Chess** The Chess Club at Wiley Middle School is holding a chess tournament in which each player earns +1 for a win, −1 point for a loss, and 0 for a draw. The chart at the right shows the records of five of the players.
   a. Determine each player's total points.
   b. Graph the total scores on a number line. Let the first letter of each player's name be the letter for each coordinate. *(Lesson 3-1)*

Player	Wins	Losses
Niko	6	0
Chuck	2	4
Rachel	1	5
Trenna	4	2
Amanda	3	3

44. **Games** Mr. Walter invents board games. In his new game *INTEGO,* you draw cards to determine how far you go on the board. Each card has an integer on it. Positive integers tell you to go forward that many spaces. Negative integers tell you to go back that many spaces. On three turns, a player drew +2, −6, and +8. Describe where the player finally landed in respect to his starting position. *(Lesson 3-4)*

45. **Education** Carmen received her graded test back during math class. The test had five sections. In each section, Carmen received a score of −3, which meant a loss of 3 points. How many points did Carmen lose on the test? *(Lesson 3-6)*

46. **Zoology** The heaviest domestic dog is the St. Bernard, which weighs up to 220 pounds. How much more does a St. Bernard weigh than the smallest domestic dog? *(Lesson 3-8)*

47. **Number Theory** When you add 5 to a certain number, then subtract −10, multiply by −4, and divide by 6, you get 12. What is the number? *(Lesson 3-7)*

Curriculum Connection Projects

- **History** Write a short report on mathematician Rene Descartes and the Cartesian coordinate system.
- **Graphic Arts** Draw a straight-line figure, such as a house, on a coordinate plane. Label the coordinates of each corner. Multiply each x- and y-coordinate by 2. Draw the new figure.

Read More About It

Arthur, Lee, Elizabeth James, and Judith B. Taylor. *Sportsmath: How It Works.*
Cresswell, Helen. *Absolute Zero.*
Mango, Karin N. *Mapmaking.*

Chapter

3 Test

Chapter Test

1. Find the absolute value of -9.

Replace each ● with >, <, or =.

2. 3 ● -20
3. $|-44|$ ● 44
4. -837 ● -164
5. $|-51|$ ● $|-14|$

6. Order the numbers in the set $\{-6, -179, 20, 134, -67, 5, -348\}$ from least to greatest.

Solve each equation.

7. $r = -582 + 68$
8. $p = -4(-16)$
9. $m = -112 \div 16$
10. $-231 - 128 = d$
11. $(8)(-10)(3) = h$
12. $3g - 14 = -50$
13. $3t = 48 - (-72)$
14. $-8 + 21 + (-12) + 15 = s$

15. Graph the set $\{5, -2, -4, 0, 3\}$ on a number line.
16. State whether the product $(-7)(-5)(-9)(-1)(-4)$ will be positive or negative.

Evaluate each expression if $c = -4$, $m = 5$, and $t = -10$.

17. $3mc^2$
18. $2ct \div m$
19. $-6c - 60$

20. Write an equation to represent the sum of a number divided by -4, plus 20, is 26. Then solve.

Name the ordered pair for the coordinates of each point shown on the graph at the right.

21. R
22. E
23. H
24. M

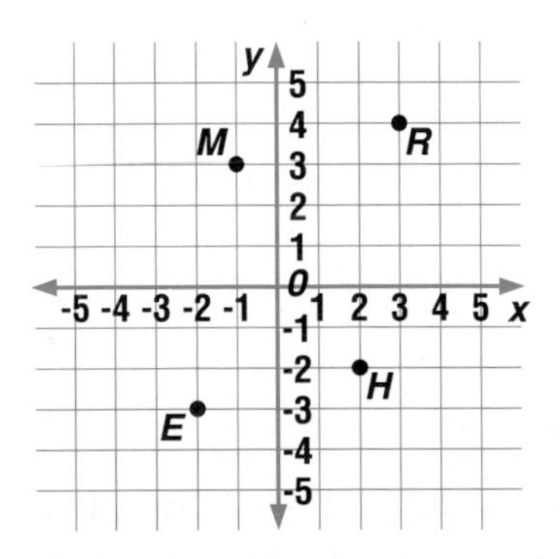

25. **Weather** Schools were closed in the district due to a winter storm. The temperature dropped 24°F over a four-hour period. If the temperature dropped at an even rate, how many degrees did the temperature fall each hour?

Bonus Find two integers x and y so that $x > y$, $xy = -12$, and $x \div y = -3$.

Chapter 3

Standard Format, Chapters 1-3

Academic Skills Test

Directions: Choose the best answer. Write A, B, C, or D.

1. Maria bought 3 blouses for $108. If each blouse cost the same amount, how much did each blouse cost?

   A. $30.24 B. $32.40
   C. $36.00 D. $54.00

2. $(3 \times 8) + (17 \times 8)$ is equivalent to

   A. 20×16
   B. $3(8 + 17)$
   C. 20×8
   D. $8(3 \times 17)$

3. John bought a stereo that cost $532.16. He paid for the stereo in 12 equal payments. Estimate: the amount of each payment was between

   A. $10 and $20
   B. $25 and $35
   C. $40 and $50
   D. $55 and $65

4. Which is a reasonable estimate of the number of meals you will eat in 10 years? (Assume 3 meals a day.)

   A. 300 B. 1,000
   C. 10,000 D. 30,000

5. 14.5 cm =

   A. 0.145 m B. 1.45 m
   C. 145 m D. 1,450 m

6. The product of a number and itself is 576. What is the number?

   A. 24 B. 30
   C. 144 D. 288

7. How is the product $4 \cdot 4 \cdot 4$ expressed using exponents?

   A. $4 \cdot 3$ B. 4^3
   C. 3^4 D. 4^4

8. If $a = 6$ and $b = 3$, what is the value of ab?

   A. 216 B. 63
   C. 18 D. 2

9. Which equation is equivalent to $x + 5 = 12$?

   A. $x = 12$
   B. $x + 5 - 5 = 12 - 5$
   C. $x + 10 = 24$
   D. $x + 5 - 5 = 12 + 5$

10. A certain number is divided by 4 and then 5 is subtracted from the result. The final answer is 25. What is the number?

    A. 5 B. 80
    C. 100 D. 120

11. What is the solution of the inequality $12 > t + 8$?

    A. $t < 4$ B. $t < 12$
    C. $t > 12$ D. $t > 20$

12. Which equation represents *three more than twice a number equals 14?*

A. $2x + 3 = 14$
B. $2x = 14 + 3$
C. $2(x + 3) = 14$
D. $3 + x + 2 = 14$

13. What is the perimeter of the rectangle shown at the right?

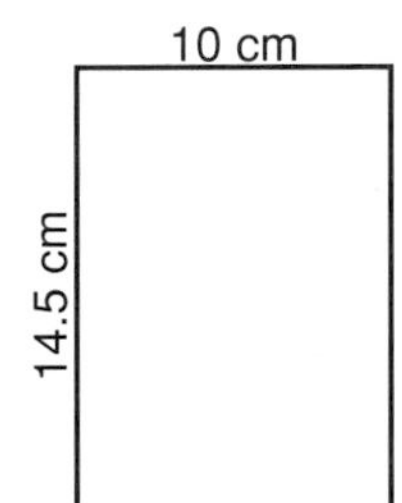

A. 24.5 cm
B. 34.5 cm
C. 49 cm
D. 145 cm

14. Which integers are graphed on the number line below?

A. {−5, −2}
B. {−5, −4, −3, ...}
C. {−5, −2, 0, 1}
D. {−5, 0, 5}

15. Which symbol replaces the ● to make a true sentence?

−4.5 ● |−5.2|

A. <
B. >
C. =
D. none of these

16. 56 + (−32) =

A. −88
B. −24
C. 24
D. 88

17. 20 · (−9) =

A. −180
B. −18
C. 18
D. 180

18. −64 ÷ 8 =

A. −12
B. −8
C. 8
D. 12

Test-Taking Tip

Most standardized tests have a time limit, so you must budget your time carefully. Some questions will be much easier than others. If you cannot answer a question within a few minutes, go on to the next one. If there is still time left when you get to the end of the test, go back and work the questions that you skipped.

You can prepare for taking standardized tests by working through practice tests like this one. The more you work with questions in a format similar to the actual test, the better you become at test taking. Do not wait until the night before taking a test to review. Allow yourself plenty of time to review the basic skills and formulas that are tested.

19. If $\frac{c}{-3} = 6$, what is the value of c?

A. −18
B. −2
C. 2
D. 18

20. Which point is in the second quadrant?

A. K
B. L
C. M
D. N

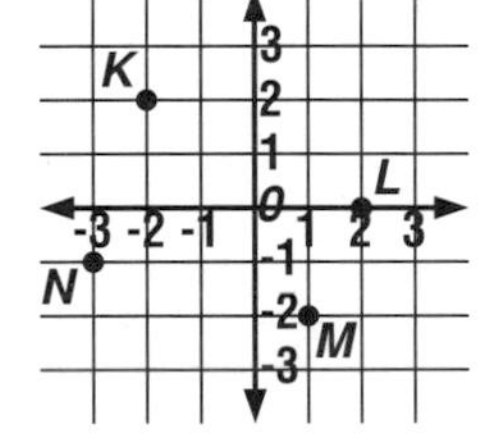

Academic Skills Test

Chapter

Statistics and Data Analysis

Spotlight on Mountain Climbing

Have You Ever Wondered...

- How many steps you would have to climb to climb as high as a mountain?
- How many laps you would have to swim in a swimming pool to equal the distance across the Atlantic Ocean?

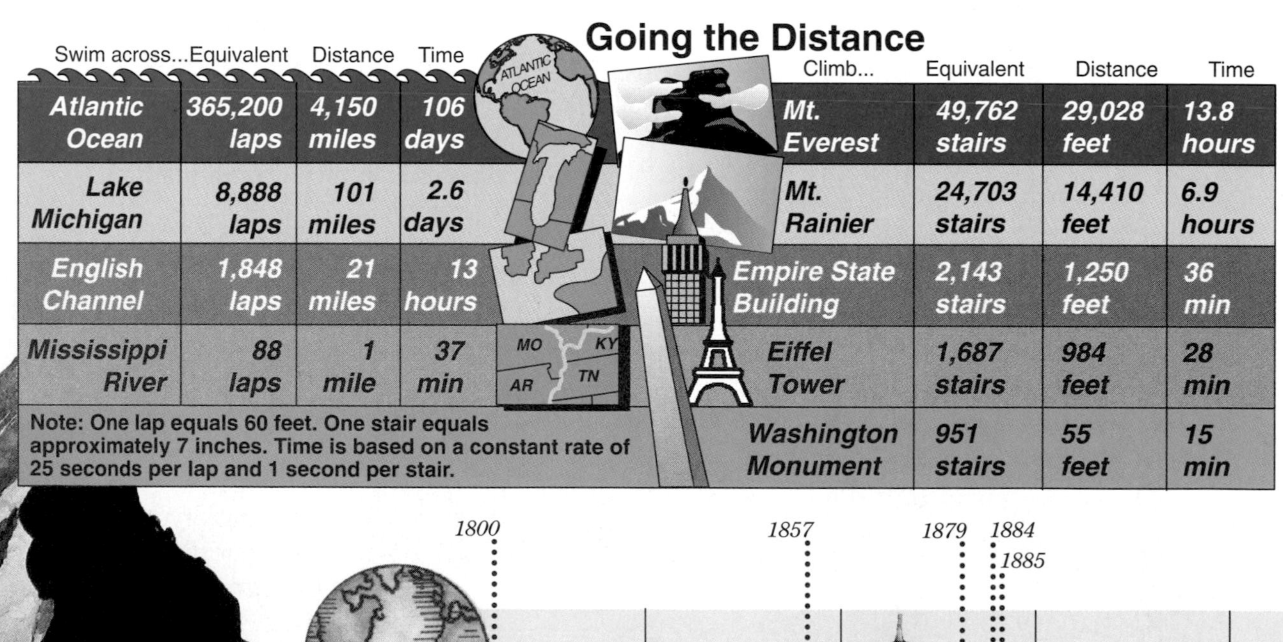

Going the Distance

Swim across...	Equivalent	Distance	Time
Atlantic Ocean	365,200 laps	4,150 miles	106 days
Lake Michigan	8,888 laps	101 miles	2.6 days
English Channel	1,848 laps	21 miles	13 hours
Mississippi River	88 laps	1 mile	37 min

Climb...	Equivalent	Distance	Time
Mt. Everest	49,762 stairs	29,028 feet	13.8 hours
Mt. Rainier	24,703 stairs	14,410 feet	6.9 hours
Empire State Building	2,143 stairs	1,250 feet	36 min
Eiffel Tower	1,687 stairs	984 feet	28 min
Washington Monument	951 stairs	55 feet	15 min

Note: One lap equals 60 feet. One stair equals approximately 7 inches. Time is based on a constant rate of 25 seconds per lap and 1 second per stair.

Chapter Project

Mountain Climbing

Work in a group.

1. Guess how many steps you climb in a day and in a week. Be sure to include the steps at home, at school, on the bus, and even at the shopping mall.
2. Keep a log of how many steps you climb each day for one week. Make a graph showing your results. Compare the actual number with your initial guess.
3. If each step is 7 inches tall, how high did you climb in a week? Is this as high as any mountain? How many weeks would it take you to climb the equivalent of Mt. McKinley?

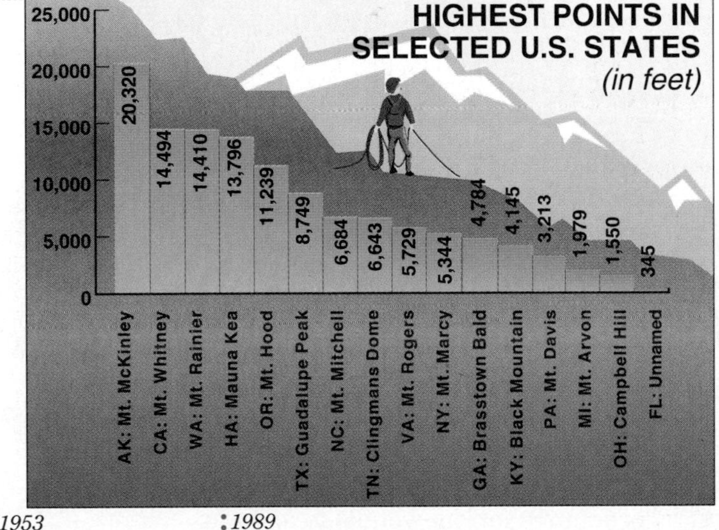

1927 — *Gertrude Ederle is the first woman to swim English Channel*

1930s — *Great Depression*

1931

1953 — *First people reach the top of Mt. Everest*

1989 — *American troops enter Panama*

1930 1960 1990

Looking Ahead

In this chapter, you will see how mathematics can be used to answer questions about mountain climbing. The major objectives of the chapter are to:

- make and interpret tables and plots
- measure the central tendency of a set of data
- represent information using frequency tables, histograms, line plots, stem-and-leaf plots, and box-and-whisker plots
- recognize misleading graphs and statistics

Cooperative Learning

4-1A Data Base

A Preview of Lesson 4-1

Objective

Explore the use of a computer data base.

A **data base** is a collection of information organized for quick search and retrieval by a computer. The information, or *data,* in a data base is referred to as a *file.* The file is organized into *fields* and *records.*

In Chapter 1, you learned that a spreadsheet is organized into rows and columns. In a data base, a field is a heading of a column. Each row in a data base is called a record. The data in each record is related.

The data base below contains data for a company's mailing list.

CUSTOMER NAME	CUSTOMER NUMBER	ADDRESS	CITY	ST	ZIP CODE
B & G INTERNATIONAL	152	113 E. FIFTH ST.	NEW YORK	NY	10003
WIRE & CABLE	350	1 INDUSTRIAL BLVD.	CLIFTON	NJ	07444
ABC CO.	210	15 EXECUTIVE WAY	TEANECK	NJ	07666

Activity One

- Look at the data base above. Name the fields.
- How many records are in this data base?

What do you think?

1. With a data base, you can ask for certain records to be retrieved that have a specific characteristic. Suppose you ask the computer to retrieve all records under the field State that are in New York. What records would it retrieve?
2. The Amex Company has a mailing list of customers with 24,000 records. They do not want to send their brochures for snowboots to all their customers.
   a. How could they use a data base to send brochures only to those customers who would be interested?
   b. Which customers do you think would be interested?

Sometimes it is necessary to alter a certain part of every record in a data base. The part can be changed for each record individually, or a quicker way is to use the REPLACE command.

Activity Two

- Study the data base shown below. Identify the fields and records.

STATE CODE	SALES DISTRICT	SALESPERSON NUMBER	SALES AMOUNT	BONUS
15	2	101	500.00	50.00
18	3	200	89.95	8.99
10	1	250	1050.00	105.00
11	4	300	925.00	92.50
15	2	101	8010.00	801.00
16	2	210	3000.00	300.00
10	1	250	2500.00	250.00
10	1	250	1750.00	175.00

- Suppose you wanted to increase everyone's bonus by 10%. This can be done by multiplying the Bonus by 1.10. Calculate the new bonus figures for each record.

What do you think?

3. Complete the command that would change the bonus values for you automatically.

   REPLACE ALL BONUS WITH BONUS* ___?___ .

4. Suppose you wanted to increase everyone's bonus by 15%. Write the command that would change this.
5. Suppose you wanted to give everyone a $200 bonus in addition to their regular bonus. Write the command that would do this.

Extension

The REPLACE command can be expanded by adding a condition for the replacement. Suppose your company wanted to increase the bonus by 10%, but only for those with bonuses less than $50. The statement to do this is

REPLACE ALL BONUS WITH BONUS*1.10 FOR BONUS < 50

6. Use the data base in Activity Two.
   a. Write the command that would give all salesperson's with sales greater than $3,000 a 20% raise in their bonus.
   b. Which records would receive the 20% raise?
   c. What would be each new bonus amount?
7. Use the data base in Activity Two.
   a. Write the command that would give all salesperson's with sales less than $1,000 a 5% raise in their bonus.
   b. Which records would receive the 5% raise?
   c. What would be each new bonus amount?

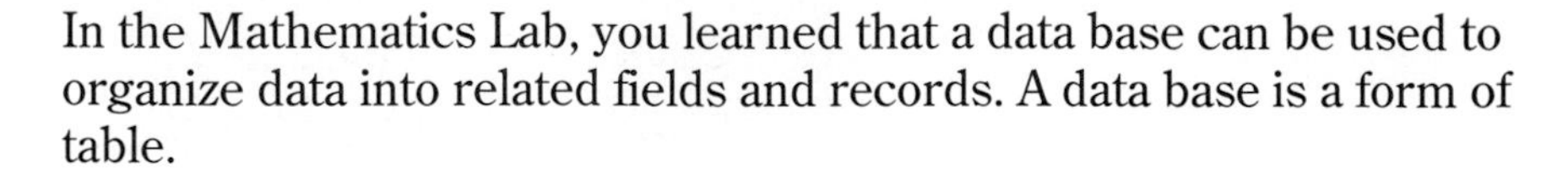

Problem-Solving Strategy

4-1 Make a Table

Objective

Solve problems by organizing data into a table.

Words to Learn

statistics
frequency table
data analysis

In the Mathematics Lab, you learned that a data base can be used to organize data into related fields and records. A data base is a form of table.

Tables are often used in **statistics.** Statistics is the branch of mathematics that deals with collecting, organizing, and analyzing data.

One type of table used in statistics is a **frequency table.** A frequency table tells how many times each piece of data occurs in a set of information. The table below shows the results of a telephone survey done by a radio station.

Favorite Music	Tally	Frequency
Classical	𝍸 𝍸 𝍸 𝍸 𝍸 𝍸 𝍸 I	36
Country	𝍸 𝍸 𝍸 𝍸 𝍸 𝍸 𝍸 𝍸 𝍸	45
Folk	𝍸 𝍸 𝍸 𝍸 𝍸	25
Instrumental	𝍸 𝍸 𝍸 𝍸 𝍸 𝍸 𝍸 III	38
Jazz	𝍸 𝍸 𝍸 𝍸 𝍸 𝍸 𝍸 𝍸	40
Rock	𝍸 𝍸 𝍸 𝍸 𝍸 𝍸 𝍸 𝍸 𝍸 𝍸 IIII	54

The statistician made a tally mark for each response in the appropriate row.

When statisticians study the data and make conclusions from the numbers they observe, they are doing **data analysis.** Sometimes, when there is a wide range of data, statisticians will group the data into intervals.

Example 1

The scores of the Amateur Charity Miniature Golf Tournament are shown below. Make a frequency table of these scores.

25	53	70	45	32
66	72	65	33	19
56	42	85	54	39
41	81	69	60	55
48	52	67	73	89

Explore None of the scores occur more than once.
What is the highest score?
What is the lowest score?
What is the range of scores?

Plan Since none of the scores occur more than once, let's use intervals to group the data. Decide on equal intervals, such as 10. Then tally the scores.

Solve Make a table with three columns. Write the intervals. Tally the scores. Total the tallies.

Scores	Tally	Frequency
11-20	I	1
21-30	I	1
31-40	III	3
41-50	IIII	4
51-60	~~IIII~~ I	6
61-70	~~IIII~~	5
71-80	II	2
81-90	III	3

Examine Check to see if you have recorded each score by finding the total number of scores in the frequency column. Then count the number of scores in the original list. If these two numbers do not match, you need to tally the scores again.

Example 2 ***Problem Solving***

Customer Service The manager of Taco Time had to report the usual time it took the customer to be served after placing an order. She made the frequency table below. What time did she report?

Serving Time(s)	Tally	Frequency
0 - 60	II	2
61 - 120	~~IIII~~ ~~IIII~~	10
121 - 180	~~IIII~~ ~~IIII~~ ~~IIII~~	15
181 - 240	~~IIII~~ ~~IIII~~ ~~IIII~~ III	18
241 - 300	~~IIII~~ ~~IIII~~ I	11

Look for the time that has the most tally marks. The usual amount of time it took to serve a customer was between 181 and 240 seconds.

Checking for Understanding

Communicating Mathematics

Read and study the lesson to answer each question.

1. **Tell** what is wrong with the frequency table at the right.
2. **Write** a sentence to tell what you conclude when you analyze the data in the frequency table on page 130.

Science Test Scores	Tally	Frequency
50-60	I	1
60-70	III	3
70-80	~~IIII~~ I	6
80-90	~~IIII~~ ~~IIII~~ ~~IIII~~	15
90-100	~~IIII~~ ~~IIII~~ II	12

Guided Practice **Make a frequency table for each set of data.**

3. To the nearest hour, how many hours did you talk on the telephone last week?

4	0	1	2	2	3	5	1	2	3
4	0	3	3	0	1	3	2	4	6

4. What was your best game in the Junior Summer Bowling League? *Bowling scores are given in terms of pins.*

150	138	110	135
89	167	133	175
169	200	203	169
133	125	109	138
145	144	189	190

Problem Solving

Practice **Solve using any strategy.**

Strategies

Look for a pattern.
Solve a simpler problem.
Act it out.
Guess and check.
Draw a diagram.
Make a chart.
Work backward.

5. Make a frequency table for this list of prices of popular video games.

\$24.99	\$16.99	\$44.99	\$50.50	\$35.99	\$32.99
\$10.99	\$29.99	\$29.99	\$43.99	\$45.99	\$37.99
\$14.99	\$18.00	\$34.89	\$55.80	\$37.90	\$40.44

   a. How many games are in this list?

   b. To the nearest \$10, which price seems most common?

6. Ninety-seven out of every 100 employees at B.Y. Chemical commute to work by car. The rest use public transportation to get to work. Approximately how many of 1,433 employees use public transportation?

7. Make a frequency table for the data.

Magazine Subscriptions Sold

70	74	12	34	23	78	45	32	55	51
89	43	32	11	25	62	43	78	70	72

8. A number divided by 0.4 equals 20. Find the number.

9. Make a frequency table for the data.

Points Scored Each Game during Basketball Season

102	78	62	98	67
88	98	101	102	89
121	66	78	102	113
120	109	88	97	88

10. Michelle spent \$56 on exactly 15 items at the grocery store. She bought nine breakfast muffins and twice as many frozen dinners as packages of vegetables. How many of each item did she buy?

4-2 Histograms

Objective

Construct and interpret histograms.

Words to Learn

histogram

Do you spend a lot of time watching television? Some experts say the average American may spend as many as 40 hours a week watching television.

One hundred eighth graders were asked how many hours they spent watching television in a week. The frequency table at the right shows the results of this survey. A special kind of bar graph, called a **histogram,** can be used to display this data.

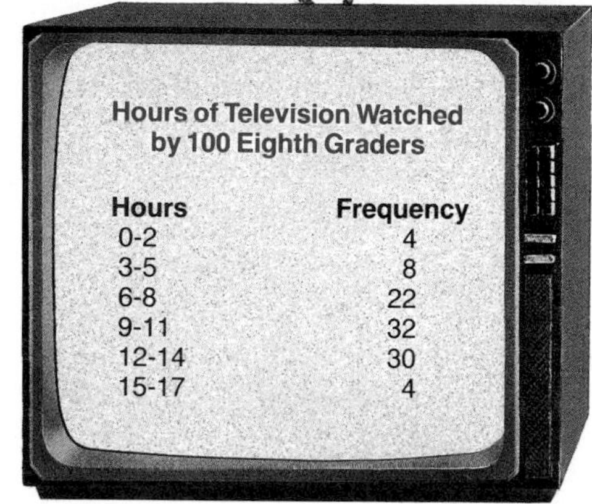

Hours of Television Watched by 100 Eighth Graders

Hours	Frequency
0-2	4
3-5	8
6-8	22
9-11	32
12-14	30
15-17	4

A histogram is a bar graph that displays the frequency of data that has been organized into equal intervals. Because the intervals cover all possible values of data, there are no spaces between the bars of the graph.

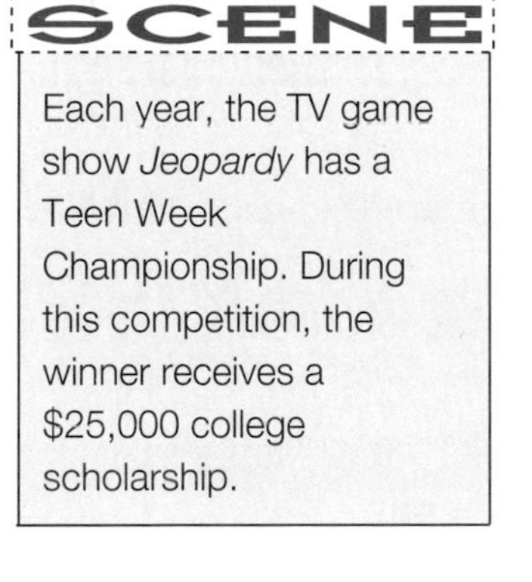

TEEN SCENE

Each year, the TV game show *Jeopardy* has a Teen Week Championship. During this competition, the winner receives a $25,000 college scholarship.

Example 1

Use the data from the frequency table above.

a. Construct a histogram of the data.

Draw a horizontal and vertical axis. Let the horizontal axis represent the time intervals and the vertical axis represent the frequency.

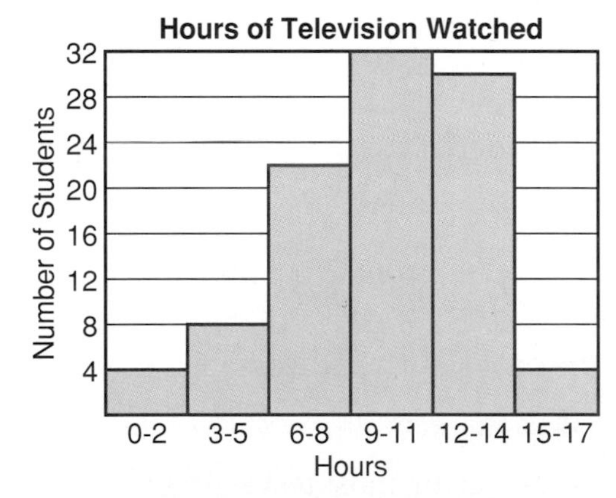

- *The equal intervals are shown on the horizontal axis.*
- *All bars have the same width.*
- *The frequency of the data in each interval is represented by the height of the bar.*
- *The vertical scale is often a factor of the greatest frequency.*

b. How does the number of students who watch TV 3-5 hours compare to the number of students who watch TV 0-2 hours?

Look at the bars representing 3-5 hours and 0-2 hours. The 3-5 bar is twice as tall as the 0-2 bar. Twice as many students watch TV 3-5 hours as those who watch TV 0-2 hours.

Example 2

The heights (to the nearest inch) of 40 teenage boys is given in the chart below. Make a histogram of this data.

Heights	60-61	62-63	64-65	66-67	68-69	70-71	72-73	74-75
Number	5	0	3	7	10	8	5	2

Notice that the interval 62-63 has a frequency of 0. This means there will be no bar at that frequency.

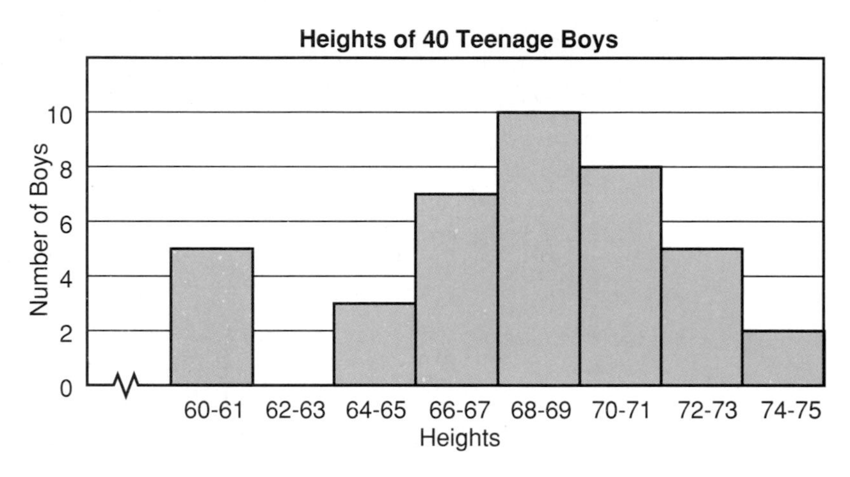

The jagged line on the horizontal axis indicates that all intervals from 0 to 59 have been omitted.

Checking for Understanding

Communicating Mathematics

Read and study the lesson to answer each question.

1. **Write** a sentence to tell how a histogram and a bar graph are the same. How are they different?
2. **Tell** why you think histograms are used to show data with a wide range rather than bar graphs.
3. **Show** why a histogram is a more effective display than a frequency table.
4. **Tell** why there are no spaces between the bars in a histogram.

Guided Practice

Use the histogram at the right to answer each question.

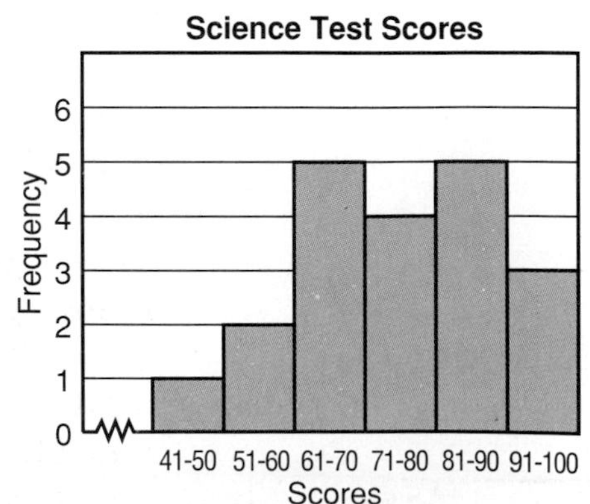

5. Describe the data shown in this graph.
6. How large is each interval?
7. Which interval has the most test scores?
8. Why is there a jagged line in the horizontal axis?
9. Make a frequency table of the same data.

Exercises

Independent Practice

10. Construct a histogram using the data in the table below.

Heights of Presidents of the United States		
Height (in)	Tally	Frequency
63-65	I	1
66-68	~~IIII~~ IIII	9
69-71	~~IIII~~ ~~IIII~~ III	13
72-74	~~IIII~~ ~~IIII~~ ~~IIII~~ III	18
75-77	I	1

a. How many men have been presidents of the United States?
b. What range of heights includes most of the presidents?
c. Compared to the total, how would you describe the number of presidents that have been 6 feet or taller?

11. A survey in the mall stopped people and asked how much change they had with them. Make a frequency table and histogram of the following results.

98¢	88¢	81¢	77¢	74¢	69¢	94¢	85¢	75¢	72¢
85¢	79¢	72¢	65¢	88¢	82¢	78¢	74¢	70¢	62¢

Mixed Review

12. **Algebra** Solve $7 = \frac{r}{15}$. Check your solution. *(Lesson 2-4)*

13. **Geometry** Graph the points $A(-3, 2)$, $B(-3, -1)$, $C(1, -1)$, and $D(1, 2)$ on the same coordinate plane. Draw AB, BC, CD, and AD. Find the perimeter of the rectangle formed. *(Lesson 3-10)*

Problem Solving and Applications

14. **Government** As of July 1990, there were 1,273 women in the United States who held office in state legislatures. New Hampshire has 136 women in legislature. The table shows the numbers of women in the state legislatures in the other 49 states.

Women in State Legislatures	
Number of Women	Number of States
1-10	7
11-20	16
21-30	15
31-40	5
41-50	4
51-60	2

a. Make a histogram of this data.
b. How does the histogram show what range of the number of women is most common?
c. How does New Hampshire compare with the other states that have the most women in legislature?

15. **Critical Thinking** Look at the data in Exercise 10.
a. How many presidents were exactly 6 feet tall? Explain your answer.
b. Why do you think so many presidents were or are tall?

16. **Data Search** Refer to pages 126 and 127.
Use the graph of mountain heights to find which region of the United States appears to have the greatest number of high mountains.

4-3 Line Plots

Objective
Construct and interpret line plots.

Words to Learn
line plot

In a taste test conducted by *Zillions* magazine, 44 readers judged potato chips according to texture, saltiness, potato flavor, and overall scrumptiousness. Twenty-six of the 29 brands of potato chips rated "good" or "very good."

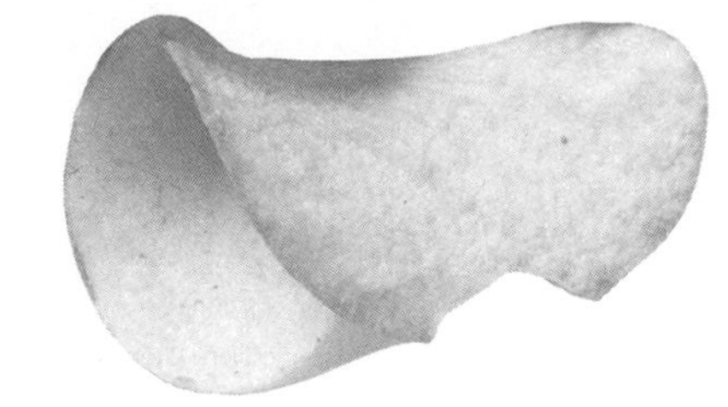

The price of all the different brands were also studied. The price per ounce of the 26 tastiest brands are listed below along with their ratings. *V = very good, G = good*

Ratings and Cost per Ounce (¢) of Potato Chip Brands

12¢-V	21¢-G	14¢-V	25¢-G	12¢-G	18¢-V
20¢-V	17¢-G	20¢-V	21¢-G	21¢-G	23¢-V
21¢-G	24¢-G	17¢-G	21¢-V	10¢-G	28¢-G
23¢-V	21¢-G	22¢-G	13¢-V	20¢-G	24¢-G
19¢-G	25¢-G				

This data could be organized into a frequency table or shown in a histogram. Another way to display the frequency of data is by using a **line plot.** A line plot is a vertical graph of the tally marks you make in creating a frequency table.

In the line plot below, the price per ounce is shown along the horizontal axis. There is no vertical scale. Instead, each x represents a brand that is priced in that category.

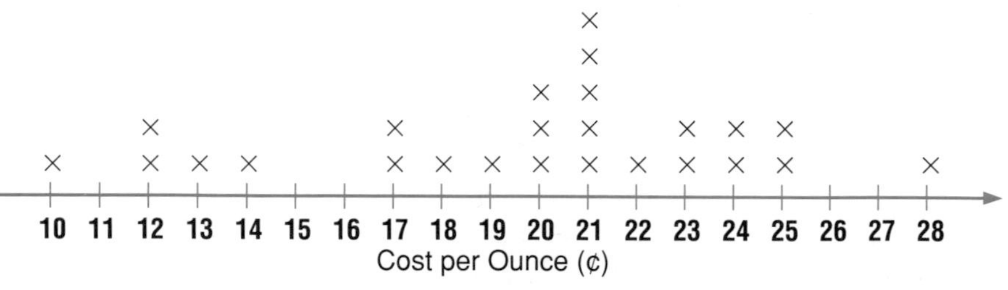

Each person in the United States eats an average of six pounds of potato chips per year.

From this representation, you can see that most potato chip brands are priced around 20-21¢ per ounce. Also you can see that 10¢ per ounce and 28¢ per ounce are extreme prices for chips.

You can alter this line plot to show more information. With this information, further data analysis can be done.

Example

Reconstruct the line plot on page 136 by replacing each x with a letter representing the quality of that brand of chip. Use V for very good and G for good. Find the best price for a chip rated very good.

Refer to the original list to find which chips are good or very good.

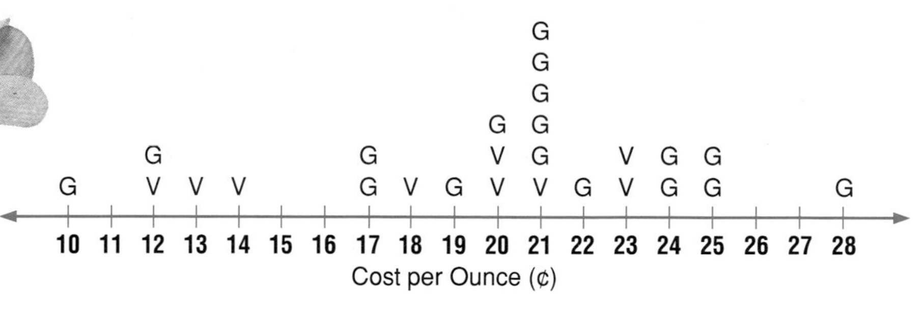

The lowest price for a chip rated very good is 12¢ per ounce.

Mini-Lab

Work in pairs.

Materials: yardstick, tape measure, or other measuring device

- Measure each other's height and record your heights on the chalkboard.
- With the help of your teacher and classmates, make a human line plot of your heights.

Talk About It

a. Which height had the most people?

b. Which heights had more girls than boys?

Checking for Understanding

Communicating Mathematics

Read and study the lesson to answer each question.

1. **Tell** how a line plot is like a frequency table.
2. **Show** how you could use a line plot to verify that the most common prices for brands of potato chips usually are chips rated as *good.*
3. **Write** what the line plot below tells you about the heights of the players on the girls' basketball team.

Heights (in inches) of the Midway Girls' Basketball Team

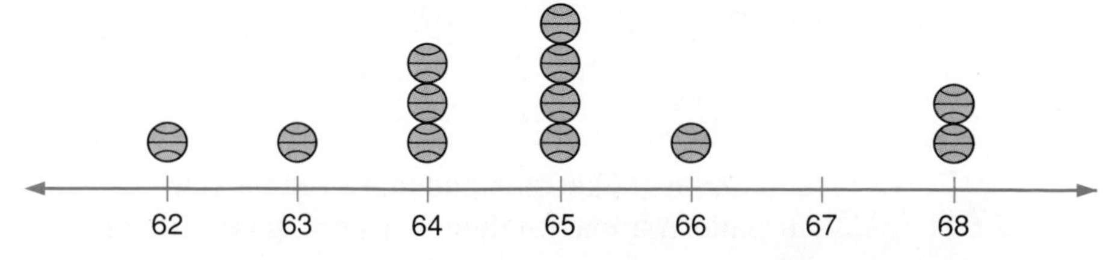

Guided Practice

4. Make a line plot for the following test scores..
   90, 92, 91, 94, 96, 98, 90, 91, 92, 90, 93
5. The weights in kilograms of the Hendricks Middle School wrestling team are shown in the table.
   a. Make a line plot of the data.
   b. Analyze the line plot. Write all conclusions you can make.

Weights (in kg)			
79	79	80	75
72	72	72	72
70	75	90	68
51	65	72	
75	72	80	
68	74	62	

Exercises

Independent Practice

6. The scores on a 50-point social studies test are given below.

45	47	38	40	41	42	50	45	47	42
34	44	41	42	39	40	33	41	45	31

   a. Make a line plot for the data.
   b. What were the highest and lowest scores?
   c. Which score occurred the most often?
   d. If 35 points is considered a passing score, how many scores were passing scores?
7. Mr. Cresky asked his students how many hours of sleep they got on the average each night for a week. He labeled the data with G for girls and B for boys. Use Gs and Bs to make a line plot of this data. Analyze your graph.

10-G	7.5-B	9-B	9-G	8-G	8.5-G	9.5-B	8.5-B	8.5-G
9.5-B	8-G	8-B	10-B	10-G	9.5-G	7.5-G	9-G	8.5-B

Mixed Review

8. Solve $n = -282 + 41$. *(Lesson 3-3)*
9. **Forestry** The ranger at Crestview Nature Preserve recorded the heights of several trees in an area of the preserve. Make a histogram of this data. *(Lesson 4-2)*

Height (in ft)	11-20	21-30	31-40	41-50	51-60	61-70
Number of trees	4	10	14	22	28	18

Problem Solving and Applications

10. **Literature** The list at the right shows the prices paid for some early editions of books written by Mark Twain.
    a. Make a line plot of this data.
    b. If a collector went to an auction of these books with $200, how many books were within the collector's budget?

Sales of Mark Twain's Books		
$150	$2,200	$160
$70	$450	$330
$110	$325	$1,600
$100	$800	$130
$60	$180	$65
$50	$420	

11. **Critical Thinking** Use the potato chip data on page 136.
    a. Select an interval and make a histogram of the data. Use different colors to divide each bar into the number of *good* brands and the number of *very good* brands.
    b. How does your histogram compare to the line plot in the Example?
12. **Journal Entry** Ask 15 friends how old they are in months. Make a line plot of your data. Write a few sentences about your data.

Cooperative Learning

4-3B Maps and Statistics

A Follow-Up of Lesson 4-3

Objective

Use maps to display United States statistical data.

Gerhardus Mercator, a Flemish geographer, was the first person to use the term *atlas* for a collection of maps. He invented the Mercator map projection. He used this type of map to produce a map of Earth similar to the ones we use today.

You can pick up almost any newspaper or magazine and see a map of the United States with statistics about the individual states displayed by shading or coloring. The map below shows the percent of change in state funding of education for 1992, grades K-12. How do geographers decide how the statistics should be displayed? Why do they use maps instead of lists?

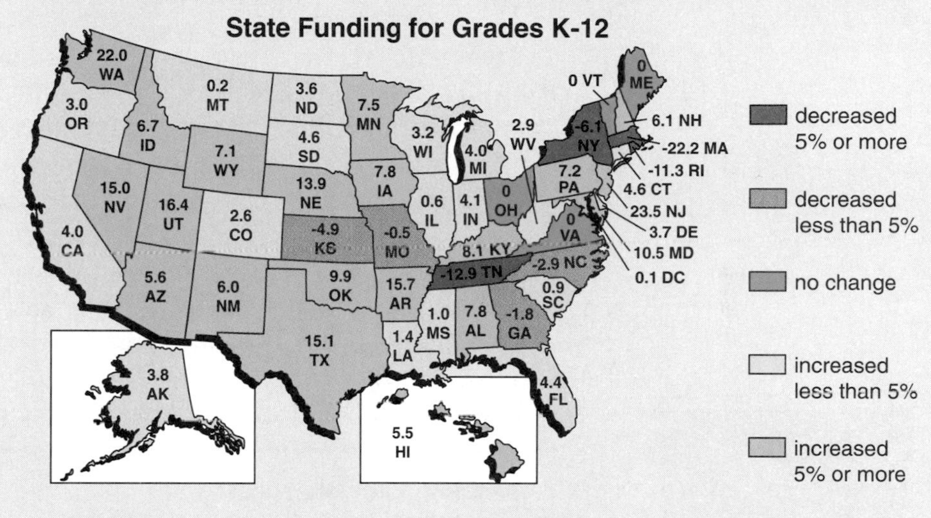

Try this!

Materials: colored pencils, outline map of the United States

State	Hispanic (Percent)	State	Hispanic (Percent)	State	Hispanic (Percent)
AL	0.6	KY	0.6	ND	0.7
AK	3.2	LA	2.2	OH	1.3
AZ	18.8	ME	0.6	OK	2.7
AR	0.8	MD	2.6	OR	4.0
CA	25.8	MA	4.8	PA	2.0
CO	12.9	MI	2.2	RI	4.6
CT	6.5	MN	1.2	SC	0.9
DE	2.4	MS	1.2	SD	0.8
DC	5.4	MO	1.2	TN	0.7
FL	12.2	MT	1.5	TX	25.5
GA	1.7	NE	2.3	UT	4.9
HI	7.3	NV	10.4	VT	0.7
ID	5.3	NH	1.0	VA	2.6
IL	7.9	NJ	9.6	WA	4.4
IN	1.8	NM	38.2	WV	0.5
IA	1.2	NY	12.3	WI	1.9
KS	3.8	NC	1.2	WY	5.7

[1990 U.S. Census, subject to update]

- The table at the left lists the 50 states and the District of Columbia with the percent of their population that is of Hispanic origin. Make a line plot of this data using the state abbreviations instead of ×s.
- Mapmakers usually like to organize the data into fewer than 7 categories. Use the ranges 0-1.99, 2-5.99, 6-11.99, 12-21.99, and 22-40 to separate the data into categories. *The intervals are selected to get a point of view across. Therefore, they are usually not equal intervals.*
- Mapmakers use colors ranging from light to dark to correspond with the ranges from least to greatest. Choose five colors of pencils. Color each state on a United States map according to its category.

What do you think?

1. Why do you think the category ranges given were chosen?
2. Why do you think mapmakers try to use less than seven colors when making a map?
3. What areas of the country have more people of Hispanic origin than others? Why do you think this is true?
4. Why is the map a more effective way to present the data than a list?

Extension

5. Depending on a point of view, someone may want a map that highlights states with a small Hispanic population. How could you change the map to emphasize this point of view?
6. Make another map using the ranges 0-3, 3-5.99, 6-11.99, 12-23.99, and 24 and over. How does the appearance of this map differ from the previous one?

Application

7. **Farming** The chart below shows the percent of land that is farmland in each state and the District of Columbia.

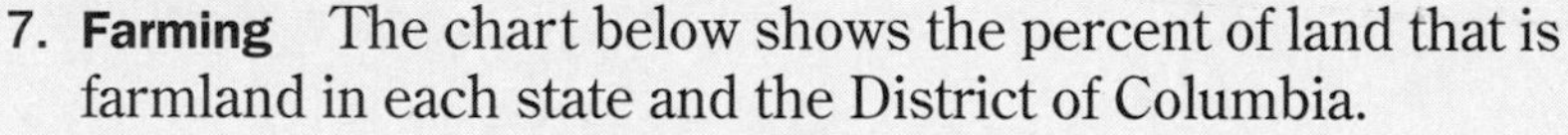

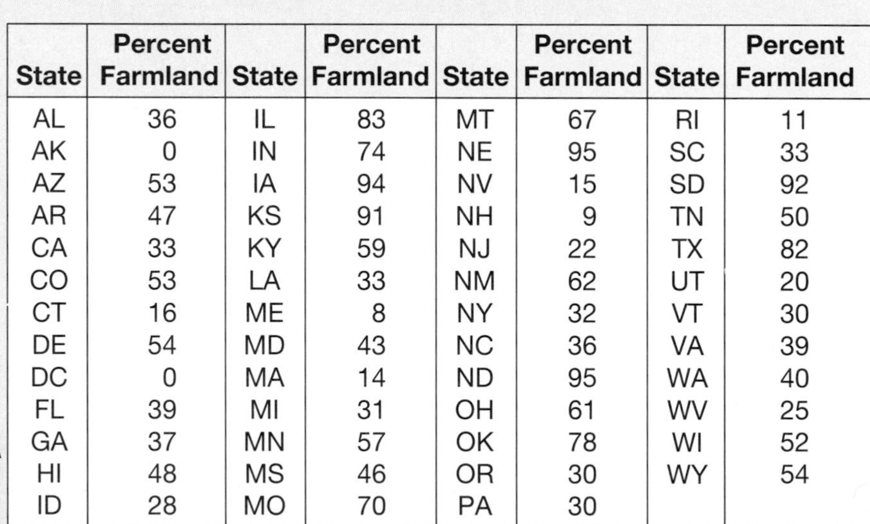

State	Percent Farmland	State	Percent Farmland	State	Percent Farmland	State	Percent Farmland
AL	36	IL	83	MT	67	RI	11
AK	0	IN	74	NE	95	SC	33
AZ	53	IA	94	NV	15	SD	92
AR	47	KS	91	NH	9	TN	50
CA	33	KY	59	NJ	22	TX	82
CO	53	LA	33	NM	62	UT	20
CT	16	ME	8	NY	32	VT	30
DE	54	MD	43	NC	36	VA	39
DC	0	MA	14	ND	95	WA	40
FL	39	MI	31	OH	61	WV	25
GA	37	MN	57	OK	78	WI	52
HI	48	MS	46	OR	30	WY	54
ID	28	MO	70	PA	30		

a. Color a map to illustrate this data.
b. How would you change your map if you were a senator trying to get more government money for agricultural programs?
c. How would you change your map if you were an opponent of the senator in Question b and were looking for funds for urban development?

4-4 Stem-and-Leaf Plots

Objective

Construct and interpret stem-and-leaf plots.

Words to Learn

stem-and-leaf plot
stem
leaf
back-to-back
stem-and-leaf plot

Growth spurts differ in boys and girls. Girls typically grow quickly around age 12, while boys grow quickly, on average, around the ages of 13 and 14.

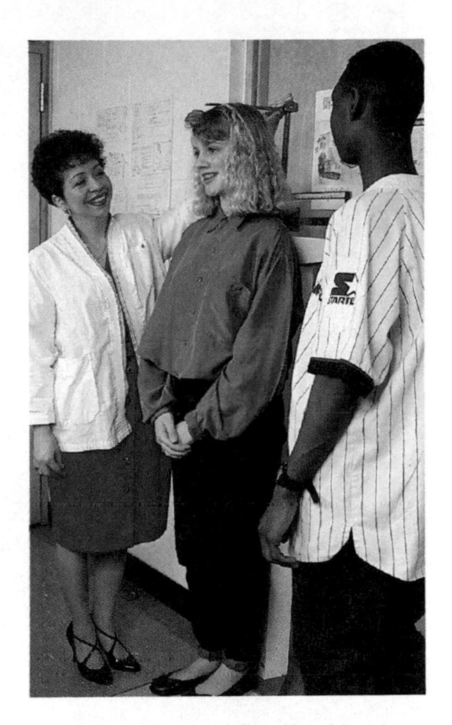

Mrs. Jeske is the school nurse at Titusville Middle School. She measured the heights in inches of 20 students in a health class.

Height of Health Class Students (in inches)

Girls				Boys			
55	59	66	64	59	63	65	62
70	68	67	63	72	71	62	60
68	58			60	61		

This data could be presented on a line plot. Another way to organize the data is to make a **stem-and-leaf plot.**

In a stem-and-leaf plot, the greatest place value of the data can be used for the **stems,** and the next greatest place value for the **leaves.**

Follow these steps to construct a stem-and-leaf plot for the data above.

- Tens place is the greatest place value. The stems will be those digits in the tens place. Use the digits only once and list them in order from least to greatest.

5	
6	
7	

- The leaves are the corresponding digits in the next greatest place value for each stem.

  For example, there are four numbers that have 5 in the tens place. They are 55, 59, 59 and 58. The 5, 9, 9, and 8 are the leaves for the stem 5. Always write every leaf even if it is a repeat of another leaf. Write the leaves in order from least to greatest.

  5|5899 5 is a stem. 5, 8, 9, and 9 are leaves.

Sometimes it is helpful to write the leaves first in any order and then to order them from least to greatest.

- Complete the stem-and-leaf plot for all stems.

5	5899
6	0012233456788
7	012

- Include a guide to the data. *6|2 means 62 inches.*

Now it is easy to see that most of the students are between 60 and 68 inches tall.

A **back-to-back stem-and-leaf plot** is used to compare two sets of data. In this type of plot, the leaves for one set of data are on one side of the stem and the leaves for the other set of data are on the other side of the stem. Two guides to the data are needed.

Example

Make a back-to-back stem-and-leaf plot of the heights of the health class, putting the girls' heights on one side and the boys' heights on the other.

Girls		Boys
985	5	9
887643	6	0012235
0	7	12

0|7 means 70 inches *7|1 means 71 inches*

Notice that the greatest data are always the outermost leaves.

Mini-Lab

Work with your classmates.

Materials: the data from the Mini-Lab on page 137, four pieces of paper labeled 4, 5, 6, and 7

- Lay the numbers on the floor in a column.
- Make a human back-to-back stem-and-leaf plot with girls on one side and boys on the other.

Talk About It

a. When everyone is in place, what can you say about the heights of the persons on the end of each row of leaves?

b. Which group appears to be taller?

Checking for Understanding

Communicating Mathematics

Read and study the lesson to answer each question.

1. **Tell** what numbers you use to form the stems in a stem-and-leaf plot.
2. **Write** what 80|9 means if the greatest place value is hundreds.
3. **Tell** when you can use a back-to-back stem-and-leaf plot. Give an example of data that you could show with one.

Guided Practice

Refer to the stem-and-leaf plot in the Example.

4. What is the height of the shortest girl?
5. How tall is the tallest boy?

6. Copy the stem-and-leaf plot on page 141.
   a. Replace each leaf with either G (girl) or B (boy) to correspond with the heights of girls or boys.
   b. What kind of information do you gain from this type of plot?
   c. What kind of information do you lose from this type of plot?
7. A park is giving prizes to the first 30 people through the gate on their anniversary date. In order to determine what type of prizes to get, the park asks the ages of the first 30 people through the gate on a day prior to their anniversary. The ages are 12, 11, 22, 32, 35, 45, 46, 14, 14, 16, 33, 30, 41, 7, 9, 25, 8, 51, 43, 18, 17, 19, 32, 34, 18, 22, 24, 56, 61, and 13.
   a. What numbers are used as stems in making a stem-and-leaf plot of this data?
   b. Make a stem-and-leaf plot of these ages.
   c. What are the ages of the youngest person and oldest person?
   d. What age group seems most representative of this group?

Exercises

Independent Practice

State the numbers you would use for the stems in a stem-and-leaf plot of each set of data. Then make the plot.

8. 17, 32, 41, 34, 42, 35
9. 9, 12, 24, 51, 33, 14, 11
10. 294, 295, 272, 253, 280, 267
11. 7.5, 5.4, 8.6, 6.3, 7.1, 5.9
12. The data below are test scores for Mr. Sopher's eighth-grade homeroom.

Mathematics	75	93	87	56	60	73	78	69
	83	89	94	97	65	73	87	85
Science	68	73	98	87	65	64	70	73
	72	78	81	83	68	57	63	75

   a. Make a back-to-back stem-and-leaf plot for the data.
   b. What is the highest mathematics score?
   c. What interval does each stem represent?
   d. In which interval do most of the science scores occur?
   e. In which subject did the class have better scores? Explain your answer.
13. Two companies list the ages of their employees by using a stem-and-leaf plot.

Bender Company

Stem	Leaf
2	34
3	5677
4	34478899
5	1266
6	58

Flexor Company

Stem	Leaf
2	2355
3	689
4	1589
5	0479
6	33489

2|2 means 22 years old.

   a. One company can estimate the average age range of their employees more quickly than the other company. Which one?
   b. Why would a company want to find out the average age?

Mixed Review

14. Find $|-20|$. *(Lesson 3-1)*
15. Solve $-48 \div 3 = c$. *(Lesson 3-7)*
16. Make a line plot for these English quiz scores:
7, 10, 6, 7, 5, 8, 2, 8, 6, 8, 9, 8, 7, 4, 9, 7, 6, 10, 5, and 7 *(Lesson 4-3)*

Problem Solving and Applications

17. **Critical Thinking** Explain how a stem-and-leaf plot is similar to a histogram.
18. **Sports** The number of home runs hit by the league leaders in professional baseball from 1921-1990 is shown in the plot below.

National League		American League
331	2	224
98887	2	
4433111100	3	22222223334
9988877766665	3	5666777799999
444433322110000	4	001112223334444
9988877776655	4	5566666788999999
42211	5	24
6	5	889
4|5 means 54	6	01 *6|0 means 60*

a. Which league's leaders tend to hit more home runs?
b. Why do you think two lines were used for each stem?

19. **Research** Find the ages of the United States presidents at their death.
a. Make a stem-and-leaf plot of the data.
b. Write a sentence to describe any patterns you notice.

Portfolio

20. **Portfolio Suggestion** Select an item from this chapter that shows your creativity and place it in your portfolio.

CULTURAL KALEIDOSCOPE

Jackie Robinson

Jackie Robinson (1919-1972) became the first African-American baseball player in modern major league history. He signed with the Brooklyn Dodgers in 1947. He played through 1956 and was both an infielder and an outfielder.

Leading the league in stolen bases, Mr. Robinson was chosen Rookie of the Year in 1947. With his help, the Dodgers won the National League Pennant that year and on September 30, 1947, he became the first African-American to compete in a World Series.

In 1949, Jackie Robinson won the batting championship with a 0.342 average and also won the Most Valuable Player award. His overall career average was 0.311. In his first year of eligibility in 1962, he was elected to the Baseball Hall of Fame.

4-5 Measures of Central Tendency

Objective

Find the mean, median, and mode of a set of data.

Words to Learn

measures of central tendency
mean
median
mode

Technology Activity

You can learn how to find the mean of a set of data with a spreadsheet in Technology Activity 2 on page 659.

Calculator Hint

You can also find the sum of a group of data by using the SUM or M+ keys. After entering each number, press SUM or M+. To retrieve the total, press RCL or MR.

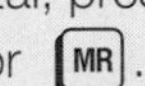

On a test covering material from their social studies project, five members of a cooperative group received the scores listed below.

Emilio	88
Nadawi	86
Sean	85
Debbie	78
Hiroshi	88

Their teacher informed them that their project score would be the average of their group's test scores. Sean told his parents that he received an 86 on his social studies project. Nadawi told her brother that she received an 88. Hiroshi told his grandmother that he received an 85. How can all three students think they are correct?

There are three common ways to describe a set of data. These ways are called **measures of central tendency**. They are the mean, the mode, and the median.

Hiroshi chose the **mean.** Most people think of the *mean* when they use the word *average.* For the scores above, the mean is

[( ] 88 [+] 86 [+] 85 [+] 78 [+] 88 [ )] [÷] 5 [=] 85

The mean score is 85.

The mean is not necessarily a member of the set of data, but here, it is.

Mean	The mean of a set of data is the sum of the data divided by the number of pieces of data.

Nadawi used the score that appears most often to describe the set. This is called the **mode.**

78 85 86 88 88

There are two 88s in this set. So, 88 is the mode.

If there was another score of 86, then 86 would also be a mode and the data would have two modes. A set of data in which no numbers appear more than once, has no mode.

If there is a mode, it is always a member of the set of data.

Mode	The mode of a set of data is the number or item that appears most often.

Sean ordered the test scores from least to greatest and chose the middle number. This is called the **median.**

78 85 86 88 88

The middle number is 86. The median score is 86.

If the number of data is even, the set has two middle numbers. In that case, the median is the mean of the two numbers. For example, in {11, 12, 15, 22, 26, and 29}, the middle numbers are 15 and 22. Their mean is $\frac{15 + 22}{2}$ or 18.5. *The median is not necessarily a member of the set of data.*

Mental Math Hint

You can find the mean of pairs like 40 and 42 without dividing. Just think "what number is halfway between?"

Median	The median is the number in the middle when the data are arranged in order. When there are two middle numbers, the median is their mean.

Example 1 *Problem Solving*

Radio One advertiser claims that the average radio station plays about 44 minutes of music per hour. The line plot below shows the minutes of music per hour for 12 radio stations. Find the mean, median, and mode of the data.

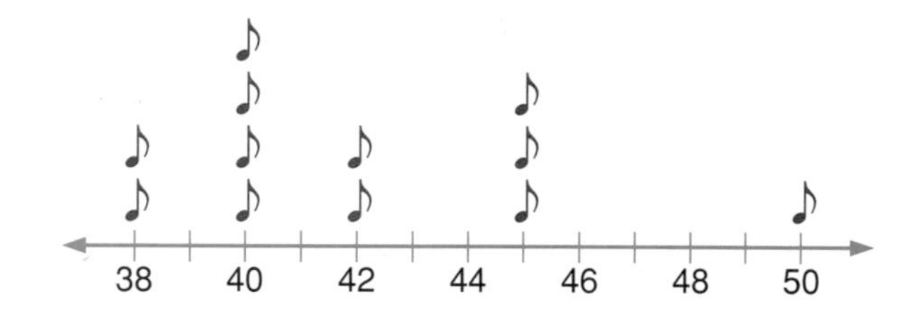

To find the mean, find the sum of the minutes and divide by 12.

$$\frac{2(38) + 4(40) + 2(42) + 3(45) + 50}{12} = \frac{505}{12} \text{ or about } 42$$

The mean is about 42 minutes.

The median is the middle number. There are 12 numbers, so the median is the mean of the 6th and 7th numbers.

$$\frac{40 + 42}{2} = 41$$ The median is 41 minutes.

The mode is the most frequent number.
The mode is 40 minutes.

Since the advertiser's claim is greater than any of these measures of central tendency, you might consider the claim to be false advertising.

Example 2

Refer to the stem-and-leaf plot on page 141. Find the mean, median, and mode of the heights.

5	5899
6	0012233456788
7	012

To find the mean, add the heights and divide by 20.

1273 [÷] 20 [=] 63.65

The mean height is 63.65 inches.

The median is the mean of the two middle numbers. Since the two middle numbers are both 63, the median height is 63 inches.

This data set has five repeated numbers, so it has five modes. They are 59, 60, 62, 63, and 68.

Checking for Understanding

Communicating Mathematics

Read and study the lesson to answer each question.

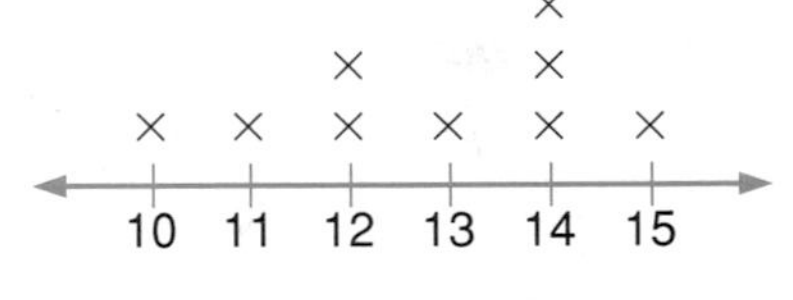

1. **Tell** how to find the mode of the data in the line plot at the right.
2. **Show** how to find the median of the set {17, 3, 15, 8, 5, 13, 7, 9, 12, 10}.
3. **Write** a sentence to tell how the mean of a set of data may differ from the median.
4. **Tell** which measure of central tendency in Example 1 best represents the number of minutes music is played? Explain.

Guided Practice

Organize the data into either a frequency table, a line plot, or a stem-and-leaf plot. Then find the mean, median, and mode.

5. 3, 5, 7, 6, 8, 2, 9, 3, 7, 7, 8
6. 22, 24, 25, 26, 25, 34, 35
7. \$49, \$49, \$50, \$50, \$52, \$52
8. 17.2, 3.5, 15.6, 8.8, 5.5, 13.1, 7, 9.6, 12.9, 10.74

Exercises

Independent Practice

Find the mean, median, and mode for each set of data. Round to the nearest tenth.

9. 2, 6, 4, 5, 6, 13, 5, 8, 13
10. 79, 84, 81, 84, 73, 75, 80, 78
11. 130, 155, 148, 184, 172
12. 21, 18, 24, 21
13. 3.4, 1.8, 2.6, 1.8, 2.3, 3.1
14. 25.98, 30, 45.36, 25, 45.36
15.

4	2
5	05
6	128
7	0355
8	178

6|1 means 61

16.

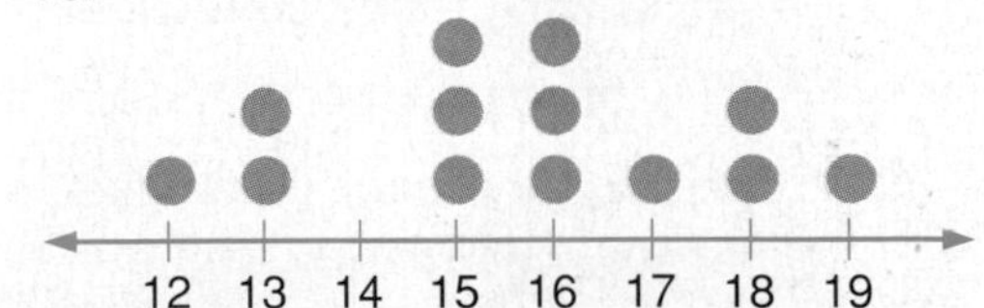

17. The growing season in Tennessee is the period from May to September. The chart at the right shows the normal rainfall for these months.
    a. Find the mean, mode, and median of this data.
    b. Suppose Tennessee received heavy rain in May totaling 8.2 inches. If this figure was used for May, how would the measures of central tendency be affected?
    c. If September is eliminated from the period, how would this affect the measures of central tendency?

Normal Rainfall for Tennessee	
May	4.8 in.
June	3.6 in.
July	3.9 in.
Aug.	3.6 in.
Sept.	3.7 in.

Mixed Review

18. Solve $5.17 = \frac{k}{2} + 3.68$. Check your solution. *(Lesson 2-7)*
19. Solve $b = 65 - (-87)$. *(Lesson 3-5)*
20. Twenty people attended a public speaking seminar. Make a stem-and-leaf plot of their ages: 15, 23, 42, 18, 34, 29, 56, 24, 36, 21, 43, 52, 35, 22, 16, 48, 32, 37, 26, and 45. *(Lesson 4-4)*

Problem Solving and Applications

21. **Critical Thinking** Mikael knows that he will have five tests this grading period and that he must have at least an 80 average to play on the school's volleyball team. His mean for the first four tests is 77. What is the least score he can get on the last test and still qualify to play volleyball?
22. **Hobbies** Meagan does one cross stitch project a month. Four of her projects required 7 colors of floss, three projects required 9 colors of floss, and five projects required 10 colors of floss.
    a. Find the mean, median, and mode of the colors.
    b. Which measure of central tendency is the best average? Explain.

23. **Journal Entry** Can you think of a time when you have heard or read about an *average*? What do you think of when you hear that word?

4

Assessment: Mid-Chapter Review

The line plot shows scores on a 25-point history test.

1. Make a frequency table for the data. *(Lesson 4-1)*
2. Construct a histogram for the data using intervals of 3 for the scores. *(Lesson 4-2)*

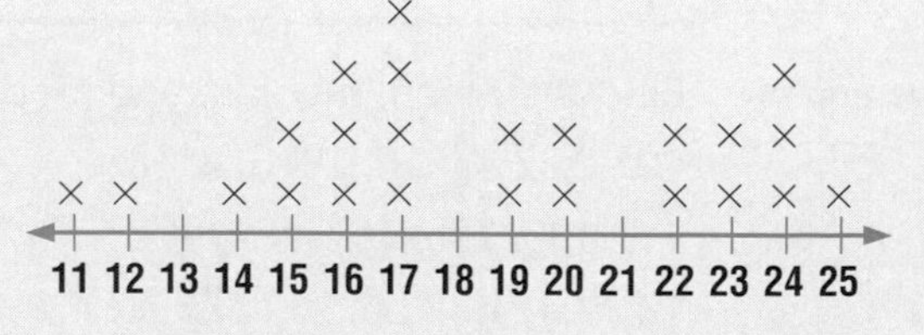

3. In what range do most of the scores lie in the line plot? *(Lesson 4-3)*
4. Construct a stem-and-leaf plot for the data. *(Lesson 4-4)*
5. Would the mean, median, or mode be the best measure of central tendency to describe the "average" score? Why? *(Lesson 4-5)*

Cooperative Learning

4-5B Making Predictions

A Follow-Up of Lesson 4-5

If you've ever watched *Wheel of Fortune* on television, you know that most contestants ask for the letters T, N, R, S, and E. That's because these are the four consonants that occur most often, and E is the vowel that occurs most often.

The frequency table below lists the average number of times each letter occurs in a sample of 100 letters in the English language.

- Are the letters T, N, S, R, and E good choices?
- What would be your next choice?

Letter	Frequency
A	8.2
B	1.4
C	2.8
D	3.8
E	13.0
F	3.0
G	2.0
H	5.3
I	6.5
J	0.1
K	0.4
L	3.4
M	2.5

Letter	Frequency
N	7.0
O	8.0
P	2.0
Q	0.1
R	6.8
S	6.0
T	10.5
U	2.5
V	0.9
W	1.5
X	0.2
Y	2.0
Z	0.07

The coded message on the next page is a poem by Robert Frost. Each letter in the code stands for a specific letter in the English language. These types of codes are often called cryptograms.

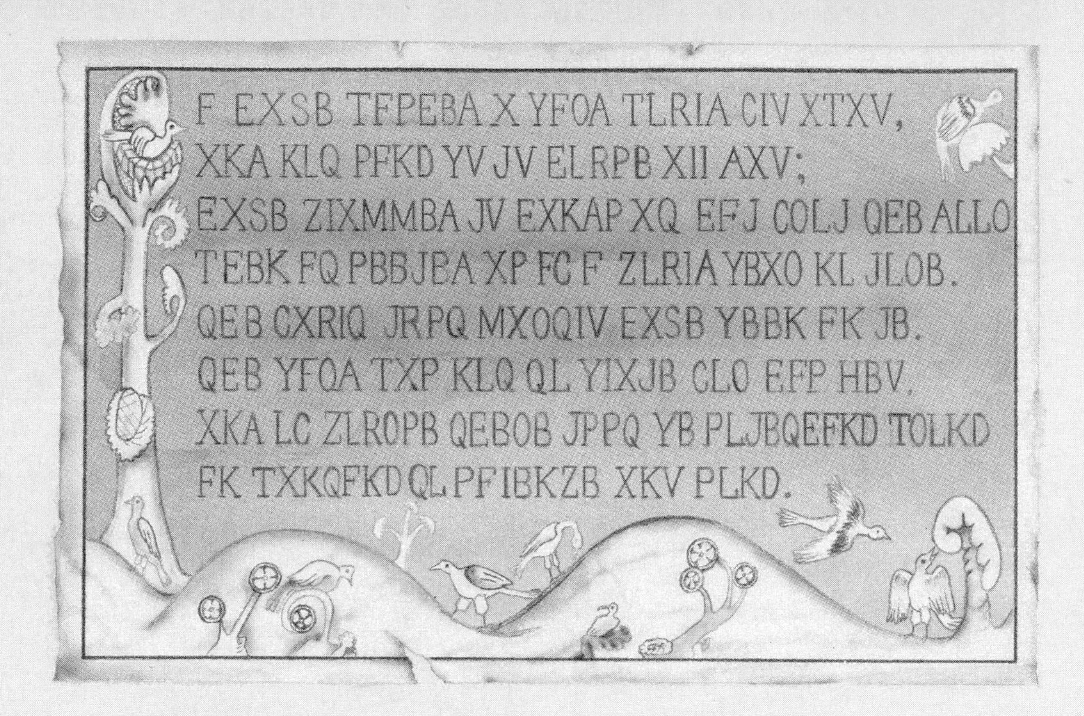

Try this!

Work with a partner to break the code.

- Make a frequency table of the number of times each letter appears in the coded message.
- Make a line plot that shows the 10 letters that occur most often in the coded message.
- Make a line plot that shows the 10 letters that occur most in the English language.
- Compare the two line plots.

What do you think?

1. Which English letter probably corresponds to the letter that occurs most often in the code?
2. Predict the English letters for the nine most frequent code letters.
3. Decode the poem.
4. Compare your predictions with the results of breaking the code. How well did you do?
5. Suppose this was a poem about zippers. What effect would this have had on the frequency table for the coded letters?

4-6 Measures of Variation

Objective

Find the range and quartiles of a set of data.

Words to Learn

variation
range
quartile
interquartile range
upper quartile
lower quartile

Tai is doing research for a health project. She wants to find the amount of sugar in cereals advertised for kids.

She visited the grocery store and tallied the number of grams of sugar per serving in 27 brands of cereal. Then she displayed her results in the line plot below.

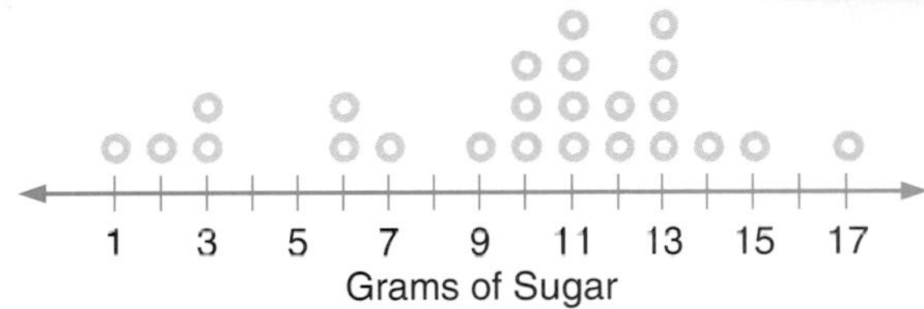

After constructing the line plot, Tai could easily see how the data extended from a low of 1 to a high of 17. The *spread* of data is called the **variation.**

One *measure of variation* is called the **range.**

Range	The range of a set of numbers is the difference between the least and the greatest number in the set.

The range of the sugar amounts is 17 - 1 or 16.

In a large set of data, such as thousands of college entrance exam test scores, it is helpful to separate the data into four equal parts called **quartiles.** Quartiles are used in another measure of variation called the **interquartile range.**

Interquartile Range	The interquartile range is the range of the middle half of the data.

To find the interquartile range, first find the middle half of the data. The steps on the following page show how you can find the interquartile range of the data displayed in the line plot above.

- Find the median of the data since the median separates the data into two halves.

1 2 3 3 6 6 7 9 10 10 10 11 11 11 11 11 11 11 12 12 13 13 13 13 14 15 17
↑
Median

- Find the median of the upper half. This number is called the **upper quartile,** indicated by UQ.

1 2 3 3 6 6 7 9 10 10 10 11 11 11 11 11 11 11 12 12 13 13 13 13 14 15 17
↑ Median ↑ UQ

- Find the median of the lower half. This number is called the **lower quartile,** indicated by LQ.

1 2 3 3 6 6 7 9 10 10 10 11 11 11 11 11 11 11 12 12 13 13 13 13 14 15 17
↑ LQ ↑ Median ↑ UQ

- The middle half of the data goes from 7 to 13. Subtract the lower quartile from the upper quartile.
  $13 - 7 = 6$ The interquartile range of Tai's data is 6. This means that the 13 middle amounts are between 7 and 13 grams.

Example 1

Refer to the stem-and-leaf plot of students' heights on page 141. Find the range and interquartile range of this data.

Since the stem-and-leaf plot is organized from the least data to the greatest, the range is the difference in the first and last data of the plot.

$72 - 55 = 17$ The range is 17.

5 | 5899
6 | 0012233456788
7 | 012

7 | 0 means 70

To find the interquartile range, first find the median. The median is 63. Use brackets to separate the data into two halves.

5 | [5899
6 | 001223][3456788
7 | 012]

Now find the upper and lower quartile by finding the median of each half. Use boxes to show the location of LQ and UQ.

5 | [5899
6 | [00] 1223][3456 [78] 8
7 | 012]

LQ is $\frac{60 + 60}{2}$ or 60.

UQ is $\frac{67 + 68}{2}$ or 67.5.

The interquartile range is $67.5 - 60$ or 7.5.

The grains used in cereals are important food crops. They belong to the grass family. The grain is the fruit or seed of the plant.

Example 2

The frequency table at the right represents the grams of sugar per serving in 31 cereals advertised for adults. Find the range and interquartile range for this set of data.

Grams	Tally	Frequency
0	𝍸	5
1		0
2	III	3
3	𝍸 I	6
4	I	1
5	𝍸	5
6	IIII	4
7	I	1
8		0
9	I	1
10	I	1
11	I	1
12	II	2
13		0
14	I	1

The greatest value is 14 and the least value is 0. The range is 14 − 0 or 14.

To find the interquartile range, find the median, LQ, and UQ.

0 0 0 0 0 2 2 2 3 3 3 3 3 3 4 5 5 5 5 5 6 6 6 6 7 9 10 11 12 12 14

↑ LQ (2) ↑ Median (5) ↑ UQ (6)

The median is 5. The lower quartile is 2 and the upper quartile is 6.

The interquartile range is 6 − 2 or 4.

Checking for Understanding

Communicating Mathematics

Read and study the lesson to answer each question.

1. **Tell** how the measures of variation differ from the measures of central tendency.
2. **Show** how a set of data is separated into quartiles.
3. **Write** how you find the two values which determine the interquartile range.
4. **Tell** what the interquartile range means.

Guided Practice

Find the range of each set of data.

5. 3, 4, 5, 7, 8, 8, 10
6. 12, 17, 16, 23, 18
7. 135, 170, 125, 174, 136, 145, 180, 156, 188

Find the median and upper and lower quartiles of each set of data.

8. 3, 4, 5, 7, 8, 8, 10
9. 12, 17, 16, 23, 18
10. 135, 170, 125, 174, 136, 145, 180, 156, 188
11. Find the interquartile range of the data in Exercise 10.

Exercises

Independent Practice

Use the data in the stem-and-leaf plot below.

12. What is the range?
13. What is the median?
14. What are the upper and lower quartiles?
15. What is the interquartile range?

Stem	Leaf
1	227
2	33344566889
3	0146
4	06

4 | 0 means 40

The gas mileages in miles per gallon (MPG) of 4-cylinder manual transmission cars are listed at the right.

16. Organize the data into a line plot or a stem-and-leaf plot.
17. What is the highest rate?
18. What is the lowest rate?
19. What is the range?
20. What is the median?
21. What are the upper and lower quartiles?
22. What is the interquartile range?

MPG of 4-Cylinder Cars			
28	32	42	37
30	25	57	38
24	32	33	44
38	34	30	44
31	28	31	29
39	29	32	29

Mixed Review

23. Find the value of $9^2 - 2^3$. *(Lesson 1-9)*
24. Solve $(-8)(12)(-10) = f$. *(Lesson 3-6)*
25. Find the mean, median, and mode of the set {6, 4, 6, 12, 10, 8, 7, 12, 11, 9}. *(Lesson 4-5)*

Problem Solving and Applications

26. **Decision Making** The Remako Company is planning a sales meeting for September 20. They have narrowed their choices to Cincinnati or Baltimore. They would like to choose a location with a milder climate. The company's research department researched the temperatures for that date for the past 14 years. The results are shown below.

Cincinnati

Stem	Leaf
3	9
4	2 5
5	3 7
6	0 0 0 4
7	1 3
8	4 6 8

3 | 9 means 39

Baltimore

Stem	Leaf
3	
4	3 5 8 9
5	2 6
6	0 0 8
7	4 7
8	1 4 5

a. Which city usually has a milder climate on September 20?
b. Which city do you think they will choose? Why?

27. **Critical Thinking** Write two different sets of data with the same range, but with different interquartile ranges.

28. **Data Search** Refer to page 667. Compare the range and the interquartile range of the record high temperatures in the U.S. What does this mean?

4-7 Box-and-Whisker Plots

Objective

Construct and interpret box-and-whisker plots.

Words to Learn

box-and-whisker plot
outlier

So far, you have made frequency tables and stem-and-leaf plots to *display* sets of data, and you have used measures of central tendency and measures of variation to *summarize* data. Now we will combine the ideas of displaying and summarizing data. A **box-and-whisker plot** summarizes data using the median, the upper and lower quartiles, and the *extreme* (greatest and least) *values.*

Here's how to construct a box-and-whisker plot to display Tai's data on sugar content in kids' cereal on page 151.

- Use a number line. Graph points above the line for the extreme values, the median, and the quartile values.

lower extreme: 1	upper extreme: 17
median: 11	LQ: 7 UQ: 13

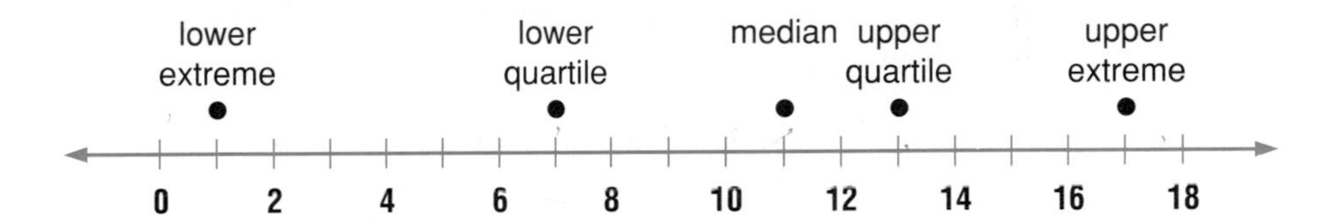

- Next, draw a *box* around the quartile values.

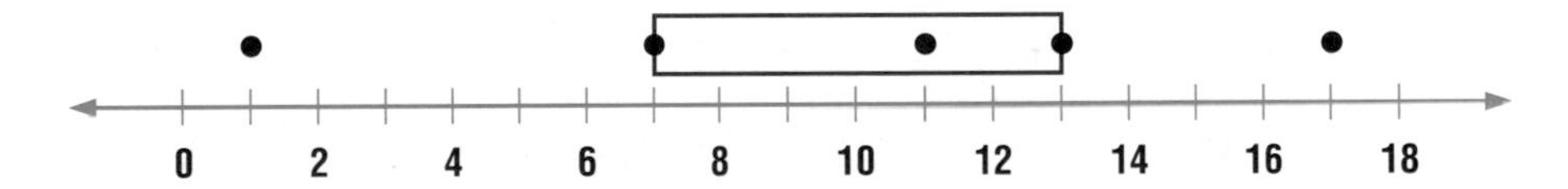

- Draw a vertical line through the median value.

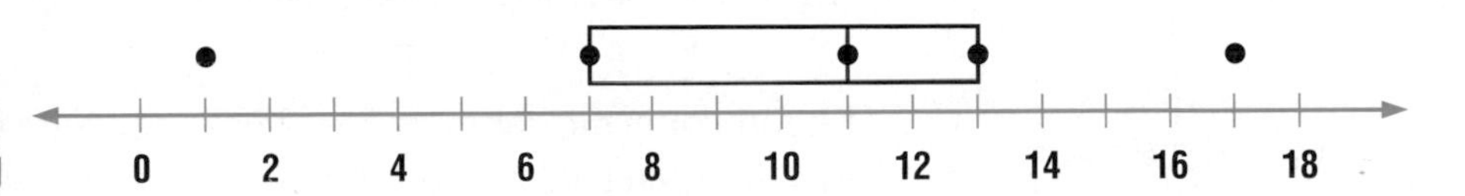

In evaluating the weekly salaries of her department, a manager can determine the median salary and the salaries in the upper and lower quartiles by using a box-and-whisker plot.

- Finally, extend *whiskers* from each quartile to the extreme data points.

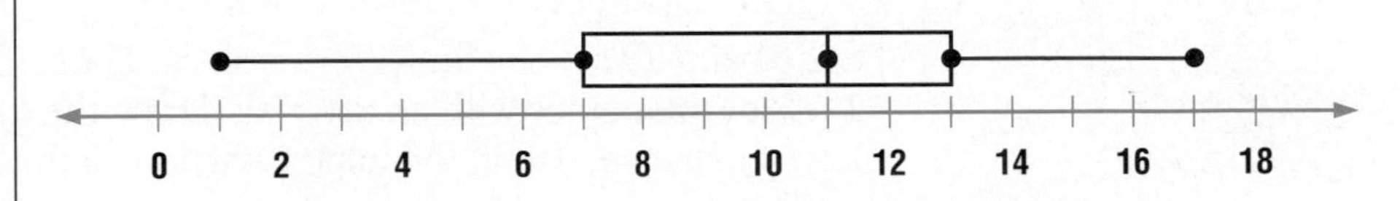

Thus, the box-and-whisker plot gives you five pieces of information about the data: lower extreme, lower quartile, median, upper quartile, and upper extreme.

Example 1

The daily high temperatures (°F) for two weeks of September in Baltimore were 43°, 45°, 48°, 49°, 52°, 56°, 60°, 60°, 68°, 74°, 77°, 81°, 84°, and 85°. Make a box-and-whisker plot of this data. How can you locate the interquartile range on the plot?

To make the plot, find the extremes, median, and quartiles.

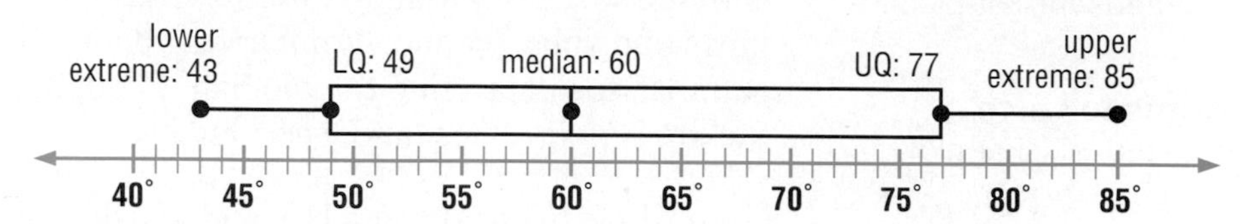

The interquartile range is the difference of the upper and lower quartiles. This is represented by the left and right sides of the box.

Sometimes the data will have such great variation that one or both of the extreme values will be far beyond the other data. Data that are more than 1.5 times the interquartile range from the upper or lower quartiles are called **outliers.**

Example 2

Draw a box-and-whisker plot for the grams of sugar per serving in cereals for adults shown below.

0 0 0 0 0 2 2 2 3 3 3 3 3 3 3 4 5 5 5 5 5 6 6 6 6 6 7 9 10 11 12 12 14

↑ LQ ↑ Median ↑ UQ

The median is 5; the upper quartile is 6, and the lower quartile is 2. Draw a box to show the median and the quartiles.

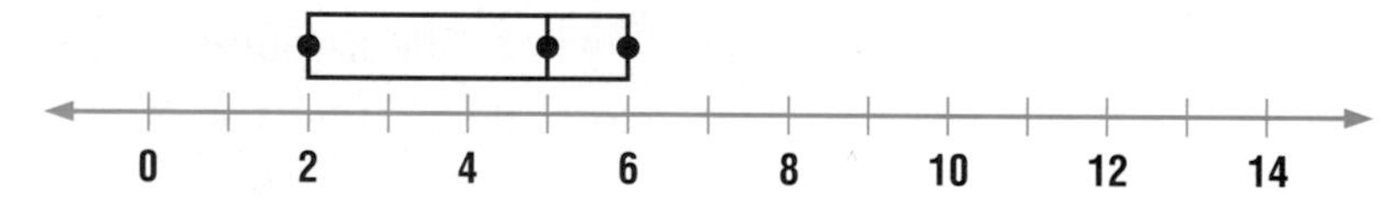

Mental Math Hint

To quickly calculate 1.5×4, think:

1.5 means $1\frac{1}{2}$.

$1 \times 4 = 4$

$\frac{1}{2} \times 4 = 2$

$4 + 2 = 6$

The interquartile range is $6 - 2$ or 4. So, data more than $1.5 \cdot 4$, or 6, from the quartiles are outliers.

Find the limits for the outliers.

Subtract 6 from the lower quartile. $2 - 6 = -4$

Add 6 to the upper quartile. $6 + 6 = 12$

So, –4 and 12 are the limits for outliers. There is one outlier in the data, 14. Plot the outlier with an asterisk. Draw the lower whisker to the lower extreme, 0, and the upper whisker to the last value that is not an outlier, 12.

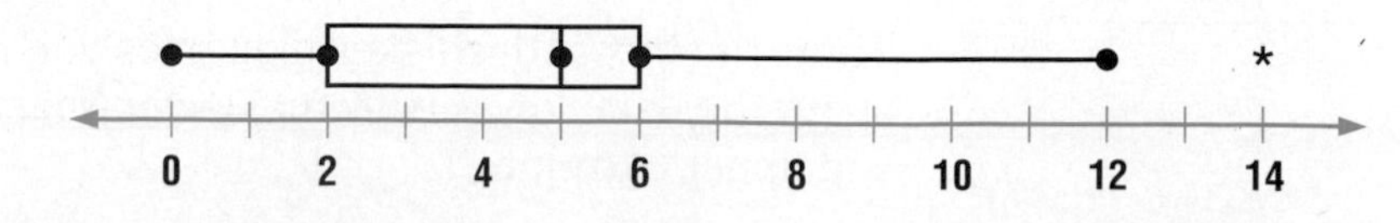

Checking for Understanding

Communicating Mathematics

Read and study the lesson to answer each question.

1. **Tell** what part of the data are enclosed by the box in a box-and-whisker plot.
2. **Write** which data are connected to the box by the whiskers in a box-and-whisker plot.
3. **Write** five pieces of information you can learn from a box-and-whisker plot.
4. **Tell** how you know if a value is an outlier.

Compare the box-and-whisker plots shown below.

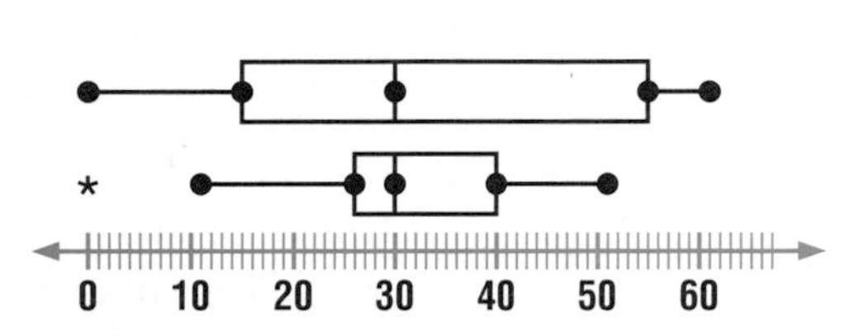

5. What is similar about the data in the two plots?
6. What is different about the data in the two plots?
7. Which set of data is more evenly spread over the range? Explain.

Refer to the potato chip prices on page 136.

8. Draw a box-and-whisker plot of the data.
9. What is the median?
10. What is the upper quartile?
11. What is the lower quartile?
12. What is the lower extreme?
13. What is the interquartile range?
14. Are there any outliers?
15. What are the limits on the outliers?
16. Compare the box-and-whisker plot to the line plot on page 136. What similarities and differences do you see?

Exercises

Independent Practice

The box-and-whisker plot below represents the heights in inches of the twenty students in the Fitness Club.

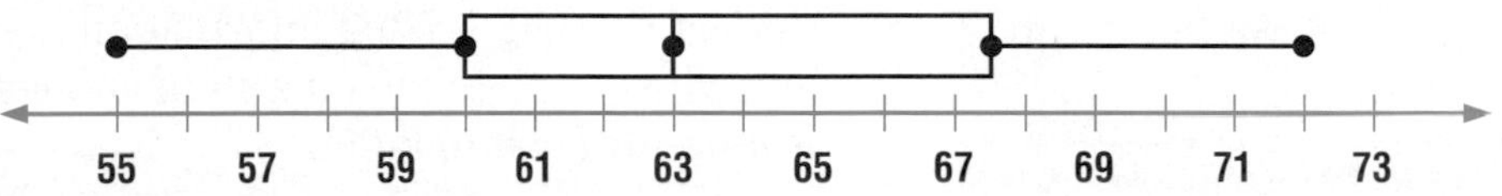

17. What is the median?
18. What is the range?
19. What is the upper quartile?
20. What is the lower quartile?
21. What is the interquartile range?
22. What are the extremes?
23. What are the limits on the outliers? Are there any outliers?
24. How many students are between 60 and 63 inches tall? Explain.
25. How many students are less than 63 inches tall? Explain.
26. How does the mean of this data relate to the box-and-whisker plot if the mean is 64?

27. Draw a box-and-whisker plot of the ages of the residents at Piney Village given in the stem-and-leaf plot shown at the right.

5	23346
6	1122233355588
7	00034447
8	5579

5 | 2 means 52 yr

Mixed Review

28. How many gallons are in 14 quarts of juice? *(Lesson 1-7)*

29. Evaluate $4^2 + 6(12 - 8) - 15 \div 3$. *(Lesson 2-1)*

30. Order the set {−219, −52, 18, 3, −24, 120, −186} from greatest to least. *(Lesson 3-2)*

31. **Music** The number of pages in *Rock Star Magazine* in the last nine issues is 196, 188, 184, 200, 168, 176, 192, 160, and 180. Find the median, upper quartile, and lower quartile of this data. *(Lesson 4-6)*

Problem Solving and Applications

32. **Consumer Math** Ten flashlight batteries were purchased from each of two different manufacturers X and Y. The batteries were tested to determine how many hours they would last. The results are given below.

    X: 15.5, 14, 14, 24, 19, 16.7, 15, 11.4, 16, 15
    Y: 18, 14, 15.8, 9, 12, 16, 20, 16, 13, 15

    a. Make a box-and-whisker plot for each set of data.
    b. Based on the plots, from which manufacturer would you buy your batteries? Explain.

33. **Critical Thinking** Name a set of data with ten numbers whose box-and-whisker plot would have only one whisker.

34. **Critical Thinking** What information is easier to find on a box-and-whisker plot than on a stem-and-leaf plot? What information is more difficult to find?

35. **School** Before teaching a unit on Mexico, the social studies teacher had his class take a test to see how much his students already knew. Then he gave a test after teaching the unit. The results are given below. Use a box-and-whisker plot to determine whether the students' scores improved from one test to the next.

    Before Unit Scores: 56, 89, 79, 70, 90, 86, 75, 68, 92, 85, 78, 87, 95, 84, 79, 80, 64, 89, 83
    After Unit Scores: 87, 89, 90, 93, 95, 97, 82, 95, 100, 88, 90, 94, 99, 93, 96, 88, 93, 89, 100

Journal

36. **Journal Entry** Write a few sentences to tell why a box-and-whisker plot might be used to display great numbers of data. What drawbacks does a box-and-whisker plot have, if any?

4-8 Scatter Plots

Objective

Construct and interpret scatter plots.

Words to Learn

scatter plot

Silvia found the results of the first state-by-state comparison of mathematics scores, measured in 1991 by the National Assessment of Educational Progress. She thinks there might be a relationship between the scores on the test and the annual average temperature for the state.

Silvia collected the temperature data for the District of Columbia and the 37 states that volunteered to report their test results. Then she formed ordered pairs in which the first number was the temperature and the second number was the score. For example, the ordered pair for North Dakota was (41, 281). After the 38 points were plotted, Silvia looked for a pattern to see if her prediction was correct.

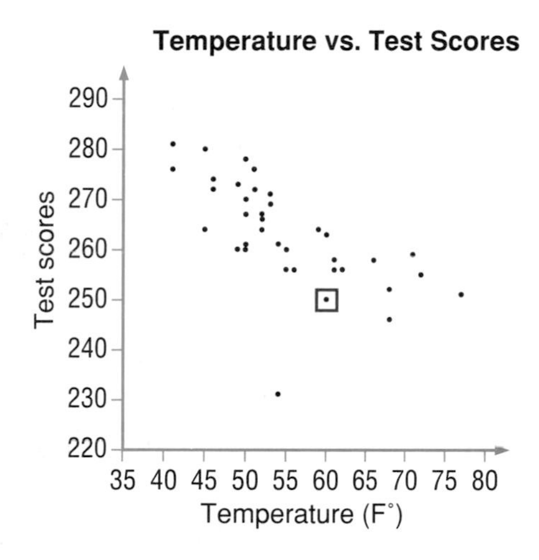

LOOKBACK

You can review graphing ordered pairs on page 117.

When you graph two sets of data as ordered pairs, you form a **scatter plot.** Unlike other displays of data you have studied, scatter plots do not measure central tendencies or variations. Instead, scatter plots can suggest whether the two sets of data are truly related.

positive relationship

negative relationship

To determine if there is a pattern for the data in a scatter plot imagine a line drawn on the scatter plot where half the points are above it and half the points are below it. If the line slants upward to the right, there is a *positive* relationship. If the line slants downward to the right, there is a *negative* relationship.

Example 1

Refer to the scatter plot for test scores versus temperature.

a. Describe the point with the box around it.

The point with a box around it is (60, 250). It represents a temperature of 60° with a test score of 250.

b. What type of relationship is shown by the scatter plot?

Since the points seem to slope downward to the right, the scatter plot shows a negative relationship. That is, as the average temperature increases, the scores seem to decrease.

Example 2

The scatter plot at the right compares the week Mike took a science quiz to the quiz score. Does there appear to be any relationship between the week he took the quiz and his quiz score?

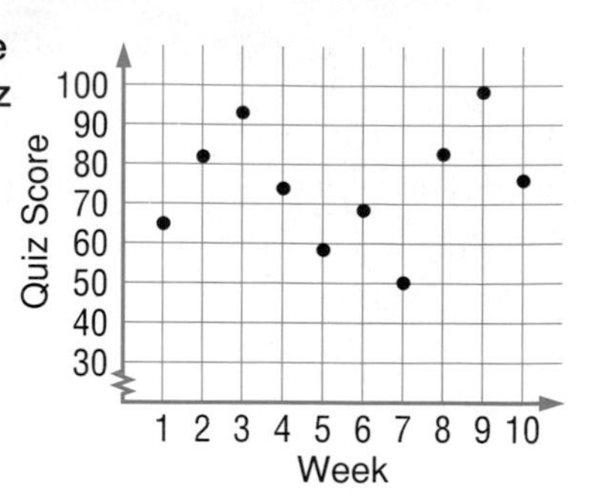

The points in the scatter plot are very spread out. There appears to be no relationship between the week Mike took the quiz and his score.

Checking for Understanding

Communicating Mathematics

Read and study the lesson to answer each question.

1. **Write** how to draw a scatter plot for two sets of data.
2. **Tell** how the scatter plot in Example 2 would look if there was a positive relationship between when Mike took the quiz and his score.

Guided Practice

Determine whether a scatter plot of the data below would show a positive, negative, or no relationship.

3. age of a used car and its value
4. heights of mothers and sons
5. hours of TV watched per week and test scores
6. income and years of school completed
7. height and month of birth
8. playing time and points scored in basketball

Determine whether each scatter plot of data shows a positive, negative, or no relationship.

9. 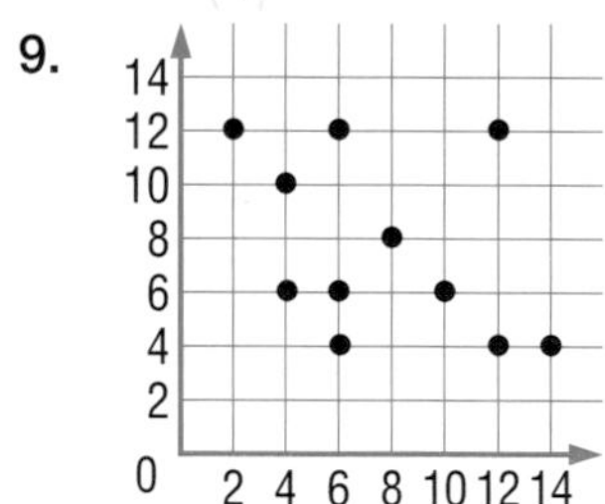

10. 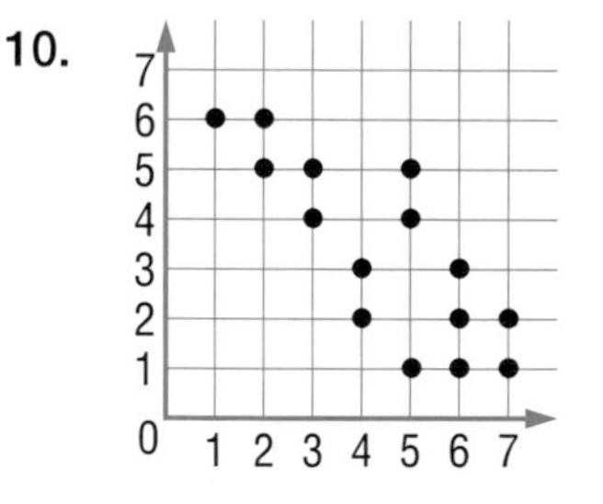

11. 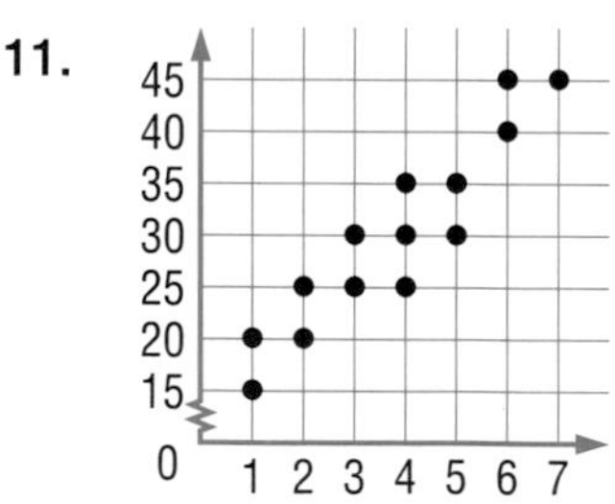

Exercises

Independent Practice

Determine whether a scatter plot of the data below would show a positive, negative, or no relationship.

12. height and weight
13. height of student and test scores
14. hair color and test scores
15. age and income
16. outside temperature and amount of the heating bill
17. miles per gallon and the weight of the car

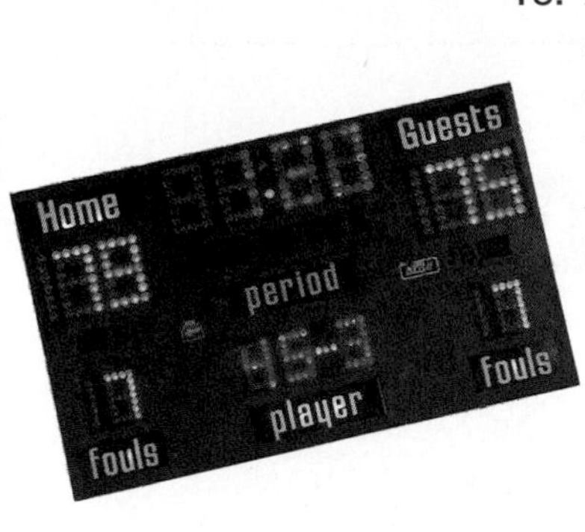

18. A scatter plot of the shots attempted and the shots made by each player in the first game of the 1991 National Basketball Association Championship series is shown at the right.
    a. Describe the data represented by the C in the upper right-hand corner.
    b. Find the point that represents the only perfect shooter.
    c. Is there a different relationship for Los Angeles and Chicago players?
    d. Suppose a player attempts 14 field goals. About how many would you expect him to make?
    e. How many players made over 10 field goals?
    f. What conclusion can you make from the scatter plot?

Mixed Review

19. Solve $158 = s - 36$. Check your solution. *(Lesson 2-3)*
20. Solve $5x + 32 < 42$. Show the solution on a number line. *(Lesson 2-10)*
21. Solve $h = -28 + 15 + 6 + (-30)$. Check your solution by solving another way. *(Lesson 3-4)*
22. Draw a box-and-whisker plot for the set {36, 26, 32, 15, 44, 29, 38, 20, 11, 33, 47, 24, 42}. *(Lesson 4-7)*

Problem Solving and Applications

23. **Finance** Make a scatter plot for the data in the table below.

Hours Worked	12	23	18	46	17	18	34	15	21	10	40	1	21
Money Earned ($)	51	82	65	199	157	68	186	75	189	42	180	5	170

    a. How will you form the ordered pairs?
    b. Does the data show a relationship between hours worked and money earned? If so, what type of relationship is it?
24. **Critical Thinking** In the scatter plot of test scores and temperatures on page 159, it appears that as the temperatures got warmer the test scores got lower.
    a. Why do you think the relationship might exist?
    b. Does this negative relationship necessarily mean that one factor caused the other? Why or why not?
    c. What are some other factors that affect test scores?
    d. How many states did not report scores from the test?
    e. What effect might the states that did not report scores have on the scatter plot?

25. **Portfolio Suggestion** Select some of your work from this chapter that shows how you used a calculator or computer. Place it in your portfolio.

DECISION MAKING

Planning a School Trip

Situation

As the editor of your school newspaper, you have been selected by the newspaper advisor to attend a conference. The 3-day conference will include sessions on utilizing word processing software for typesetting, graphic design, and using windows to make layout of copy more efficient. The advisor has set aside $600 from the budget for your expenses. The question is, how many days can you afford to stay and which sessions should you attend?

Hidden Data

Cost of travel: Is the conference local or will you need to travel by plane, bus, or train to get there?
Cost of lodging: Will you need to stay in a hotel?
Cancellation charges: Will you get your deposit refunded if you are unable to attend the conference?

Analyzing the Data

1. Which conference day(s) seem better suited to the newspaper editor?
2. Are any sessions offered more than once?

Tuesday, Januaary 28

8:30 - 10:00
- Yearbook Editor Issues
- Product Promises vs. Reality
- Photography in the Yearbook
- School Issues
- Word Processing into Type
- Multimedia

10:30 - 11:30

KEY NOTE ADDRESS AND PANEL DISCUSSION

12:00 - 1:30
- Multimedia
- What Fonts Can Do
- Developing Solutions: Case Studies
- PC or Mac?
- Software Design Principles
- Windows: What are They?

2:00 - 3:00

EXHIBITS VIEWING

3:30 - 5:00
- Training Your Staff
- Computer Editing: Is it for You?
- Microsoft Strategy Briefing
- Yearbook Advisors Tell New Trends
- Graphiics: Import or use PMTs
- 4-Color Yearbook Production

Wednesday, January 29

8:30 - 10:00
- Sharpening Your Skills
- Window Presentation Sofware
- Style, Design, and Graphics
- What's Hot and What's Not in Newspapers
- Training Your Staff
- Desktop Publishing

10:30 - 12:00
- Clip Art
- Utilities for the PC
- State of the Art Publishing
- Computer Editing
- Beyond B & W Photos
- Layout and Design

12:30 - 1:30

COLUMNISTS PANEL DISCUSSION

2:00 - 3:30

Utilities for the PC
Which Pagemaking Program?
Sharpening Your Skills
Style, Design, and Graphics
Distributing Your Newspaper
Layout and Design

4:00 - 5:00

EXHIBIITS VIEWING

Thursday, January 30

8:30 - 9:30

DEVELOPER'S KEYNOTE

10:00 - 11:30
- Desktop Publishing II
- Data Bases for Circulation
- Spreadsheets in the Real World
- Computer Editing
- Developing Macros
- Logos

12:00 - 1:00

SPECIAL INTERESTS SESSION

1:30 - 3:00
- Distributing Your Newspaper
- Managing Styles
- State of the Art Publishing
- Mail-Order Software
- Developing Skills for Writers
- Computer-Shy Writers and You

3:30 - 4:30

EXHIBITS VIEWING

PROFESSIONAL SEMINARS

Making a Decision

3. **Are you able** to attend more than one day of the conference?
4. **What if** the conference is not compatible with the methods you currently use for newspaper production?
5. **Is the cost** of the full conference program less expensive than the number of days you plan to attend?

STUDENT PUBLISHING CONFERENCE

REGISTRATION

Full Conference Program $295.00

Conference registration includes admission to all educational sessions, fast track and keynote presentations, all conference materials, unlimited access to the exhibit hall and all special events.

Single Day Conference Program $175.00

Conference registration includes admission to all educational sessions, keynote presentations, conference materials, and all special events on any one of the three days. Includes unlimited access to the exhibit hall on all three days. Please be sure to indicate which day you plan to attend.

Making Decisions in the Real World

6. Investigate the seminars or conferences available to student journalists in your area. Share this information with your newspaper advisor.

4-9 Misleading Graphs and Statistics

Objective

Recognize when graphs and statistics are misleading.

Words to Learn

sample

An advertising agency conducted a test at the Mill Creek Mall for Mrs. Field's cookies. They prepared their finding in two different graphs.

- Do both graphs show the same information?
- Which graph might influence your choice of cookie?

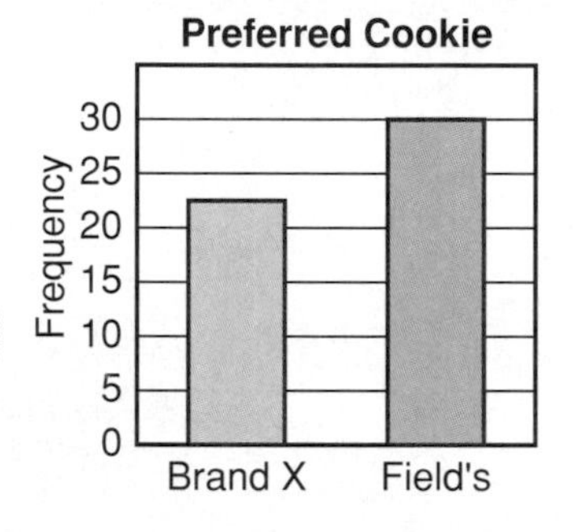

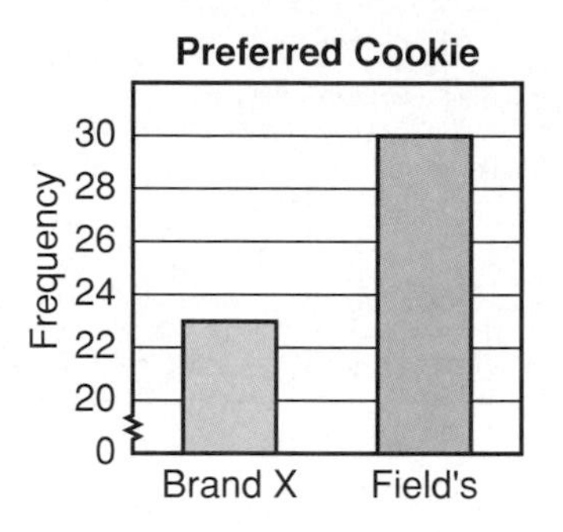

Sometimes using a different scale helps to make a graph easier to read. However, scales may also be chosen to make the data appear to support a particular point of view.

The graph at the right does not begin at 0 and has a different scale than the graph at the left. The difference in the number of votes is the same in each graph. However, the bars in the graph at the right make it appear that Mrs. Field's is the overwhelming choice.

Example 1

The graph shows the number of students in each grade that had a B average or better.

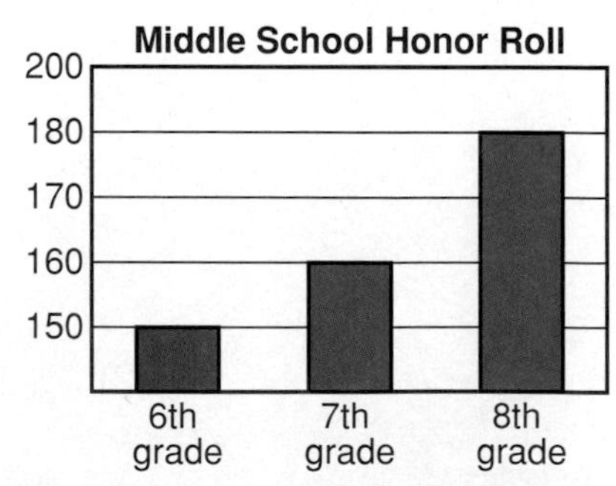

a. **How many more eighth graders are on the honor roll than sixth graders?**

   There are 30 more eighth graders on the honor roll than sixth graders.

b. **What impression does the height of the bars give you about comparing the number of seventh graders to the number of eighth graders?**

   The height of the bars make it look as if the eighth grade has twice as many on the honor roll as the seventh grade.

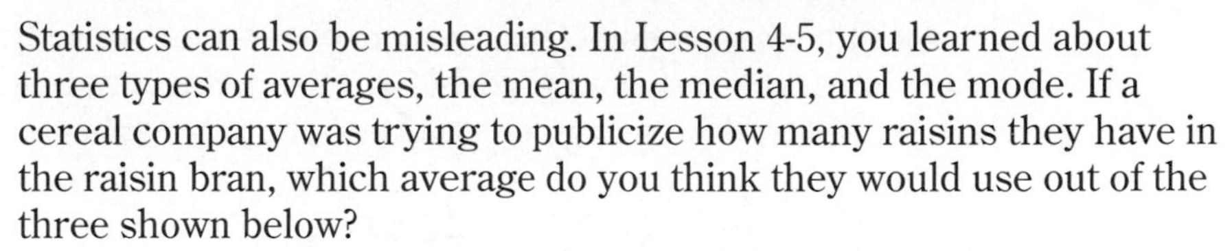

LOOKBACK

You can review measures of central tendency on pages 145 and 146.

Statistics can also be misleading. In Lesson 4-5, you learned about three types of averages, the mean, the median, and the mode. If a cereal company was trying to publicize how many raisins they have in the raisin bran, which average do you think they would use out of the three shown below?

mean = 186 raisins

median = 184 raisins

mode = 214 raisins

They probably would use the mode, since it is the greatest number.

When a survey of 100 people is taken, the 100 responses to the survey are called a **sample.** Statisticians often use samples to represent larger groups of data. However, if the sample is not typical of all the data, the wrong conclusions could be drawn from the results of the sample.

Example 2

Alicia asked 20 adults how many times they went to the movies in the past month. Then she asked 20 of her classmates the same question. How did her choice of sample affect her results?

The results of asking 20 adults is probably a number less than the results of asking 20 classmates, since young people usually attend movies more often than adults.

Checking for Understanding

Communicating Mathematics

Read and study the lesson to answer each question.

1. **Tell** two ways that a graph might be misleading.
2. **Write** some things you might question when reading the results of a survey on your favorite radio station.
3. **Draw** a graph that is misleading and tell why it is misleading.

Guided Practice

4. Look at the graph at the right.
   a. How did the number of house sales change from December 1991 to July 1992?
   b. How does this change compare with the change in the sizes of the houses on the graph?
   c. Is this a misleading graph? Explain.

5. The test scores of five students on a project were Emilio 88, Nadawi 86, Dean 85, Debbie 78, and Hiroshi 88. Their actual grade will be the average of these grades. Which average do they hope the teacher will choose and why?

Exercises

Independent Practice

6. The two graphs below show the increases in Marty's allowance over the last six years.

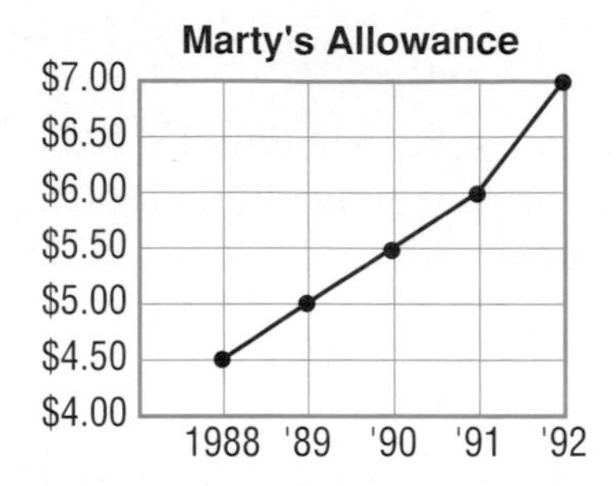

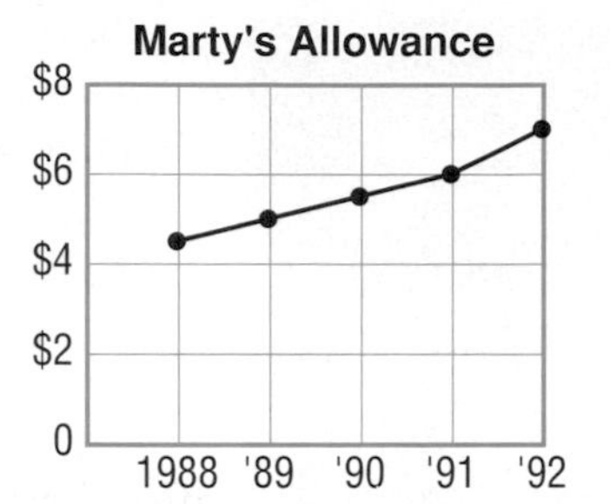

a. Tell if either of the graphs are misleading.
b. If Marty wants an increase in his allowance, which graph would he probably show his parents? Explain your answer.

7. The number of calories in a sample of six cookies from the Sugar-Lite company are 45, 60, 63, 75, 75, and 90. Which average would they use to promote the low calorie count in their cookies?

Decide whether each location given is a good place to find representative samples for the selected survey. Justify each answer.

8. Survey the number of books read in a month in a shopping mall
9. Survey the favorite kind of entertainment in a concert hall
10. Survey the number of dogs owned in an apartment complex
11. Survey the favorite fast food restaurant in a school

12. Roosevelt Junior High School sold candy bars to raise money for the parent-teacher association. The sixth graders sold 300 candy bars. The seventh graders sold 385 candy bars. The eighth graders sold 500 candy bars.
   a. Draw a bar graph that is not misleading.
   b. Draw a bar graph that makes it appear that the eighth graders sold three times as many candy bars as the sixth graders.

Mixed Review

13. Use mental math to find $12 \cdot 15$. *(Lesson 1-2)*
14. Solve $-460 = -16a + 52$. *(Lesson 3-9)*
15. Determine whether a scatter plot of hours spent fishing and fish caught would show a positive, negative, or no relationship. Explain your answer. *(Lesson 4-8)*

Problem Solving and Applications

16. **Health** A magazine says "Four out of five doctors recommend Multi-Zip vitamins."
    a. How could the number of doctors in the survey affect the truth of this ad?
    b. Dr. Mentos was part of this survey. He says that he also recommends other brands of vitamins in addition to Multi-Zip. Does the ad reflect this?
    c. How might the results of this survey be different if this survey was taken in Lincoln, Nebraska instead of Los Angeles, California?
17. **Critical Thinking** Suppose your grades on the last six history tests are 74, 75, 80, 88, 90, and 94.
    a. You have been grounded because your grades have not been good. Draw a graph of your test scores to show your parents how well you have improved your grades.
    b. If your parents ask what your average test score is, what score might you tell them and why?

18. **Sports** Mr. Otashi loves soccer and wants the local cable television company to carry more stations that cover this sport. He wants to get 500 people to sign a petition to increase the cable company's soccer coverage. What would be a good way to get 500 signatures of people genuinely interested in seeing more soccer on cable television?
19. **Mathematics and Food Processing** Read the following paragraphs.

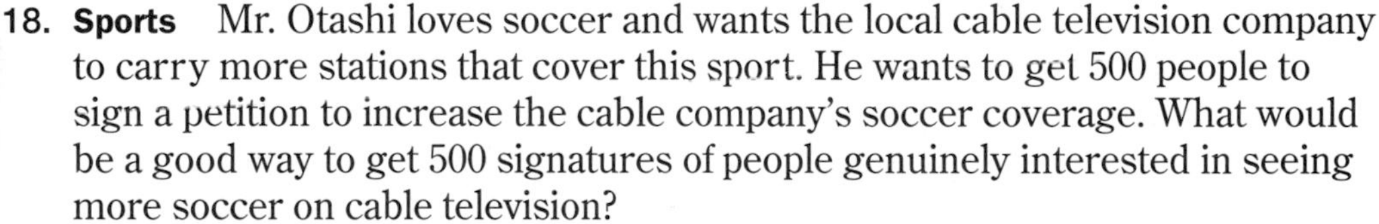

Statistical methods are used in all stages of food manufacturing. The information from samples are used as the measure of the quality of the product. Samples are also used in testing how many of the products fall below government or company standards.

The rate at which samples are taken is determined by previous experience with the product. The cost of taking samples and testing is also considered. New products are often tested more frequently to make sure the characteristics desired are presented in all samples.

A manufacturer produces 1 million applesauce cups per week. They test the first 100 cups for contamination.

a. Is this a representative sample?
b. What does the manufacturer need to consider in the evaluation of the results of this sample?

Chapter

4 Study Guide and Review

Communicating Mathematics

Choose the correct term to complete each sentence.

1. In a stem-and-leaf plot, the greatest place value of the data can be used for the (stem, leaf).
2. A (scatter plot, line plot) is a graph whose ordered pairs consist of two sets of related data.
3. A (frequency table, histogram) is a bar graph that shows the frequency of data organized in intervals.
4. The (interquartile range, range) of a set of numbers is the difference between the least and greatest number in the set.
5. Data that are more than 1.5 times the interquartile range from the quartiles are called (whiskers, outliers).
6. A (mode, sample) is a small group representative of a larger group.
7. The (mean, median) of a set of data is the sum of the data divided by the number of pieces of data.
8. Tell what information a box-and-whisker plot shows that a line plot does not.
9. Explain the difference between the mean and the median of a set of data.

Self Assessment

Objectives and Examples

Upon completing this chapter, you should be able to:

Review Exercises

Use these exercises to review and prepare for the chapter test.

- construct and interpret histograms *(Lesson 4-2)*

Construct a histogram for {1, 2, 2, 3, 4, 5, 5, 6, 6, 6, 7, 8, 9, 9, 10}

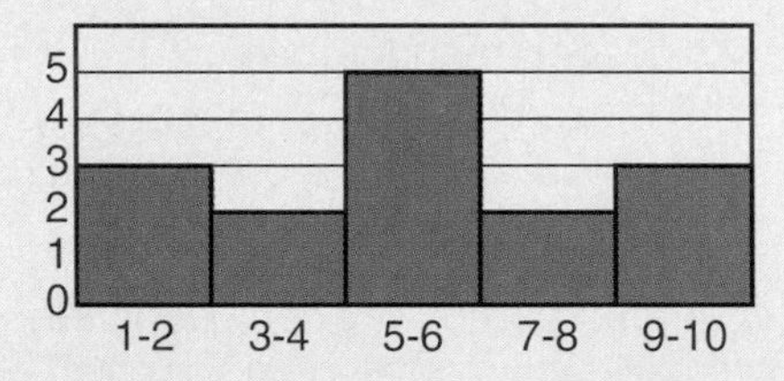

10. **Construct a histogram for the following science test scores:**

70, 75, 90, 83, 72, 77, 93, 80, 67, 84, 70, 68, 86, 71, 56, 87, 94, 79, 91

- construct and interpret line plots *(Lesson 4-3)*

Construct a line plot for the data above

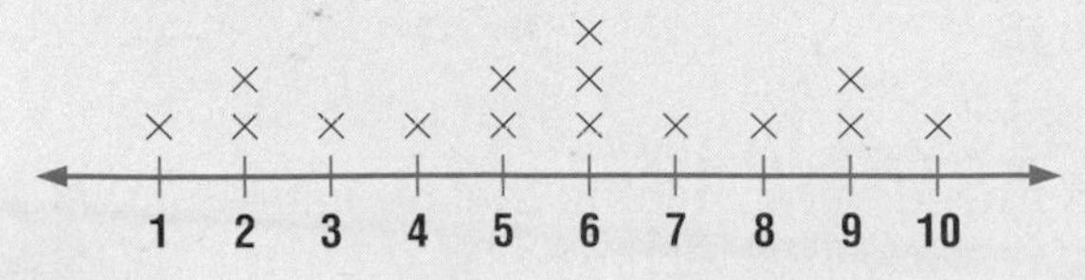

11. **Construct a line plot for the following ages:**

72, 73, 68, 78, 76, 69, 65, 70, 77, 80, 74, 63, 73, 69, 66, 75, 70, 77, 75, 73

Objectives and Examples

- construct and interpret stem-and-leaf plots *(Lesson 4-4)*

Construct a stem-and-leaf plot for {13, 15, 17, 20, 21, 22, 22, 26, 29, 34, 53}

1	3 5 7
2	0 1 2 2 6 9
3	4
4	
5	3

5 | 3 means 53

Review Exercises

12. Make a stem-and-leaf plot for the following temperatures:
    60, 53, 68, 72, 66, 80, 73, 51, 62, 48, 56, 84, 77, 45, 79, 65

- find the mean, median, and mode of a set of data *(Lesson 4-5)*

Find the mean, median, and mode of {53, 53, 57, 62, 65}.

The mean is $\frac{53+53+57+62+65}{5}$ or 58.

The median is 57. The mode is 53.

Find the mean, median, and mode for each set of data. Round to the nearest tenth.

13. 8.2, 6.8, 7.4, 8.5, 7.1, 8.3
14. 41, 26, 35, 30, 20, 41, 21
15. 173, 132, 157, 191, 118
16. 4, 6, 11, 2, 4, 8, 12, 3

- find the range and quartiles of a set of data *(Lesson 4-6)*

Find the range and quartiles of {1, 2, 2, 2, 3, 3, 4, 4, 5, 6, 6}.

The range is $6 - 1 = 5$. The median is 3. The lower quartile is 2 and the upper quartile is 5.

Find the range and quartiles of each set of data.

17. 149, 137, 144, 126, 151, 132, 142, 120, 166
18. 4, 3, 1, 6, 2, 6, 5, 1, 5, 3, 4, 2, 1, 3, 6
19. 25, 30, 28, 22, 24

- construct and interpret box-and-whisker plots *(Lesson 4-7)*

Construct a box-and-whisker plot of {1, 2, 2, 2, 3, 3, 4, 4, 5, 6, 6}

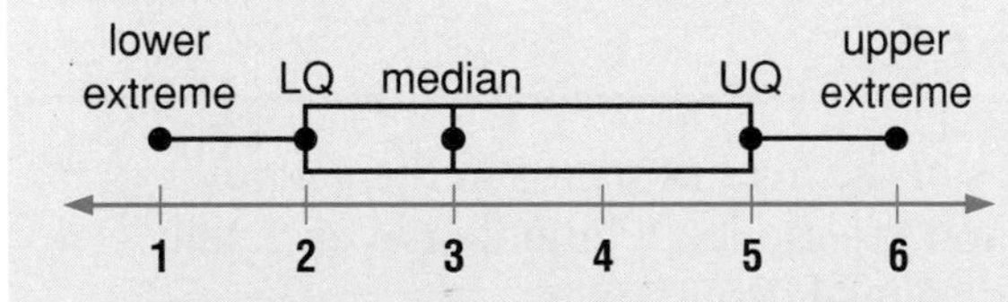

Draw a box-and-whisker plot for each set of data.

20. 12, 15, 14, 18, 20, 17
21. 41, 36, 45, 20, 31, 54, 22, 35, 49, 23, 51, 37, 42

- construct and interpret scatter plots *(Lesson 4-8)*

A scatter plot is a graph of points whose ordered pairs consist of two sets of data.

Determine whether a scatter plot of the data below would show a positive, negative, or no relationship.

22. height and income
23. temperature and snow

Objectives and Examples	Review Exercises
• recognize when graphs and statistics are misleading *(Lesson 4-9)* Statistics can be misleading if the data comes from a sample that is not representative of the larger group. A graph can be misleading if the scale is inappropriate or parts of the scale are missing.	24. Tell which graph below might be misleading and tell why.

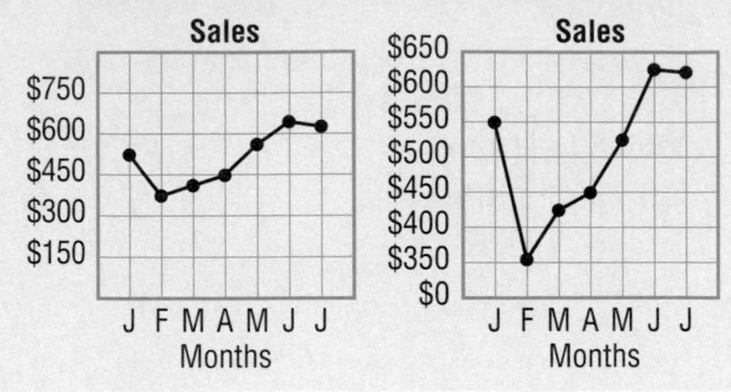

Applications and Problem Solving

25. The cost of air fares from Denver to Newark are shown below.

$248	$756	$350	$298	$650	$987	$326	$211	$345
$378	$888	$458	$298	$321	$789	$667	$703	$498

a. Make a frequency table for the data. Use reasonable intervals. *(Lesson 4-1)*
b. Make a histogram of the data. *(Lesson 4-2)*

26. **Grades** In order to receive a B in Mr. Carmona's math class, the average of Ling's five test scores must be 80 or above. His test scores are 78, 85, 83, 76, and 80. Will he receive a B in Mr. Carmona's class? Explain. *(Lesson 4-5)*

27. **Wildlife** Miss Kaiser is a biologist working for the state of Ohio. She is studying fish in Leesville Lake. These are the lengths in inches of large-mouth bass that she captured and released: 12, 9, 8, 15, 7, 8, 13, 10, 6, 14, 17, 8, 14, 12, and 7. Draw a box-and-whisker plot of the data. *(Lesson 4-7)*

Curriculum Connection Projects

- **Automotive** Make a tally of all the first numbers of the license plates in the school parking lot, and another of all the first letters. Find the mode and median of each tally.
- **Government** Construct stem-and-leaf plots of the ages of your state representatives and the ages of your state senators.

Read More About It

McWhirter, Norris, and Ross. *Guinness Sports Record Book.*
Janeczko, Paul B. *Loads of Codes and Ciphers.*
Brooks, Bruce. *The Moves Make the Man.*

Chapter

4 Test

Use the histogram at the right to answer each question.

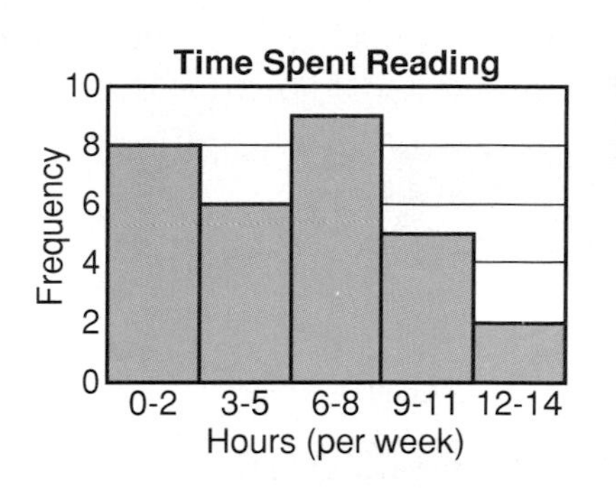

1. For which interval is the frequency the greatest?
2. How many people spend 3-5 hours reading per week?
3. How many people were surveyed?
4. How many people spend more than 5 hours per week reading?

The scores on a 50-point history test are given at the right.

43	48	40	36	50
42	49	33	37	41
45	32	50	44	47
38	41	49	33	42
45	35	38	42	43

5. Make a frequency table for the data.
6. Make a line plot for the data.
7. Make a stem-and-leaf plot for the data.
8. What is the frequency of the score that occurred most often?
9. How many students received a score of 38?
10. How many students received a score of 45 or above?

11. What numbers would you use for stems in a stem-and-leaf plot of {319, 334, 321, 325, 312, 346, and 338}?
12. Find the mean, median, and mode for the following set of data. 30, 14, 23, 20, 12, 18, 14, 19

The ages of the customers at Bob's Ice Cream Palace are 20, 34, 26, 40, 13, 23, 15, 41, 45, 35, 31, 12, 38, 18, 25, 48, 29.

13. What is the range?
14. What is the median?
15. What are the upper and lower quartiles?
16. What is the interquartile range?
17. What are the limits on the outliers? Are there any outliers?
18. Draw a box-and-whisker plot for the data.

19. Make a scatter plot for the data in the table below. Does the data show a positive or negative relationship between gasoline used and miles driven?

Gas Used (gal)	12	20	15	18	10	22	19	11	17	8	13
Miles Driven	215	336	247	320	181	356	336	172	312	149	201

20. Make a frequency table for these points scored in a basketball game.

102	78	62	98	67	88	98	101	102	89
121	66	78	102	113	120	109	88	97	98

Bonus Write six numbers whose mean, median, and mode is the same number.

Chapter 5

Investigations in Geometry

Spotlight on Rivers and Waterfalls

Have You Ever Wondered...

- What the longest river in the world is?
- Where the tallest waterfall in the world is?

River	Outflow	Length (mi)
Amazon	Atlantic Ocean	4,000
Amur	Tatar Strait	2,744
Chang Jiang	East China Sea	3,964
Congo	Atlantic Ocean	2,718
Huang	Yellow Sea	2,903
Lena	Laptev Sea	2,734
Mississippi	Gulf of Mexico	2,340
Missouri	Mississippi River	2,540
Nile	Mediterranean Sea	4,160
Ob-Irtysh	Gulf of Ob	3,362

THE FAR SIDE

By GARY LARSON

© Chronicle Features, 1980

3-11

LARSON

. . . and then the second group comes in—"row, row, row your boat" . . .

Chapter Project

Rivers and Waterfalls

Work in a group.

1. Construct a detailed model (or map) of an area of land that includes a river with a waterfall.
2. Illustrate the source of the river and the direction in which it flows.
3. Study the shape and path of the river. Show the best place to get to the other side.

Looking Ahead

In this chapter, you will see how mathematics can be used to answer questions about rivers and waterfalls. The major objectives of the chapter are :

- describe and construct parallel lines
- solve problems by using a Venn diagram
- classify triangles and quadrilaterals by their characteristics
- identify and draw translations, reflections, rotations, and dilations

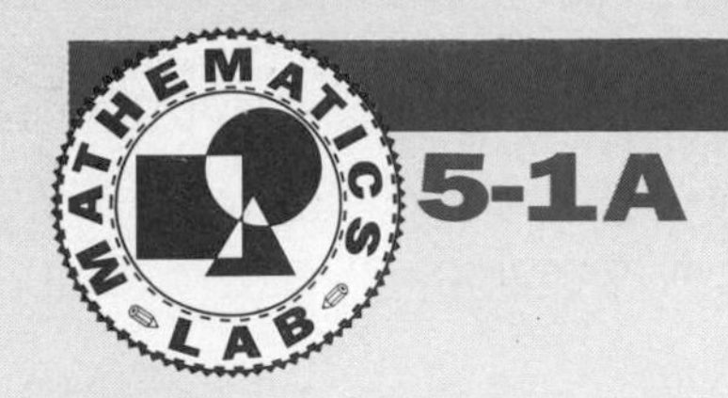

Cooperative Learning

5-1A Measurement in Geometry

A Preview of Lesson 5-1

Objective

Define congruence in terms of angle measurement and length.

Materials

straightedge (ruler)
compass
protractor

In Chapter 1, you learned two systems of measurement, the metric system and the customary system. In geometry we use both of these systems for measuring line segments.

Activity One

Work with a partner.

- Each segment below can be named by its endpoints. For example, a segment with endpoints W and Z can be named as WZ or as ZW. Name each segment shown below.

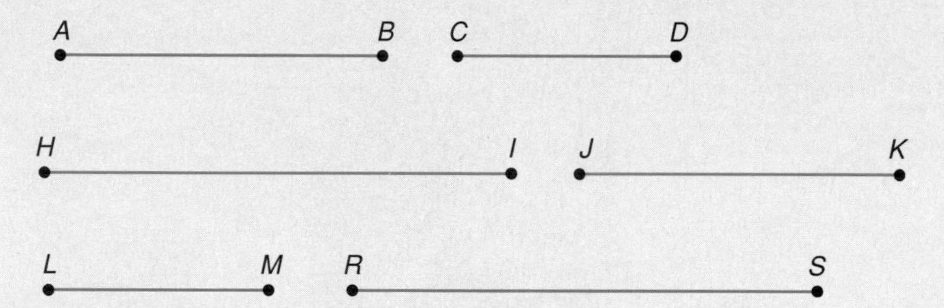

- Measure each segment to the nearest centimeter. Record your measurements.

What do you think?

1. Which segments had the same length?
2. Segments with the same length are said to be *congruent.* To say *segment WZ is congruent to segment OP,* we write $\overline{WZ} \cong \overline{OP}$. Use the proper symbols to write all pairs of congruent segments shown above.

Extension

3. A compass and straightedge are used to construct congruent segments. Construct a segment congruent to $\overline{AB}$ by following these steps.
   a. Draw a long line with a straightedge. Put a point on the line. Call it F.
   b. Open your compass to the same width as the length of $\overline{AB}$.
   c. Put the point of the compass at F and draw an arc to intersect the line. Label this intersection G. Segment FG should be congruent to segment AB, or $\overline{FG} \cong \overline{AB}$.

Angles are measured in units called *degrees.* You can use a protractor to measure an angle or to draw an angle of a given measurement.

Activity Two

Use a protractor to measure $\angle ABC$.

$\overrightarrow{BC}$ means ray BC. A ray is part of a line. It has one endpoint, B, and goes on continuously in one direction from that point.

- Trace $\angle ABC$ on your paper and extend sides $\overrightarrow{BA}$ and $\overrightarrow{BC}$.
- Place a protractor over $\angle ABC$ with the center point on vertex B.
- Align the horizontal line with side $\overrightarrow{BC}$.
- Locate the point on the protractor where $\overrightarrow{BA}$ intersects the edge of the protractor.
- Read the measurement from the scale that has 0° along $\overrightarrow{BC}$. To say, *the measure of angle ABC is 110 degrees,* we write $m\angle ABC = 110°$.

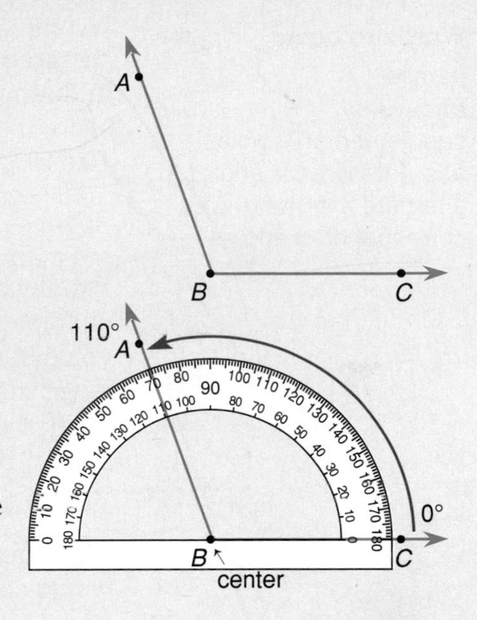

What do you think?

4. How do you know which scale to use when measuring an angle?
5. Two angles that have the same measurement are congruent. Suppose $m\angle XYZ = 110°$. Write a sentence relating $\angle XYZ$ and $\angle ABC$.

Extension

6. You can use a protractor to draw an angle with a specific measure. Draw $\angle MNO$ if $m\angle MNO = 45°$.
   a. Draw a ray. Label the endpoint N and put point M on the ray.
   b. Align your protractor so that the center is at N and the horizontal line aligns with $\overrightarrow{MN}$.
   c. Find the scale containing 0° along $\overrightarrow{MN}$. Follow that scale until you find 45°. Label this point O.
   d. Draw $\overrightarrow{NO}$. $m\angle MNO = 45°$.

5-1 Parallel Lines

Objective

Identify lines that are parallel.

Words to Learn

parallel
transversal
supplementary angles
alternate interior angles
alternate exterior angles
corresponding angles

In the Olympics, two gymnastic events are men's parallel bars and women's uneven parallel bars. Gymnasts are judged on required moves and difficulty in their routines. The highest score they can receive is a 10.0.

Parallel bars have some of the same characteristics as **parallel lines.** That is, if you extended them indefinitely, they would never meet. In order for two lines to never meet, they must be the same distance apart. To say *line ℓ is parallel to line m,* we write $\ell \parallel m$.

Many geometric figures contain parallel segments. If the lines that would contain those segments are parallel, the segments are parallel.

In this book, when two lines look parallel, you can assume they are.

Example 1

Name the parallel segments in each figure.

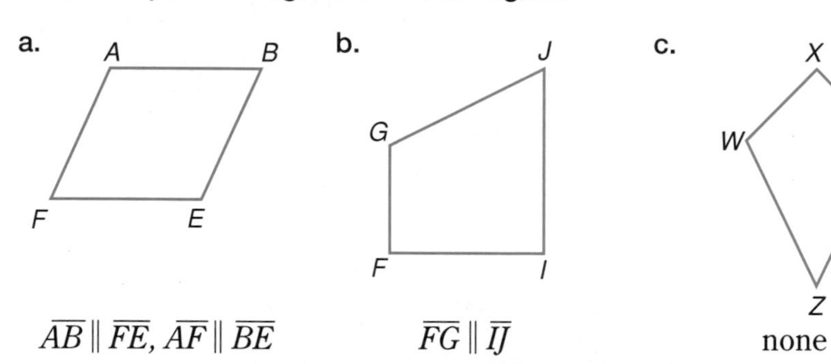

a. $\overline{AB} \parallel \overline{FE}$, $\overline{AF} \parallel \overline{BE}$

b. $\overline{FG} \parallel \overline{IJ}$

c. none

A line that intersects two other lines is called a **transversal.** In the figure, $m \parallel n$. Line p is the transversal. Eight angles are formed when a transversal intersects two parallel lines.

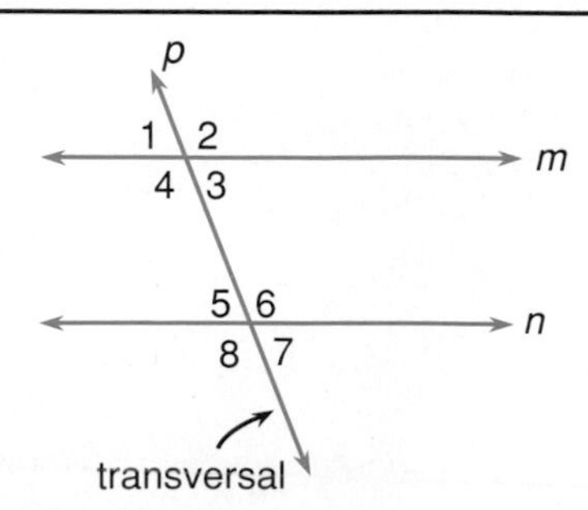

In the 1972 Olympics in Munich, Soviet gymnast Olga Korbut, age 16, won gold medals in balance beam, floor exercises, and as a member of the USSR team, which had the most team points. She also won a silver medal in parallel bars.

Mini-Lab

Work with a partner.

Materials: notebook paper, colored pencils, ruler, protractor

- Draw two parallel lines using the lines on your notebook paper. Draw any line to intersect these two lines.

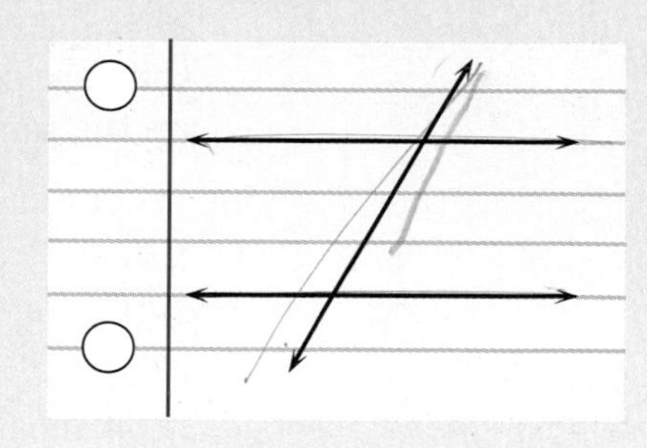

- Label the angles formed using the numbers 1 through 8. Measure each angle and record its measurement.
- Use a different colored pencil to circle the numbers of all of the angles that are congruent. *Congruent angles have the same measure.*

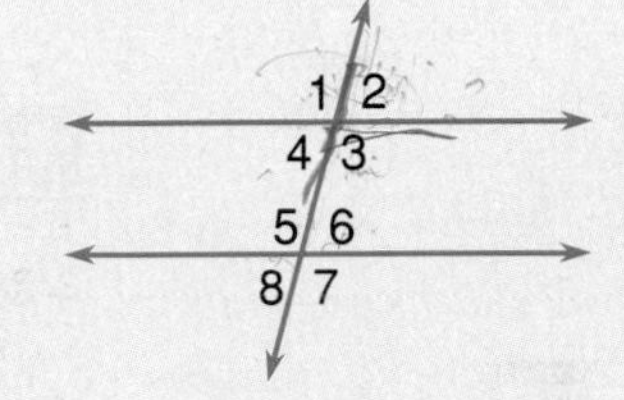

Talk About It

a. List the pairs of congruent angles in which one angle is formed by the transversal and one of the parallel lines and the other angle is formed by the transversal and the other parallel line.

b. Which pairs of congruent angles are on the same side of the transversal?

c. Which pairs of congruent angles are on opposite sides of the transversal?

d. How do $m\angle 2$ and $m\angle 3$ relate? $\angle 2$ and $\angle 3$ are **supplementary angles.** Name other pairs of supplementary angles.

If the sum of two angle measures is 180°, the angles are supplementary.

The congruent angles formed by parallel lines and a transversal have special names. Remember that the symbol $\cong$ means *is congruent to.*

Why do you think $\angle 1$ and $\angle 5$ are called corresponding angles?

Congruent Angles with Parallel Lines

If a pair of parallel lines is intersected by a transversal, these pairs of angles are congruent:

alternate interior angles:
$\angle 4 \cong \angle 6$, $\angle 3 \cong \angle 5$

alternate exterior angles:
$\angle 1 \cong \angle 7$, $\angle 2 \cong \angle 8$

corresponding angles:
$\angle 1 \cong \angle 5$, $\angle 2 \cong \angle 6$, $\angle 3 \cong \angle 7$, $\angle 4 \cong \angle 8$.

1 2
4 3
5 6
8 7

Example 2

In the figure, $m \parallel n$ and $m\angle 7 = 100°$.

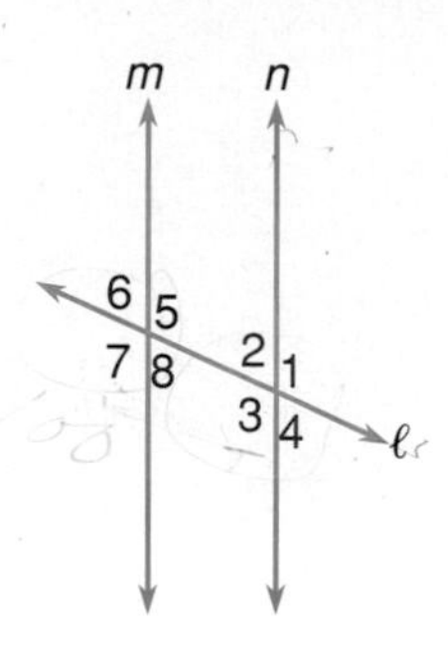

a. Find $m\angle 3$.
$\angle 7$ and $\angle 3$ are corresponding angles. They are congruent so their measures are the same.

$$m\angle 3 = m\angle 7$$
$$m\angle 3 = 100°$$

b. Find $m\angle 5$.
$\angle 3$ and $\angle 5$ are alternate interior angles. Their measures are the same, so $m\angle 5 = 100°$.

Example 3 *Connection*

Algebra In the figure, $m \parallel n$. Find the value of x, if $m\angle 5 = 110°$ and $m\angle 3 = 2x + 10°$.

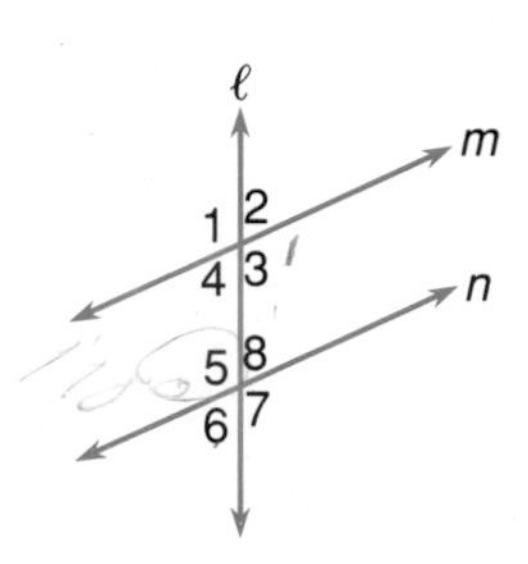

$\angle 5$ and $\angle 3$ are corresponding angles. They are congruent, so their measures are equal. You can write an equation.

$$m\angle 5 = m\angle 3$$
$$110 = 2x + 10$$
$$110 - 10 = 2x + 10 - 10$$
$$100 = 2x$$
$$\frac{100}{2} = \frac{2x}{2}$$
$$50 = x$$

The value of x is 50.

Checking for Understanding

Communicating Mathematics

Read and study the lesson to answer each question.

1. **Write** a definition for parallel lines.
2. **Tell** how you think a parallelogram got its name.
3. **Show** two examples of parallel segments in your classroom.
4. **Draw** two parallel lines p and q. Then draw a transversal r. Measure one angle. Then, without measuring, put the measures of all angles on your drawing.
5. **Tell** what it means when two angles are supplementary.

Guided Practice

Name all parallel segments in each figure.

6. 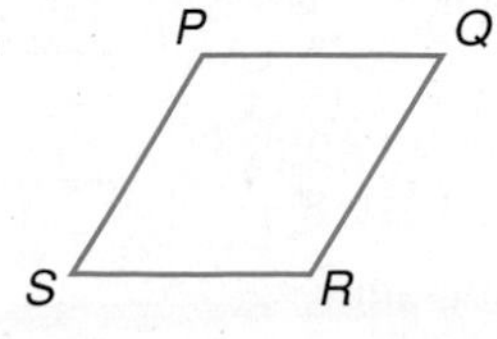

7. 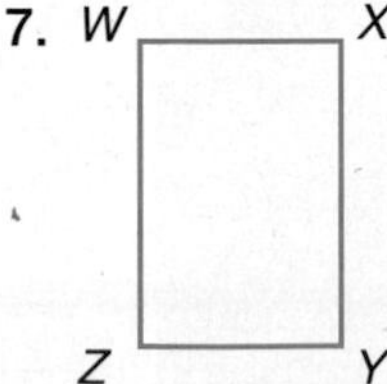

8. 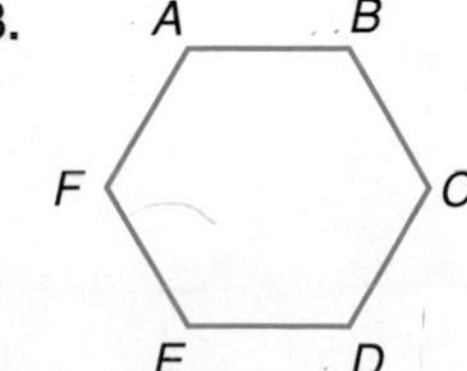

9. Find the value of each variable in the figure at the right if $p \parallel q$ and $h = 60°$.

a. a b. b c. c d. d

e. e f. f g. g

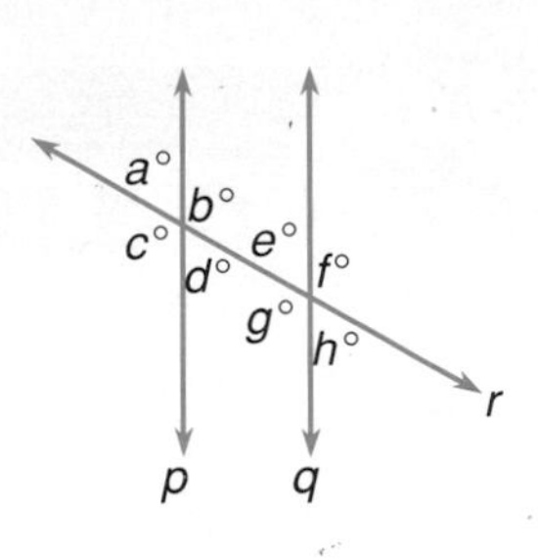

Exercises

Independent Practice

Name the parallel segments, if any, in each figure.

10. 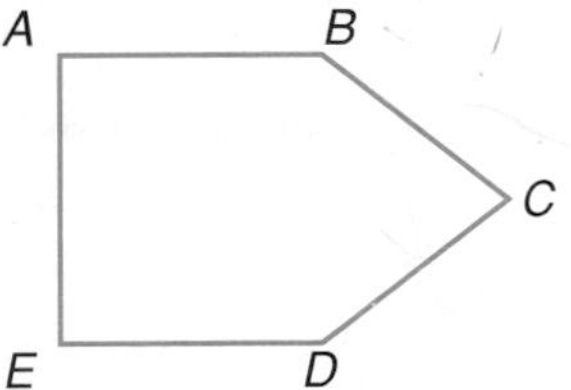

11. 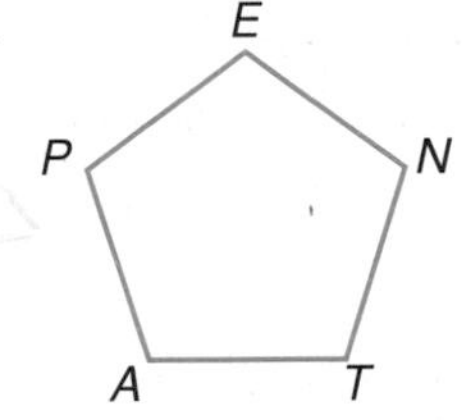

12. 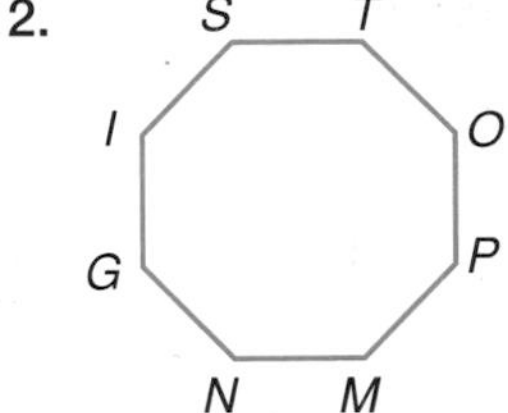

Use the figure at the right for Exercises 13-16.

13. Find $m\angle 6$, if $m\angle 2 = 35°$.
14. Find $m\angle 3$, if $m\angle 5 = 77°$.
15. Find $m\angle 4$, if $m\angle 8 = 122°$.
16. Find $m\angle 7$, if $m\angle 1 = 68°$.

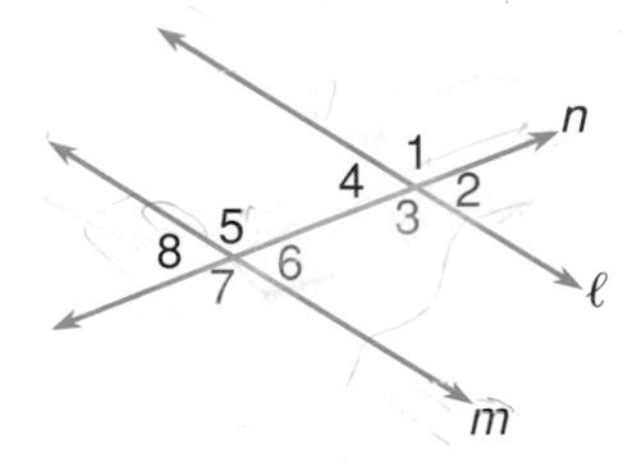

Find the value of x in each figure.

17. 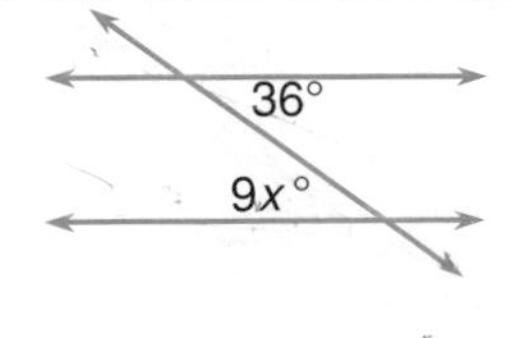

18. 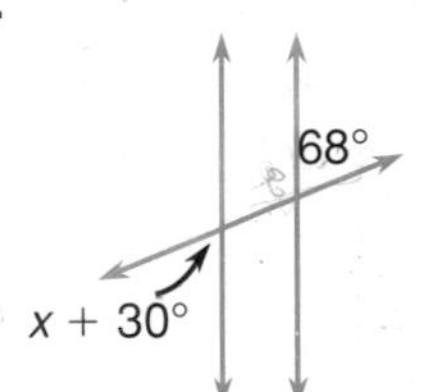

19. 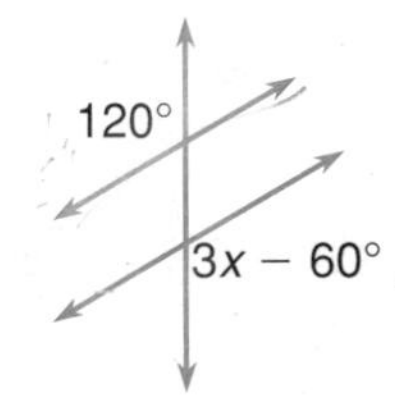

Mixed Review

20. Use mental math to find \$5.00 − \$2.49. *(Lesson 1-2)*
21. Find the absolute value of 10. *(Lesson 3-1)*
22. **Statistics** Which average (mean, median, or mode) would best describe the height of eighth grade students at Dunbar Junior High? *(Lesson 4-8)*

Problem Solving and Applications

23. **Engineering** In Birmingham, Alabama, the city blocks are arranged like a grid. The roads running north and south are called streets and those running east and west are called avenues.
    a. What is true of all the avenues?
    b. What is true of all the streets?
    c. What is the measure of the angle formed when a street intersects an avenue?
24. **Critical Thinking** Find the value of x in the figure at the right if $m\angle 1 = 30°$. Explain how you found this value.

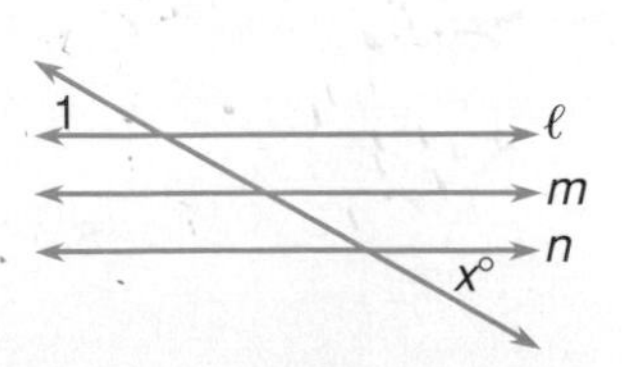

25. **Journal Entry** Write how the statement *If $\ell \parallel m$ and $m \parallel n$, then $\ell \parallel n$* has to be true.

Cooperative Learning

5-1B Constructing Parallel Lines

A Follow-Up of Lesson 5-1

Objective

Construct a line parallel to a given line.

Materials

compass
straightedge

You can use a compass and straightedge to construct a line parallel to a given line.

Try this!

- Draw a line and label it ℓ.
- Choose any point P, not on ℓ.
- Draw a line through P that intersects ℓ. Label this point Q.

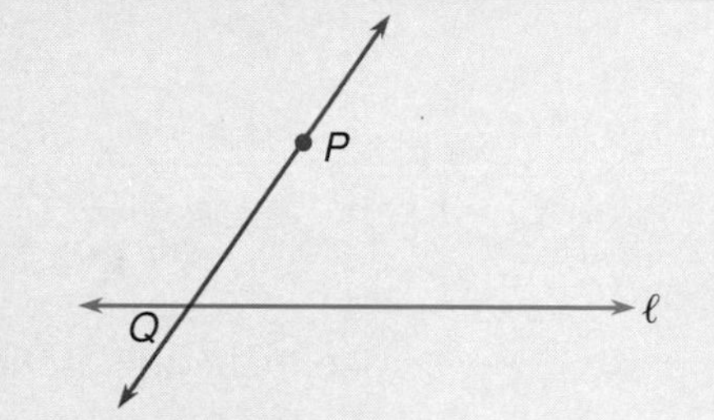

- Place your compass point at Q and draw a large arc. Label points R and S.
- With the same setting, place the compass point at P and draw a large arc. Label point T.

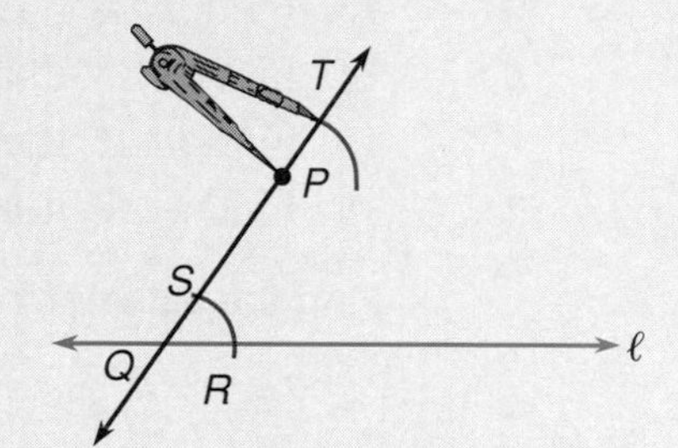

- Use your compass to measure the distance from R to S.
- With the same setting, place your compass at T and draw an arc to intersect the one already drawn. Label this point U.

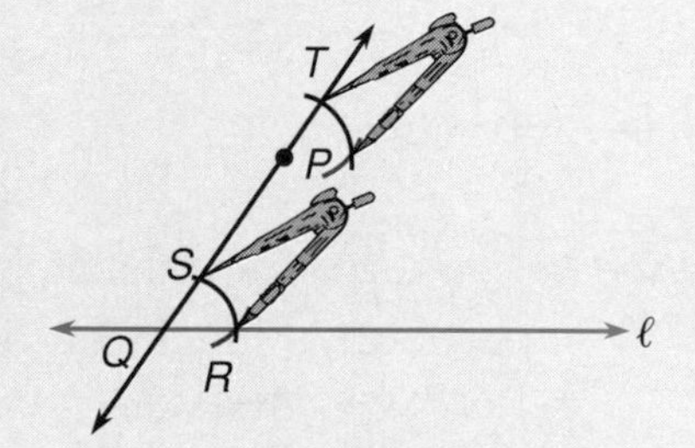

- Draw a line through P and U. Label it line m.
  By construction, $\ell \parallel m$.

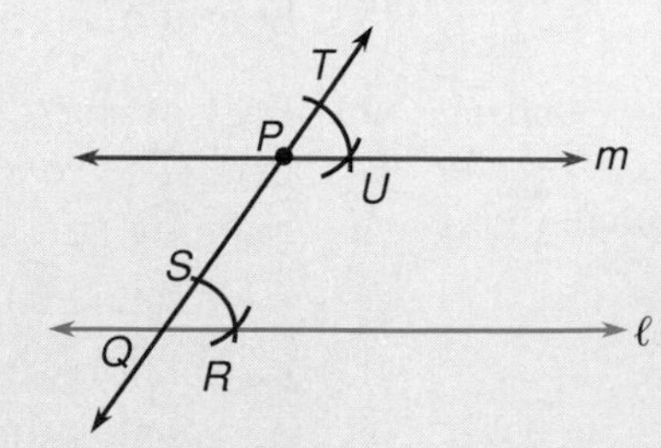

What do you think?

1. Look at your completed construction. What type of angles did you use to create your parallel lines?
2. Measure the angles in your construction. The angles you constructed may not be congruent. Explain how this could happen.

Problem-Solving Strategy

5-2 Use a Venn Diagram

Objective

Use a Venn diagram to solve problems.

Words to Learn

Venn diagram

John Venn (1834-1923) was an English logician. He used diagrams to solve problems. Overlapping circles indicated sets that shared some of the same elements. These elements were placed in the overlapping part of the two circles. These diagrams are known as **Venn diagrams.**

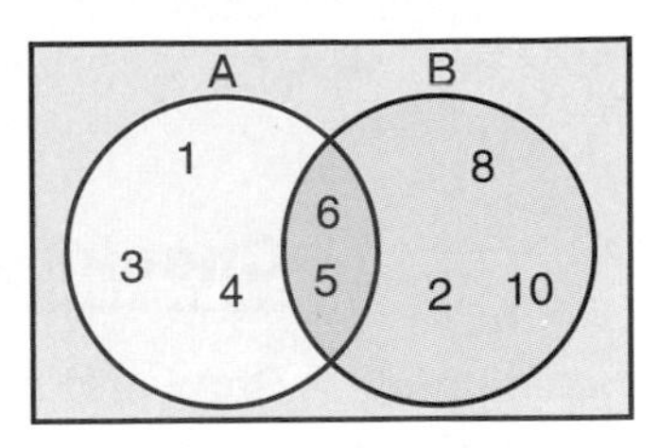

set A = {1, 3, 4, 5, 6}
set B = {2, 5, 6, 8, 10}

Example

The table below shows the states that produce over 100 million bushels of corn, wheat, or soybeans per year. Make a Venn diagram of this information and determine how many states produce over 100 million bushels of either corn or wheat.

Crop	State
corn	MI, NE, SD, WI, TX, IL, IA, IN, MN, OH
wheat	CO, KS, MT, ND, OK, WY, TX
soybeans	MO, IL, IA, IN, MN, OH

Explore *What do you know?*
You know which states produce over 100 million bushels of corn, wheat, or soybeans.

What are you trying to find?
How many states produce over 100 million bushels of either corn or wheat?

Plan Draw a Venn diagram to organize the states. Make a note of which states appear in more than one category.

Solve Draw three intersecting circles in a rectangle to represent corn, wheat, and soybeans. List the state abbreviations in the appropriate sections. Count the number of states in both the corn circle and the wheat circle.

corn only		*corn and soybeans*		*corn and wheat*		*wheat only*		
4	+	5	+	1	+	6	=	16 states

Examine Compare the lists of states for corn and wheat. How many states are in each list? How many states are in both lists? Subtract the number of states in both lists from the total.

corn	+	*wheat*	−	*both*	=	?
10	+	7	−	1	=	16

This matches the count from the Venn diagram.

Checking for Understanding

Communicating Mathematics

Read and study the lesson to answer each question.

1. **Write** what the section in the Venn diagram where all three circles overlap represents in the Example.
2. **Tell** how many states produce over 100 million bushels of either corn or soybeans.

Guided Practice

Solve using a Venn diagram.

3. Refer to the Example.
   a. How many states produce over 100 million bushels of either wheat or soybeans?
   b. How many states produce over 100 million bushels of all three grains?
4. The results of a supermarket survey showed that 83 customers chose wheat cereal, 83 chose rice, and 20 chose corn. Of those customers who bought two boxes, six bought corn and wheat, 10 bought rice and corn, and 12 bought rice and wheat. Four customers bought all three. Make a Venn diagram of this information.

Problem Solving

Practice

Solve using any strategy.

Strategies

Look for a pattern.
Solve a simpler problem.
Act it out.
Guess and check.
Draw a diagram.
Make a chart.
Work backward.

5. Of the 30 members in a cooking club, 20 like to mix salads, 17 prefer baking desserts, and 8 like to do both.
   a. Make a Venn diagram of this information.
   b. How many like to mix salads, but not bake desserts?
   c. How many do not like either baking desserts or mixing salads?
6. Luisa put some counters into a beaker holding 500 cm^3 of water. The water level rose to a reading of 625 cm^3. What is the volume of the counters?
7. A number increased by 29 is 6 less than twice the number. What is the number?
8. Refer to the Venn diagram at the right.
   a. How many eighth grade students are in the band and the chorus?
   b. How many are in the band or the chorus?
   c. Are there more in the orchestra than in the band? Explain.

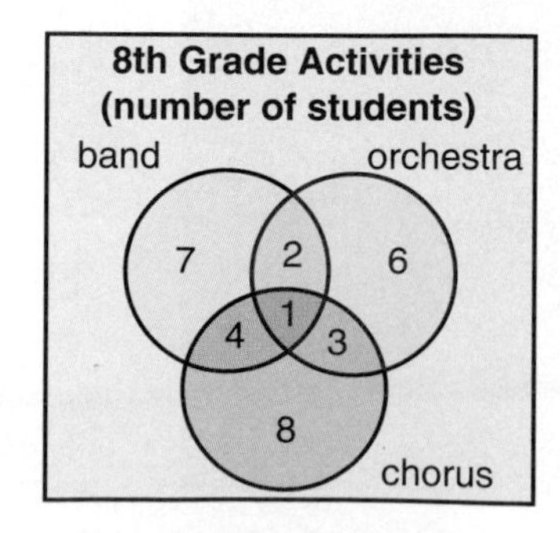

5-3 Classifying Triangles

Objective
Classify triangles.

Words to Learn
scalene
isosceles
equilateral
acute
right
obtuse
perpendicular

Have you ever looked under the bleachers at the football field? If you look at the end of the bleachers, you will probably notice that a triangular frame was used to support them.

Triangles are named by using letters at their vertices. The triangle at the right can be named as ΔXYZ. *A vertex is the point where two sides meet.*

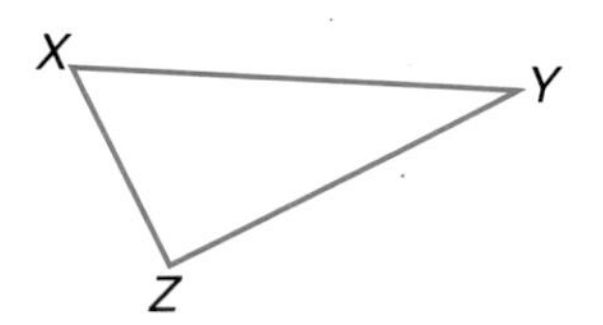

Triangles are often classified by how many congruent sides they have.

Slashes are often used to show which sides of figures are congruent.

Triangles Classified by Sides	scalene	isosceles	equilateral
	no sides congruent	two sides congruent	three sides congruent

Triangles can also be classified by the type of angles they have. An angle is classified by its measure.

Acute angles have measures less than 90°.
Right angles have measures equal to 90°.
Obtuse angles have measures greater than 90°, but less than 180°.
Straight angles have measures equal to 180°.

Mental Math Hint

Use the corner of your notebook paper to estimate whether an angle has a measure greater than 90°, equal to 90°, or less than 90°.

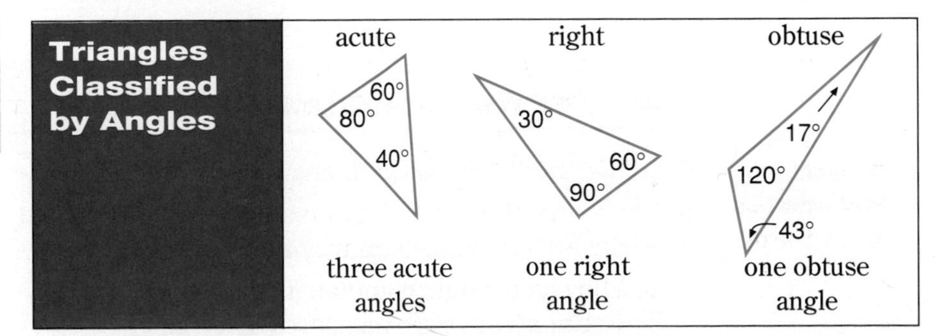

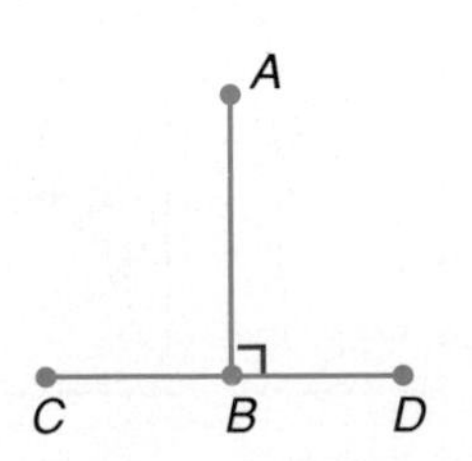

The ⌝ symbol in the figure at the left, indicates that an angle is a right angle. When segments meet to form right angles, they are **perpendicular.** To say $\overline{AB}$ is perpendicular to $\overline{CD}$, we write $\overline{AB} \perp \overline{CD}$.

Examples

Classify each triangle by its sides and by its angles.

Equilateral triangles are also equiangular. *That is, all its angles have the same measure.*

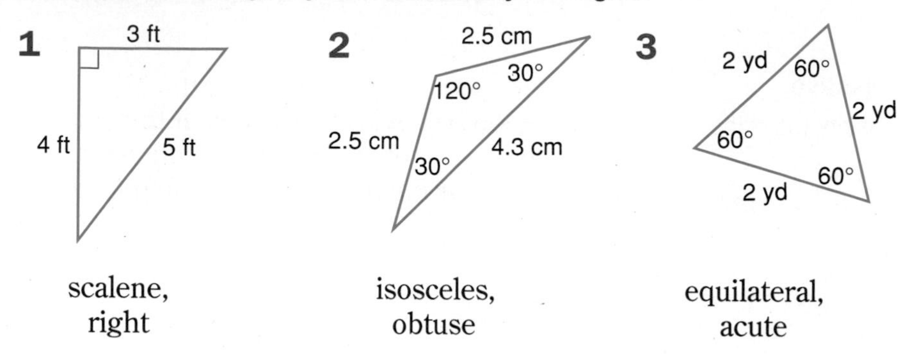

1. scalene, right
2. isosceles, obtuse
3. equilateral, acute

Mini-Lab

Work with a partner.

Materials: scissors, straightedge

- Draw any large triangle and cut it out. Label the angles A, B, and C.

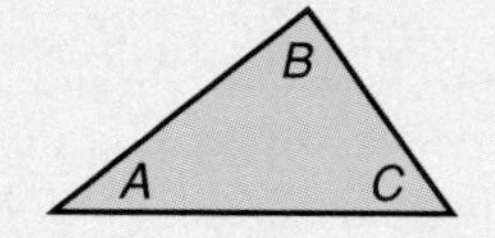

- Tear off the angles and arrange as shown.

Talk About It

a. When the angles were put together, what type of angle did they form? What is the measure of this type of angle?

b. Measure each of the angles and find the sum of their measures. Compare this sum to your result in question a.

c. Try this activity with another triangle. What do you think is true about the measures of the angles of *any* triangle?

Checking for Understanding

Communicating Mathematics

Read and study the lesson to answer each question.

1. **Write** what the Venn diagram at the right says about isosceles and equilateral triangles.
2. **Tell** how you would complete this sentence. The sum of the angle measures of a triangle is ___?___°.
3. **Draw** a right isosceles triangle.
4. **Tell** what the symbol $\perp$ means.

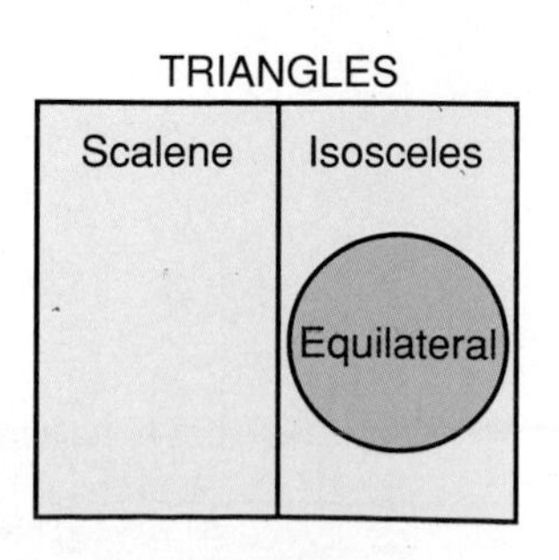

5. Show how you can use the results of the Mini-Lab to find the value of x in ΔPQR.

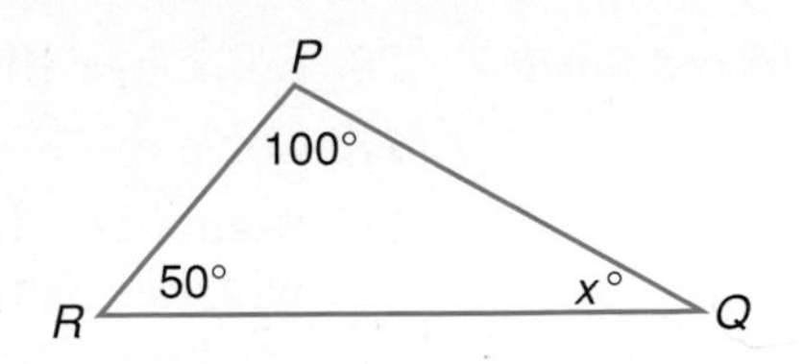

Guided Practice

Classify each triangle by its sides and by its angles.

6.

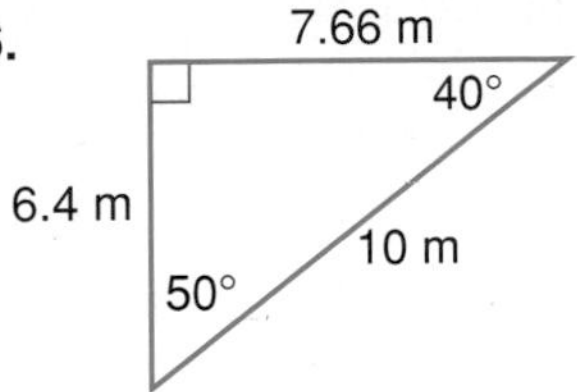

7.

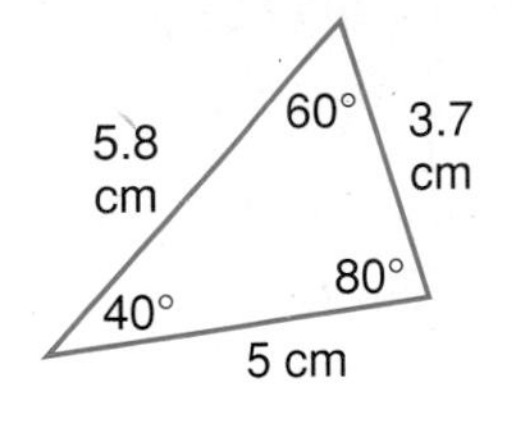

8.

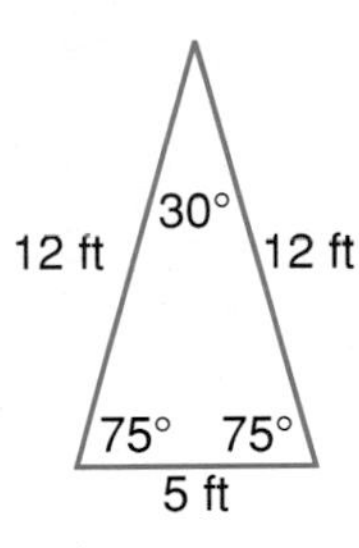

9.

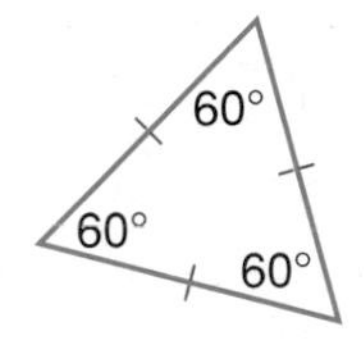

10.

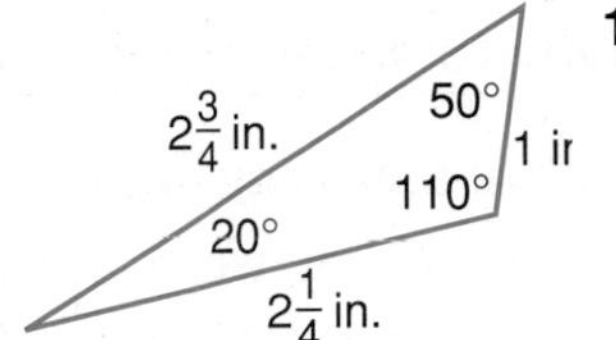

11.

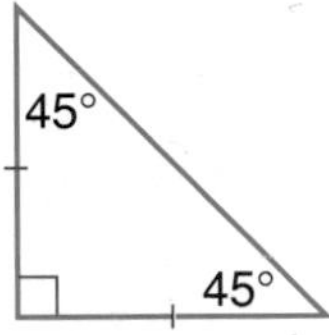

Exercises

Independent Practice

Classify each triangle by its sides and by its angles.

12.

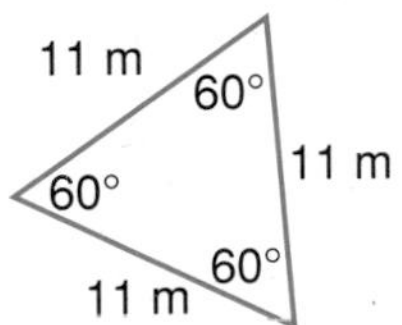

13.

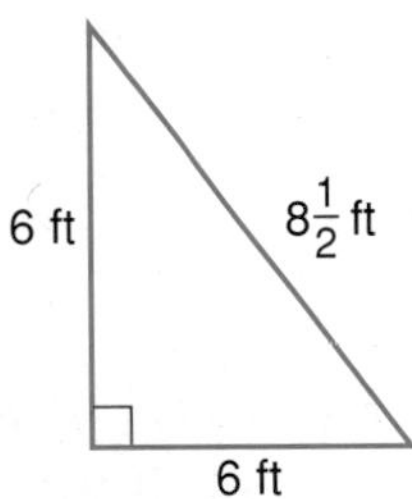

14.

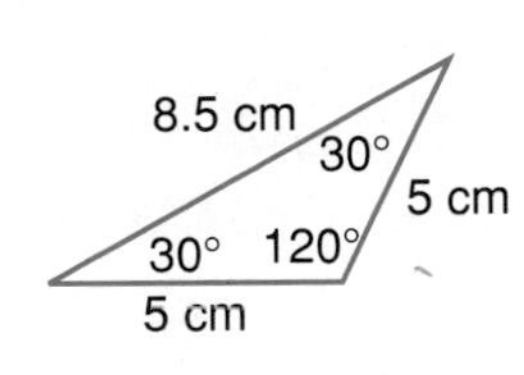

15.

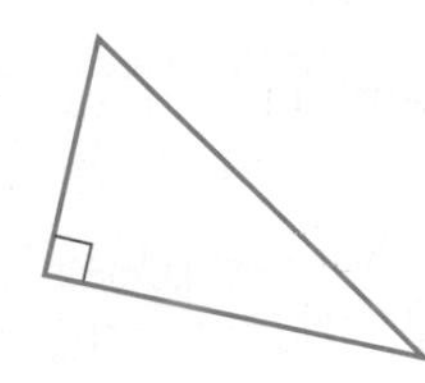

16.

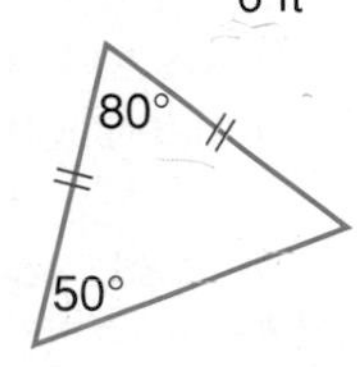

17.

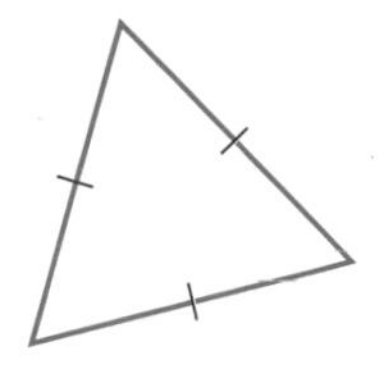

Tell if each statement is *true* or *false*. Then draw a figure to justify your answer.

18. A triangle can be isosceles and acute.
19. A triangle can be obtuse and scalene.
20. A right triangle can also have an obtuse angle.
21. A triangle can contain two obtuse angles.
22. An equilateral triangle can never be a right triangle.
23. An equilateral triangle is also an isosceles triangle.

24. **Algebra** Find the value of x in ΔABC if $m\angle A = 90°$, $m\angle B = 32°$, and $m\angle C = x°$.
25. **Algebra** Find the value of x in ΔXYZ if $m\angle X = 2x°$, $m\angle Y = 64°$, and $m\angle Z = 36°$.

Mixed Review

26. Solve $b - 19 = 73$. *(Lesson 2-3)*
27. Solve $e = 8(-3)(-4)^2$. *(Lesson 3-6)*
28. **Statistics** Refer to the temperature box-and-whisker plot in Example 1 on page 156. What fraction of the days had temperatures above 49°? *(Lesson 4-7)*

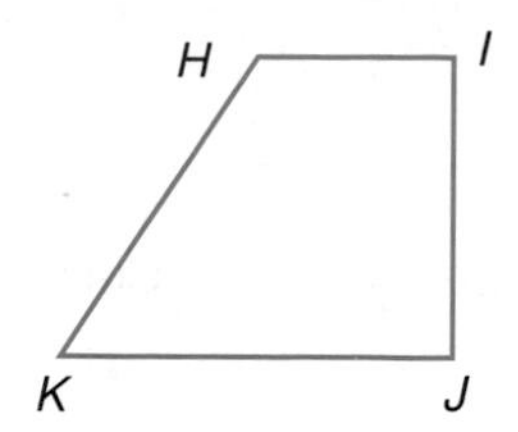

29. Name the parallel segments in the figure at the right. *(Lesson 5-1)*

Problem Solving and Applications

30. **Critical Thinking** Use the number of toothpicks as units to form the sides of as many triangles as possible and complete the chart.

Number of toothpicks	3	4	5	6	7	9	12
Number of triangles formed	?	?	?	?	?	?	?
Types of triangles formed	?	?	?	?	?	?	?

a. Were there any numbers for which you could not build a triangle?

b. Suppose you were told the lengths of the sides of a scalene triangle were 5 meters, 2 meters, and 7 meters. Is this possible? Explain your answer.

c. Write what you can conclude about the relationship between the sum of the lengths of two sides of a triangle and the length of the third side.

31. **Architecture** Triangles are often used in structural designs.

a. Name all the types of triangles in the structure shown at the left.

b. Why do you think architects use triangles in structures rather than other figures?

32. **Critical Thinking** Build the figure at the right with straws. Take away four straws to leave exactly four congruent equilateral triangles.

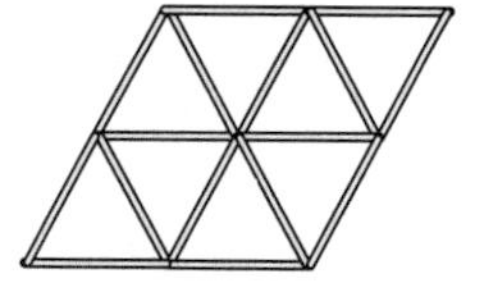

33. **Journal Entry** Write all possible types of triangles if you classify them both by angles and by sides.

34. **Mathematics and History** Read the following paragraph.

> In 2600 B.C., the Egyptians used a detailed system of geometry to build the pyramid of Snefu. The builders measured the base of the pyramid with a rope that had 12 equal sections knotted in it. It was stretched around three pegs to form sides of three, four, and five sections. This formed a right triangle.

a. The measures of complementary angles have a sum of 90°. If two angles of a triangle are complementary, what is the measure of the third angle?

b. Find the area of a triangle with a 7-centimeter base and a height of 9 centimeters. Use the formula $A = \frac{1}{2}bh$.

5-4 Classifying Quadrilaterals

Objective

Classify quadrilaterals.

Words to Learn

quadrilateral
rhombus
trapezoid

What characteristics do a baseball diamond, a football field, a basketball court, and a long jump pit have in common? You may have guessed that all are four-sided figures. Any four-sided figure is a **quadrilateral.**

When you say the word *cat,* different people picture different animals. Each breed of cat has its own set of characteristics. Some breeds share characteristics, or attributes. The same is true of quadrilaterals.

You are already familiar with three types of quadrilaterals.

LOOKBACK

You can review parallelograms, rectangles, and squares on page 73.

parallelogram

opposite sides parallel
opposite sides congruent

rectangle

opposite sides parallel
opposite sides congruent
four right angles

square

opposite sides parallel
four sides congruent
four right angles

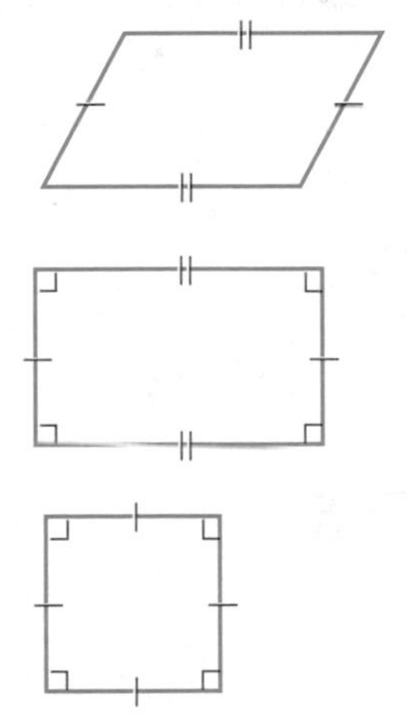

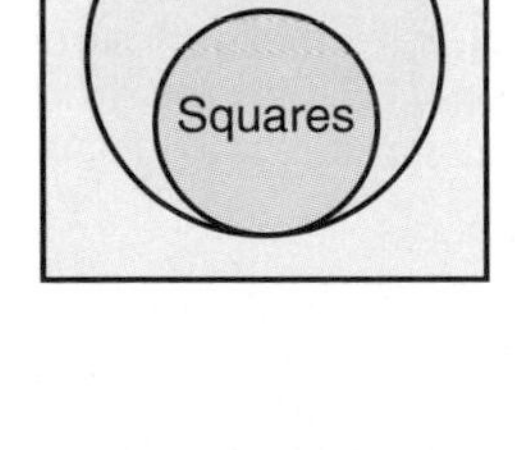

The Venn diagram relates these quadrilaterals by their attributes. It shows that rectangles and squares are types of parallelograms. A square is also a special type of rectangle.

Another type of parallelogram is a **rhombus.** A rhombus is a parallelogram that has four congruent sides. Two *rhombi* are shown at the right. A rhombus with four right angles is also a square. A quadrilateral that is not in the parallelogram family is the **trapezoid.** A trapezoid has *only one* pair of parallel sides. The other pair of sides are not parallel.

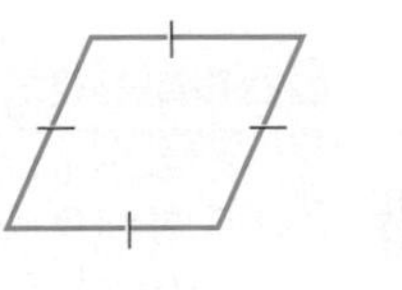

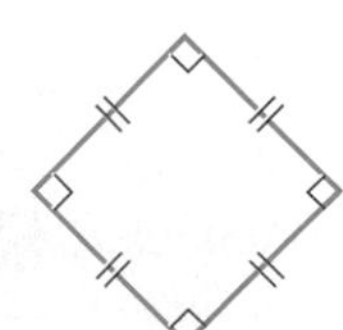

Quadrilaterals are named by the letters of their four vertices.

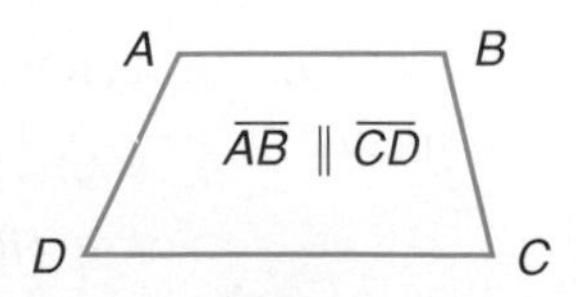

Examples

Identify all names that describe each quadrilateral.

1

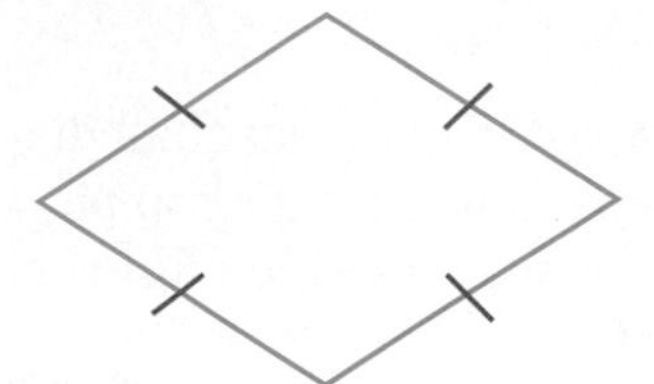

parallelogram, rhombus

2

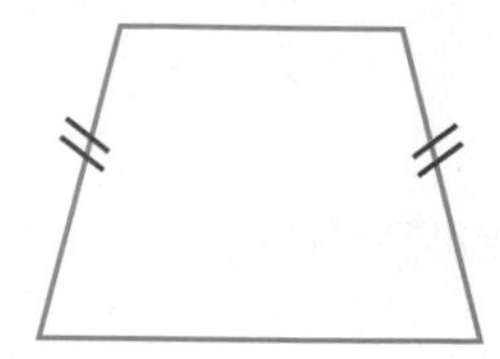

trapezoid

If the two nonparallel sides of a trapezoid are congruent, the trapezoid is called an *<u>isosceles trapezoid</u>.*

Mini-Lab

Work with a partner.

Materials: straightedge, protractor

- Draw a parallelogram. Use your protractor to measure all four angles. Record your measurements.
- Draw a trapezoid. Measure all four angles. Record your measurements.
- Draw a quadrilateral that has no sides parallel. Measure all four angles. Record your measurements.

Talk About It

a. What is the sum of the angle measures for each figure?

b. What is the sum of the angle measures in a rectangle?

c. Write a sentence to describe the sum of the angle measures in any quadrilateral.

Checking for Understanding

Communicating Mathematics

Read and study the lesson to answer each question.

1. **Tell** what attribute all quadrilaterals have.
2. **Draw** a rhombus that isn't a square.
3. **Tell** why a parallelogram is not a trapezoid.
4. **Write** the number that completes this statement.
   The sum of the angle measures of any quadrilateral is ___?___ °.

Guided Practice **Sketch each figure on your paper. Let Q = quadrilateral, P = parallelogram, R = rectangle, S = square, RH = rhombus, and T = trapezoid. Write all letters inside the figure that describe it.**

5. 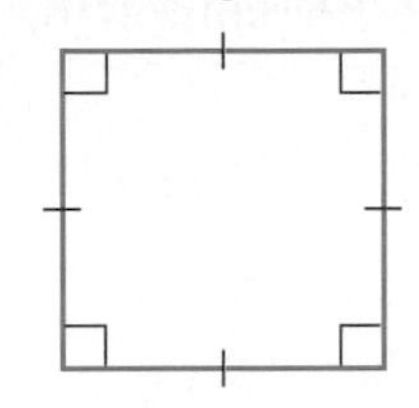

6. 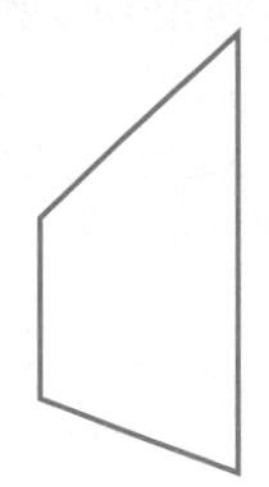

7. 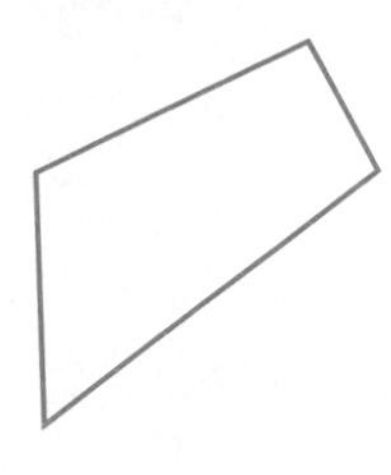

8. In quadrilateral $ABCD$, $m\angle A = 90°$, $m\angle B = 90°$, and $m\angle C = 120°$. Find $m\angle D$.

Exercises

Independent Practice **Sketch each figure on your paper. Let Q = quadrilateral, P = parallelogram, R = rectangle, S = square, RH = rhombus, and T = trapezoid. Write all letters inside the figure that describe it.**

9. 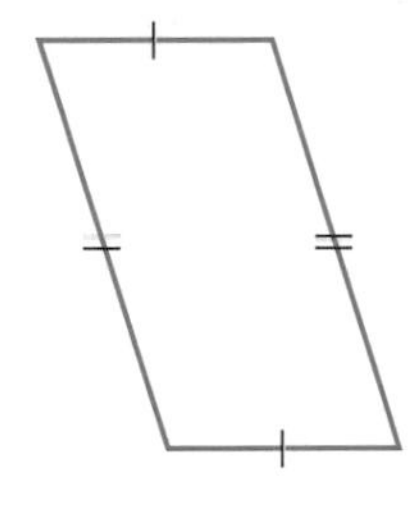

10. 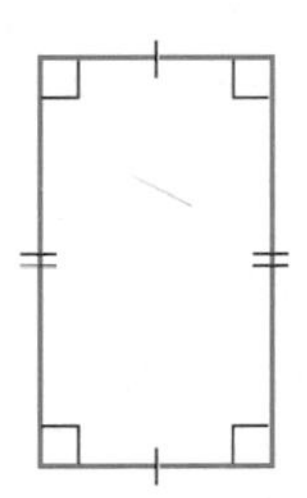

11. 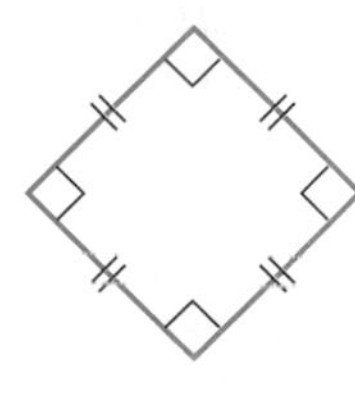

12. 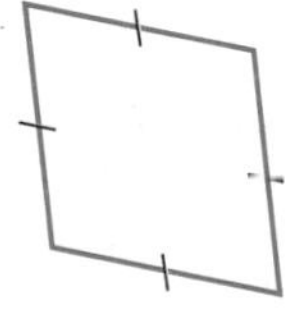

13. 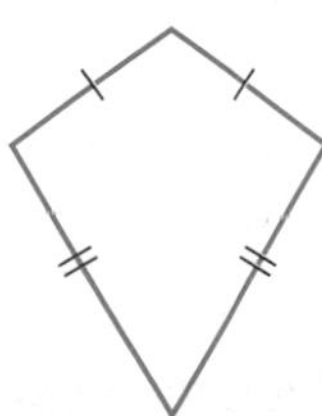

14. 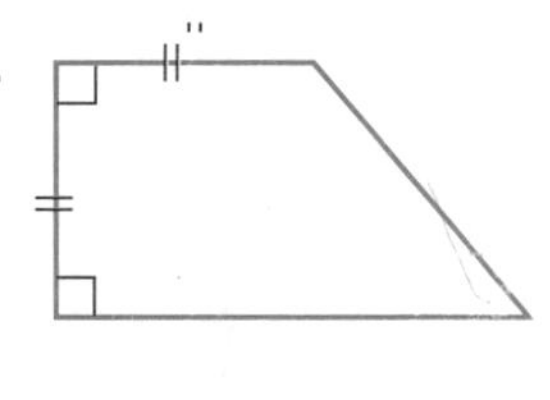

15. Name all quadrilaterals that have both pairs of opposite sides parallel.
16. Name all quadrilaterals that have four right angles.
17. Name all quadrilaterals that have four congruent sides.

Tell if each statement is *true* or *false*. Then draw a figure to justify your answer.

18. Every square is a parallelogram.
19. A rhombus is a square.
20. A trapezoid can have only one right angle.

21. **Algebra** In trapezoid $WXYZ$, $m\angle W = 2a°$, $m\angle X = 40°$, $m\angle Y = 110°$, and $m\angle Z = 70°$. Find the value of a.

22. **Algebra** In rhombus $PQRS$, $m\angle P = 60°$, $m\angle Q = x°$, $m\angle R = 60°$, and $m\angle S = x°$. Find the value of x.

Mixed Review

23. Find the perimeter and area of a rectangle that has a length of 3 centimeters and a width of 5.5 centimeters. *(Lesson 2-9)*
24. Without graphing, tell in which quadrant the point $M(6, -4)$ lies. *(Lesson 3-10)*
25. **Algebra** Find the value of x in ΔDEF if $m\angle D = 4x°$, $m\angle E = 60°$, and $m\angle F = 72°$. *(Lesson 5-3)*

Problem Solving and Applications

26. Draw a Venn diagram to represent the relationships between all quadrilaterals.

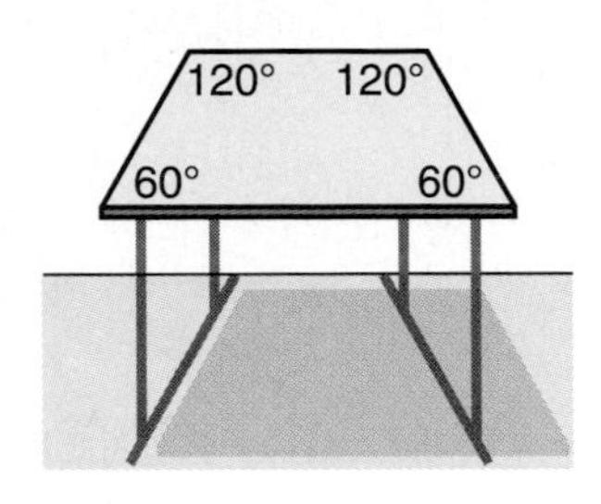

27. **Teaching** A furniture company manufactures tables whose tops are shaped like trapezoids for use in the classroom. A teacher wants to arrange six of these tables so that 12 students can sit in a group. Make a sketch of how these tables could be arranged.

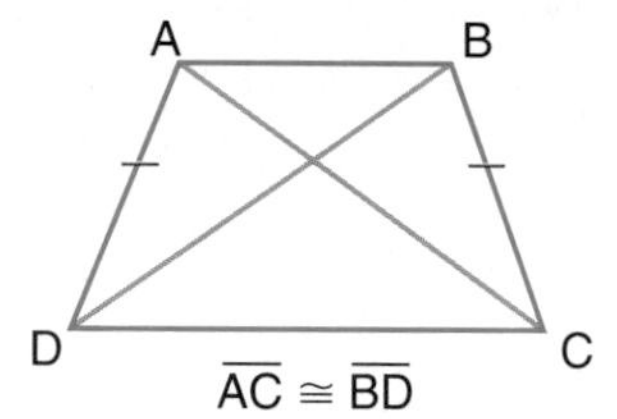

28. **Critical Thinking** A diagonal is a segment that connects any two nonconsecutive vertices. An isosceles trapezoid has two congruent diagonals. What other types of quadrilaterals have two congruent diagonals?

29. **Data Search** Refer to pages 172 and 173. Suppose you were 100 feet from the bottom of Angel Falls. Assume the waterfall forms a 90° angle with the river.
    a. Draw the triangle formed by connecting the points representing the top of the falls, the bottom of the falls, and where you are standing.
    b. Would you draw a different triangle if the falls were Niagara Falls? Explain.

5 Assessment: Mid-Chapter Review

Use the figure for Exercises 1-3.

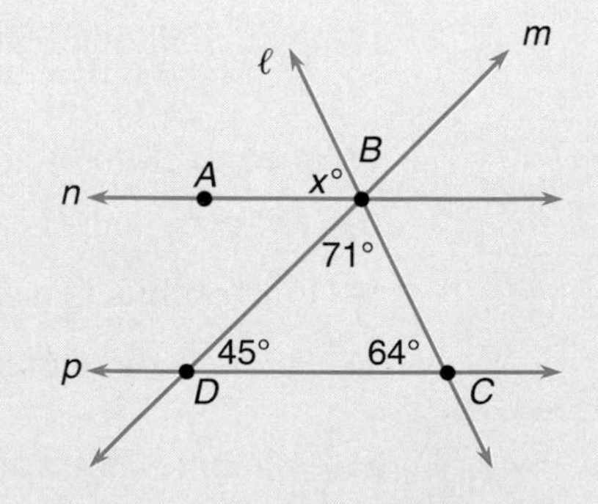

1. Name all parallel segments. *(Lesson 5-1)*
2. What is the value of x? *(Lesson 5-1)*
3. Classify ΔBCD by its sides and by its angles. *(Lesson 5-3)*
4. Draw a Venn diagram to determine how many odd numbers from 1 to 50 are multiples of both 3 and 5. *(Lesson 5-2)*
5. Draw a rhombus that has at least one right angle. Describe your rhombus. *(Lesson 5-4)*

Cooperative Learning

5-5A Reflections

A Preview of Lesson 5-5

Objective

Construct a reflection using a geoboard.

Materials

geoboards
rubber bands

When you see your reflection in a mirror, it seems that your image is as far back from the mirror as you are in front of the mirror. In mathematics, we create reflections of figures using a line instead of a mirror.

Try this!

Work with a partner.

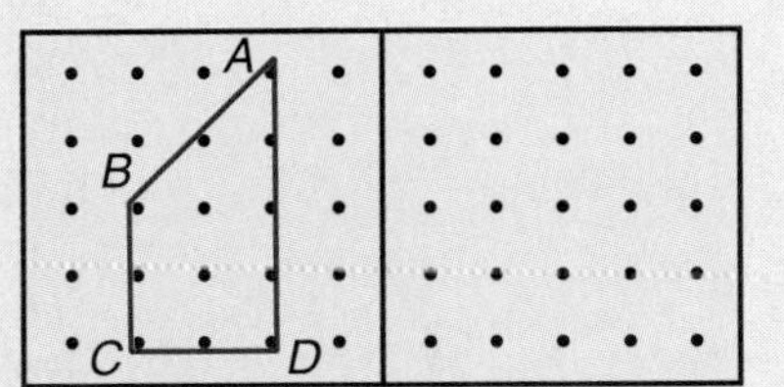

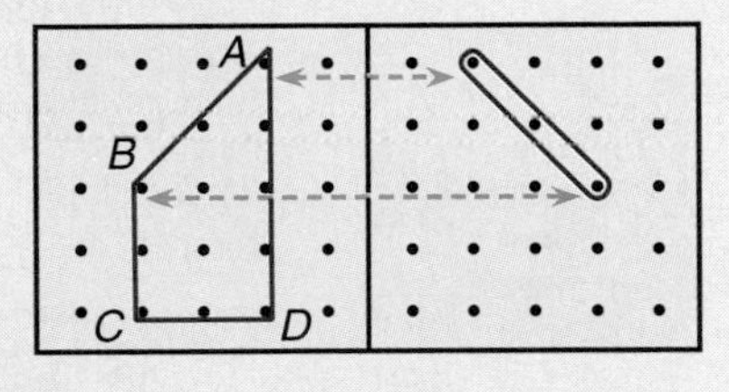

- Place two geoboards side by side. Use a rubber band to create the figure shown.
- Let the edge where the two boards meet be the line of reflection. Vertex A is 2 pegs left of the line. Find a peg on the same row that is 2 pegs right of the line. Place a rubber band around this peg.
- Vertex B is 4 pegs left of the line. Find a peg on the same row that is 4 pegs right of the line. Stretch the rubber band to fit around this peg.
- Find the corresponding pegs for vertices C and D. Complete the figure with the rubber band. The figure on the right-hand board is the reflection of the image on the left-hand board.

What do you think?

1. Compare the two figures. Are their corresponding sides congruent?
2. What type of quadrilateral are these figures?
3. Write a sentence to explain how this reflection is like a reflection in a mirror.

Use side-by-side geoboards to reflect each figure.

4. 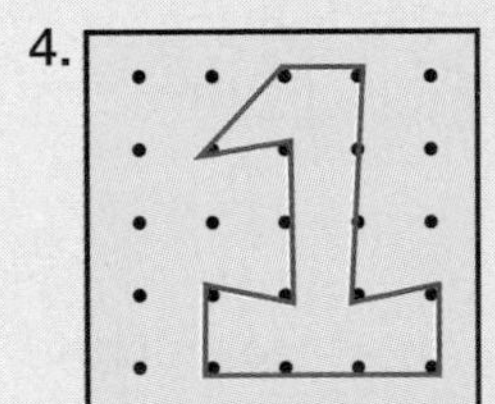

5. 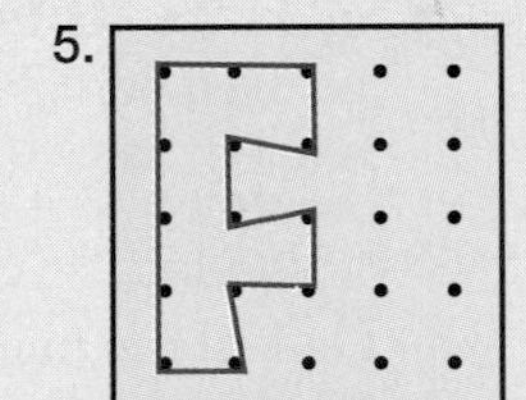

6. 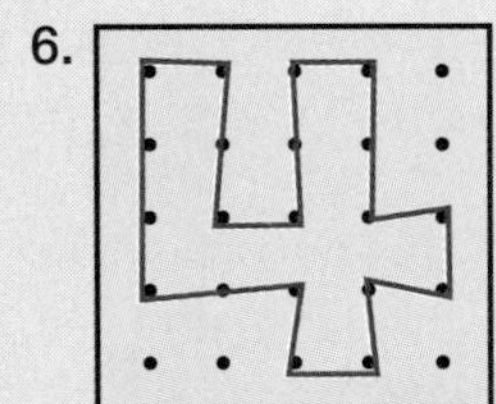

7. Suppose the second geoboard is placed below the first. Repeat Exercises 4-6 using this type of reflection. How do they compare with the first reflections?

Did You Know
Mirrors and mirror arrangements with reducing convex mirrors or concave lenses were used in the 17th and 18th centuries as drawing aids in the preparation of reproductions.

5-5 Symmetry

Objective

Explore line symmetry and rotational symmetry.

Words to Learn

line symmetry
reflection
rotational symmetry

Yoshizawa Akira of Tokyo is one of the world's best known origamist. Origami is the Japanese art of paper folding. A 15 centimeter by 25 centimeter sheet of thin colored paper is folded to form birds, fish, and flowers without cutting or gluing.

One of the first steps in many origami figures is to fold the rectangle in half. When you do this, each point on one half of the rectangle is matched with a point on the other half. If you can fold a figure exactly in half, it is said to have **line symmetry.**

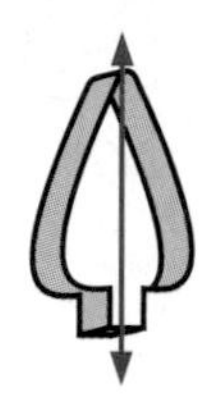

The cookie cutter at the right has line symmetry. The right half is a **reflection** of the left half and the center line is the *line of reflection.* The line of reflection is also called the *line of symmetry.* Some figures have more than one line of symmetry.

DID YOU KNOW

Origami is divided into two categories, figures used in ceremonial etiquette and objects such as birds, animals, insects, and flowers. Some figures have movable parts.

Mini-Lab

Work with a partner.
Materials: MIRA®, colored pencil

- Draw a large square on a piece of paper.
- Place the MIRA® on the square. Look at one side of the MIRA®. Notice the position of the reflection.
- Move the MIRA® until the reflection exactly matches the other half of the square you see through the pane.
- Hold the MIRA® steady. Use a colored pencil to draw along the edge of the MIRA®.

Talk About It

a. Look at the line you drew. How do the parts on each side of the line compare?
b. There are four lines of symmetry in a square. Use the MIRA® to find all of them.

Examples

Trace each figure. Draw all lines of symmetry.

1 

2 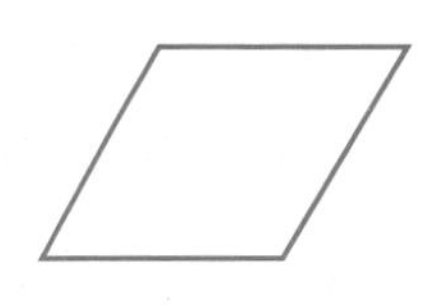

3 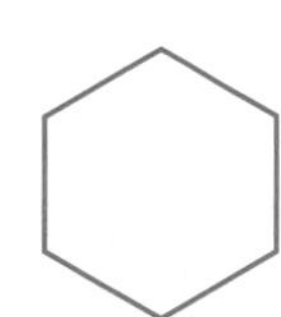

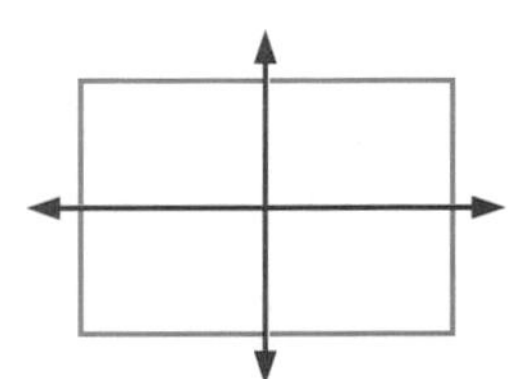

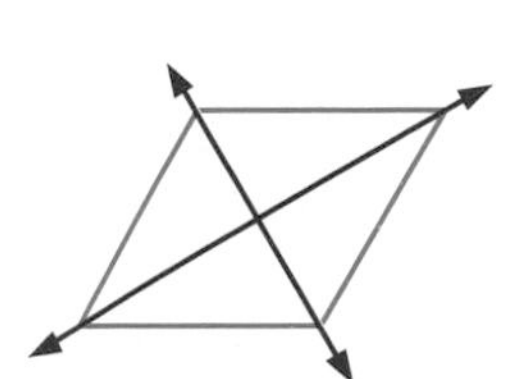

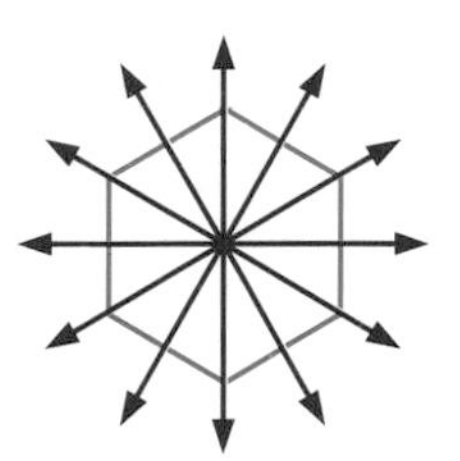

You can also use a MIRA® or paper folding to find the reflection.

Example 4 *Problem Solving*

Design The art club is making reflection neckties to sell to raise money for art supplies. They make a pattern for each design. Create a reflection design for the figure at the right.

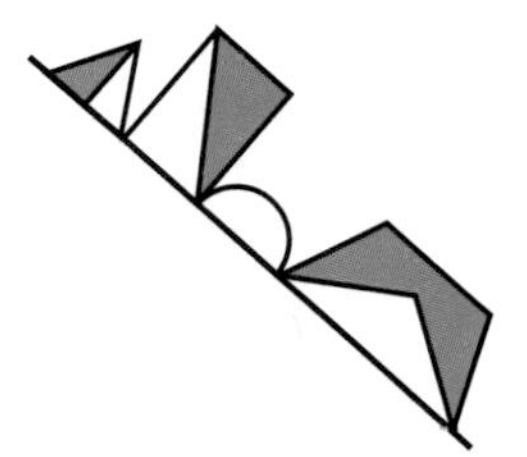

- Copy the design on tracing paper. Fold the paper along the vertical line so the figure is on the outside.
- Use another color to trace all parts you see through the paper.
- Open the paper. The colored parts are the reflection of the original design.

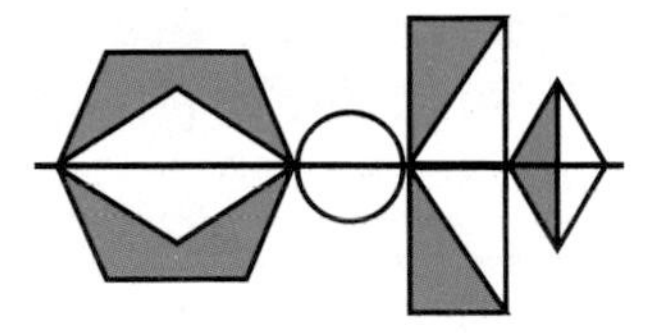

When you turn your compass completely around, it forms a circle. A circle contains 360°. If a figure can be turned less than 360° about its center and it looks like the original, then the figure has **rotational symmetry.**

original

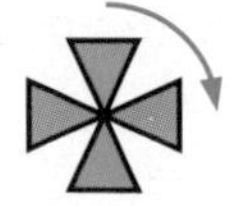

90° turn

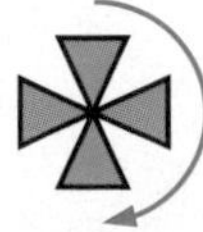

180° turn

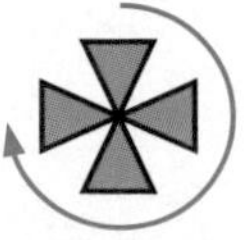

270° turn

360° turn

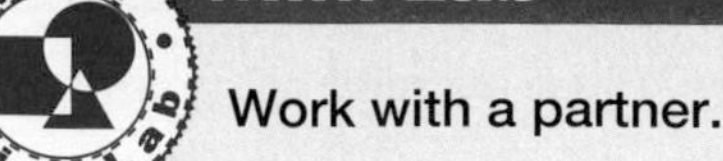

Mini-Lab

Work with a partner.

Materials: straightedge, pin, tracing paper

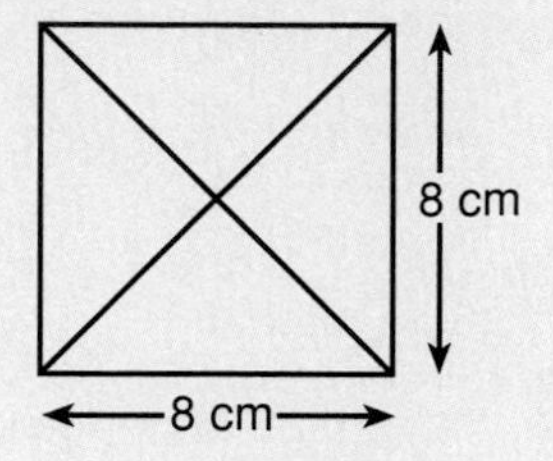

- Draw a square that is 8 centimeters on each side. Then draw its diagonals.
- Place a piece of tracing paper over the square and trace it.
- The point where the diagonals meet is the center point. Place a pin through the center points to hold the two figures together.
- Turn the top figure. Make a note of how many times the top matches the bottom.

Talk About It

a. How many times did the top figure match the bottom in a full turn? *Caution: Don't count the original position twice.*

b. Divide 360° by the number of times it matched. This is the first angle of rotation. What are the other angles of rotation?

Checking for Understanding

Communicating Mathematics

Read and study the lesson to answer each question.

1. **Write** a sentence to tell how a reflection is related to line symmetry.
2. **Draw** a shape that has both line symmetry and rotational symmetry.
3. **Tell** how many degrees are in a full turn.

Guided Practice

Trace each figure. Determine if the figure has line symmetry. If so, draw the lines of reflection.

4. 

5. 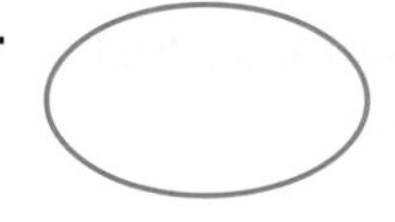

6. 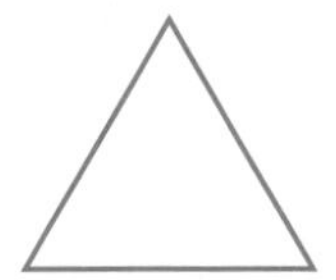

7. Which of the figures in Exercises 4-6 have rotational symmetry?

Exercises

Independent Practice

Trace each figure. Determine if the figure has line symmetry. If so, draw the lines of reflection.

8. 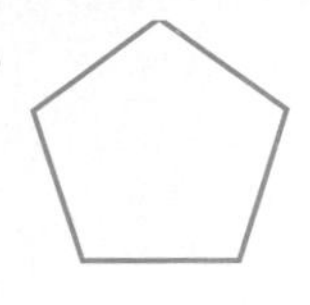

9. 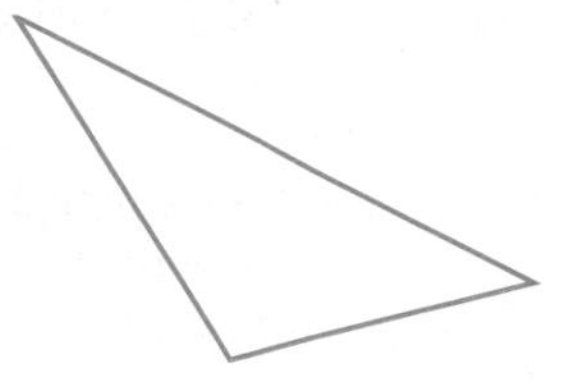

10. 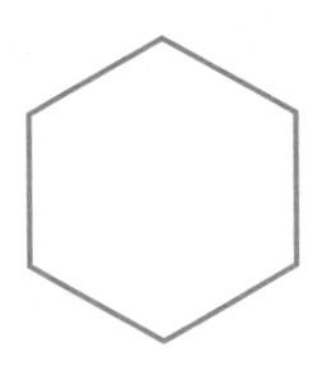

Determine if each figure has rotational symmetry.

11. 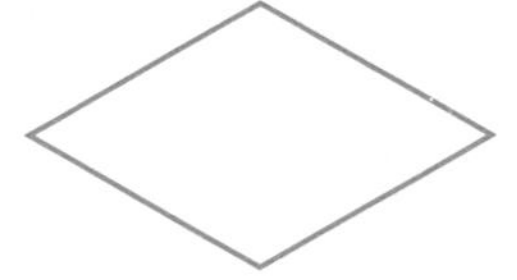

12. 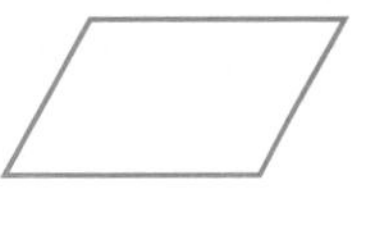

13. 

14. Which of the quadrilaterals you have studied have line symmetry? Draw them.
15. Which quadrilaterals have rotational symmetry? Draw them.
16. Are there any quadrilaterals that have both rotational and line symmetry?

Mixed Review

17. Solve $-281 - (-52) = j$. *(Lesson 3-5)*
18. **Statistics** Determine whether a scatter plot that related the amount of candy eaten with the number of cavities would show a positive, negative, or no relationship. *(Lesson 4-8)*
19. **Algebra** In trapezoid $LMNO$, $m\angle L = 120°$, $m\angle M = 120°$, $m\angle N = 2x°$, and $m\angle O = 60°$. Find the value of x. *(Lesson 5-4)*

Problem Solving and Applications

20. **Science** List five objects in nature that have symmetry. Describe the type of symmetry they have.
21. **Data Search** Refer to page 667.
A font is the type style used in printing. Which letters of the alphabet in the Helvetica font have line symmetry?

22. **Critical Thinking** Copy each pattern on graph paper. Shade in additional squares so that the patterns have the given lines of symmetry.
    a. vertical line of symmetry
    b. two diagonal lines of symmetry
    c. vertical and horizontal lines of symmetry

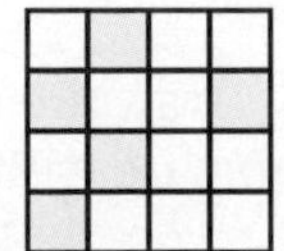

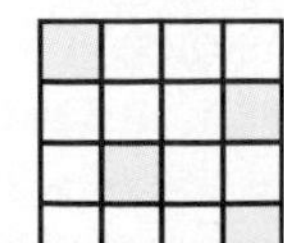

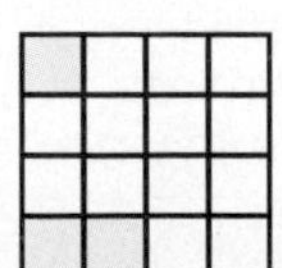

Cooperative Learning

5-6A Dilations

A Preview of Lesson 5-6

Objective

Draw an enlargement of a figure.

Materials

graph paper
colored pencils

When you get a picture developed, you have to specify the size you want. All the pictures come from a small piece of film, so for any size picture the image on the film must be enlarged. In mathematics, an enlargement or reduction of an image is called a **dilation.**

Try this!

Work with a partner.

- Place a piece of graph paper over a cartoon or picture you want to enlarge. Trace the picture. *It may help to put the paper against a window pane to see the picture more clearly.*
- On another piece of graph paper, use a colored pencil to draw horizontal lines every 3 squares. Then draw vertical lines every 3 squares.
- Now sketch the parts of the figure contained in each small square of your original picture onto each large square of the grid you created.

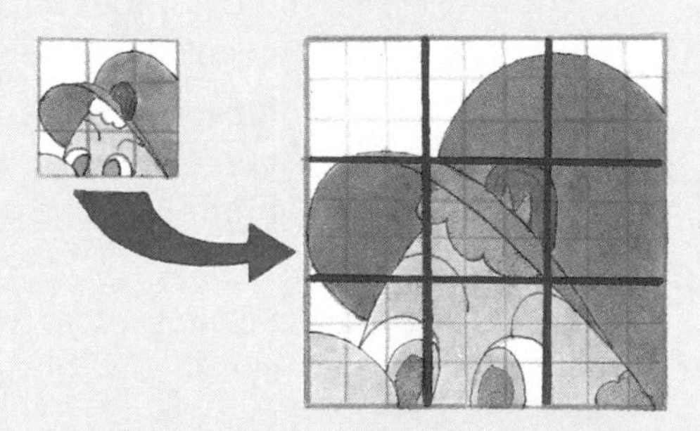

What do you think?

1. Measure the heights of your original figure and your enlargement. How do they compare?
2. What type of grid do you think you would use to reduce a picture?
3. When you enlarge or reduce a figure, does the shape of the figure change? Explain.
4. On graph paper, draw a rectangle that is 12 units by 6 units with these dimensions.
   a. Divide each measurement by 3. Draw the new rectangle.
   b. How does the new rectangle compare with the original in size and shape?

A photojournalist's ability to perceive the significant in a fraction of a second and to use the camera with such speed and precision is a great creative gift. To be a good photographer, you also need to have good mathematical skills to estimate distance, calculate F-stop settings, and crop photos.

For more information, contact:
Associated Press
50 Rockefeller Plaza
New York, NY 10020

5-6 Congruence and Similarity

Objective

Identify congruent figures and similar figures.

Words to Learn

congruent
similar

Have you ever looked into the back of a clock or watch? All of the gears fit finely together to keep the clock running smoothly. How can hundreds of these be made every day so that each one will work?

Each part in the clock is die cast by a computerized machine so that thousands of that part come out exactly the same size and shape. When all parts are made this way, they fit together creating a clock guaranteed to work.

In geometry, when two figures are exactly the same size and shape, we say they are **congruent.** That is, when you cut one figure out, it can fit exactly on top of the other. Congruent figures have corresponding sides and angles that are congruent.

Example 1

Determine if $\triangle ABC$ is congruent to $\triangle DEF$.

LOOKBACK

You can review congruent angles and congruent segments on pages 174-175.

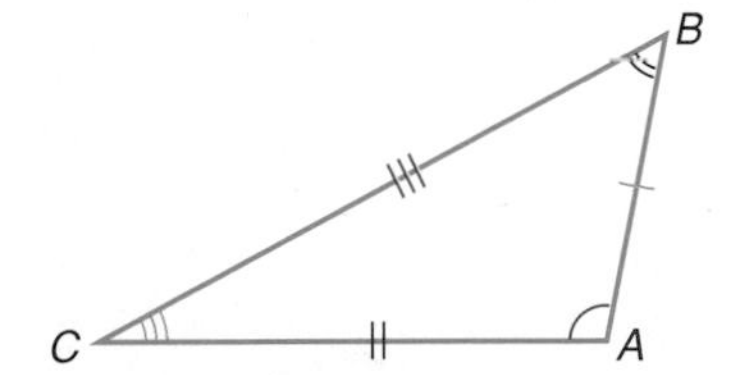

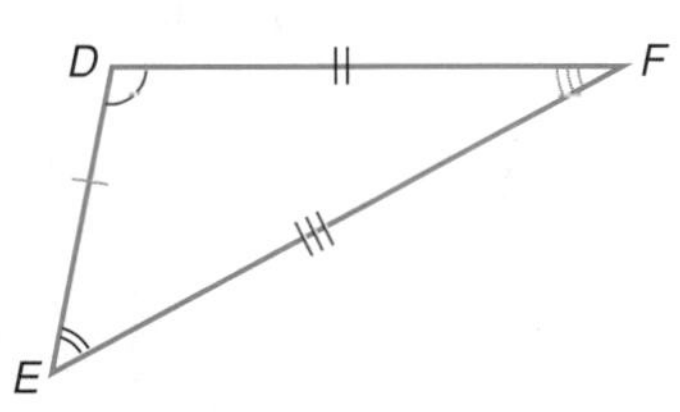

Compare the measures of the sides and angles.

The slashes tell us which sides are congruent.	*The arcs tell us which angles are congruent.*
$\overline{AB} \cong \overline{DE}$	$\angle ABC \cong \angle DEF$
$\overline{BC} \cong \overline{EF}$	$\angle BCA \cong \angle EFD$
$\overline{AC} \cong \overline{DF}$	$\angle CAB \cong \angle FDE$

All corresponding parts are congruent. So, $\triangle ABC \cong \triangle DEF$.

Check: Trace $\triangle DEF$ and cut it out. Does it fit exactly on top of $\triangle ABC$? *You may have to turn or flip it to get it to fit.*

You may wonder if you always have to look at both the sides and the angles to see if two figures are congruent. Wouldn't looking at one set of corresponding parts be enough?

Mental Math Hint

Which sides are corresponding? Find two pairs of congruent angles and compare the sides they include. If $\angle X \cong \angle P$ and $\angle Y \cong \angle Q$, then compare $\overline{XY}$ and $\overline{PQ}$.

Congruent figures are also similar figures.

Examples

Determine if each pair of figures is congruent.

2

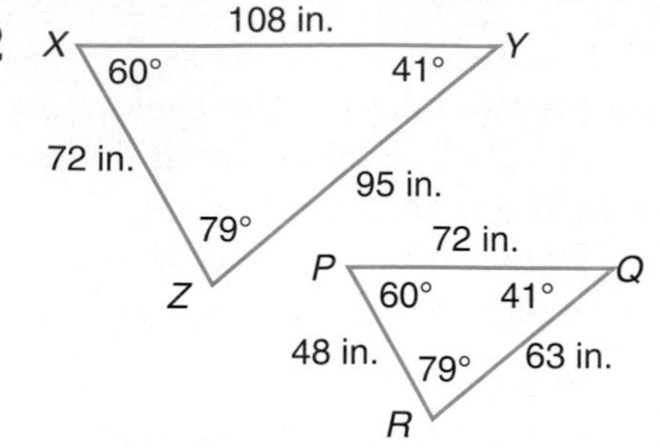

The angles are congruent.
The sides are *not* congruent.
The triangles are *not* congruent.

3

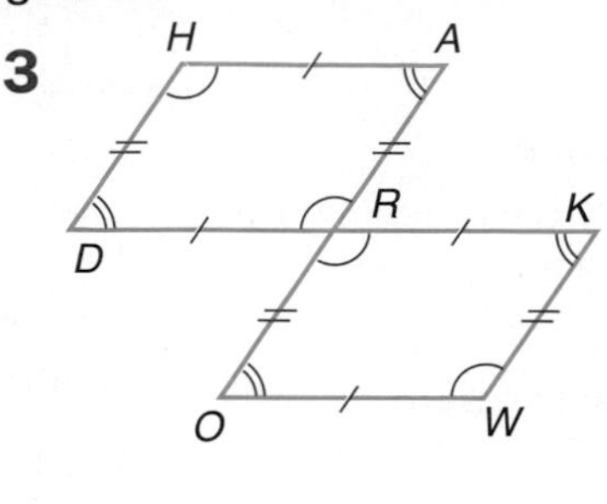

The sides are congruent.
The angles are congruent.
$\square HARD \cong \square WORK$
$\square$ is the symbol for parallelogram.

In Example 2, the triangles have the same shape, but are not congruent. Figures that have the same shape, but may differ in size are **similar** figures. To say ΔXYZ *is similar to* ΔPQR, we write $\Delta XYZ \sim \Delta PQR$.

Congruent figures can be cut out and matched exactly. When you cut out similar triangles, other relationships can be observed.

Mini-Lab

Work with a partner.
Materials: scissors, tracing paper

- Trace each pair of figures below and cut them out.

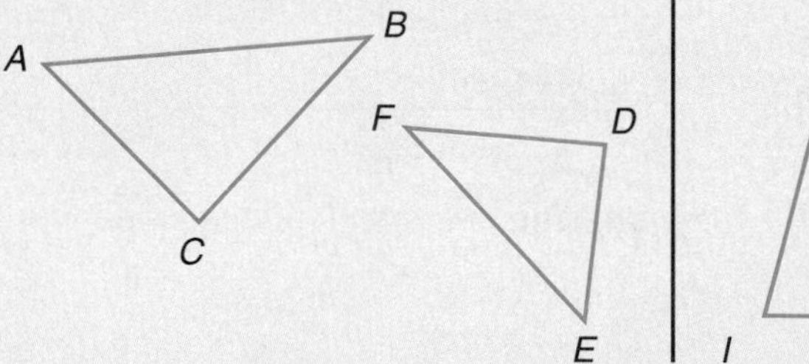

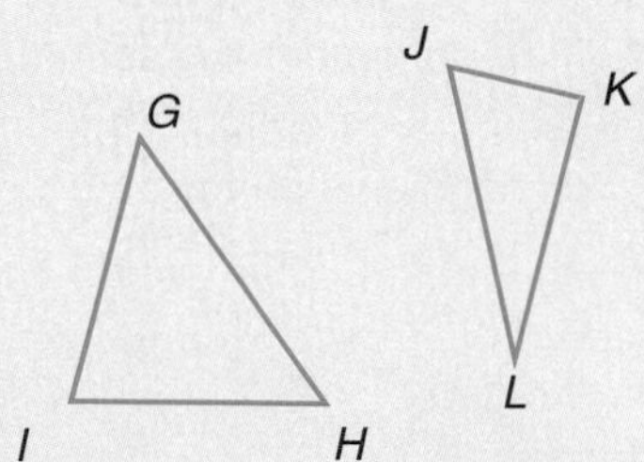

- Compare the angles in each pair of figures.
- For each pair of triangles, see if you can place one figure inside the other with an equal amount of space separating the corresponding sides in every case.

Talk About It

a. What did you find when you compared the angles?
b. If you can fit one triangle inside another with equal space separating all corresponding sides, the triangles are similar. Name the similar figures. Is this true for all figures?

You can use the relationships between the angles and sides of congruent and similar figures to find missing measures.

Example 4 ***Connection***

Algebra Find the value of x in each pair of figures.

a. $\Delta ROB \cong \Delta STL$

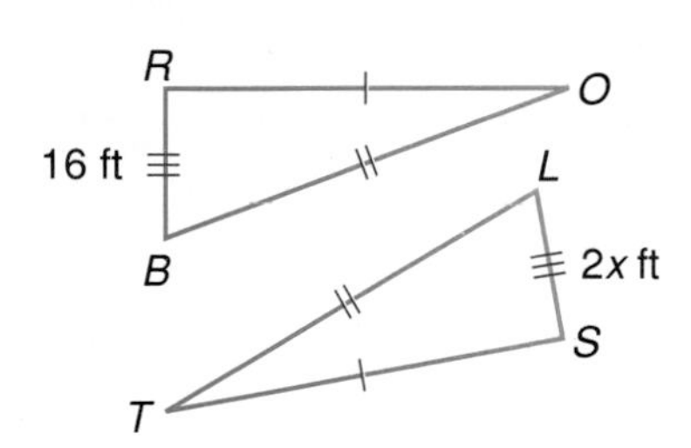

The corresponding sides are congruent.

$RB = 2x$

$16 = 2x$ *$\overline{RB}$ is 16 feet long.*

$8 = x$ *Divide each side by 2.*

b. $\square ABCD \sim \square EFGH$

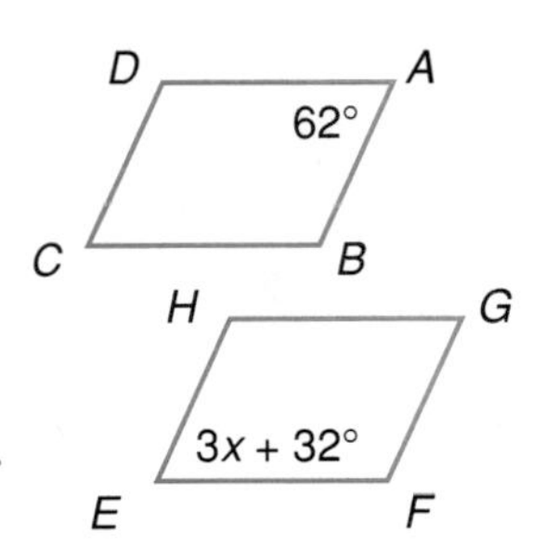

The corresponding angles are congruent, so $m\angle A = m\angle E$.

$m\angle A = m\angle E$

$62 = 3x + 32$ *Replace $m\angle A$ with 62 and $m\angle E$ with $3x + 32$.*

$30 = 3x$ *Subtract 32 from each side.*

$10 = x$ *Divide each side by 3.*

Checking for Understanding

Communicating Mathematics

Read and study the lesson to answer each question.

1. **Tell** what must be true for two figures to be congruent.
2. **Show** one way that you can tell if two figures are similar.
3. **Draw** two congruent parallelograms.
4. **Draw** two similar trapezoids.
5. **Write** why two congruent figures are also similar.

Guided Practice

Tell if each pair of figures is congruent, similar, or neither. Justify your answer.

6.

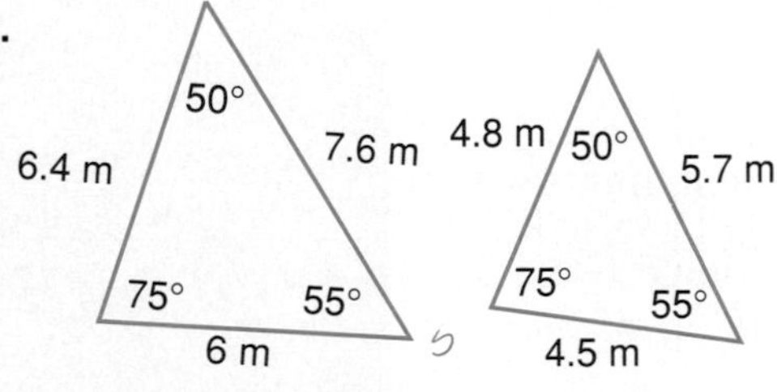

7.

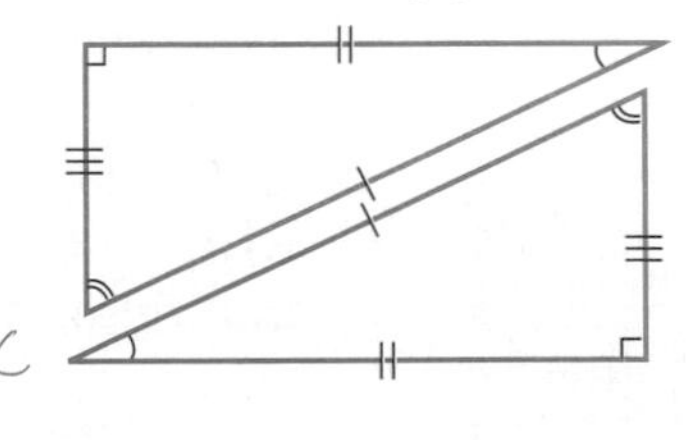

8.

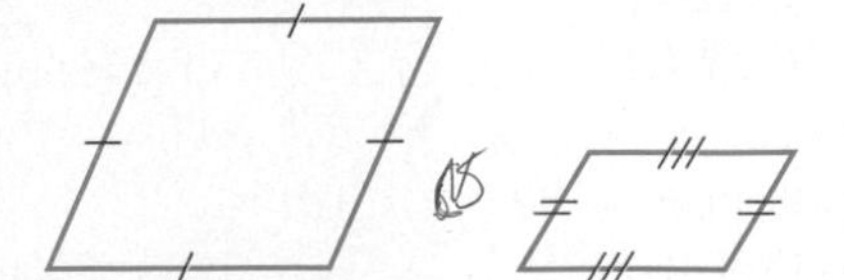

9.

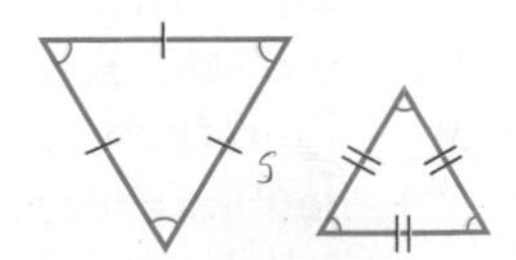

Exercises

Independent Practice

Tell if each pair of figures is congruent, similar, or neither. Justify your answer.

10. 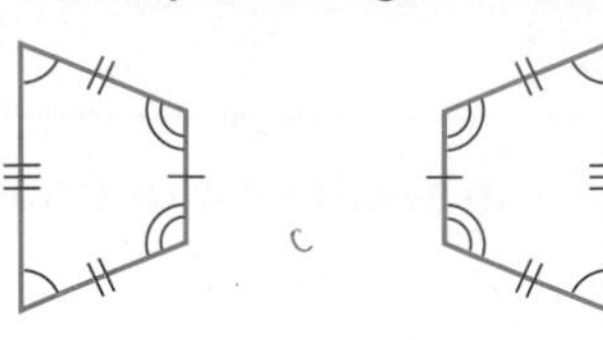

11. 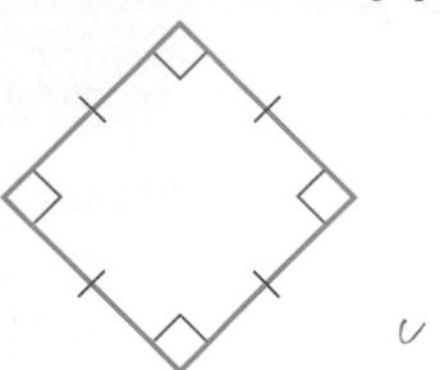 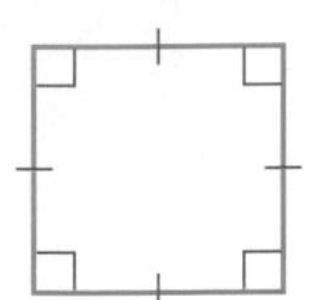

12. 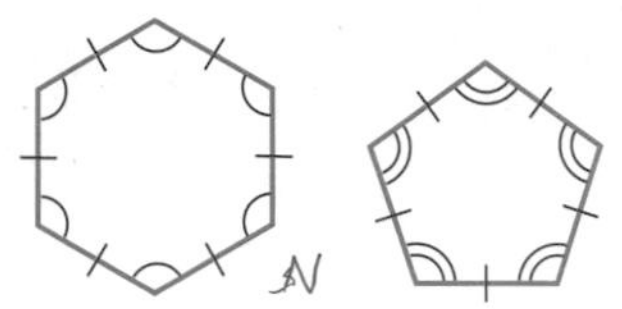

13. 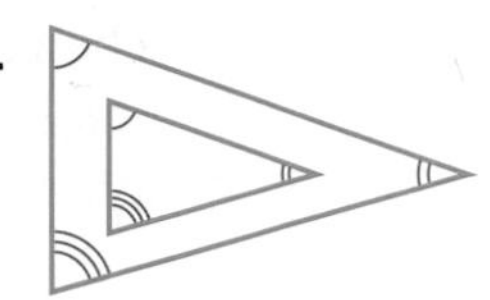

Find the value of x in each pair of figures.

14. $\Delta PQR \cong \Delta MNO$

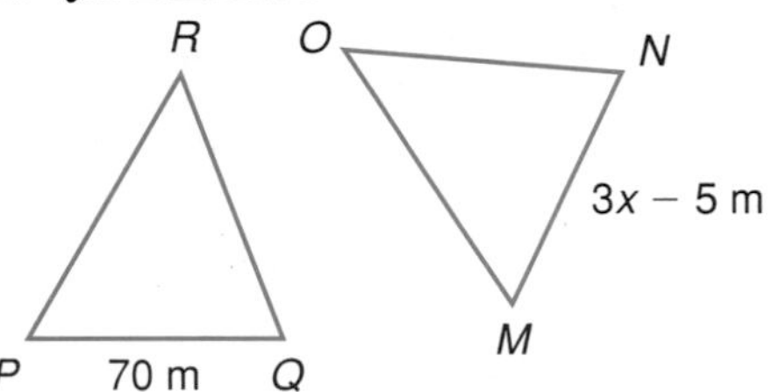

15. $\Delta VFW \sim \Delta PDR$

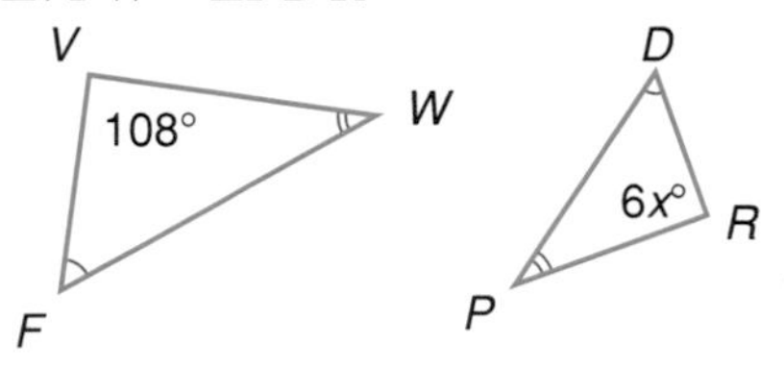

Mixed Review

16. Estimate 5,381 + 3,416 first by rounding and then by front-end estimation. *(Lesson 1-3)*
17. Evaluate $5a^2 - (b + c)$ if $a = 3$, $b = 8$, and $c = 10$. *(Lesson 2-1)*
18. Determine if the figure at the right has rotational symmetry. *(Lesson 5-5)*

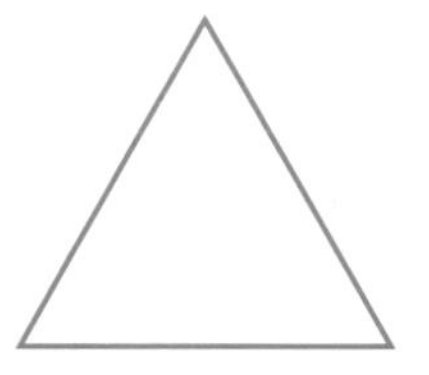

Problem Solving and Applications

19. **Entertainment** The graph shows an estimated cost of a night out for a family of four.
    a. If the charge for adults and children is the same, what is the average cost for a rock concert ticket?
    b. If a family of four spent about $100 on entertainment, how do you think they spent it?

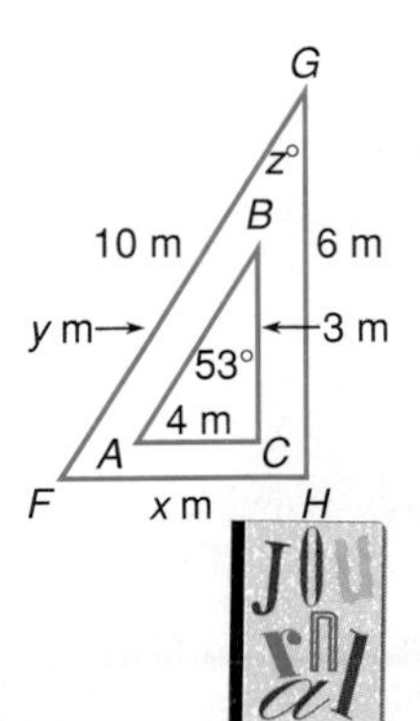

20. **Critical Thinking** In the figure at the left, ΔABC is similar to ΔFGH. Find the values of x, y, and z.
21. **Journal Entry** Write a sentence telling some occupations that would use congruent or similar figures.

Cooperative Learning

5-6B Constructing Congruent Triangles

A Follow-Up of Lesson 5-6

Objective

Construct a triangle congruent to a given triangle.

Materials

compass
straightedge

Mapmakers, or cartographers, often need to copy figures from one place to another. They often use a compass and straightedge to do this.

Try this!

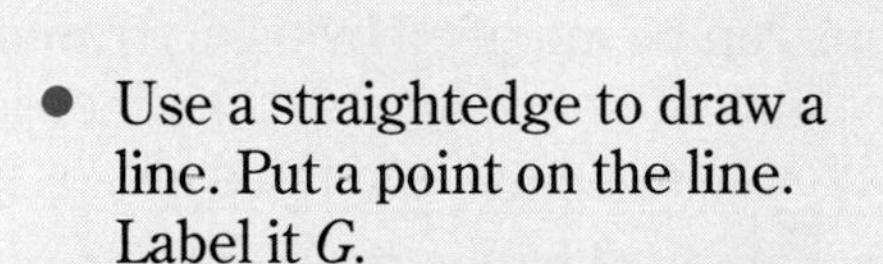

Work with a partner. Construct a triangle congruent to $\triangle ABC$.

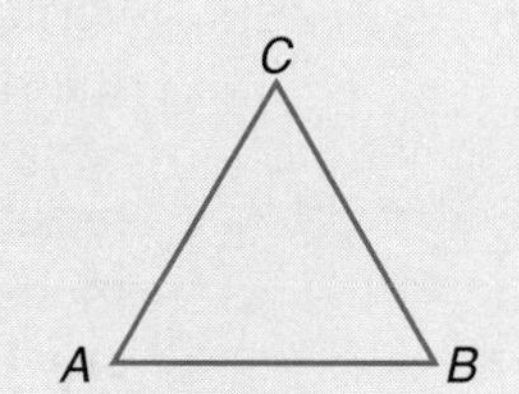

- Use a straightedge to draw a line. Put a point on the line. Label it *G*.
- Set your compass to the same width as the length of $\overline{AB}$. Put the compass point at *G*. Draw an arc to intersect the line. Call this intersection *H*.

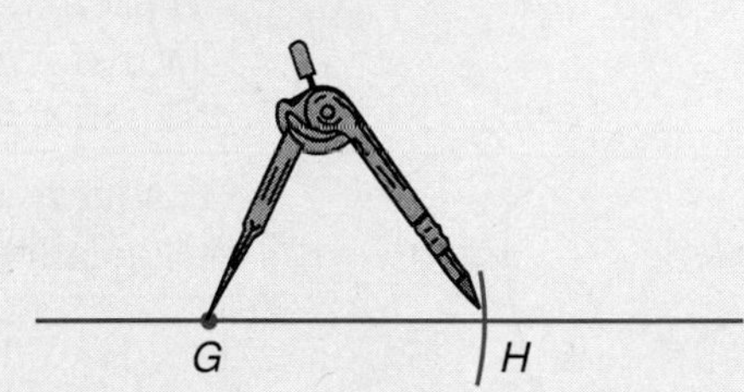

- Set your compass to the same width as the length of $\overline{AC}$. Put the compass at *G* and draw an arc above the line.

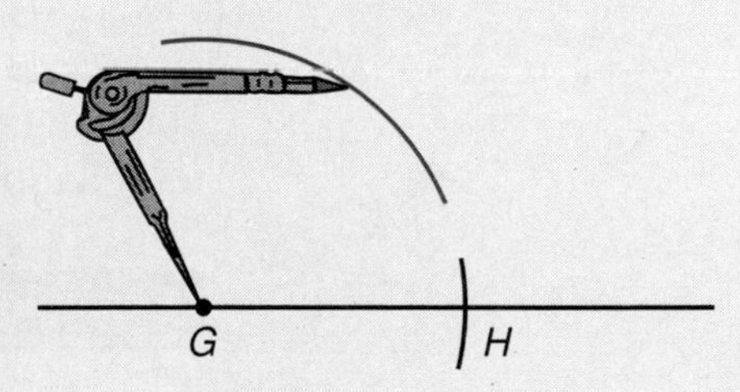

- Set your compass to the same width as the length of $\overline{BC}$. Put the compass at *H* and draw an arc above the line to intersect the arc you just drew. Label this point *I*.
- Draw $\overline{HI}$ and $\overline{IG}$.

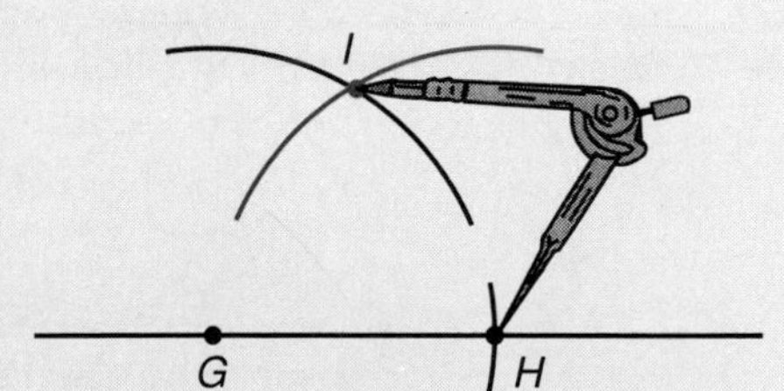

What do you think?

1. How do the lengths of $\overline{AB}$, $\overline{BC}$, and $\overline{CA}$ compare to the lengths of $\overline{GH}$, $\overline{HI}$, and $\overline{IG}$?
2. How do the corresponding angles of the two triangles compa
3. What is the relationship of $\triangle ABC$ and $\triangle GHI$?

Art Connection

5-7 Transformations and M. C. Escher

Objective

Create Escher-like drawings by using translations and rotations.

Words to Learn

tessellation
transformation
translation
rotation

Maurits Cornelis Escher (1898-1972) was a Dutch artist famous for his repetitive, interlocking patterns. His works look like paintings but were done by woodcarving and lithographs.

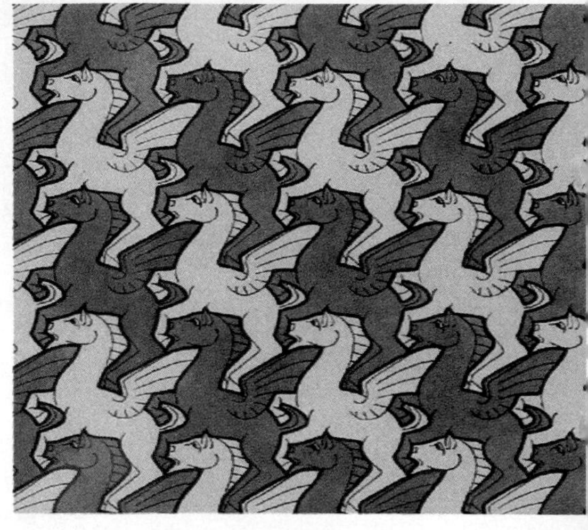

Escher's designs are made from variations on tiling patterns called **tessellations.** A floor covered by square tiles is an example of a tessellation of squares. By modifying these squares, you can create an Escher-like drawing.

Tessellations can be modified by using **transformations.** Transformations are movements of geometric figures. One transformation commonly used is a slide, or **translation,** of a figure.

A square can be modified by making a change on one side and then translating that change to the opposite side.

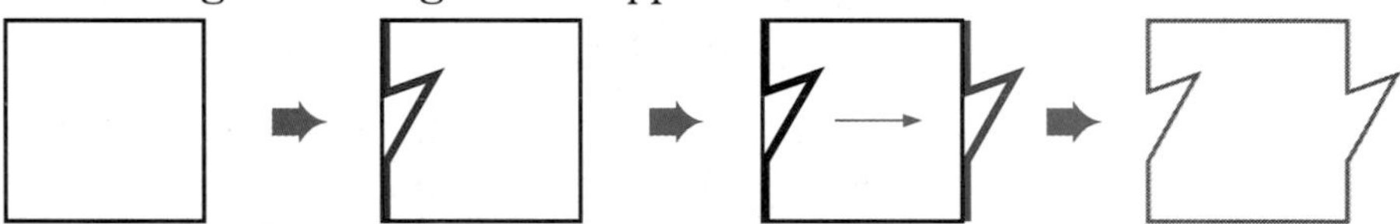

You can create more complex designs starting with square tessellations and making changes on both pairs of sides.

Example 1

Make an Escher-like drawing using the modification shown at the right.

Draw a square. Copy the pattern shown. Translate the change shown on the top side to the bottom side.

Now translate the change shown on the left side to the right side to complete the pattern unit.

Repeat this pattern on a tessellation of squares. *It is sometimes helpful to complete one pattern unit, cut it out, and trace it for the other units.*

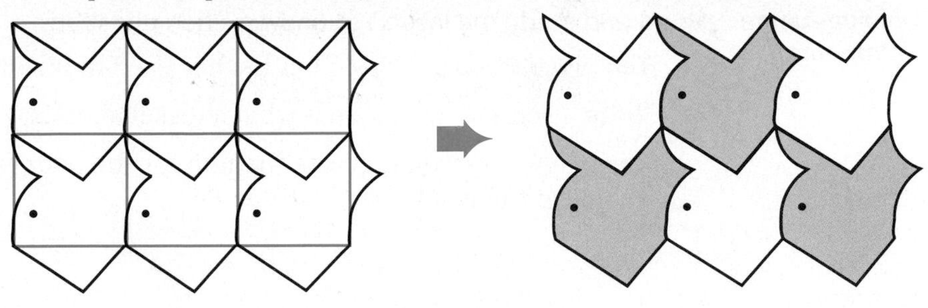

Other tessellations can be modified by using changes with a **rotation.** Rotations are transformations that involve a turn about the vertices of the base figure.

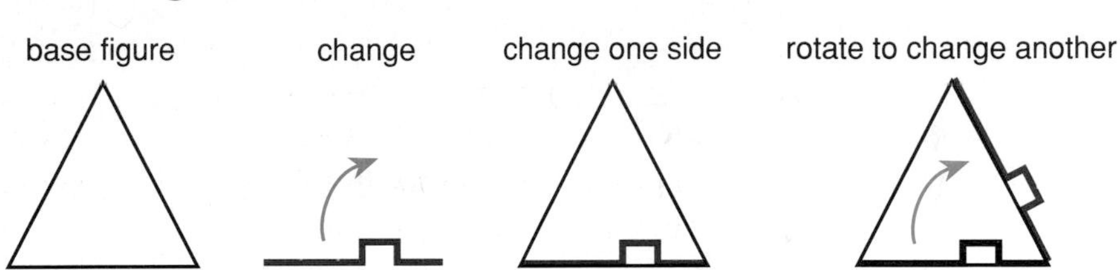

Example 2

Make an Escher-like drawing using a tessellation of equilateral triangles and the rotation shown at the right.

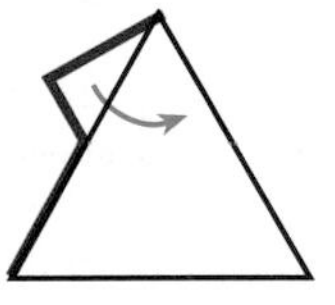

This involves one rotation.
Draw an equilateral triangle.
Copy the change shown.

Rotate the triangle so you can copy the change on the side indicated.

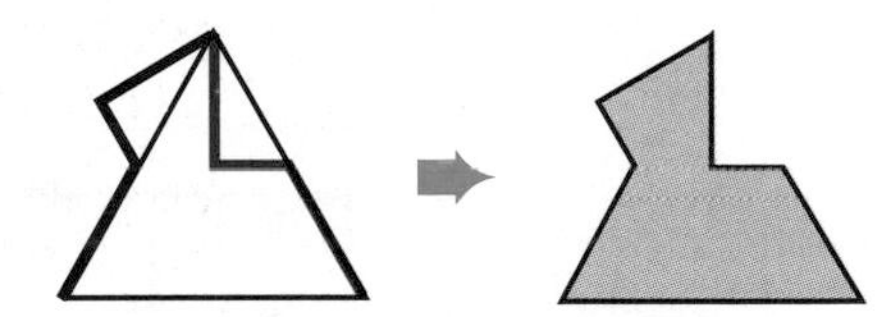

Now repeat this pattern unit on a tessellation of equilateral triangles.

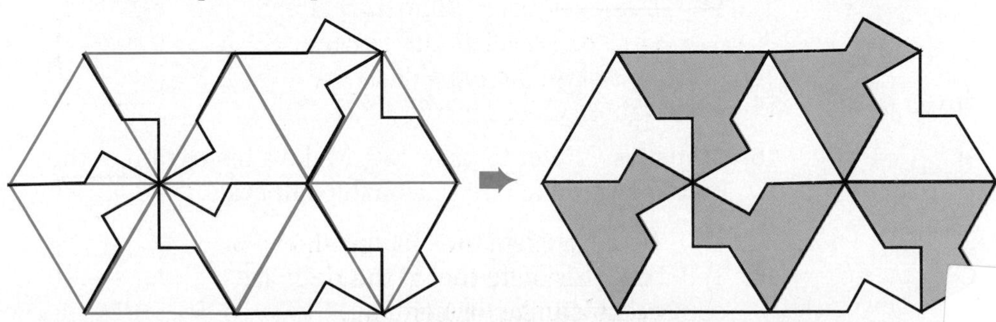

Checking for Understanding

Communicating Mathematics

Read and study the lesson to answer each question.

1. **Tell** two transformations that Escher used in his drawings.
2. **Write** a sentence to explain what a tessellation is.
3. **Show** how you would create the unit for the pattern shown at the right.

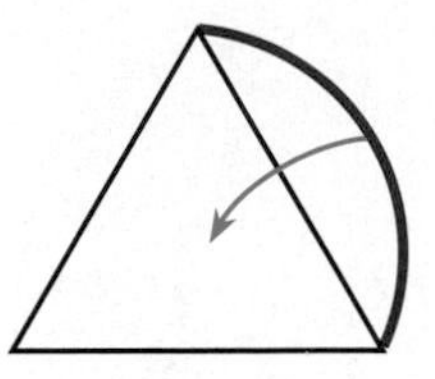

Guided Practice

Make an Escher-like drawing for each pattern described. Use a tessellation of two rows of three squares as your base.

4. 

5. 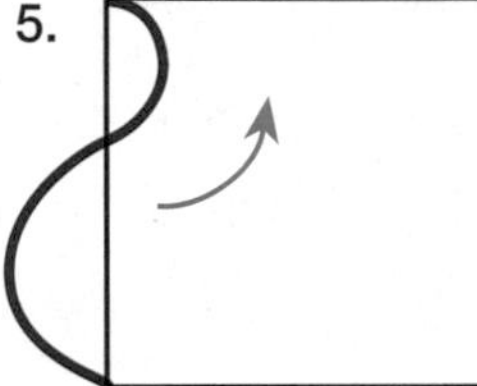

6. 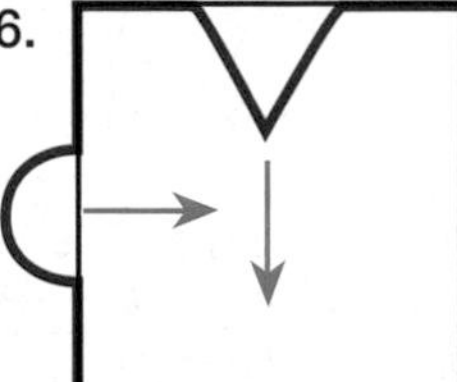

7. Tell whether each pattern in Exercises 4-6 involves translations or rotations.

Exercises

Independent Practice

Make an Escher-like drawing for each pattern described. For squares, use a tessellation of two rows of three squares as your base. For triangles, use a tessellation of two rows of five equilateral triangles as your base.

8. 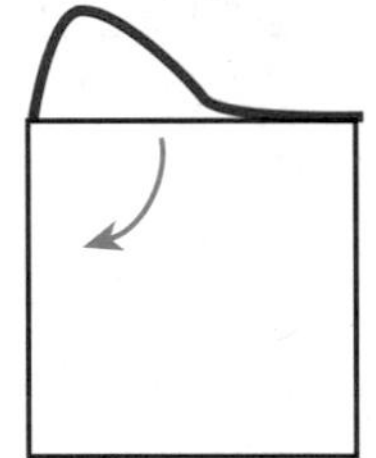

9. 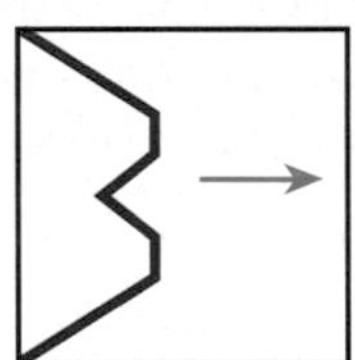

10. 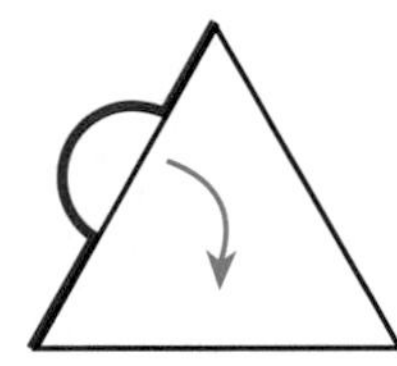

11. 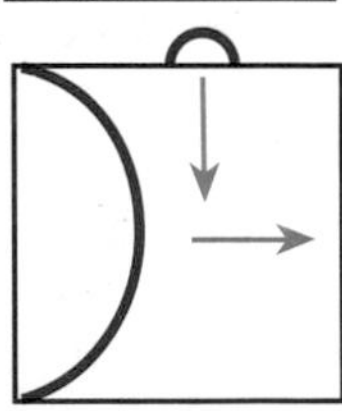

12. 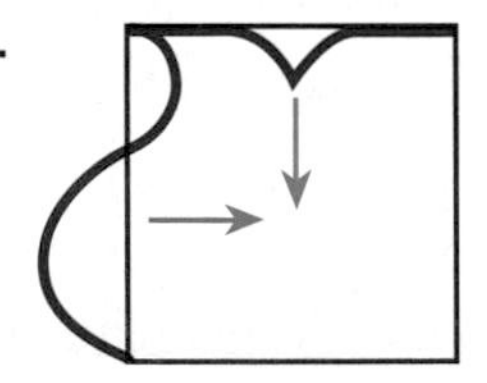

13. 

Mixed Review

14. Solve $\frac{a}{15} - 5 = 1$. *(Lesson 2-7)*

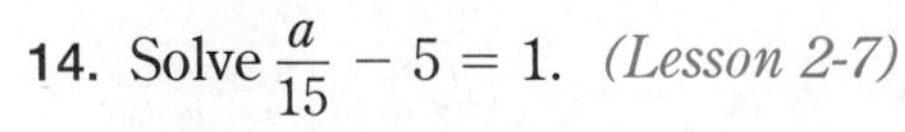

15. **Statistics** Refer to page 136. Make a histogram of the potato chip data. Use reasonable intervals. *(Lesson 4-2)*
16. Tell if the pair of figures at the right are congruent, similar, or neither. *(Lesson 5-6)*

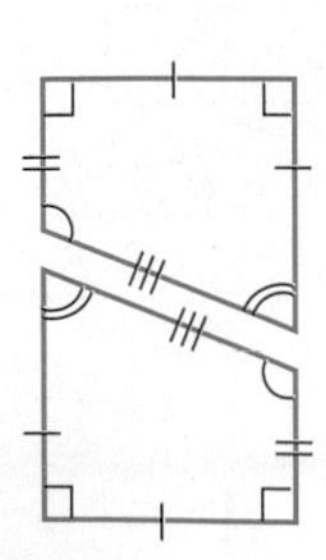

Problem Solving and Applications

17. **Critical Thinking** Look at the tessellation at the right.

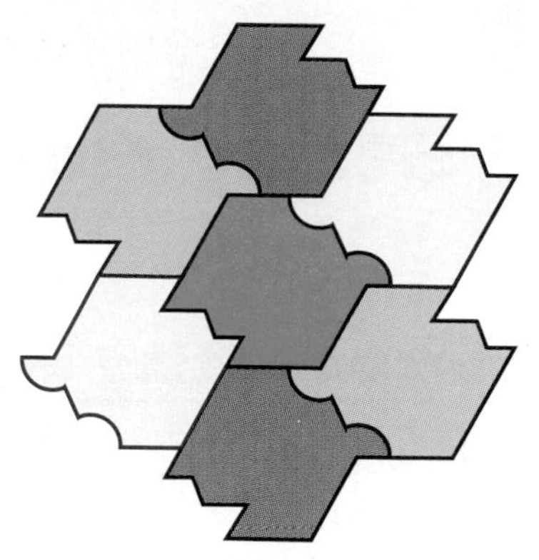

   a. What type of base figure was modified to produce the pattern unit?
   b. What transformation was used to create the pattern unit?
   c. Describe each change used to create the pattern unit.

18. **Art** The Escher lithograph below, entitled *Study of Regular Division of the Plane with Reptiles,* was created in 1939. Study the figures carefully.

   a. Trace the outside edge of one pattern unit.
   b. What geometric shape do you think was used for this tessellation?
   c. What transformation did this use?

Portfolio

19. **Portfolio Suggestion** Select your favorite word problem from this chapter and place your solution to it in your portfolio. Attach a note explaining why it is your favorite.

20. **Computer Connection** The LOGO program below will tessellate a square design to fill the computer screen. The commands that create the design unit are in blue.

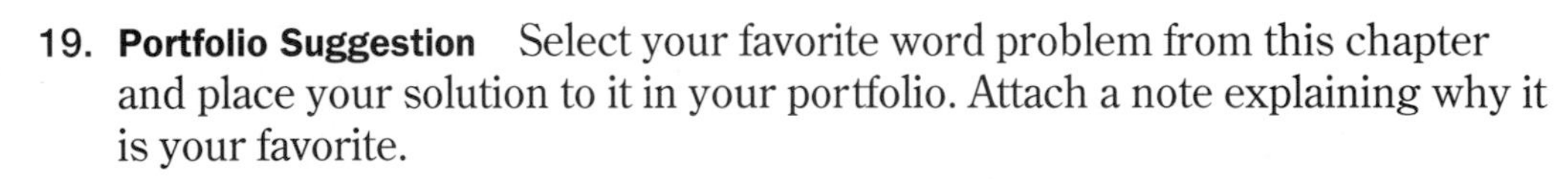

```
PU SETXY -100 100 PD HT
REPEAT 10[REPEAT 10[REPEAT 4[FD 20 RT 90]
FD 10 RT 45 REPEAT 4[FD 14.14 RT 90] LT 45 BK 10 RT 90
FD 20 LT 90] PU LT 90 FD 200 RT 90 BK 20 PD]
```

   a. Sketch the pattern unit.
   b. Sketch a portion of the tessellation.

Chapter 5

Study Guide and Review

Communicating Mathematics

Choose a word or symbol from the list at the right to correctly complete each statement.

rhombus
trapezoid
rotational symmetry
line symmetry
transversal
acute
obtuse
rotations
⊥
~
∥

1. A(n) ___?___ is a line that intersects two other lines.
2. ___?___ are transformations that involve a turn about a given point.
3. A parallelogram that has four congruent sides is a(n) ___?___.
4. A figure that can be folded exactly in half is said to have ___?___.
5. ___?___ angles have measures greater than 90° and less than 180°.
6. The symbol ___?___ means *is parallel to.*
7. The symbol ___?___ means *is perpendicular to.*
8. The symbol ___?___ means *is similar to.*
9. Tell how to determine if a figure has rotational symmetry.
10. Draw a scalene obtuse triangle.

Self Assessment

Objectives and Examples

Upon completing this chapter, you should be able to:

Review Exercises

Use these exercises to review and prepare for the chapter test.

- identify lines that are parallel *(Lesson 5-1)*

Name the parallel segments in the figure.

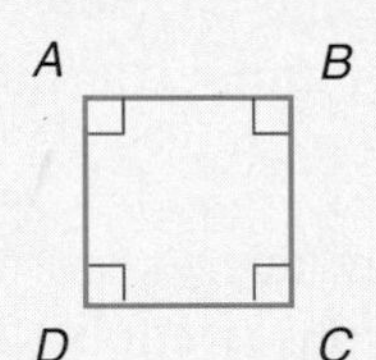

$\overline{AB} \parallel \overline{DC}$, $\overline{AD} \parallel \overline{BC}$

Name the parallel segments in each figure.

11.

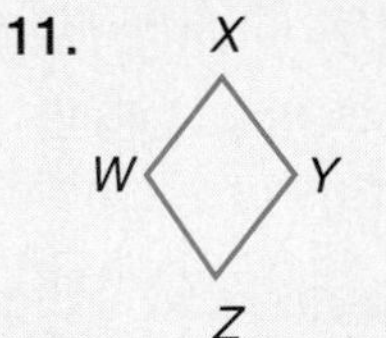

12.

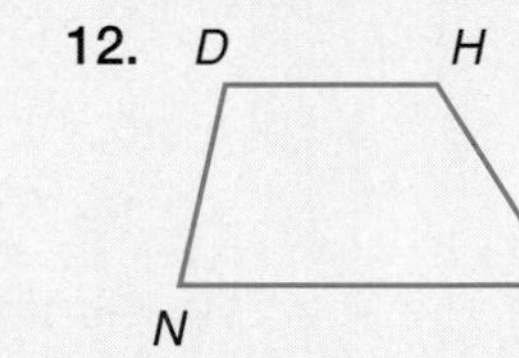

- classify triangles *(Lesson 5-3)*

Classify the triangle by its sides and by its angles.

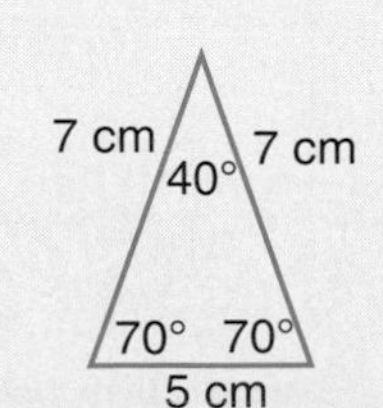

The triangle is isosceles and acute.

Classify each triangle by its sides and by its angles.

13. **14.**

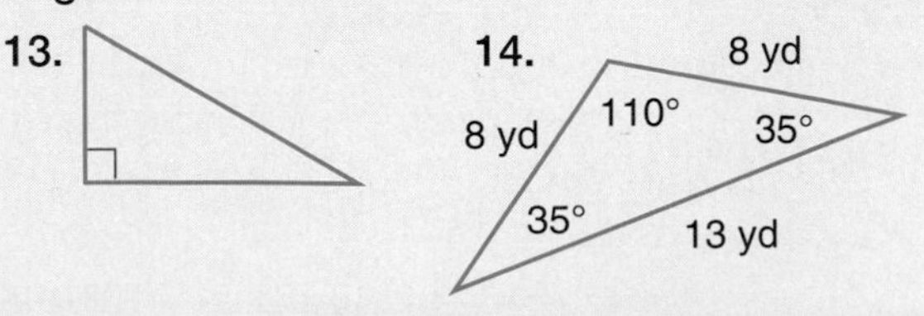

Objectives and Examples	Review Exercises
• classify quadrilaterals *(Lesson 5-4)* Identify all names that describe the quadrilateral. 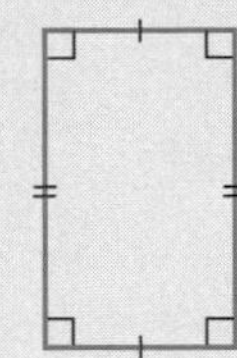 parallelogram, rectangle	Sketch each figure on your paper. Let Q=quadrilateral, P=parallelogram, R=rectangle, S=square, RH=rhombus, and T=trapezoid. Write all letters inside the figure that describe it. 15. 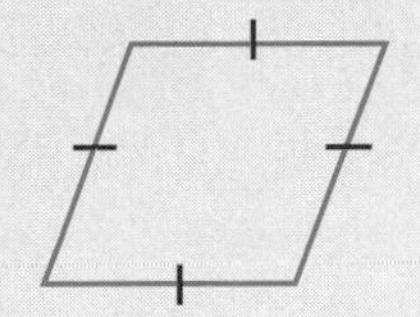 16. 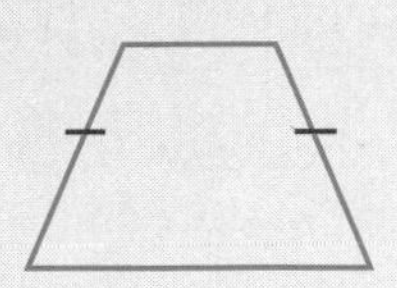
• explore line symmetry *(Lesson 5-5)* Draw all lines of symmetry for a rectangle. 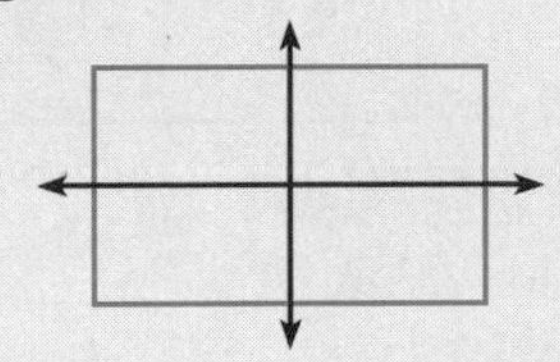	Copy each figure. Determine if the figure has line symmetry. If so, draw the lines of symmetry. 17. 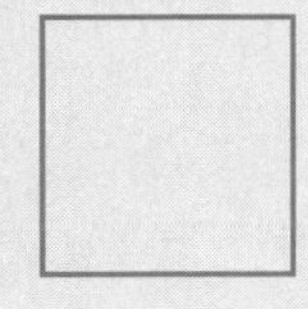 18. 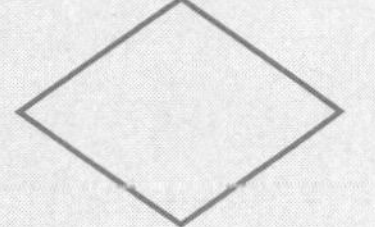
• explore rotational symmetry *(Lesson 5-5)* A figure has rotational symmetry if it can be turned less than 360° about its center and looks like the original.	Determine if each figure has rotational symmetry. 19. 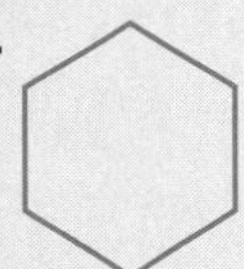 20. 
• identify congruent and similar figures *(Lesson 5-6)* Figures that have the same size and shape are congruent. Figures that have the same shape, but *not* the same size are similar.	Tell if each pair of figures are congruent, similar, or neither. 21. 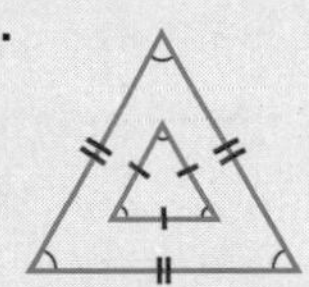 22. 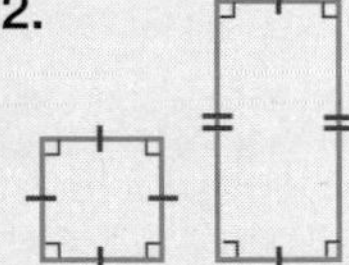
• create Escher-like drawings by using translations and rotations *(Lesson 5-7)* An example of a tessellation of squares is a floor covered by tiles. By modifying these squares, an Escher-like drawing can be made.	Make an Escher-like drawing for each pattern described. Use a tessellation of two rows of three squares as your base. 23. 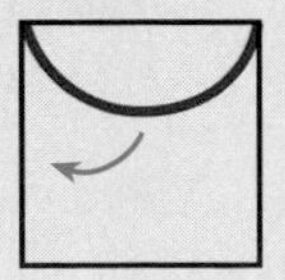 24. 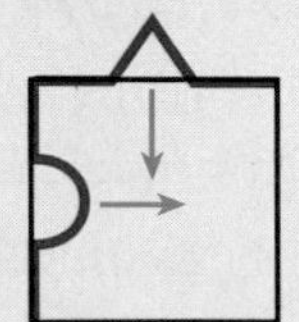

Applications and Problem Solving

25. **Traffic** The shape of a STOP sign is an octagon. How many pairs of edges in a STOP sign are parallel? *(Lesson 5-1)*

26. A 24-foot ladder is leaning against a house. The base of the ladder is 8 feet from the house. *(Lesson 5-3)*
    a. Draw a sketch of this situation.
    b. What type of triangle is formed by the house, ground, and ladder?

27. The 48 members of the senior class voted on their choice of a class trip. They could choose Atlantic City (A), Barnegat Bay (B), or Cape May (C). They could vote for as many as all three choices. The results were displayed as the Venn diagram shown at the right. Use the diagram to find out which location received the most total votes. *(Lesson 5-2)*

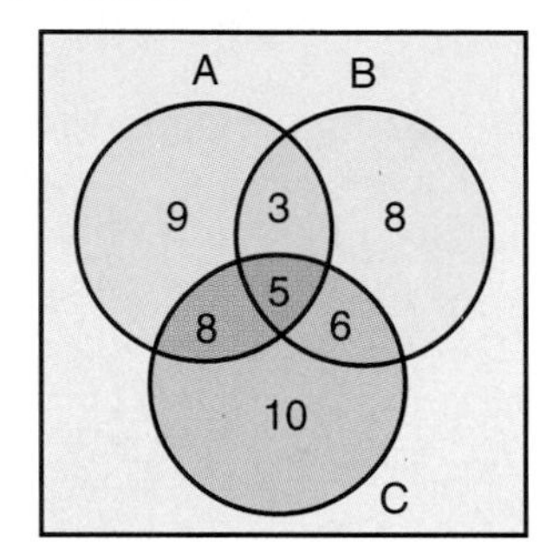

28. **Hobbies** Olivedale Senior Center offers different activities each quarter for their members. Twenty-one members have signed up for aerobics classes. Thirty members have signed up for healthful cooking classes. Of the members that have signed up for classes, six have signed up for both activities. *(Lesson 5-2)*
    a. Make a Venn diagram to show how many have signed up for activities at the Olivedale Senior Center.
    b. What is the total number of members that have signed up for these two classes?

Curriculum Connection Projects

- **Music and Entertainment** Find how many students in your class have a VCR, a CD player, a Walkman-type cassette player, or none of these. Draw a Venn diagram of your results.
- **Language Arts** Survey the adults in your school to find how many regularly read novels, poetry, magazines, newspapers, or none of these. Draw a Venn diagram of your results.

Read More About It

- DeClements, Barthe, and Greimes, Christopher, *Double Trouble*
- White, Laurence B., Jr. and Broehel, Ray, *Optical Illusions*
- Demi, *Demi's Reflective Fables*

5 Test

Find the measure of each angle in the figure at the right if $m \parallel n$ and $m\angle 4 = 40°$.

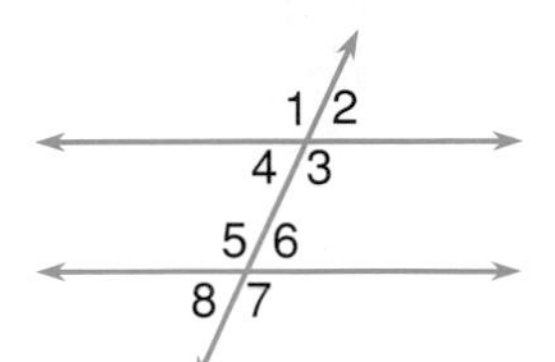

1. $\angle 1$
2. $\angle 2$
3. $\angle 3$
4. $\angle 5$
5. Can a triangle be isosceles and obtuse? Draw a figure to justify your answer.

Classify each triangle by its sides and by its angles.

6. 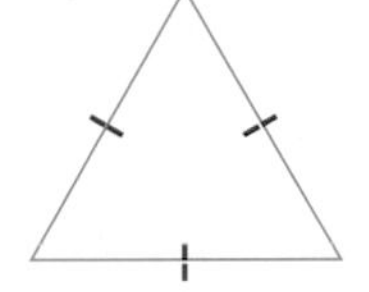

7. 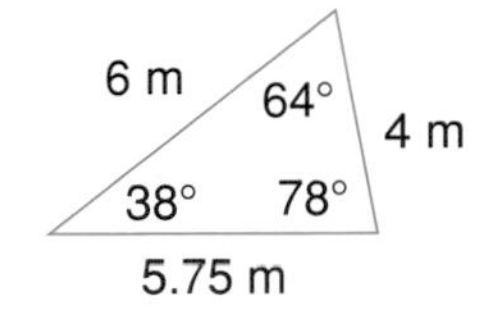

8. 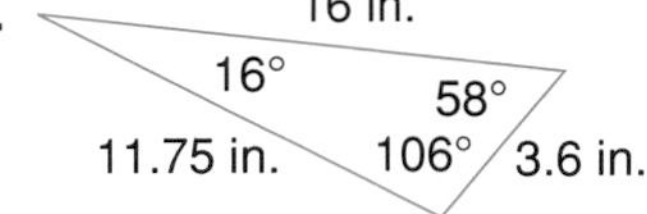

9. Draw an isosceles right triangle.
10. In rhombus $DEFG$, $m\angle D = 66°$, $m\angle E = 114°$, $m\angle F = 66°$, and $m\angle G = 2x°$. Find the value of x.

Let Q = quadrilateral, P = parallelogram, R = rectangle, S = square, RH = rhombus, and T = trapezoid. Write all letters that describe each figure.

11. 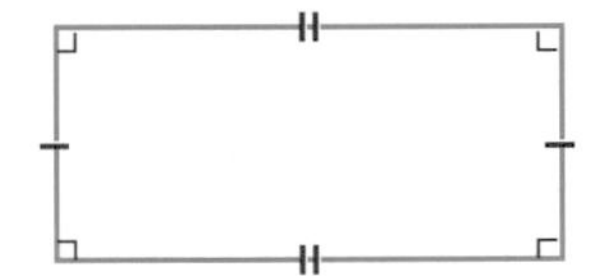

12. 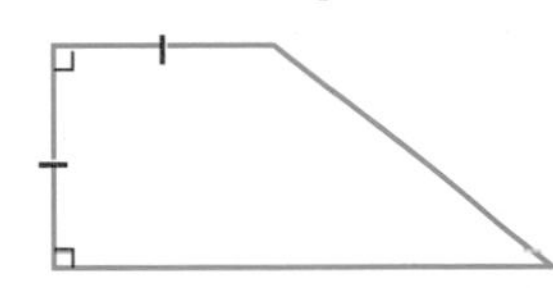

13. 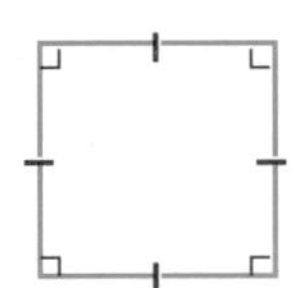

14. Draw a parallelogram that is not a rectangle, a square, or a rhombus.
15. Does an equilateral triangle have line symmetry? If so, draw the lines of symmetry.
16. Does a rectangle have rotational symmetry?

Find the value of x for each pair of figures.

17. $\triangle ABC \cong \triangle DEF$

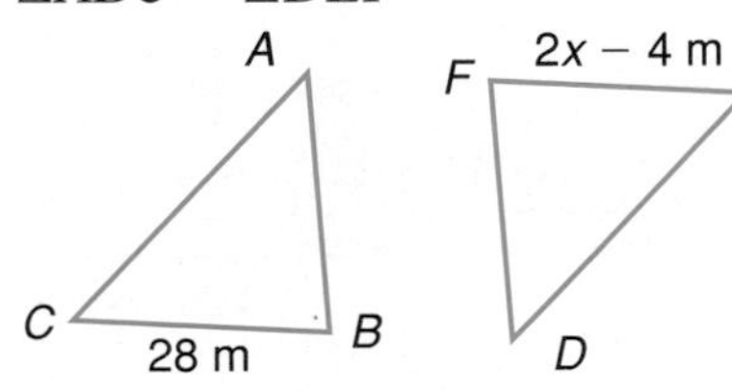

18. $\triangle HMB \sim \triangle TKR$

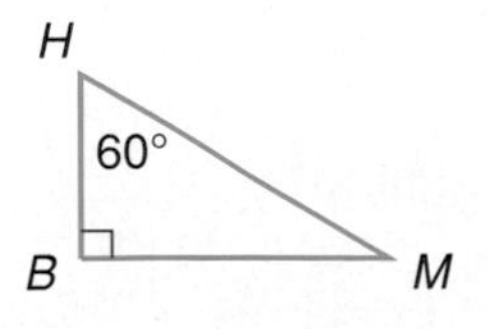

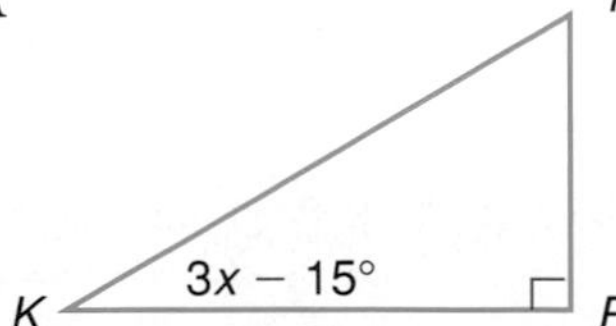

19. Does the pattern shown at the right involve a translation or a rotation?

20. Draw a Venn diagram to show the set of odd numbers from 1 through 15 and the set of counting numbers from 1 through 10.

Bonus Draw a figure that has line symmetry but not rotational symmetry.

Chapter

6

Patterns and Number Sense

Spotlight on the Planets

Have You Ever Wondered...

- How far each of the planets is from the sun?
- How many planets the size of Earth could fit into a planet the size of Jupiter?

Average Distance from the Sun

Planet	Miles
Mercury	3.59×10^7
Venus	6.7×10^7
Earth	9.29×10^7
Mars	1.42×10^8
Jupiter	4.83×10^8
Saturn	9.14×10^8
Uranus	1.78×10^9
Neptune	2.79×10^9
Pluto	3.68×10^9

Mass of Each Planet

Planet	Mass
Mercury	7.283×10^{23} lb
Venus	1.07×10^{25} lb
Earth	1.32×10^{25} lb
Mars	1.42×10^{24} lb
Jupiter	4.187×10^{27} lb
Saturn	1.2583×10^{27} lb
Uranus	1.909×10^{26} lb
Neptune	2.271×10^{26} lb
Pluto	1.45×10^{22} lb

Chapter Project

The Planets

Work in a group.

1. Make a model or draw a picture of our solar system. Be sure to illustrate the relative distances between the planets as well as their relative sizes.
2. Write how each planet's distance from the sun relates to the distance of the farthest planet by using fractions.
3. Make a chart showing how the distance from the sun affects the temperature on the surface of each planet.

"Dear Henry: Where were you? We waited and waited but finally decided that . . ."

Looking Ahead

In this chapter, you will see how mathematics can be used to answer questions about the planets in our solar system. The major objectives of the chapter are to:

- find the prime factorization of a composite number
- find the greatest common factor or the least common multiple of a set of numbers
- simplify, compare, and order rational numbers
- express rational numbers as decimals and vice versa
- express numbers in scientific notation

6-1 Divisibility Patterns

Objective
Use divisibility rules for 2, 3, 4, 5, 6, 8, 9, and 10.

Words to Learn
divisible

You can review factors on page 34.

Michael DeSoto received a $40 check from his grandmother for his birthday. When he went to the bank to cash the check, the teller asked, "How would you like this?" He said, "All the bills the same." Then the teller asked, "What kind of bills do you want?"

What kind of bills could the teller give Michael? This is another way of asking for all the factors of 40. You can divide to find factors. The factors of a whole number divide that number with a remainder of zero.

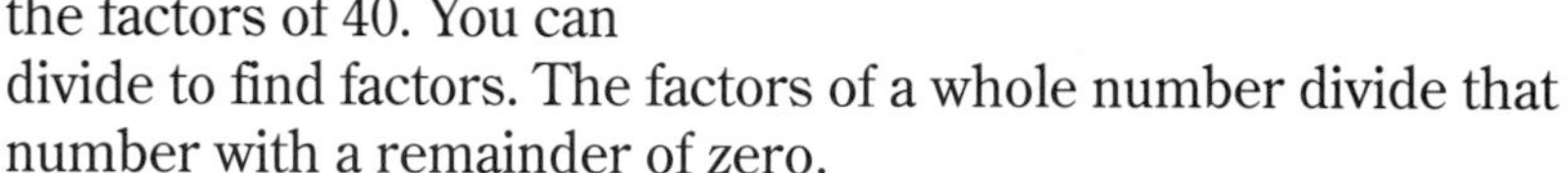

In order to determine the kind of bills the teller can give Michael, divide 40 by 1, 2, 5, 10, and 20. These are the possible types of bills.

$40 \div 1 = 40$ *The remainder is 0 so 1 and 40 are factors.*
$40 \div 2 = 20$ *The remainder is 0 so 2 and 20 are factors.*
$40 \div 5 = 8$ *The remainder is 0 so 5 and 8 are factors.*
$40 \div 10 = 4$ *The remainder is 0 so 10 and 4 are factors.*
$40 \div 20 = 2$ *The remainder is 0 so 20 and 2 are factors.*

Michael could receive 40 $1-bills, 20 $2-bills, 8 $5-bills, 4 $10-bills, or 2 $20-bills.

If a number is a factor of a given number, we can also say that the given number is **divisible** by the factor. For example, since 4 is a factor of 40, 40 is *divisible* by 4. The following rules will help you determine whether a number is divisible by 2, 3, 4, 5, 6, 8, 9, or 10.

Mental Math Hint

You can use the divisibility rule for 2 to determine whether a number is divisible by 4 and 8. Mentally divide the number by 2. If the quotient is even, the original number is divisible by 4. Divide the quotient by 2 again. If the new quotient is even, the original number is divisible by 8.

A number is divisible by:

- 2 if the ones digit is divisible by 2.
- 3 if the sum of the digits is divisible by 3.
- 4 if the number formed by the last two digits is divisible by 4.
- 5 if the ones digit is 0 or 5.
- 6 if the number is divisible by 2 *and* 3.
- 8 if the number formed by the last three digits is divisible by 8.
- 9 if the sum of the digits is divisible by 9.
- 10 if the ones digit is 0.

Examples

1 **Determine whether 104 is divisible by 2, 3, 4, 5, 6, 8, 9, or 10.**

2: The ones digit is divisible by 2. So 104 is divisible by 2.
3: The sum of the digits, $1 + 0 + 4 = 5$, is *not* divisible by 3. So 104 is *not* divisible by 3.
4: The number formed by the last two digits, 04, is divisible by 4. So 104 is divisible by 4.
5: The ones digit is *not* 0 or 5. So 104 is *not* divisible by 5.
6: The number is divisible by 2 but *not* by 3. So 104 is *not* divisible by 6.
8: The number formed by the last 3 digits, 104, is divisible by 8. So 104 is divisible by 8.
9: The sum of the digits, $1 + 0 + 4 = 5$, is *not* divisible by 9. So 104 is not divisible by 9.
10: The ones digit is not 0, so 104 is *not* divisible by 10.

Therefore, 104 is divisible by 2, 4, and 8, but not 3, 5, 6, 9, or 10.

2 **Find two different pairs of factors of 3,267.**

The sum of the digits $3 + 2 + 6 + 7$, or 18, is divisible by 3 and 9. So 3,267 is divisible by both 3 and 9.

3,267  3 [=] 1089   3,267 [÷] 9 [=] 363

Two pairs of factors of 3,267 are 3 and 1,089, and 9 and 363.

Checking for Understanding

Communicating Mathematics

Read and study the lesson to answer each question.

1. **Tell** why 7 is a factor of 42.
2. **Write** two factors of 78 that are greater than 1 and less than 78.
3. **Tell** why an odd number cannot be divisible by 6.
4. **Write** one or two sentences to explain why a number that is divisible by 9 is also divisible by 3.

Guided Practice

Using divisibility rules, determine whether each number is divisible by 2, 3, 4, 5, 6, 8, 9, or 10.

5. 58
6. 153
7. −330
8. 881
9. 12,345

10. Is 6 a factor of 228?
11. Is 5 a factor of 523?
12. Is 231 divisible by 9?
13. Is 432 divisible by 4?
14. Find two ways to write 126 as a product of two factors.
15. Find two numbers that are divisible by 2 and 3.

Exercises

Independent Practice

Use divisibility rules to determine if the first number is divisible by the second number. Write *yes* or *no*.

16. 7,774; 2 **17.** 335; 5 **18.** 991; 3 **19.** 902; 6

20. 112; 8 **21.** 2,034; 9 **22.** 5,142; 4 **23.** 10,105; 10

24. Use divisibility rules to find two different pairs of factors of 984.

25. Use divisibility rules and a calculator to find two different pairs of factors of 43,210.

Use mental math, paper and pencil, or a calculator to find a number that satisfies the given conditions.

26. a number divisible by both 4 and 9

27. a three-digit number divisible by 3, 6, and 9

28. a four-digit number divisible by 4, but *not* 8

29. a five-digit number *not* divisible by 4, 5, 9, or 10

Mixed Review

30. Fund Raising Kim is participating in a walk-a-thon for the Lung Association. Jamal is sponsoring her at 10¢ for each mile she walks and Andreina is sponsoring her at 11¢ per mile. If Kim walks 9 miles, what is the total amount she will collect from Jamal and Andreina? *(Lesson 1-1)*

31. Solve $p = -654 - 175$. *(Lesson 3-5)*

32. Statistics Determine whether a scatterplot of height and eye color would show a positive, negative, or no relationship. *(Lesson 4-8)*

33. Geometry Make an Escher-like drawing for the pattern shown at the right. Use a tessellation of two rows of five equilateral triangles as your base. *(Lesson 5-7)*

Problem Solving and Applications

34. Entertainment There are 144 students in the school band. They need to march in a rectangular formation. List all the formations that are possible.

35. Critical Thinking What is the greatest six-digit number that is *not* divisible by 4 or 9?

COMPUTER CONNECTION

36. Computer Connection The BASIC program shown below will print the factors of a given number.

```
10  INPUT N
20  FOR D = 1 TO N
30  IF INT(N/D) = N/D THEN 50
40  GOTO 60
50  PRINT D
60  NEXT D
```

Use the program to find the number, less than 100, that has the greatest number of factors.

6-2 Prime Factorization

Objective

Find the prime factorization of a composite number.

Words to Learn

prime
composite
prime factorization
factor tree
Fundamental Theorem of Arithmetic

Did you ever wonder how water changes to steam and then the steam seems to disappear? As water is heated, the molecules begin to move more rapidly. The molecules move farther apart forming a vapor. The vapor bubbles rise and escape into the air.

If you separate a water molecule into smaller parts, you would find 2 atoms of hydrogen and 1 atom of oxygen. These are the basic elements of water.

In mathematics, we use factoring to separate a number into smaller parts. The basic elements of a number are its factors. When a whole number *greater than 1* has *exactly* two factors, 1 and itself, it is called a **prime number.** For example, 5 is a prime number since it has two factors, 1 and 5.

Any whole number, except 0 and 1, that is not prime can be written as a product of prime numbers. When a whole number *greater than 1* has more than two factors, it is called a **composite number.** For example, 6 is a composite number since it has four factors, 1, 2, 3, and 6.

The numbers 0 and 1 are *neither* prime *nor* composite. Notice that 0 has an endless number of factors and that 1 has only one factor, itself.

To find the prime factors of any composite number, begin by expressing the number as a product of two factors. Then continue to factor until all the factors are prime. When a number is expressed as a product of factors that are all prime, the expression is called the **prime factorization** of the number.

The diagrams below each show a different way to find the prime factorization of 24. These diagrams are called **factor trees.**

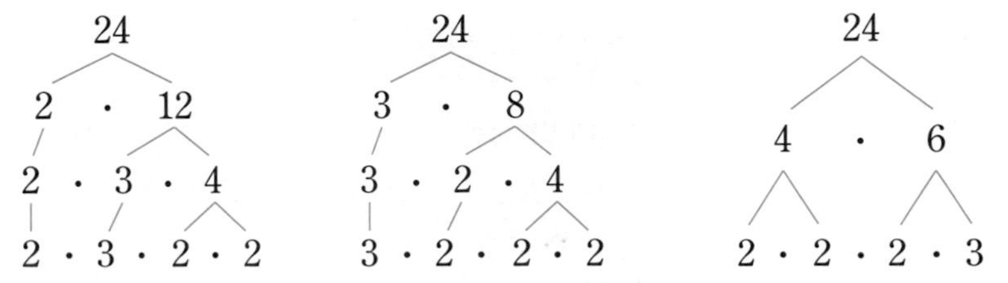

Every number has a unique set of prime factors. Notice that the bottom row of "branches" in each factor tree is the same except for the order in which the factors are written. This property of numbers is called the **Fundamental Theorem of Arithmetic.**

You can use a calculator and the divisibility rules as tools to find the prime factorization of a number.

Calculator Hint

When you use a calculator to find prime factors, it is helpful to record them on paper as you find each one.

Example 1

Find the prime factorization of 315.

Divide by prime factors until the quotient is prime.

prime factors ↓

315 [÷] 5 [=] 63 — *The ones digit is 5. So 315 is divisible by 5.*

63 [÷] 3 [=] 21 — *The sum of the digits, 6 + 3 = 9, is divisible by 3. So 63 is divisible by 3.*

21 [÷] 3 [=] 7 — *7 is a prime number.*

$315 = 3^2 \cdot 5 \cdot 7$

For centuries, mathematicians have tried to find an expression for calculating every prime number. While an expression may yield some prime numbers, none has been found to yield all prime numbers.

Example 2 *Connection*

Algebra **Show that the expression $n^2 - n + 41$ yields a prime number for $n = 1, 2,$ and 3.**

$n^2 - n + 41 = 1^2 - 1 + 41$ — *Replace n with 1.*
$= 41$ — *41 has only two factors, 1 and 41. 41 is a prime number.*

$n^2 - n + 41 = 2^2 - 2 + 41$ — *Replace n with 2.*
$= 43$ — *43 has only two factors, 1 and 43. So 43 is a prime number.*

$n^2 - n + 41 = 3^2 - 3 + 41$ — *Replace n with 3.*
$= 47$ — *47 has only two factors, 1 and 47. So 47 is a prime number.*

Sometimes you may need to factor a negative integer. Any negative integer may be written as the product of −1 and a whole number.

Example 3

Find the prime factorization of −140.

First, write −140 as the product $-1 \cdot 140$.

$-140 = -1 \cdot 140$
$= -1 \cdot 2 \cdot 70$ — *140 is divisible by 2.*
$= -1 \cdot 2 \cdot 2 \cdot 35$ — *70 is divisible by 2.*
$= -1 \cdot 2 \cdot 2 \cdot 5 \cdot 7$ — *35 is divisible by 5. 7 is a prime number.*

$-140 = -1 \cdot 2^2 \cdot 5 \cdot 7$

Checking for Understanding

Communicating Mathematics

Read and study the lesson to answer each question.

1. **Tell** the difference between prime and composite numbers.
2. **Draw** two different factor trees to find the prime factorization of 60.
3. **Write** the number whose prime factorization is given by $2^3 \cdot 3^2$.

Guided Practice

Determine whether each number is *prime* or *composite*.

4. 52
5. 13
6. 29
7. 51

Draw a factor tree to find the prime factorization of each number.

8. 57
9. 36
10. 90
11. 180

Exercises

Independent Practice

Determine whether each number is *prime, composite,* or *neither.*

12. 23
13. 93
14. 77
15. 68
16. 1
17. 453
18. 43,000
19. –97
20. State the least prime number that is a factor of the number 48.

Find the prime factorization of each number.

21. 81
22. 605
23. 64
24. 31
25. 400
26. –144
27. –2,700
28. 20,310

Mixed Review

29. Solve $4m < 24$. Show the solution on a number line. *(Lesson 2-10)*
30. **Statistics** Make a line plot for the data below. *(Lesson 4-3)*
    27, 22, 23, 26, 22, 20, 25, 28, 23, 20, 21, 22, 25, 23, 22
31. **Geometry** Determine if the figure at the right has rotational symmetry. *(Lesson 5-5)*
32. Use divisibility rules to determine if 672 is divisible by 3. *(Lesson 6-1)*

Problem Solving and Applications

33. **Number Sense** Primes that differ by two are called *twin primes.* One such pair is 11 and 13.
    a. Find all the twin primes less than 100.
    b. List the numbers that are between twin primes for primes greater than 3 and less than 37. What do these numbers have in common? Is this true for all twin primes? Why or why not?
34. **Critical Thinking** Find the prime factorization of the perfect squares 9, 36, and 100.
    a. What characteristics do the prime factorizations have?
    b. Is this true for all perfect squares? Explain.
35. **Algebra** Is the value of $3a + 4b$ prime or composite if $a = 5$ and $b = 4$?
36. **Number Sense** Two primes that differ by one are 2 and 3. Find all such pairs of primes less than 1,000.
37. **Algebra** Find the least whole number, n, for which the expression $n^2 - n + 41$ is not prime.

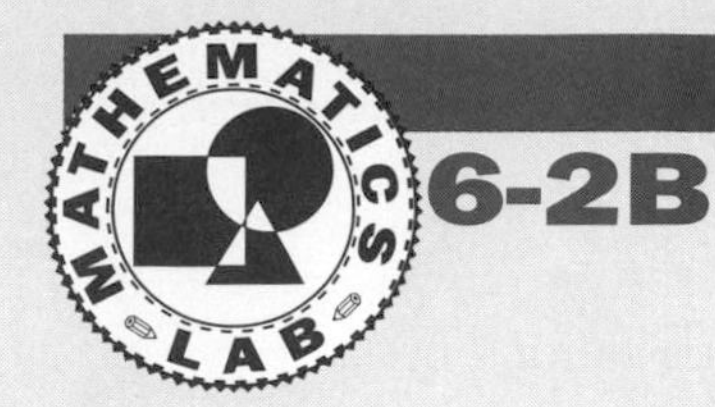

Cooperative Learning

6-2B Basketball Math

A Follow-Up of Lesson 6-2

Objective

Discover the relationship of relatively prime numbers.

Materials

colored pencils
tracing paper

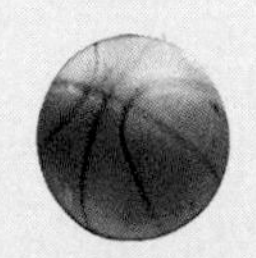

Ms. Carozza coaches the eighth grade girls' basketball team. She is also a mathematics teacher. Sometimes she uses drills at basketball practice that reinforce concepts she covered in class.

Try this!

- The figure at the right represents the members of the team standing around a circle for a passing drill. The drill starts when the player with the ball (B) calls out a number, such as 4, and passes the ball to the fourth player to her right. The person catching the ball continues the pattern by passing the ball to the fourth person to her right and so on.

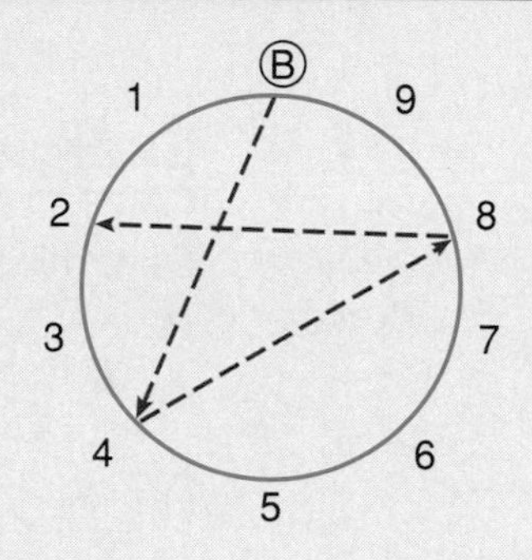

- Sometimes the player who starts the drill calls out a number that results in every player around the circle touching the ball once before it returns to her. When this happens, the team is given a water break. Use colored pencils to record the different paths the ball takes when each of the numbers 1 through 9 is called.
- Copy and complete the table at the right to determine when the team gets a water break.

Number of Players	Number Called	Water Break?
10	1	
10	2	
10	3	
10	4	
10	5	
10	6	
10	7	
10	8	
10	9	

What do you think?

1. What numbers can be called so the team can take a break?
2. Make a conjecture about the relationship between 10 and the numbers in the answer to Exercise 1.
3. Find the greatest factor common to 10 and each number in the answer to Exercise 1.
4. Describe the pattern that exists between the pairs of numbers you used in Exercise 3.
5. When two whole numbers have 1 as their greatest common factor, the numbers are **relatively prime.** How did the coach use relatively prime numbers in the passing drill?

Problem-Solving Strategy

6-3 Make a List

Objective

Solve problems by making an organized list.

You know that prime numbers have exactly two factors. What numbers have exactly three factors? or four factors?

Explore

What do you know?
Prime numbers have exactly two factors.
Composite numbers are not prime.

What are you trying to find?
The composite numbers that have exactly three factors
The composite numbers that have exactly four factors

Plan

Find the factors of the numbers from 2 through 15. List them according to the number of factors each has. Describe the numbers in the 3-factor list. Then describe the numbers in the 4-factor list.

Solve

Exactly 2 Factors	Exactly 3 Factors	Exactly 4 Factors	Exactly 5 Factors
2: 1, 2	4: 1, 2, 4	6: 1, 2, 3, 6	16: 1, 2, 4, 8, 16
3: 1, 3	9: 1, 3, 9	8: 1, 2, 4, 8	
5: 1, 5		10: 1, 2, 5, 10	
7: 1, 7		14: 1, 2, 7, 14	
11: 1, 11		15: 1, 3, 5, 15	
13: 1, 13			

Study the lists for 3 and 4 factors. Look for a pattern to help describe the numbers in terms of their factors.

3-Factor List
The numbers are squares of their prime factor.

4-Factor List
Except for 8, each number is the product of its two prime factors.

Examine

Determine the next number in each list to test its description.

25: 1, 5, 25 — *25 is the square of its prime factor and has three factors. The description is accurate.*

21: 1, 3, 7, 21 — *21 is the product of its two prime factors and has four factors. The description is accurate.*

Checking for Understanding

Communicating Mathematics

Read and study the lesson to answer each question.

1. **Tell** what you can do to a prime number to get a number that will have exactly three factors.
2. **Show** that the description *a double or triple of a prime number* does not describe the numbers with exactly four factors.

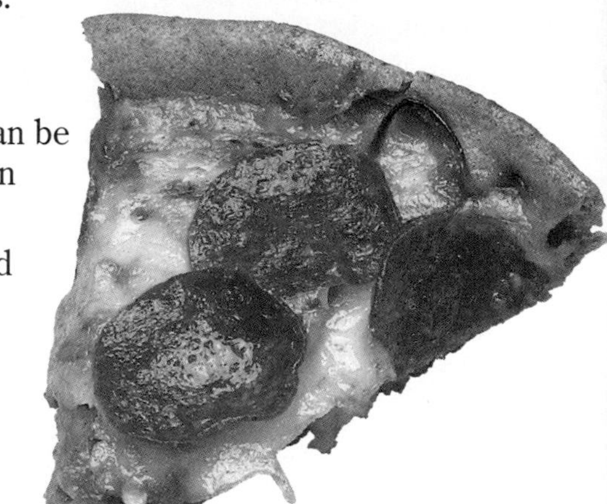

Guided Practice

Solve by making a list.

3. Predict a number from 50 through 150 that can be placed in each list. Use the BASIC program on page 214 to check your predictions.
4. How many three-digit numbers can be formed by the digits 1, 2, and 3 if no digit is repeated? What are the numbers?
5. Extend the list of numbers that have exactly 5 factors. Write a description of the list.

Problem Solving

Practice

Solve using any strategy.

Strategies

Look for a pattern.
Solve a simpler problem.
Act it out.
Guess and check.
Draw a diagram.
Make a chart.
Work backward.

6. Marty and some friends are sharing a pizza. Marty starts with 9 slices of pizza. She takes the first slice, then passes the pizza around the table. Each friend takes one slice and continues to pass the pizza around the table. Marty takes the last slice. Counting Marty, how many people could be sharing the pizza?
7. There are 108 students signed up for the basketball league. They have 18 sponsors. How many teams of 9 players can be formed?
8. Paper cups can be purchased in packages of 40 or 75. Alexis buys 7 packages and gets 350 cups. How many packages of 75 does she buy?
9. Neal is using a beaker containing water for a science experiment. He uses half of the water and gives half of the remaining amount to Juanita for her experiment. At the end of class, Neal has 225 mL left in the beaker. How much water was in the beaker originally?
10. Alishanee bought 15 posters to give to friends. She wants to know how many friends she can equally share them with if she keeps one for herself. How can she determine all the possibilities without making lists?
11. Mary has a basket containing 21 muffins to share with friends. She takes the first one and each friend takes one as they continue around the table. There is one muffin left when the basket returns to Mary. How many people could have shared the muffins if the basket can go around the table more than once?
12. **Journal Entry** Write a problem that can be solved by making a list.

6-4 Greatest Common Factor

Objective

Find the greatest common factor of two or more integers.

Words to Learn

greatest common factor (GCF)

In July of 1976, Dr. Robert B. Craven of the U.S. Center for Disease Control in Atlanta began to receive calls from physicians in Pennsylvania. The doctors described a rare form of pneumonia which they had never seen before. The disease was often fatal. Upon further investigation, the Center found that all of the people struck with this disease had been in or near the Bellevue-Stratford Hotel in Philadelphia during a convention of the American Legion. This disease became known as Legionnaire's disease.

Doctors unraveled the mystery of Legionnaire's disease by looking for *common factors* in those who were ill.

Two or more numbers may also have some common factors. The greatest of the factors common to two or more numbers is called the **greatest common factor** (GCF) of the numbers. You can use prime factorization to find the GCF. Consider the prime factorization of 84 and 90 shown below.

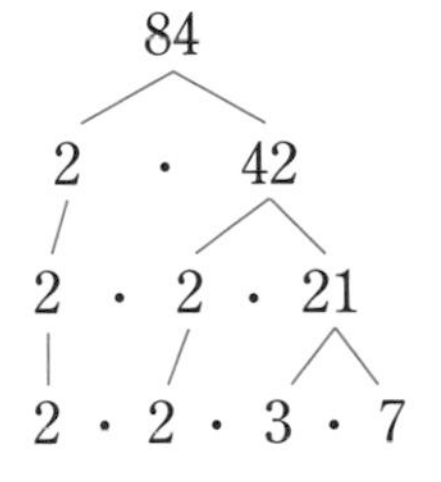

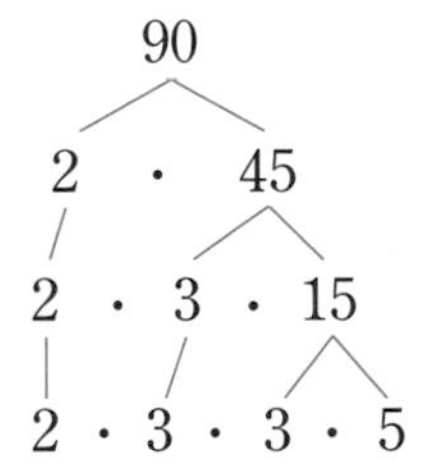

The integers 84 and 90 have 2 and 3 as common prime factors. The product of these prime factors, $2 \cdot 3$ or 6, is the GCF of 84 and 90.

Example 1

Use prime factorization to find the GCF of 84, 126, and 210.

Write each number as a product of prime factors.

$84 = \mathbf{2} \cdot 2 \cdot \mathbf{3} \cdot \mathbf{7}$

$126 = \mathbf{2} \cdot \mathbf{3} \cdot 3 \cdot \mathbf{7}$

$210 = \mathbf{2} \cdot \mathbf{3} \cdot 5 \cdot \mathbf{7}$ *The common prime factors are 2, 3, and 7.*

Thus, the GCF is $2 \cdot 3 \cdot 7$ or 42.

Another way to find the GCF is to list all the factors of each number. Then find the greatest number that is in both lists.

Problem-Solving Hint

Use the *make a list* strategy.

Example 2

Find the GCF of 24 and 60.

factors of 24: 1, 2, 3, 4, 6, 8, 12, 24
factors of 60: 1, 2, 3, 4, 5, 6, 10, 12, 15, 20, 30, 60

The *common factors* of 24 and 60, shown in blue, are 1, 2, 3, 4, 6, and 12. The greatest common factor of 24 and 60 is 12.

Checking for Understanding

Communicating Mathematics

Read and study the lesson to answer each question.

1. **Tell** how to find the GCF of two or more numbers.
2. **Draw** factor trees to find the prime factorization of 120 and 168. Then circle the common factors.
3. **Tell** what the GCF is for the two numbers whose prime factorizations are shown at the right.
4. **Write** two numbers whose GCF is 15.

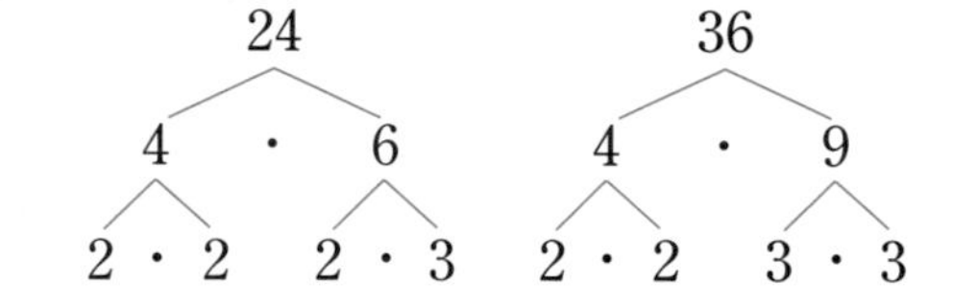

Guided Practice

Find the GCF for each set of numbers.

5. $8 = 2 \cdot 2 \cdot 2$
   $12 = 2 \cdot 2 \cdot 3$
6. $35 = 5 \cdot 7$
   $20 = 2 \cdot 2 \cdot 5$
7. $28 = 2 \cdot 2 \cdot 7$
   $70 = 2 \cdot 5 \cdot 7$
8. 60, 105
9. 9, 15, 24
10. 15, 45
11. 18, 27
12. 7, 11
13. 26, 34, 64

Exercises

Independent Practice

Find the GCF for each set of numbers.

14. 8, 34
15. 16, 24
16. 36, 27
17. 125, 100
18. 28, 56
19. 96, 108
20. 12, 18, 30
21. 10, 15, 30
22. 45, 105, 75
23. 210, 330, 150
24. 510, 714, 306
25. How do you know just by looking that two numbers will have 5 as a common factor?
26. What is the GCF of $3^2 \cdot 5$ and $2 \cdot 3 \cdot 5^2$?

Mixed Review

27. How many liters are in 864 milliliters? *(Lesson 1-6)*
28. Find the absolute value of −21. *(Lesson 3-1)*
29. **Geometry** Classify the triangle at the right according to its sides and its angles. *(Lesson 5-3)*

56 cm, 29°, 21°, 27 cm, 130°, 36 cm

30. Find the prime factorization of 56. *(Lesson 6-2)*

Problem Solving and Applications

31. **Industrial Technology** Charles wants to resurface the top of a table with ceramic tiles. The table is 30 inches long and 24 inches wide.
    a. What is the largest square tile that he can use without having to use any partial squares?
    b. How many of these tiles will Charles need?

32. **Data Search** Refer to page 665.
    Suppose Venus and Earth are aligned with the sun, that is, at the same point in their orbit relative to the sun. How many years will pass before they are aligned again?
33. **Number Sense** Write three numbers that have a GCF of 7.
34. **Critical Thinking** Let p represent a prime number.
    a. What is the GCF of $8p$ and $16p$?
    b. What is the GCF of $15p$ and $27p$?
35. **Critical Thinking** Numbers that have a GCF of 1 are **relatively prime.** Use this definition to determine if each statement is *true* or *false* and tell why.
    a. Any two prime numbers are relatively prime.
    b. If two numbers are relatively prime, one of them must be prime.

SAVE THE PLANET • SAVE THE PLANET •

Save Planet Earth

Environmentally Safe Cars Mercedes Benz's "S-Class" cars are free of asbestos and other toxic materials. They contain only recyclable plastics and have a sophisticated emissions-control system. Their air-conditioning system does not contain CFCs. None of these compounds, which can destroy the ozone, are used in the making of foam in seats and interior panels. In 1994, all of the "S" cars will be painted with a water-based paint rather than petroleum-based paint.

While the price of these cars is out of reach for most consumers, the technology will eventually make its way into less-expensive vehicles.

How You Can Help

- Write to major car manufacturers and ask them what environmentally safe technology they plan to use in their new models. Consumer pressure could convince many car manufacturers to start making environmentally safe vehicles much sooner.

6-5 Rational Numbers

Objective

Identify and simplify rational numbers.

Words to Learn

rational number
simplest form

Edison's light was a glass bulb containing a carbon filament in a vacuum. It burned for $13\frac{1}{2}$ hours.

In 1879, Thomas Alva Edison invented the electric light. His light was an incandescent bulb. That is, light was produced by running electricity through a filament to produce the light. Fluorescent bulbs produce light by running current through mercury vapor.

Fluorescent bulbs save energy. They last longer and use about $\frac{1}{4}$ the energy of an incandescent bulb. A 60-watt incandescent bulb costs around \$0.53, while a compact fluorescent bulb costs about \$12.

The numbers $\frac{1}{4}$, 0.53, and 12 are all **rational numbers.** Rational numbers can also be negative.

Rational Numbers	Any number that can be expressed in the form $\frac{a}{b}$, where a and b are integers and $b \neq 0$, is called a rational number.

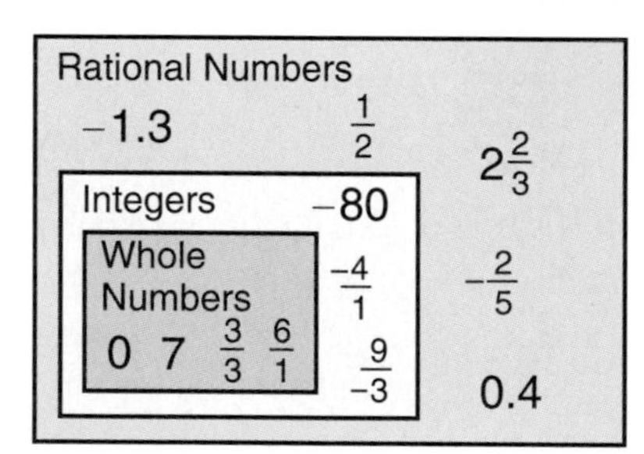

You can use a Venn diagram to show how sets within the rational numbers are related.

In the Venn diagram at the left, the smallest rectangle contains the set of *whole numbers.* The middle rectangle contains the set of *integers,* and the largest rectangle contains the set of *rational numbers.*

Notice that whole numbers and integers are included in the set of rational numbers because integers, such as 6 and −4, can be written as $\frac{6}{1}$ and $\frac{-4}{1}$.

The set of rational numbers also includes decimals because they can be written as fractions with a power of ten for the denominator. For example, −0.17 can be written as $-\frac{17}{100}$.

`The number line below is separated into eighths to show the graphs of some common fractions and decimals.

0 0.125 0.25 0.375 0.5 0.625 0.75 0.875 1

$\frac{0}{8}$ $\frac{1}{8}$ $\frac{1}{4}$ $\frac{3}{8}$ $\frac{1}{2}$ $\frac{5}{8}$ $\frac{3}{4}$ $\frac{7}{8}$ $\frac{8}{8}$

Mixed numbers are also included in the set of rational numbers because any mixed number can be written as an improper fraction. For example, $2\frac{2}{3}$ can be written as $\frac{8}{3}$.

Examples

Name all the sets of numbers to which each number belongs.

1 18 — 18 is a whole number and an integer. Since it can be written as $\frac{18}{1}$, it is also a rational number.

2 −7 — −7 is an integer and a rational number. −7 can be written as $\frac{-7}{1}$ or $\frac{7}{-1}$.

3 $-4\frac{3}{5}$ — Since $-4\frac{3}{5}$ can be written as $-\frac{23}{5}$, it is a rational number.

4 0.261 — Since 0.261 can be written as $\frac{261}{1{,}000}$, it is a rational number.

When a rational number is represented as a fraction, it is commonly expressed in **simplest form.** A fraction is in simplest form when the GCF of the numerator and denominator is 1.

Example 5

Write $\frac{12}{30}$ in simplest form.

Method 1

$12 = 2 \cdot 2 \cdot 3$ *The GCF of 12 and 30 is 2 · 3 or 6.*

$30 = 2 \cdot 3 \cdot 5$

$\frac{12}{30} \Rightarrow \frac{12 \div 6}{30 \div 6} = \frac{2}{5}$

Method 2

$\frac{12}{30} = \frac{\overset{1}{\cancel{2}} \cdot 2 \cdot \overset{1}{\cancel{3}}}{\underset{1}{\cancel{2}} \cdot \underset{1}{\cancel{3}} \cdot 5} = \frac{2}{5}$

The slashes indicate that the numerator and denominator are divided by 2 · 3, the GCF.

Since the GCF of 2 and 5 is 1, the fraction $\frac{2}{5}$ is in simplest form.

Checking for Understanding

Communicating Mathematics

Read and study the lesson to answer each question.

1. **Tell** how you know if a number is a rational number.
2. **Write** an example of a rational number that is not an integer.
3. **Tell** how you know whether a fraction is in simplest form.
4. **Write** directions telling how to express $\frac{20}{25}$ in simplest form.

Guided Practice Name all the sets of numbers to which each number belongs.

5. $1\frac{2}{3}$ 6. -11.6 7. $\frac{15}{32}$ 8. 0

Determine whether each fraction is in simplest form. If it is not in simplest form, write it in simplest form.

9. $-\frac{42}{48}$ 10. $\frac{4}{5}$ 11. $\frac{8}{16}$ 12. $\frac{13}{52}$ 13. $-\frac{27}{49}$

Exercises

Independent Practice Name all the sets of numbers to which each number belongs.

14. 1.5 15. $16\frac{8}{9}$ 16. -10 17. 625

Write each fraction in simplest form.

18. $-\frac{2}{8}$ 19. $\frac{-3}{12}$ 20. $\frac{6}{21}$ 21. $\frac{24}{54}$ 22. $-\frac{5}{8}$

23. $\frac{3}{51}$ 24. $-\frac{10}{16}$ 25. $-\frac{45}{72}$ 26. $\frac{81}{99}$ 27. $\frac{14}{66}$

28. Write two other names for -10.

Mixed Review

29. Evaluate $3[14 - (8 - 5)^2] + 20$. *(Lesson 2-1)*
30. Solve $s = -\frac{132}{11}$. *(Lesson 3-7)*
31. **Statistics** Find the interquartile range of the set {239, 226, 232, 212, 243, 216, 250}. *(Lesson 4-6)*
32. Find the GCF of 36, 108, and 180. *(Lesson 6-4)*

Problem Solving and Applications

33. **Critical Thinking** Does $\frac{8}{3.2}$ name a rational number? If so, what one? If not, why not?
34. **Number Sense** Both the numerator and denominator of a fraction are even. Can you tell if the fraction is in simplest form? Explain.
35. **School** Vanesa correctly answered 16 of 25 history test questions. How would you describe her success as a fraction in simplest form?

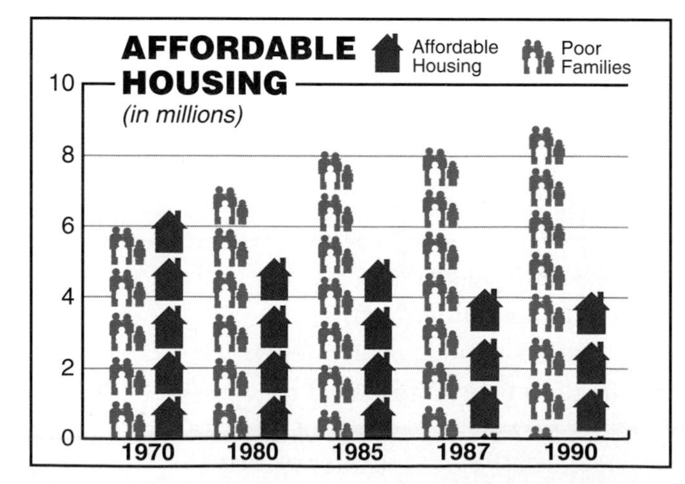

36. **Housing** The graph at the left shows the number of poor families and the number of affordable housing.
    a. How does the number of poor families in 1970 compare to the number of poor families in 1985?
    b. Suppose p represents the number of poor families and h represents the number of affordable houses. Describe the value of $\frac{p}{h}$ for each year shown on the graph.

6-6 Rational Numbers and Decimals

Objective

Express rational numbers as decimals and terminating decimals as fractions.

Words to Learn

terminating decimal

In 1952, A. Pilkington developed the float glass process, the main method of producing glass today. Molten glass is poured onto the surface of molten tin. This makes the surface of the glass very smooth. A ribbon of glass passes over rollers into a tunnel where it is cooled and cut into sheets.

Scott Kuns and Richard Buchanan install glass in office buildings. Each pane of glass they installed in the 13th and 14th floors of the Beggs Building in Columbus, Ohio was $1\frac{3}{16}$ inch thick and weighed about 400 pounds each. The mixed number $1\frac{3}{16}$ is equivalent to $\frac{19}{16}$.

It is sometimes more convenient to write rational numbers as decimals instead of fractions. One reason is that it is easier to compare decimals. It may also be more convenient to use decimals when computing with a calculator or computer.

Consider the fraction $\frac{19}{16}$. Remember that a fraction is another way of writing a division problem. So, $\frac{19}{16}$ means $19 \div 16$. Any fraction can be expressed as a decimal by dividing the numerator by the denominator.

Use a calculator.

19 [÷] 16 [=] 1.1875

Use pencil and paper.

$$\begin{array}{r} 1.1875 \\ 16\overline{)19.0000} \\ \underline{-16} \\ 3\,0 \\ \underline{-1\,6} \\ 1\,40 \\ \underline{-1\,28} \\ 120 \\ \underline{-112} \\ 80 \\ \underline{-80} \\ 0 \end{array}$$

Annex zeros to the numerator: 19 = 19.0000.

Mental Math Hint

Here are some commonly used fraction-decimal equivalencies. It is helpful to know them by memory.

$\frac{1}{2} = 0.5$ $\quad \frac{1}{5} = 0.2$

$\frac{1}{4} = 0.25$ $\quad \frac{1}{10} = 0.1$

$\frac{1}{8} = 0.125$

The fraction $\frac{19}{16}$ can be expressed as the decimal 1.1875. Remember that a decimal like 1.1875 is called a **terminating decimal** because the division ends or terminates when the remainder is zero.

Examples

1 **Express $\frac{3}{5}$ as a decimal.**

3 [÷] 5 [=] 0.6 So, $\frac{3}{5} = 0.6$.

2 **Express $4\frac{9}{16}$ as a decimal.**

$4\frac{9}{16} = 4 + \frac{9}{16}$

4 [+] 9 [÷] 16 [=] 4.5625 So, $4\frac{9}{16} = 4.5625$.

Every terminating decimal can be expressed as a fraction with a denominator of 10, 100, 1,000, and so on. Thus, terminating decimals are rational numbers.

Examples

Express each decimal as a fraction or mixed number in simplest form.

3 **0.65**

$0.65 = \frac{65}{100}$

$= \frac{13}{20}$ *Simplify. The GCF of 65 and 100 is 5.*

4 **−2.625**

$-2.625 = -2\frac{625}{1,000}$

$= -2\frac{5}{8}$ *Simplify.*

Checking for Understanding

Communicating Mathematics

Read and study the lesson to answer each question.

1. **Tell** how to express a fraction as a decimal.
2. **Write** two examples of fractions that can be expressed as terminating decimals.
3. **Show** why 0.37 is a rational number.
4. **Write** sixteen thousandths as a decimal and then as a fraction.

Guided Practice

Express each fraction or mixed number as a decimal.

5. $\frac{2}{5}$ 6. $-\frac{7}{8}$ 7. $\frac{13}{4}$ 8. $-\frac{7}{20}$ 9. $3\frac{14}{25}$

Express each decimal as a fraction or mixed number in simplest form.

10. −0.4 11. 0.75 12. 0.17 13. 3.12 14. −5.375

Exercises

Independent Practice

Express each fraction or mixed number as a decimal.

15. $-\frac{4}{5}$ 16. $\frac{9}{10}$ 17. $-\frac{7}{25}$ 18. $\frac{11}{4}$ 19. $-5\frac{3}{8}$

20. $\frac{5}{16}$ 21. $7\frac{1}{4}$ 22. $-\frac{17}{20}$ 23. $-\frac{9}{32}$ 24. $\frac{71}{40}$

Express each decimal as a fraction or mixed number in simplest form.

25. 0.05	**26.** −1.3	**27.** 0.64	**28.** −8.52
29. 3.85	**30.** 4.105	**31.** −0.075	**32.** −20.35

33. When $\frac{131}{200}$ is expressed as a decimal, what kind of decimal is it?

34. How would you use a calculator to express $9\frac{11}{15}$ as a decimal?

Mixed Review

35. Algebra Write an algebraic expression to represent *the quotient of y and 3 decreased by 15.* *(Lesson 2-6)*

36. Replace ● with >, <, or = in 10 ● −10. *(Lesson 3-2)*

37. Statistics What numbers would you use as stems to make a stem-and-leaf plot of 11.4, 9.7, 10.8, 11.2, 9.5, 12.4, and 10.3? *(Lesson 4-4)*

38. Name all the sets of numbers to which −38 belongs. *(Lesson 6-5)*

Problem Solving and Applications

39. Critical Thinking A unit fraction is a fraction that has one as its numerator. Determine the six greatest unit fractions that are terminating decimals.

40. Measurement A micrometer measured the thickness of the wall of a plastic pipe as 0.084 inches. What fraction of an inch is this?

41. Portfolio Suggestion Select one of the assignments from this chapter that you found particularly challenging. Place it in your portfolio.

6 Assessment: Mid-Chapter Review

Use divisibility rules to determine if the first number is divisible by the second number. Write *yes* or *no*. *(Lesson 6-1)*

1. 582; 6	**2.** 838; 4	**3.** 342; 9	**4.** 10,290; 5

Find the prime factorization of each number. *(Lesson 6-2)*

5. 36	**6.** −45	**7.** 128	**8.** 200

9. Weather The high temperatures during your vacation were: 31°F, 26°F, 30°F, 29°F, 28°F, 27°F, 30°F, 32°F, 31°F, 28°F, 30°F, 29°F, 26°F, 27°F. Find the mode. *(Lesson 6-3)*

Find the GCF for each group of numbers. *(Lesson 6-4)*

10. 25, 30	**11.** 64, 48	**12.** 12, 72, 24

Express each fraction in simplest form. *(Lesson 6-5)*

13. $\frac{8}{32}$	**14.** $-\frac{45}{60}$	**15.** $\frac{35}{175}$	**16.** $\frac{27}{15}$

Express each decimal as a fraction in simplest form and each fraction as a decimal. *(Lesson 6-6)*

17. −0.8	**18.** $\frac{13}{25}$	**19.** $-\frac{5}{8}$	**20.** 4.85

6-7 Repeating Decimals

Objective

Express repeating decimals as fractions.

Words to Learn

repeating decimals
bar notation

How many students in your class wear glasses or contacts? Did you know that $\frac{2}{3}$ of all adults in the United States wear glasses or contacts at some time in their lives?

Calculator Hint

Most calculators round answers, but some truncate answers. *Truncate* means to cut-off at a certain place-value position, ignoring the digits that follow. How can you tell whether your calculator rounds or truncates?

When you divide 2 by 3 to find its decimal equivalent, you find that the digit 6 is repeated.

$3\overline{)2.0000}$ = 0.6666... or 2 [÷] 3 [=] 0.6666667

Decimals like 0.666666... are called **repeating decimals.** Since it is awkward to write all of these digits, you can use **bar notation** to show that the 6 repeats.

$$0.666666... = 0.\overline{6}$$

Examples

Express each decimal using bar notation.

1 0.363636...

The digits 36 repeat.

$0.363636... = 0.\overline{36}$

2 10.0456456...

The digits 456 repeat.

$10.0456456... = 10.0\overline{456}$

The following examples show how to rename repeating decimals as fractions.

Example 3 *Connection*

Algebra Express $0.\overline{2}$ as a fraction.

Let $N = 0.\overline{2}$, or 0.222.... Then $10N = 2.222....$ *Multiply N by 10, because 1 digit repeats.*

Subtract N = 0.222 to eliminate the repeating part, 0.222....

$$\begin{aligned} 10N &= 2.222... \\ -1N &= 0.222... \\ 9N &= 2 \\ N &= \frac{2}{9} \end{aligned}$$

N = 1N

So, $0.\overline{2} = \frac{2}{9}$.

Check: 2 [÷] 9 [=] 0.2222222 ✓

TEEN SCENE

In 1938, Orbig Co., a U.S. corporation, produced the first plastic contact lenses. In the late 1970s, manufacturers began selling different-colored contact lenses for cosmetic purposes.

Example 4

Express $6.\overline{30}$ as a fraction.

Let $N = 6.\overline{30}$. Then $100N = 630.\overline{30}$. *Multiply N by 100, because 2 digits repeat.*

$$\begin{aligned} 100N &= 630.\overline{30} \\ -\ N &= 6.\overline{30} \\ \hline 99N &= 624 \\ N &= \frac{624}{99} \text{ or } 6\frac{10}{33} \end{aligned}$$

Subtracting eliminates the repeating part, $0.\overline{30}$.

So, $6.\overline{30} = \frac{624}{99}$ or $6\frac{10}{33}$. **Check:** 624 ÷ 99  6.3030303 ✓

Since repeating decimals can be expressed as fractions, repeating decimals are included in the set of rational numbers.

Checking for Understanding

Communicating Mathematics

Read and study the lesson to answer each question.

1. **Tell** how to express $2\frac{3}{11}$ as a repeating decimal.
2. **Write** two examples of fractions that can be expressed as repeating decimals.
3. **Tell** how you would choose a multiplier to express $1.\overline{13}$ as a fraction.
4. **Tell** why one of the steps in expressing a repeating decimal as a fraction involves multiplying by a power of 10.

Guided Practice

Write the first ten decimal places of each decimal.

5. $0.\overline{264}$
6. $0.9\overline{2}$
7. $0.\overline{5082}$
8. $0.5\overline{082}$
9. $0.2\overline{16}$

What multiplier, 10 or 100, would you use to express each decimal as a fraction?

10. 0.454545...
11. $8.\overline{7}$
12. 10.622222...
13. $-9.\overline{81}$

Exercises

Independent Practice

Express each decimal using bar notation.

14. 0.166666...
15. 29.272727...
16. 0.42857142...
17. –2.454545...
18. 98.666666...
19. –7.074747...

20. If $N = 0.\overline{894}$, what power of ten should you multiply by in order to express the decimal as a fraction?
21. Which number is greater, 8.32 or $8.\overline{31}$?

Express each repeating decimal as a fraction.

22. $0.\overline{4}$
23. $-1.\overline{7}$
24. $2.\overline{6}$
25. $0.\overline{54}$
26. $-4.\overline{01}$
27. $2.\overline{5}$
28. $0.6\overline{2}$
29. $0.\overline{24}$
30. $7.\overline{52}$
31. $0.\overline{345}$

Mixed Review

32. Write $4 \cdot 4 \cdot 5 \cdot 5 \cdot 4$ using exponents. *(Lesson 1-9)*

33. Solve $63.45 = 4.23g$. Check your solution. *(Lesson 2-4)*

34. **Geometry** Find the value of x in the figure below. *(Lesson 5-1)*

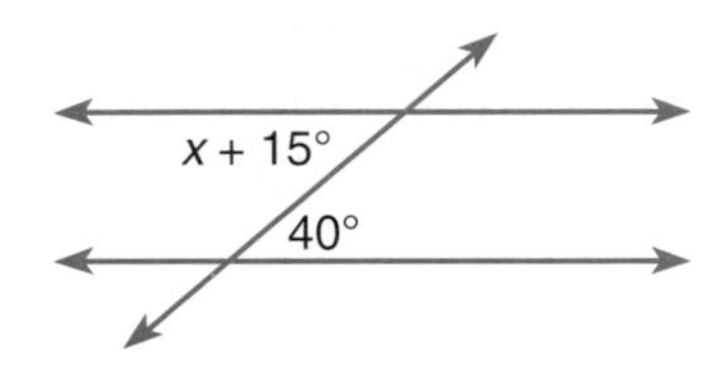

35. Draw a box-and-whisker plot of the following bowling scores: 122, 158, 163, 193, 111, 124, 194, 133, 193, 175, 166, 135, 164. What are the extremes? *(Lesson 4-7)*

36. Express 8.68 as a mixed number in simplest form. *(Lesson 6-6)*

Problem Solving and Applications

37. **Geometry** In the figure at the right, what part of the perimeter is represented by the length of $\overline{AB}$? Express it as a decimal.

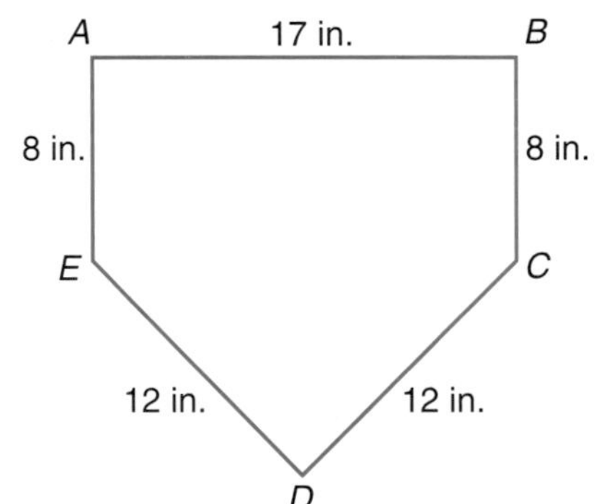

38. **Technology** A spreadsheet's display and two calculators' displays for $5 \div 9$ are listed below. How would each display the answer to $3 \div 11$?

    a. 0.6

    b. 0.5555556

    c. 0.555555

39. **Critical Thinking** List the factors of the denominators of several fractions in simplest form that can be expressed as repeating decimals. List the factors of the denominators of several fractions that can be expressed as terminating decimals. How can you tell by looking at the factors of the denominator of a fraction whether the fraction can be expressed as a repeating decimal?

40. **Number Sense** Write a number that is between the pair of numbers shown in Exercise 21.

41. **Journal Entry** What concept in this chapter have you found most challenging? What do you think made it more difficult for you?

Probability Connection

6-8 Simple Events

Objective

Find the probability of a simple event.

Words to Learn

outcome
random
event
probability

Mr. Cummins teaches mathematics at Lyons Township Middle School. When his class works in cooperative groups, he tosses a number cube to determine which group will demonstrate how they solved a problem. What is the chance that the number cube lands on 4?

When Mr. Cummins tosses the number cube, there are six equally-likely results or **outcomes.** The cube can show either 1, 2, 3, 4, 5, or 6. Each outcome has the same chance of occurring. When all outcomes have an *equally likely* chance of happening, we say that the outcomes happen at **random.**

An **event** is a specific outcome or type of outcome. In this case, the event is tossing a 4. **Probability** is the chance that an event will happen.

Definition of Probability	Probability = $\frac{\text{number of ways that an event can occur}}{\text{number of possible outcomes}}$

"When am I ever going to use this?"

Suppose you play tennis after school once a week with 5 friends. Each week, your opponent is selected by picking names from a hat. To determine the probability that you will get a particular opponent, you can make a sample space of all the possibilities.

Use the scale below when you consider the probability of an event. When it is *impossible* for an outcome to happen, its probability is 0. An outcome that is *certain* to happen has a probability of 1.

impossible	50-50 chance	certain
0	$\frac{1}{2}$	1

The probability of an outcome is written as a number from 0 to 1.

Mini-Lab

Work with a partner.
Materials: dice or number cubes

- Roll two dice 50 times and find the sum of the numbers on the dice. Record each sum as being either even or odd.
- Roll two dice 50 times and find the product of the numbers on the dice. Record each product as being either even or odd.

Talk About It

a. Would you say that an even sum is more likely or less likely to occur than an odd sum? Why?

b. Would you say that an even product is more likely or less likely to occur than an odd product? Why?

Example Connection

Geometry A pin is dropped onto the square at the right. The point of the pin lands in one of the small regions.

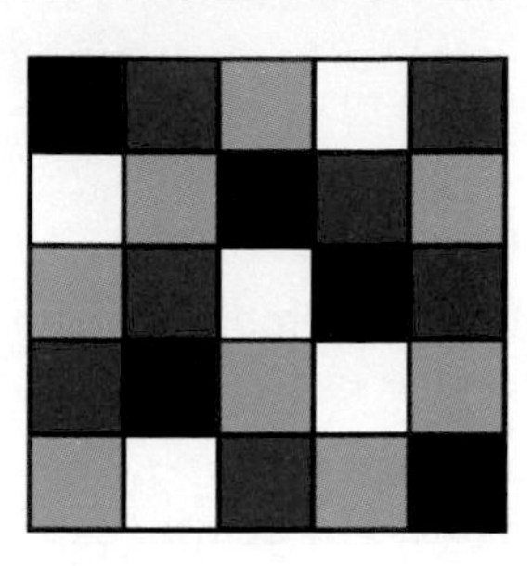

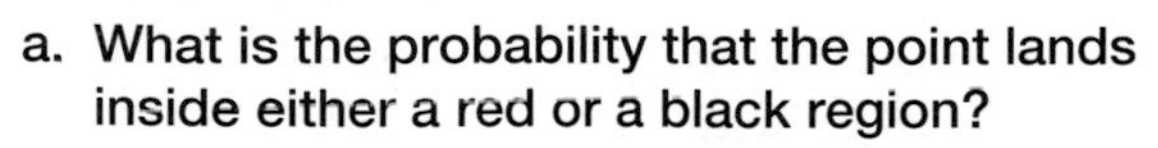

a. What is the probability that the point lands inside either a red or a black region?

$$P(\text{red or black}) = \frac{\text{red squares} + \text{black squares}}{\text{total number of squares}}$$

$$= \frac{7 + 5}{25}$$

$$= \frac{12}{25} \text{ or } 0.48$$

b. What is the probability that the point lands inside a blue region?

Since there are no blue regions, the probability is 0.

Checking for Understanding

Communicating Mathematics

Read and study the lesson to answer each question.

1. **Tell** what probability means.
2. **Tell** what it means for an outcome to have a probability of 1.
3. **Write** an example of an outcome with a probability of 1.
4. **Tell** why the probability of spinning a number divisible by three on the spinner at the right is 0.
5. **Draw** a spinner where the probability of an outcome of one is $\frac{1}{2}$.

Guided Practice

State the probability of each outcome as a fraction and a decimal.

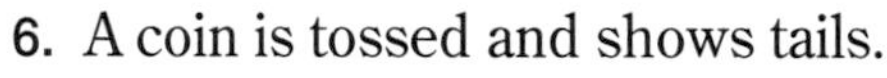

6. A coin is tossed and shows tails.
7. Your friend will live to be 500 years old.
8. A die is rolled and it shows a number divisible by three.
9. A date picked at random is Tuesday.
10. This is a mathematics book.

A bag contains three black, seven red, four orange, and six green jelly beans. A blindfolded student draws one jelly bean. Find the probability of each outcome.

11. It is a white jelly bean.
12. It is a white or a green jelly bean.
13. It is red.
14. It is not orange.
15. It is not purple.
16. It is black, red, or green.

Exercises

Independent Practice

The letters of the word "commutative" are written one each on 11 identical slips of paper and shuffled in a hat. A blindfolded student draws one slip of paper. Find the probability of each outcome.

17. P(t)
18. P(vowel)
19. P(n)
20. P(not c)
21. P(m or v)
22. P(not m or t)

Numbers from 1 to 49 are printed on table tennis balls, mixed by a machine, and drawn at random. One ball is drawn.

23. What is P(an even number)?
24. What is P(a two-digit number)?
25. What is P(a positive number)?
26. What is P(a perfect square)?
27. What is P(a number divisible by 10)?
28. What is P(a negative number)?
29. What is P (a number that is the product of four different prime numbers)?

Mixed Review

30. **Geometry** Find the perimeter and area of the figure at the right. *(Lesson 2-9)*

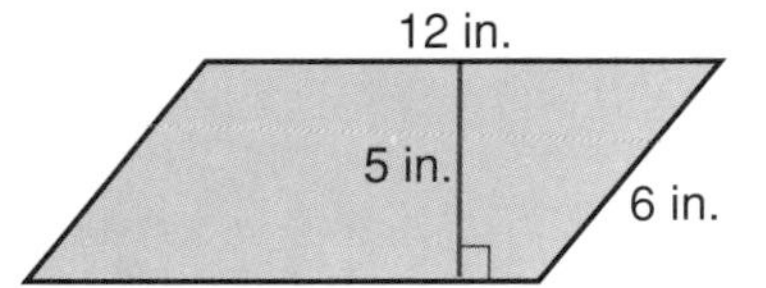

31. Solve $s = -72 + 59$. *(Lesson 3-3)*
32. **Geometry** Tell if the pair of figures at the right are congruent, similar, or neither. *(Lesson 5-6)*

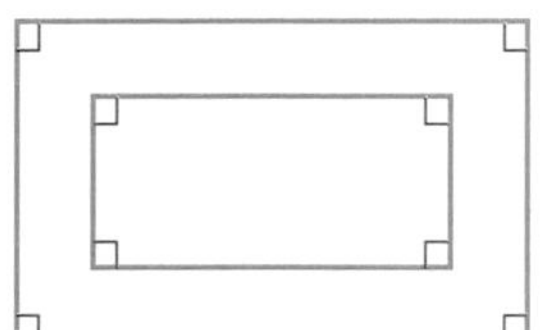

33. Express $8.\overline{72}$ as a mixed number. *(Lesson 6-7)*

Problem Solving and Applications

34. **Make Up a Problem** Write a problem in which the answer will be a probability of $0.\overline{3}$.
35. **Marketing** Pricey's Drive Thru is promoting their new sausage sandwich by distributing cards with a chance to win a free sandwich, a free soft drink, or a free dessert. They want the probability of winning a sandwich to be 0.015, of winning a drink to be 0.2, and of winning a dessert to be 0.04. They will distribute 10,000 cards.
    a. How many winning cards will there be for a free sandwich?
    b. How many winning cards will there be for a free drink?
    c. How many winning cards will there be for a free dessert?
    d. How many cards will say "Better luck next time!"?
36. **Number Sense** Can a probability ever be greater than 1? Why?
37. **Number Sense** Can a probability ever be less than 0? Why?
38. **Critical Thinking** A spider lowers itself onto one of the six squares of the T-shaped figure shown and then randomly moves to an adjacent square. Out of these moves, what is the probability that the spider ends up on a red square?

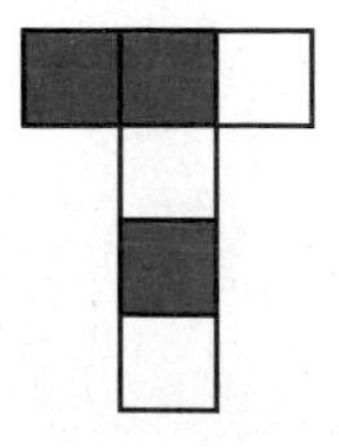

6-9 Least Common Multiple

Objective

Find the least common multiple of two or more integers.

Words to Learn

multiple
least common multiple (LCM)

In the United States, a president is elected every four years. Members of the House of Representatives are elected every two years, and Senators are elected every six years. If a voter has the opportunity to vote for a president, a representative, and a senator in the same year, how long will it be until the same situation occurs again?

You can use multiples and the problem-solving strategy *make a list* to answer this question. A **multiple** of a number is the product of that number and any whole number.

Mini-Lab

Work with a partner.

Materials: colored pencils

- List the numbers from 1 to 30 on a sheet of paper. Use divisibility rules to cross out all the multiples of 2. Using a different color, cross out all the multiples of 4. Repeat the procedure with a third color to cross out all the multiples of 6.

Talk About It

a. Which numbers were crossed out by all three colors?
b. How would you describe these numbers?
c. What is the least number crossed out by all three colors?

In the Mini-Lab, you discovered some common multiples of 2, 4, and 6. The least of the nonzero common multiples of two or more numbers is called the **least common multiple (LCM)** of the numbers. The LCM of 2, 4, and 6 is 12. So, a voter will vote for all three positions in the same year 12 years from now.

Problem Solving Hint

Find multiples by making a list.

Examples

1 List the first four multiples of 6.

multiples of 6 $\rightarrow$ $0 \cdot 6 = 0$
$1 \cdot 6 = 6$
$2 \cdot 6 = 12$
$3 \cdot 6 = 18$

2 List the first four multiples of x.

multiples of x $\rightarrow$ $0 \cdot x = 0$
$1 \cdot x = x$
$2 \cdot x = 2x$
$3 \cdot x = 3x$

Calculator Hint

You can use a calculator to find the LCM of a pair of numbers. Divide multiples of the greater number by the lesser number until you get a whole number quotient.

Example 3

Find the LCM of 6, 10, and 15.

multiples of 6: 0, 6, 12, 18, 24, 30, 36, 42, . . .
multiples of 10: 0, 10, 20, 30, 40, 50, 60, . . .
multiples of 15: 0, 15, 30, 45, 60, 75, 90, . . .

The LCM is 30.

Prime factorization can be used to find the LCM of a set of numbers. A common multiple contains *all* the prime factors of each number in the set. The LCM contains *each* factor the greatest number of times it appears in the set.

Examples

Use prime factorization to find the LCM for each set of numbers.

4 **8, 20**

$8 = 2 \cdot 2 \cdot 2$

$20 = 2 \cdot 2 \cdot 5$

Express each common factor and all other factors.

$2 \cdot 2 \cdot 2 \cdot 5$

Multiply all the factors, using the common factors only once.

The LCM of 8 and 20 is $2^3 \cdot 5$ or 40.

5 **21, 25, 9**

$21 = 3 \cdot 7$ — *The greatest power of 3 is 3^2.*

$25 = 5 \cdot 5$ or 5^2 — *The greatest power of 5 is 5^2.*

$9 = 3 \cdot 3$ or 3^2 — *The greatest power of 7 is 7^1.*

The LCM of 21, 25, and 9 is $3^2 \cdot 5^2 \cdot 7$ or 1,575.

Checking for Understanding

Communicating Mathematics

Read and study the lesson to answer each question.

1. **Tell** how to find the first six multiples of any number.
2. **Write** a three-step procedure for finding the LCM of a set of numbers based on prime factorization.
3. **Tell** what is always true about the first multiple of any number.
4. **Tell** how to find the LCM of a set of numbers if you know the greatest power of each prime factor.

Guided Practice

List the first six multiples of each number.

5. 5 6. 18 7. 20 8. n

Use lists to find the LCM for each set of numbers.

9. 1, 3, 5 10. 7, 21, 5 11. 15, 45, 60 12. 8, 28, 30

Use prime factorization to find the LCM for each set of numbers.

13. 12, 16 14. 10, 15, 20 15. 35, 25, 49 16. 18, 24

Exercises

Independent Practice

Find the LCM for each set of numbers.

17. 12, 15 18. 16, 88 19. 12, 35 20. 16, 24
21. 20, 50 22. 7, 12 23. 4, 8, 12 24. 10, 12, 14
25. 24, 12, 6 26. 45, 10, 6 27. 71, 17, 7 28. 68, 170, 4
29. Determine whether 116 is a multiple of 6.

Mixed Review

30. Use mental math to find 320 + 159 + 80. *(Lesson 1-2)*
31. Solve $12 = \frac{a}{14} + 10$. Check your solution. *(Lesson 2-7)*

32. Solve $r + (-125) = 483$. Check your solution. *(Lesson 3-9)*
33. **Statistics** Would a pet store be a good location to find a representative sample for a survey of number of pets owned? *(Lesson 4-9)*
34. *True* or *false:* Every square is a rhombus. Make a drawing to justify your answer. *(Lesson 5-4)*
35. **Probability** A die is rolled. What is the probability it shows an odd number? *(Lesson 6-8)*

Problem Solving and Applications

36. **Packaging** Brass cabinet hinges are shipped in cartons of individually boxed hinges. Each hinge box is 6 cm wide, 2 cm tall, and either 9 cm, 12 cm, or 18 cm long. A carton is long enough to hold 3 rows of boxes.

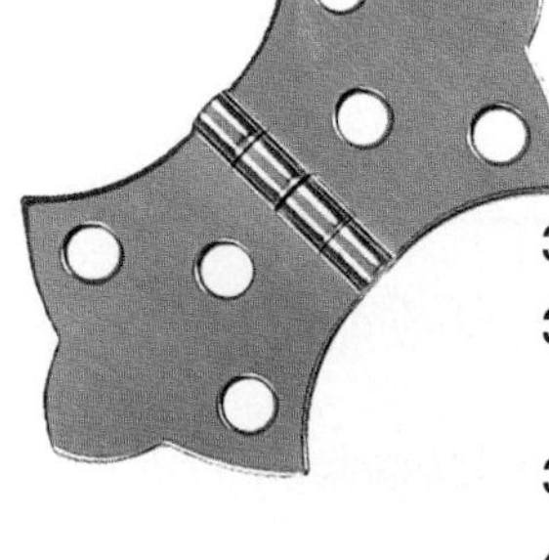

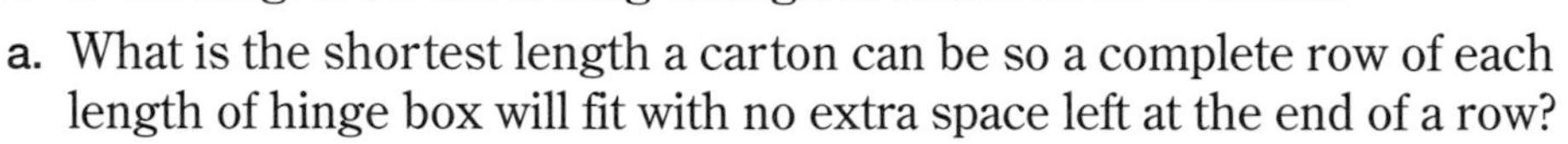

   a. What is the shortest length a carton can be so a complete row of each length of hinge box will fit with no extra space left at the end of a row?
   b. If the carton in part **a** is 12 cm tall, how many of each kind of hinge box will the carton hold?

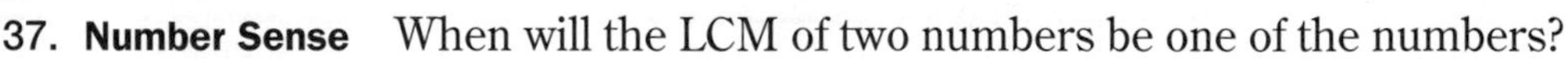

37. **Number Sense** When will the LCM of two numbers be one of the numbers?
38. **Number Sense** When will the LCM of two numbers be the product of the two numbers?
39. **Algebra** What is the LCM of $3n$, $6n^2$, and 8?
40. **Critical Thinking** Find three numbers whose LCM is the product of the numbers.
41. **Journal Entry** Write a sentence or two describing how you can find the LCM of a set of numbers if you have the prime factorization of each number.

Cooperative Learning

6-10A Density Property

A Preview of Lesson 6-10

Objective

Use customary units of measurement.

Materials

12-inch ruler
typing paper
calculator

How many numbers are there between $-1\frac{7}{8}$ and $-1\frac{3}{4}$? Would you guess a *few, many,* or *none?* The **density property** states that between any two rational numbers, no matter how close they may seem, there is at least one other rational number.

Activity One

Work with a partner.

- Near the middle of a piece of typing paper draw an eight-inch line segment parallel to the long side of the paper. Mark the ends and each one-inch interval with a vertical dash. Below the line, label the left endpoint 0, the right endpoint 2, and the middle point 1.

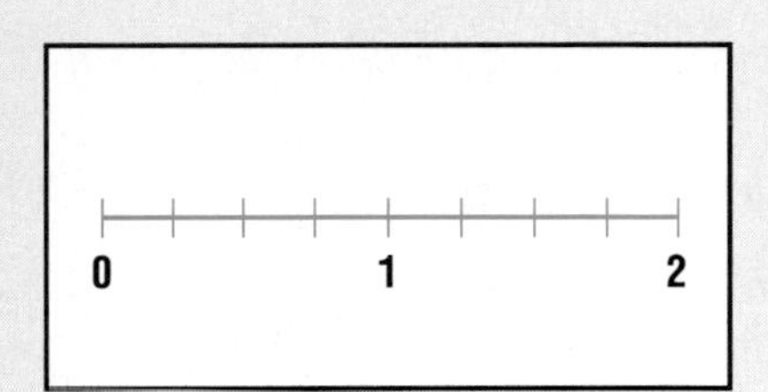

Endpoints	Midpoint	Mean
0 and 2	1	
1 and 2		
1 and $1\frac{1}{2}$		
$1\frac{1}{2}$ and $1\frac{1}{4}$		
$1\frac{3}{8}$ and $1\frac{1}{4}$		

- Copy and complete the *Midpoint* column of the chart at the left to find the midpoint between each pair of endpoints.
- Express the numbers in the *Endpoints* column as decimals. Find the mean of the two decimals. Record each result in the *Mean* column.

What do you think?

1. What is the relationship between the numbers in the *Midpoint* column and the numbers in the *Endpoints* column?
2. What is the relationship between the numbers in the *Midpoint* column and the numbers in the *Mean* column?
3. Tell how you would find the midpoint between $\frac{5}{8}$ and $\frac{3}{4}$.
4. Tell how you would find a rational number between any two rational numbers.

Activity Two

Work with a partner.

- List any two decimals with all but the last digits the same; for example, 3.14 and 3.15. Add a digit other than 0 to the end of the lesser number.
- Repeat this procedure for four more pairs of decimals.

What do you think?

5. What is the relationship between the original numbers and the number formed by adding a digit?
6. Explain why this procedure works for finding a number between any two rational numbers.

Extension

7. List any two fractions.
   a. Add the numerators and add the denominators to form a new fraction. For example, $\frac{3}{7}, \frac{4}{9} \rightarrow \frac{3+4}{7+9} \rightarrow \frac{7}{16}$
   b. Express the three fractions as decimals and order the decimals from least to greatest. Use a calculator.
   c. Repeat this procedure for four other pairs of fractions.
8. Describe any patterns you notice when you order the decimal representations.

6-10 Comparing and Ordering Rational Numbers

Objective

Compare and order rational numbers expressed as fractions and/or decimals.

Words to Learn

least common denominator (LCD)

The eighth grade ensemble at Ferris Middle School was assembled in the gym to have their picture taken. The first task was to line everyone up from shortest to tallest. The teachers helped students do this by comparing one student to another to determine who was taller.

You can follow a similar procedure to compare and arrange any set of rational numbers. Sometimes a number line is used in comparing numbers. However, when numbers are very small, very large, or the difference between them is very small, a number line is not convenient.

Suppose you correctly answered 17 of 20 questions on a history test. On a science test, you answered 22 of 25 questions correctly. On which test did you do better?

To solve this problem, compare $\frac{17}{20}$ and $\frac{22}{25}$. One way to compare these fractions is to express them as decimals and then compare the decimals.

$\frac{17}{20}$ → 17 [÷] 20 [=] 0.85 $\qquad$ $\frac{22}{25}$ → 22 [÷] 25 [=] 0.88

In the hundredths place, 5 < 8.

Since $0.85 < 0.88$, $\frac{17}{20} < \frac{22}{25}$. You did better on the science test.

Another way to compare two rational numbers is to express them as equivalent fractions with like denominators. Any common denominator may be used, but the computation may be easier if the *least common denominator* is used. The **least common denominator (LCD)** is the LCM of the denominators.

Example 1 *Problem Solving*

Sports After playing 18 basketball games, the New York Knicks had won 12 games. During the same time, the Portland Trailblazers won 13 games out of 20. Which team had the better record?

The LCM of 18 and 20 is 180.

$\frac{12}{18} = \frac{■}{180} \rightarrow \frac{12 \times 10}{18 \times 10} = \frac{120}{180} \qquad \frac{13}{20} = \frac{■}{180} \rightarrow \frac{13 \times 9}{20 \times 9} = \frac{117}{180}$

Since $\frac{120}{180} > \frac{117}{180}$, then $\frac{12}{18} > \frac{13}{20}$. The Knicks had the better record.

Examples

Mental Math Hint

Any negative number is always less than any positive number.

Replace each ● with <, >, or = to make a true sentence.

2 $-\frac{1}{2}$ ● 0.12

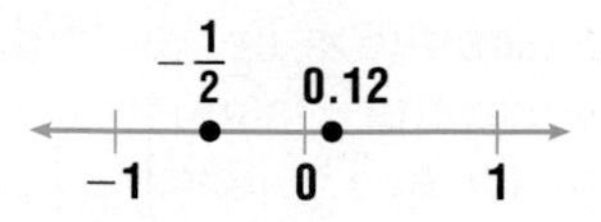

When two numbers are graphed on the same number line, the number on the left is less.

Therefore, $-\frac{1}{2} < 0.12$.

3 −0.7 ● $-\frac{5}{6}$

−0.7 ● −0.8333 ... *Express $-\frac{5}{6}$ as a decimal.*

5 [+/−] [÷] 6 [=] −0.8333333

In the tenths place, −7 > −8. Therefore, $-0.7 > -\frac{5}{6}$.

You have seen rational numbers compared by expressing them in the same form. When there are several numbers, it is usually easier and faster if all the numbers are decimals.

Example 4

Order 0.6, $\frac{6}{11}$, $\frac{1}{2}$, and $0.\overline{63}$ from least to greatest.

Change the fractions to decimals.

$\frac{6}{11}$ → 6 [÷] 11 [=] 0.5454546 $\frac{1}{2} = 0.5$

The order from least to greatest is $\frac{1}{2}$, $\frac{6}{11}$, 0.6, and $0.\overline{63}$.

Example 5

LOOKBACK

You can review median on page 146.

Statistics Find the median of the set of data below.

21.4, 19, $21\frac{1}{2}$, $21\frac{3}{8}$, and 20.9

To find the median, the numbers must be in order.

$21\frac{1}{2}$ → 21.5 $21\frac{3}{8}$ → 21 [+] 3 [÷] 8 [=] 21.375

The order from least to greatest is 19, 20.9, $21\frac{3}{8}$, 21.4, and $21\frac{1}{2}$. The median is $21\frac{3}{8}$.

Checking for Understanding

Communicating Mathematics

Read and study the lesson to answer each question.

1. **Tell** two different ways to compare 0.625 and $\frac{1}{2}$.
2. **Draw** a 10-by-10 square on graph paper. Let the square represent 1. Use the square to explain why $0.2 > 0.09$.

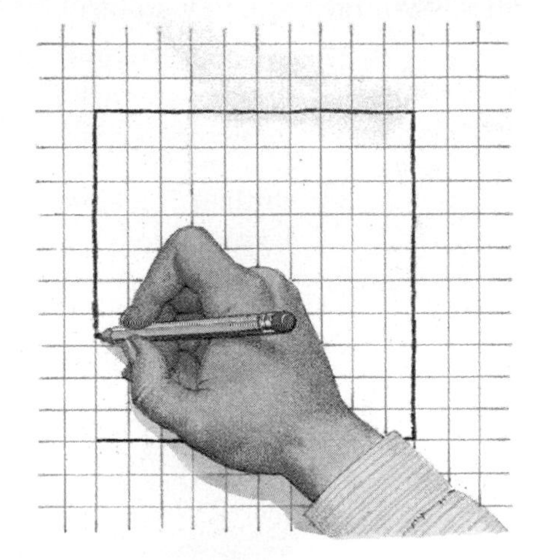

Guided Practice

Find the LCD for each pair of fractions.

3. $\frac{1}{2}, \frac{1}{3}$
4. $\frac{2}{5}, \frac{3}{8}$
5. $\frac{5}{6}, \frac{7}{9}$
6. $\frac{7}{25}, -\frac{3}{4}$

Replace each ● with <, >, or = to make a true statement.

7. $3\frac{3}{7}$ ● $3\frac{4}{9}$
8. $\frac{1}{6}$ ● $0.1\overline{6}$
9. $-\frac{7}{2}$ ● $-\frac{3}{4}$
10. 1.4 ● 1.403
11. −1.808 ● −1.858
12. $10\frac{2}{5}$ ● 10.4
13. 5.92 ● $5\frac{23}{25}$

Order each set of rational numbers from least to greatest.

14. −5, −1, 2, −12, 5
15. $\frac{1}{2}, \frac{4}{5}, \frac{2}{5}, 0$
16. $\frac{3}{8}$, 0.376, 0.367, $\frac{2}{5}$

Exercises

Independent Practice

Replace each ● with <, >, or = to make a true sentence.

17. −4.6 ● −4.58
18. 1.5 ● −1.52
19. 0.88 ● $0.\overline{8}$
20. $\frac{5}{7}$ ● $\frac{9}{21}$
21. $-9\frac{2}{3}$ ● $8\frac{7}{8}$
22. $\frac{4}{5}$ ● $\frac{8}{10}$
23. $11\frac{1}{8}$ ● 11.26
24. 7.47 ● $7\frac{47}{100}$
25. −5.2 ● $5\frac{1}{5}$
26. Using a number line, explain why $-7.4 < -7$.
27. What is the LCD of $\frac{3}{8}$ and $\frac{13}{25}$?

Order each set of rational numbers from least to greatest.

28. 0.056, 0.56, 0.5, 0.06
29. $\frac{1}{9}, \frac{1}{10}, -\frac{1}{3}, -\frac{1}{4}$
30. 1.8, 1.07, $\frac{17}{9}, \frac{18}{9}$
31. 0.182, $0.182\overline{5}$, $0.18\overline{2}$, $0.\overline{18}$

Mixed Review

32. How many quarts are in $3\frac{1}{2}$ gallons? *(Lesson 1-7)*
33. Solve $t = -62 + 47 + (-18) + 22$. Check by solving another way. *(Lesson 3-4)*
34. **Statistics** Find the mean, median, and mode for the data set 14, 3, 6, 8, 11, 9, 3, 2, 7. *(Lesson 4-5)*
35. Find the LCM of 8, 15, and 12. *(Lesson 6-9)*

Problem Solving and Applications

36. **Travel** Sharon has traveled 5 miles of her 16-mile trip, and Bob has traveled 21 miles of his 64-mile trip. Who has completed the greater part of their journey so far?

37. **Portfolio Suggestion** Select an item from this chapter that you feel shows your best work and place it in your portfolio. Attach a note explaining why you selected it.

38. **Critical Thinking** Are there any rational numbers between $0.\overline{4}$ and $\frac{4}{9}$? Explain.

39. **Meteorology** The chart below shows the Monthly Normal Precipitation in inches for several cities in the United States.

City	Feb	Mar	Apr	May	June
Asheville, NC	3.6	5.1	3.8	4.2	4.2
Burlington, VT	1.7	2.2	2.8	3.0	3.6
Columbus, OH	2.2	3.2	3.4	3.8	4.0
Honolulu, HI	2.7	3.5	1.5	1.2	0.5
Kansas City, MO	1.0	2.1	2.7	3.4	4.1
Los Angeles, CA	3.0	2.4	1.2	0.2	0.0
Phoenix, AZ	0.6	0.8	0.3	0.1	0.2
Seattle, WA	4.2	3.6	2.4	1.6	1.4
San Antonio, TX	1.9	1.3	2.7	3.7	3.0

a. What is the median rainfall for April for the cities in the table?

b. Which of the cities listed has the highest median precipitation?

40. **Mathematics and Science** Read the following paragraphs.

Ions are atoms or groups of atoms with positive or negative electrical charges. British physicist Michael Faraday demonstrated that it takes a definite amount of charge to convert an ion of an element into an atom of the element and that the amount of charge depends on the element used.

Faraday's work was the first to imply the electrical nature of matter and the existence of subatomic particles and a fundamental unit of charge. Faraday wrote: "The atoms of matter are in some way endowed or associated with electrical powers, to which they owe their most striking qualities, and amongst them their mutual chemical affinity." Faraday did not, however, conclude that atoms cause electricity.

Which has the greater electrical charge, 3 magnesium and 1 phosphate ion or 2 aluminum and 3 oxide ions?

Electrical Charge of Various Ions

Ion	Charge	Ion	Charge
chloride	−1	magnesium	+2
silver	+1	phosphate	−3
oxide	−2	aluminum	+3

6-11 Scientific Notation

Objective

Express numbers in scientific notation.

Words to Learn

scientific notation

Have you ever been to a ballgame where the national anthem was being played live and also broadcast over a radio system? You could hear it on your radio sooner than you did in person. This is because radio waves travel at 3.0×10^8 meters per second (m/s) while sound waves travel slower at 3.4×10^2 m/s. The numbers 3.0×10^8 and 3.4×10^2 are written in **scientific notation.**

In standard form, the speed of radio waves is written 300,000,000 m/s. So, $3.0 \times 10^8 = 300{,}000{,}000$. Large numbers like 300,000,000 are written in scientific notation to lessen the chance of omitting a zero or misplacing the decimal point.

Notice that when a number is expressed in scientific notation, it is written as a product of a factor and a power of 10. The factor must be greater than or equal to 1 and less than 10.

Mental Math Hint

Multiplying by a positive power of 10 moves the decimal point to the right the same number of places as the exponent.

Example 1

Express 7.821×10^6 in standard form.

$7.821 \times 10^6 = 7.821 \times 1{,}000{,}000$ 7.821000

$= 7{,}821{,}000$ *6 places*

Scientific notation is also used to express very small numbers. Study the patterns in the chart to see when a negative power of 10 is used.

$3.45 \times 10^1 = 34.5$
$3.45 \times 10^0 = 3.45$
$3.45 \times 10^{-1} = 0.345$
$3.45 \times 10^{-2} = 0.0345$
$3.45 \times 10^{-3} = 0.00345$

Mental Math Hint

Multiplying by a negative power of 10 moves the decimal point to the left the same number of places as the absolute value of the exponent.

Example 2

Express 8.3×10^{-5} in standard form.

$8.3 \times 10^{-5} = 8.3 \times \frac{1}{10^5}$

$= 8.3 \times \frac{1}{100{,}000}$

$= 8.3 \times 0.00001$ 00008.3

$= 0.000083$ *5 places*

The Mini-Lab below shows how to enter a number in standard form and have the calculator express it in scientific notation.

Mini-Lab

Work with a partner.

Materials: scientific calculator

- Copy the chart below and use a calculator to express the numbers in scientific notation. Record each display as the product of the first three digits and 10 to the power of the last two digits. For example, 3.45 06 means 3.45×10^6. Record the absolute value of the exponent in the third column.

ENTER	Scientific Notation	Absolute Value of Exponent
3450 EE =		
345 EE =		
34.5 EE =		
3.45 EE =		
0.345 EE =		
0.0345 EE =		

Talk About It

What is the relationship between the number of places the decimal point moved and the absolute value of the exponent?

The relationship you discovered in the Mini-Lab allows you to use mental math to express any number in scientific notation.

Examples

Express each number in scientific notation.

3 12,345,000

$12{,}345{,}000 = 1.2345 \times 10^7$

The decimal point moves 7 places to the left. Divide by 10^7.

4 0.0000375

$0.0000375 = 3.75 \times 10^{-5}$

The decimal point moves 5 places to the right. Divide by 10^{-5}.

Example 5 shows how to enter a number in a calculator with too many digits to fit on the display screen in standard form.

Example 5

Enter 0.000000825 into a calculator.

First write the number in scientific notation. $0.000000825 = 8.25 \times 10^{-7}$

Then enter the number. 8.25 EE 7 +/− The display shows 8.25 −0.7.

Checking for Understanding

Communicating Mathematics

Read and study the lesson to answer each question.

1. **Tell** why 45.6×10^2 and 0.456×10^{-6} are not written in scientific notation.
2. **Write** an example of a number that is written in scientific notation.
3. **Tell** why very large numbers and very small numbers are written in scientific notation.

Guided Practice

Express each number in standard form.

4. 3.45×10^7
5. 8.9×10^{-5}
6. 3.777×10^4

State where the decimal point should be placed in order to express each number in scientific notation. State the power of ten by which you should multiply. Then express the number in scientific notation.

7. 12,300,000
8. 1,230,000
9. 0.000123
10. 12.3
11. 0.0056789
12. 829
13. 0.000007
14. 0.001^2

Exercises

Independent Practice

Express each number in standard form.

15. -9.99×10^{-8}
16. 4.2×10^6
17. 4.2×10^{-6}
18. 2.54×10^3
19. 9.6×10^{-2}
20. 3.853×10^4
21. the distance to the sun, 9.3×10^7 miles

Express each number in scientific notation.

22. 9,700,000
23. 85,420,000
24. 0.0000635
25. 0.000056
26. 3,478
27. 0.0002^2
28. the product of 7,000,000 and 800
29. the product of 0.00008 and 0.0009

Mixed Review

30. Solve $y - 8.3 = 20.9$. Check your solution. *(Lesson 2-3)*
31. Solve $\frac{c}{4} + 8 > 10$. Show the solution on a number line. *(Lesson 2-10)*
32. Solve $-319 - (-98) = w$. *(Lesson 3-5)*
33. Find the LCD of $-\frac{5}{6}$ and $\frac{3}{8}$. *(Lesson 6-10)*

Problem Solving and Applications

34. **Critical Thinking** The galaxy NGC1232 is over 65 million light years from the earth. A light year is 9.46×10^{12} kilometers. Use scientific notation to express the distance to the galaxy in kilometers.

35. **Data Search** Refer to pages 210 and 211.
    a. Which planet is the largest?
    b. How far is this planet from the Sun?
    c. How far is this planet from Earth?

Chapter 6

Study Guide and Review

Communicating Mathematics

Choose the correct term to complete each sentence.

1. The (factors, multiples) of a whole number divide that number with a remainder of zero.
2. A whole number greater than 1 that has exactly two factors is called a (composite, prime) number.
3. A fraction is in (bar notation, simplest form) when the GCF of the numerator and denominator is 1.
4. The least common denominator of two fractions is the (LCM, GCF) of the denominators.
5. An outcome that is certain to happen has a probability of (1, 0).
6. Explain how to express a terminating decimal as a fraction.

Self Assessment

Objectives and Examples	***Review Exercises***
Upon completing this chapter, you should be able to:	*Use these exercises to review and prepare for the chapter test.*

- use divisibility rules for 2, 3, 4, 5, 6, 8, 9, and 10 *(Lesson 6-1)*

  Determine whether 738 is divisible by 9.

  The sum of the digits, 18, is divisible by 9. So 738 is divisible by 9.

Use divisibility rules to determine if the first number is divisible by the second number. Write *yes* or *no*.

7. 523; 3
8. 1,895; 5
9. 328; 4
10. 4,291; 8
11. 16,542; 6
12. 1,001; 10

- find the prime factorization of a composite number *(Lesson 6-2)*

  Factor 60 completely.

$$60 = 2 \cdot 30$$
$$= 2 \cdot 2 \cdot 15$$
$$= 2 \cdot 2 \cdot 3 \cdot 5 \quad \text{or} \quad 2^2 \cdot 3 \cdot 5$$

Find the prime factorization of each number.

13. 48
14. −56
15. 175
16. −252
17. 33
18. −27

- find the greatest common factor of two or more numbers *(Lesson 6-4)*

  Find the GCF of 20 and 32.

  factors of 20: 1, 2, 4, 5, 10, 20
  factors of 32: 1, 2, 4, 8, 16, 32
  The GCF of 20 and 32 is 4.

Find the GCF for each set of numbers.

19. 18, 54
20. 15, 45
21. 14, 28, 49
22. 36, 84, 108
23. 120, 440, 360

Objectives and Examples	Review Exercises
• identify and simplify rational numbers *(Lesson 6-5)* Write $\frac{60}{150}$ in simplest form. $\frac{60}{150} = \frac{\cancel{2} \cdot 2 \cdot \cancel{3} \cdot \cancel{5}}{\cancel{2} \cdot \cancel{3} \cdot \cancel{5} \cdot 5} = \frac{2}{5}$	**Write each fraction in simplest form.** **24.** $-\frac{15}{18}$ **25.** $\frac{12}{16}$ **26.** $\frac{63}{72}$ **27.** $-\frac{42}{63}$
• express fractions as decimals *(Lesson 6-6)* Write $\frac{3}{11}$ as a decimal. $\frac{3}{11} = 3 \div 11$ $= 0.2727272$ or $0.\overline{27}$	**Express each fraction or mixed number as a decimal. Use bar notation if necessary.** **28.** $\frac{3}{8}$ **29.** $\frac{4}{22}$ **30.** $1\frac{2}{5}$ **31.** $6\frac{8}{12}$
• express terminating decimals as fractions *(Lesson 6-6)* Express 0.75 as a fraction. $0.75 = \frac{75}{100} = \frac{3}{4}$	**Express each decimal as a fraction or mixed number in simplest form.** **32.** 0.45 **33.** −0.028 **34.** −11.375 **35.** 4.8125
• express repeating decimals as fractions *(Lesson 6-7)* Express $2.\overline{84}$ as a fraction. Let $N = 2.\overline{84}$. Then $100N = 284.\overline{84}$. $100N = 284.\overline{84}$ $-\ N = 2.\overline{84}$ $99N = 282$ $N = \frac{282}{99}$ or $2\frac{28}{33}$ So, $2.\overline{84} = 2\frac{28}{33}$.	**Express each repeating decimal as a fraction.** **36.** $0.\overline{7}$ **37.** $-5.\overline{28}$ **38.** $-0.\overline{318}$ **39.** $6.\overline{630}$ **40.** $0.\overline{48}$ **41.** $-0.\overline{03}$
• find the probability of a simple event *(Lesson 6-8)* Probability = $\frac{\text{number of ways an event can occur}}{\text{number of possible outcomes}}$	**A bag contains five red, seven blue, and eight white marbles. A blindfolded student draws a marble. Find the probability of each outcome.** **42.** It is white. **43.** It is not red. **44.** It is red or blue.
• find the least common multiple of two or more integers *(Lesson 6-9)* Find the LCM of 6 and 8. multiples of 6: 0, 6, 12, 18, **24**, . . . multiples of 8: 0, 8, 16, **24**, 32, . . . The LCM of 6 and 8 is 24.	**Find the LCM for each set of numbers.** **45.** 15, 20 **46.** 18, 24 **47.** 54, 72 **48.** 8, 20, 24 **49.** 5, 174, 30

Objectives and Examples

- compare and order rational numbers expressed as fractions and/or decimals *(Lesson 6-10)*

Is $0.65 < \frac{2}{3}$?

$\frac{2}{3} = 0.\overline{66}$, and $0.65 < 0.\overline{66}$

So, yes, $0.65 < \frac{2}{3}$.

Review Exercises

Replace each ● with <, >, or = to make a true sentence.

50. -10.29 ● -10.3

51. $\frac{11}{15}$ ● $\frac{4}{5}$

52. $\frac{3}{8}$ ● 0.375

53. $0.\overline{4}$ ● 0.4

- express numbers in scientific notation *(Lesson 6-11)*

Express 0.000294 in scientific notation.

$0.000294 \rightarrow 2.94 \times 10^{-4}$

4 places

Express each number in scientific notation.

54. 5,830,000
55. 0.0000735
56. 12,500
57. 0.00068
58. 95,700,000

Applications and Problem Solving

59. Eva and Karl are saving money to attend an art show at the museum. Travel and tickets cost $31.50 in all. Eva starts with $10.15 and saves $1.60 a week. Karl starts with $8.30 and saves $1.75 a week. When will they save enough? *(Lesson 6-3)*

60. **Science** Radio waves travel at 3.0×10^8 m/s. Sound waves travel at 3.4×10^2 m/s. Write each number in standard form and find the difference in the two speeds. *(Lesson 6-11)*

Curriculum Connection Projects

- **Geography** Consult an atlas or almanac and express the populations of ten countries in scientific notation.
- **Life Science** Take a survey of shoe sizes in your class. Find the probability of a new student having the same size as the most frequent size of your classmates.

Read More About It

Gordon, A. C. *Solve-a-Crime.*
Wells, H. G. *The Invisible Man.*
White, Lawrence B. Jr. and Ray Broekel *Math-a-Magic: Number Tricks for Magicians.*

Chapter

6 Test

1. Is 718 divisible by four? Why or why not?

Find the GCF for each set of numbers.

2. 40, 24
3. 56, 98
4. 108, 234, 30
5. 320, 16, 176

Write each fraction in simplest form.

6. $-\frac{10}{16}$
7. $\frac{42}{72}$
8. $-\frac{18}{81}$
9. $\frac{90}{21}$

10. Write the prime factorization of 360.

Write each decimal as a fraction or mixed number and each fraction as a decimal.

11. -3.45
12. $\frac{5}{16}$
13. 0.4
14. $-\frac{3}{8}$
15. 20.8125

Express each repeating decimal as a fraction.

16. $-3.\overline{2}$
17. $0.\overline{621}$
18. $5.\overline{28}$

The letters of the word "composite" are written one each on nine identical slips of paper and shuffled in a bag. A blindfolded student draws one slip of paper. Find the probability of each outcome.

19. P(c)
20. P(o or p)
21. P(vowel)
22. P(not o or t)

23. Express 18.42727... using bar notation.
24. Use prime factorization to find the LCM of 24 and 28.

Find the LCM for each set of numbers.

25. 9, 21
26. 12, 15, 18
27. 81, 34, 54

Replace each ● with <, >, or = to make a true sentence.

28. $\frac{4}{9}$ ● $\frac{11}{27}$
29. $5\frac{31}{100}$ ● 5.31
30. -4.68 ● -4.7

31. Express 3.7×10^{-4} in standard form.
32. Express 58,930,000 in scientific notation.
33. The CLV company supplies CDs, LPs, and Vs (videos) to five stores. Records show that 426 CDs, 152 LPs, and 102 Vs are in the stores. CLV delivers 509 CDs, 209 LPs, and 205 Vs. Later the stores report 453 CDs, 183 LPs, 198 Vs in stock. How much of each product was sold?

Bonus Is it possible for three different numbers to have 1 as their GCF? If so, name three such numbers.

Chapter Test

Chapter 6

Standard Format, Chapters 1-6

Academic Skills Test

Academic Skills Test

Directions: Choose the best answer. Write A, B, C, or D.

1. To find the difference of 3.4 and 1.8,

   A subtract 2 from 3.2.
   B subtract 2 from 3.6.
   C subtract 2.2 from 3.
   D subtract 1.2 from 4.

2. Three friends will equally share the cost of a $28.95 board game. Which is *not* a reasonable estimate of each person's share?

   A $9.90
   B $9.65
   C $8.80
   D All are reasonable.

3. 96 yd = __?__ ft

   A 960 B 288
   C 96 D 32

4. If $b = 5$ and $c = 8$, what is the value of $b(10 - c)$?

   A 10 B 42
   C 50 D 52

5. If $4.2 = \frac{x}{3}$, what is the value of x?

   A 1.2 B 1.26
   C 1.4 D 12.6

6. On a math test, Lee scored 10 points less than twice the lowest score. If his score was 96, what was the lowest score?

   A 86 B 53
   C 48 D 43

7. $-3 + 8 + (-3) + 1 =$

   A −15 B 3
   C 9 D 15

8. $-145 - 86 =$

   A 231 B 59
   C −59 D −231

9. An airplane descended 250 feet in 5 minutes. How can you find its average change in altitude per minute?

   A Divide −250 by 5.
   B Multiply 250 by 5.
   C Subtract −25 from 250.
   D Not enough information given to solve.

10. The line plot shows the heights in inches of the Washington Middle School girls' basketball team.

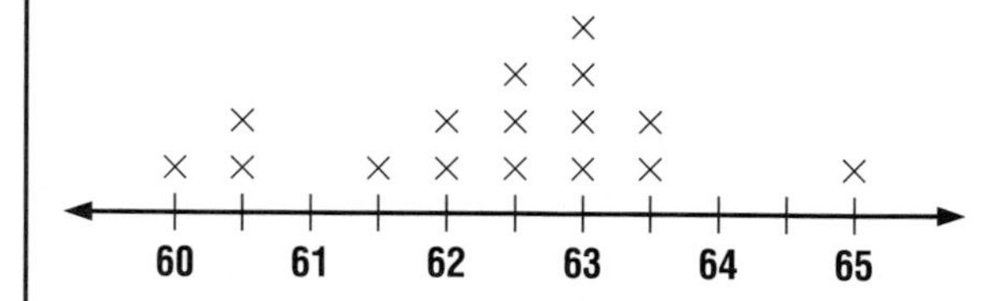

Which sentence best describes the data?

   A The range is 6 inches.
   B Most of the players are 63 inches tall.
   C There are 8 players.
   D Most of the players are between 62 and 64 inches tall.

11. These are the daily low temperatures (°F) for two weeks in April.
36, 42, 38, 50, 48, 44, 46, 50, 52, 49, 48, 45, 46, 48
What is the median temperature?

A 44° B 46°
C 47° D 48°

12. What is the interquartile range of the data in Exercise 11?

A 2 B 5
C 12 D 16

13. Which figure could contain exactly one right angle?

A acute triangle
B isosceles triangle
C obtuse triangle
D rectangle

14. How many lines of symmetry does a square have?

A none B 1
C 2 D 4

15. How many pairs of congruent triangles are formed by the diagonals of rectangle *ABCD?*

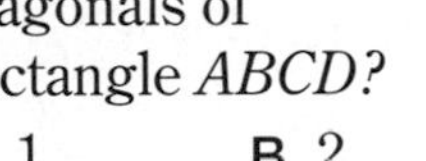

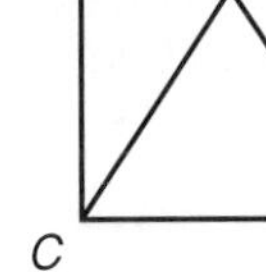

A 1 B 2
C 3 D 4

16. If ΔJKL is similar to ΔQRP, what is the value of x?

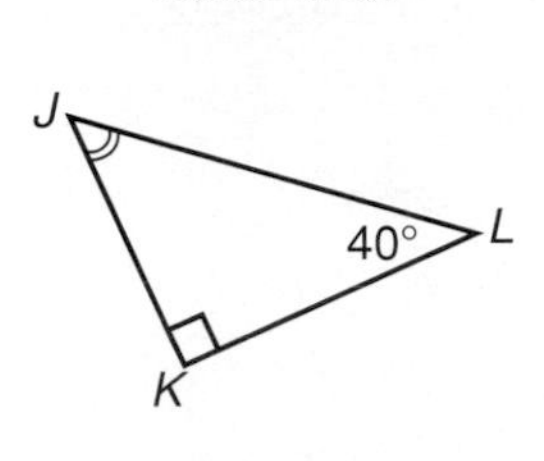

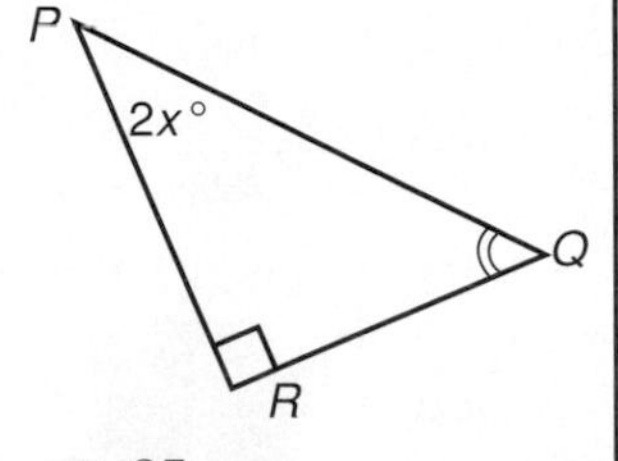

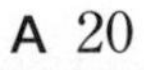

A 20 B 25
C 40 D 50

Test-Taking Tip

When you prepare for a standardized test, review basic definitions such as the ones below.

- A number is *divisible* by another number if it can be divided with 0 as a remainder. A number is divisible by each of its *factors.*
- A number is *prime* if it has exactly two factors, itself and 1. 5 is a prime number. 12 is not prime since it has factors other than itself and 1. 2, 3, 4, and 6 are also factors of 12.

17. What is the least prime factor of 54?

A 1 B 2
C 3 D 6

18. The number −13.6 belongs to what set(s) of numbers?

A rational numbers
B whole numbers and rational numbers
C integers and rational numbers
D none of these

19. What is the probability of the spinner landing on a prime number?

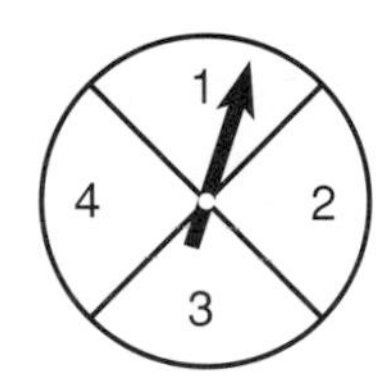

A 0.25 B 0.5
C 0.75 D 1

20. A red blood cell is about 0.00075 cm long. How is this measure expressed in scientific notation?

A 0.75×10^{-3}
B 7.5×10^{-4}
C 7.5×10^{-5}
D 75×10^{-5}

Chapter

Rational Numbers

Spotlight on Baseball

Have You Ever Wondered...

- How a player's batting average is figured?
- What the fraction $\frac{11}{31}$ means when it describes the number of left-handed Cy Young Award winners?

LEADING BATTERS IN BASEBALL HALL OF FAME

Player	Year Inducted	Years Played	At Bats	Hits	Batting Avg.
Ty Cobb	1936	24	11,436	4,190	.366
Rogers Hornsby	1942	23	8,173	2,930	.358
Ed Delhanty	1945	16	7,493	1,593	.346
Willie Keeler	1939	19	8,570	2,955	.345
Billy Hamilton	1961	14	6,262	2,157	.344
Ted Williams	1966	19	7,706	2,654	.344
Tris Speaker	1937	22	10,208	3,515	.344
Dan Brouthers	1945	19	6,682	2,288	.342
Jessee Burkett	1946	16	8,389	2,872	.342
Babe Ruth	1936	22	8,399	2,873	.342
Harry Heilmann	1952	17	7,787	2,660	.342
Bill Terry	1954	14	6,428	2,193	.341
George Sisler	1939	15	8,267	2,812	.340
Lou Gehrig	1939	17	8,001	2,721	.340

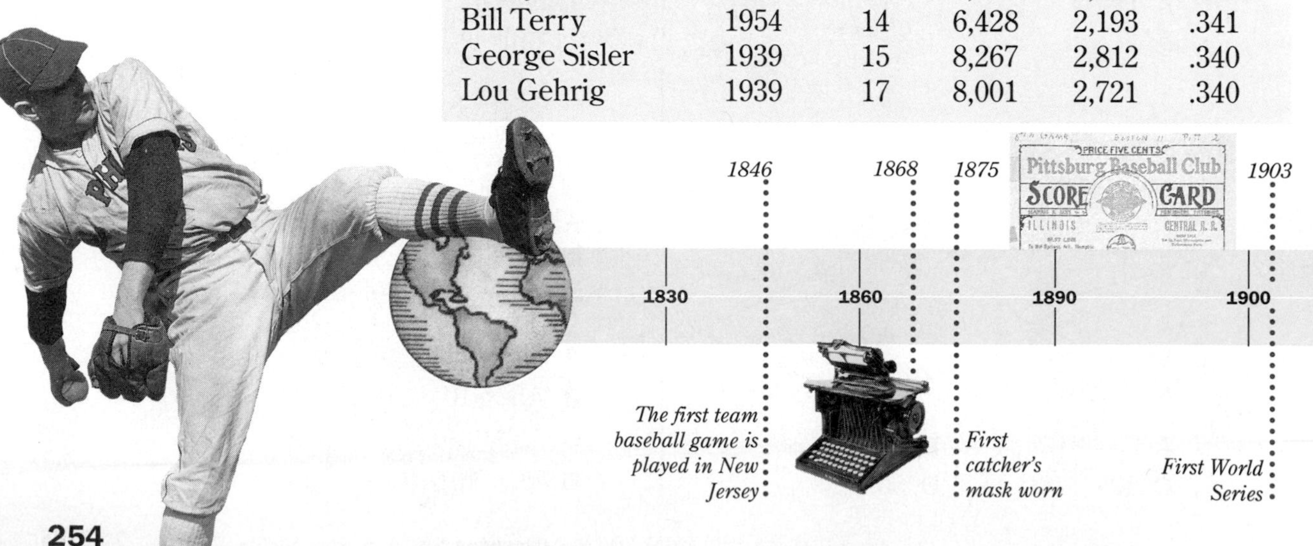

Looking Ahead

In this chapter, you will see how mathematics can be used to answer questions about baseball. The major objectives of the chapter are to:

- add, subtract, multiply, and divide fractions
- solve problems by finding and extending a pattern
- find the areas of triangles, trapezoids, and circles
- solve equations involving rational numbers

Chapter Project

Baseball

Work in a group.

1. Choose several athletes currently playing a sport.
2. Follow their statistics for the next month and present them in a creative chart.
3. Use fractions to describe each athlete's performance as a part of the season.

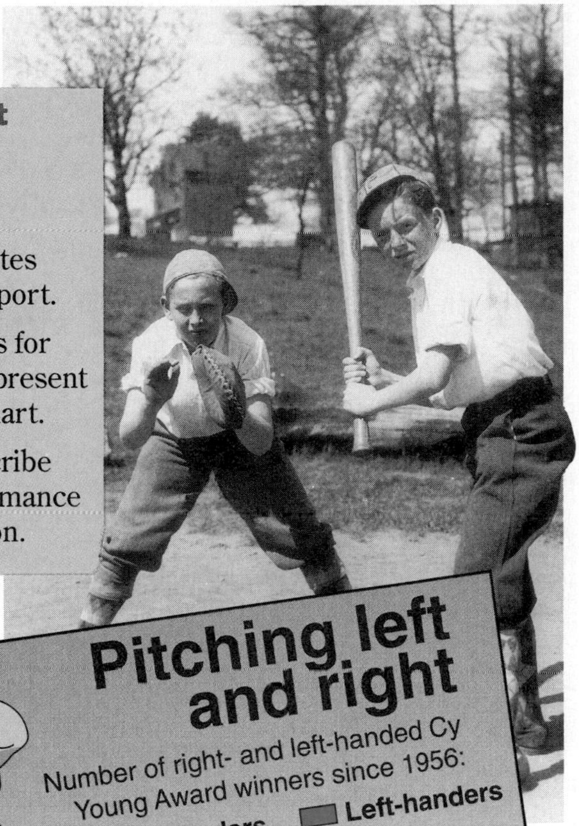

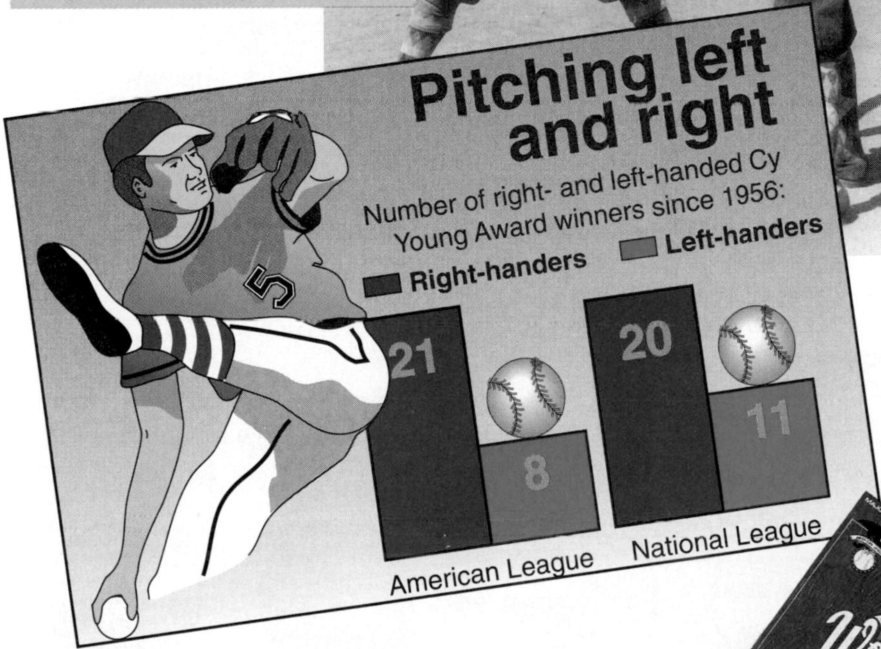

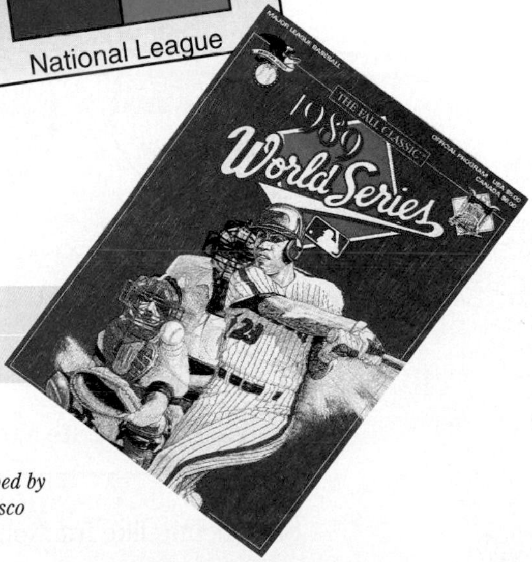

Pitching left and right

Number of right- and left-handed Cy Young Award winners since 1956:

	Right-handers	Left-handers
American League	21	8
National League	20	11

1930 — 1960 — 1990

1937 Golden Gate Bridge completed in San Francisco

1974 Hank Aaron beats Babe Ruth's lifetime home-run record

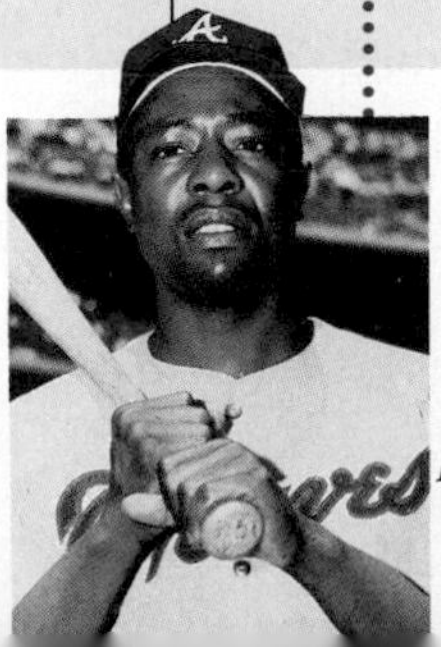

1989 World Series stopped by major San Francisco earthquake

7-1 Adding and Subtracting Like Fractions

Objective

Add and subtract fractions with like denominators.

Words to Learn

mixed number

The electric eel, found in Venezuela and Brazil, is really not an eel, but a greenish-black fish 2 to 4 feet long. The inner organs of the eel are located in the first $\frac{1}{5}$ of its body. The remaining $\frac{4}{5}$ contain the organs that produce an electric current.

Notice that the rational numbers $\frac{1}{5}$ and $\frac{4}{5}$ have like denominators. Fractions with like denominators are also called *like fractions*.

Adding Like Fractions	**In words:** To add fractions with like denominators, add the numerators.	
	Arithmetic	**Algebra**
	$\frac{2}{7} + \frac{3}{7} = \frac{5}{7}$	$\frac{a}{c} + \frac{b}{c} = \frac{a+b}{c}, c \neq 0$

LOOKBACK

You can review simplifying fractions on page 225.

Sometimes the sum of two fractions is greater than 1. When this happens, we usually write the sum as a mixed number in simplest form. A **mixed number** is the sum of a whole number and a fraction.

Example 1

John ate $\frac{3}{4}$ of a pizza and Jeanne ate $\frac{3}{4}$ of a pizza. How much pizza did they eat together?

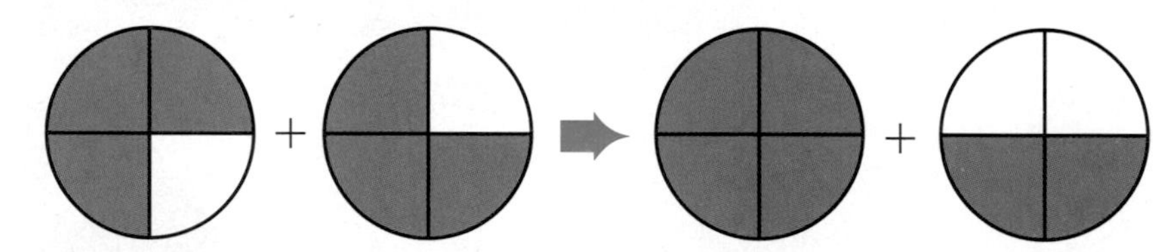

$\frac{3}{4} + \frac{3}{4} = \frac{6}{4}$ *Add the numerators.*

$= 1\frac{2}{4}$ *Rename $\frac{6}{4}$ as a mixed number, $1\frac{2}{4}$.*

$= 1\frac{1}{2}$ *Write the mixed number in simplest form.*

John and Jeanne ate a total of $1\frac{1}{2}$ pizzas.

Subtracting like fractions is similar to adding them.

Subtracting Like Fractions	**In words:** To subtract fractions with like denominators, subtract the numerators.
	Arithmetic $\frac{3}{5} - \frac{2}{5} = \frac{1}{5}$ **Algebra** $\frac{a}{c} - \frac{b}{c} = \frac{a-b}{c}, c \neq 0$

Rational numbers include positive and negative fractions. Use the rules for adding and subtracting integers to determine the sign of the sum or difference of two rational numbers.

Example 2

Solve $a = -\frac{5}{6} - \left(-\frac{7}{6}\right)$.

$-\frac{5}{6}$, $\frac{-5}{6}$, and $\frac{5}{-6}$ all name the same rational number.

$a = -\frac{5}{6} - \left(-\frac{7}{6}\right)$ *Since the denominators are the same, subtract the numerators.*

$a = \frac{-5 - (-7)}{6}$

$a = \frac{2}{6}$ or $\frac{1}{3}$ *Simplify.*

Example 3 ***Connection***

Algebra Simplify $3\frac{1}{4}x + \frac{3}{4}x - 2\frac{3}{4}x$.

$3\frac{1}{4}x + \frac{3}{4}x - 2\frac{3}{4}x = \left(3\frac{1}{4} + \frac{3}{4} - 2\frac{3}{4}\right)x$ *Distributive Property*

$= \left(3\frac{4}{4} - 2\frac{3}{4}\right)x$ *Add $3\frac{1}{4}$ and $\frac{3}{4}$.*

$= 1\frac{1}{4}x$

Checking for Understanding

Communicating Mathematics

Read and study the lesson to answer each question.

1. **Write** the addition sentence shown by the following model.

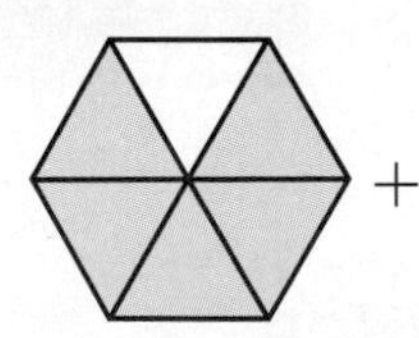 + 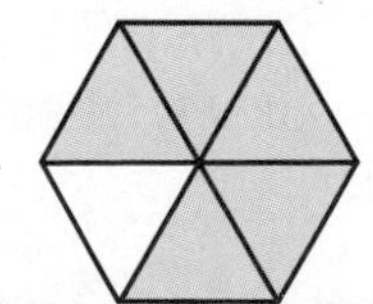 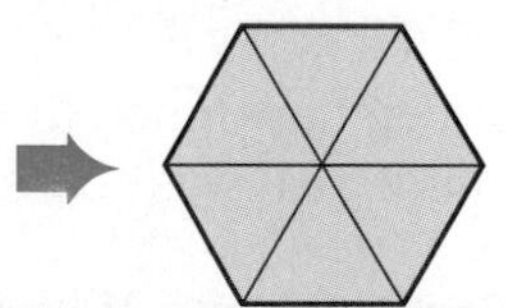 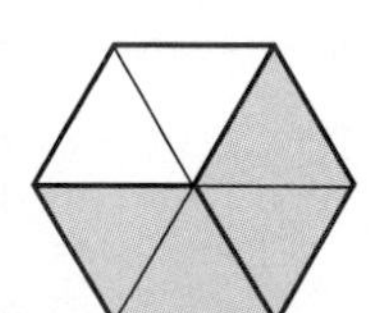

2. **Draw** a model to show the sum of $\frac{5}{8}$ and $\frac{5}{8}$. Write the sum as a mixed number.
3. **Tell** a simple rule for adding and subtracting like fractions.

Guided Practice

Solve each equation. Write each solution in simplest form.

4. $\frac{3}{8} + \frac{1}{8} = d$
5. $\frac{3}{5} + -\frac{2}{5} = x$
6. $\frac{36}{21} - \frac{8}{21} = y$
7. $\frac{3}{16} + \frac{15}{16} = m$
8. $s = -\frac{4}{5} - \left(-\frac{3}{5}\right)$
9. $r = \frac{9}{7} - \frac{5}{7}$

Exercises

Independent Practice

Solve each equation. Write each solution in simplest form.

10. $\frac{5}{8} + \frac{1}{8} = n$
11. $\frac{17}{9} + \left(-\frac{1}{9}\right) = y$
12. $b = -\frac{7}{12} - \frac{5}{12}$
13. $z = \frac{1}{18} - \frac{7}{18}$
14. $5\frac{3}{4} + 2\frac{3}{4} = g$
15. $c = \frac{21}{8} - \frac{49}{8}$
16. $m = -\frac{3}{5} - \left(-\frac{4}{5}\right)$
17. $-2\frac{5}{9} - \frac{5}{9} = r$
18. $k = -\frac{2}{3} - \frac{1}{3}$
19. Find the sum of $\frac{5}{6}$ and $-\frac{1}{6}$.
20. Find the sum of $-\frac{5}{12}$ and $-\frac{1}{12}$.
21. Subtract $-\frac{5}{8}$ from $-\frac{1}{8}$.
22. Subtract $-2\frac{1}{2}$ from $5\frac{1}{2}$.
23. Find the sum of $-\frac{25}{6}$ and $\frac{7}{6}$.
24. Subtract $-\frac{7}{9}$ from $\frac{4}{9}$.

Evaluate each expression if $c = -\frac{3}{5}$ and $d = \frac{12}{5}$.

25. $d - c$
26. $c + d$
27. $c - d$

Simplify each expression.

28. $4\frac{1}{2}y + \frac{1}{2}y - 2\frac{1}{2}y$
29. $-2\frac{1}{3}n + \left(-\frac{2}{3}n\right) + 4n$
30. $\frac{9}{5}r + \left(-\frac{1}{5}r\right) - \frac{3}{5}r$

Mixed Review

31. How many tons are in 7,000 pounds? *(Lesson 1-7)*
32. Solve $-42 \div (-14) = p$. *(Lesson 3-7)*
33. **Geometry** Determine if the figure at the right has rotational symmetry. *(Lesson 5-5)*

34. Express 52,380,000 in scientific notation. *(Lesson 6-11)*

Problem Solving and Applications

35. **Football** The first touchdown in a football game was made $3\frac{1}{4}$ minutes after the opening kickoff. Another $5\frac{3}{4}$ minutes passed before a field goal was made. If the clock started counting down from 15 minutes, how much time was left when the field goal was scored?

36. **Critical Thinking** Which is greater, $\frac{x}{1}$ or $\frac{1}{x}$? Use examples to defend your answer.
37. **Probability** A number cube is rolled.
    a. What is the probability of rolling a 3 or a 6?
    b. What is the probability of *not* rolling a 3 or a 6?
    c. Find the sum of the answers to Exercises **a** and **b**.

38. **Journal Entry** Write a few sentences that would explain to a younger student how to add and subtract fractions with like denominators.

7-2 Adding and Subtracting Unlike Fractions

Objective

Add and subtract fractions with unlike denominators.

Johann Strauss, Jr. (1825-1899) is known as the Waltz King. A waltz is written in $\frac{3}{4}$ time. This means there are three beats in a measure and the quarter note gets one beat. The value of the notes can be expressed as fractions and the total value of each measure is $\frac{3}{4}$. The first few measures of *The Laughing Song* are shown below with the values of each note. What type of note must be used to finish the last measure?

DID YOU KNOW

The word *orchestra* means dancing place. In ancient Greek theaters, dancers and musicians performed on a space between the audience and the stage.

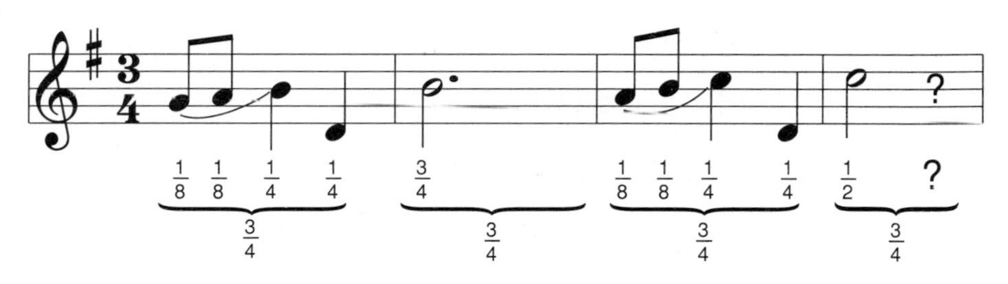

To find the missing note, subtract $\frac{1}{2}$ from $\frac{3}{4}$. The fractions do not have the same denominator, but $\frac{1}{2}$ can be renamed with a denominator of 4.

$\frac{3}{4} - \frac{1}{2} = \frac{3}{4} - \frac{2}{4}$ *Rename $\frac{1}{2}$ as $\frac{2}{4}$.*

$= \frac{1}{4}$ *The difference, $\frac{1}{4}$, can be read as "one quarter."*

A note with a value of $\frac{1}{4}$ could be used to finish the last measure.

Adding and Subtracting Unlike Fractions	To find the sum or difference of two fractions or mixed numbers with unlike denominators, rename the fractions with a common denominator. Then add or subtract and simplify.

Example 1

The least common denominator (LCD) is the least common multiple of the denominators. The LCD is helpful in renaming fractions for adding and subtracting.

Solve $a = \frac{5}{6} + \frac{7}{9}$. Write the solution in simplest form.

$a = \frac{5}{6} + \frac{7}{9}$ *$6 = 2 \cdot 3$ and $9 = 3^2$. The LCM of 6 and 9 is $2 \cdot 3^2$ or 18.*

$a = \frac{15}{18} + \frac{14}{18}$ *$\frac{5 \cdot 3}{6 \cdot 3} = \frac{15}{18}$ and $\frac{7 \cdot 2}{9 \cdot 2} = \frac{14}{18}$*

$a = \frac{29}{18}$ or $1\frac{11}{18}$ *Rename $\frac{29}{18}$ as $1\frac{11}{18}$.*

Example 2

Solve $y = \frac{1}{4} - \left(-\frac{5}{8}\right)$.

$y = \frac{1}{4} - \left(-\frac{5}{8}\right)$

$y = \frac{2}{8} - \left(-\frac{5}{8}\right)$ *Use the LCM of 4 and 8 to rename $\frac{1}{4}$ as $\frac{2}{8}$.*

$y = \frac{2}{8} + \frac{5}{8}$ *Subtract $-\frac{5}{8}$ by adding its inverse, $\frac{5}{8}$.*

$y = \frac{2+5}{8}$ or $\frac{7}{8}$ *Add the numerators.*

To add or subtract mixed numbers with unlike denominators, first rename the fractions with a common denominator.

Examples

3 Solve $d = 3\frac{7}{8} + 5\frac{5}{24}$. *Estimate: 4 + 5 = 9*

$d = 3\frac{7}{8} + 5\frac{5}{24}$

$d = 3\frac{21}{24} + 5\frac{5}{24}$ *Use the LCM of 8 and 24 to rename $\frac{7}{8}$ as $\frac{21}{24}$.*

$d = 8\frac{26}{24}$ *Add the whole numbers. Then add the fractions.*

$d = 9\frac{2}{24}$ or $9\frac{1}{12}$ *Simplify.*

4 Solve $5\frac{1}{3} - 3\frac{3}{4} = m$. *Estimate: 5 − 4 = 1*

In Example 4, why is it necessary to rename $5\frac{4}{12}$ as $4\frac{16}{12}$?

$$\begin{array}{r} 5\frac{1}{3} \\ -3\frac{3}{4} \\ \hline \end{array} \rightarrow \begin{array}{r} 5\frac{4}{12} \\ -3\frac{9}{12} \\ \hline \end{array} \rightarrow \begin{array}{r} 4\frac{16}{12} \\ -3\frac{9}{12} \\ \hline 1\frac{7}{12} \end{array}$$

Rename $5\frac{4}{12}$ as $4\frac{16}{12}$.

So, $1\frac{7}{12} = m$.

Checking for Understanding

Communicating Mathematics

Read and study the lesson to answer each question.

1. **Tell** the first step you should take when adding and subtracting fractions with unlike denominators.
2. **Tell** why you might rename $3\frac{5}{8}$ as $2\frac{13}{8}$ in a subtraction problem.

Guided Practice

Complete.

3. $7\frac{2}{5} = 6\frac{\blacksquare}{5}$
4. $3\frac{1}{4} = 2\frac{\blacksquare}{4}$
5. $4\frac{5}{12} = 3\frac{\blacksquare}{12}$
6. $9\frac{3}{8} = 8\frac{\blacksquare}{8}$

Solve each equation. Write each solution in simplest form.

7. $\frac{2}{3} - \frac{3}{4} = a$
8. $x = \frac{1}{2} + \frac{2}{3}$
9. $\frac{1}{2} + \left(-\frac{7}{8}\right) = y$
10. $h = 3\frac{1}{2} - 2\frac{2}{9}$
11. $5 - 3\frac{1}{3} = k$
12. $g = 9\frac{1}{3} - 2\frac{1}{2}$

Exercises

Independent Practice

Solve each equation. Write each solution in simplest form.

13. $h = \frac{1}{2} + \frac{4}{5}$
14. $-\frac{7}{10} + \frac{1}{5} = f$
15. $j = -\frac{3}{4} + \left(-\frac{1}{3}\right)$
16. $5\frac{1}{5} - \left(-2\frac{7}{10}\right) = z$
17. $w = -4\frac{1}{8} - 5\frac{1}{4}$
18. $9 - 6\frac{1}{4} = x$
19. $7\frac{3}{4} - \left(-1\frac{1}{8}\right) = y$
20. $t = 3\frac{2}{3} + \left(-5\frac{3}{4}\right)$
21. $-8\frac{5}{9} - 2\frac{1}{6} = m$
22. Find the sum of $\frac{7}{6}$ and $\frac{5}{18}$.
23. What is $2\frac{1}{2}$ less than $-8\frac{1}{5}$?

Evaluate each expression if $a = \frac{5}{8}$, $b = 3\frac{11}{12}$, and $c = -\frac{5}{9}$.

24. $c - b$
25. $a + b + c$
26. $a - c$
27. $b - (-c)$

Mixed Review

28. Solve $16 + n = 43$. *(Lesson 2-3)*
29. **Statistics** Find the median and upper and lower quartiles for the following set of data. 128, 140, 132, 146, 120 *(Lesson 4-6)*
30. Write $\frac{24}{32}$ in simplest form. *(Lesson 6-5)*
31. Find the sum of $-5\frac{1}{6}$ and $2\frac{5}{6}$. *(Lesson 7-1)*

Problem Solving and Applications

32. **Carpentry** The Cabinet Shop made a desktop by gluing a sheet of $\frac{1}{16}$-inch oak veneer to a sheet of $\frac{3}{4}$-inch plywood. What was the total thickness of the desktop?
33. **Critical Thinking** A drainpipe is cut in half and one-half is used. Then one-fifth of the remaining pipe is cut off and used. The piece left is 12 feet long. How long was the pipe originally?
34. **Publishing** The width of a page of a newspaper is $13\frac{3}{4}$ inches. The left margin is $\frac{7}{16}$ inch and the right margin is $\frac{1}{2}$ inch. What is the width of the page inside the margins?

35. **Journal Entry** Write a sentence or two explaining how to subtract fractions that have different denominators.

7-3 Multiplying Fractions

Objective
Multiply fractions.

The length of a flag is called the *fly*, and the width is called the *hoist*. The blue rectangle in the United States flag is called the *union*. The length of the union is $\frac{2}{5}$ of the fly and the width is $\frac{7}{13}$ of the hoist. If the fly of a United States flag is 3 feet, how long is the union?

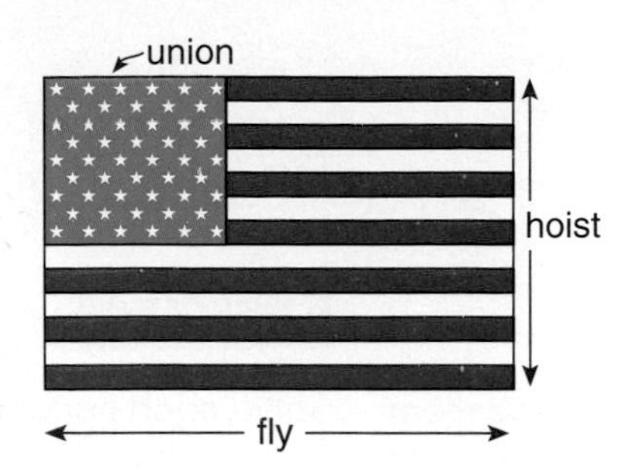

The largest United States flag measures 64 by 125 meters. It is kept in the White House and brought out for display each Flag Day.

The length of the union would be $\frac{2}{5}$ of 3 feet. You can multiply to find the actual length.

$$\frac{2}{5} \cdot 3 = \frac{2}{5} \cdot \frac{3}{1}$$ *Rewrite the whole number as a fraction.*

$$= \frac{2 \cdot 3}{5 \cdot 1}$$ *Multiply the numerators. Multiply the denominators.*

$$= \frac{6}{5}$$

$$= 1\frac{1}{5}$$ *Rename as a mixed number.*

The width of the union is $1\frac{1}{5}$ feet.

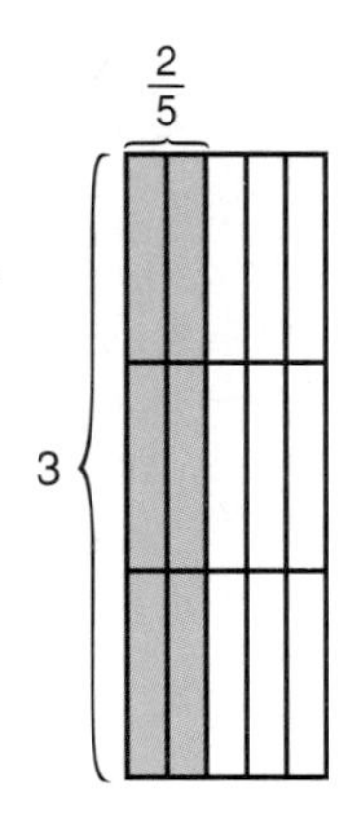

Multiplying Fractions

In words: To multiply fractions, multiply the numerators and multiply the denominators.

Arithmetic	**Algebra**
$\frac{2}{5} \cdot \frac{1}{3} = \frac{2}{15}$	$\frac{a}{b} \cdot \frac{c}{d} = \frac{ac}{bd}$; $b \neq 0, d \neq 0$

Example

1 Solve $m = \frac{3}{5} \cdot \frac{4}{9}$.

Method 1

Multiply first.
Then simplify.

$m = \frac{3}{5} \cdot \frac{4}{9}$

$m = \frac{3 \cdot 4}{5 \cdot 9}$ *Multiply.*

$m = \frac{12}{45}$ or $\frac{4}{15}$ *Simplify.*

Method 2

Divide common factors.
Then multiply.

$m = \frac{3}{5} \cdot \frac{4}{9}$

$m = \frac{\overset{1}{\cancel{3}}}{5} \cdot \frac{4}{\underset{3}{\cancel{9}}}$ *The GCF of 3 and 9 is 3. Divide 3 and 9 by 3.*

$m = \frac{1 \cdot 4}{5 \cdot 3}$ or $\frac{4}{15}$ *Multiply.*

Use the rules of signs for multiplying integers when you multiply rational numbers.

Example 2

Solve $-5\frac{1}{2} \cdot 2\frac{2}{3} = y.$

$-5\frac{1}{2} \cdot 2\frac{2}{3} = y$ *Estimate: $-5 \cdot 3 = -15$*

$\frac{-11}{2} \cdot \frac{8}{3} = y$ *Rename $-5\frac{1}{2}$ as $\frac{-11}{2}$ and $2\frac{2}{3}$ as $\frac{8}{3}$.*

$\frac{-11}{\cancel{2}_1} \cdot \frac{\cancel{8}^4}{3} = y$ *Divide 8 and 2 by their GCF, 2.*

$\frac{-44}{3} = y$ *The product of two rational numbers with different signs is negative.*

$-14\frac{2}{3} = y$ *Compare with the estimate.*

LOOKBACK

You can review exponents on page 35.

The rules for exponents that were stated for whole numbers also hold for rational numbers. For example, the expression $\left(\frac{2}{3}\right)^2$ means $\frac{2}{3} \cdot \frac{2}{3}$.

Example 3 ***Connection***

Algebra Evaluate $(ab)^2$ if $a = -\frac{1}{2}$ and $b = \frac{5}{6}$.

$(ab)^2 = \left(-\frac{1}{2} \cdot \frac{5}{6}\right)^2$

$(ab)^2 = \left(-\frac{5}{12}\right)^2$

$(ab)^2 = \left(-\frac{5}{12}\right)\left(-\frac{5}{12}\right)$

$(ab)^2 = \frac{25}{144}$

Checking for Understanding

Communicating Mathematics

Read and study the lesson to answer each question.

1. **Write** the product of $\frac{1}{3}$ and $\frac{3}{4}$.
2. **Tell** how the model at the right shows the product of $\frac{1}{3}$ and $\frac{3}{4}$.

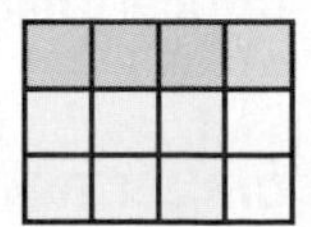

3. **Draw** a model that shows the product of $\frac{3}{5}$ and $\frac{2}{3}$.

Guided Practice Solve each equation. Write each solution in simplest form.

4. $\frac{3}{5} \cdot \frac{5}{8} = a$
5. $\frac{5}{9} \cdot \frac{8}{15} = r$
6. $y = \frac{4}{9} \cdot 2$
7. $-4\frac{1}{2}\left(-\frac{2}{3}\right) = j$
8. $5\frac{1}{3} \cdot \frac{5}{12} = g$
9. $d = -2\left(\frac{3}{8}\right)$
10. $\frac{15}{16} \cdot 3\frac{3}{5} = k$
11. $-2\frac{1}{4} \cdot \frac{2}{3} = h$
12. $-2\frac{2}{5}\left(-1\frac{3}{4}\right) = x$
13. Find the product of $-\frac{1}{3}$ and $2\frac{1}{4}$.

Exercises

Independent Practice Solve each equation. Write each solution in simplest form.

14. $\frac{2}{7} \cdot \frac{3}{4} = p$
15. $-\frac{5}{12} \cdot \frac{8}{9} = n$
16. $\frac{1}{19}\left(-\frac{15}{16}\right) = x$
17. $-3\frac{3}{8} \cdot -\frac{5}{6} = y$
18. $z = \frac{1}{6} \cdot 1\frac{3}{5}$
19. $k = (-6)\left(2\frac{1}{4}\right)$
20. $-3\frac{1}{3}\left(-1\frac{1}{5}\right) = d$
21. $h = 8\frac{1}{4} \cdot 3\frac{1}{3}$
22. $6\left(-7\frac{1}{2}\right) = m$
23. $x = \left(\frac{2}{3}\right)^2$
24. $b = \left(-\frac{8}{13}\right)^2$
25. $5 \cdot \left(\frac{4}{5}\right)^2 = t$

Evaluate each expression if $a = \frac{4}{5}$, $d = -3\frac{3}{4}$, $r = \frac{1}{2}$ and $p = -1\frac{1}{3}$.

26. ar
27. $2d$
28. p^2
29. $r^2(-d)$

Mixed Review

30. Find the perimeter and area of the figure at the right. *(Lesson 2-9)*

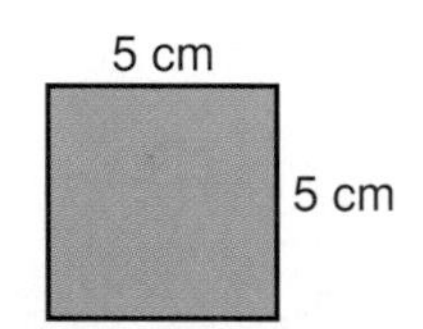

31. Solve $d = 56 + (-18)$. *(Lesson 3-3)*
32. Order the set of rationals $\left\{\frac{1}{2}, -\frac{1}{5}, \frac{1}{3}, \frac{1}{8}, -\frac{1}{10}\right\}$ from least to greatest. *(Lesson 6-10)*
33. Find the sum of $-1\frac{3}{8}$ and $-3\frac{5}{12}$. *(Lesson 7-2)*

Problem Solving and Applications

34. **Geometry** In the United States flag, the fly is $1\frac{9}{10}$ times the hoist.
    a. What is the area of the flag if the hoist is 18 meters?
    b. What is the area of the union if the hoist is $9\frac{1}{2}$ feet?

35. **Critical Thinking** How is $\left(-\frac{3}{4}\right)^2$ different from $-\left(\frac{3}{4}\right)^2$?
36. **Critical Thinking** In multiplying a negative number by $\frac{2}{3}$, will the product be greater or less than the original number? Explain.

7-4 Properties of Rational Numbers

Objective

Identify and use rational number properties.

Words to Learn

multiplicative inverse
reciprocal

The Alaskan brown bear is one of the largest bears in the world. The average Alaskan brown bear is about $1\frac{1}{8}$ times as long as a grizzly bear. If the average grizzly bear is 8 feet long, how long is an Alaskan brown bear? *You will solve the bear problem in Example 1.*

You can use fraction properties to find the length mentally. All of the properties that were true for addition and multiplication of integers are also true for addition and multiplication of rationals.

Property	Arithmetic	Algebra
Commutative	$\frac{2}{3} + \frac{1}{2} = \frac{1}{2} + \frac{2}{3}$ $\frac{1}{4} \cdot \frac{3}{5} = \frac{3}{5} \cdot \frac{1}{4}$	$a + b = b + a$ $a \cdot b = b \cdot a$
Associative	$\left(-\frac{1}{4} + \frac{2}{3}\right) + \frac{1}{2} = -\frac{1}{4} + \left(\frac{2}{3} + \frac{1}{2}\right)$ $\left(\frac{1}{3} \cdot \frac{5}{6}\right) \cdot \frac{3}{8} = \frac{1}{3} \cdot \left(\frac{5}{6} \cdot \frac{3}{8}\right)$	$(a + b) + c = a + (b + c)$ $(a \cdot b) \cdot c = a \cdot (b \cdot c)$
Identity	$\frac{2}{3} + 0 = \frac{2}{3}$ $\quad \frac{4}{5} \cdot 1 = \frac{4}{5}$	$a + 0 = a$ $a \cdot 1 = a$

In Chapter 3, you learned about the inverse property of addition. A similar property that applies to multiplication of rational numbers is called the **inverse property of multiplication.** Two numbers whose product is 1 are **multiplicative inverses,** or **reciprocals,** of each other. For example, $-\frac{4}{3}$ and $-\frac{3}{4}$ are multiplicative inverses because $-\frac{4}{3} \cdot \left(-\frac{3}{4}\right) = 1$.

Inverse Property of Multiplication

In words: The product of a rational number and its multiplicative inverse is 1.

Arithmetic	Algebra
$\frac{1}{3} \cdot \frac{3}{1} = 1$	$\frac{a}{b} \cdot \frac{b}{a} = 1$, where $a, b, \neq 0$.

The distributive property involves two operations, multiplication and addition.

Distributive Property	**In words:** The product of a number and a sum of numbers is the same as the sum of the products of the number and each addend. **Arithmetic** $9\left(5 + \frac{1}{3}\right) = 9(5) + 9\left(\frac{1}{3}\right)$ **Algebra** $x \cdot (y + z) = x \cdot y + x \cdot z$

You can use the multiplicative inverse property and the distributive property to find products mentally.

Example 1 *Problem Solving*

Zoology Use the information at the beginning of the lesson to find the length of an average Alaskan brown bear.

$1\frac{1}{8} \cdot 8 = 8 \cdot 1\frac{1}{8}$ *Commutative property*

$= 8\left(1 + \frac{1}{8}\right)$

$= (8 \cdot 1) + \left(8 \cdot \frac{1}{8}\right)$ *Distributive property*

$= 8 + 1 \text{ or } 9$ *Identity and Inverse Properties*

The average Alaskan brown bear is 9 feet long.

Examples

Find each product.

2 $6 \cdot 3\frac{5}{6}$

$$6 \cdot 3\frac{5}{6} = 6 \cdot \left(3 + \frac{5}{6}\right)$$
$$= (6 \cdot 3) + \left(6 \cdot \frac{5}{6}\right)$$
$$= 18 + 5 \text{ or } 23$$

3 $\frac{3}{4} \cdot 12\frac{1}{3}$

$$\frac{3}{4} \cdot 12\frac{1}{3} = \frac{3}{4}\left(12 + \frac{1}{3}\right)$$
$$= \frac{3}{4}(12) + \frac{3}{4}\left(\frac{1}{3}\right)$$
$$= 9 + \frac{1}{4} \text{ or } 9\frac{1}{4}$$

Checking for Understanding

Communicating Mathematics

Read and study the lesson to answer each question.

1. **Write** the multiplicative inverse of $\frac{5}{7}$.
2. **Tell** how you would find the multiplicative inverse of $-4\frac{2}{5}$.
3. **Write** a number sentence that shows the commutative property of multiplication for rational numbers.
4. **Tell** what property allows you to compute $\frac{1}{3} \cdot \left(6 \cdot \frac{4}{3}\right)$ as $\left(\frac{1}{3} \cdot 6\right) \cdot \frac{4}{3}$.

Guided Practice

State which pairs of numbers are multiplicative inverses. Write yes or no.

5. $8, \frac{1}{8}$
6. $-\frac{6}{5}, \frac{5}{6}$
7. $0.75, \frac{3}{4}$
8. $\frac{5}{6}, 1\frac{1}{5}$

Name the multiplicative inverse of each number.

9. 5
10. $-\frac{2}{3}$
11. 0.2
12. $2\frac{4}{5}$

Solve using mental math.

13. $m = \left(-\frac{1}{4} \cdot \frac{2}{3}\right) \cdot \frac{1}{2}$
14. $n = \frac{1}{2} \cdot 12\frac{4}{5}$
15. $4 \cdot 5\frac{1}{2} = k$

Exercises

Independent Practice

Name the multiplicative inverse of each of the following.

16. 10
17. $-\frac{3}{5}$
18. 0.4
19. $2\frac{8}{9}$
20. $\frac{7}{8}$
21. -1
22. $\frac{c}{d}$
23. $-x$
24. Is $\frac{8}{9}$ the multiplicative inverse of $-1\frac{1}{8}$? Why or why not?
25. Is 0.3 the multiplicative inverse of $3\frac{1}{3}$? Why or why not?

Evaluate each expression if $a = -\frac{1}{2}$, $b = \frac{2}{3}$, $x = -2\frac{1}{4}$, and $y = 1\frac{5}{6}$.

26. by
27. $2x$
28. a^2
29. $ax + \frac{1}{2}$
30. $3b - 4a$
31. $\frac{1}{2} + a$
32. $b^2(y + 5)$
33. $a + b + x$

Mixed Review

34. How many millimeters are in 2.75 meters? *(Lesson 1-6)*
35. **Statistics** What is the interquartile range of the data in the box-and-whisker plot at the right? *(Lesson 4-7)*

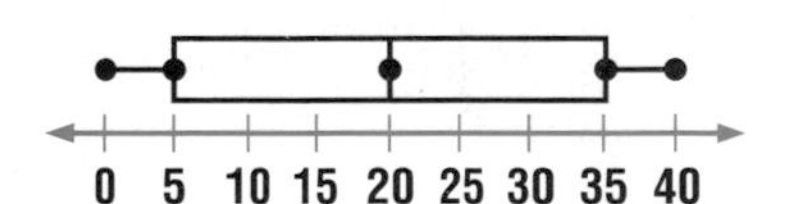

36. Find the GCF of 24 and 64. *(Lesson 6-4)*
37. Solve $-\frac{7}{16} \cdot \frac{4}{9} = j$. Write the solution in simplest form. *(Lesson 7-3)*

Problem Solving and Applications

38. **Home Economics** Marcie wants to make enough custard to serve 16 friends. The recipe that serves 4 requires $1\frac{3}{4}$ cups of milk. How much milk will she need?

39. **Critical Thinking** Complete the table shown below.

n	1	2	3	4
n^2				
$\frac{1}{n^2}$				

a. What pattern do you notice as you move across each row to the right?

b. What relationship do you see between n^2 and $\frac{1}{n^2}$?

DECISION MAKING

Choosing Employment

Situation

You and three other teens have formed a band called Sweet Tooth. Your band usually plays for free, but sells T-shirts and cassettes of the music after the performance to earn money. You buy the T-shirts wholesale, have them printed locally, and sell them for three times the wholesale price of one shirt. You send an original recording to a manufacturer to have duplicate tapes produced and sell them at three times the manufacturer's cost for one tape.

Suppose Sweet Tooth is offered two possible engagements for the same evening. One offer is to play at a school dance. The other offer is from the Park and Recreation Board to play at a community picnic. However, the Board wants to pay each of the band members $10 per hour instead of allowing them to sell shirts and cassettes. Which engagement should you accept?

Hidden Data

- attendance at each event: Will there be enough students at the dance to sell a lot of shirts and cassettes?
- length of performance: How long does the Board want the band to play?
- shipping costs of T-shirts and cassettes: How much extra will this add to the order?

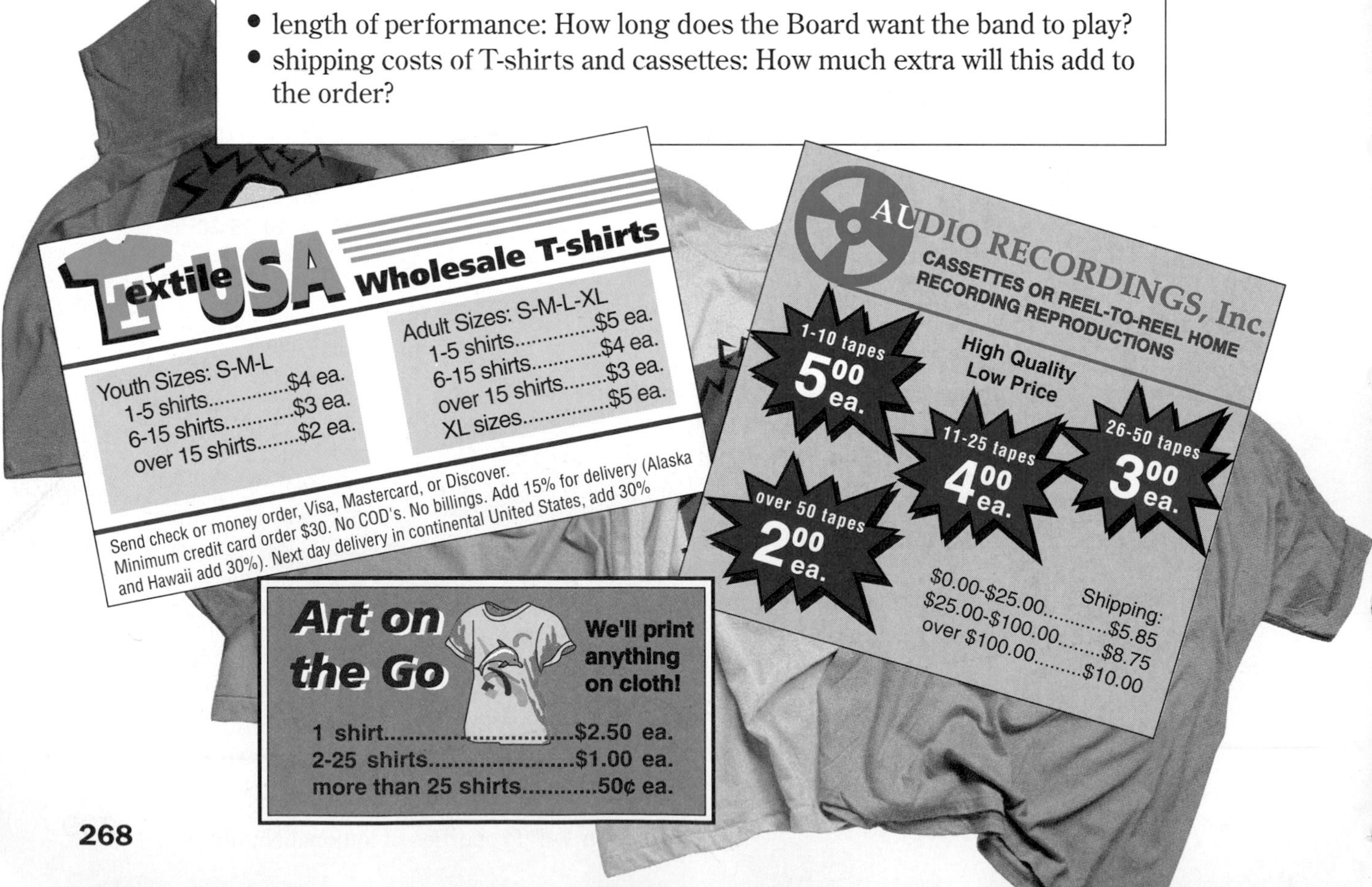

Analyzing the Data

1. How much profit can Sweet Tooth make on each medium adult shirt?
2. Including shipping, what is the actual cost per tape if you order 15 tapes? What if you order 25 tapes?
3. What is the total cost of ordering and printing 5 youth shirts, size L, 5 adult shirts, size M, and 5 adults shirts, size, L?

Making a Decision

4. **How many items** must you sell to beat the Park and Recreation Board's offer?
5. **If you play** at the school
   a. **How many** shirts and cassettes should you order?
   b. **How many** of each size shirt should you order?
6. **What if** you talk the board into letting you sell shirts and cassettes instead of being paid cash? Does this affect your decision?

Making Decisions in the Real World

7. Take a poll of students who have been to a recent concert. Compare the types of souvenirs offered for sale and the prices of these items.

Problem-Solving Strategy

7-5 Find a Pattern

Objective

Solve problems by finding and extending a pattern.

Have you ever heard of a wayward goose winging its way to Tallahassee, Florida? It started from Akimiske Island in the Hudson Bay one sunny morning to catch up with friends 1,600 miles away. It flew halfway and stopped for a rest. It could only fly half the remaining distance each time. Did it ever get to Tallahassee?

Explore *What do you know?*

The distance is 1,600 miles.

The goose could fly half the remaining distance on each flight.

What are you trying to find out?

Whether the goose ever gets to Tallahassee.

Plan Visualize the trip with a model. Let the 1,600-mile journey equal 1 unit. Then each flight is a fraction of 1.

Solve The goose traveled half the distance on its first flight. After the second flight it traveled half the distance left: $\frac{1}{2}$ of $\frac{1}{2}$, or $\frac{1}{4}$. After two flights, the goose had traveled $\frac{1}{2} + \frac{1}{4}$, or $\frac{3}{4}$, of the total distance. Copy the drawing below. Draw and label a few more flights.

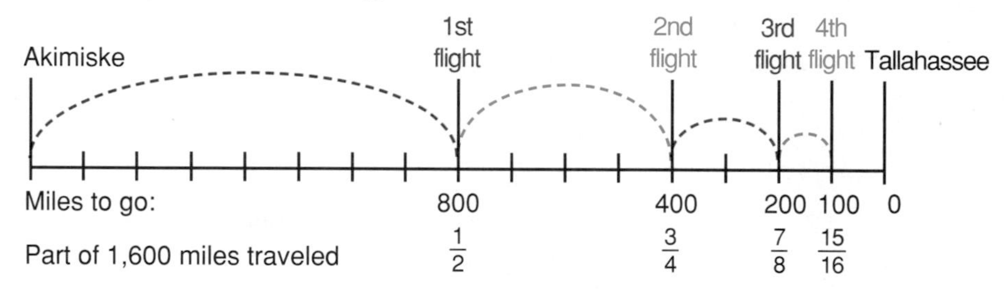

According to this pattern the goose will have 50 miles to go after the fifth flight. The distance left would continue as follows.

$$50 \cdot \frac{1}{2} = 25 \quad \rightarrow \quad 25 \cdot \frac{1}{2} = 12\frac{1}{2} \quad \rightarrow \quad 12\frac{1}{2} \cdot \frac{1}{2} = 6\frac{1}{4}$$

In theory, the goose continues to get closer, but will never travel the entire 1,600 miles. In reality, the distance will get so small that the goose will eventually travel to Tallahassee.

Examine Let n represent the number of flights. The distance remaining can be represented by the following list.

$$\frac{1}{2}, \frac{1}{4}, \frac{1}{8}, \frac{1}{16}, \cdots \frac{1}{2^n}$$

As n increases, the distance remaining approaches 0, but it can never equal 0.

Checking for Understanding

Communicating Mathematics

Read and study the lesson to answer each question.

1. **Tell** where 2^n comes from in the list of the remaining distances in the *Examine* step.
2. **Write** the distance the goose traveled after eight flights as a fraction of 1,600.

Solve by finding a pattern.

Guided Practice

3. What is the total number of rectangles in the figure below?

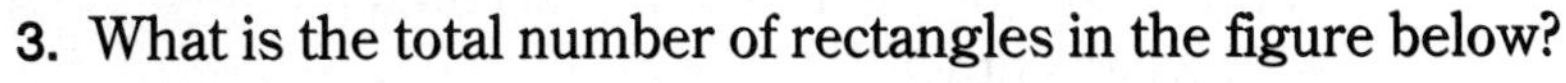

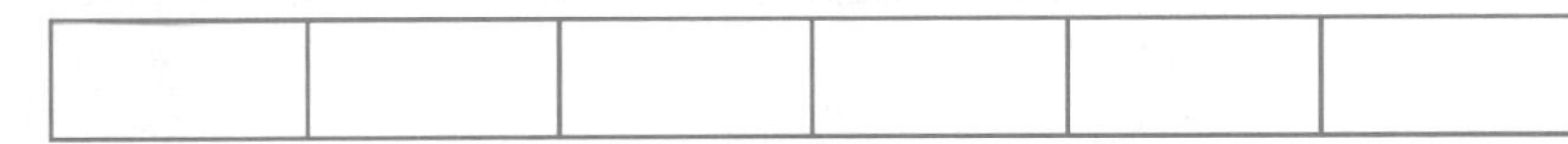

4. How many diagonals does a 7-sided polygon have?
5. Find the next number in the set {100, 95, 85, 70, __?__}.

Problem Solving

Practice

Solve using any strategy.

Strategies

Look for a pattern.
Solve a simpler problem.
Act it out.
Guess and check.
Draw a diagram.
Make a chart.
Work backward.

6. Jackson Middle School's football team scored seven times for a total of 34 points. They only scored touchdowns and field goals. Only once did they succeed in getting the extra point after a touchdown. How many touchdowns did the team make?
7. José was assigned some math exercises for homework. He answered half of them in study hall. After school he answered 7 more exercises. If he still has 11 exercises to do, how many exercises were assigned?
8. Marla Guerrero needs to buy numbers to put on the doors of each apartment in a 48-unit apartment building. The apartments are numbered 1 through 48. How many of each digit 0, 1, 2, 3, 4, 5, 6, 7, 8, and 9 should she order?

9. **Data Search** Refer to page 667.
How many Olympic swimming pools would fit on a soccer field? Draw a diagram to verify your answer.

10. **Computer Connection** Suppose your science teacher wants to investigate how many times a ball will bounce, when it is dropped from a height of 25 feet. You know that each time the ball bounces, it returns to a height that is 0.40 times the previous height.

BALL BOUNCE

	A	B
1	INITIAL HIT (FT)	25
2	BOUNCE FACTOR	0.40
3	HIT NUMBER	RETURN HT
4	0	= B1
5	= A4 + 1	= B4*B2
6	= A5 + 1	= B5*B2

Instead of performing tedious calculations, you can organize the data into a spreadsheet like the one shown at the left.

a. What value will be in cell B5?

b. A different ball will return to a height that is 0.5 times the previous height. How would you modify the spreadsheet?

Algebra Connection

7-6 Sequences

Objective

Recognize and extend arithmetic and geometric sequences.

Words to Learn

sequence
term
arithmetic sequence
common difference
geometric sequence
common ratio

Joanne won by spelling "antipyretic," which is an agent used to reduce fever.

In 1991, Joanne Lagatta of Clintonville, Wisconsin, won the National Spelling Bee. For winning, she received $5,000. Suppose she opened a savings account with it and adds $10 to the account every month.

The chart below shows her balance after each month during the first year.

Month	0	1	2	3	4	5	6
Total Deposits	5,000	5,010	5,020	5,030	5,040	5,050	5,060

Month	7	8	9	10	11	12
Total Deposits	5,070	5,080	5,090	5,100	5,110	5,120

A list of numbers in a certain order, such as 0, 1, 2, 3, ... or 5,000, 5,010, 5,020, 5,030, ... is called a **sequence.** Each number is called a **term** of the sequence. When the difference between any two consecutive terms is the same, the sequence is called an **arithmetic sequence.** The difference is called the **common difference.**

Examples

State whether each sequence is arithmetic. Then write the next three terms of each sequence.

1 $4, 7\frac{1}{3}, 10\frac{2}{3}, 14, \ldots$

$$4 \xrightarrow{+3\frac{1}{3}} 7\frac{1}{3} \xrightarrow{+3\frac{1}{3}} 10\frac{2}{3} \xrightarrow{+3\frac{1}{3}} 14$$

The difference between any two consecutive terms is $3\frac{1}{3}$. So, the sequence is arithmetic. Add $3\frac{1}{3}$ to the last term of the sequence, and continue adding until the next three terms are found. The next three terms are $17\frac{1}{3}$, $20\frac{2}{3}$, and 24.

2 11, 4, –2, –7, ...

$$11 \xrightarrow{-7} 4 \xrightarrow{-6} -2 \xrightarrow{-5} -7$$

Since there is no common difference, the sequence is *not* arithmetic. Add –4 to the last term of the sequence, and continue adding the next greater integer until the next three terms are found. The next three terms are –11, –14, and –16.

Calculator Hint

Some calculators have a constant function. That is, the calculator will repeat the last entry and operation by pressing the [=] key repeatedly. By using this feature, you save time when calculating other terms in a sequence.

When consecutive terms of a sequence are formed by multiplying by a constant factor, the sequence is called a **geometric sequence.** The factor is called the **common ratio.**

Examples

State whether each sequence is geometric. Then write the next three terms of each sequence.

3 2, –6, 18, –54, ...

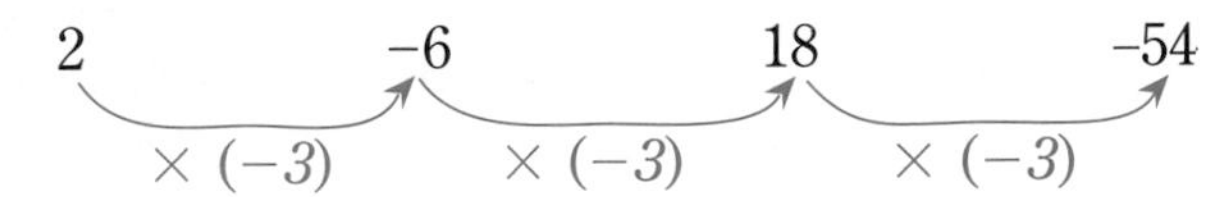

Since there is a common ratio, –3, the sequence is geometric. Multiply the last term of the sequence by –3, and continue multiplying until the next three terms are found.

The next three terms are 162, –486, and 1,458.

Problem-Solving Hint

Look for a pattern when deciding whether a sequence is geometric. Does each consecutive term result from multiplying by the same number each time?

4 24, 12, 4, 1, ...

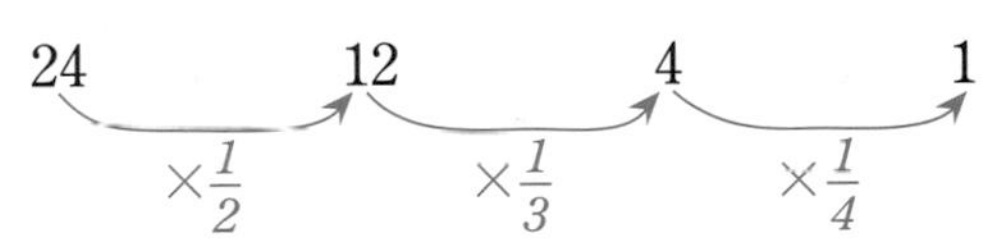

Since there is no common ratio, the sequence is *not* geometric. Multiply the last term by $\frac{1}{5}$, and continue multiplying by $\frac{1}{6}$ and then $\frac{1}{7}$ to find the next three terms. The next three terms are $\frac{1}{5}$, $\frac{1}{30}$, and $\frac{1}{210}$.

Example 5 *Connection*

Algebra **The first term of a sequence is represented by a, the second term a_2, and so on up to the nth term, a_n. Find the twentieth term, a_{20}, in the sequence 18, 14, 10, 6, ...**

This sequence is arithmetic, and the common difference is –4. To find the twentieth term, you can add –4 to 18 nineteen times, or you can use the formula $a_n = a + (n - 1)d$, where n is the number of the term you want to find and d is the common difference.

$a_n = a + (n - 1)d$

$a_{20} = 18 + (20 - 1)(-4)$ *Replace n with 20, d with –4, and a with 18.*

$a_{20} = 18 + (19)(-4)$

$a_{20} = 18 + (-76)$

$a_{20} = -58$

So, the twentieth term is –58.

Checking for Understanding

Communicating Mathematics

Read and study the lesson to answer each question.

1. **Write** an arithmetic sequence with a common difference of 4.
2. **Tell** how you can find the next term of a geometric sequence if you know one term and the common ratio.
3. **Tell** the sixth term of the sequence 0, 3, 6, 9, ...
4. **Write** the first five terms in a geometric sequence with a common ratio of 0.5. The first term is 20.

Guided Practice

State whether each sequence is arithmetic, geometric, or neither. Then write the next three terms of each sequence.

5. 1, 4, 9, 16, ...
6. $-5, 1, -\frac{1}{5}, \frac{1}{25}, \ldots$
7. 2, 4, 8, 16, ...
8. 98.6, 98.2, 97.8, ...
9. 9, 3, −3, −9, ...
10. 20, 24, 28, 32, ...
11. 99, 88, 77, 66, ...
12. 1, −3, 9, −27, ...
13. 1.5, 3, 4.5, 6, ...
14. 89, 89, 89, 89, ...
15. −6, −4, −2, 0, ...
16. −256, 128, −64, ...

Exercises

Independent Practice

Write the next three terms of each sequence.

17. 100, 91, 82, 73, ...
18. $25, 5, 1, \frac{1}{5}, \ldots$
19. 9, 6, 3, 0, ...
20. 0.3, −0.9, 2.7, −8.1, ...
21. 1,256, −628, 314, ...
22. 97, 85, 110, 98, 123, 111, 136, ...
23. 5, 6.5, 8, 9.5, ...
24. 54, 60, 66, 72, ...
25. 1, 2, 5, 10, 17, ...
26. −3, 12, −48, 192, ...
27. $\frac{1}{4}, \frac{1}{12}, \frac{1}{36}, \ldots$
28. $4\frac{1}{2}, 4\frac{1}{6}, 3\frac{5}{6}, 3\frac{1}{2}, \ldots$
29. Name the tenth term in the sequence 100, 95, 90, 85, ...
30. Write the first four terms in a geometric sequence with a common ratio of $\frac{3}{4}$. The first term is 16.
31. The fifth term of an arithmetic sequence is 14. The common difference is −2. Find the first four terms.
32. Name the eighth term in the sequence 16, 8, 4, ...

Mixed Review

33. Use mental math to find 700 − 365. *(Lesson 1-2)*
34. Solve $x = -43 - (-89)$. *(Lesson 3-5)*
35. **Statistics** Construct a line plot for the following English quiz scores. 21, 24, 17, 19, 16, 22, 25, 19, 18, 21, 23, 18, 24, 19, 25. *(Lesson 4-3)*
36. Express $-\frac{5}{8}$ as a decimal. *(Lesson 6-6)*
37. Compute $\frac{5}{6} \cdot 3\frac{3}{7}$ mentally. *(Lesson 7-4)*

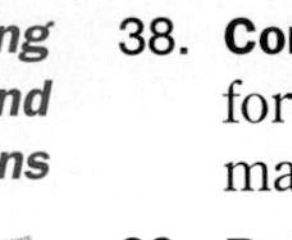

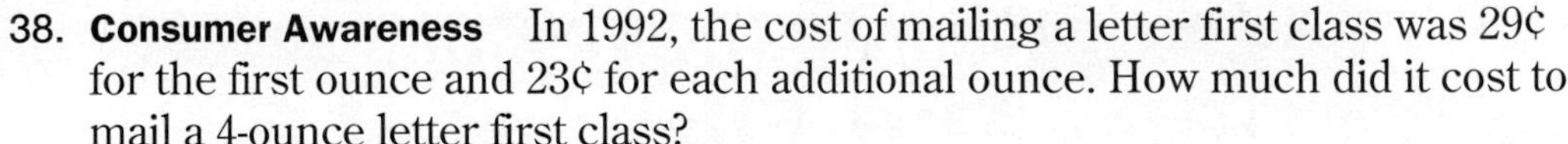

Problem Solving and Applications

38. **Consumer Awareness** In 1992, the cost of mailing a letter first class was 29¢ for the first ounce and 23¢ for each additional ounce. How much did it cost to mail a 4-ounce letter first class?

39. **Business** Lisa bought a car for \$12,200. If it loses $\frac{1}{5}$ of its value every year, what is the value at the end of three years?

40. **Critical Thinking** Can a geometric sequence contain zero as a term? Explain.

41. **Critical Thinking** The second term of an arithmetic sequence is 6 and the sixth term is 38.
    a. What is the common difference? How did you find it?
    b. Explain how to find the common difference of an arithmetic sequence if you know any two terms.

42. **Geometry** Study the pattern of circles.

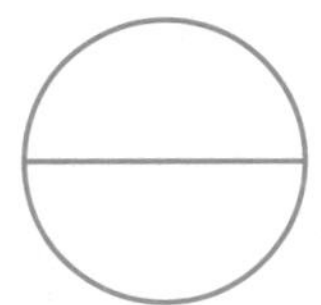
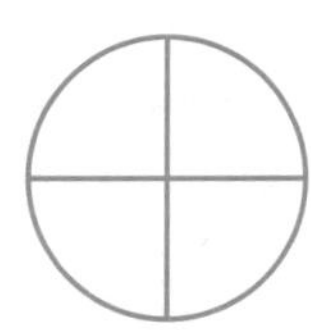

    a. Draw the next two figures in the sequence.
    b. Write a sequence for the number of pieces in each circle.
    c. What kind of sequence does this illustrate? Explain.
    d. Write a sequence for the number of cuts across each circle.
    e. What is the relationship between the number of cuts and the number of pieces?

7 Assessment: Mid-Chapter Review

Solve each equation. Write each solution in simplest form. *(Lessons 7-1, 7-2, 7-3)*

1. $n = -\frac{3}{4} + \frac{1}{4}$
2. $\frac{7}{8} - \frac{3}{8} = m$
3. $2\frac{8}{11} + 3\frac{6}{11} = d$
4. $x = \frac{3}{4} + \frac{4}{5}$
5. $y = -1\frac{5}{6} + \left(-8\frac{3}{8}\right)$
6. $g = 12 - 5\frac{3}{7}$
7. $a = 2\frac{4}{5}(-10)$
8. $\left(-\frac{4}{9}\right)^2 = h$
9. Explain how to compute $12 \times 3\frac{3}{4}$ using the distributive property. *(Lesson 7-4)*
10. Marianne is conditioning for a 10-kilometer race. On the first day she ran 2 laps around the track. The second day she ran 4 laps. The third day she ran 6 laps. On what day will she reach her goal of 14 laps? *(Lesson 7-5)*

State whether each sequence is arithmetic or geometric. Then find the next three numbers in the sequence. *(Lesson 7-6)*

11. 20, 23, 26, 29, ...
12. 768, 192, 48, ...

Cooperative Learning

7-6B The Fibonacci Sequence

A Follow-Up of Lesson 7-6

Objective

Discover the numbers that make up the Fibonacci Sequence.

Materials

colored pencils

Leonardo was born about A.D. 1170 in the city of Pisa, Italy. This famous mathematician was also known by the name Fibonacci. He wrote many works, but his *Liber Abaci* was the most famous. He is credited with introducing the Arabic numbers we use today to the European scholars who, at the time, were still using Roman numerals.

Leonardo loved to create interesting story problems. Many of the ones he created centered around a series of numbers that became known as the Fibonacci Sequence.

Try this!

Work in groups of three.

Study the honeycomb shown below. The bees go from cell to cell in the comb to store honey and other food stuffs. Suppose the path they take must have the letters in alphabetical order.

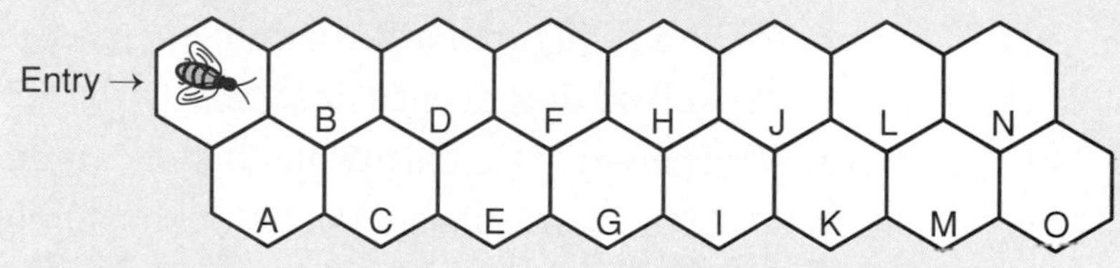

- Suppose the bee wanted to go from the entry to cell A. There is only one path possible.

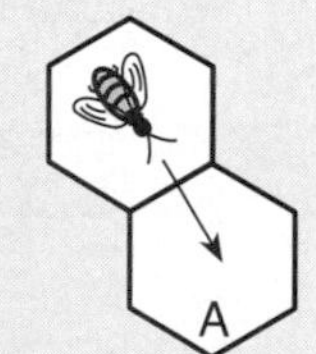

- Suppose the bee wanted to go from the entry to cell B. There are two paths possible.

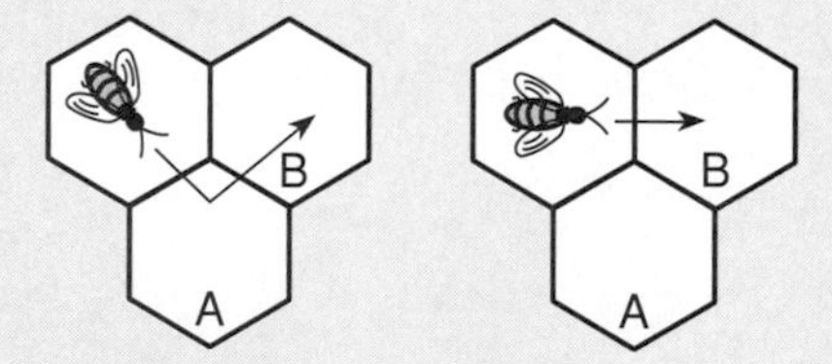

- Suppose the bee wanted to go from the entry to cell C. There are three paths possible.

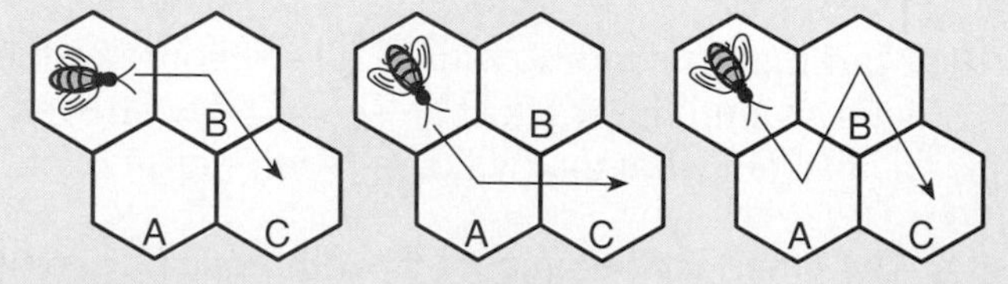

- Make a drawing of the honeycomb up to cell D. Use colored pencils to draw all the possible paths from the entry to cell D, making each path a different color. How many paths are there?

- Make another drawing of the honeycomb up to cell E. Use colored pencils to draw all the possible paths to cell E.
- Copy and complete the chart.

Cell	A	B	C	D	E	F
Number of paths to the cell	1	2	3			

What do you think?

1. Look for a pattern in the numbers. How is each number related to the previous numbers?
2. Without making a drawing, how many paths are there to cell G? How did you arrive at this number?
3. The numbers in the chart are the first numbers of the Fibonacci Sequence. Find the next five numbers in the sequence.
4. Is the Fibonacci Sequence arithmetic, geometric, or neither? Explain.

Applications

The pinball machine, one of the earliest coin-activated electromechanical games was originated in its modern form in about 1930. The early machines used marbles and cost a penny to play.

5. **Money** Tokens for the arcade machines cost 25¢ each. The token machine at *Laser One Arcade* will accept only half-dollars and quarters. For example, if you wanted to buy two tokens, you could use two quarters (QQ) or one half-dollar (H). There are two ways to buy two tokens.
   a. Use Qs and Hs to list the ways to insert the coins to buy each number of tokens. Then copy and complete the chart below.

Number of Tokens	1	2	3	4	5	6	7
Ways to Buy							

   b. How do your results relate to the Fibonacci Sequence?

6. **Nature** The pine cone, pineapple, daisy, and sunflower all have characteristics of the Fibonacci Sequence. For example, look at the spirals on the bottom of a pine cone.
   a. The blue tint shows a clockwise spiral. How many clockwise spirals are there?
   b. The red tint shows a counterclockwise spiral. How many counterclockwise spirals are there?
   c. Why do we say that the pine cone is an example of Fibonacci's numbers?

Geometry Connection

7-7 Area of Triangles and Trapezoids

Objective

Find the areas of triangles and trapezoids.

Words to Learn

trapezoid
base
altitude
height

A hydrologist studies the movement, or stream flow, of water. The first step in determining stream flow is to find the area of a cross section of the stream. The stream bed shown at the right approximates the shape of a trapezoid. The area of this trapezoid is a good estimate of the area of the cross section. *You will find the area of the trapezoid in Example 1.*

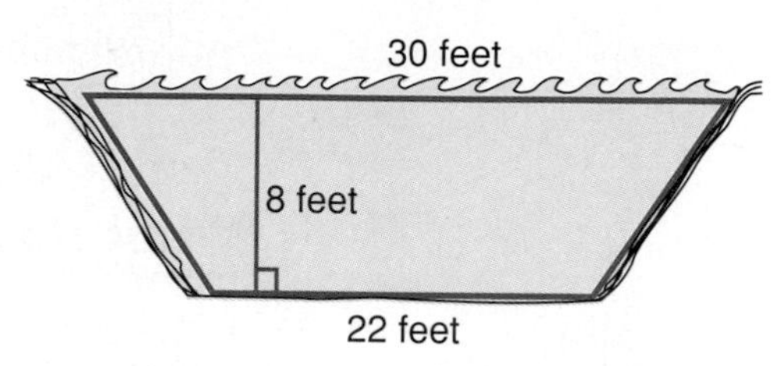

Remember that a **trapezoid** is a quadrilateral with exactly one pair of parallel sides called **bases.** A segment perpendicular to both bases, with endpoints on the base lines, is called the **altitude.** The length of the altitude is called the **height.**

In the trapezoid at the right, sides $\overline{GH}$ and $\overline{JK}$ are the bases. $\overline{EF}$ is an altitude.

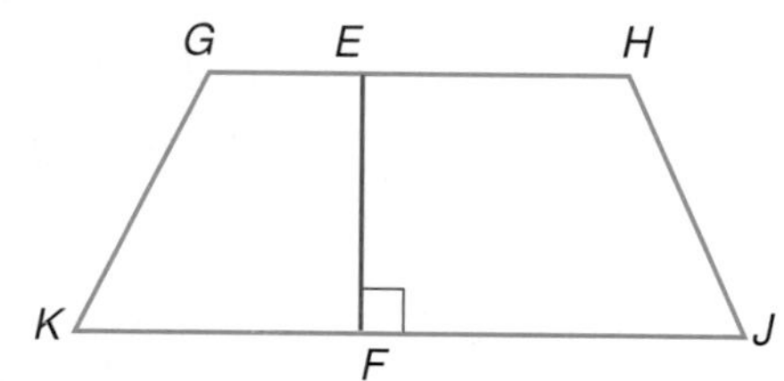

Mini-Lab

Work with a partner.

Materials: graph paper, scissors

- Copy the trapezoid at the right on a piece of graph paper. Cut it out.

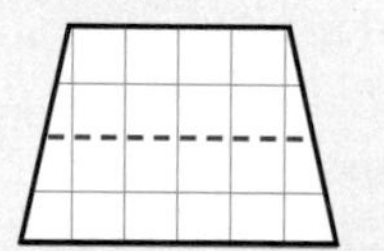

- Make a cut along the dashed line. Move the parts so they form a parallelogram.

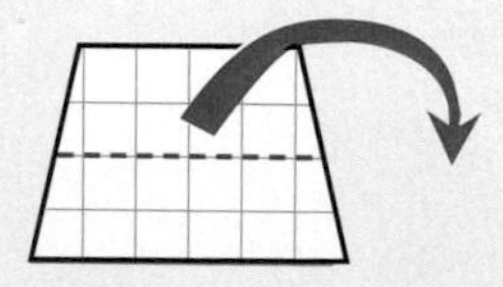

Talk About It

a. What is the length of the base and the height of the parallelogram?

b. What is the area of the parallelogram?

c. Explain how the height and length of the base of the parallelogram are related to the height and sum of the lengths of the bases of the original trapezoid.

You can review area of parallelograms on page 75.

The Mini-Lab suggests the following rule for finding the area of a trapezoid.

Area of a Trapezoid	**In words:** The area of a trapezoid is equal to the product of half the height and the sum of the bases. **In symbols:** If a trapezoid has bases of a units and b units and a height of h units, $A = \frac{1}{2}h(a + b)$.

Example 1 *Problem Solving*

Hydrology A hydrologist needs to find the area of the cross section of the stream shown at the beginning of the lesson to complete a flood control study. What is the area of the trapezoid representing the cross section?

$A = \frac{1}{2}h(a + b)$

$A = \frac{1}{2} \cdot 8(30 + 22)$ *Replace a with 30, b with 22, and h with 8.*

$A = \frac{1}{2} \cdot 8 \cdot 52$ or 208 The area is 208 square feet.

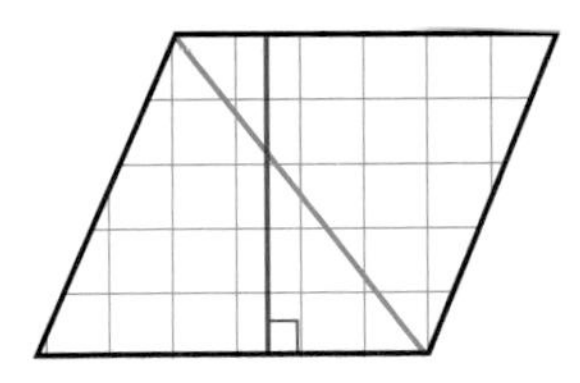

A parallelogram can also be used to find the area of a triangle. The diagonal separates a parallelogram into two congruent triangles.

The area of the parallelogram at the left is $6 \cdot 5$ or 30 square units. The area of each triangle is one half the area of the parallelogram. So, each triangle has an area of 15 square units.

Area of a Triangle	**In words:** The area of a triangle is equal to half the product of its base and height. **In symbols:** If a triangle has a base of b units and a height of h units, $A = \frac{1}{2}bh$.

Any side of a triangle can be used as a base. The height is the length of the corresponding altitude. The altitude is a line segment perpendicular to the base from the opposite vertex.

Example 2

Find the area of the triangle.

3.75 inches

7.5 inches

$A = \frac{1}{2}bh$

$A = 0.5(7.5)(3.75)$ *Replace b with 7.5 and h with 3.75.*

0.5 [×] 7.5 [×] 3.75 [=] 14.0625

The area is 14.0625 square inches.

Checking for Understanding

Communicating Mathematics

Read and study the lesson to answer each question.

1. **Tell** which two sides of a trapezoid are the bases.
2. **Draw** a triangle similar to the one shown at the right. Sketch an altitude.
3. **Tell** how to find the area of a triangle.
4. **Draw** a trapezoid on graph paper. Label its height and bases. Then find its area.

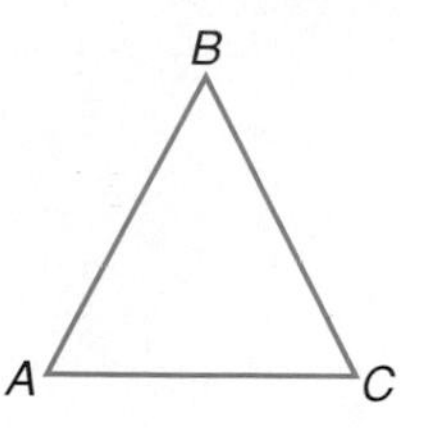

Guided Practice

State the measures of the base(s) and the height of each triangle or trapezoid. Then find the area.

5. 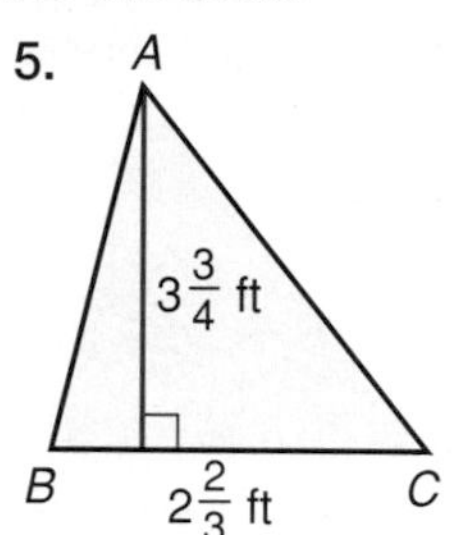

6. 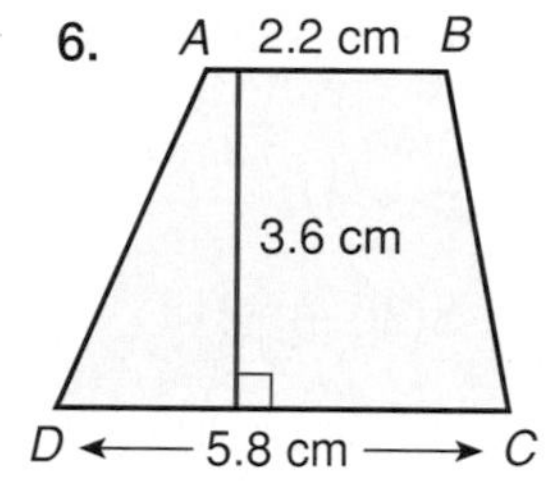

7. 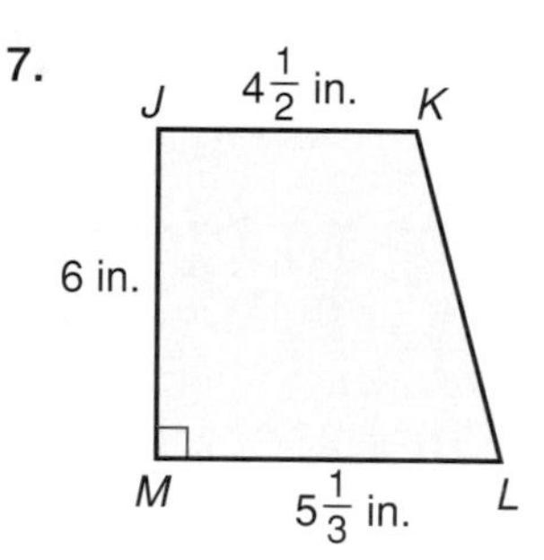

Find the area of each figure described below.

8. triangle: base, $3\frac{3}{4}$ ft; height, $4\frac{1}{2}$ ft
9. triangle: base, 9 cm; height, 2.6 cm
10. triangle: base, 10 in.; height, 7 in.
11. trapezoid: bases, 12 m and 8 m; height, 9 m
12. trapezoid: bases, 0.3 km and 0.5 km; height, 0.2 km
13. trapezoid: bases $2\frac{1}{3}$ yd and $4\frac{1}{6}$ yd; height, 3 yd

Exercises

Independent Practice

State the measures of the base(s) and the height of each triangle or trapezoid. Then find the area.

14. 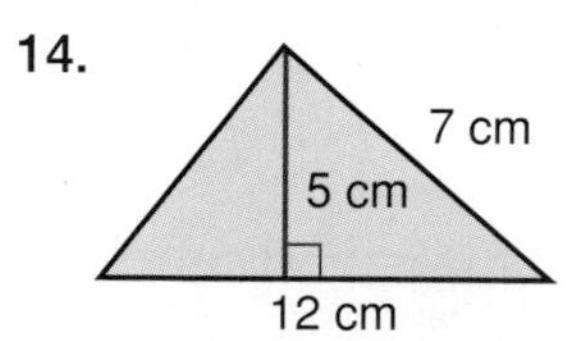

15. 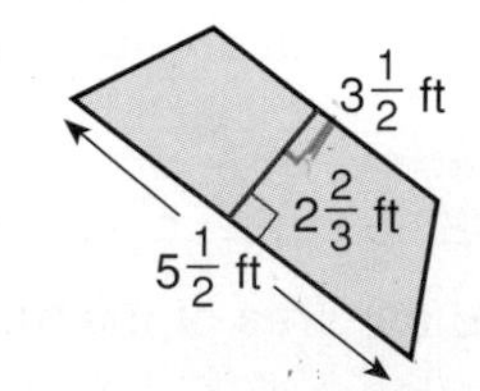

16. 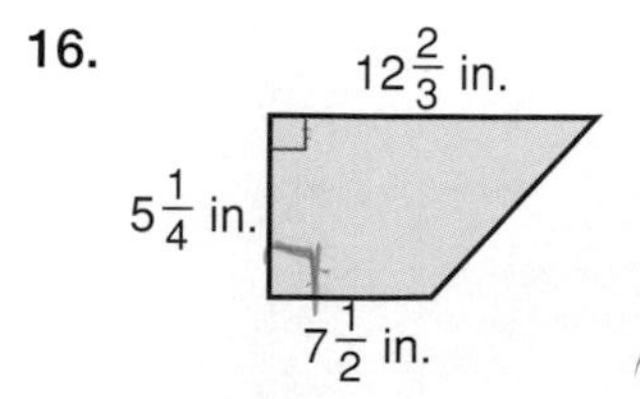

17. 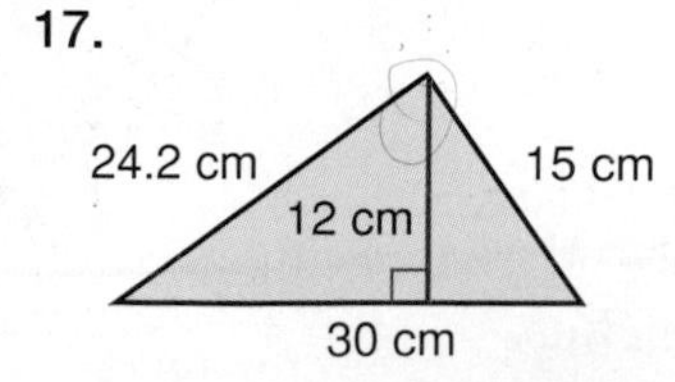

18. 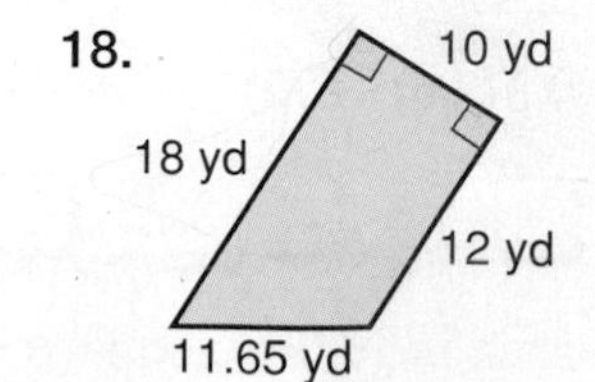

19. 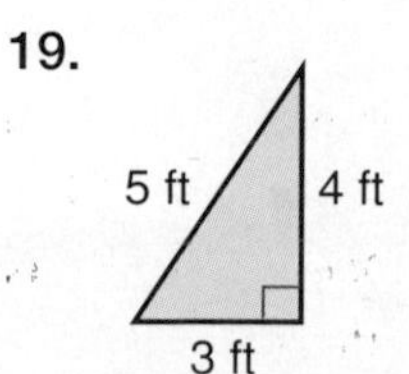

Find the area of each triangle described below.

	base	height
20.	8 cm	3.8 cm
22.	$2\frac{2}{3}$ ft	$3\frac{1}{12}$ ft
24.	20 m	17 m

	base	height
21.	22 yd	27 yd
23.	$1\frac{3}{4}$ in.	$5\frac{1}{8}$ in.
25.	16 mm	14 mm

Find the area of each trapezoid described below.

	base *(a)*	base *(b)*	height
26.	19 cm	24 cm	12 cm
28.	26 in.	24 in.	15 in.
30.	4.7 cm	5.9 cm	2.2 cm

	base *(a)*	base *(b)*	height
27.	12 ft	18 ft	17 ft
29.	8.3 km	8.5 km	8.2 km
31.	0.3 m	0.72 m	0.4 m

Mixed Review

32. Write an equation to represent *three more than two times the number of cars is fifteen.* *(Lesson 2-6)*

33. Solve $\frac{w}{4} = -125$. *(Lesson 3-9)*

34. **Statistics** Find the mean, median, and mode of the set {40, 27, 19, 31, 35, 23, 40, 39}. *(Lesson 4-5)*

35. **Geometry** Tell if the pair of figures at the right are congruent, similar, or neither. *(Lesson 5-6)*

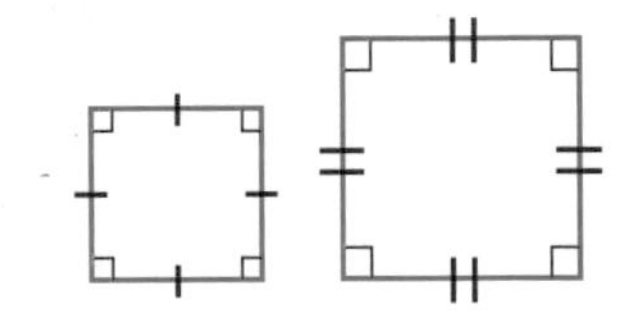

36. Write the next three terms of the sequence 80, 76, 72, 68, ... *(Lesson 7-6)*

Problem Solving and Applications

37. **Hydrology** To get a more accurate picture of the cross section of a stream, a hydrologist takes depth readings every 5 feet and draws a sketch of the cross section. Estimate the area of the cross section. *Hint: Find the area of each small section.*

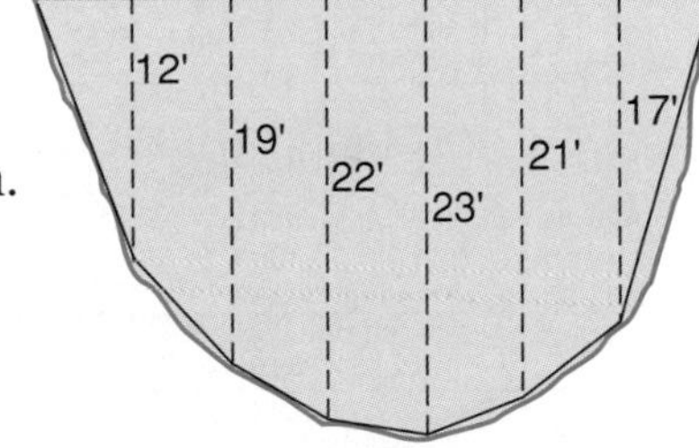

38. **Critical Thinking** What is the effect on the area of a trapezoid if the length of each base is doubled and the height is doubled?

39. **Geography** The state of South Carolina is shaped something like a triangle.
    a. Describe how to find an estimate of the area of South Carolina.
    b. Estimate the area of South Carolina.

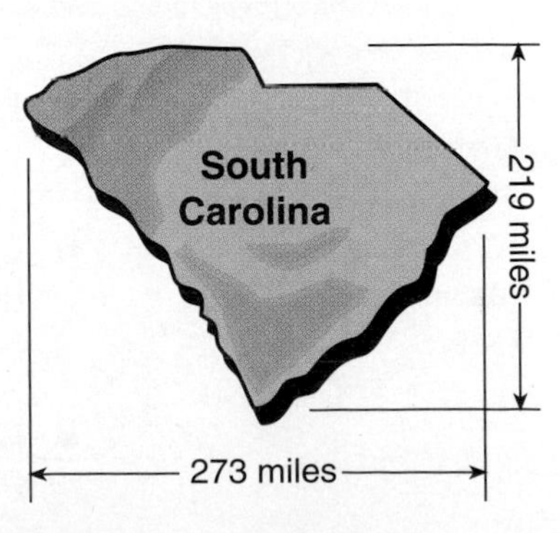

40. **Journal Entry** Write a few sentences explaining how the area of a triangle and a trapezoid are related.

Cooperative Learning

7-7B Area and Pick's Theorem

A Follow-Up of Lesson 7-7

Objective

Connect algebra and geometry to find the area of a triangle.

Materials

dot paper or geoboards

Pick's Theorem is a formula that uses information from dot paper to find the area of a triangle. Let's investigate this formula on the following triangles.

Try this!

Work in groups of four.

- Make these triangles on your geoboard. *If you don't have a geoboard, draw them on dot paper.*

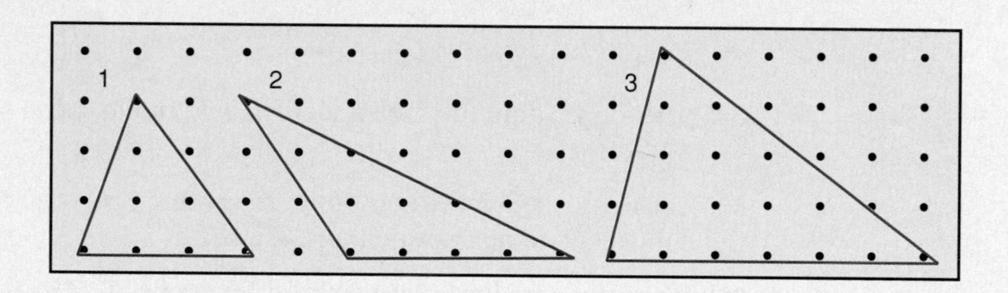

- Copy the spreadsheet below.

	A	B	C	D	E	F
1	TRIANGLE	AREA USING A = 1/2 bh	NUMBER OF DOTS ON TRIANGLE	NUMBER OF DOTS INSIDE TRIANGLE	PICK'S THEOREM	AREA USING PICK'S THEOREM
2	1				C2/2 + D2 - 1	
3	2				C3/2 + D3 - 1	
4	3				C4/2 + D4 - 1	

Computer Hint

C2/2 means divide the value in cell C2 by 2.

- Find the area of each triangle using the formula $A = \frac{1}{2}bh$. Record your findings in column B of the spreadsheet.
- Count the number of dots on the sides of the triangle. Then count the number of dots inside the triangle. Record these numbers in columns C and D of the spreadsheet.
- Use the spreadsheet formula in column E to find the result of using Pick's Theorem. Record each result in column F.

What do you think?

1. Describe triangles 1, 2, and 3 shown on the geoboard.
2. Compare your results in column B with your results in column F. What do you notice?
3. If x represents the number of dots on the figure, y represents the number of dots inside the figure, and A represents the area of the figure, write a formula for Pick's Theorem.
4. Do you think Pick's Theorem would work for a rectangle? Show several rectangles on your geoboard to support your answer.
5. Would Pick's Theorem work for trapezoids? Show examples to support your answer.

Extension

6. Make each of the figures below on your geoboard or dot paper.
   a. How could you find the area of each figure without using Pick's Theorem?
   b. Investigate Pick's Theorem with each figure. With which types of figures can you use Pick's Theorem?

I.

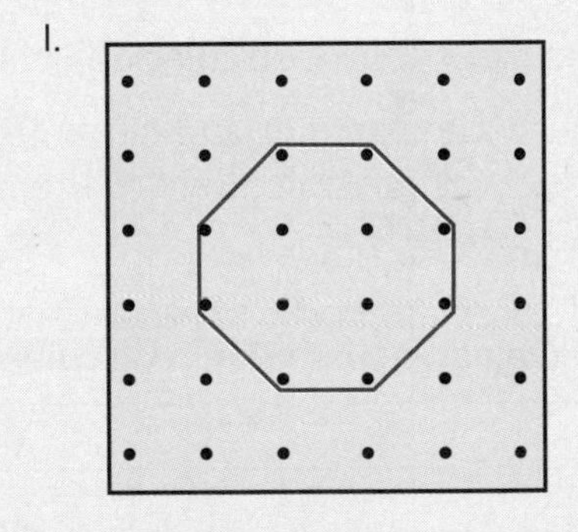

II.

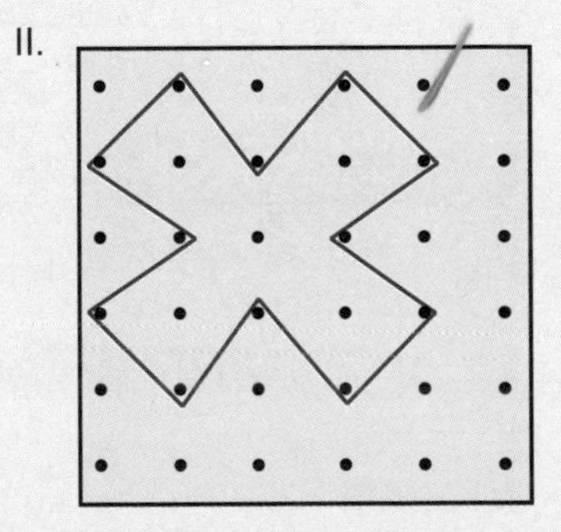

III.

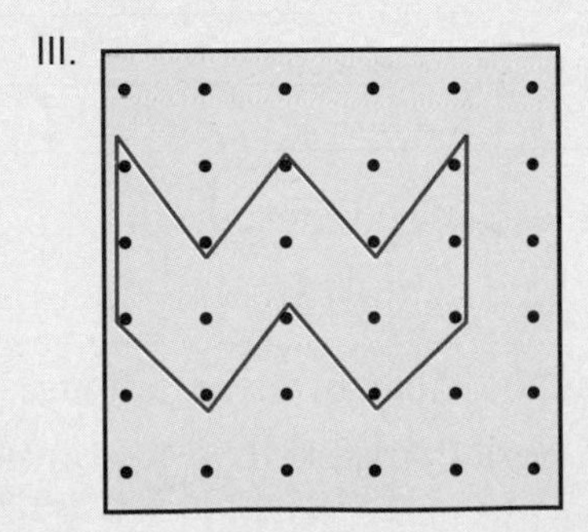

IV.

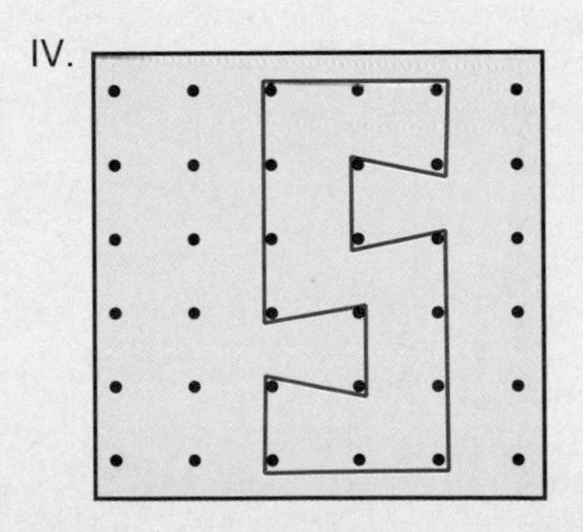

Geometry Connection

7-8 Circles and Circumference

Objective

Find the circumference of circles.

Words to Learn

circle
center
radius
diameter
circumference

What is displayed on your calculator when you press [π]? Is it possible to generate the same number without using this specially-marked key?

Ancient mathematicians used a variety of numbers to approximate π in their work with circles. Before you learn about π, let's first study some terms and properties related to circles.

A **circle** is a set of points in a plane that are the same distance from a given point in the plane. The given point is called the **center.** The distance from the center to any point on the circle is called the **radius** (r). The distance across the circle through the center is its **diameter** (d). The **circumference** (C) of a circle is the distance around the circle. The diameter of a circle is twice its radius, or $d = 2r$.

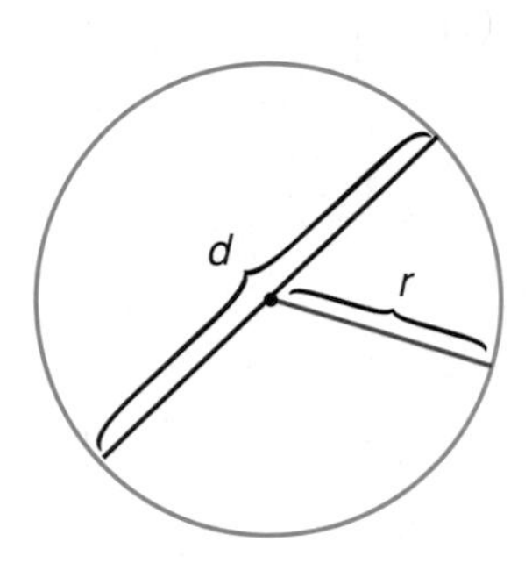

Mini-Lab

Work in groups of three or four.

Materials: tape measure, circular objects, calculator

- Collect three to five circular objects.
- Measure the diameter and circumference of each as accurately as possible. Record the measurements in a table like the one below.

You can review mean on page 145.

	Object	Diameter	Circumference	$\frac{\text{Circumference}}{\text{Diameter}}$
#1				
#2				
#3				
#4				
#5				

- Divide each circumference by the corresponding diameter.

Talk About It

a. What is the mean of the quotients?

b. How does it compare to [π]?

The relationship you discovered in the Mini-Lab is true for all circles. The circumference of a circle divided by its diameter is always 3.1415926.... The Greek letter π (pi) represents this number. Although π is not a rational number, the rational numbers 3.14 and $\frac{22}{7}$ are often used as approximations for π.

Circumference	**In words:** The circumference of a circle is equal to its diameter times π, or 2 times its radius times π. **In symbols:** $C = \pi d$ or $C = 2\pi r$

Example 1 *Connection*

Geometry The "spokes" of a Ferris wheel extend 21 feet from the center to the outside of the wheel.

a. How far does a car that is connected to the outside travel in one trip around?

Using $\frac{22}{7}$ for π.

$C = 2\pi r$

$C \approx 2 \cdot \frac{22}{7} \cdot 21$

$C \approx 2 \cdot \frac{22}{\cancel{7}_1} \cdot \frac{\cancel{21}^3}{1}$

$C \approx 132$

The car travels about 132 feet.

Using 3.14 for π.

$C = 2\pi r$

$C \approx 2 \cdot 3.14 \cdot 21$

$C \approx 131.88$

b. How many turns of the Ferris wheel would it take to travel 1 mile?

Divide 5,280 by 132 to find how many turns of the Ferris wheel it takes to equal 1 mile.

5280  132  40

It would take the Ferris wheel about 40 turns to travel 1 mile.

Example 2

Find the circumference of a 26-inch bicycle tire.

$C = \pi d$

$C = \pi \cdot 26$ *Replace d with 26.*

  26 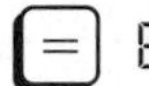 81.681409

The circumference is about 81.7 inches

Checking for Understanding

Communicating Mathematics

Read and study the lesson to answer each question.

1. **Tell** how you can find the diameter of a circle if you know the radius.
2. **Write** the formula for the circumference of a circle if you know the radius.
3. **Tell** how to estimate the circumference of a circle if you know the diameter.

Guided Practice

Find the circumference of each circle described below.

4. 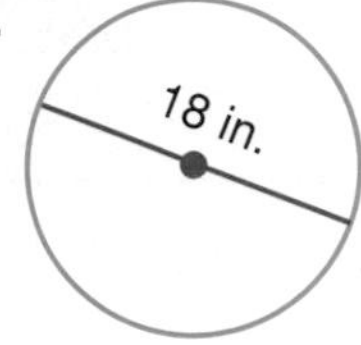

5. 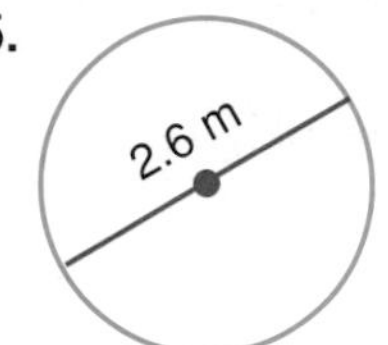

6. 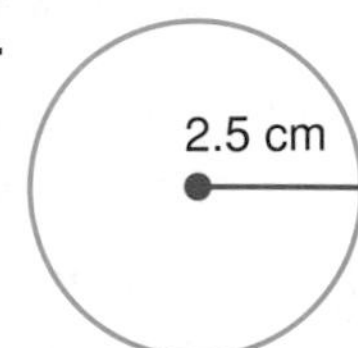

7. 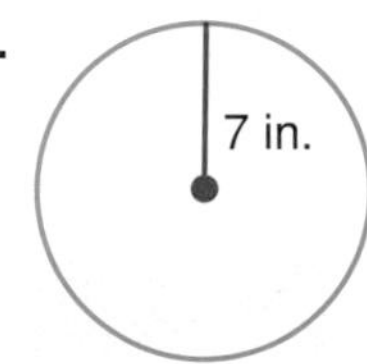

8. The diameter is 13.6 meters.
9. The diameter is $5\frac{1}{4}$ inches.
10. The radius is 3.5 kilometers.
11. The radius is $\frac{1}{2}$ foot.

Exercises

Independent Practice

Find the circumference of each circle described below.

12. 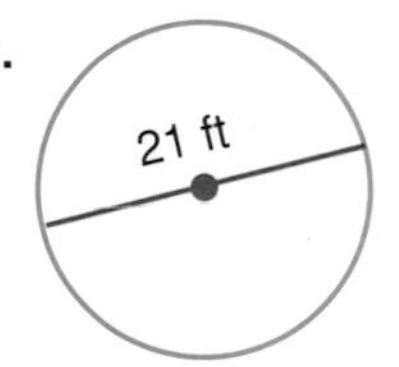

13. 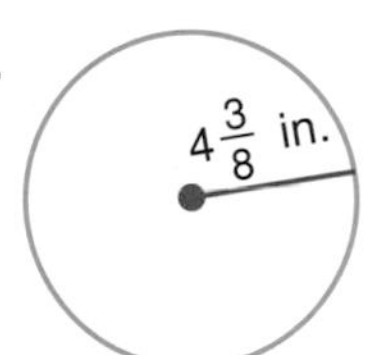

14. 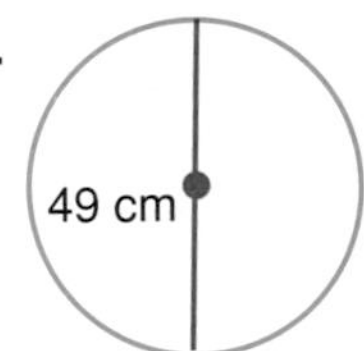

15. 

16. The diameter is 36 inches.
17. The radius is 8 feet.
18. The radius is 3.4 centimeters.
19. The diameter is 8.8 meters.
20. The diameter is $2\frac{1}{3}$ feet.
21. The radius is $4\frac{1}{2}$ yards.

Mixed Review

22. Find the value of the expression 2^6. *(Lesson 1-9)*
23. **Statistics** The chart below shows history test scores. *(Lesson 4-4)*

71	83	67	91	73	75	84	94	71	65
58	76	85	82	93	72	88	97	79	62

a. Make a stem-and-leaf plot of these scores.
b. In what interval do most of the scores lie?

24. State the measure of the bases and the height of the trapezoid at the right. Then find the area of the trapezoid. *(Lesson 7-7)*

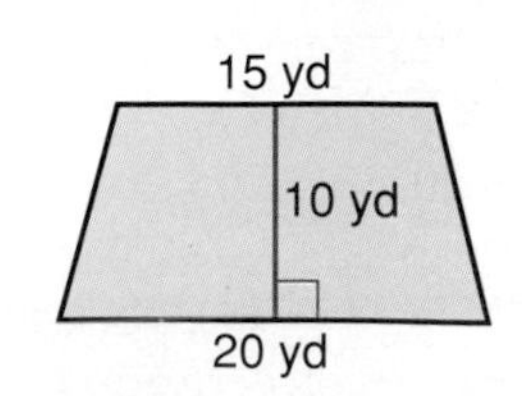

Problem Solving and Applications

25. **Manufacturing** A label is to be placed on a can. The can has a diameter of 9.1 centimeters and a height of 7.5 centimeters. What shape and size must the label be in order to fit the can exactly?

26. **Critical Thinking** Three tennis balls are packaged one on top of the other in a can. Which measure is greater, the height or the circumference of the can? Explain.

27. **Astronomy** The diameter of Saturn at its equator is about 75,100 miles. Find the approximate circumference of Saturn at its equator.

28. **Portfolio Suggestion** Review the items in your portfolio. Make a table of contents of the items, noting why each item was chosen. Replace any items that are no longer appropriate.

29. **Sports** The diameter of a basketball rim is 18 inches.
    a. Find the circumference of the basketball rim.
    b. The circumference of a standard-size basketball is 30 inches. About how much room is there between the rim and the ball if it goes exactly in the center of the rim?

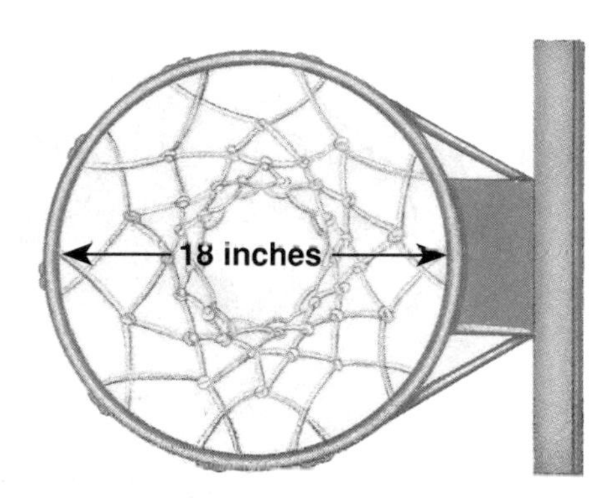

30. **Data Search** Refer to pages 254 and 255. What fractional part of the total number of Cy Young Award winners are left-handed?

CULTURAL KALEIDOSCOPE

An Wang

An Wang (1920-1990) was born in Shanghai, China and later came to the United States to study technology. He graduated from Harvard University with a Ph.D. in physics. His first major invention in computer technology was the magnetic-core computer memory, which he sold to IBM. With the profits from that sale, he financed his own business and Wang Laboratories was born.

Wang soon became the front-runner in the computer industry, and his inventions are used for data and text processing, telecommunications, and network processing.

In 1986, *Forbes* magazine listed Wang as one of the wealthiest people in America. On July 4, 1986, at the 100th anniversary celebration of the Statue of Liberty, President Ronald Reagan awarded Mr. Wang the Medal of Liberty—an award to honor naturalized citizens who have made significant contributions to society.

7-9 Dividing Fractions

Objective
Divide fractions.

In 1980, only $\frac{2}{5}$ of the radio stations in the United States were FM stations. Today, nearly $\frac{1}{2}$ of the 9,356 radio stations are FM stations. Approximately how many stations are FM stations?

You could find this number in two ways.

1. Divide by 2.

   9356 [÷] 2 [=] 4678

2. Multiply by $\frac{1}{2}$.

   9356 [×] 0.5 [=] 4678 *$\frac{1}{2} = 0.5$*

There are about 4,678 FM stations.

In the example above, notice that dividing by 2 and multiplying by $\frac{1}{2}$ give you the same result.

"When am I ever going to use this?"

Suppose it takes you $\frac{3}{4}$ of a minute to answer each multiple choice question on a test. How many questions can you complete in 15 minutes?

Dividing Fractions	**In words:** To divide by a fraction, multiply by its multiplicative inverse.
	Arithmetic $3 \div \frac{3}{4} = 3 \cdot \frac{4}{3}$
	Algebra $\frac{a}{b} \div \frac{c}{d} = \frac{a}{b} \cdot \frac{d}{c}$, where $b, c, d \neq 0$

Example 1 *Problem Solving*

Food **Mrs. Rodriguez had $\frac{1}{2}$ of a square cake left for lunch. She divided it into 6 equal parts for her family. What part of the cake will each person receive?**

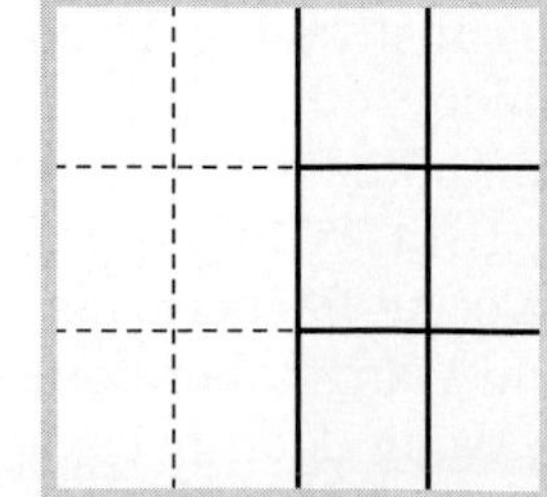

$\frac{1}{2} \div 6 = \frac{1}{2} \div \frac{6}{1}$ *Rename 6 as $\frac{6}{1}$.*

$= \frac{1}{2} \cdot \frac{1}{6}$ *Dividing by $\frac{6}{1}$ is the same as multiplying by $\frac{1}{6}$.*

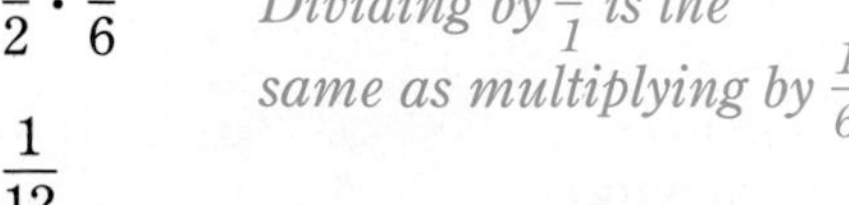

$= \frac{1}{12}$

Each serving is $\frac{1}{12}$ of the cake. Compare with the model.

Use the rules of signs for dividing integers when you divide rational numbers.

Examples

Solve each equation.

Fractions like $\frac{12}{-3\frac{1}{3}}$ are called complex fractions.

2 $y = \frac{12}{-3\frac{1}{3}}$ — *Estimate: 12 ÷ −3 = −4*

$y = 12 \div \left(-3\frac{1}{3}\right)$ — *Remember that the fraction bar means division.*

$y = \frac{12}{1} \cdot \left(-\frac{3}{10}\right)$ — *Dividing by $-3\frac{1}{3}$ or $-\frac{10}{3}$ is the same as multiplying by $-\frac{3}{10}$.*

$y = \frac{\overset{6}{\cancel{12}}}{1} \cdot \left(-\frac{3}{\underset{5}{\cancel{10}}}\right)$

$y = -\frac{18}{5}$ or $-3\frac{3}{5}$ — *Rename as a mixed number in simplest form.*

3 $x = -\frac{9}{4} \div \left(-\frac{3}{8}\right)$

$x = -\frac{9}{4} \cdot \left(-\frac{8}{3}\right)$

$x = -\frac{\overset{3}{\cancel{9}}}{\underset{1}{\cancel{4}}} \cdot \left(-\frac{\overset{2}{\cancel{8}}}{\underset{1}{\cancel{3}}}\right)$

$x = 6$

4 $p = -3\frac{3}{4} \div 6\frac{2}{3}$

$p = -\frac{15}{4} \div \frac{20}{3}$

$p = -\frac{15}{4} \cdot \frac{3}{20}$

$p = -\frac{\overset{3}{\cancel{15}}}{4} \cdot \frac{3}{\underset{4}{\cancel{20}}}$

$p = -\frac{9}{16}$

Checking for Understanding

Communicating Mathematics

Read and study the lesson to answer each question.

1. **Tell** why b, c, or d cannot equal zero in the expression $\frac{a}{b} \div \frac{c}{d}$.
2. **Write** the quotient of 3 and $\frac{3}{4}$.
3. **Tell** how the model at the right shows the quotient of 3 and $\frac{3}{4}$.
4. **Draw** a model that shows how to find the quotient of $2\frac{1}{2} \div \frac{5}{8}$.
5. **Tell** whether $10 \div \frac{1}{2}$ is greater or less than 10. Explain.

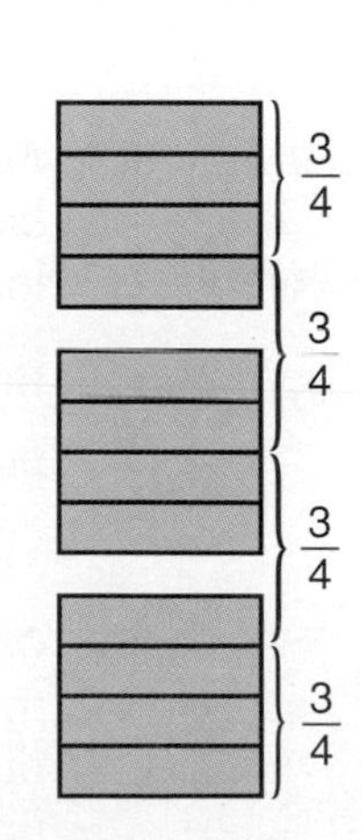

Guided Practice State a multiplication expression for each division expression. Then compute.

6. $\frac{2}{3} \div \frac{5}{6}$
7. $\frac{3}{8} \div \frac{9}{10}$
8. $-\frac{5}{6} \div \frac{2}{9}$
9. $\frac{4}{9} \div 6$
10. $-12 \div \left(-3\frac{3}{8}\right)$
11. $1\frac{1}{3} \div 2\frac{2}{9}$
12. $15 \div 2\frac{8}{11}$
13. $7\frac{5}{9} \div (-8)$
14. What is the value of $\frac{3\frac{3}{5}}{10}$?
15. Find the quotient of -6 divided by $-1\frac{1}{2}$.

Exercises

Independent Practice Solve each equation. Write each solution in simplest form.

16. $a = \frac{3}{4} \div \frac{5}{6}$
17. $10 \div (-2) = t$
18. $b = \frac{8}{9} \div \frac{6}{7}$
19. $7\frac{1}{3} \div 1\frac{2}{9} = d$
20. $n = \frac{3}{8} \div (-6)$
21. $4\frac{1}{2} \div \frac{3}{4} = h$
22. $j = \frac{9}{10} \div 6$
23. $2\frac{2}{3} \div 4 = y$
24. $g = -3\frac{3}{4} \div \left(-2\frac{1}{2}\right)$

Evaluate each expression.

25. $c \div d$ if $c = \frac{1}{2}$ and $d = 2\frac{1}{3}$
26. $a^2 \div b^2$ if $a = -\frac{2}{3}$ and $b = \frac{4}{5}$
27. $x + y \div z$ if $x = \frac{3}{4}$, $y = 0.5$, and $z = \frac{1}{4}$

Mixed Review

28. Solve $\frac{b}{2} + 7 > 3$. Show the solution on the number line. *(Lesson 2-10)*
29. Solve $r = (-8) + 12 + 3 + 9$. Check by solving another way. *(Lesson 3-4)*
30. Express 15.363636... using bar notation. *(Lesson 6-7)*
31. **Geometry** Find the circumference of a circle with a radius of 5 millimeters. *(Lesson 7-8)*

Problem Solving and Applications

32. **Advertising** A page of sale items is to be folded into thirds before it is stapled and mailed. If the page is 11 inches long, how wide is each section?
33. **Home Economics** How many slices of pepperoni, each $\frac{1}{16}$ inch thick, can be cut from a stick 8 inches long?
34. **Critical Thinking** A positive number is both multiplied and divided by the same rational number n, where $0 < n < 1$. Which is greater, the product or the quotient? Explain your reasoning.

Algebra Connection

7-10 Solving Equations

Objective

Solve equations with rational number solutions.

The West Montpelier track team qualified to compete in an international track meet to be held in Montreal. One of the track members is Terry O'Malley. He read that the temperature there would be about 20 degrees Celsius (°C). His coach said they could find out the temperature in degrees Fahrenheit (°F) using the formula, $C = \frac{5}{9}(F - 32)$. What is the temperature in degrees Fahrenheit for Montreal?

You can apply the skills you have learned for rational numbers to solve equations containing rational numbers, such as $C = \frac{5}{9}(F - 32)$.

$$C = \frac{5}{9}(F - 32)$$

$$20 = \frac{5}{9}(F - 32)$$ *Replace C with 20.*

$$\frac{9}{5} \cdot 20 = \frac{9}{5} \cdot \frac{5}{9}(F - 32)$$ *Multiply each side by $\frac{9}{5}$, the multiplicative inverse of $\frac{5}{9}$.*

$$36 = F - 32$$

$$36 + 32 = F - 32 + 32$$ *Add 32 to each side.*

$$68 = F$$

The equivalent Fahrenheit temperature is 68°.

Example 1

Solve $t + 0.25 = -4.125$. Check your solution.

$$t + 0.25 = -4.125$$

$$t + 0.25 - 0.25 = -4.125 - 0.25$$ *Subtract 0.25 from each side.*

4.125 [+/–] [–] 0.25 [=] -4.375

$$t = -4.375$$

Check: 4.375 [+/–] [+] 0.25 [=] -4.125 ✓

The solution is –4.375.

Examples

Solve each equation. Check each solution.

2 $\frac{3}{4}y = -\frac{7}{8}$

$$\frac{4}{3} \cdot \frac{3}{4}y = \left(\frac{4}{3}\right)\left(-\frac{7}{8}\right)$$ *Multiply each side by $\frac{4}{3}$.*

$$y = -\frac{7}{6} \text{ or } -1\frac{1}{6}$$

Check: $\frac{3}{4}y = -\frac{7}{8}$

$$\frac{3}{4}\left(-\frac{7}{6}\right) \stackrel{?}{=} -\frac{7}{8}$$

$$-\frac{7}{8} = -\frac{7}{8} \checkmark$$

3 $7m + \left(-\frac{2}{9}\right) = \frac{5}{9}$

$$7m + \left(-\frac{2}{9}\right) = \frac{5}{9}$$

$$7m + \left(-\frac{2}{9}\right) + \frac{2}{9} = \frac{5}{9} + \frac{2}{9}$$ *Add $\frac{2}{9}$ to each side.*

$$7m = \frac{7}{9}$$

$$7m \cdot \frac{1}{7} = \frac{7}{9} \cdot \frac{1}{7}$$ *Multiply each side by $\frac{1}{7}$, the multiplicative inverse of 7.*

$$m = \frac{1}{9}$$

Check: $7m + \left(-\frac{2}{9}\right) = \frac{5}{9}$

$$7\left(\frac{1}{9}\right) + \left(-\frac{2}{9}\right) \stackrel{?}{=} \frac{5}{9}$$

$$\frac{5}{9} = \frac{5}{9} \checkmark$$ The solution is $\frac{1}{9}$.

Checking for Understanding

Communicating Mathematics

Read and study the lesson to answer each question.

1. **Tell** how you would solve $\frac{2}{3}x - 4 = 8$.
2. **Write** an equation that you would use the multiplicative inverse of $-1\frac{5}{8}$ to solve.

Guided Practice

Solve each equation. Check your solution.

3. $1.1 + y = -4.4$
4. $-\frac{5}{8}x + \frac{1}{6} = \frac{3}{5}$
5. $3.6 = \frac{c}{0.9}$
6. $\frac{2}{3}h - (-3) = 6$
7. $2\frac{1}{2}d = 5\frac{3}{4}$
8. $-\frac{8t}{5} = 4$
9. $\frac{b}{1.5} - 13 = 2.2$
10. $-12 = -\frac{z}{7}$
11. $-11g + 15 = 12.5$

Exercises

Independent Practice

Solve each equation. Check your solution.

12. $2x = -12$
13. $\frac{t}{3} = -6$
14. $a - (-0.03) = 3.2$
15. $-\frac{1}{4}c = 3.8$
16. $\frac{y}{3.2} = -4.5$
17. $-1.6w + 3.5 = 0.48$
18. $\frac{n}{2} = -1.6$
19. $\frac{6d}{2} = -0.36$
20. $\frac{m}{2.3} - 1.3 = -5.2$
21. $k + \frac{2}{3} = -\frac{4}{9}$
22. $-\frac{3}{5}h = \frac{2}{3}$
23. $4p - \frac{1}{5} = -7\frac{2}{5}$
24. What is the solution of $z - \frac{2}{5} = -2$?

Mixed Review

25. Estimate 618 + 182 + 375 by rounding. *(Lesson 1-3)*
26. **Statistics** Determine whether a scatter plot of the number of base hits to the number of runs scored would show a positive, negative, or no relationship. *(Lesson 4-8)*
27. Solve $46 = 8k - 2$. Check your solution. *(Lesson 2-7)*
28. Find the LCM of 9 and 30. *(Lesson 6-9)*
29. Solve $g = \frac{5}{6} \div \frac{4}{3}$. *(Lesson 7-9)*

Problem Solving and Applications

30. **Critical Thinking** Using the formula at the beginning of this lesson, explain how you could find the Fahrenheit temperature when the Celsius temperature is 0°.
31. **Safe Driving** The graph below shows the distance needed to stop an automobile traveling at various speeds. Each of the given formulas approximates the stopping distance, d (in feet), based on the speed, s, in miles per hour. Determine which speed is the closest match for each of the given formulas.

a. $d = 3s$ b. $d = 4s$ c. $d = 2\frac{1}{2}s$

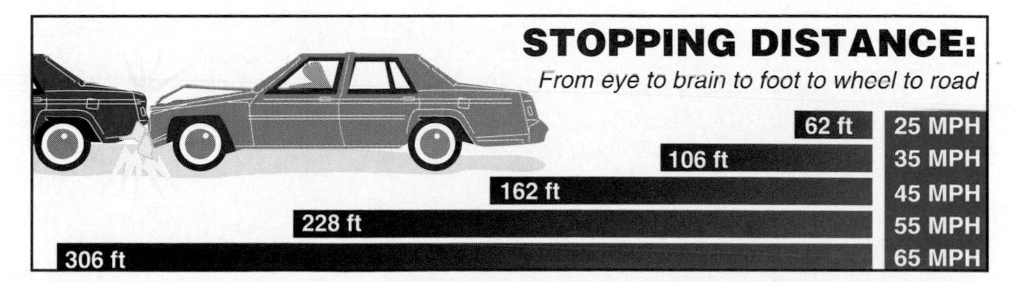

d. Look for a pattern in the graph above. What conclusion can you make about the stopping distance and the speed of a car?

Chapter 7 Study Guide and Review

Communicating Mathematics

Choose the letter that best matches each phrase.

1. the sum of a whole number and a fraction
2. the LCM of 2 and 6
3. a sequence whose terms increase or decrease by a constant factor
4. a sequence having the same difference between any two consecutive terms
5. a quadrilateral with exactly one pair of parallel sides
6. the distance across a circle through the center
7. the distance around a circle
8. the number you would multiply each side of the equation $\frac{3}{4}t = -\frac{5}{8}$ by to solve it

a. $\frac{3}{4}$
b. $\frac{4}{3}$
c. 6
d. 12
e. diameter
f. circumference
g. mixed number
h. geometric sequence
i. arithmetic sequence
j. trapezoid

9. In your own words, describe how to multiply two mixed numbers.

Self Assessment

Objectives and Examples

Upon completing this chapter, you should be able to:

Review Exercises

Use these exercises to review and prepare for the chapter test.

- add and subtract fractions with like denominators *(Lesson 7-1)*

Solve $b = \frac{1}{5} - \frac{4}{5}$.

$b = \frac{1-4}{5}$ or $-\frac{3}{5}$

Solve each equation. Write each solution in simplest form.

10. $\frac{2}{7} + \frac{3}{7} = n$
11. $w = -\frac{1}{8} - \frac{5}{8}$
12. $x = \frac{5}{12} + \frac{7}{12}$

- add and subtract fractions with unlike denominators *(Lesson 7-2)*

Solve $h = \frac{1}{2} + \frac{2}{3}$.

$h = \frac{3}{6} + \frac{4}{6}$

$h = \frac{7}{6}$ or $1\frac{1}{6}$

Solve each equation. Write each solution in simplest form.

13. $m = -\frac{3}{5} + \frac{1}{3}$
14. $z = 4 - 2\frac{3}{5}$
15. $t = -4\frac{2}{3} + \left(-6\frac{3}{4}\right)$

Objectives and Examples	Review Exercises
• multiply fractions *(Lesson 7-3)* Solve $f = -\frac{7}{8} \cdot \frac{1}{2}$. $f = \frac{-7 \cdot 1}{8 \cdot 2}$ $f = -\frac{7}{16}$	**Solve each equation. Write each solution in simplest form.** **16.** $p = \left(-\frac{1}{6}\right)\left(-\frac{3}{5}\right)$ **17.** $2\frac{2}{5}\left(-4\frac{3}{8}\right) = s$ **18.** $-\frac{7}{10} \cdot \frac{4}{7} = k$ **19.** $g = \frac{4}{9} \cdot 5\frac{1}{4}$
• identify and use rational number properties *(Lesson 7-4)* Find $3\frac{1}{4} \cdot 4$. $3\frac{1}{4} \cdot 4 = 4 \cdot 3\frac{1}{4}$ $= 4\left(3 + \frac{1}{4}\right)$ $= 4(3) + 4\left(\frac{1}{4}\right)$ $= 12 + 1$ or 13	**Name the multiplicative inverse of each of the following.** **20.** $\frac{5}{7}$ **21.** $-6\frac{1}{3}$ **Find each product.** **22.** $5 \cdot 7\frac{2}{5}$ **23.** $\left(-\frac{4}{5}\right)\left(-3\frac{1}{2}\right)$
• recognize and extend arithmetic and geometric sequences *(Lesson 7-6)* State whether the sequence 3, 6, 12, 24, ... is geometric. Then write the next three terms. Since there is a common ratio, 2, the sequence is geometric. The next three terms are 48, 96, and 192.	**State whether each sequence is arithmetic, geometric, or neither. Then write the next three terms of each sequence.** **24.** −10, −7, −4, −1, ... **25.** 27, 9, 3, 1, ... **26.** 6, 13, 19, 24, ... **27.** 60, 56, 52, 48, ...
• find the areas of triangles and trapezoids *(Lesson 7-7)* Find the area of the triangle. 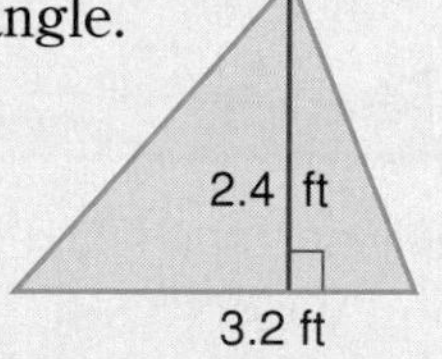 $A = \frac{1}{2}bh$ $A = \frac{1}{2}(3.2)(2.4)$ $A = 3.84$ The area is 3.84 square feet.	**Find the area of each figure described below.** **28.** triangle; base, $2\frac{1}{2}$ cm; height, $3\frac{1}{4}$ cm **29.** trapezoid; bases, 10 m and 7 m; height, 6 m
• find the circumference of circles *(Lesson 7-8)* Find the circumference of the circle. 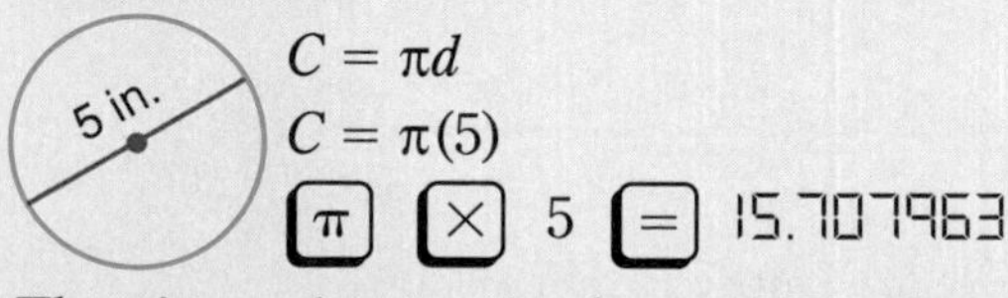 $C = \pi d$ $C = \pi(5)$ The circumference is about 15.7 inches.	**Find the circumference of each circle described below.** **30.** The diameter is $3\frac{1}{3}$ feet. **31.** The radius is 2.4 meters. **32.** The radius is 19 yards. **33.** The diameter is 5.5 inches.

Study Guide and Review

Objectives and Examples	***Review Exercises***
• divide fractions *(Lesson 7-9)* Solve $r = \frac{5}{8} \div \left(-\frac{3}{4}\right)$. $r = \frac{5}{8} \cdot \left(-\frac{4}{3}\right)$ $r = -\frac{5}{6}$	Solve each equation. Write each solution in simplest form. **34.** $i = -\frac{7}{9} \div \left(-\frac{2}{3}\right)$ **35.** $2\frac{2}{5} \div 4 = f$ **36.** $3\frac{1}{7} \div \left(-2\frac{1}{5}\right) = d$ **37.** $q = \frac{3}{4} \div 3\frac{3}{5}$
• solve equations with rational number solutions *(Lesson 7-10)* Solve $\frac{2}{3}n = \frac{5}{12}$. $\frac{3}{2} \cdot \frac{2}{3}n = \frac{3}{2} \cdot \frac{5}{12}$ $n = \frac{5}{8}$	Solve each equation. **38.** $\frac{x}{6} = -4.3$ **39.** $-6.2 = \frac{e}{1.7} + 4$ **40.** $2b - \frac{4}{7} = 6\frac{1}{7}$ **41.** $t - (-0.9) = 5$

Applications and Problem Solving

42. Bob rides in the March Bike Marathon. He rides 12 miles the first day, 18 the second, 27 the third. If he continues this pattern, how many miles will Bob have ridden in all by the end of the fourth day? *(Lesson 7-5)*

43. The length of the minute hand on a clock is 9 centimeters. How far does the end of the minute hand travel as it moves from 12 to 4? *(Lesson 7-8)*

Curriculum Connection Projects

- **Business** From the Business section of today's newspaper, record the "volume" and "net change" for the "Top Percent Losers" and "Top Percent Gainers" and find total gain or loss for each stock.
- **Sports** Find the circumference of the three-point circle on a basketball court.
- **Sports** Use triangles to find the area of the infield of a baseball field.

Read More About It

Jordan, Sheryl. *A Time of Darkness.*
Keightley, Moy. *Investigating Art: A Practical Guide for Young People.*
Miller, Marvin. *You Be the Jury.*

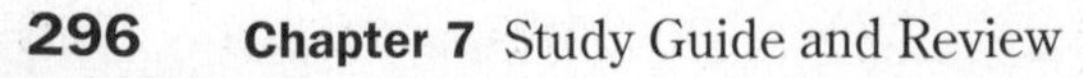

Chapter 7

Test

1. What is the multiplicative inverse of $1\frac{1}{9}$?
2. Find the product of $5\frac{1}{3}$ and $-2\frac{1}{4}$.

Solve each equation. Write each solution in simplest form.

3. $-6\frac{3}{7} + 4\frac{6}{7} = m$
4. $\frac{8}{15} + \frac{8}{15} = a$
5. $\frac{5}{8} - \frac{4}{5} = g$
6. $\frac{33}{40} \cdot \frac{8}{9} = t$
7. $c = -\frac{7}{8} \div 2\frac{4}{5}$
8. $5\frac{5}{6} \div \left(-1\frac{2}{3}\right) = x$
9. Write the first four terms in an arithmetic sequence with a common difference of 2. The first term is 20.
10. The fifth term of a sequence is 6. The common ratio is $\frac{1}{2}$. Find the first four terms.

Find the area of each triangle described below.

	base	height
11.	12 m	19 m
12.	$4\frac{1}{2}$ ft	$6\frac{3}{4}$ ft

Find the area of each trapezoid described below.

	base *(a)*	base *(b)*	height
13.	15 in.	26 in.	$9\frac{1}{2}$ in.
14.	3.7 mm	5.4 mm	8.2 mm

Find the circumference of each circle described below.

15. The diameter is 6.3 yards.
16. The radius is $2\frac{5}{8}$ meters.

Solve each equation. Check your solution.

17. $2\frac{1}{2}w = 4\frac{3}{8}$
18. $\frac{a}{2.8} - 6.8 = 12$
19. $\frac{1}{4} = -\frac{5}{6}x + \frac{9}{16}$
20. **Archeology** Arrowheads from a prehistoric site in Montana have been arranged in rows in a museum showcase. The first row contains 12 arrowheads. The second row contains 10 arrowheads. The third row has 13, the fourth row has 11, and so on, until the last row contains 13 arrowheads. Find the pattern and determine how many rows of arrowheads there were in the showcase.

Bonus Does zero have a multiplicative inverse? Why or why not?

Chapter Test

Chapter 8

Real Numbers

Spotlight on Animals

Have You Ever Wondered...

- What animal can move the quickest?
- How long different animals live?

Length of Life for Animals

Mammals (Average life span in years)		Birds (Maximum life span in years)	
Chimpanzee	40-50	Blue jay	4
Elephant	60	Canada goose	32
Grizzly Bear	20	Canary	24
Horse	20-30	Cardinal	22
Lion	20-25	Ostrich (African)	50
Mouse	1-2	Penguin (king)	26
Squirrel	9	Raven	69
Tiger	11	Robin	12
Fish (Maximum life span in years)		**Reptiles and amphibians (Maximum life span in years)**	
Electric Eel	11	Alligator	56
Flounder	10	Bullfrog	15
Goldfish	25	Crocodile	13
Perch	11	Garter snake	6
Sea horse	4	Gila monster	20
Sturgeon	50	Rattlesnake	18
Trout (rainbow)	4	Turtle (box)	123

1870 | 1890 | 1910 | 1930

1872 — *First National Park declared at Yellowstone*

1916 — *National Park System Formed*

1935 — *Game of Monopoly designed*

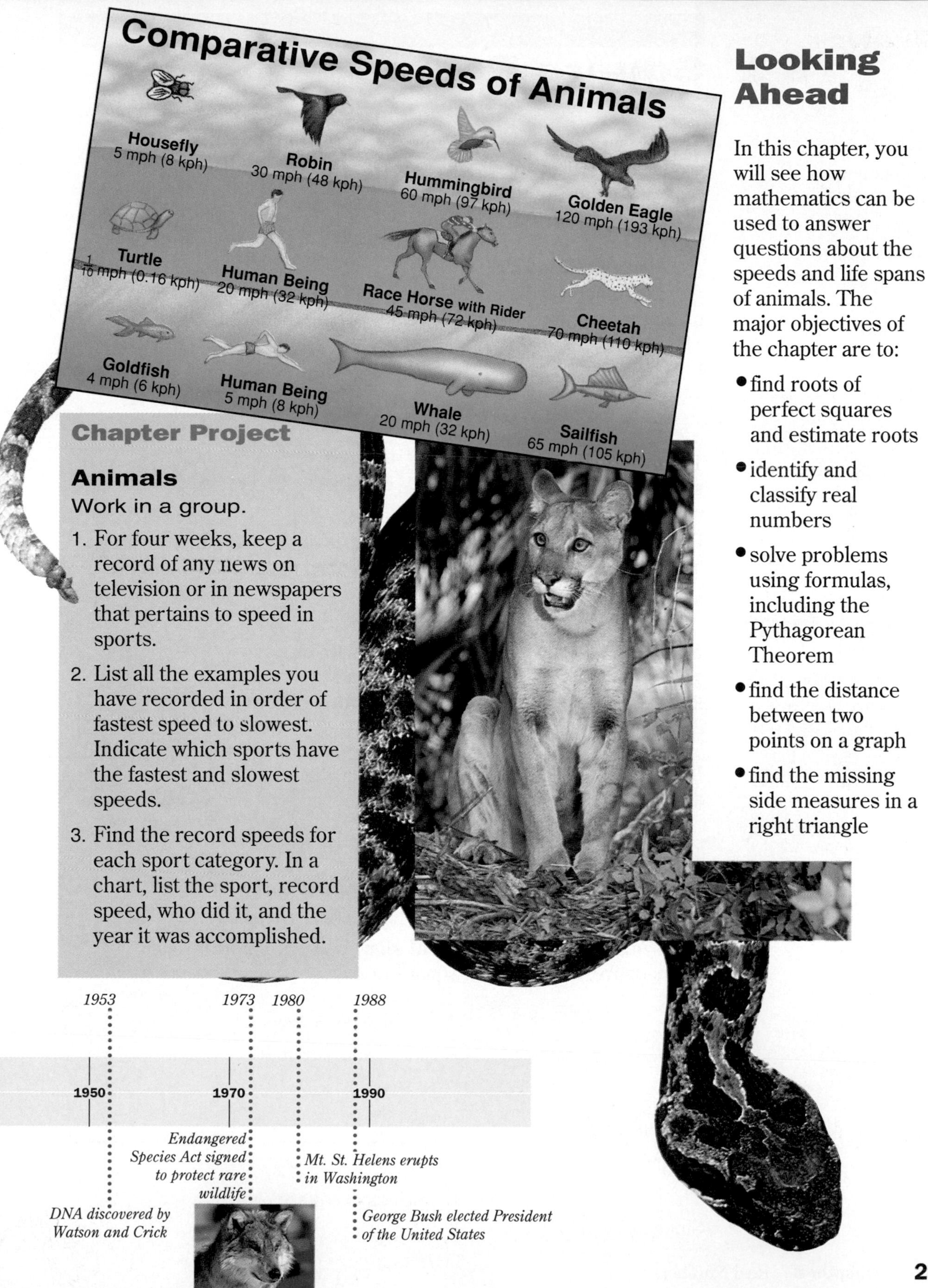

Chapter Project

Animals

Work in a group.

1. For four weeks, keep a record of any news on television or in newspapers that pertains to speed in sports.
2. List all the examples you have recorded in order of fastest speed to slowest. Indicate which sports have the fastest and slowest speeds.
3. Find the record speeds for each sport category. In a chart, list the sport, record speed, who did it, and the year it was accomplished.

Looking Ahead

In this chapter, you will see how mathematics can be used to answer questions about the speeds and life spans of animals. The major objectives of the chapter are to:

- find roots of perfect squares and estimate roots
- identify and classify real numbers
- solve problems using formulas, including the Pythagorean Theorem
- find the distance between two points on a graph
- find the missing side measures in a right triangle

8-1 Square Roots

Objective

Find square roots of perfect squares.

Words to Learn

perfect square
square root
radical sign
principal square root

The Great Pyramid at Giza, built around 2600 B.C., is one of the "Seven Wonders of the Ancient World." It measures 776 feet on each side of the square base. The area covered by the base of the Great Pyramid can be found by using the formula $A = s^2$, where s feet is the length of the side of the base.

776 $\boxed{x^2}$ 602176

The area of the base is 602,176 square feet.

Products such as 602,176 that are squares of rational numbers are called **perfect squares.** Some other perfect squares are 25, 0.04, and $\frac{4}{9}$.

More than 2 million stone blocks, each weighing between 2 tons and 150 tons, were used in the construction of the Great Pyramid of Gizeh.

Notice the relationship that exists between the area of the square shown at the right and the length of its side. We say that 5 is a **square root** of 25, because $5^2 = 25$.

It is also true that $(-5)^2 = 25$. This suggests that another square root of 25 is -5.

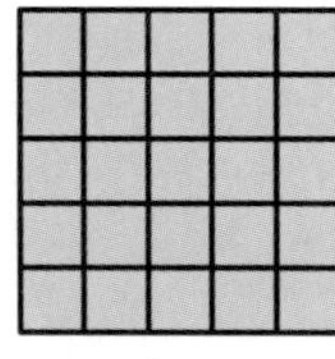

$5^2 = 25$

Square Root If $x^2 = y$, then x is a square root of y.

The symbol $\sqrt{}$, called a **radical sign,** is used to indicate a nonnegative, or **principal,** square root.

$\sqrt{25} = 5$ *principal square root*

$-\sqrt{25} = -5$ *negative square root*

Example 1

Find $\sqrt{144}$.

The symbol $\sqrt{144}$ indicates the principal square root.
Since $12^2 = 144$, $\sqrt{144} = 12$

Calculator Hint

$\sqrt{x}$ is the square root key. When this key is pressed, the calculator replaces the number in the display with its nonnegative square root.

Example 2

Find $-\sqrt{2.25}$.

Use your calculator. 2.25 $\sqrt{x}$ 1.5

The symbol $-\sqrt{}$ indicates the negative square root.

Since $1.5^2 = 2.25$, $-\sqrt{2.25} = -1.5$.

Example 3 *Connection*

Geometry **The area of a square is 256 square inches. Find its perimeter.**

256 square inches

First find the length of each side.

256 $\sqrt{x}$ 16

The length of each side is 16 inches.

$P = 4s$

$P = 4 \cdot 16$ *Replace s with 16.*

$P = 64$ The perimeter is 64 inches.

You can review area and perimeter of squares on page 75.

Checking for Understanding

Communicating Mathematics

Read and study the lesson to answer each question.

1. **Tell** why 36 is a perfect square.
2. **Write** the symbol for the principal square root of 100.
3. **Draw** and label a square that has an area of 16 square centimeters.

Guided Practice

Find each square root.

4. $\sqrt{49}$ 5. $\sqrt{81}$ 6. $\sqrt{121}$ 7. $-\sqrt{64}$

Exercises

Independent Practice

Find each square root.

8. $\sqrt{25}$ 9. $\sqrt{400}$ 10. $\sqrt{225}$ 11. $-\sqrt{9}$

12. $\sqrt{196}$ 13. $\sqrt{625}$ 14. $-\sqrt{289}$ 15. $-\sqrt{100}$

16. $\sqrt{\frac{4}{9}}$ 17. $\sqrt{0.16}$ 18. $-\sqrt{2.89}$ 19. $\sqrt{\frac{64}{100}}$

20. Find two square roots of 225. Explain which is the principal square root.
21. **Geometry** If the area of a square is 1.69 square meters, what is the length of its side?

Mixed Review

22. How many yards are in 270 inches? *(Lesson 1-7)*
23. Graph the points $C(2, 5)$, $H(-3, 3)$, $E(4, 1)$, and $G(-1, -2)$ on the same coordinate plane. *(Lesson 3-10)*
24. Find the GCF of 12 and 63. *(Lesson 6-4)*
25. Solve $3x - 5 = -6$. *(Lesson 7-11)*

Problem Solving and Applications

26. **Critical Thinking** Is the product of two perfect squares always a perfect square? Explain why or why not.

27. **Gardening** The area of a square garden is 289 square feet. How much will it cost to fence the garden if fencing costs $0.35 per foot?

28. **Journal Entry** In your own words, write a definition of square root.

SAVE THE PLANET • SAVE THE PLANET •

Save Planet Earth

Disposable Batteries Do you have a game that requires batteries? Did you know that household batteries contain heavy metals such as mercury and cadmium? When these batteries are thrown out with the trash and taken to landfills, they break apart and release the metals into the soil. When batteries are incinerated, the toxic substances are released into the air.

Americans use 2 billion disposable batteries every year. The annual use of mercury in batteries exceeds the federal limits for trash by 400%. Prolonged exposure to mercury can make people sick and affect their behavior.

How You Can Help

- Use rechargeable batteries. Even though they contain cadmium, they last longer and do not require frequent replacement.
- If possible, recycle alkaline batteries.

Cooperative Learning

8-2A Estimating Square Roots

A Preview of Lesson 8-2

Objective

Use models to estimate square roots.

Materials

tiles or
base-10 blocks

Suppose you have 50 tiles and want to arrange them into a square. Can you do it? No, a square cannot be built with 50 tiles. This suggests that 50 is not a perfect square.

In this Lab, you will estimate the square root of numbers that are not perfect squares.

Try this!

Work with a partner.

- Arrange 50 tiles into the largest square possible.

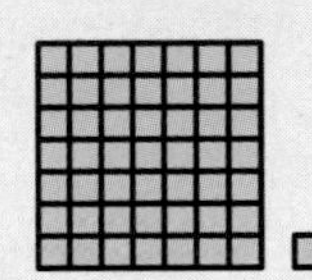

The square has 49 tiles, with one left over.

- Add tiles until you have the next larger square.

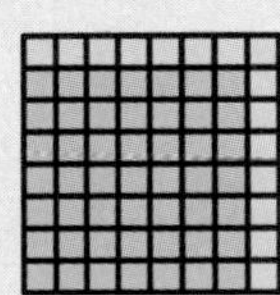

You need to add 14 tiles. This square has 64 tiles.

What do you think?

1. What is the square root of 49?
2. What is the square root of 64?
3. Between what two whole numbers is the square root of 50?
4. Is $\sqrt{50}$ closer to 7 or 8? Explain your reasoning.

Applications

For each given number, arrange tiles or base-10 blocks into the largest square possible. Then add tiles until you have the next larger square. To the nearest whole number, estimate the square root of each number.

5. 20
6. 76
7. 133
8. 150
9. 200
10. 2

8-2 Estimating Square Roots

Objective
Estimate square roots.

Have you ever had one of those days when everything went your way? You aced your geography test, got the last chocolate milk at lunch, and the experiment in science lab worked! More often than not, our days are not perfect and we have to make adjustments.

Similarly, most numbers are not perfect squares, and we have to approximate their square roots. For example, the number 160 is not a perfect square. However, we know that 160 is between two perfect squares, 144 and 169. So the square root of 160 is between 12 and 13.

$144 < 160 < 169$

$12^2 < 160 < 13^2$

$12 < \sqrt{160} < 13$

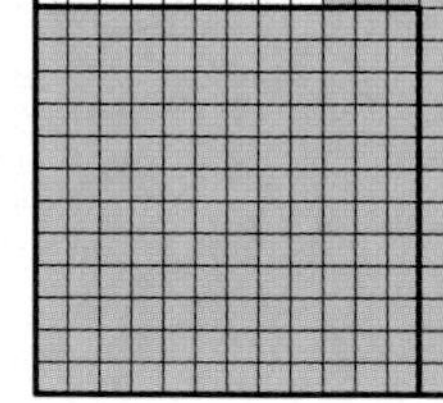

Since 160 is closer to 169 than 144, the best whole number estimate for the square root of 160 is 13.

Example 1

Estimate $\sqrt{90}$.

$81 < 90 < 100$ *81 and 100 are perfect squares.*

$9^2 < 90 < 10^2$

$9 < \sqrt{90} < 10$ Since 90 is closer to 81 than to 100, the best whole number estimate for $\sqrt{90}$ is 9.

Example 2 *Problem Solving*

Physics You can estimate the distance you can see to the horizon by using the formula $d = 1.22 \times \sqrt{h}$. In the formula d represents the distance you can see, in miles, and h represents the height your eyes are from the ground, in feet. Suppose your eyes are 5 feet from the ground. About how far can you see to the horizon?

$d = 1.22 \times \sqrt{5}$ *Replace h with 5.*

$\approx 1.22 \times 2$ *$\sqrt{5}$ is between 2 and 3.*

≈ 2.44

If your eyes are 5 feet from the ground, you can see about 2.4 miles to the horizon.

Checking for Understanding

Communicating Mathematics

Read and study the lesson to answer each question.

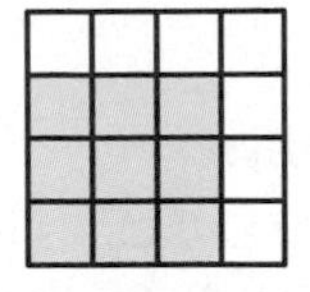

1. **Tell** how the drawing at the right can be used to estimate $\sqrt{12}$.
2. **Draw** a figure that can be used to explain why the square root of 75 is between 8 and 9.

Guided Practice

Estimate to the nearest whole number.

3. $\sqrt{50}$
4. $\sqrt{135}$
5. $\sqrt{29}$
6. $\sqrt{11}$

Exercises

Independent Practice

Estimate to the nearest whole number.

7. $\sqrt{23}$
8. $\sqrt{44}$
9. $\sqrt{56}$
10. $\sqrt{17.5}$
11. $\sqrt{47}$
12. $\sqrt{113}$
13. $\sqrt{175}$
14. $\sqrt{200}$
15. $\sqrt{408}$
16. $\sqrt{17.25}$
17. $\sqrt{957}$
18. $\sqrt{30.8}$

Mixed Review

19. Solve $s = 3.6 \div 0.6$. *(Lesson 2-4)*
20. Solve $4\frac{2}{3} - 6\frac{1}{4} = m$. *(Lesson 7-2)*
21. Find the principal square root of 900. *(Lesson 8-1)*

Problem Solving and Applications

22. **Traffic Safety** Police officers can estimate how fast a car was going by measuring the length of its skid marks. The formula $s = \sqrt{24d}$ can be used to estimate the speed on a dry, concrete road. In the formula, s is the speed in miles per hour, and d is the distance in feet the car skidded after its brakes were applied. What was the approximate speed of a car that left skid marks for 20 feet?

23. **Physics** Suppose you are in a hot air balloon that is flying at an altitude of 900 feet. About how far can you see to the horizon? Use the formula in Example 2 on page 304.

24. **Critical Thinking** A square has an area of 20 square units. Explain why the length of its side is between 4.4 and 4.5 units.

8-3 The Real Number System

Objective
Identify and classify numbers in the real number system.

Words to Learn
irrational number
real number

Swedish botanist Carolus Linnaeus (1707—1778) developed the system used today to classify every kind of living thing according to common characteristics. For example, a guinea pig is classified as a rodent, but it is also a mammal.

You can review rational numbers on page 264.

In mathematics, we classify numbers that have common characteristics. So far in this text, we have classified numbers into the following sets.

Natural Numbers	{1, 2, 3, 4, ...}
Whole Numbers	{0, 1, 2, 3, 4, ...}
Integers	{..., -2, -1, 0, 1, 2, ...}
Rational Numbers	{all numbers that can be expressed in the form $\frac{a}{b}$, where a and b are integers and $b \neq 0$}

Remember that terminating or repeating decimals are rational numbers since they can be expressed as fractions. Also, the square roots of perfect squares are rational numbers. For example, $\sqrt{0.09}$ is a rational number because $\sqrt{0.09} = 0.3$, a rational number.

We can summarize the classification of rational numbers in a Venn diagram.

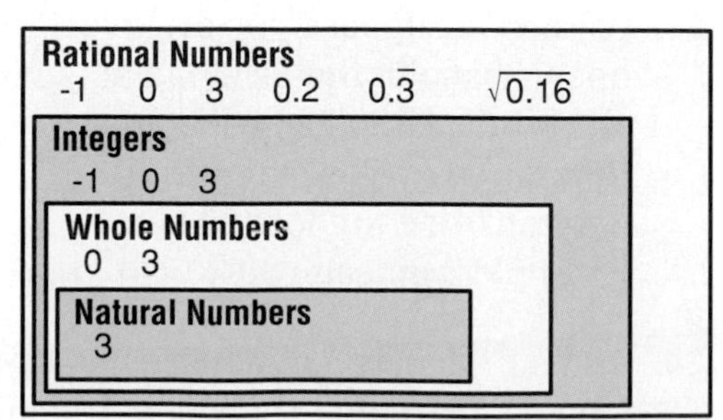

Numbers like $\sqrt{2}$ and $\sqrt{5}$ are the square roots of numbers that are *not* perfect squares. Notice what happens when you find these square roots with your calculator.

2 [$\sqrt{x}$] 1.4142136 . . .   5 [$\sqrt{x}$] 2.236068 . . .

The numbers continue forever without any pattern of repeating digits. These numbers are not rational numbers since they are not terminating or repeating decimals. Numbers like $\sqrt{2}$ and $\sqrt{5}$ are called **irrational numbers.**

Definition of Irrational Number	An irrational number is a number that cannot be expressed as $\frac{a}{b}$, where a and b are integers and b does not equal 0.

Examples

Determine whether each number is rational or irrational.

1 0.66666 . . .

This repeating number is a rational number since it can be expressed as $\frac{2}{3}$.

2 0.141141114 . . .

This decimal does not terminate and it does not repeat. It is an irrational number.

3 π

$\pi = 3.1415926\ldots$ It is an irrational number.

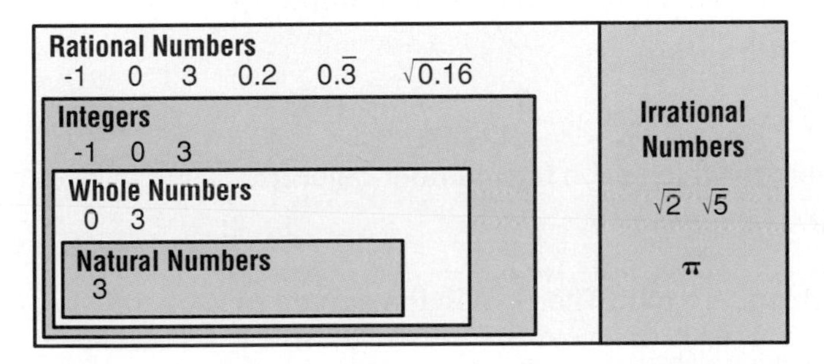

You have graphed rational numbers on a number line. But if you graphed all of the rational numbers, you would still have some "holes" in the number line. The irrational numbers "fill in" the number line. The set of rational numbers and the set of irrational numbers combine to form the set of **real numbers**.

The graph of all real numbers is the entire number line.

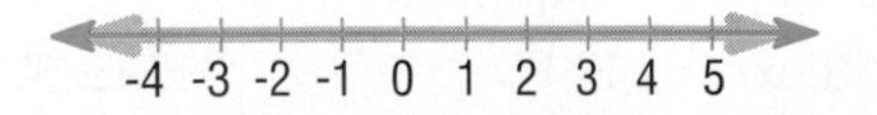

To graph irrational numbers, you can use a calculator or a table of squares and square roots to find approximate square roots in decimal form.

Example

4 **Graph $\sqrt{2}$, $\sqrt{5}$, and π on the number line.**

2 [$\sqrt{x}$] 1.4142136

5 [$\sqrt{x}$] 2.236068

[π] 3.1415927

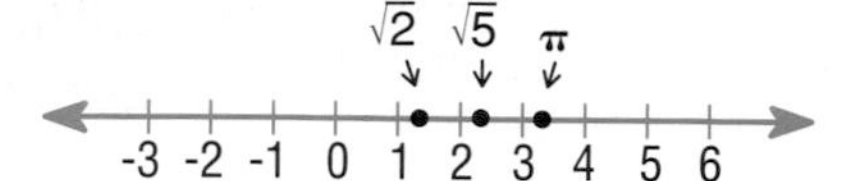

Throughout this text you have solved equations that have rational number solutions. Some equations have solutions that are irrational numbers. You can solve some equations that involve squares by taking the square root of each side.

Examples

5 **Solve $x^2 = 36$.**

$x^2 = 36$

$x = \sqrt{36}$ or $x = -\sqrt{36}$

$x = 6$ or $x = -6$

6 **Solve $x^2 = 180$.**

$x^2 = 180$

$x = \sqrt{180}$ or $-\sqrt{180}$

180 [$\sqrt{x}$] 13.416408

$x \approx 13.4$ or $x \approx -13.4$

Estimation Hint

In Example 6, THINK

$100 < 180 < 225$

$10 < \sqrt{180} < 15$

Checking for Understanding

Communicating Mathematics

Read and study the lesson to answer each question.

1. **Write** a definition of real numbers.
2. **Tell** whether 0.010010001 . . . is a rational or irrational number.
3. **Draw** a number line and graph $\sqrt{10}$.
4. **Tell** the solutions of $x^2 = 25$.

Guided Practice

Name the set or sets of numbers to which each real number belongs.

5. $\sqrt{5}$
6. $0.\overline{27}$
7. $-\sqrt{9}$
8. -2.5

Find an approximation for each square root. Then graph the square root on the number line.

9. $\sqrt{7}$
10. $\sqrt{8}$
11. $\sqrt{20}$
12. $-\sqrt{2}$

Solve each equation.

13. $x^2 = 144$
14. $x^2 = 900$
15. $y^2 = 50$

Exercises

Independent Practice

Name the set or sets of numbers to which each real number belongs.

16. 7
17. $\sqrt{11}$
18. $-\sqrt{36}$
19. $0.4545\ldots$
20. $\frac{5}{8}$
21. 6.06060606
22. $0.121121112\ldots$

Find an approximation for each square root. Then graph the square root on the number line.

23. $\sqrt{6}$
24. $\sqrt{50}$
25. $\sqrt{27}$
26. $\sqrt{108}$

Solve each equation. Round decimal answers to the nearest tenth.

27. $x^2 = 64$
28. $m^2 = 12$
29. $y^2 = 360$
30. $n^2 = 17$
31. $p^2 = 1.44$
32. $t^2 = 1$

Mixed Review

33. Solve $33 = 4x - 15$. *(Lesson 2-7)*
34. Solve $d = 28(-12)$. *(Lesson 3-6)*
35. **Statistics** Find the mean, median, and mode for the following set of data: 64, 52, 57, 65, 59, 61, 55, 50, 68. *(Lesson 4-5)*
36. Find the product of $-\frac{5}{8}$ and $-3\frac{2}{5}$. *(Lesson 7-3)*
37. Estimate $\sqrt{300}$ to the nearest whole number. *(Lesson 8-2)*

Problem Solving and Applications

38. **Physics** The formula $d = 16t^2$ represents the distance, d, in feet that an object falls in t seconds. Suppose a ball is rolled off a platform that is 25 feet above the ground. How long does it take for the ball to hit the ground?

39. **Algebra** In the geometric sequence 4, 12, __?__, 108, 324, the missing number is called the *geometric mean* of 12 and 108. It can be found by simplifying $\sqrt{ab}$ where a and b are the numbers on either side of the geometric mean. Find the missing number.

40. **Critical Thinking** Name all whole numbers whose square roots are between 3 and 4.

41. **Journal Entry** How can you remember the classifications of numbers? Can you think of a method for recalling the sets?

Problem-Solving Strategy

8-4 Use a Formula

Objective

Solve problems by using a formula.

Motion is all around us. People walk, run, ride bicycles and skateboards, and travel from place to place in automobiles, airplanes, and trains.

The Amtrak Commuter Train #620 runs from Philadelphia to Penn Station in New York City. It leaves 30th Street Station in Philadelphia at 5:41 A.M. and arrives at Penn Station at 7:23 A.M., traveling a distance of 104 miles. Along the way, the train makes some stops to pick up more passengers. These stops take about 12 minutes in all.

You may have used the formula $d = rt$ when solving problems that involve motion. In the formula, d is the distance the object has moved, r is the rate or speed of the object, and t is the time. Use the formula $d = rt$ to find the average running speed of the train.

Explore We know the starting time and arrival time of the train. We also know that the train was not moving for about 12 minutes. The distance between stations is 104 miles. The problem asks for the average running speed of the train.

Plan Find the actual time that the train runs. Use the formula $d = rt$.

Solve Subtract to find the time between the starting and arrival times.

1 hour = 60 minutes

$7:23 - 5:41 \Rightarrow 6:83 - 5:41 = 1:42$

Subtract the time when the train was not moving.

$1:42 - 0:12 = 1:30$

The train runs for 1 hour and 30 minutes. Since 30 minutes is 0.5 hour, another way to express the time is 1.5 hours.

Now, substitute values into the formula $d = rt$.

$$d = rt$$

$$104 = r(1.5)$$ *Replace d with 104 and t with 1.5.*

$$\frac{104}{1.5} = \frac{r(1.5)}{1.5}$$ *Divide each side by 1.5.*

104 [÷] 1.5 [=] 69.3333333

$$69.3 \approx r$$

The average running speed of the train is about 69.3 miles per hour. *Examine this solution.*

Checking for Understanding

Communicating Mathematics

Read and study the lesson to answer each question.

1. **Tell** what the variables d, r, and t represent in the formula $d = rt$.
2. **Write** the formula for the area of a rectangle.

Guided Practice

Solve. Use a formula.

3. Find the distance you travel if you drive your car at an average speed of 45 miles per hour for 3 hours.
4. The formula $P = I - E$ is used to find the profit (P) when income (I) and expenses (E) are known. Find the profit if $I = \$12,995$ and $E = \$15,000$.
5. Amtrak Commuter Train #624 leaves Philadelphia at 6:20 A.M. and arrives at Penn Station at 8:08 A.M. If this train makes the same stops as #620, find the average running speed of the train.

Problem Solving

Practice

Strategies

Look for a pattern.
Solve a simpler problem.
Act it out.
Guess and check.
Draw a diagram.
Make a chart.
Work backward.

Solve using any strategy.

6. The difference between two whole numbers is 9. Their product is 360. Find the two numbers.
7. A commuter-train car is 60 feet long and 12 feet wide. Find the maximum amount of carpeting needed to cover the car's floor.
8. Draw the next figure in the sequence. Write a sentence that describes the sequence.

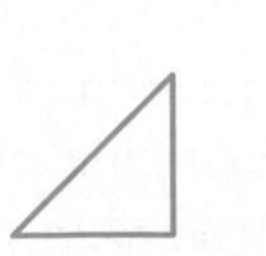  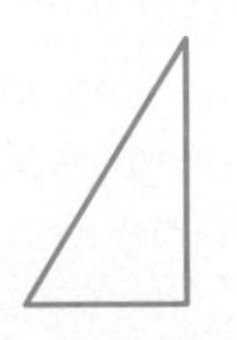 

9. **Mathematics and Science** Read the following paragraphs.

A flash of lightning is a huge spark of electricity that travels between a cloud and the ground or between two clouds. Lightning also causes thunder. Light from the flash travels almost instantly to your eyes. The sound of the thunder travels more slowly and arrives a few seconds later.

You can use the formula $d = 0.2t$ to find the distance between you and a thunderstorm. In the formula, d is the distance, in miles, and t is the time, in seconds. The speed of sound is about 0.2 miles per second.

a. On her way home from work, Darlene saw a flash of lightning. About 20 seconds later, she heard the thunder. How far away was the storm?

b. Sam was watching the Weather Up-Date on television, when the weather forecaster announced that a severe thunderstorm was 4 miles west of the city. About how many seconds should Sam expect between the lightning and the sound of the thunder?

8 Assessment: Mid-Chapter Review

Find each square root. *(Lesson 8-1)*

1. $\sqrt{36}$ 2. $\sqrt{225}$ 3. $-\sqrt{25}$ 4. $-\sqrt{0.16}$

Estimate to the nearest whole number. *(Lesson 8-2)*

5. $\sqrt{90}$ 6. $\sqrt{2}$ 7. $\sqrt{28}$ 8. $\sqrt{200}$

Name the set or sets of numbers to which each real number belongs. *(Lesson 8-3)*

9. 10 10. $\sqrt{4}$

11. 0.121212... 12. $\sqrt{3}$

Solve each equation. Round decimal answers to the nearest tenth. *(Lesson 8-3)*

13. $x^2 = 49$ 14. $y^2 = 50$

15. What is the average speed of a baseball if it takes 0.5 seconds for the ball to reach home plate, 60 feet away? Use the formula $d = rt$. *(Lesson 8-4)*

Cooperative Learning

8-5A The Pythagorean Theorem

A Preview of Lesson 8-5

Objective

Explore the relationships in a right triangle.

Materials

geoboard or dot paper

In the previous lessons, you learned about the relationship between the area of a square and the length of its side. You will use this relationship when you investigate a famous rule about right triangles. This rule is called the Pythagorean Theorem.

Try this!

Work with a partner.

- Build a triangle like the one shown on the geoboard. Remember it is called a *right triangle* because it has one right angle.

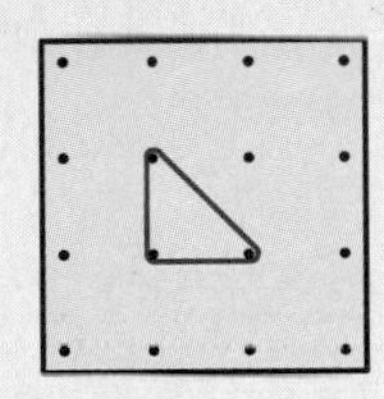

- Using the longest side of the triangle, build a square. The side of the square has the same length as the longest side of the triangle.

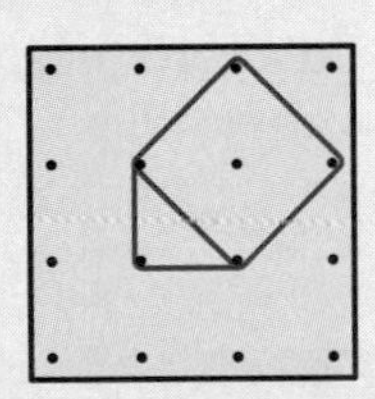

- The area of this square is 2 square units.

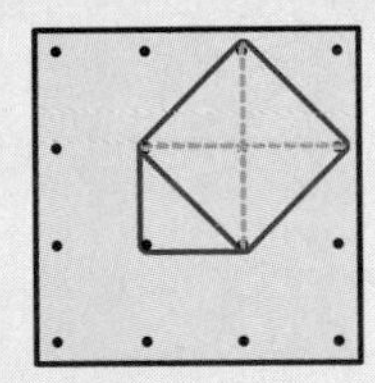

- Now build squares on the two shorter sides. Each square has an area of 1 square unit.

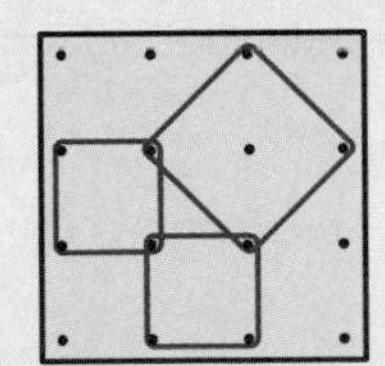

- The areas of the squares are 1 square unit, 1 square unit, and 2 square units.

What do you think?

1. Build squares on each side of the triangles shown below. Record the areas of the squares.

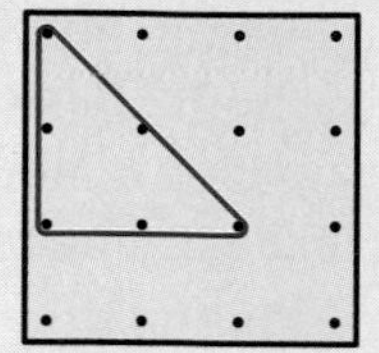

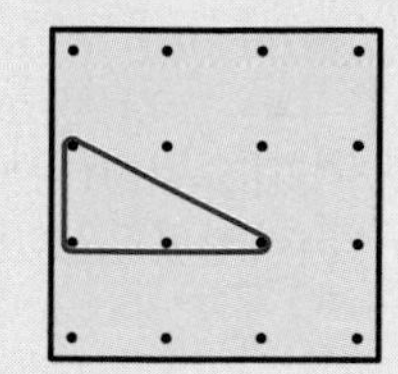

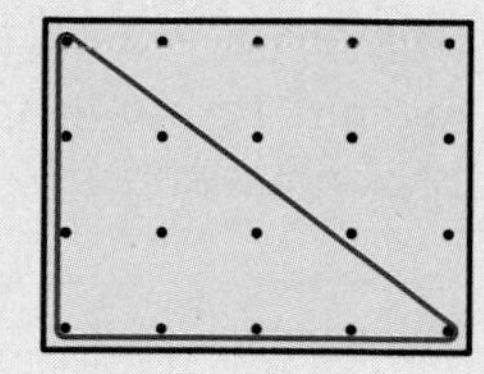

2. For each triangle, how does the area of the two smaller squares compare to the area of the larger square?
3. Build at least four different right triangles on your geoboard. Then build squares on each side. Look for a relationship between the area of the two smaller squares and the area of the larger square. Write a statement that describes the relationship.

Applications

4. In a right triangle, the two shorter sides are 2 units and 3 units long. Without making a drawing, what is the area of the square built on the longest side?
5. In a right triangle, the longest side is 6 units long, and one of the other sides is 2 units long. What is the area of the square built on the remaining side?

Extension

Can you build a square with an area of 10 square meters? If you think of 10 as the sum of 1 and 9, you can build squares like the ones shown below. The square that is built on the longest side is 10 square units.

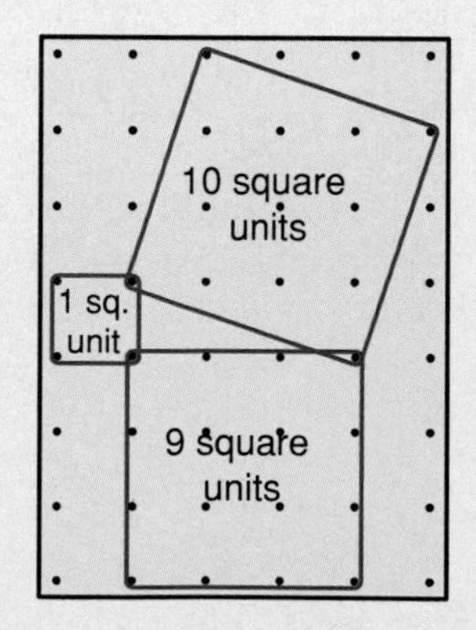

6. Build squares with areas of 5, 8, 13, 17, 32, 40, and 50 square units.

8-5 The Pythagorean Theorem

Objective

Use the Pythagorean Theorem.

Words to Learn

hypotenuse
legs
Pythagorean Theorem
converse

One of the earliest uses of mathematics was for measurement. Engineers and builders in ancient Egypt pioneered many aspects of mathematics. Every year, the Nile River flooded, covering their farmland for weeks. After the waters drained away, it was necessary to remeasure the boundaries.

About 2000 B.C., Egyptians discovered a 3-4-5 triangle. Taking a piece of rope knotted into 12 equal spaces, they stretched the rope around three stakes to form a triangle. The sides of the triangle had lengths of 3, 4, and 5 spaces. The longest side was opposite a right angle. Today, we call the longest side of a right triangle the **hypotenuse**. The sides that form the right angle are called **legs**.

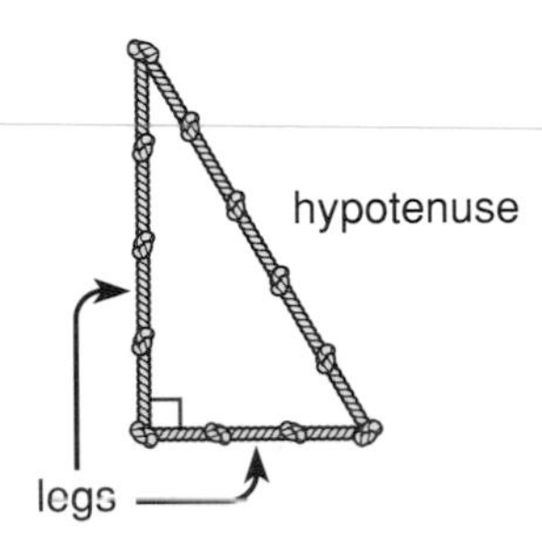

Many years later, in the fifth century B.C., a Greek mathematician, Pythagoras, and his followers learned about this 3-4-5 triangle. They noticed that if they built a square on each of the sides, the area of the two smaller squares was equal to the area of the large square.

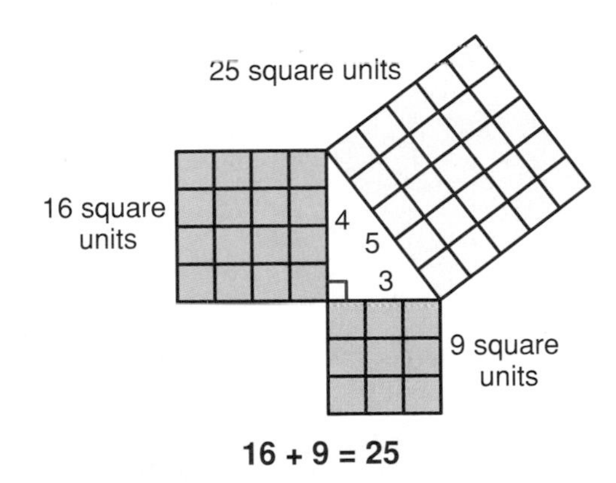

Today, we call this relationship the **Pythagorean Theorem.** It is true for *any* right triangle.

Pythagorean Theorem	**In words:** In a right triangle, the square of the length of the hypotenuse is equal to the sum of the squares of the lengths of the legs. **In symbols:** $c^2 = a^2 + b^2$

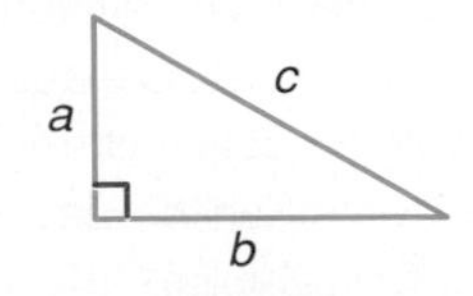

Examples

1 **Find the length of the hypotenuse in the triangle shown at the right.**

$c^2 = a^2 + b^2$ *Pythagorean Theorem*

$c^2 = 9^2 + 12^2$ *Replace a with 9 and b with 12.*

$c^2 = 81 + 144$

$c^2 = 225$

$c = \sqrt{225}$

$c = 15$ *You can ignore $-\sqrt{225}$ because it is not reasonable to have a negative length.*

The length of the hypotenuse is 15 feet.

2 **The hypotenuse of a right triangle is 30 meters long, and one of its legs is 12 meters long. Find the length of the other leg.**

$c^2 = a^2 + b^2$

$30^2 = a^2 + 12^2$

$900 = a^2 + 144$

$900 - 144 = a^2 + 144 - 144$

$756 = a^2$

$\sqrt{756} = a$

756 [√x] 27.495454

The length of the leg is about 27.5 meters.

Estimation Hint

THINK:
$25 \times 25 = 625$
$30 \times 30 = 900$
$\sqrt{756}$ is between 25 and 30.

Example 3 *Problem Solving*

Baseball **A baseball diamond is actually a square. The distance between bases is 90 feet. When the catcher throws the ball from home plate to second base, how far does the ball travel?**

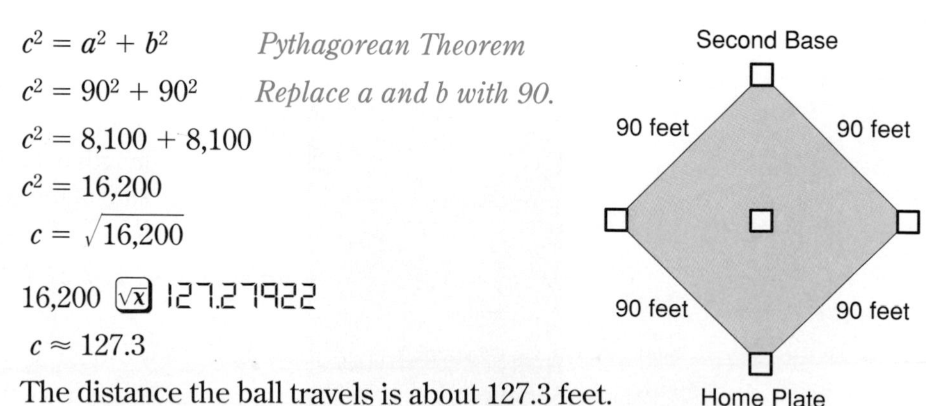

$c^2 = a^2 + b^2$ *Pythagorean Theorem*

$c^2 = 90^2 + 90^2$ *Replace a and b with 90.*

$c^2 = 8{,}100 + 8{,}100$

$c^2 = 16{,}200$

$c = \sqrt{16{,}200}$

16,200 [√x] 127.27922

$c \approx 127.3$

The distance the ball travels is about 127.3 feet.

Each year the Continental Amateur Baseball Association conducts a World Series for 14-year olds. There are 19 teams from 15 states, Canada, Puerto Rico, and Brazil. In 1991, the Series was played in Dublin, Ohio.

Some triangles may look like right triangles, but in reality they are not. The **converse** of the Pythagorean Theorem can be used to test whether a triangle is a right triangle.

Converse of Pythagorean Theorem	If the sides of a triangle have lengths a, b, and c units, such that $c^2 = a^2 + b^2$, then the triangle is a right triangle.

Examples

Determine whether each triangle with sides of given length is a right triangle. *Remember, the hypotenuse is the longest side.*

4 **8 cm, 13 cm, 17 cm**

$c^2 = a^2 + b^2$

$17^2 \stackrel{?}{=} 8^2 + 13^2$

$289 \stackrel{?}{=} 64 + 169$

$289 \neq 233$

The triangle is *not* a right triangle.

5 **5 ft, 12 ft, 13 ft**

$c^2 = a^2 + b^2$

$13^2 \stackrel{?}{=} 5^2 + 12^2$

$169 \stackrel{?}{=} 25 + 144$

$169 = 169$

The triangle is a right triangle.

Checking for Understanding

Communicating Mathematics

Read and study the lesson to answer each question.

1. **Tell** the area of the shaded square shown at the right.
2. **Draw** a right triangle and label the hypotenuse and the legs.
3. **Tell** whether a triangle with sides of 4, 5, and 6 inches is a right triangle.

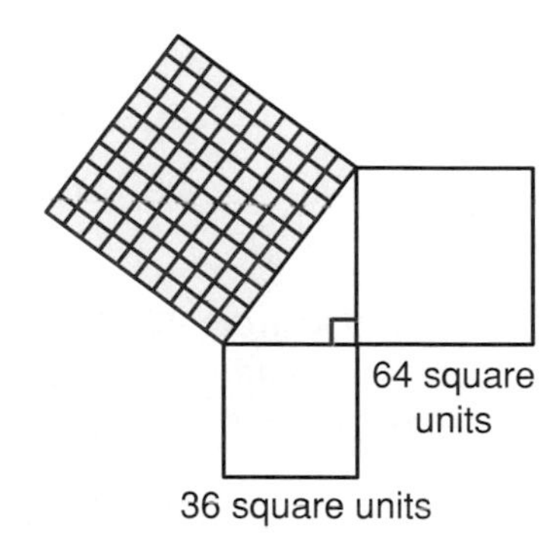

Guided Practice

State an equation you could use to find the length of the missing side of each right triangle. Then find the missing length.

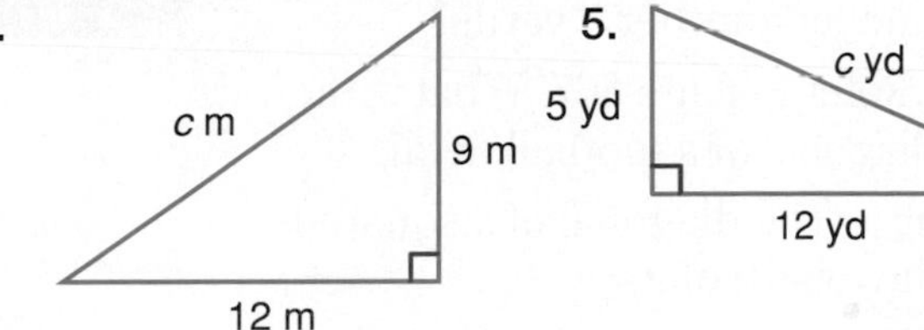

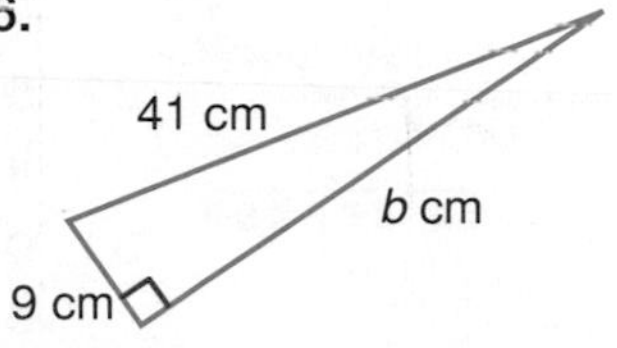

Determine whether each triangle with sides of given length is a right triangle.

7. 10 cm, 12 cm, 15 cm
8. 18 ft, 24 ft, 30 ft

Exercises

Independent Practice

Find the missing measure for each right triangle. Round answers to the nearest tenth.

9. a, 9 ft; c, 12 ft
10. a, 5 in.; b, 5 in.
11. a, 3 m; c, 8 m
12. b, 99 mm; c, 101 mm
13. b, 12 cm; c, 22 cm
14. a, 48 yd; b, 55 yd
15. a, 40 in.; c, 41 in.
16. a, 3.5 m; b, 12.5 m

Write an equation to solve for x. Then solve. Round answers to the nearest tenth.

17. 10 in.; x in.; 10 in.

18. 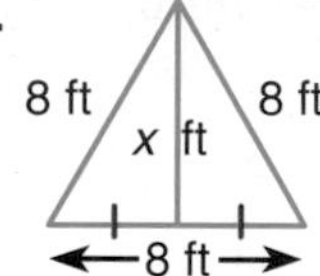

19. 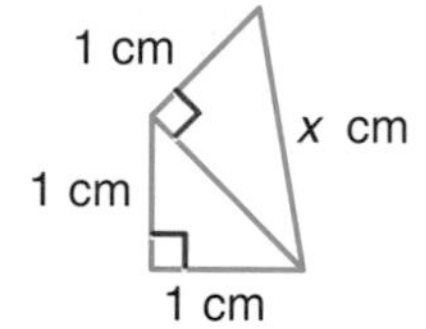

Determine whether each triangle with sides of given length is a right triangle.

20. 5 in., 10 in., 12 in.
21. 1 mi, 1 mi, $\sqrt{2}$ mi
22. 4 m, 7 m, 5 m
23. 28 cm, 197 cm, 195 cm
24. 9 in., 40 in., 41 in.
25. 24 mm, 143 mm, 145 mm

Mixed Review

26. How many grams are in 1.25 kilograms? *(Lesson 1-6)*
27. **Statistics** Name the stems you would use to plot the following set of data: 328, 351, 336, 357, 348, 324. *(Lesson 4-4)*
28. Express -3.47×10^{-5} in standard form. *(Lesson 6-11)*
29. Solve $t^2 = 144$. *(Lesson 8-3)*

Problem Solving and Applications

30. **Geometry** Find the length of a diagonal of a rectangle with a length of 8 units and a width of 5 units.
31. **Sports** A popular pass play in football is the "down and out." A receiver runs down the field, parallel to the sideline, then turns sharply toward the sideline to receive the ball. How far from his original position is a receiver who runs downfield 15 yards, turns and runs toward a sideline for another 8 yards?

32. **Data Search** Refer to page 667. What is the length of the diagonal of a football field?
33. **Critical Thinking** The diagonal of a square is 8 units. Find the length of its side.
34. **Geometry** Find the area and perimeter of the triangle shown at the right.

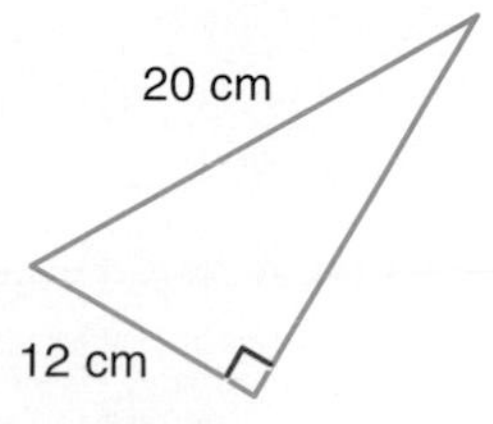

8-6 Using the Pythagorean Theorem

Objective

Solve problems using the Pythagorean Theorem.

Words to Learn

Pythagorean triple

The illustration at the right is from a Chinese work called *Chóu-peï Suan-king*, written sometime during 2000–1000 B.C. The block print shows a principle we know as the Pythagorean Theorem. Can you see several 3–4–5 right triangles in the illustration? However, the work contains no explanation relating the measures of the figures.

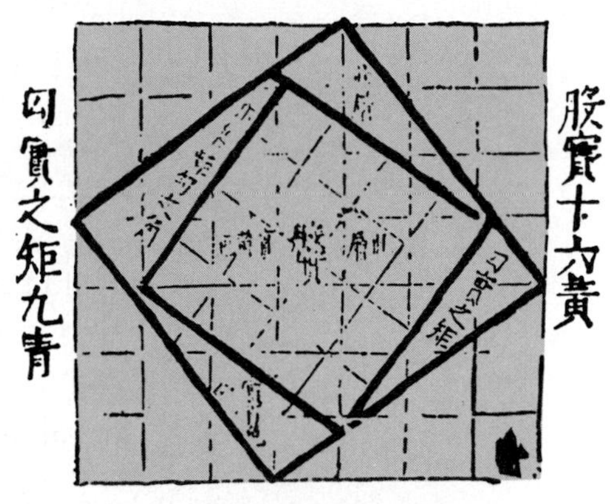

CHÓU-PEÏ SUAN-KING

You can use your knowledge of the Pythagorean Theorem to find the areas of the figures in the illustration.

Example 1 ***Connection***

Geometry **Find the total area of the four small triangles in the illustration above.**

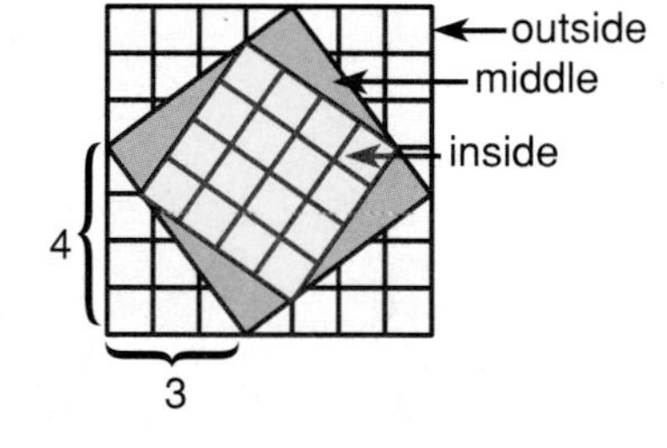

Explore First, let's re-draw the figure to show the inside, middle, and outside squares. You need to find the area of the shaded triangles.

Plan To find the area of the shaded triangles, you can find the area of the middle square and subtract the area of the inside square from it.

Solve The middle square is built on the hypotenuse of the 3-4-5 right triangle. Its area is the square of the hypotenuse.

$c^2 = a^2 + b^2$ *Pythagorean Theorem*

$c^2 = 3^2 + 4^2$

$c^2 = 9 + 16$ or 25

The area of the middle square is 25 square units.

Now subtract the area of the inside square. It is 4^2 or 16 square units.

$25 - 16 = 9$

The total area of the four small triangles is 9 square units.

Examine Estimate the area of each small triangle by counting squares. Then multiply by 4 to find the total area.

$2 \cdot 4 = 8$

The answer is reasonable.

The Pythagorean Theorem can be applied in many situations involving measurement.

Example 2 *Problem Solving*

Home Maintenance For safety reasons the base of a 24-foot ladder should be at least 8 feet from the wall. How high can a 24-foot ladder safely reach?

Let b represent the height the ladder will reach.

$$c^2 = a^2 + b^2$$
$$24^2 = 8^2 + b^2$$
$$576 = 64 + b^2$$
$$576 - 64 = 64 + b^2 - 64$$
$$512 = b^2$$
$$\sqrt{512} = b$$

512 $\boxed{\sqrt{x}}$ 22.627417

The ladder can safely reach a height of about 22.6 feet.

By now you can recognize 3–4–5 as integers that satisfy the Pythagorean Theorem. Such numbers are called **Pythagorean triples.** You can use multiples to find other Pythagorean triples that are based on 3–4–5.

Mental Math Hint

You can mentally find a Pythagorean triple by finding a multiple of a Pythagorean triple you already know.

Example 3 *Connection*

Number Theory The chart below shows several Pythagorean triples. Study the pattern in the chart to find the next triple in the set.

a	b	c
3	4	5
6	8	10
9	12	15
■	■	■

a	b	c
3	4	5
6	8	10
9	12	15
12	16	20

The next triple in the set is 12–16–20.

The triple 3–4–5 is called a *primitive* Pythagorean triple because the numbers are relatively prime. The triples 6–8–10, 9–12–15, and so on are in the 3–4–5 family.

Checking for Understanding

Communicating Mathematics **Read and study the lesson to answer each question.**

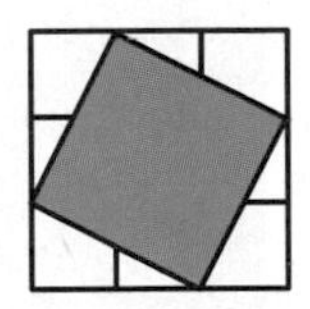

1. **Tell** the area of the shaded square shown at the right.
2. **Write** two Pythagorean triples in the same family as 5–12–13.

Guided Practice **State an equation that can be used to answer each question. Then solve. Round answers to the nearest tenth.**

3. How long is each rafter?

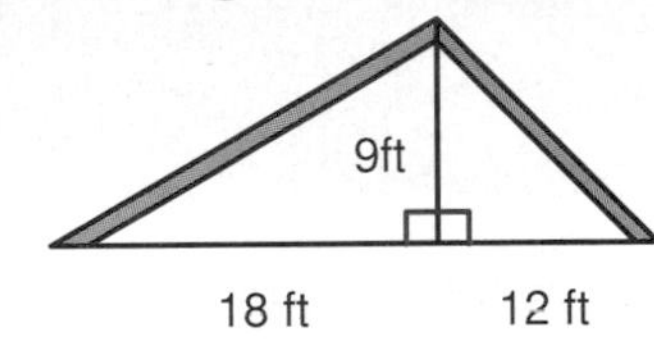

4. How long is the lake?

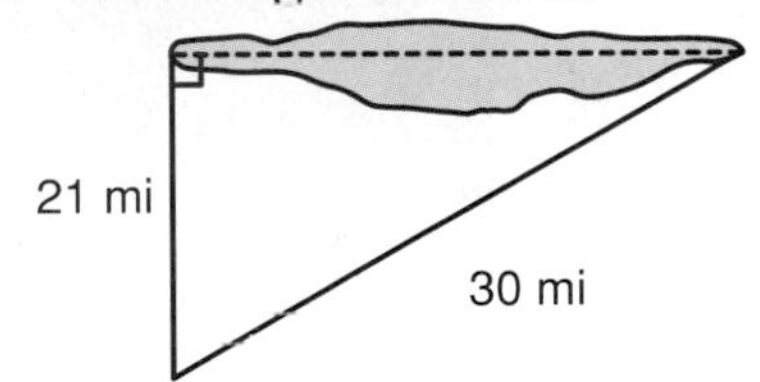

Exercises

Independent Practice **Solve. Round answers to the nearest tenth.**

5. How high does the ladder reach?

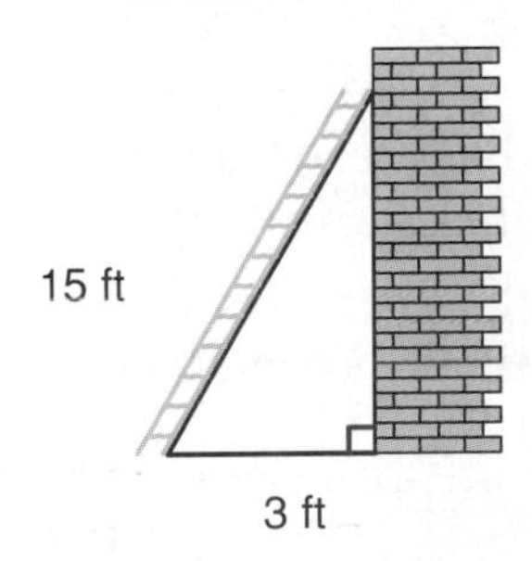

6. How far apart are the planes?

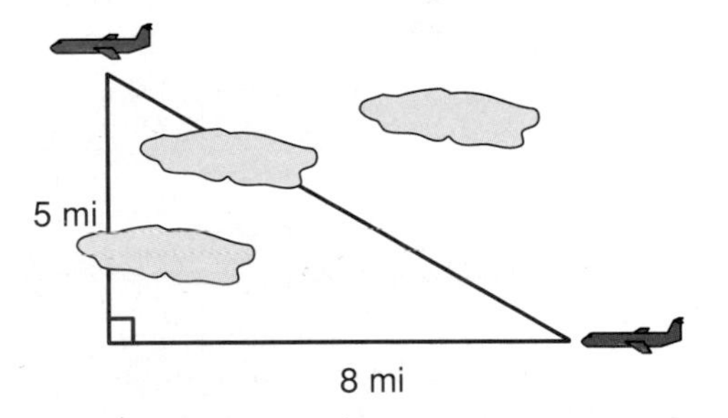

For each Pythagorean triple, find two triples in the same family.

7. 8–15–17
8. 7–24–25
9. 9–40–41

10. Name the family to which the triple 36–48–60 belongs.

Mixed Review

11. Evaluate $4^2 - 2^3$. *(Lesson 1-9)*

12. The lengths of three sides of a triangle are 12 meters, 20 meters, and 13 meters. Determine whether the triangle is a right triangle. *(Lesson 8-5)*

Problem Solving and Applications

For Exercises 13-16, make a drawing of each situation. Then solve. Round answers to the nearest tenth.

13. **Skateboarding** The acceleration ramp for the skateboard competition is 20 meters long and extends 15 meters from the base of the starting point. How high is the ramp?

14. **Hiking** A hiker walked 11 kilometers north, then walked 3 kilometers west. How far was she from the starting point?

15. **Landscaping** A newly-planted tree needs to be staked with three wires. Each wire is staked 3 feet from the base of the tree and extends 5 feet high on the trunk. How much wire is needed to stake four trees?

16. **Portfolio Suggestion** Select your favorite word problem from this chapter and place your solution to it in your portfolio. Attach a note explaining why it is your favorite.

17. **Critical Thinking** Several Pythagorean triples are listed below. Study the pattern in the table.

a	b	c
3	4	5
5	12	13
7	24	25
9	40	41

a. Find the Pythagorean triple that has 11 as the measure of the shortest leg.

b. Find the Pythagorean triple that has 85 as the measure of the hypotenuse.

18. **Computer Connection** You can also find Pythagorean triples by inputing positive integers x and y into the following program. Note that x must be greater than y.

```
10 INPUT X, Y
20 A =X^2 - Y^2: B = 2*X*Y: C = X^2 + Y^2
30 PRINT A, B, C
```

a. Write the expressions that are used to find values of a, b, and c.

b. Run the program for several values of x and y.

Cooperative Learning

8-6B Graphing Irrational Numbers

A Follow-Up of Lesson 8-6

Objective

Use the Pythagorean Theorem to graph irrational numbers on a number line.

Materials

compass
straightedge

Most people know that Alexander Graham Bell invented the telephone. But did you know that the drawings needed to get a patent for the telephone were done by Lewis Howard Latimer (1848–1928), an African-American engineer and draftsman?

Draftsmen, inventors, and engineers often use a compass and straightedge to copy measurements accurately from one location to another. You already know how to graph integers and rational numbers on a number line. How would you graph an irrational number like $\sqrt{5}$?

Activity One

LOOKBACK

You can review perpendicular segments on page 183.

- Draw a number line.
- At 2, construct a perpendicular line segment 1 unit in length. Draw the line segment shown in color. Label it *c*.

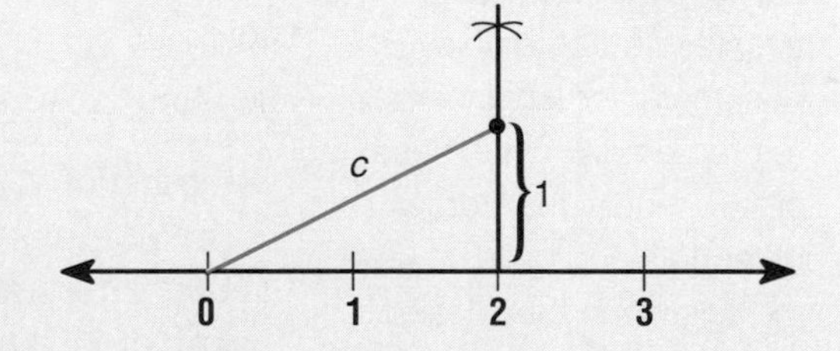

- The Pythagorean Theorem can be used to show that *c* is $\sqrt{5}$ units long.

$c^2 = a^2 + b^2$

$c^2 = 1^2 + 2^2$ *Replace a with 1 and b with 2.*

$c^2 = 5$

$c = \sqrt{5}$

You can construct a perpendicular segment by folding the paper line at 2, making sure the two parts of the number line align when you hold the paper up to the light. The fold is a segment perpendicular to the number line.

- Open the compass to the length of *c*. With the tip of the compass at 0, draw an arc that intersects the number line at *B*. The distance from 0 to *B* is $\sqrt{5}$ units.

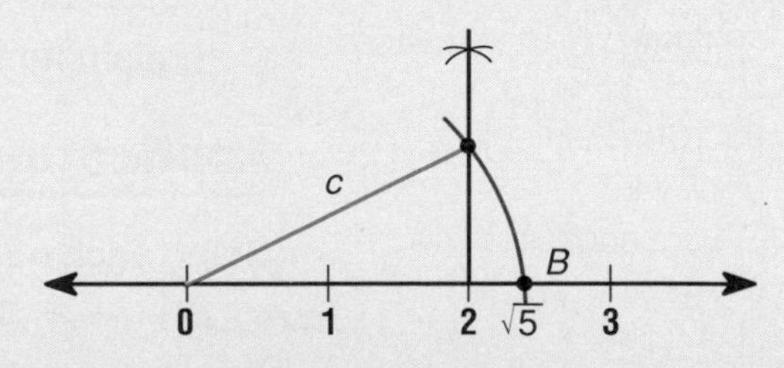

In Activity One, the irrational number that was graphed is the sum of two squares and is represented by the hypotenuse. Sometimes, the irrational number to be graphed will be the difference of two squares and can be represented by a leg of a triangle. Consider the following method for graphing $\sqrt{7}$.

Activity Two

- Draw a number line.
- At 3, construct a perpendicular line segment. Put the tip of the compass at 0. With the compass set at 4 units, construct an arc that intersects the perpendicular line segment. Label the perpendicular leg *a*.

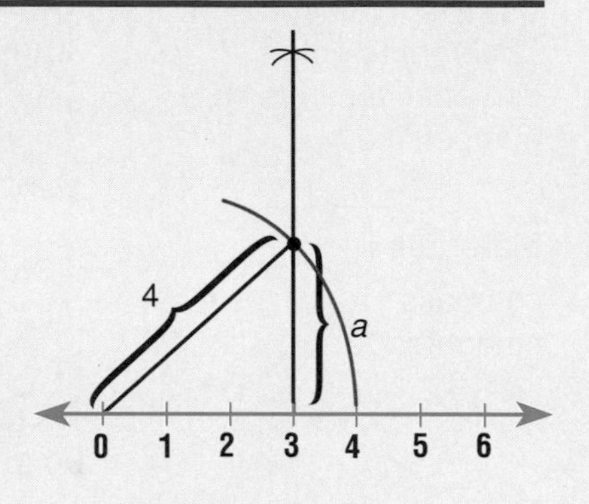

- The Pythagorean Theorem can be used to show that *a* is $\sqrt{7}$ units long.

$$
\begin{aligned}
c^2 &= a^2 + b^2 \\
4^2 &= a^2 + 3^2 \\
16 &= a^2 + 9 \quad \textit{Replace c with 4 and b with 3.} \\
7 &= a^2 \\
\sqrt{7} &= a
\end{aligned}
$$

- Open the compass to the length of *a*. With the tip of the compass at 0, draw an arc that intersects the number line at *D*. The distance from 0 to *D* is $\sqrt{7}$ units.

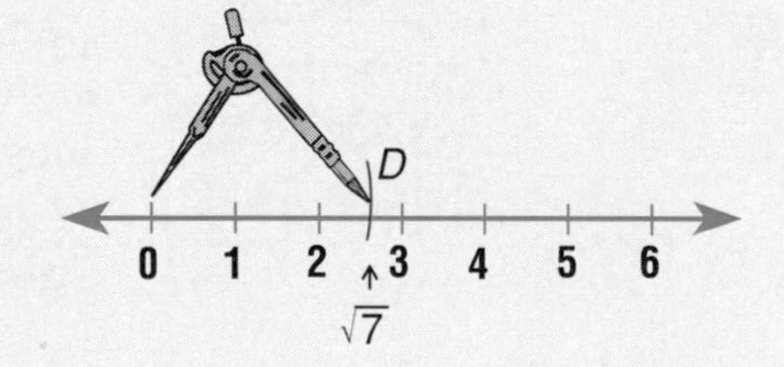

“When am I ever going to use this?”

Large toy companies, such as Mattel, employ designers and engineers to create new toys. These creative individuals draw blueprints and build models when they design new games, toys, and cars.

A background in mathematics and drafting is essential in the study of mechanical engineering and industrial design.

For more information contact:
Mattel
333 Continental Boulevard
El Segundo, California 90245

What do you think?

1. Explain how to graph $\sqrt{2}$.
2. Describe two different ways to graph $\sqrt{8}$.
3. Explain how the graph of $\sqrt{2}$ can be used to locate the point that represents $\sqrt{3}$.
4. Explain how to graph $-\sqrt{5}$.

Applications

Graph each number on a number line.

5. $\sqrt{2}$
6. $\sqrt{3}$
7. $\sqrt{8}$
8. $\sqrt{10}$

Measurement Connection

8-7 Distance on the Coordinate Plane

Objective

Find the distance between points in the coordinate plane.

One day Jim and his brother David were arguing. Their parents sent them to their room to calm down. Since they share the same room, they wanted to be as far away from each other as possible. Jim wanted to know how far he was from David. He drew a graph of their room on a grid like the one shown at the right. *The side of each square represents 1 foot.*

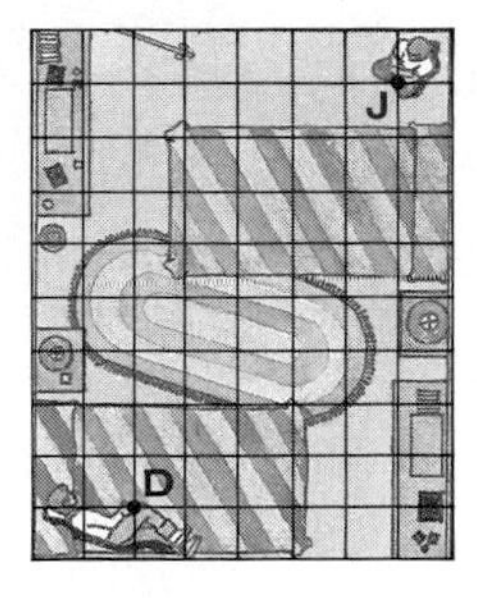

LOOK BACK

You can review the coordinate plane on page 117.

The graph looks like a coordinate plane. If we draw the *x*- and *y*- axes, David's location is represented by the ordered pair (2, 1). Jim's location is represented by the ordered pair (7, 9).

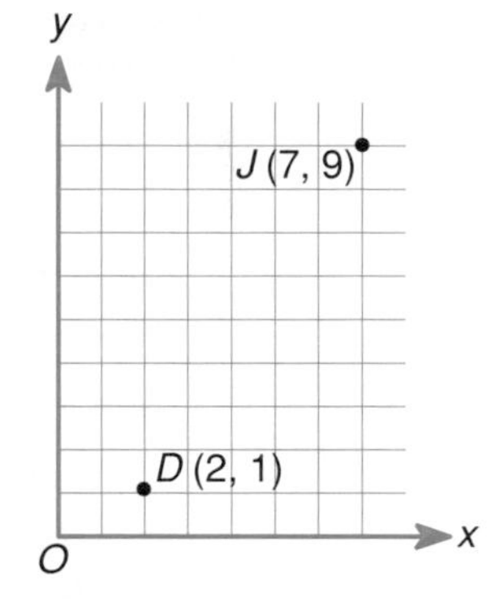

Let's see how these ordered pairs can help determine how far Jim is from David.

Mini-Lab

Work with a partner.

Materials: graph paper

- Graph points D and J on a coordinate plane and connect them with a line segment.
- Draw a horizontal line through D and a vertical line through J. Call the point of intersection T.

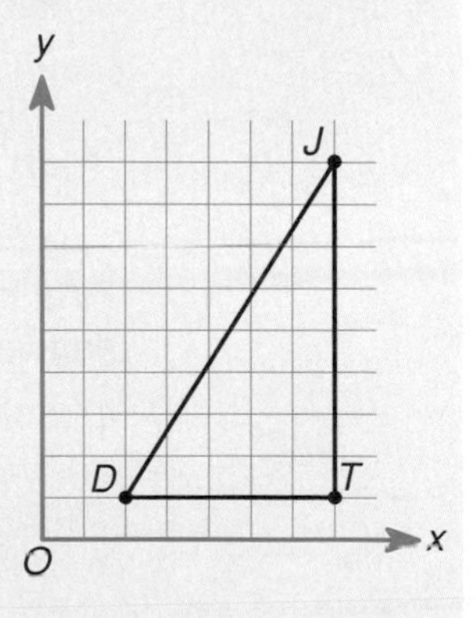

Talk About It

a. What is the length of $\overline{DT}$?

b. What is the length of $\overline{JT}$?

c. Explain how you can find the length of $\overline{DJ}$.

DID YOU KNOW

AT&T has a reference book listing coordinates for all locations in North America. The phone company then calculates the distance between two cities, and from this, the long distance telephone charge is determined.

As you found in the Mini-Lab, the distance between Jim and David can be related to the hypotenuse of a right triangle.

Example

Find the distance between Jim and David.

Problem Solving Hint

Use a formula.

Technology Activity

You can learn how to use a graphing calculator to find the distance between two points in Technology Activity 3 on page 660.

Let c = the distance between Jim and David.
Let $a = 5$.
Let $b = 8$.

$c^2 = a^2 + b^2$
$c^2 = 5^2 + 8^2$
$c^2 = 25 + 64$
$c^2 = 89$
$c = \sqrt{89}$

89 [√x] 9.4339811

Jim and David are about 9.4 feet apart.

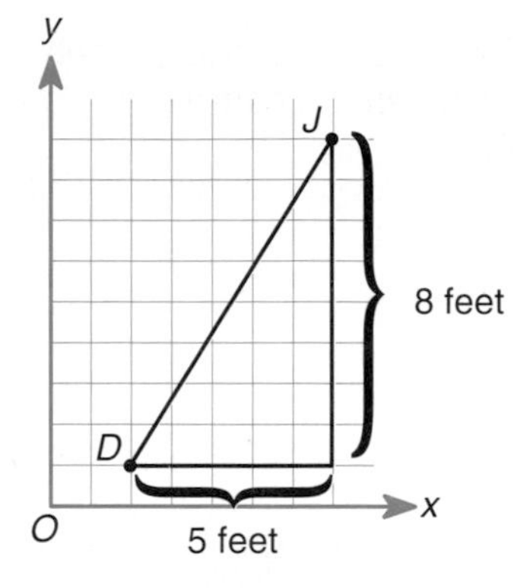

Checking for Understanding

Communicating Mathematics

Read and study the lesson to answer each question.

1. **Tell** the length of $\overline{BC}$ and $\overline{AC}$ in the figure at the right.
2. **Write** the steps you could use to find the distance between (5, 5) and (2, 2).
3. **Draw** the triangle that you can use to find the distance between (-2, 1) and (-5, -3) on the coordinate plane.

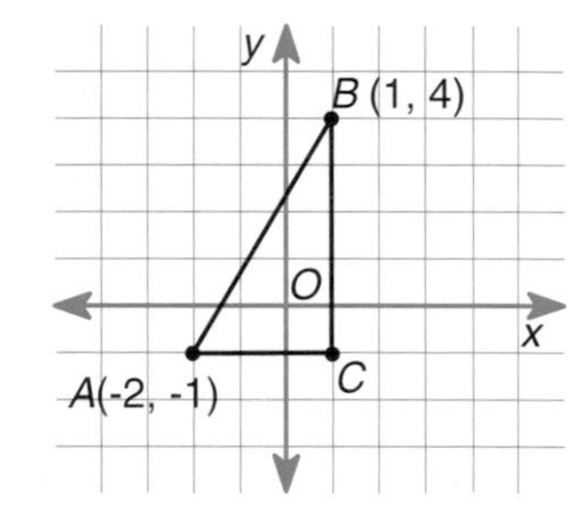

Guided Practice

Find the distance between each pair of points whose coordinates are given. Round answers to the nearest tenth.

4.

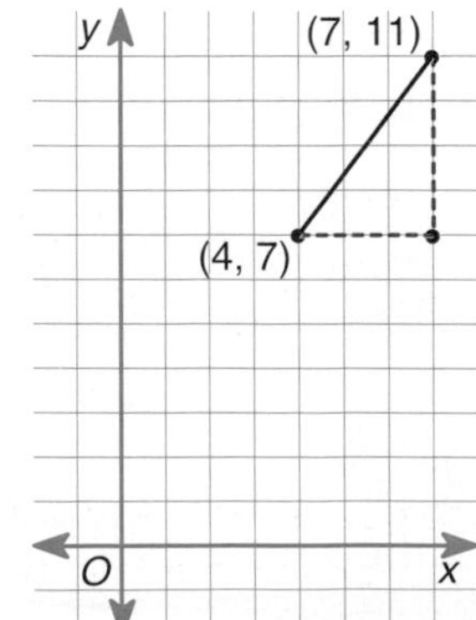

5.

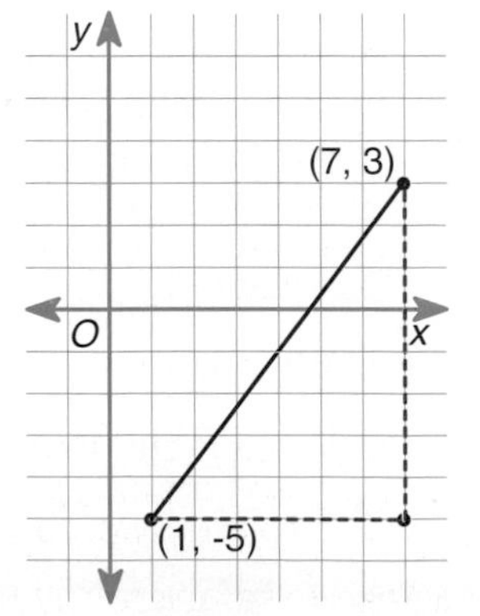

6.

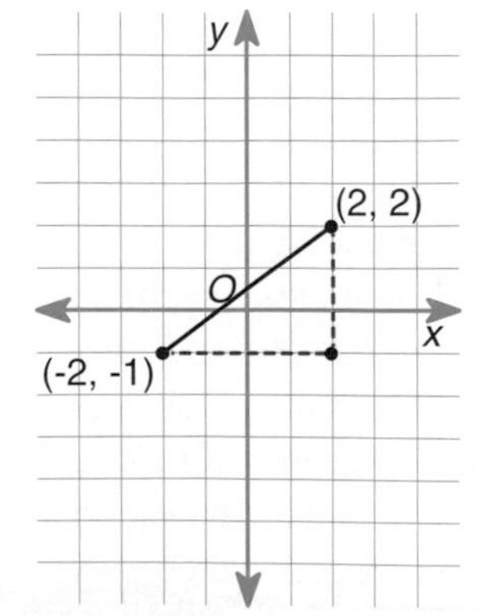

Exercises

Independent Practice

Find the distance between each pair of points whose coordinates are given. Round answers to the nearest tenth.

7. 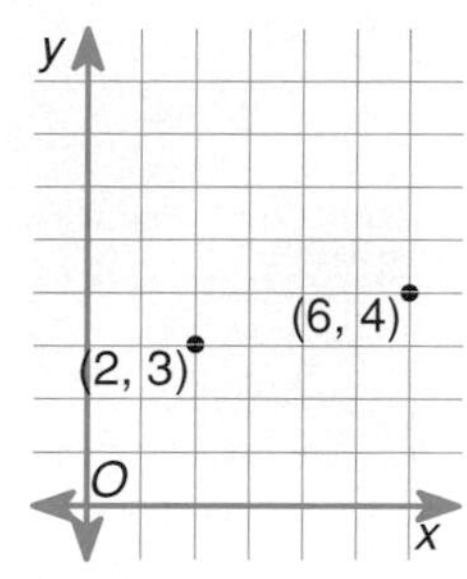

8. 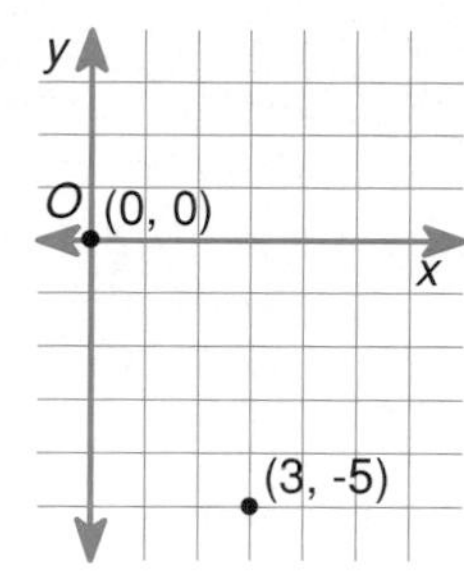

9. 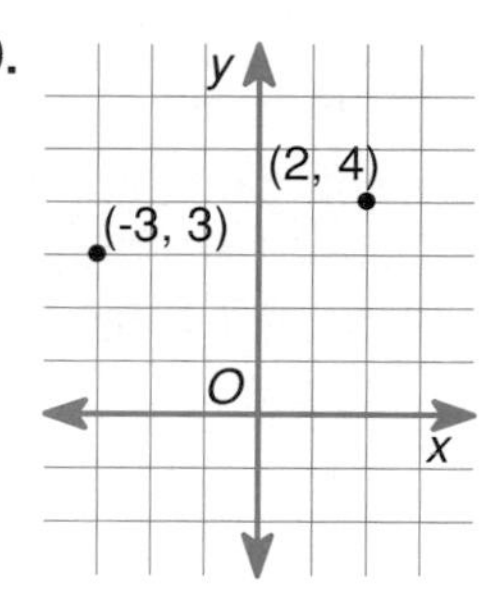

Graph each pair of ordered pairs. Then find the distance between the points. Round answers to the nearest tenth.

10. (3, 5); (3, -2)
11. (-1, 0); (2, 7)
12. (1, 5); (3, 1)
13. (-2, 4); (3, -5)
14. Find the distance between $A(-5, -2)$ and $C(1, -2)$.
15. The coordinates of point R and S are (4, 3) and (1, 6). What is the distance between the points?

Mixed Review

16. Solve $h - 15 = 27$. *(Lesson 2-3)*
17. Solve $\frac{n}{10} = -8$. *(Lesson 7-10)*
18. **Measurement** Find the height of the tower shown at the right. *(Lesson 8-6)*

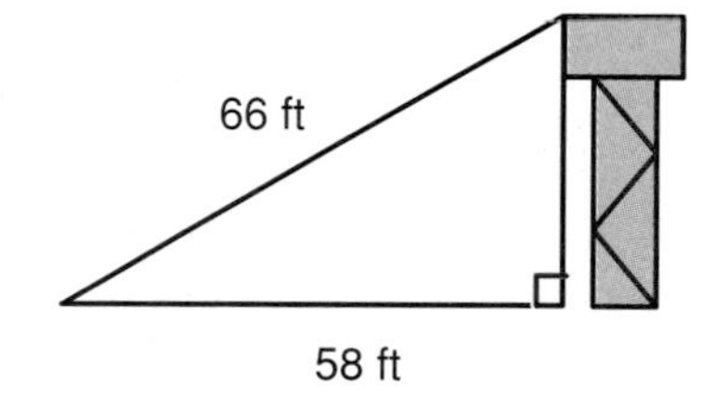

Problem Solving and Applications

19. **Geometry** A triangle on the coordinate plane has vertices $A(2, -1)$, $B(-2, 2)$ and $C(-6, 14)$.
    a. Draw the triangle.
    b. Find the perimeter of the triangle.
20. **Critical Thinking** The distance between points A and B is 17 units. Find the value of x if the coordinates of A and B are $(-3, x)$ and $(5, 2)$.
21. **Map Making** On a scaled street map, where the side of each square represents 1 mile, the Ball Park is located at (1, 2) and the Rollerdome is located at (6, 10). A diagonal street runs directly between the two locations. Approximately how far is it from the Ball Park to the Rollerdome?

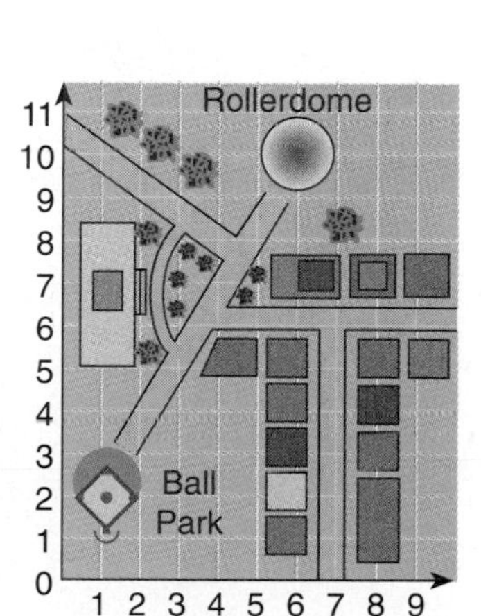

DATA SEARCH

22. **Data Search** Refer to pages 298 and 299.
    You can use the formula $d = rt$ to find the average speed (r) that an object travels, given its distance (d) and time (t). On average, how many seconds does it take for a dragonfly to travel 5 miles?

Geometry Connection

8-8 Special Right Triangles

Objective

Find missing measures in 30°-60° right triangles and 45°-45° right triangles.

In 1987, the United States regained the America's Cup trophy in yachting from Australia. *Stars and Stripes* won over *Kookaburra III* in four one-sided races. The sails on boats such as these are in the shape of right triangles.

The other two angles of the right triangle often have measurements of 30° and 60°. In a 30°–60° right triangle, the lengths of the sides are related in a special way.

Mini-Lab

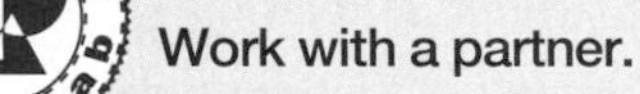

Work with a partner.

Materials: compass, protractor, scissors, ruler

- Construct and cut out an equilateral triangle.
- Fold the triangle in half and cut along the fold line.
- Measure each side and each angle.

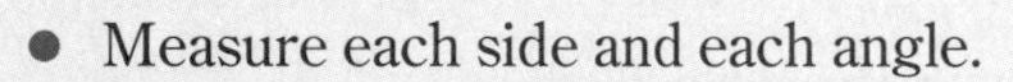

- Repeat the above steps for several other equilateral triangles.

Talk About It

What is the relationship between the length of the hypotenuse and the length of the side opposite the 30° angle?

The relationship you discovered in the Mini-Lab is always true in a 30°–60° right triangle. The length of the side opposite the 30° angle is one-half the length of the hypotenuse.

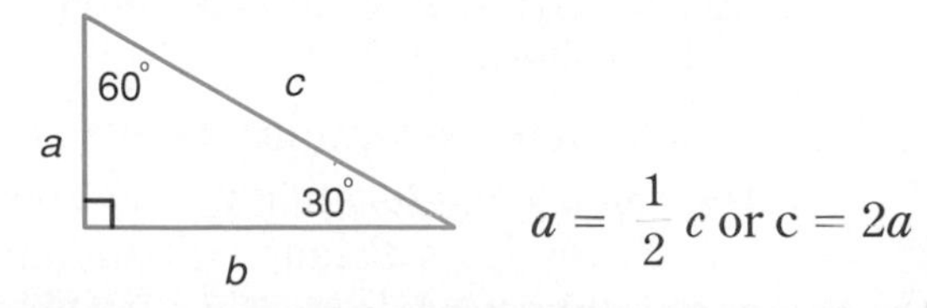

$a = \frac{1}{2}c$ or $c = 2a$

Example 1

Find the missing lengths in $\triangle PQR$.

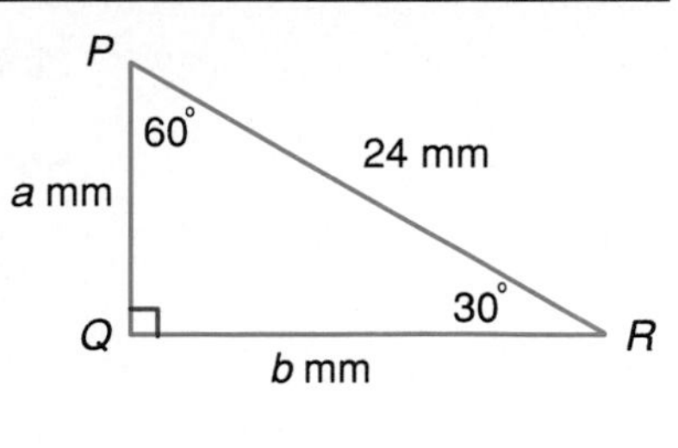

Step 1: Find a.

$$a = \frac{1}{2}c$$

$$a = \frac{1}{2}(24) \quad \textit{Replace c with 24.}$$

$$a = 12$$

Step 2: Find b.

$$c^2 = a^2 + b^2 \quad \textit{Pythagorean Theorem}$$

$$24^2 = 12^2 + b^2 \quad \textit{Replace c with 24 and a with 12.}$$

$$576 = 144 + b^2$$

$$432 = b^2 \quad 432 \ \boxed{\sqrt{x}}\ 20.78461$$

$$b \approx 20.8$$

The length of $\overline{PQ}$ is 12 millimeters, and the length of $\overline{PR}$ is about 20.8 millimeters.

Sometimes the other two angles of a right triangle have measurements of 45° and 45°. The lengths of the sides are also related in a special way.

Mini-Lab

Work with a partner.

Materials: protractor, ruler, scissors

- Construct and cut out a square.
- Fold the square in half and cut along the diagonal.
- Measure each side and each angle.

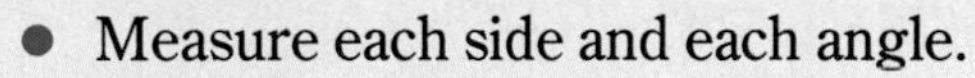

- Repeat the above steps for several other squares.

Talk About It

What is the relationship between the length of the legs in the 45°–45° right triangles?

In a 45 °–45° right triangle, the lengths of the legs are equal. $a = b$

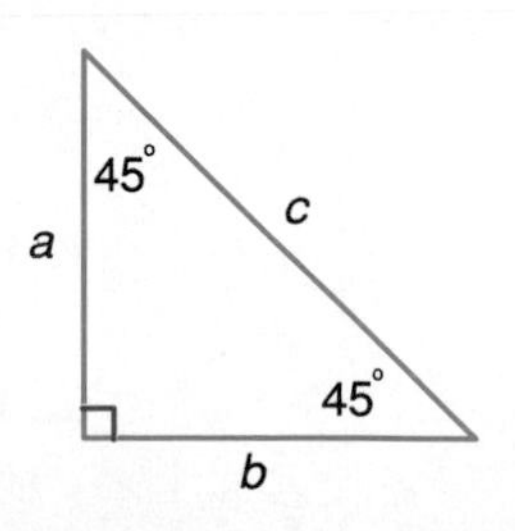

Example 2

Find the missing lengths of $\triangle QRS$.

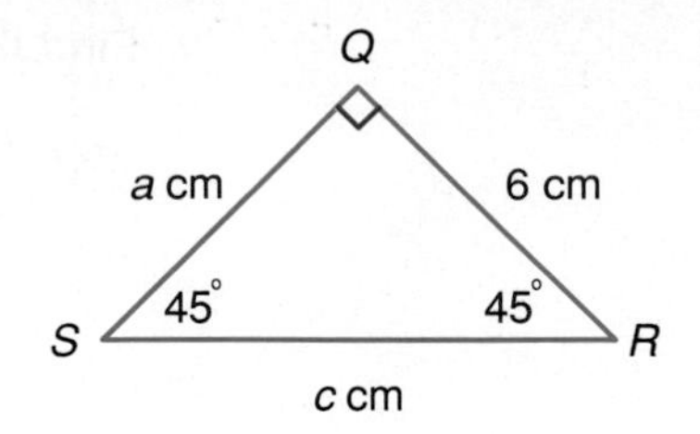

Step 1 Find a.

$a = b$

$b = 6$

$a = 6$

Step 2 Find c.

$c^2 = a^2 + b^2$ *The Pythagorean Theorem*

$c^2 = 6^2 + 6^2$ *Replace a and b with 6.*

$c^2 = 36 + 36$

$c^2 = 72$

$c = \sqrt{72}$

72 [$\sqrt{x}$] 8.4852814

$c \approx 8.5$

The length of $\overline{QS}$ is 6 centimeters, and the length of $\overline{SR}$ is about 8.5 centimeters.

Checking for Understanding

Communicating Mathematics

Read and study the lesson to answer each question.

1. **Draw** a 30°–60° right triangle. Indicate which side is opposite the 30° angle.
2. **Tell** why a 45°–45° right triangle is also called an *isosceles* right triangle.
3. **Write** a sentence describing the relationship between the hypotenuse of a 30°–60° right triangle and the leg opposite the 30° angle.

Guided Practice

Find the lengths of the missing sides. Round answers to the nearest tenth.

4.

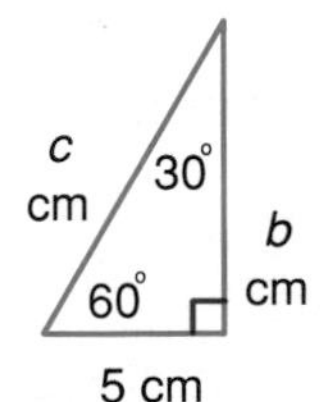

5.

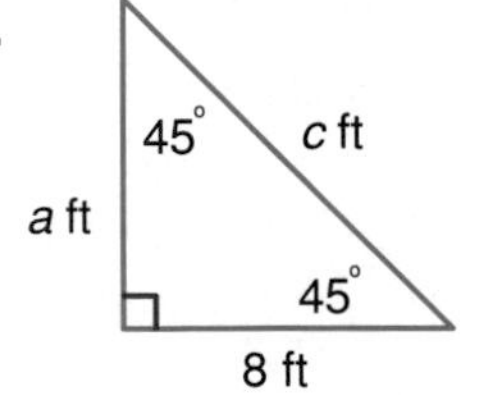

6.

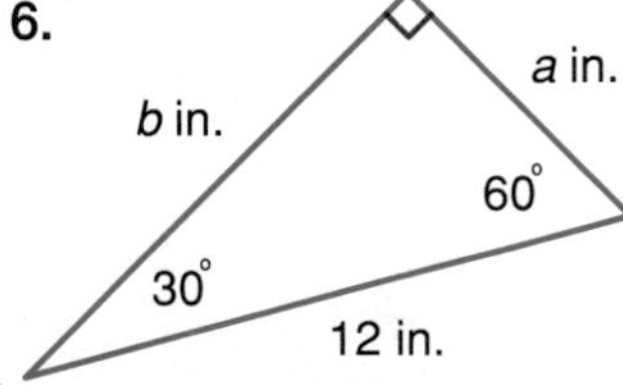

7. One leg of a 45°–45° right triangle is 15 centimeters long. What is the length of the other leg?
8. The shorter leg in a 30°–60° right triangle is 3 inches long. What is the length of the hypotenuse?

Exercises

Independent Practice

Find the lengths of the missing sides. Round decimals to the nearest tenth.

9. 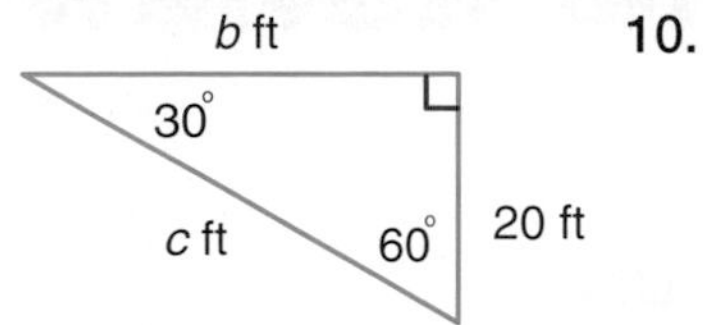

10.

3.2 m
45°
b m
45°
c m

11. 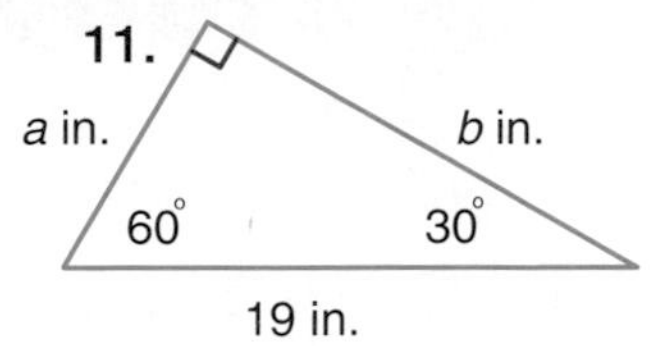

12. One leg of a 45°–45° right triangle is 21.5 inches long. Find the length of the hypotenuse.
13. The length of the hypotenuse of a 30°–60° right triangle is 7.5 inches. Find the length of the side opposite the 30° angle.

Mixed Review

14. **Statistics** Find the median and upper and lower quartiles for the following lengths, in inches, of rope: 56, 62, 48, 71, 60, 50, 64. *(Lesson 4-6)*
15. Write $\frac{54}{81}$ in simplest form. *(Lesson 6-5)*
16. Find the distance between $P(-3, 4)$ and $Q(5, -2)$. *(Lesson 8-7)*

Problem Solving and Applications

17. **Construction** Ms. Gomez wants to add an access ramp to the rear entrance of her store. The ramp makes a 30° angle with the ground. If the rear entrance is 6 feet above the ground, how long is the ramp?

18. **Portfolio Suggestion** Select an item from this chapter that you feel shows your best work and place it in your portfolio. Explain why you selected it.

19. **Critical Thinking** The area of a square is 400 square meters. Find the length of each of its diagonals.

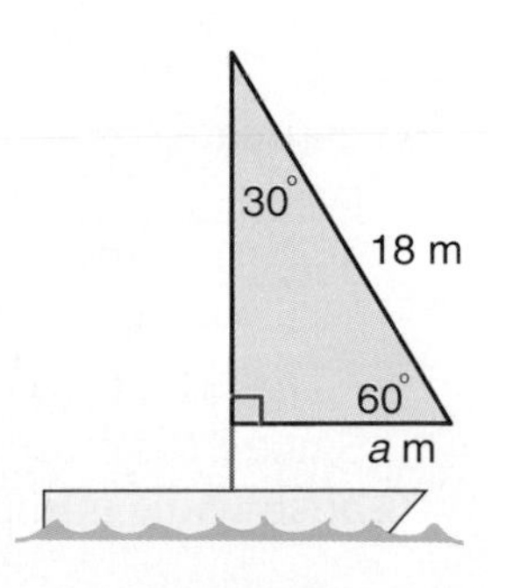

20. **Journal Entry** Write a detailed solution of the following problem. How much material is needed to make the sail shown at the right?

Chapter

8 Study Guide and Review

Communicating Mathematics

Choose the letter of the correct word or words to complete each statement.

1. The symbol $\sqrt{\ }$ is used to indicate a(n) ______ square root.
2. A(n)______ is the square of a rational number.
3. The set of rational numbers and the set of irrational numbers combine to form the set of ______.
4. A(n)______ can always be expressed as a terminating or repeating decimal.
5. In a right triangle, the ______ is the side opposite the right angle.
6. The Pythagorean Theorem is true for any ______.
7. In a 30°–60° right triangle, the length of the side opposite the 30° angle is one-half the length of the ______.
8. Tell how you can use the Pythagorean Theorem to determine if a triangle is a right triangle.
9. Explain the difference between a rational number and an irrational number. Give an example of each.

a. triangle
b. principal
c. real numbers
d. rational number
e. irrational number
f. square root
g. perfect square
h. right triangle
i. hypotenuse

Self Assessment

Objectives and Examples

Upon completing this chapter, you should be able to:

Review Exercises

Use these exercises to review and prepare for the chapter test.

- find square roots of perfect squares *(Lesson 8-1)*

Find $\sqrt{81}$.
Since $9^2 = 81$, $\sqrt{81} = 9$.

Find each square root.

10. $\sqrt{36}$
11. $-\sqrt{2.25}$
12. $-\sqrt{\frac{9}{16}}$
13. $\sqrt{\frac{49}{100}}$
14. $-\sqrt{169}$
15. $\sqrt{5.29}$

- estimate square roots *(Lesson 8-2)*

Estimate $\sqrt{31}$.
$25 < 31 < 36$
$5^2 < 31 < 6^2$
$5 < \sqrt{31} < 6$
Since 31 is closer to 36 than to 25, the best whole number estimate is 6.

Estimate to the nearest whole number.

16. $\sqrt{136}$
17. $\sqrt{50.2}$
18. $\sqrt{725}$
19. $\sqrt{372}$
20. $\sqrt{19.33}$
21. $\sqrt{250}$

Objectives and Examples

- identify and classify numbers in the real number system *(Lesson 8-3)*

Determine whether 8.41 is a rational or irrational number.

8.41 is a terminating decimal. So it is a rational number.

Review Exercises

Name the set or sets of numbers to which each real number belongs.

22. $\sqrt{33}$

23. -21

24. -0.525225 . . .

25. 0.686868 . . .

- find the length of the side of a right triangle using the Pythagorean Theorem *(Lesson 8-5)*

Find the length of the hypotenuse.

$c^2 = a^2 + b^2$

$c^2 = 6^2 + 8^2$

$c^2 = 36 + 64$

$c^2 = 100$

$c = 10$

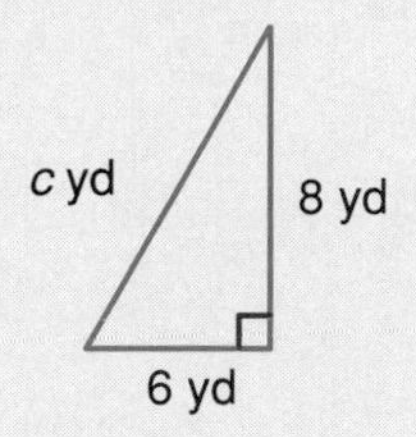

The hypotenuse is 10 yards long.

Find the missing measure for each right triangle. Round decimal answers to the nearest tenth.

26. $a = 16$ ft; $c = 20$ ft

27. $a = 5$ cm; $b = 7$ cm

28. $b = 28$ mm; $c = 55$ mm

29. $a = 7.5$ in.; $c = 8.5$ in.

30. $a = 18$ m; $b = 30$ m

- solve problems using the Pythagorean Theorem *(Lesson 8-6)*

How tall is the tree? Let b represent the height of the tree.

$c^2 = a^2 + b^2$

$63^2 = 40^2 + b^2$

$3{,}969 = 1{,}600 + b^2$

$2{,}369 = b^2$

$48.7 \approx b$

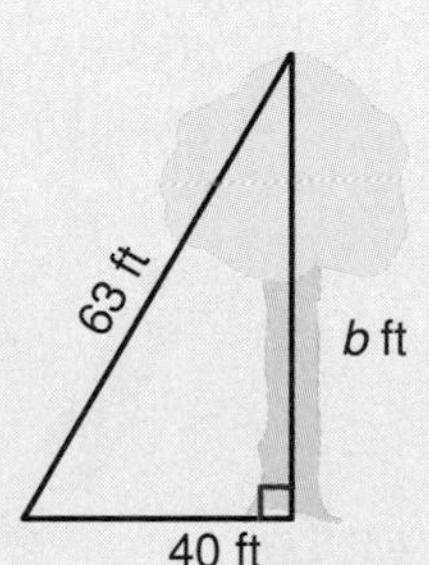

The tree is about 48.7 feet tall.

Solve. Round decimal answers to the nearest tenth.

31. How wide is the door?

32. How far is the airplane from the airport?

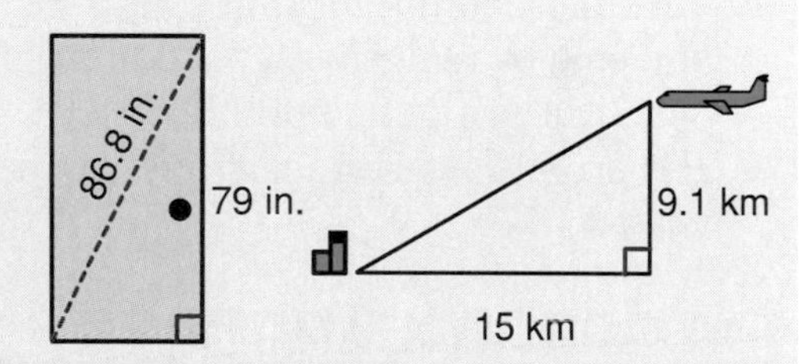

- find the distance between two points on the coordinate plane. *(Lesson 8-7)*

Find c.

$c^2 = a^2 + b^2$

$c^2 = 5^2 + 3^2$

$c^2 = 34$

$c = \sqrt{34}$

$c \approx 5.8$

The distance is about 5.8 units.

Graph each pair of ordered pairs. Find the distance between the points. Round decimal answers to the nearest tenth.

33. (3, 3) and (1, 6)

34. (-2, 1) and (3, 7)

35. (-2, -4) and (5, 4)

36. (-5, -3) and (4, -6)

Objectives and Examples

- find the missing measures in 30°-60° right triangles *(Lesson 8-8)*

Find the length of $\overline{AB}$ in $\triangle ABC$.

$a = \frac{1}{2}C$

$a = \frac{1}{2}(16)$

$a = 8$

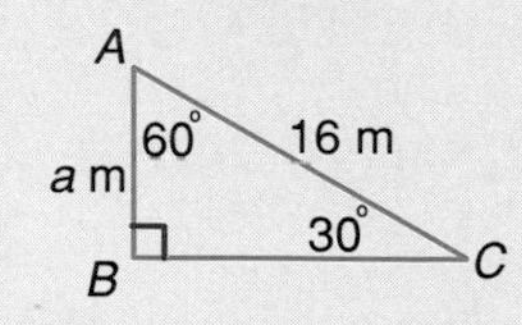

The length of $\overline{AB}$ is 8 meters.

Review Exercises

Find the missing lengths.

37.

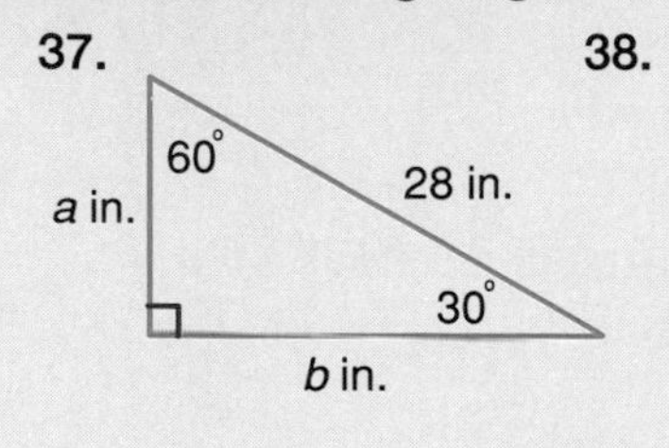

38.

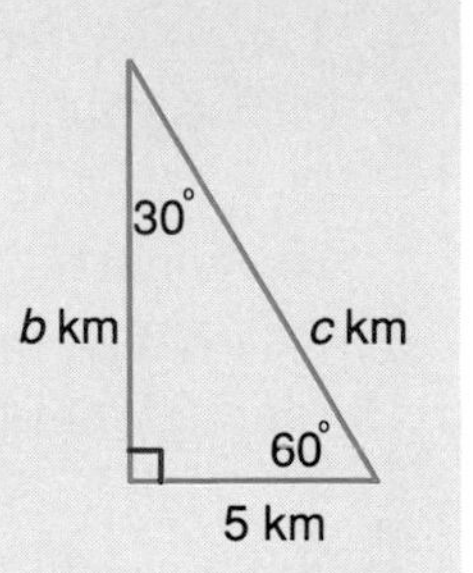

- find the missing measures in 45°-45° right triangles *(Lesson 8-8)*

Find the length of $\overline{YZ}$ in $\triangle XYZ$.

$a = b$

$8 = b$

The length of $\overline{YZ}$ is 8 feet.

Find the missing lengths.

39.

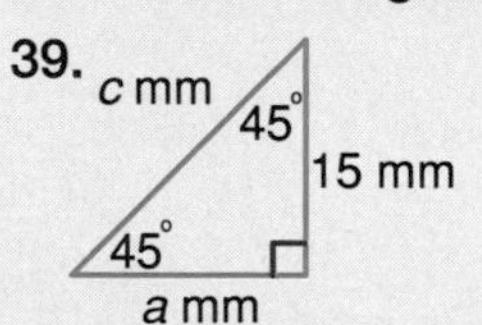

40.

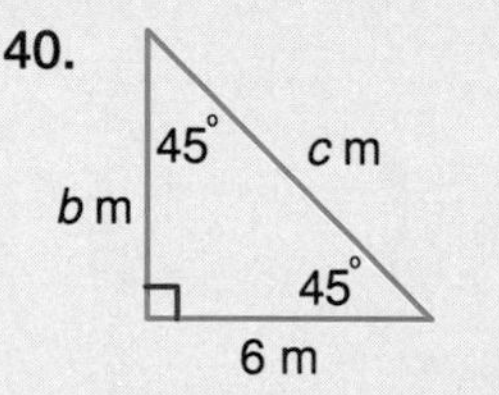

Applications and Problem Solving

41. **Skateboarding** Suppose a skateboarder travels a distance of 20 meters in 16 seconds. Find the average speed. Use the formula $d = rt$. *(Lesson 8-4)*

42. **Sales** How much profit does the owner of The Computer Connection make on a microcomputer that sells for \$599.99 if the computer cost her \$390.29, and she paid the salesperson a commission of \$49.72? Use the formula $P = I - E$, where P is profit, I is income, and E is expense. *(Lesson 8-4)*

Curriculum Connection Projects

- **Science** Step off the length and width of a rectangular area of your school grounds. Calculate the diagonal, in steps. Then step off the diagonal to see if your answer is correct.
- **Sports** Research the dimensions of a tennis court, volleyball court, and a basketball court. Find the length of each diagonal.

Read More About It

Galen, Laura. *Out of this World: Science Fiction and Fantasy.*
Froman, Robert. *Angles Are Easy as Pie.*
Chrisman, Arthur Bowie. *Shen of the Sea.*

Chapter

8 Test

1. Find the distance between $R(-3, -3)$ and $S(4, 5)$. Round to the nearest tenth.

Find each square root.

2. $-\sqrt{144}$
3. $\sqrt{\frac{49}{64}}$
4. $-\sqrt{0.25}$

Estimate to the nearest whole number.

5. $\sqrt{66}$
6. $\sqrt{605}$
7. $\sqrt{137.8}$
8. State the Pythagorean Theorem.

Name the set or sets of numbers to which each real number belongs.

9. $\sqrt{13}$
10. $-\sqrt{25}$
11. 8.1212 . . .
12. $\sqrt{28.347}$

Solve each equation.

13. $y^2 = 80$
14. $x^2 = 225$.

Find the missing measure for each right triangle. Round decimal answers to the nearest tenth.

15. $a = 1.5$ km; $b = 2$ km
16. $b = 20$ in.; $c = 33$ in.

Determine whether each triangle with sides of given measure is a right triangle.

17. 16 cm, 34 cm, 30 cm
18. 12 ft, 18 ft, 23 ft
19. A ladder is leaning against a house. The top of the ladder is 20 feet from the ground and the base of the ladder is 15 feet from the side of the house. How long is the ladder?
20. Find the perimeter of a right triangle with legs of 9 inches and 8 inches.

Find the missing lengths.

21.
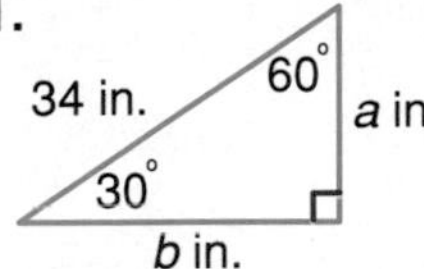

22.
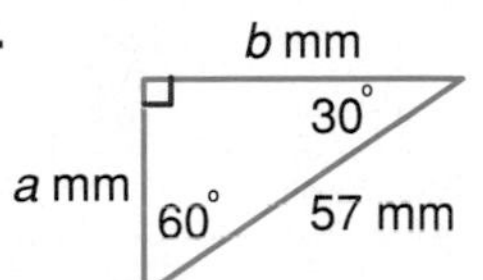

23.
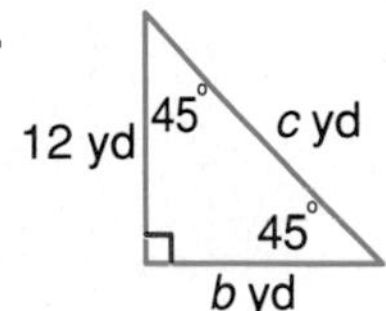

24.
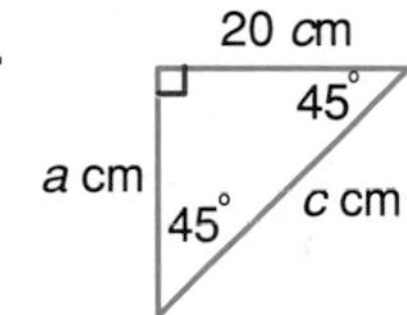

25. The formula for finding gas mileage is $m = \frac{d}{g}$, where m is miles per gallon, d is the distance traveled, and g is the number of gallons of gasoline used. Suppose your car averages 28 miles per gallon on the highway, and your car's gas tank holds 13 gallons of gasoline. Can you drive 400 miles on one tank of gas? Explain why or why not.

Bonus In a rectangular container, all pairs of intersecting edges are perpendicular. What is the length of the diagonal of the container shown at the right?

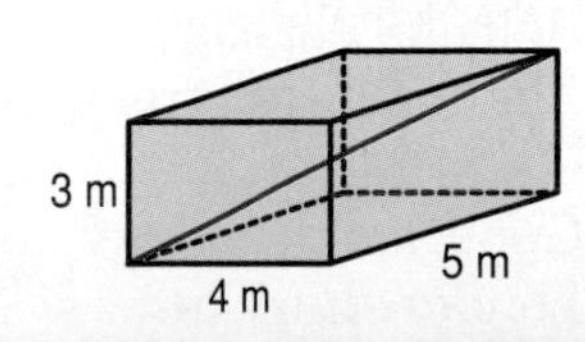

Chapter Test

Chapter

9

Applications with Proportion

Spotlight on Travel and Money

Have You Ever Wondered...

- What it means when the United States dollar has an exchange rate of 1:19 with Jamaican dollars?
- What it means when the cost of living in Paris, France, is 1:1 with the cost of living in New York City?

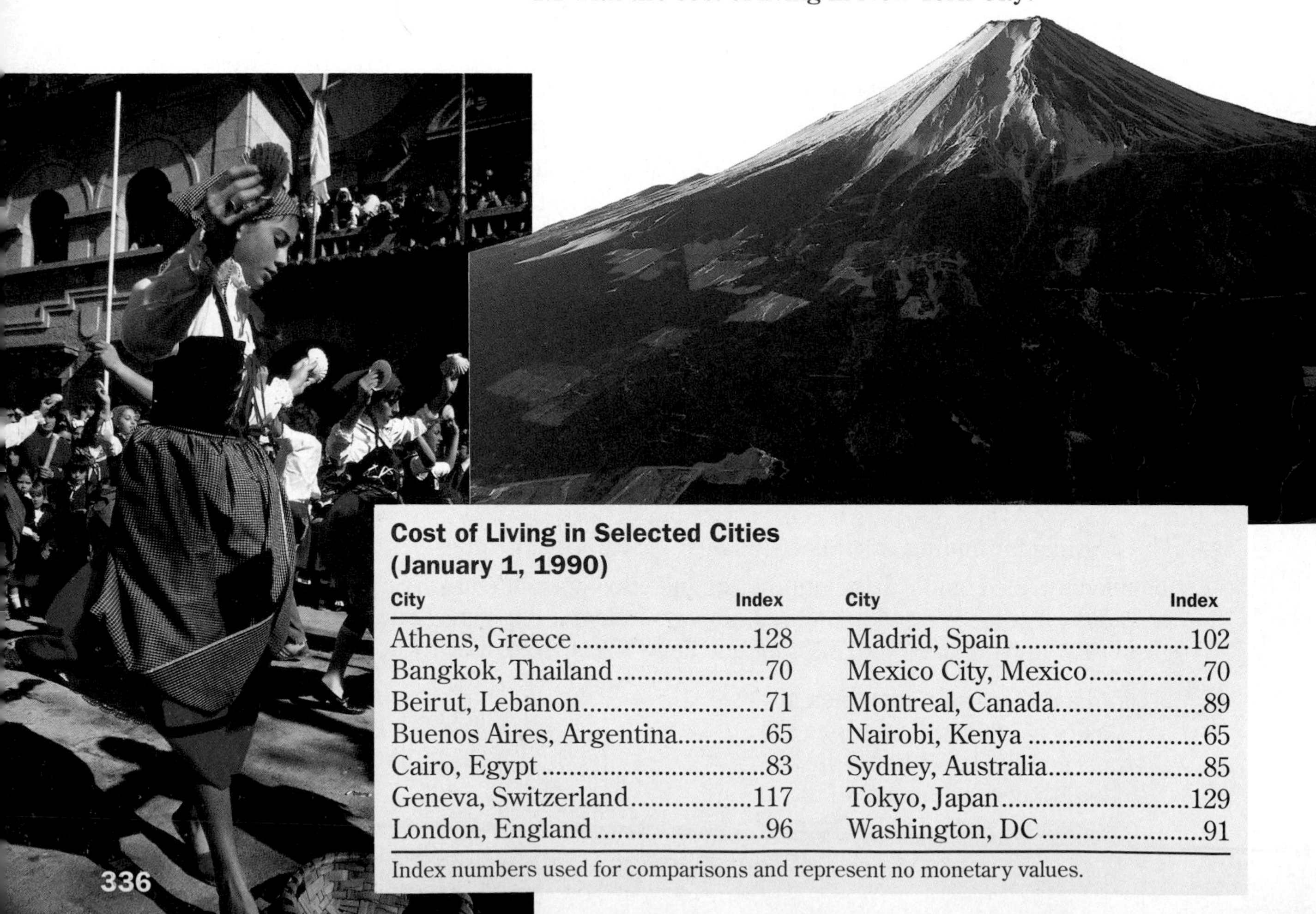

Cost of Living in Selected Cities (January 1, 1990)

City	Index	City	Index
Athens, Greece	128	Madrid, Spain	102
Bangkok, Thailand	70	Mexico City, Mexico	70
Beirut, Lebanon	71	Montreal, Canada	89
Buenos Aires, Argentina	65	Nairobi, Kenya	65
Cairo, Egypt	83	Sydney, Australia	85
Geneva, Switzerland	117	Tokyo, Japan	129
London, England	96	Washington, DC	91

Index numbers used for comparisons and represent no monetary values.

Chapter Project

Travel and Money

Working a group.

1. Design a questionnaire for your friends to find out whether they have traveled or lived in another country or state. Ask them what items, such as food, cab fare, or movies, were cheaper or more expensive than what they are at home.
2. Organize your data. Make a bar graph to compare the differentcosts of one item in different states and/or countries.

Exchange Rates (January 3, 1994)

Country	Amount Equivalent to $1 U.S. Currency
Australia	1.47 dollars
Britain	0.68 pounds
Canada	1.324 dollars
France	5.91 francs
India	31.13 rupee
Japan	111.61 yen
Mexico	3.105 pesos
Poland	20,438 zloty

Looking Ahead

In this chapter, you will see how mathematics can be used to answer questions about travel and money. The major objectives of the chapter are to:

- express ratios as fractions and determine unit rates
- solve problems by using proportions and drawing diagrams
- solve problems involving similar triangles and scale drawings
- graph dilations on a coordinate plane
- find the tangent, sine, and cosine of an angle

9-1 Ratios and Rates

Objectives

Express ratios as fractions in simplest form and determine unit rates.

Words to Learn

ratio
rate
unit rate

At the time of the 1991 All-Star Game, the Texas Rangers led the American League West division by winning 44 of their first 77 games. You can compare these two numbers using a **ratio.**

Ratio	A ratio is a comparison of two numbers by division.

The ratio that compares 44 to 77 can be written as follows.

44 to 77   44:77   44 out of 77 $\frac{44}{77}$

Since a ratio can be written as a fraction, it can also be simplified. Since 44 and 77 have a common factor of 11, the ratio $\frac{44}{77}$ is equivalent to $\frac{4}{7}$.

Examples

Express each ratio in simplest form.

1 **16 brown-eyed students: 20 blue-eyed students**

$\frac{16}{20} = \frac{4}{5}$ *Divide the numerator and denominator by their GCF, 4.*

The ratio in simplest form is $\frac{4}{5}$ or 4:5.

LOOKBACK

You can review GCF on page 221.

2 **6 inches out of 1 foot**

$\frac{6 \text{ inches}}{1 \text{ foot}} = \frac{6 \text{ inches}}{12 \text{ inches}}$ *1 foot = 12 inches*

$\frac{6}{12} = \frac{1}{2}$ *The GCF of 6 and 12 is 6.*

The ratio in simplest form is $\frac{1}{2}$ or 1 out of 2.

The ratios in Examples 1 and 2 compare quantities with the same unit or quantities that can be rewritten so that the units are the same. Often it is necessary to compare two quantities with different units. For example, $\frac{\$10.00}{8.7 \text{ gallons}}$ compares the number of dollars spent for gasoline to the number of gallons purchased. This type of ratio is called a **rate.**

Rate	A rate is a ratio of two measurements with different units.

When a rate has a denominator of 1, it is called a **unit rate.** This type of rate is frequently used when comparing statistics.

Example 3 *Problem Solving*

Geography In 1990, Texas had a population of 16,991,000 and covers an area of 266,807 square miles. Washington had a population of 4,761,000 with an area of 68,139 square miles. Which state was more densely populated?

First, find the rate of people per square mile for each state. Then make a comparison.

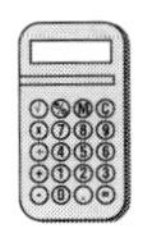

To find the unit rate, divide both the numerator and the denominator by the denominator. Use your calculator.

> **Problem Solving Hint**
>
> You can also find the unit rate by dividing the numerator by the denominator.

Texas

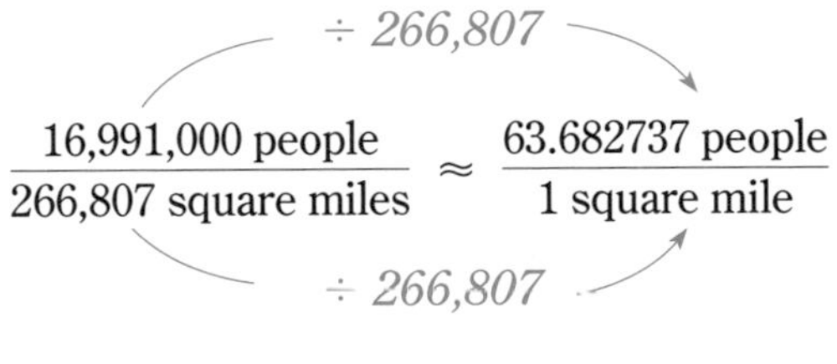

Texas has about 63.7 people per square mile.

Washington

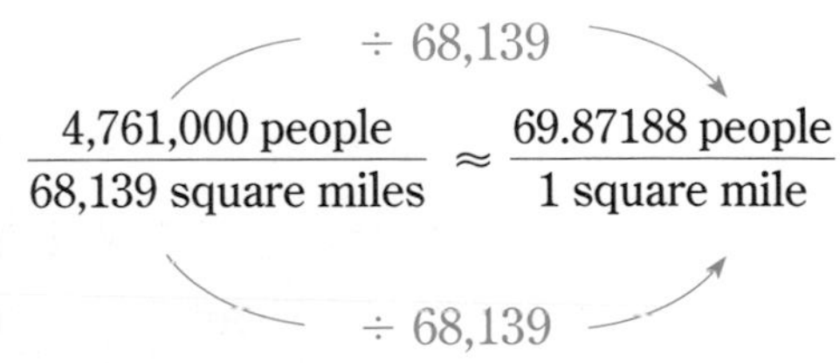

Washington has about 69.9 people per square mile.

Washington has more people per square mile than Texas. So Washington is more densely populated.

Checking for Understanding

Communicating Mathematics

Read and study the lesson to answer each question.

1. **Tell** the difference between a ratio and a rate.
2. **Tell** why unit rates are helpful. Give two examples.
3. **Write** a ratio about the students in your class.
   a. Write your ratio in simplest form.
   b. Is your ratio also a rate? Why?

Guided Practice

Express each ratio or rate as a fraction in simplest form.

4. 5 out of 7 people
5. 2 cups to 16 cups
6. 20 out of 25 free throws
7. 35 wins in 55 games
8. 18 brown-eyed student : 12 blue-eyed students
9. 1 foot per 1 yard

Express each comparison as a rate in simplest form. Tell whether the simplest form is a unit rate.

10. 100 miles in 4 hours
11. 24 pounds lost in 8 weeks
12. 4 inches of rain in 30 days
13. 102 passengers in 9 minivans

Exercises

Independent Practice

Express each ratio or rate as a fraction in simplest form.

14. 11 out of 12
15. 49 tiles:77 tiles
16. 27 is to 15
17. 99 wins:99 losses
18. 18 boys:27 students
19. 65 out of 105
20. 17 out of 51
21. 64 to 16
22. 165:200
23. 144 to 96
24. 188 students:$354
25. 3 inches per foot
26. 6 absences in 180 school days
27. 20 minutes per hour

Express each as a unit rate.

28. $1.75 for 5 minutes
29. $25 for 10 disks
30. 300 students to 20 teachers
31. $420 for 15 tickets
32. $8.80 for 11 pounds
33. 96¢ per dozen
34. $25,000 fifth-place prize money among 100 winners

Mixed Review

35. Solve $2x - 8 = 12$. *(Lesson 2-7)*
36. **Statistics** What is meant by the mean, median, and mode? *(Lesson 4-5)*

37. **Probability** A cereal box stated that it may contain a prize. A consumer protection agency bought 50 boxes of the cereal. Five of them contained a prize. Estimate the probability of winning a prize when you buy a box of that brand of cereal. *(Lesson 6-8)*

38. Order 7.35, $7\frac{2}{7}$, and $\frac{37}{5}$ from least to greatest. *(Lesson 6-10)*

39. Find the next number in the sequence 3, 7, 11, __?__. *(Lesson 7-6)*

40. Name all the sets of numbers to which $\sqrt{100}$ belongs. *(Lesson 8-3)*

Problem Solving and Applications

41. **Critical Thinking** Ms. Al-Khwarizmi works for a law firm from a computer in her home. The computer communicates with the law office by use of a telephone modem. The telephone bill for the modem was \$7.50 for 18 minutes of use. Find the unit rate for one hour of use.

42. **Sports** In the first ten games of the season, the Pittsburgh Pirates won 6 games. What is their win:loss ratio in simplest terms?

43. **School** Eisenhower Middle School has 24 computers and 432 students. Thales Central Middle School has 567 students and 27 computers. In which school would you have a better chance of getting computer time?

44. **Data Search** Refer to page 666. Enrique Perez can drive from Charlotte, North Carolina to Birmingham, Alabama in about 7 hours.
    a. Find the unit rate of miles per hour or average speed of the car.
    b. If the speed limit is 55 miles per hour, should the State Trooper give him a ticket for speeding?

45. **Mathematics and Sports** Read the following paragraphs.

> The modern game of baseball was invented by Abner Doubleday at Cooperstown, New York in 1839. In 1845, Alexander J. Cartwright established standard rules for the game and organized the Knickerbocker Baseball Club of New York.
>
> Since the beginning of baseball, statisticians have kept track of players' performances. For example, a player's batting average is really a ratio.
>
> $$\frac{\text{number of hits}}{\text{number of times at bat}} = \text{batting average}$$

In the 1991 baseball season, Jeff Treadway led the Atlanta Braves in individual batting with 98 hits in 306 times at bat. To the nearest thousandth, what was his batting average?

Cooperative Learning

9-1B The Golden Ratio

A Follow-Up of Lesson 9-1

Objective

Find the value of the golden radio.

Materials

graph paper
scissors
calculator
drawing paper

The ancient Greeks were concerned with aesthetics. That means they wanted to know what made some things more pleasing to look at than other things. For example, many Greek philosophers and mathematicians searched for the perfect, beautiful rectangle. In these rectangles, the ratio of the length to the width always resulted in the same ratio. This ratio became known as the **golden ratio,** and the rectangles were called **golden rectangles.**

Try this!

Work in groups of three.

- Cut a rectangle out of graph paper that is 34 squares by 21 squares. Find the ratio of length to width and record your result. Express the ratio as a decimal.
- **Step 1** Draw a square so that the edge along the width of the square is one side of the square. Cut this square from the rectangle.

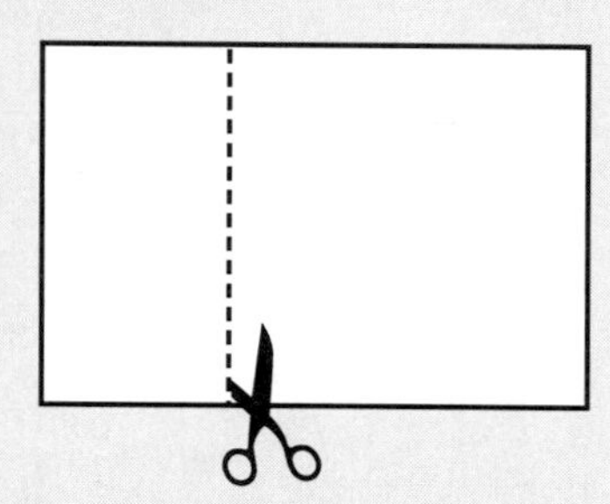

Step 2 Measure the new rectangle that remains. Record its length and width. Express the ratio of length to width as a decimal. Record your result.

- Repeat Steps 1 and 2 with your new rectangle. Record the measurements and ratio.
- Continue repeating the steps and recording your information until the remaining rectangle is 1 square by 2 squares.

What do you think?

1. Make a conjecture about the ratios you recorded.
2. Does your conjecture hold true for any rectangle? Draw several rectangles to support your answer.
3. If the rectangle above were described as a golden rectangle, what do you think is the value of the golden ratio?

Applications

Investigate the ratio of the two measurements in each situation. Which ones are close to the golden ratio?

4. Find a painting of a landscape. Measure the height of the painting and the distance from the horizon line to the bottom of the painting.
5. Measure the height of each person in your group. Then measure the distance from the floor to each person's navel.
   a. Find the ratio for each person.
   b. Find the average ratio for the group.
6. Measure the height of your face and the width of your face at your cheekbone. Compare your ratio with those in your group.

The ancient Greeks considered the value of the golden ratio to be 1.618. In Exercises 7–9, a rectangle has been drawn. Measure each rectangle and determine how closely it resembles a golden rectangle.

7. The Parthenon (Greece, 450 B.C.)
8. Bill Clinton (1994)
9. Augustus (Rome, 20 B.C.)

Extension

10. The spiral of the chambered nautilus follows the pattern of the golden rectangle.
    a. Trace the figure at the right. It is a smaller version of the golden rectangle you cut apart at the beginning of this lesson.
    b. An arc connects the opposite corners of the large square. Sketch an arc that connects the opposite corners of the next largest square. Continue the process.

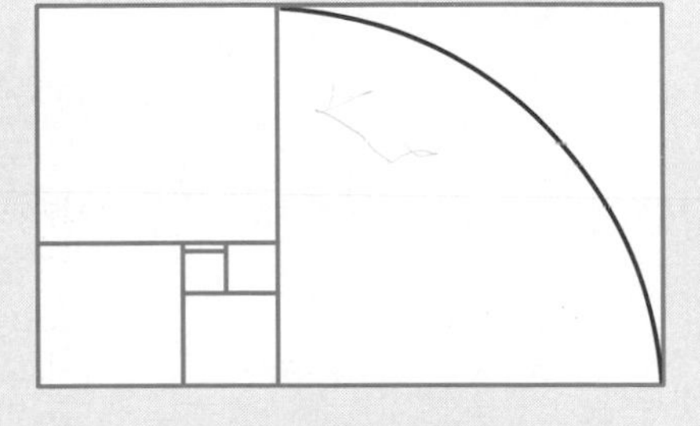

9-2 Proportions

Objectives

Detemine if a pair of ratios form a proportion and solve proportions.

Words to Learn

proportion
cross product

A recent survey reported that 3 out of 50 eighth-graders in Montana and North Dakota watch television six or more hours a day.

The ratio *3 out of 50* can be expressed as the fraction $\frac{3}{50}$. The report means that $\frac{3}{50}$ of the eighth-graders surveyed watch television six or more hours a day. There are about 20,000 eighth-graders in Montana and North Dakota. You can estimate how many of them watch television six or more hours a day by finding equivalent fractions.

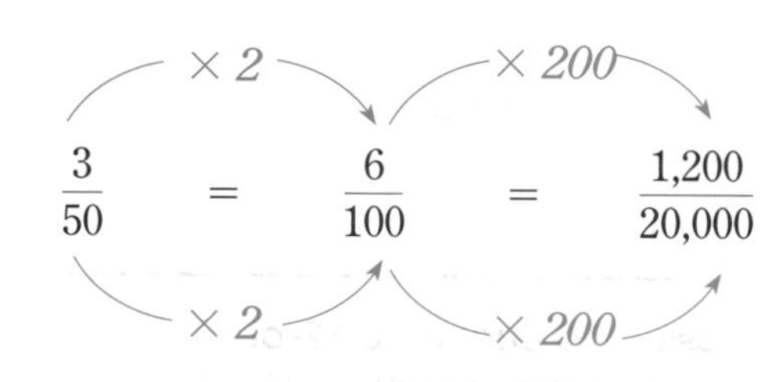

TEEN SCENE

If you are an "average" teenager, you might watch television about 22 hours per week. Of those 22 hours, advertisements take 3 to 4 hours. That's about 15% of the time you spend watching television.

This means that 3 out of 50 is equivalent to 1,200 out of 20,000. Therefore, about 1,200 eighth-graders in Montana and North Dakota watch television six or more hours a day. The equation $\frac{3}{50} = \frac{1,200}{20,000}$ states that the two ratios are equivalent. This is an example of a **proportion.**

Proportion	**In words:** A proportion is an equation that shows that two ratios are equivalent.	
	Arithmetic $\frac{3}{50} = \frac{1,200}{20,000}$	**Algebra** $\frac{a}{b} = \frac{c}{d}, b \neq 0, d \neq 0$

There are two ways to determine if a pair of ratios form a proportion. The following example shows one way.

Example 1

Jane made 9 out of 12 free throws, and Alesha made 18 out of 24. Do these ratios form a proportion?

You can write each ratio in simplest form and compare them.

$\frac{9}{12} = \frac{3}{4}$ $\quad\frac{18}{24} = \frac{3}{4}$

Since both ratios simplify to $\frac{3}{4}$, they are equivalent.
The ratios do form a proportion.

The proportion in Example 1 can also be verified by using **cross products.** In the proportion shown at the right, $9 \cdot 24$ and $12 \cdot 18$ are cross products.

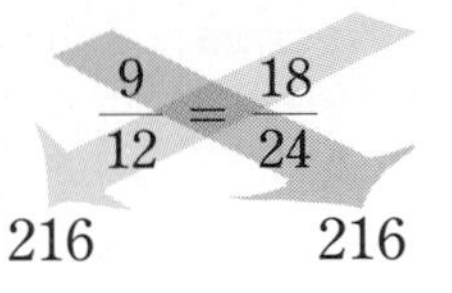

Since the products are equal, the ratios form a proportion.

Property of Proportions	**In words:** The cross products of a proportion are equal. **In symbols:** If $\frac{a}{b} = \frac{c}{d}$, then $ad = bc$. If $ad = bc$, then $\frac{a}{b} = \frac{c}{d}$.

If one value in a proportion is not known, you can use cross products to solve the proportion.

Example 2 *Problem Solving*

Fitness To burn off a 12-ounce can of regular cola, you have to cycle for 24 minutes at 13 mph. How long would you have to cycle at 13 mph to burn off a 16-ounce bottle of cola?

Write a proportion. Let m represent the number of minutes.

$$\begin{matrix} \textit{ounces} \rightarrow \\ \textit{minutes} \rightarrow \end{matrix} \frac{12}{24} = \frac{16}{m} \begin{matrix} \leftarrow \textit{ounces} \\ \leftarrow \textit{minutes} \end{matrix}$$

Find the cross products. Then solve for m.

$$\frac{12}{24} = \frac{16}{m}$$
$$12 \cdot m = 24 \cdot 16$$
$$12m = 384$$
$$\frac{12m}{12} = \frac{384}{12}$$

24 [×] 16 [÷] 12 [=] 32

$$m = 32$$

You would have to cycle for 32 minutes at 13 mph to burn off a 16-ounce bottle of cola.

Calculator Hint

To solve a proportion using a calculator, find one cross product. Then divide by the remaining number. Do not press [=] until after you divide.

Checking for Understanding

Communicating Mathematics

Read and study the lesson to answer each question.

1. **Tell** two methods for determining if two ratios form a proportion.
2. **Write** two ratios that do not form a proportion. Explain why they do not.

Guided Practice Tell whether each pair of ratios form a proportion.

3. $\frac{16}{12}, \frac{12}{9}$
4. $\frac{7}{6}, \frac{6}{7}$
5. $\frac{75}{100}, \frac{3}{4}$
6. $\frac{3}{11}, \frac{55}{200}$

Solve each proportion.

7. $\frac{3}{5} = \frac{c}{10}$
8. $\frac{a}{0.9} = \frac{0.6}{2.7}$
9. $\frac{120}{b} = \frac{24}{60}$
10. $\frac{3}{7} = \frac{2.1}{d}$

Exercises

Independent Practice Tell whether each pair of ratios form a proportion.

11. $\frac{8}{12}, \frac{10}{15}$
12. $\frac{3}{2}, \frac{4}{6}$
13. $\frac{10}{12}, \frac{5}{6}$
14. $\frac{2}{7}, \frac{6}{21}$
15. $\frac{6}{9}, \frac{8}{12}$
16. $\frac{10}{16}, \frac{25}{45}$
17. $\frac{75}{100}, \frac{4}{3}$
18. $\frac{18}{14}, \frac{54}{42}$

Solve each proportion.

19. $\frac{5}{4} = \frac{y}{12}$
20. $\frac{4}{100} = \frac{12}{n}$
21. $\frac{2}{34} = \frac{5}{x}$
22. $\frac{9}{1} = \frac{n}{14}$
23. $\frac{2}{3} = \frac{7}{y}$
24. $\frac{0.35}{3} = \frac{c}{18}$
25. $\frac{n}{2} = \frac{7}{4}$
26. $\frac{10}{6} = \frac{y}{26}$

Mixed Review

27. Solve $480 \div (-40) = t$. *(Lesson 3-7)*
28. **Geometry** Find the value of x in the figure at the right. *(Lesson 5-1)*

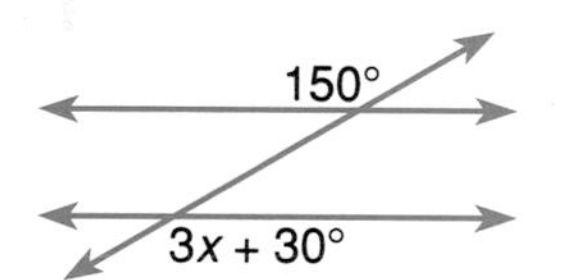

29. Solve $\left(-2\frac{6}{7}\right)\left(-5\frac{3}{5}\right) = y$. *(Lesson 7-3)*
30. Express *240 shrimp in 6 pounds* as a unit rate. *(Lesson 9-1)*

Problem Solving and Applications

31. **Critical Thinking** Solve $\frac{9}{x} = \frac{x}{16}$.
32. **Science** Light travels approximately 1,860,000 miles in 10 seconds. How long will it take light to travel the 93,000,000 miles from the sun to Earth?
33. **Health** Take your pulse to count the number of times your heart beats in 15 seconds. Let c = this count. Substitute your value for c into the proportion $\frac{c}{15} = \frac{p}{60}$. Then solve for p and you will know your pulse rate per minute.

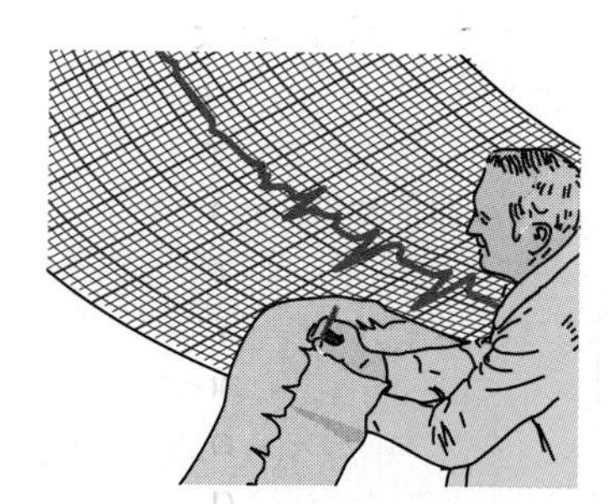

34. **Data Search** Refer to pages 336 and 337. How many U.S. dollars are equivalent to 2,500 Japanese yen?

9-3 Using Proportions

Objective

Solve problems by using proportions.

In Home Economics class, Wai Lui used $1\frac{1}{4}$ cups of sugar in a recipe that made 25 cookies. If he wanted to make 60 cookies, how much sugar should he use?

This problem can be solved using a proportion. In the proportion below, c represents the number of cups of sugar needed for the 60 cookies.

$$\begin{matrix} \textit{cups} \rightarrow \\ \textit{cookies} \rightarrow \end{matrix} \frac{1\frac{1}{4}}{25} = \frac{c}{60} \begin{matrix} \leftarrow \textit{cups} \\ \leftarrow \textit{cookies} \end{matrix}$$

You will solve this problem in Exercise 5.

This proportion involves two rates. Remember you must always write the proportions so that the units correspond.

Technology Activity

You can learn how to use a spreadsheet to solve proportions in Technology Activity 4 on page 661.

Example 1 *Problem Solving*

Health **Out of 60 blood cells on a microscope slide, there are 35 red blood cells. How many red blood cells would be expected in a blood sample of 75,000 blood cells?**

Explore $\frac{35}{60}$ is the ratio of red cells to all blood cells. The sample has 75,000 blood cells.

Plan Let r represent the number of red cells. Write a proportion.

$$\begin{matrix} \textit{red blood cells} \rightarrow \\ \textit{all blood cells} \rightarrow \end{matrix} \frac{35}{60} = \frac{r}{75{,}000} \begin{matrix} \leftarrow \textit{red blood cells} \\ \leftarrow \textit{all blood cells} \end{matrix}$$

Solve Solve the proportion.

$$\frac{35}{60} = \frac{r}{75{,}000}$$

35 [×] 75000 [÷] 60 [=] 43750

$r = 43{,}750$

43,750 red blood cells would be expected in a sample of 75,000 blood cells.

Suppose you want to triple a recipe for banana bread. The ingredients must stay in the same proportion for the bread to taste the same. When you multiply the amount of each ingredient by three, you are using proportions.

Examine Check your solution by substituting 43,750 for r in your original ratios. Simplify each ratio.

$\frac{35}{60} = \frac{7}{12}$ $\frac{43{,}750}{75{,}000} = \frac{7}{12}$

Since both ratios are equivalent to the same fraction, they form a proportion. So, 43,750 is correct.

Example 2 ***Connection***

Statistics **The Student Council at Walnut Springs Middle School decided to sell school sweatshirts as a fundraiser. They surveyed 75 students. Of those students, 53 said they would buy a sweatshirt. If there are 1,023 students in the school, about how many will buy a sweatshirt?**

Let s represent the number of sweatshirts. Write a proportion.

$\frac{53}{75} = \frac{s}{1{,}023}$

53 [×] 1023 [÷] 75 [=] 722.92

$s = 722.92$

About 723 students will buy a sweatshirt.

Checking for Understanding

Communicating Mathematics

Read and study the lesson to answer each question.

1. **Show** how to use a calculator to solve $\frac{25}{x} = \frac{15}{9}$.
2. **Tell** which proportions from the list below could be used to solve the problem. Be prepared to defend your answer.

   If 12 eggs cost 96¢, how much will 6 eggs cost?

   a. $\frac{12}{0.96} = \frac{6}{x}$ b. $\frac{0.96}{12} = \frac{x}{6}$ c. $\frac{12}{6} = \frac{0.96}{x}$

   d. $\frac{12}{x} = \frac{0.96}{6}$ e. $\frac{x}{12} = \frac{6}{0.96}$ f. $\frac{6}{12} = \frac{x}{0.96}$

Guided Practice

Write a proportion to solve each problem. Then solve.

3. 40 ounces costs \$43. 25 ounces costs \$$x$.
4. 10 pounds cooks in 4 hours. y pounds cooks in 5.6 hours.
5. Refer to the cookie problem at the beginning of this lesson. How much sugar will Wai need to make 60 cookies?

Exercises

Independent Practice

Use a proportion to solve each problem.

6. 60 lines can be printed in 10 seconds. k lines can be printed in 15 seconds.
7. Sue can type 2 pages in 15 minutes. How many minutes will it take her to type 35 pages?
8. The average heart beats 72 times in 60 seconds. How many times does it beat in 15 seconds?
9. 24 cans of soft drink cost \$6.24. x cans cost \$4.68.

Mixed Review

10. Express $9\frac{3}{5}$ as a decimal. *(Lesson 6-6)*
11. What is the negative square root of 64? *(Lesson 8-1)*
12. Solve $\frac{3}{7} = \frac{n}{28}$. *(Lesson 9-2)*

Problem Solving and Applications

13. **Science** The ratio of a person's weight on the moon to his or her weight on Earth is 1:6. If you weigh 126 pounds on Earth, how much would you weigh on the moon?
14. **Smart Shopping** The Winn-Dixie store sells 8 cans of Cool Cola for \$2.59. Super-Duper sells 6 cans of Cool Cola for \$1.79. Which is the better buy?

15. **Statistics** A political advisor surveyed 150 voters before election day to find their preference for mayor. The results are shown in the table below.

Candidate	Number
Alvarez	45
Cruz	54
Hoffman	33
Newton	18

Suppose 250,000 voters are expected to vote in the election. Predict the number of votes each candidate will receive.

16. **Critical Thinking** Three friends, Arturo, Beth, and Carmen, invested \$200, \$300, and \$500 respectively in a business. Profits will be divided in the same ratio as the investment. If the business made a profit of \$1,500, what is each investor's share in the profit?

Problem-Solving Strategy

9-4 Draw a Diagram

Objective

Solve problems by drawing a diagram.

The famous mathematician, Fibonacci, liked to invent problems to solve. One concerns a pair of rabbits.

Suppose you place a pair of adult rabbits in an enclosure on January 1. Every month, each adult pair of rabbits produces a pair of baby rabbits. These baby rabbits begin to give birth two months after their own birth. How many pairs of rabbits will there be on the following January 1?

One way to solve this problem is to draw a diagram. Diagrams are useful because they can help you decide how to solve the problem. They can also help you organize your information.

Explore Every adult pair of rabbits produces a pair of baby rabbits each month. In two months, the babies begin to reproduce.

Plan Organize the information in a drawing. Then look for a pattern.

Solve In the drawings, *R* stands for a pair of adult rabbits, *b* stands for a pair of baby rabbits, and *r* stands for a pair of rabbits that are not old enough to reproduce.

Date		Number of Pairs of Rabbits
January 1	R	1
February 1	R → b	2
March 1	b ← R r	3
April 1	r b ← R R → b	5
May 1	b ← R r b ← R R → b r	8
June 1	r b ← R b ← R r R → b b ← R r R → b	13

It is clear that the drawing is becoming unmanageable. So, let's look for a pattern in the number of pairs of rabbits.

1 2 3 5 8 13

LOOKBACK

You can review the Fibonacci Sequence on page 276.

You may recognize this sequence of numbers as the Fibonacci Sequence. Each number is the sum of the previous two numbers. *The Fibonacci Sequence actually begins with two 1s.*

Extend the sequence to find the number of pairs of rabbits on January 1.

1	2	3	5	8	13	21	34	55	89	144	233	377
J	F	M	A	M	J	J	A	S	O	N	D	J

On January 1, you will have 377 pairs of rabbits!

Examine Make a model using different color counters to picture each generation.

Checking for Understanding

Communicating Mathematics

Read and study the lesson to answer each question.

1. **Tell** two reasons why a diagram is useful in solving problems.
2. **Tell** how many pairs of rabbits you would have on the following February 1.

Guided Practice

Solve. Make a drawing.

3. A section of a theater is arranged so that each row has the same number of seats. Andrew is seated in the fifth row from the front and the third row from the back. His seat is sixth from the left and second from the right. How many seats are in this section?
4. Sixteen players are competing in a bowling tournament. Each player in the competition bowls against another opponent and is eliminated after one loss. How many games does the winner play?

Problem Solving

Practice Solve using any strategy.

Strategies

- Look for a pattern.
- Solve a simpler problem.
- Act it out.
- Guess and check.
- Draw a diagram.
- Make a chart.
- Work backward.

5. When a number is decreased by four, the result is -23. What is the number?
6. How many different three-member teams can be formed from six players?
7. There are eight people at a business meeting. Each person shakes hands with everyone else exactly once. How many handshakes occur?
8. Ms. Straub's dance class is standing evenly spaced in a circle. If the sixth person is directly opposite the sixteenth person, how many people are in the circle?
9. Two sides of a triangle have the same length. The third side is 2 meters long. If the perimeter of the triangle is 20 meters, find the lengths of the sides.

COMPUTER CONNECTION

10. **Computer Connection** A Fibonacci ratio is found by dividing each term of the Fibonacci sequence by the next term. The BASIC program at the right can be used to print Fibonacci ratios.
    a. Run the program to determine the value of the Fibonacci ratio.
    b. Compare the Fibonacci ratio to the golden ratio found on page 342. How are they the same?

```
10 LET X = 1
20 LET Y = 1
30 LET Z = X/Y
40 PRINT Z
50 LET W = X + Y
60 LET X = Y
70 LET Y = W
80 LET N = N + 1
85 IF N = 15 THEN
    95
90 GOTO 30
95 END
```

9 Assessment: Mid-Chapter Review

Express each ratio or rate as a fraction in simplest form. *(Lesson 9-1)*

1. 15 out of 20
2. $500 from 320 students
3. 80¢ per dozen

Solve each proportion. *(Lesson 9-2)*

4. $\frac{2}{5} = \frac{x}{250}$
5. $\frac{7.5}{a} = \frac{3}{4}$
6. $\frac{6.5}{19.5} = \frac{13}{y}$

Solve.

7. A 10-acre field produced 750 bushels of corn. At that rate, how much corn can be produced from a 14-acre field? *(Lesson 9-3)*
8. Patrice lives nine blocks west of Julio. Julio lives four blocks east of Margie. Where does Margie live in relationship to Patrice? *(Lesson 9-4)*

Geometry Connection

9-5 Similar Polygons

Objective

Identify corresponding parts of similar polygons.

Words to Learn

polygon
pentagon
similar polygons

There are 17 miles of corridors in the Pentagon, and there are over 7,700 windows to clean.

LOOKBACK

You can review similar figures on page 198.

The largest office building in the world is the Pentagon in Washington, D.C. It has a floor area of 6,500,000 square feet. Notice how the figure outlining the courtyard has the same shape as the outer walls of the Pentagon.

In mathematics, a **polygon** is a simple, closed figure in a plane formed by three or more sides. A **pentagon** is a polygon with five sides. The pentagons shown below have the same shape, but differ in size. These pentagons are **similar polygons.**

pentagon $ABCDE \sim$ pentagon $VWXYZ$

The symbol $\sim$ means similar to.

Notice that the corresponding angles are congruent.

$\angle A \cong \angle V \quad \angle B \cong \angle W \quad \angle C \cong \angle X \quad \angle D \cong \angle Y \quad \angle E \cong \angle Z$

A special relationship also exists among the corresponding sides of the polygons. Let's compare the ratios of the lengths of the corresponding sides.

$\frac{AB}{VW} = \frac{7.5}{5}$ or $\frac{3}{2}$ $\quad \frac{BC}{WX} = \frac{7.5}{5}$ or $\frac{3}{2}$ $\quad \frac{CD}{XY} = \frac{9}{6}$ or $\frac{3}{2}$

$\frac{DE}{YZ} = \frac{6}{4}$ or $\frac{3}{2}$ $\quad \frac{AE}{VZ} = \frac{6}{4}$ or $\frac{3}{2}$

As you can see, the ratios of the lengths of the corresponding sides all equal $\frac{3}{2}$. Since they are all equivalent, you can form proportions using corresponding sides.

Similar Polygons	Two polygons are similar if their corresponding angles are congruent and their corresponding sides are in proportion.

Estimation Hint

Often you can tell which angles are corresponding by estimating their angle measure instead of measuring each angle.

Proportions are useful in finding the missing length of a side in any pair of similar polygons.

Example

The two quadrilaterals below are similar. Find the length of side $\overline{AB}$.

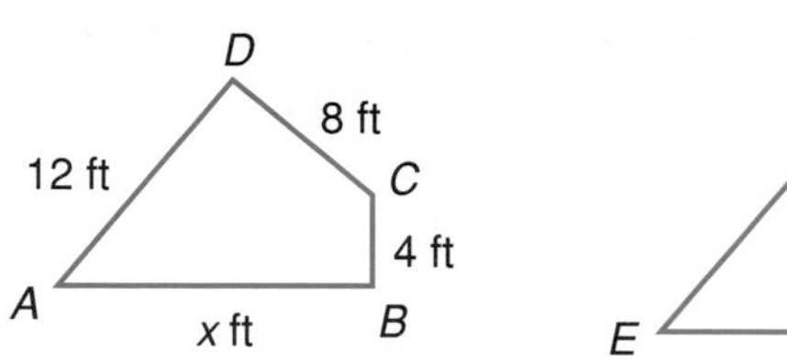

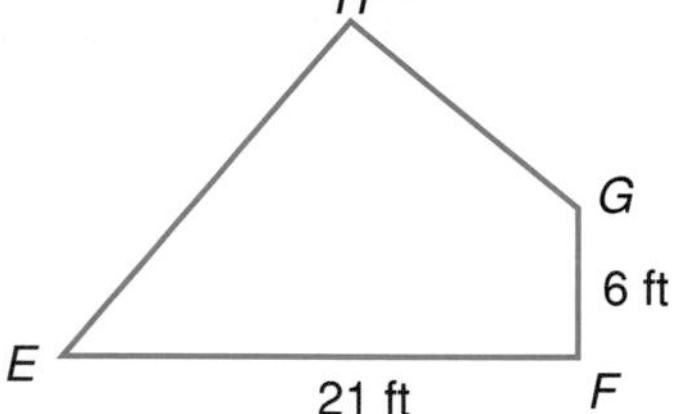

$\overline{BC}$ corresponds to $\overline{FG}$ and $\overline{AB}$ corresponds to $\overline{EF}$, so you can write a proportion. Then use the values from the quadrilaterals to find the missing measure.

$$\frac{BC}{FG} = \frac{AB}{EF}$$

$$\frac{4}{6} = \frac{x}{21}$$ *Substitute values from the quadrilaterals.*

$$4 \cdot 21 = 6 \cdot x$$ *Find the cross products.*

$$84 = 6x$$

$$14 = x$$

The length of $\overline{AB}$ is 14 feet.

Checking for Understanding

Communicating Mathematics

Read and study the lesson to answer each question.

1. **Write** a definition of similar polygons.
2. **Draw** two similar polygons and explain how you know which parts are corresponding.
3. **Tell** the difference between similar polygons and congruent polygons.

Guided Practice

Tell whether each pair of polygons are similar.

4.

5.

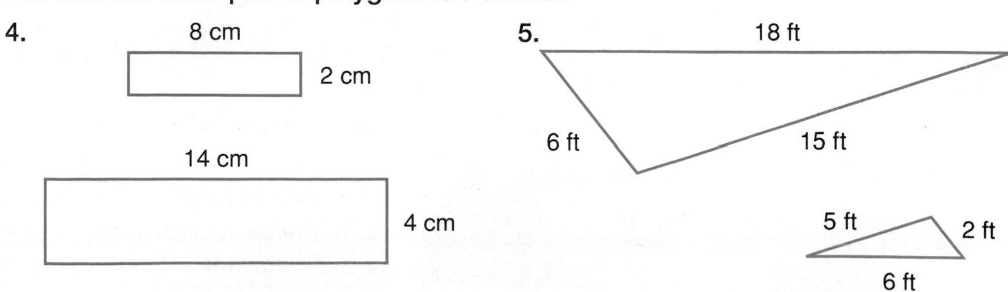

Refer to the Example on page 354.

6. Find the length of $\overline{GH}$.

7. Find the length of $\overline{EH}$.

8. If the length of $\overline{BC}$ is 8 units, how would your answers to Exercises 6 and 7 change?

Exercises

Independent Practice

Tell whether each pair of polygons are similar.

9.

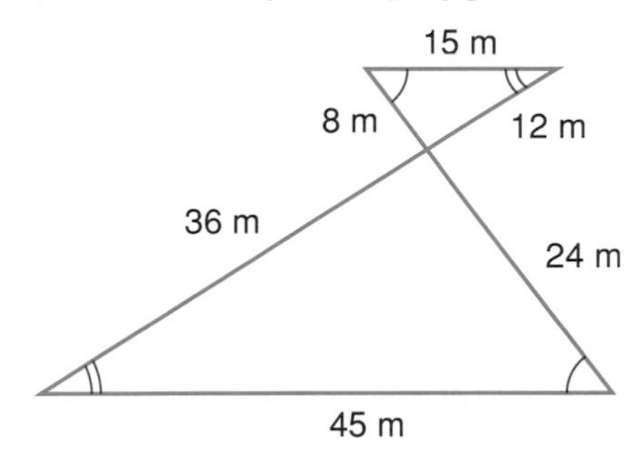

10.

3 in.
3 in.
3 in.
3 in.

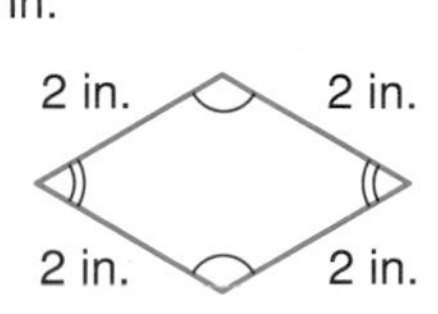

In the figure at the right, $\triangle ABC \sim \triangle ADE$. Use this information to answer Exercises 11-14.

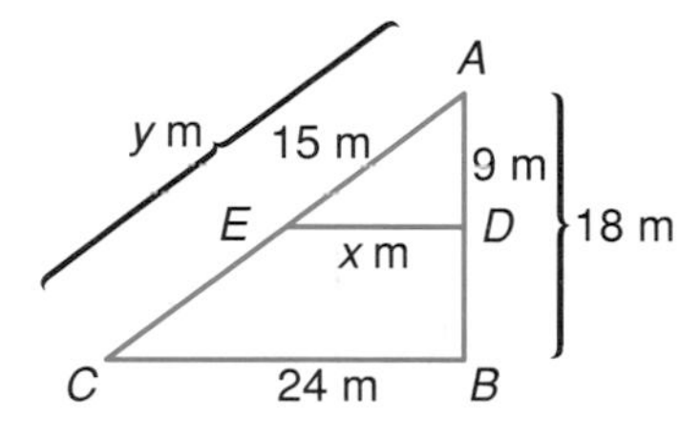

11. List all pairs of corresponding angles.

12. Writc three ratios relating the corresponding sides.

13. Write a proportion and solve for x.

14. Write a proportion and solve for y.

Mixed Review

15. Write $4 \cdot 4 \cdot 8 \cdot 8 \cdot 4$ as an expression using exponents. *(Lesson 1-9)*

16. Order 23, -8, 0, -16, 51, -51, and -30 from greatest to least. *(Lesson 3-2)*

17. Solve $a = \frac{7}{12} \div \frac{3}{4}$. *(Lesson 7-9)*

18. **Cooking** Nancy used 6 tablespoons of ground coffee to make 3 cups of coffee. If she wanted to make 5 cups of coffee, how many tablespoons of ground coffee should she use? *(Lesson 9-3)*

Problem Solving and Applications

19. **Critical Thinking** The symbol for similar is $\sim$ and the symbol for congruent is $\cong$. The symbols are related in the same way that the two words are related. What is this relationship?

20. **Photography** The 4 in. by 5 in. portrait proofs for the school yearbook must be reduced to fit three across on a page. The ratio of the original to the reduced print needed to do this is 8:5. Find the dimensions of the pictures as they will appear in the yearbook.

21. **Hobbies** Craig likes to build models of antique cars. At a recent car show, he took photographs of a 1934 Buick coupe, which was 174 inches long and 66 inches high. He wants his model to be 6 inches long. To the nearest tenth of an inch, what will be the height of his model?

9-6 Indirect Measurement

Objective

Solve problems involving similar triangles.

Words to Learn

indirect measurement

For geography class, Marcia had to draw a map of the United States on a poster. In the atlas she has, the distance on the map from Buffalo to Tampa Bay is 5.4 cm. The distance from Buffalo to Houston is 6.7 cm. The distance from Tampa Bay to Houston is 4.2 cm. On her poster, the distance from Buffalo to Tampa Bay is 21.6 cm. Without measuring, what should the other two distances be?

You will be asked to solve the map problem in Exercise 8.

If Marcia measures the distances between the other two cities on her poster, she makes a direct measurement. When we use proportions to find a measurement, we use **indirect measurement.**

Example 1 ***Problem Solving***

Forestry **The park ranger wanted to estimate the height of a tree that had been planted five years ago. At 2:00 P.M., her shadow was 3 feet long. The shadow of the tree was 19 feet long. If the ranger is $5\frac{1}{2}$ feet tall, how tall is the tree?**

Explore Make a drawing and fill in the information you know. Draw similar triangles.

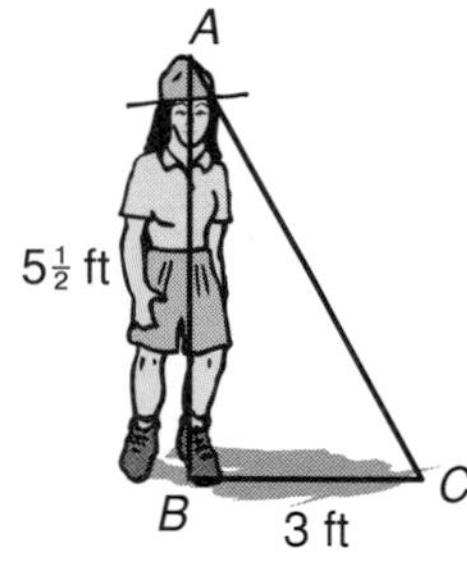

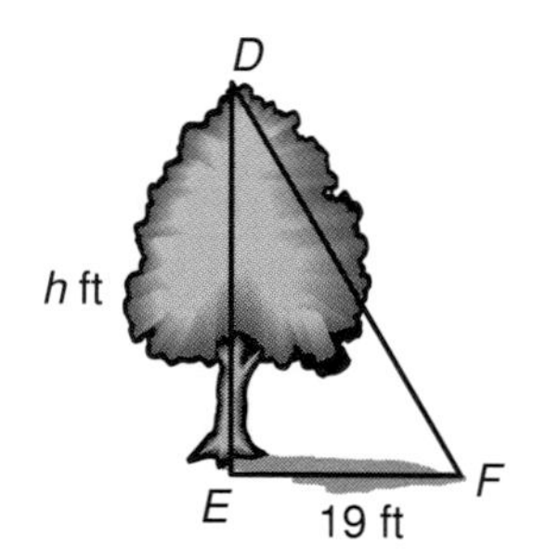

Plan Write a proportion from the information in the drawing. Let h = the tree's height.

ranger's shadow → $\frac{3}{19} = \frac{5\frac{1}{2}}{h}$ ← *ranger's height*
tree's shadow → ← *tree's height*

Solve

$$\frac{3}{19} = \frac{5\frac{1}{2}}{h}$$

$$3 \cdot h = 19 \cdot 5\frac{1}{2}$$ *Find the cross product.*

$$3h = 104\frac{1}{2}$$ *Divide each side by 3.*

$$h = 34\frac{5}{6}$$ The tree is about 35 feet high.

Mental Math Hint

Sometimes you can solve a proportion mentally by using equivalent fractions.

$$\frac{3}{18} = \frac{6}{h}$$

THINK: $3 \cdot 2 = 6$

$18 \cdot 2 = 36$

So, $h = 36$

Example 2 *Problem Solving*

Geography Use the diagram below to find how far is it from Tybert to Crawford by boat. The triangles are similar triangles.

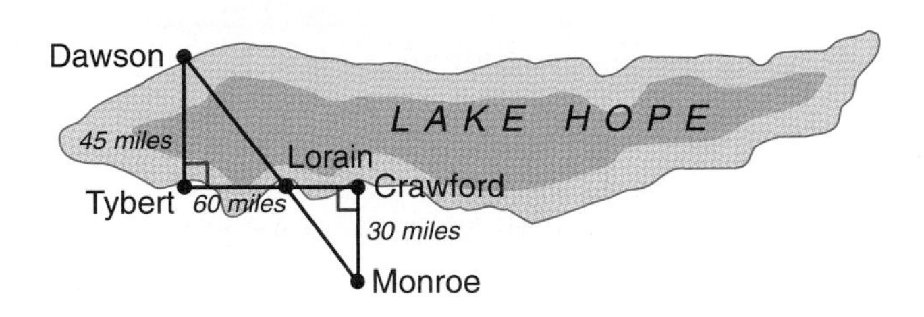

Explore The distance from Tybert (T) to Crawford (C) is equal to the distance from Tybert to Lorain (L), 60 miles, plus the distance from Lorain to Crawford (C). Similar triangles can help us find the distance from Lorain to Crawford.

Plan Let the first letters of the cities represent the vertices of the triangles. So, $\triangle LTD \sim \triangle LCM$. Now write a proportion.

$$\frac{MC}{CL} = \frac{DT}{TL} \rightarrow \frac{30}{x} = \frac{45}{60}$$

Solve $\frac{30}{x} = \frac{45}{60}$

30 [×] 60 [÷] 45 [=] 40 So, $40 = x$.

The distance from Tybert to Crawford by boat is $60 + 40$ or 100 miles.

Examine Estimate the distance by comparing it to those on the map. Does it seem reasonable?

Checking for Understanding

Communicating Mathematics Read and study the lesson to answer each question.

1. **Tell** what is meant by indirect measurement.
2. **Write** a proportion that you can solve mentally.
3. **Draw** a diagram like the one in Example 1, in which the ranger's shadow is 9 feet, the tree's shadow is 15 feet, and the ranger is 6 feet tall.
4. **Tell** a situation where you could use indirect measurement.

Guided Practice Draw a diagram of the situation. Then write a proportion and solve it.

5. A flagpole casts a shadow 8.75 feet long at the same time that a 5-story office building, 60 feet tall, casts a 15-foot shadow. How tall is the flagpole?
6. A guy wire is attached to the top of a telephone pole and goes to the ground 9 feet from its base. When Jarem stands under the guy wire so that his head touches it, he is 2 feet 3 inches from where the wire goes into the ground. If Jarem is 5 feet tall, how tall is the telephone pole?

Exercises

Independent Practice

Write a proportion to solve each problem. Assume the triangles are similar.

7. Find the distance across Clarence Lake if Town Road and Park Lane are perpendicular.

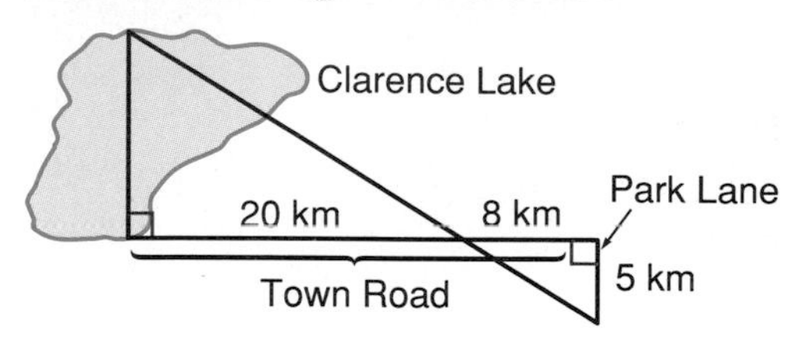

8. Refer to the poster problem at the beginning of the lesson. What is the length from Tampa Bay to Houston and the length of Buffalo to Houston on Marcia's poster?

9. The state highway department is investigating the possibility of building a tunnel through the mountain from Picketon to Skylight. The surveyors provided the map shown at the right. How long would the tunnel be?

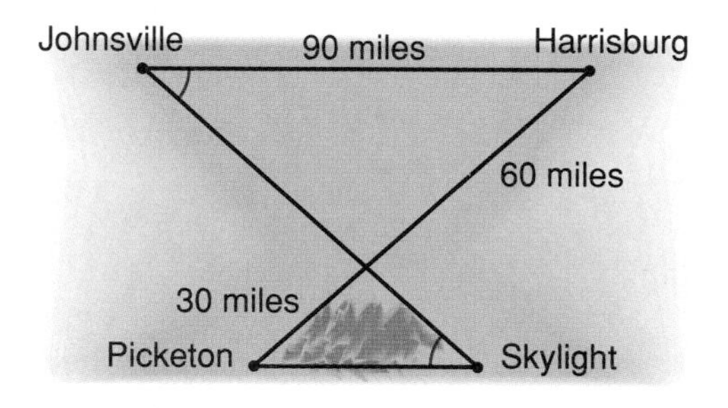

Mixed Review

10. Solve $n = -28 + 73$ *(Lesson 3-3)*

11. Find the LCM of 48 and 60. *(Lesson 6-9)*

12. **Geometry** In the figures at the right, $\triangle ABC \sim \triangle DEF$. Write three ratios relating the corresponding sides. *(Lesson 9-5)*

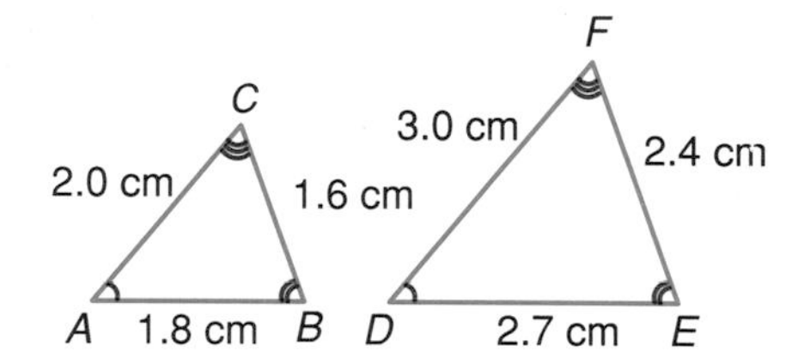

Problem Solving and Applications

13. **Sailing** A tugboat known to be 33.85 feet tall is tied up next to the *Velsheda,* the world's tallest single-masted yacht. The tugboat casts a 5-foot shadow while the *Velsheda* casts a 25-foot shadow. How tall is the *Velsheda?*

14. **History** The Pyramid of the Sun at Teotihuacán in northeastern Mexico casts a shadow 13.3 meters long. At the same time, a 1.83-meter tall man casts a shadow 0.4 meters long. How tall is the Pyramid of the Sun?

15. **Architecture** At 1:00 P.M. in Paris, the shadow of the Eiffel Tower is 123.24 feet long. When the sun is at the same angle in Chicago, the Sears Tower, the world's tallest building, casts a 181.75-foot shadow. If the Sears Tower is 1,454 feet tall, how tall is the Eiffel Tower?

16. **Critical Thinking** What factors might cause you to get an incorrect measurement when using the indirect measurement method?

17. **Journal Entry** Write a plan for measuring the height of your school building by the indirect measurement method. Include a drawing with your plan.

9-7 Scale Drawings

Objective

Solve problems involving scale drawings.

A **scale drawing** is used to represent something that is too large or too small to be conveniently drawn actual size. The scale is determined by the ratio of a given length on the drawing to its corresponding length in reality. Blueprints and maps are commonly-used scale drawings.

Example ***Problem Solving***

Aviation Denise Cortez works for American Airlines in Dallas, Texas. She makes blueprints for the design layouts of new planes. On a blueprint of a Boeing 747, the scale is $\frac{1}{8}$ in. = 1 ft, or $\frac{1}{8}$ in.:1 ft. The drawing of the plane has a length of $18\frac{1}{8}$ in., a wingspan of 15 in., and a tail height of 2 in. What is the actual length of the plane?

Use the scale and the length on the blueprint to form a proportion. Then solve the proportion.

$$\frac{\frac{1}{8}\text{ in.}}{1\text{ ft}} = \frac{18\frac{1}{8}\text{ in.}}{x\text{ ft}}$$

$\frac{1}{8}x = 18\frac{1}{8}$ *Find the cross products.*

$8\left(\frac{x}{8}\right) = 8\left(18\frac{1}{8}\right)$ *Multiply each side by 8.*

$x = 145$ The Boeing 747 is 145 feet long.

Mini-Lab

Work in groups of eight.

Materials: a map of your state, scissors, large pieces of drawing paper, tape

- Cut the map into 8 congruent rectangles. Each member of the group receives a rectangle of the map.
- Using a scale of 1:3, draw an enlargement of your piece of the map. Include large cities and important geographical features.
- Have your group tape together their enlargements to form the new map. Then, tape together the original map.

Talk About It

a. Compare the dimensions of the new map to the original one. What do you find?

b. Measure the distance between the same two cities on each map. Compare these measurements to the scale.

Checking for Understanding

Communicating Mathematics

Read and study the lesson to answer each question.

1. **Tell** what is meant by a scale drawing.
2. **Tell** how scale drawings relate to proportions.
3. **Write** two situations in real life where scale drawings are used.

Guided Practice

4. Refer to the Example on page 359. Find the actual length of the wingspan and the tail.

The distance on a map is given. Find the actual distance, if the scale on the map is 1 in.:60 mi.

5. 3 in.
6. $2\frac{1}{4}$ in.
7. $4\frac{1}{8}$ in.
8. $\frac{5}{8}$ in.

Exercises

Independent Practice

The figure at the right is a scale drawing of a playhouse. In the drawing, the side of each square represents 2 feet. Find the actual size of each segment.

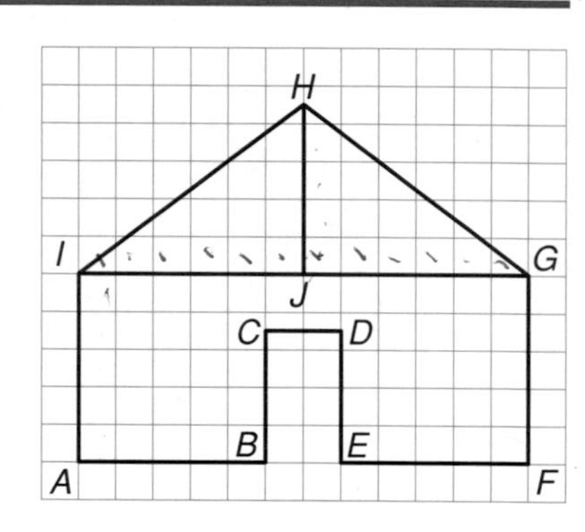

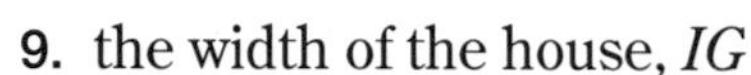

9. the width of the house, *IG*
10. the width of the door, *CD*
11. the height of the wall, *AI*
12. the height of the roof, *JH*

Mixed Review

13. **Statistics** Find the interquartile range for 79, 84, 81, 84, 73, 75, 80, and 78. *(Lesson 4-6)*
14. In the afternoon, the Empire State Building in New York City casts a shadow 156.25 feet long. At the same time, a nearby four-story office building, 84 feet high, casts a shadow 10.5 feet long. How tall is the Empire State Building? *(Lesson 9-6)*

Problem Solving and Applications

15. **Movies** The model of the bee used in the movie *Honey, I Shrunk the Kids* was built from a scale drawing. If the length of the drawing of the bee is 32 cm, what is the length of the model? *scale: 1 cm = 12 cm*

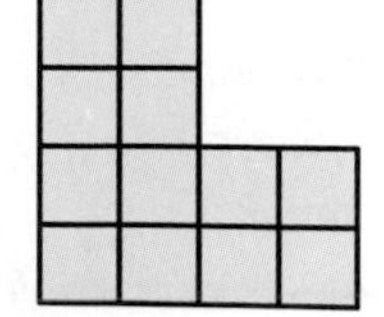

16. **Critical Thinking** In the figure at the right, the scale is 1 unit = 4 feet. Find the area of the scale drawing. Then find the area of the real figure. Compare the two numbers. Make a conjecture about your findings.

17. **Journal Entry** Make a scale drawing of your bedroom. Be sure to include the scale.

Geometry Connection

9-8 Dilations

Objective

Graph dilations on a coordinate plane.

Words to Learn

dilation
scale factor

Have you ever wondered how movies appear on the screen in a movie theater? Light from a movie projector passes through each 35-mm frame of the film and projects a larger image onto the screen, which might be more than 100 feet wide. The process of enlarging or reducing an image in mathematics is called a **dilation.**

Because a dilated image is the same shape as the original, the two images are similar. The ratio of the new image to the original is called the **scale factor.**

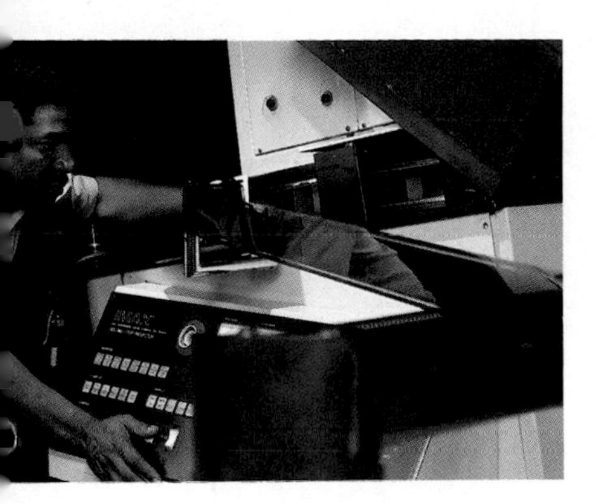

A dilated image is usually named using the same letters as the original figure, but with primes, as in $\triangle ABC \sim \triangle A'B'C'$.

Example

Graph $\triangle ABC$ with vertices $A(4, 6)$, $B(10, 2)$, and $C(14, 10)$. Graph its dilation with a scale factor of 1.5.

Graph the three vertices and connect them to form the triangle. Label the vertices.

To find the vertices of the dilation, multiply each coordinate in the ordered pairs by 1.5.

$A(4, 6) \rightarrow (4 \cdot 1.5, 6 \cdot 1.5) \rightarrow A'(6, 9)$
$B(10, 2) \rightarrow (10 \cdot 1.5, 2 \cdot 1.5) \rightarrow B'(15, 3)$
$C(14, 10) \rightarrow (14 \cdot 1.5, 10 \cdot 1.5) \rightarrow C'(21, 15)$

Now graph A', B', and C'. Connect them to form $\triangle A'B'C'$. $\triangle A'B'C'$ is the dilation of $\triangle ABC$ with a scale factor of 1.5.

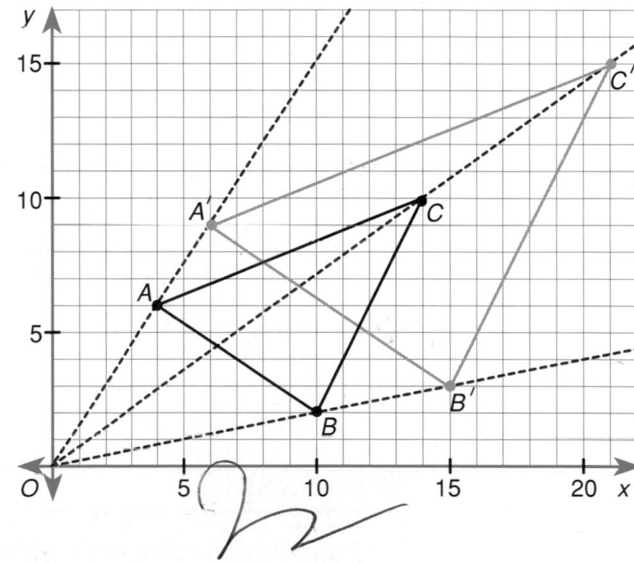

To check your graph, draw lines through the origin and each of the vertices of the original figure. If the vertices of the dilated figure don't lie on those same lines, you've made a mistake.

DID YOU KNOW

The film in movie theaters is wound on large reels that will hold 2,000 feet of film. The average movie uses 5 or 6 reels.

What happens if the scale factor of the dilation is less than 1, or perhaps equal to 1? Try this Mini-Lab to find the answer.

Mini-Lab

Work with a partner.
Materials: graph paper, ruler

- Graph $\triangle ABC$ from the Example on page 361 on a coordinate grid.
- Find the coordinates for a dilation with a scale factor of $\frac{1}{2}$.
- Graph $\triangle A'B'C'$.

Talk About It

a. How does the dilated image relate in size to the original?

b. Graph a dilation of $\triangle ABC$ with a scale factor of 1. How do the two triangles compare?

Checking for Understanding

Communicating Mathematics

Read and study the lesson to answer each question.

1. **Tell** what a dilation is.
2. **Write** a general rule for finding the new coordinates of any ordered pair (x, y) for a dilation with a scale factor of k. $(x, y) \rightarrow (\underline{\ ?\ }, \underline{\ ?\ })$
3. **Draw** a figure on a coordinate plane. Then draw the dilations of the figure if the scale factor is greater than 1, less than 1, and equal to 1.

Guided Practice

In the figure, $\triangle A'OB'$ is a dilation of $\triangle AOB$.

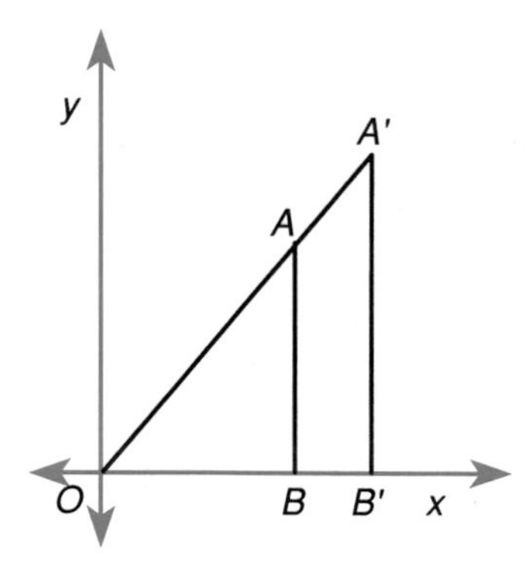

4. Find the image of $A(3, 4)$ for a dilation with a scale factor of 2.
5. Find the image of $A(3, 4)$ for a dilation with a scale factor of 5.
6. Find the scale factor if $OA = 16$, $OB = 20$ and $OA' = 48$.
7. Find the length of $\overline{A'B'}$ if $AB = 20$ and the scale factor is $\frac{5}{4}$.

Exercises

Independent Practice

8. Graph the segment GH with $G(-2, -2)$ and $H(3, -3)$. Then graph its image for a dilation with a scale factor of 6.
9. Graph segment QR with $Q(-6, -6)$ and $R(12, -12)$. Then graph its image for a dilation with a scale factor of $\frac{3}{4}$.

Triangle *PQR* has vertices *P*(-4, 12), *R*(-2, -4), and Q(8, 6). Find the coordinates of its image for a dilation with each given scale factor. Graph △*PQR* and its dilation.

10. 2 **11.** $\frac{1}{4}$ **12.** 1

Mixed Review

13. Solve $1.29b = 5.16$ *(Lesson 2-4)*

14. Use divisibility rules to determine whether 284 is divisible by 6. Explain why or why not. *(Lesson 6-1)*

15. The distance between two cities on a map is $2\frac{3}{8}$ inches. Find the actual distance between the cities if the scale on the map is 1 inch:300 miles. *(Lesson 9-7)*

Problem Solving and Applications

16. Critical Thinking Graph △*PQR* with vertices *P*(1, 1), *Q*(1, 4), and *R*(5, 1). Then graph its image for a dilation with a scale factor of -1. Make a conjecture about dilations with negative scale factors.

17. Geometry Graph polygon *ABCD* with vertices *A*(3, 3), *B*(15, 3), *C*(15, 12), and *D*(3, 12).

a. What type of polygon is *ABCD?*

b. Find its perimeter and area.

c. Graph the image of polygon *ABCD* for a dilation with a scale factor of 3.

d. Find the perimeter and area of polygon *A'B'C'D'*.

e. Write a ratio in simplest form comparing the perimeter of *A'B'C'D'* to the perimeter of *ABCD*.

f. Write a ratio in simplest form comparing the area of *A'B'C'D'* to the area of *ABCD*.

g. Make a conjecture about your findings.

CULTURAL KALEIDOSCOPE

Faith Ringgold

As a child, Faith Ringgold was a chronic asthmatic who spent much of her childhood at home learning from her parents. She learned an appreciation for art and creating designs from cloth.

In 1948, she tried to enroll in college as a liberal arts student, but was rejected because she was a woman. Instead, she enrolled in the school of education and taught for the next twenty years in New York City Public Schools. However, she continued to study art. Through her students, she learned new ways to see and appreciate fabric and patterns and to express her African-American heritage in her work.

Ms. Ringgold is known for the quilting in her artwork. She uses bright colors and bold patterns to tell stories. Each piece fits into the story and expresses ideas and feelings through color, shape, and form.

Cooperative Learning

9-9A Right Triangles

A Preview of Lesson 9-9 and Lesson 9-10

Objective

Discover ratios among the sides of a right triangle.

Materials

protractor
metric ruler
calculator

Any triangle that contains a right angle is called a right triangle. Mathematicians have been studying the relationships among the sides of right triangles since before 2000 B.C.

Try this!

Work in groups of three.

- Each person should draw a right triangle ABC in which $m\angle A = 30°$, $m\angle B = 60°$, and $m\angle C = 90°$.
- Use your ruler to measure the side opposite the 30° angle. Record the measurement to the nearest millimeter.
- The leg adjacent to an angle is the side of the angle that is not the hypotenuse. Measure the leg adjacent to the 30° angle and record the measurement.
- Measure the hypotenuse and record the measurement.
- Use your measurements and your calculator to find each ratio for the 30° angle.

$$\text{ratio 1: } \frac{\text{opposite leg}}{\text{adjacent leg}}$$

$$\text{ratio 2: } \frac{\text{opposite leg}}{\text{hypotenuse}}$$

$$\text{ratio 3: } \frac{\text{adjacent leg}}{\text{hypotenuse}}$$

- Repeat the above procedure for the 60° angle.

What do you think?

1. Compare your ratios with the others in your group. How do they compare?
2. Make a conjecture about the ratio of the sides of any 30°-60° right triangle.
3. Repeat the activity for 45°–45° right triangles. What do you find?

Extension

4. For each triangle in your group find the value of $(\text{ratio } 2)^2 + (\text{ratio } 3)^2$. What do you discover?

9-9 The Tangent Ratio

Objectives

Find the tangent of an angle and find the measure of an angle using the tangent.

Words to Learn

tangent

The industrial technology class plans to add a wheelchair ramp to the emergency exit of the auditorium as a class project. They know that the landing is 3 feet high and that the angle the ramp makes with the ground cannot be greater than 6°. What is the minimum distance from the landing that the ramp should start? *This problem will be solved in Example 1.*

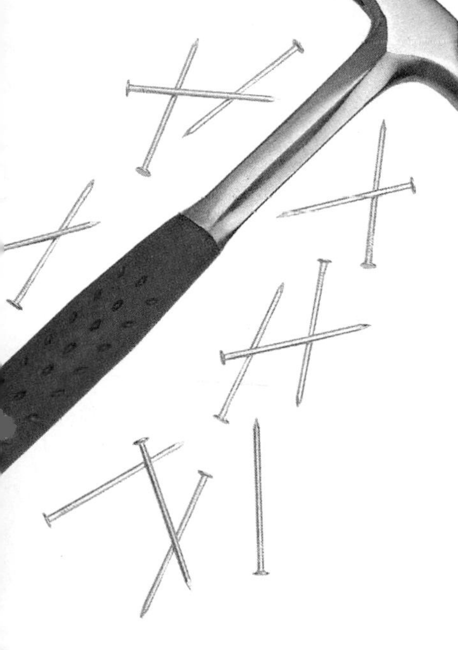

Problems like the one above involve a right triangle. Right triangles and the relationships among their sides have been studied for thousands of years. One ratio, called the **tangent,** compares the measure of leg opposite an angle with the measure of the leg adjacent to that angle. The symbol for the tangent of angle A is tan A.

Tangent	If A is an acute angle of a right triangle, $\tan A = \dfrac{\text{measure of the leg opposite } \angle A}{\text{measure of the leg adjacent to } \angle A}$.

You can also use the symbol for tangent to write the tangent of an angle measure. The tangent of a 60° angle is written as tan 60°. If you know the measures of one leg and an acute angle of a right triangle, you can use the tangent to solve for the measure of the other leg.

Example 1 ***Problem Solving***

Safety Solve the problem about the wheelchair ramp.

First draw a diagram.

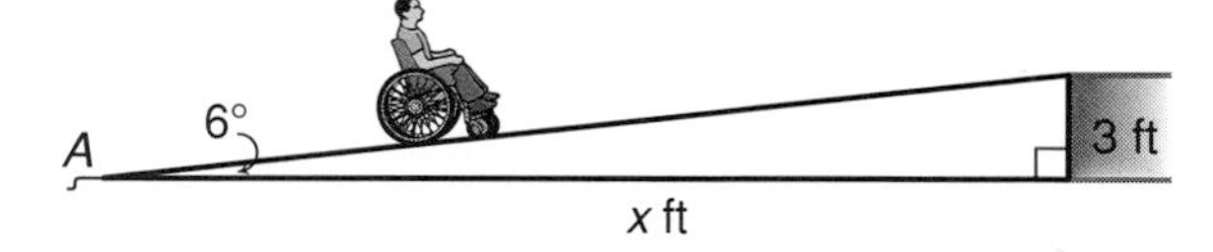

$m\angle A = 6°$
leg adjacent to $\angle A = x$ feet
leg opposite $\angle A = 3$ feet

Now substitute these values into the definition of tangent.

$\tan 6° = \dfrac{3}{x}$ ← *opposite leg* / ← *adjacent leg*

$(\tan 6°)(x) = 3$ *Multiply each side by x.*

$x = \dfrac{3}{\tan 6°}$ *Divide each side by tan 6°.*

If your calculator does not have a TAN key, you can use the table on page 583.

You can now use your calculator to find the value of x.

3 [÷] 6 [TAN] [=] 28.543093

The ramp must begin about 28.5 feet from the landing.

You can also use your calculator to find the measure of an acute angle of a right triangle when you know the measures of the two legs.

Calculator Hint

Some calculators do not have an INV key. Instead they may have TAN^{-1} written above the TAN key. This symbol means the inverse of the tangent. You have to press the 2nd key and then TAN to use TAN^{-1}.

Example 2

Find the measure of ∠A.

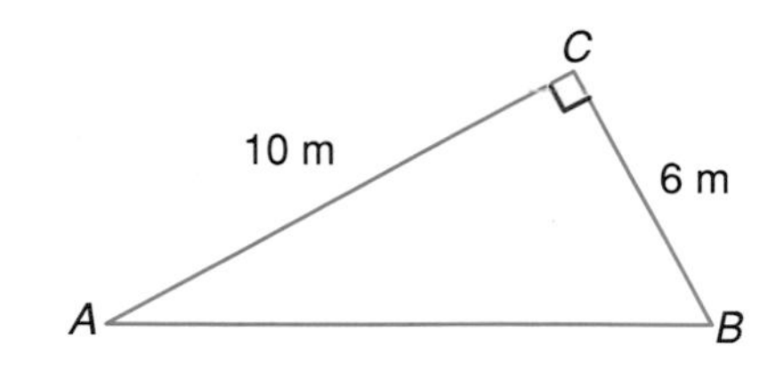

From the figure, you know the values of the two legs. Use the definition of tangent.

$$\tan A = \frac{\text{opposite leg}}{\text{adjacent leg}}$$

$$\tan A = \frac{6}{10}$$

Now use your calculator.

6 [÷] 10 [=] [INV] [TAN] 30.963757

The measure of ∠A is about 31°.

Checking for Understanding

Communicating Mathematics

Read and study the lesson to answer each question.

1. **Write** a definition of the tangent ratio.
2. **Tell** how to use the tangent ratio to find the measure of a leg of a right triangle.
3. **Tell** how to find the measure of an angle in a right triangle when you know the measures of the two legs.

Guided Practice

Use the figures at the right for Exercises 4–11. Write the ratios in simplest form. Find angle measures to the nearest degree.

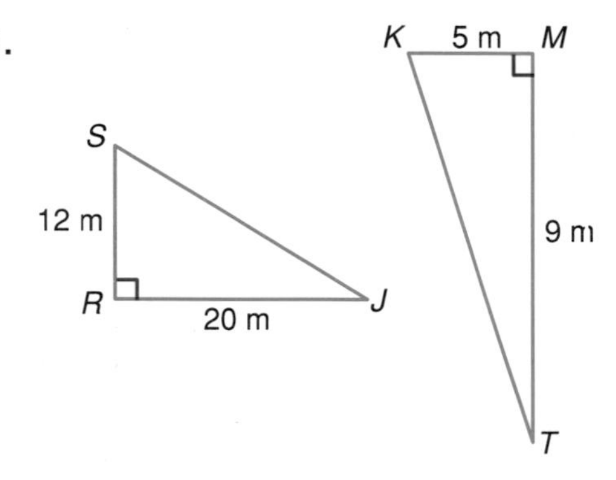

4. Find tan J.
5. Find tan S.
6. Find tan K.
7. Find tan T.
8. Find $m\angle J$.
9. Find $m\angle S$.
10. Find $m\angle K$.
11. Find $m\angle T$.
12. If the leg opposite the 53° angle in a right triangle is 4 inches long, how long is the other leg to the nearest tenth of an inch?
13. If the leg adjacent to a 29° angle in a right triangle is 9 feet long, what is the measure of the other leg to the nearest foot?

Exercises

Independent Practice

Complete each exercise using the information in the figure. Find angle measures to the nearest degree.

14.

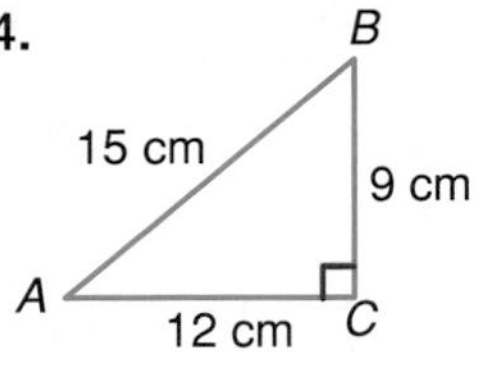

$\tan A = \underline{?}$
$\tan B = \underline{?}$
$m\angle A = \underline{?}$
$m\angle B = \underline{?}$

15.

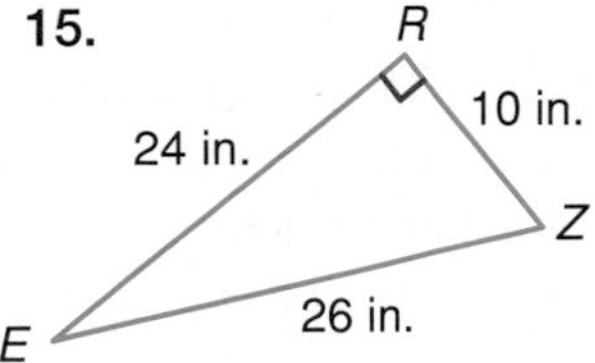

$\tan Z = \underline{?}$
$\tan E = \underline{?}$
$m\angle Z = \underline{?}$
$m\angle E = \underline{?}$

16.

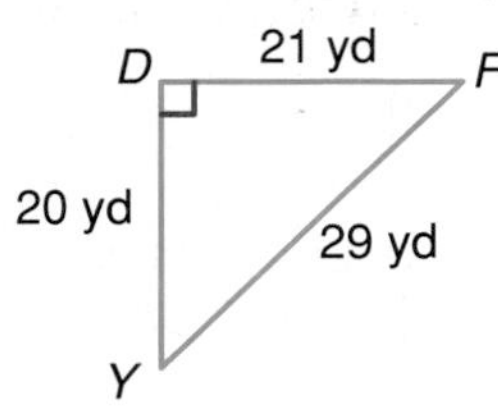

$\tan F = \underline{?}$
$\tan Y = \underline{?}$
$m\angle F = \underline{?}$
$m\angle Y = \underline{?}$

Find the value of x to the nearest tenth or nearest degree.

17.

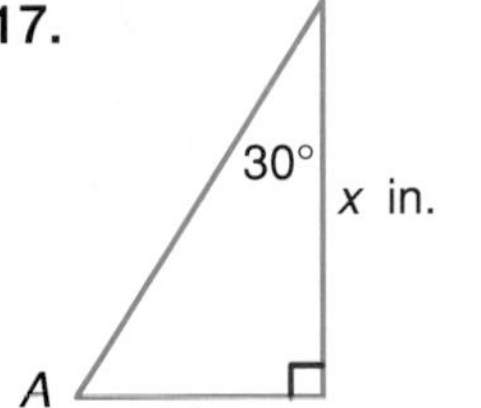

18.

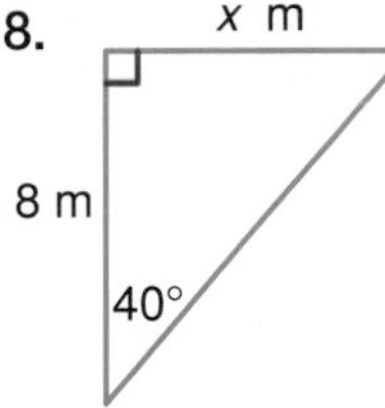

19.

20.

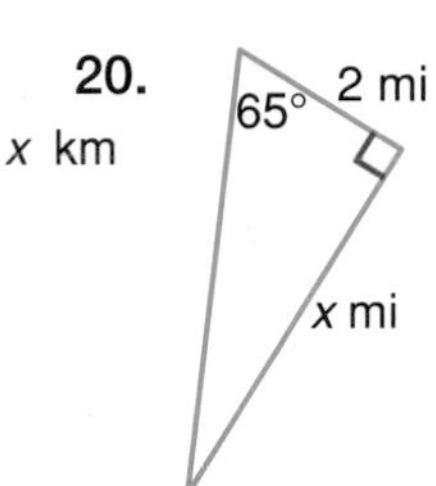

Mixed Review

21. **Geometry** Draw an isosceles, right triangle. *(Lesson 5-3)*

22. Solve $7\frac{3}{8} = 2\frac{9}{16} + j$. *(Lesson 7-2)*

23. Graph line segment EF with endpoints $E(2, 6)$ and $F(4, -4)$. Then graph its image for a dilation with a scale factor of $\frac{1}{2}$. *(Lesson 9-8)*

Problem Solving and Applications

24. **Critical Thinking** In this lesson, you have learned about the tangent of acute angles. Use your calculator to investigate the tangent of the following angle measures. What do you discover?

90°	120°	150°	180°	210°
240°	270°	300°	330°	360°

25. **Surveying** A surveyor is measuring the width of a river for a proposed bridge. A theodolite is used by the surveyor to measure angles. The distance from where the surveyor is standing to the proposed bridge site is 40 feet. The angle from where the surveyor is to the bridge site across the river is 50°. Find the length of the bridge.

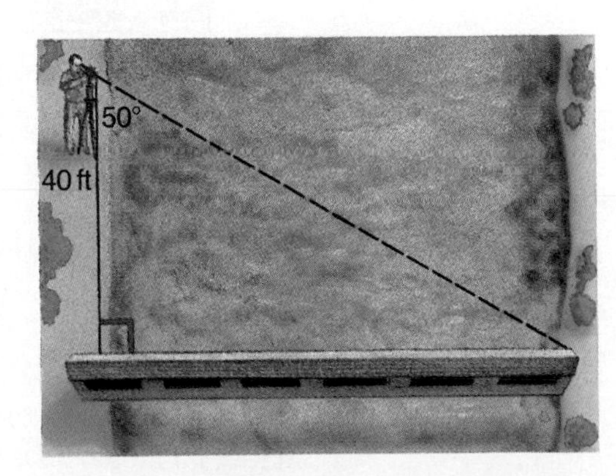

Measurement Connection

9-10 The Sine and Cosine Ratios

Objective

Find the sine and cosine of an angle and find the measure of an angle using sine or cosine.

Words to Learn

trigonometry
sine
cosine

Toni decided to make a scale drawing of the Leaning Tower of Pisa for her project in art class. She knows the tower is 177 feet tall and tilts $16\frac{1}{2}$ feet off the perpendicular. First, she wants to draw the angle representing the tilt of the tower. What should be the measure of the angle? *This problem will be solved in Example 3.*

When you look at the diagram, you see that you know the measures of one leg and the hypotenuse. These are not the measures you need to use the tangent ratio you learned about in Lesson 9-9. The tangent ratio is only one of several ratios used in the study of **trigonometry.**

Two other ratios in trigonometry are the **sine** ratio and the **cosine** ratio. They can be written as sin A and cos A. They are defined below.

	If A is an acute angle of a right triangle,
Sine	$\sin A = \frac{\text{measure of the leg opposite } \angle A}{\text{measure of the hypotenuse}}$
Cosine	$\cos A = \frac{\text{measure of the leg adjacent to } \angle A}{\text{measure of the hypotenuse}}$

Example 1

Use $\triangle ABC$ to find sin A, cos A, sin B, and cos B.

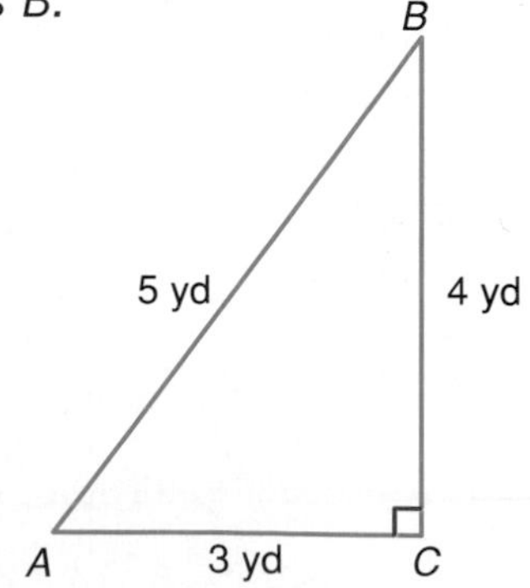

$\sin A = \frac{BC}{AB}$

$= \frac{4}{5}$ or 0.8

$\sin B = \frac{AC}{AB}$

$= \frac{3}{5}$ or 0.6

$\cos A = \frac{AC}{AB}$

$= \frac{3}{5}$ or 0.6

$\cos B = \frac{BC}{AB}$

$= \frac{4}{5}$ or 0.8

You can also find the sine and cosine of an angle if you know its measure by using your calculator.

sin 63° → 63 [SIN] 0.8910065 cos 63° → 63 [COS] 0.4539905

$\sin 63° \approx 0.891$ $\cos 63° \approx 0.454$

In Lesson 9-9, you learned you could use the tangent ratio to find missing lengths of sides or angle measures in a right triangle. The same is true of the sine and cosine ratios.

Example 2

Find the length of $\overline{XY}$ in $\triangle XYZ$.

You know the measure of $\angle X$ and the length of the hypotenuse. You can use the cosine ratio.

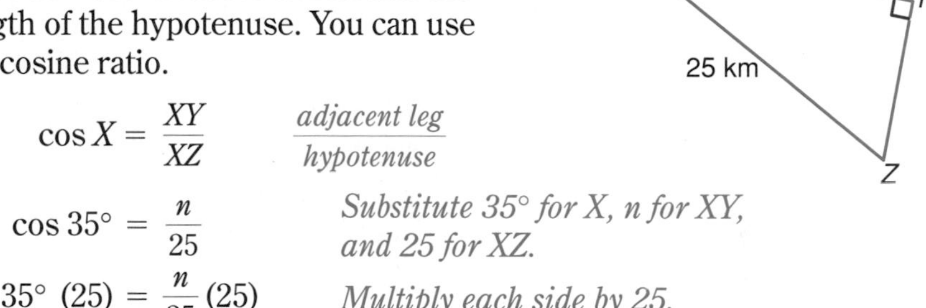

$$\cos X = \frac{XY}{XZ} \quad \textit{adjacent leg / hypotenuse}$$

$$\cos 35° = \frac{n}{25} \quad \textit{Substitute 35° for X, n for XY, and 25 for XZ.}$$

$$\cos 35°\ (25) = \frac{n}{25}(25) \quad \textit{Multiply each side by 25.}$$

$$25 \cdot \cos 35° = n$$

Now use your calculator.

25 [×] 35 [COS] [=] 20.478801

The length of $\overline{XY}$ is about 20.48 kilometers.

Example 3 *Problem Solving*

DID YOU KNOW

Did You Know?

The Leaning Tower of Pisa took 199 years to build. It was completed in 1372. The ground beneath the tower started to sink after the first three stories were built. Since 1911, the tower has increased its lean by an average of 0.05 inch per year.

Find the angle that Toni needs to draw for her scale drawing of the Leaning Tower of Pisa.

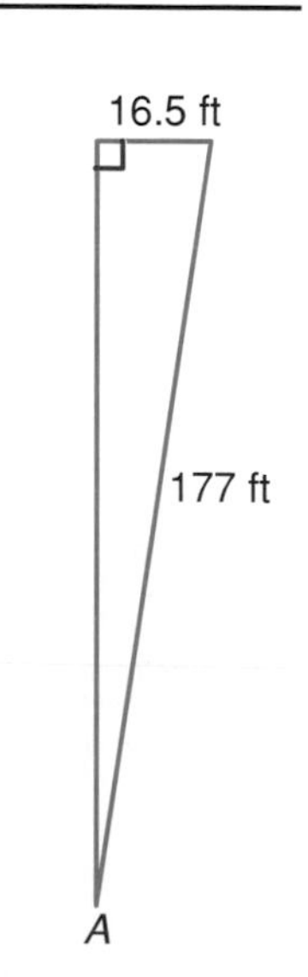

Explore We know the leg opposite the angle and the hypotenuse. We can use the sin A.

Plan Substitute the known values in the definition.

$$\sin A = \frac{16.5}{177}$$

Solve

$$\sin A = \frac{16.5}{177}$$

Use your calculator.

16.5 [÷] 177 = [INV] [SIN] 5.3488982

Toni must draw an angle of about 5°.

Examine Toni knows the angle in her drawing will be very narrow. 5° is a very small angle, so it is a reasonable answer.

Checking for Understanding

Communicating Mathematics **Read and study the lesson to answer each question.**

1. **Write** what the sine and cosine ratios are.
2. **Tell** which branch of mathematics uses the sine, cosine, and tangent ratios.
3. **Show** how you could use the sine or cosine to find the missing measure of one of the legs if you know the hypotenuse and an acute angle.

Guided Practice **Use the figures at the right for Exercises 4–13. Write the ratios in simplest form. Find angle measures to the nearest degree.**

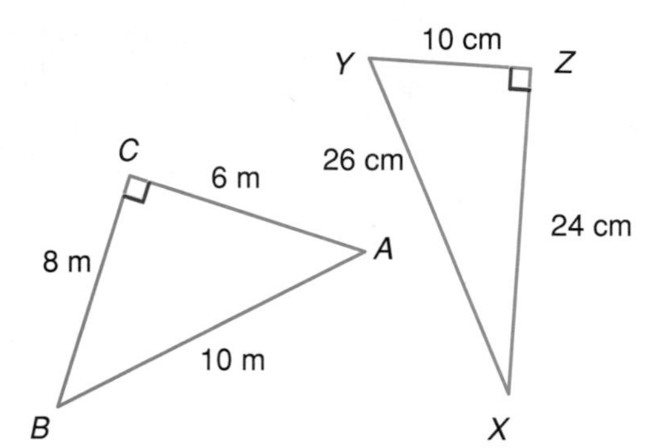

4. Find $\cos A$.
5. Find $\sin A$.
6. Find $m\angle A$.
7. Find $\sin B$.
8. Find $\cos B$.
9. Find $m\angle B$.
10. Find $\cos Y$.
11. Find $\sin X$.
12. Find $m\angle X$.
13. Find $m\angle Y$.

Exercises

Independent Practice **Complete each exercise using the information in the figure. Round angles to the nearest degree.**

14.

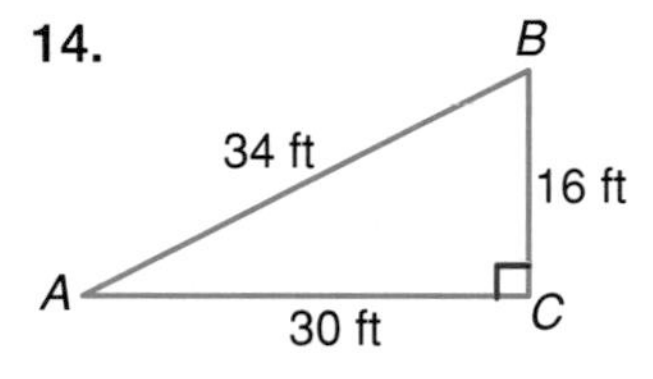

$\sin A =$?
$\sin B =$?
$\cos A =$?
$\cos B =$?
$m\angle A =$?
$m\angle B =$?

15.

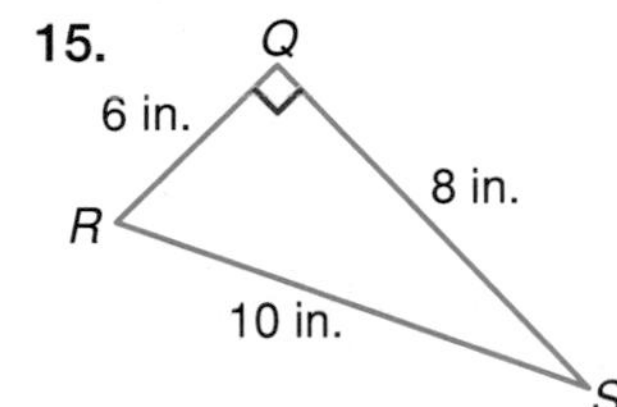

$\sin R =$?
$\cos R =$?
$\sin S =$?
$\cos S =$?
$m\angle R =$?
$m\angle S =$?

16.

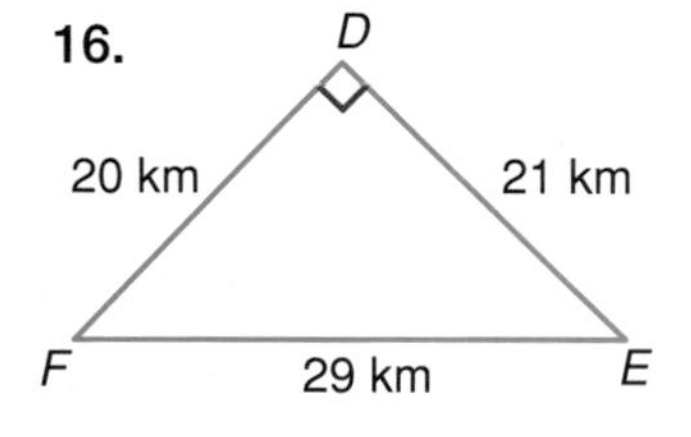

$\sin E =$?
$\cos F =$?
$m\angle E =$?
$\cos E =$?
$\tan F =$?
$m\angle F =$?

Find the value of x to the nearest tenth or to the nearest degree.

17.

x yd
13 yd
67°

18.

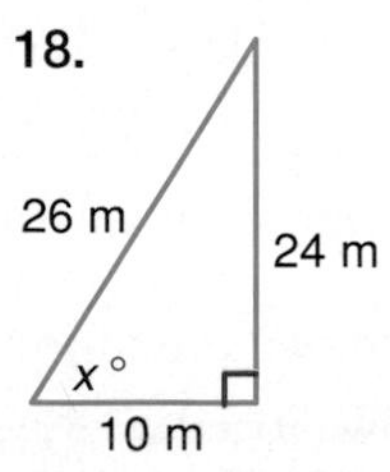

19.

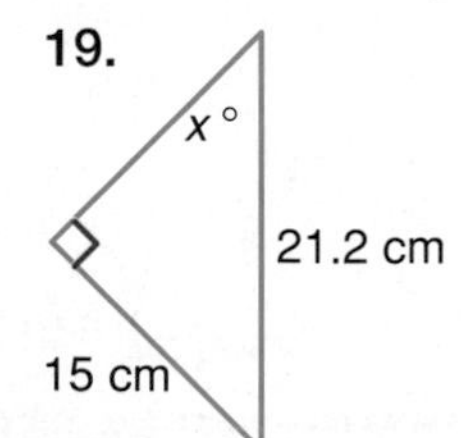

20.

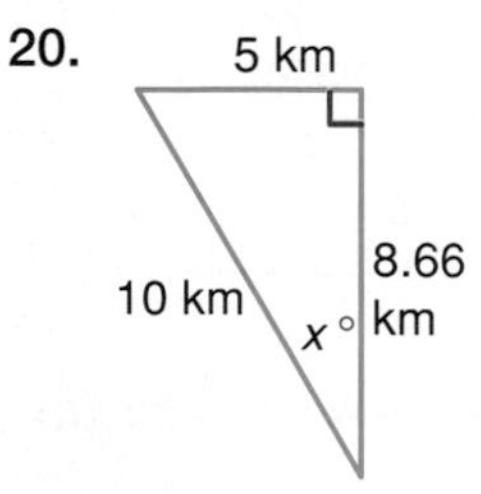

Mixed Review

21. Solve $g = -5(12)(3)$. *(Lesson 3-6)*
22. Express 9.2×10^{-4} in standard form. *(Lesson 6-11)*
23. Use the triangle at the right to find tan A. *(Lesson 9-9)*

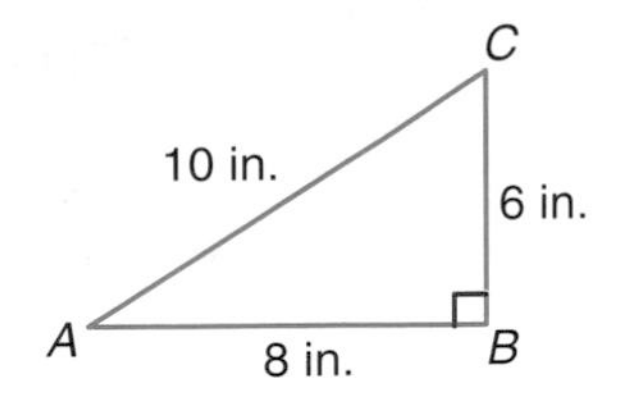

Problem Solving and Applications

24. **Critical Thinking** Look at your results in Exercises 14–16. Make a conjecture about the sine and cosine of complementary angles.

25. **School Clubs** The Student Government wants to hang banners from all of the school clubs on a wire for the Club Appreciation Assembly. They propose stringing a wire from the ceiling at the front of the auditorium to a spot on the floor so that it makes a 50° angle with the floor. If the ceiling is 30 feet high, at least how long must the wire be?

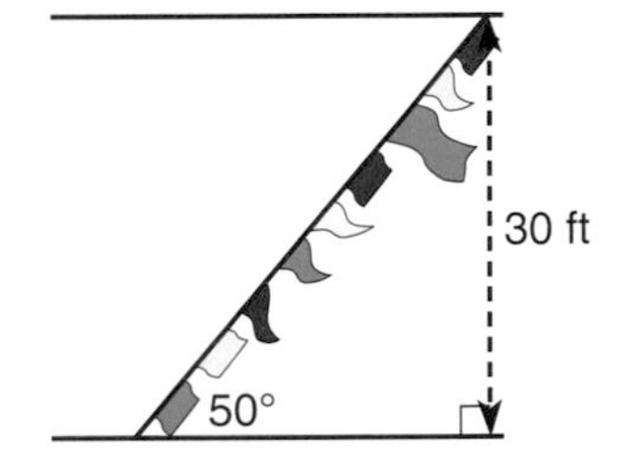

26. **Guess and Check** Suppose A represents the tangent ratio, B represents the cosine ratio, and C represents the sine ratio. Which of the following proportions is true? Be prepared to defend your answers.

a. $\frac{A}{B} = C$ b. $\frac{A}{C} = B$ c. $\frac{B}{C} = A$

d. $\frac{B}{A} = C$ e. $\frac{C}{A} = B$ f. $\frac{C}{B} = A$

27. **Travel** Many tourists travel to New York City during the holidays to see the decorated tree placed at Rockefeller Center near the skating rink. One year the tree was 63.2 feet tall. A guy wire was attached to the top of the tree and made a 72° angle with the ground. To the nearest foot, how long was that wire?

28. **Journal Entry** Write and solve a problem in which a trigonometric ratio can be used. Include a drawing as part of the solution.

29. **Portfolio Suggestion** Select one of the graphs from this chapter that you found especially challenging. Place it in your portfolio.

Chapter

9 Study Guide and Review

Communicating Mathematics

Choose the letter that best matches each phrase.

1. an example of a ratio
2. an example of a proportion
3. an example of a rate
4. figures that have the same shape but may differ in size
5. finding a measurement using proportions
6. the process of enlarging or reducing an image in mathematics
7. a branch of mathematics that uses the sine, cosine, and tangent ratios

a. direct measurement
b. indirect measurement
c. similar figures
d. congruent figures
e. dilation
f. 3 miles in 7 hours
g. trigonometry
h. $\frac{3}{7}$
i. $\frac{3}{7} = \frac{9}{21}$

8. In your own words, explain how the scale of a scale drawing is determined.

Self Assessment

Objectives and Examples

Upon completing this chapter, you should be able to:

- express ratios as fractions in simplest form *(Lesson 9-1)*

Express the ratio 8 out of 12 in simplest form.

$\frac{8}{12} = \frac{2}{3}$

Review Exercises

Use these exercises to review and prepare for the chapter test.

Express each ratio or rate as a fraction in simplest form.

9. 10 green apples:20 red apples
10. 20 centimeters per meter
11. 36 to 5
12. 4 men:20 people

- solve proportions *(Lesson 9-2)*

Solve $\frac{x}{3} = \frac{10}{15}$.

$15x = 3(10)$ *Cross products.*
$15x = 30$
$x = 2$

Solve each proportion.

13. $\frac{3}{8} = \frac{n}{40}$
14. $\frac{6}{r} = \frac{15}{20}$
15. $\frac{d}{18} = \frac{9}{4}$
16. $\frac{0.5}{30} = \frac{0.25}{b}$

Objectives and Examples	Review Exercises

Objectives and Examples

- solve problems by using proportions *(Lesson 9-3)*

If 3 cassettes cost $26.97, how much will 1 cassette cost?

$$\frac{3}{26.97} = \frac{1}{x}$$
$$3x = 26.97$$
$$x = 8.99$$

One cassette costs $8.99.

Review Exercises

Use a proportion to solve each problem.

17. Vicki earns $70 in 8 hours. How much will she earn in 20 hours?
18. A snail moves 6 inches in 3 minutes. How far does the snail move in 5 minutes?
19. It costs $1.20 to copy 10 pages. How many pages can be copied for $3.00?

- identify corresponding parts of similar polygons *(Lesson 9-5)*

Two polygons are similar if their corresponding angles are congruent and their corresponding sides are in proportion.

Tell whether each pair of polygons are similar.

20.

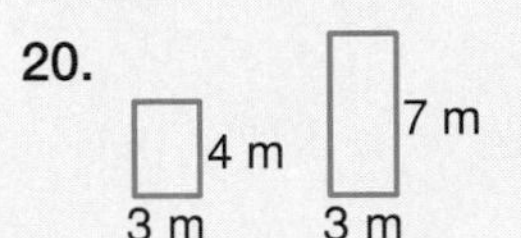

21.

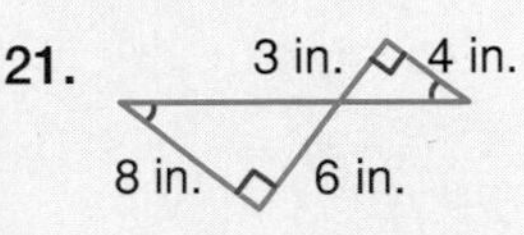

- solve problems involving similar triangles *(Lesson 9-6)*

How far is Malden from Bristow?

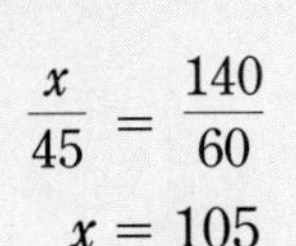

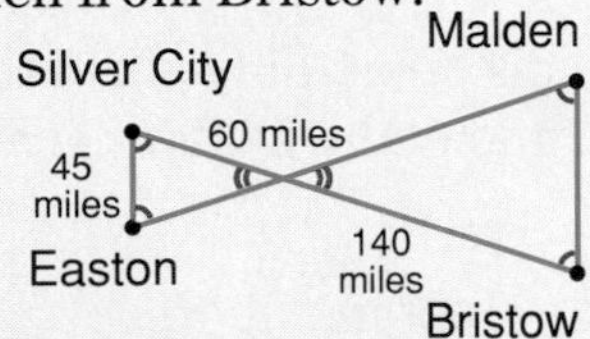

$$\frac{x}{45} = \frac{140}{60}$$
$$x = 105$$

Malden is 105 miles from Bristow.

Write a proportion to solve the following problem.

22. Sonia is standing next to a flagpole. She casts a shadow 6 feet 8 inches long, while the flagpole casts a shadow 26 feet 8 inches long. If Sonia is 5 feet tall, how tall is the flagpole?

- solve problems involving scale drawings *(Lesson 9-7)*

The scale of a model car is $\frac{1}{2}$ in.:1 ft. If the model is 7 inches long, how long is the actual car?

$$\frac{\frac{1}{2}\text{ in.}}{7\text{ in.}} = \frac{1\text{ ft}}{x\text{ ft}}$$
$$\frac{1}{2}x = 7$$
$$x = 14$$

The car is 14 feet long.

The distance on a map is given. Find the actual distance, if the scale on the map is 2 cm:35 km.

23. 6 cm
24. 9 cm
25. 3.2 cm
26. 4.6 cm
27. 8.4 cm

- graph dilations on a coordinate plane *(Lesson 9-8)*

In a dilation with a scale factor of k, the ordered pair (x, y) becomes the ordered pair (kx, ky).

Graph $\triangle HIJ$ with vertices $H(2, 4)$, $I(-6, -4)$, and $J(6, -8)$. Find the coordinates for a dilation with each scale factor. Graph the dilation.

28. $\frac{1}{2}$
29. 2
30. $\frac{1}{4}$

Objectives and Examples	Review Exercises

- find the tangent of an angle *(Lesson 9-9)*

Find the tangent of $\angle Y$.

$$\tan Y = \frac{3}{4} = 0.75$$

Find each ratio using the information in $\triangle ABC$.

31. $\tan A$

32. $\tan B$

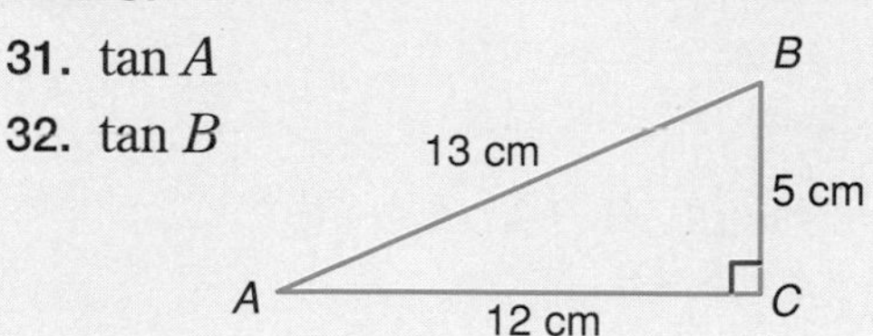

- find the sine and cosine of an angle *(Lesson 9-10)*

Find the sine and cosine of $\angle D$.

$$\sin D = \frac{6}{10} = 0.6$$

$$\cos D = \frac{8}{10} = 0.8$$

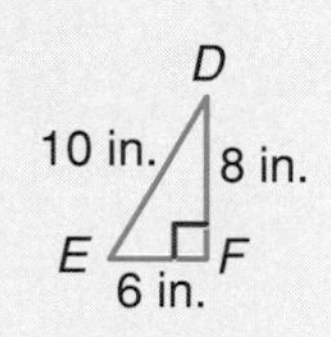

Find each ratio using the information in $\triangle RQS$.

33. $\sin Q$

34. $\sin R$

35. $\cos Q$

36. $\cos R$

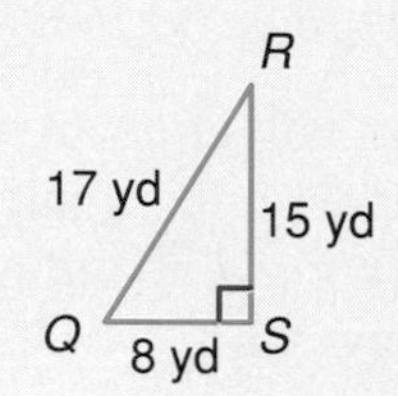

Applications and Problem Solving

37. Hobbies Eddie has 128 baseball cards in his collection. Fifty-six of the cards are of Los Angeles Dodgers players. Find the ratio of the Dodgers cards to all of the cards. *(Lesson 9-1)*

38. The Fort Couch Middle School Chess Club has eight members. Recently, each member agreed to play every other member exactly once. How many games will be played in all? *(Lesson 9-4)*

Curriculum Connection Projects

- **Astronomy** Find the ratio of Earth's distance from the sun to each of the other eight planets' distances from the sun.
- **History** Find the ratio of the length to the height of the Greek Parthenon. Compare this ratio to the golden ratio.
- **Music** Find the ratio of the frequency of middle C on the musical scale to the frequency of the other notes of the scale.

Read More About It

Crichton, Michael. *Jurassic Park.*
McCauley, David. *Castle.*
Wood, Paul W. *Starting with Stained Glass.*

Chapter

9 Test

Express each ratio or rate as a fraction in simplest form.

1. 8 misses in 20 tries
2. 9 out of 14
3. 12 wins:4 losses
4. 38 to 24

5. Express *290 miles in 5 hours* as a unit rate.

Solve each proportion.

6. $\frac{5}{6} = \frac{11}{x}$
7. $\frac{n}{30} = \frac{5}{10}$
8. $\frac{9}{33} = \frac{b}{11}$
9. $\frac{2}{y} = \frac{8}{5}$

10. Do $\frac{32}{88}$ and $\frac{12}{36}$ form a proportion?
11. A car uses 4 gallons to travel 96 miles. How many gallons are used to travel 60 miles?

Determine if each pair of polygons below are similar.

12.

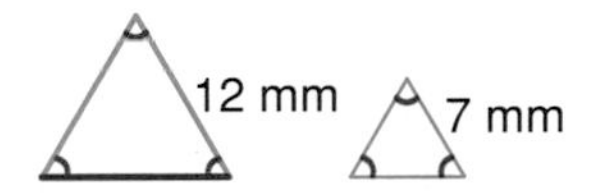

13.

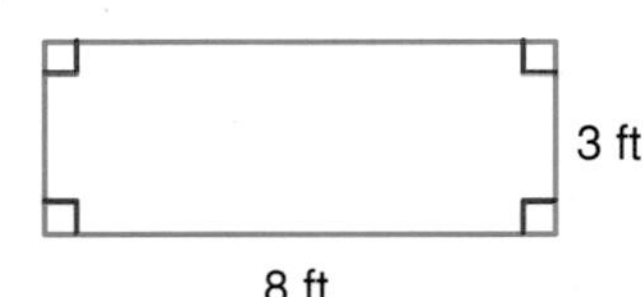

The distance on a map is given. Find the actual distance, if the scale on the map is 1 in.:100 mi.

14. 4 in.
15. $\frac{1}{4}$ in.
16. $3\frac{1}{8}$ in.

17. Polygon $ABCD$ has vertices $A(4, 3)$, $B(-3, 2)$, $C(-1, -4)$, and $D(5, -5)$. Find the vertices for a dilation with a scale factor of 3.

Complete each exercise using the information in $\triangle ABC$. Round angle measures to the nearest degree.

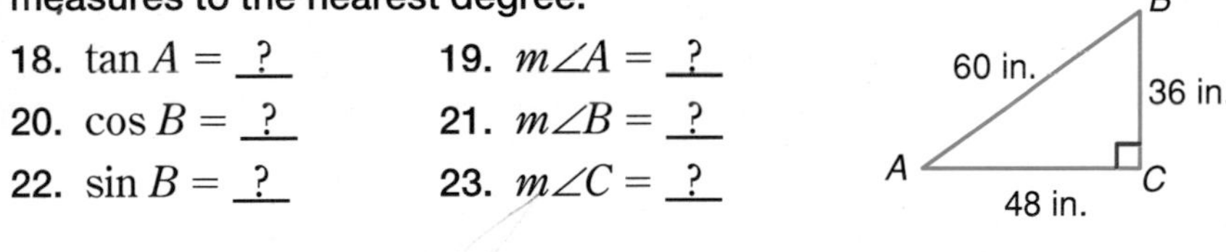

18. $\tan A = \underline{?}$
19. $m\angle A = \underline{?}$
20. $\cos B = \underline{?}$
21. $m\angle B = \underline{?}$
22. $\sin B = \underline{?}$
23. $m\angle C = \underline{?}$

24. At 2:00 P.M., the shadow of Mr. Harris' house is 20 feet long. At the same time, the shadow of a tree in his backyard is 30 feet long. If Mr. Harris' house is 24 feet tall, how tall is the tree?
25. A photographer hangs her prints to dry on a cord using clothespins. If she has 10 clothespins, what is the greatest number of photos she can hang? Assume that each photo needs two clothespins.

Bonus If A is an acute angle of a right triangle, explain why $\tan A = \frac{\sin A}{\cos A}$.

Chapter 9

Standardized Format, Chapters 1–9

Academic Skills Test

Directions: Choose the best answer. Write A, B, C, or D.

1. Sue bought grocery items for the following prices: \$1.39, \$2.89, 58¢, and \$1.19. The best estimate of the total cost is

   A \$4 B \$5
   C \$6 D \$9

2. Which equation is equivalent to $\frac{c}{5} = 2.5$?

   A $\frac{c}{5} = \frac{2.5}{5}$

   B $\frac{c}{5} = 2.5 \cdot 5$

   C $\frac{c}{5} \cdot 5 = 2.5 \cdot 5$

   D $\frac{c}{5} = 2.5 \div 5$

3. If $\frac{x}{5} + 4 = 32$, what is the value of x?

   A 5.6 B 7.2
   C 28 D 140

4. $(-8)^3 =$

   A -512 B -24
   C 24 D 512

5.

Quiz Score	Number of Students
9-11	3
12-14	4
15-17	9
18-20	6

   How many students had scores less than 15?

   A 4 B 7
   C 16 D none of these

6. 8th Graders Monthly Allowances (\$)

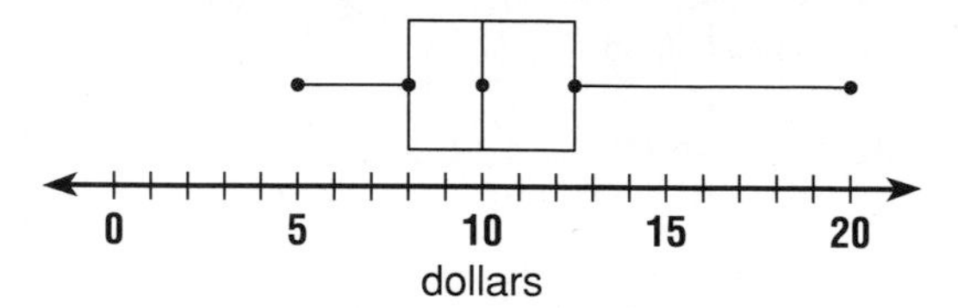

   If the box-and-whisker plot represents 30 students, about how many students receive \$10 or less?

   A 15 B 10
   C 5 D 3

7. Use the box-and-whisker plot in problem 6. Which average could be misleading?

   A mean B median
   C mode D none of these

8. In the figure below, $x \parallel y$. If $d = 72°$, what is the value of e?

   A 180
   B 108
   C 90
   D 72

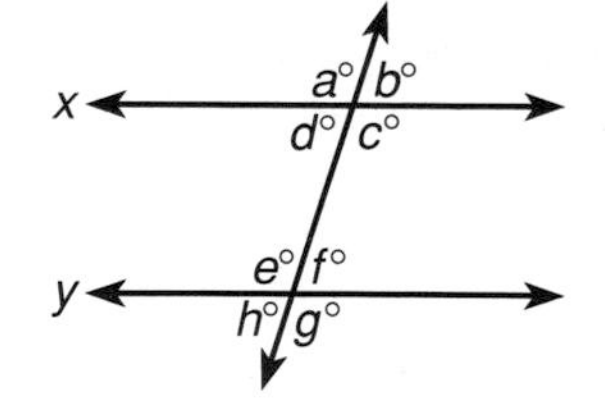

9. A four-sided figure with only one pair of parallel sides is a

   A rectangle B rhombus
   C parallelogram D trapezoid

10. What is the greatest common factor of 14 and 50?

   A 1 B 2
   C 50 D 700

11. To use a calculator to divide 6 by $2\frac{1}{8}$, what decimal number should you enter for $2\frac{1}{8}$?

A 12.75 B 2.18
C 2.125 D 0.125

12. If $x + 2\frac{2}{3} = -4\frac{1}{3}$, what is the value of x?

A $1\frac{2}{3}$ B $-1\frac{2}{3}$
C -6 D -7

13. What are the next two terms in the sequence 2, 5, 10, 17, . . . ?

A 22, 27 B 24, 31
C 26, 35 D 26, 37

14. What is the area of this figure?

A 40 sq ft
B 64 sq ft
C 88 sq ft
D 110 sq ft

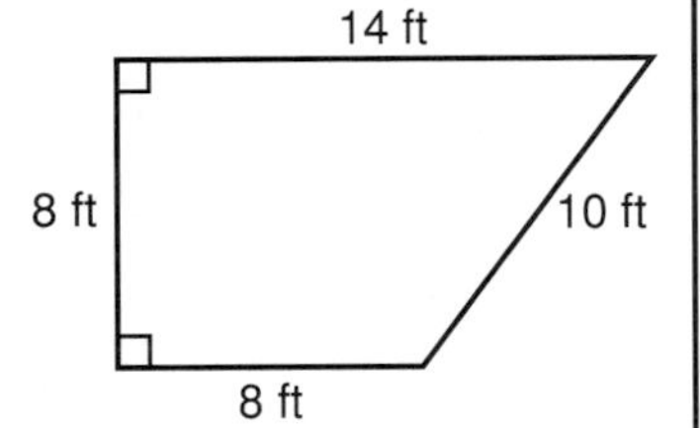

15. $-\sqrt{0.81} =$

A -0.09 B -0.9
C 0.09 D 0.9

16. To the nearest inch, how long is the diagonal of a 20 in. by 50 in. window?

A 100 in. B 70 in.
C 54 in. D 29 in.

17. To the nearest foot, what is the value of c?

A 25
B 50
C 100
D 150

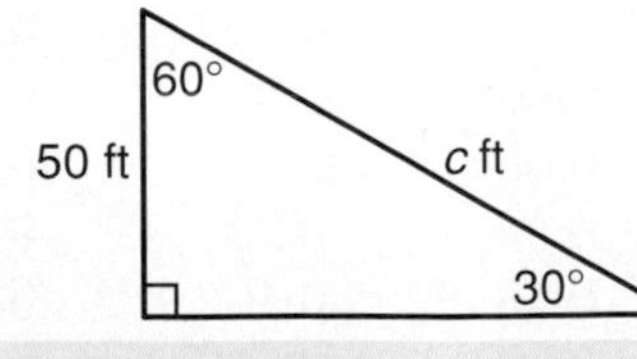

Test-Taking Tip

Most of the basic formulas that you need to answer the questions on a standardized test are usually given to you in the test booklet. However, it is a good idea to review the formulas before the test.

The area of a circle with radius r is πr^2.

The area of a triangle is $\frac{1}{2}$ the product of the base and the height.

The area of a trapezoid is $\frac{1}{2}$ the product of the altitude and the sum of the bases.

The perimeter of a rectangle is twice the sum of its length and width.

The circumference of a circle with radius r is $2\pi r$.

18. On a trip, the Browns drive 102 miles in 2 hours. If they continue at the same rate, which proportion will give h, the total number of hours, for a 450-mile trip?

A $\frac{102}{2} = \frac{h}{450}$ B $\frac{102}{2} = \frac{450}{h}$
C $\frac{2}{h} = \frac{450}{102}$ D $\frac{h}{2} = \frac{102}{450}$

19. $\triangle MNO$ is similar to $\triangle PQR$. What side corresponds to $\overline{NO}$?

A $\overline{NM}$
B $\overline{PQ}$
C $\overline{QR}$
D $\overline{RP}$

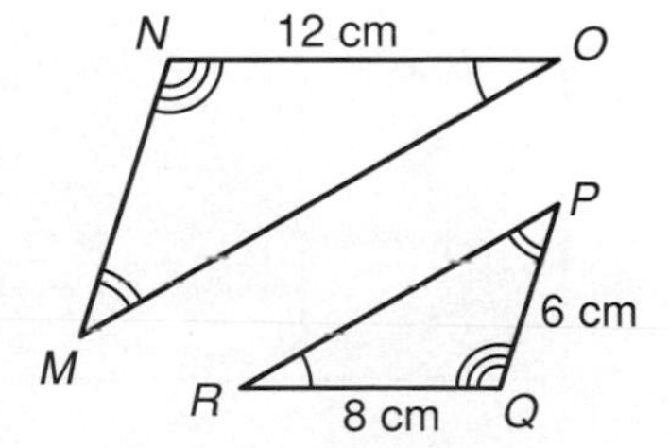

20. Use the similar triangles shown in problem 19. What is the length of $\overline{MN}$?

A 8 cm B 9 cm
C 10 cm D 11 cm

Chapter 10

Applications with Percent

Spotlight on Recycling

Have You Ever Wondered...

- If more containers are being recycled each year?
- If the amount of waste disposed of has increased or decreased over the past 10 years?

WHERE DOES YOUR WASTE GO?
(in pounds per person per day)

Year	Waste Generated	Waste Recycled	Combustion for Energy	Combustion (no Energy)	Landfill Deposits
1960	2.67	0.18	—	0.82	1.67
1965	3.00	0.19	0.01	0.75	2.05
1970	3.27	0.23	0.02	0.66	2.37
1975	3.26	0.25	0.02	0.45	2.54
1980	3.61	0.35	0.06	0.27	2.93
1985	3.71	0.38	0.17	0.10	3.06
1986	3.80	0.42	0.22	0.07	3.09
1987	3.92	0.45	0.36	0.05	3.06
1988	4.00	0.52	0.59	0.02	2.87

Chapter Project

Recycling

Working a group.

1. For two weeks, keep track of what you throw away each day.
2. Make a chart showing the type and amount of trash you throw away each day.
3. Indicate what can be recycled or reused. Include information about what kinds of recycling programs are available in your area.
4. Suggest ways that you can reduce the amount of waste you contribute to landfills

Looking Ahead

In this chapter, you will see how mathematics can be used to answer questions about recycling. The major objectives of the chapter are to:

- express fractions and decimals as percents and vice versa
- solve problems using percents
- construct circle graphs
- find the percent of increase or decrease
- solve problems involving discounts or simple interest

10-1 The Percent Proportion

Objectives

Express fractions as percents. Solve problems using the percent proportion.

Words to Learn

percent
percentage
base
rate
percent proportion

The word *percent* and the symbol % are so common in magazines and newspapers, radio and TV, that you may take percents for granted. From an 87% on your health test and shooting 74% from the foul line, to buying in-line skates at 15% off and paying 6% sales tax, you are constantly bombarded with percents.

A **percent** is a ratio that compares a number to 100. Percent also means *hundredths,* or *per hundred.* The models below represent 50%, 25%, and 10%.

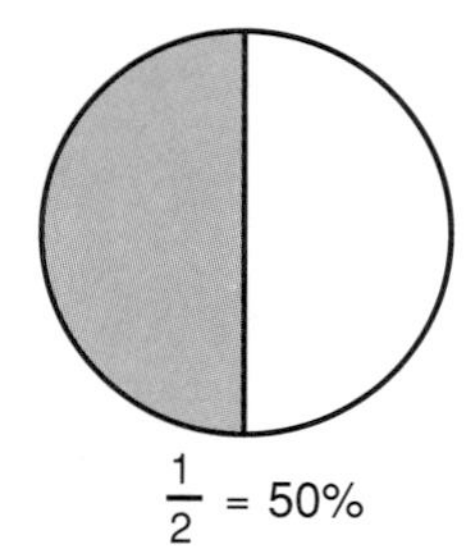

$\frac{1}{2}$ = 50%

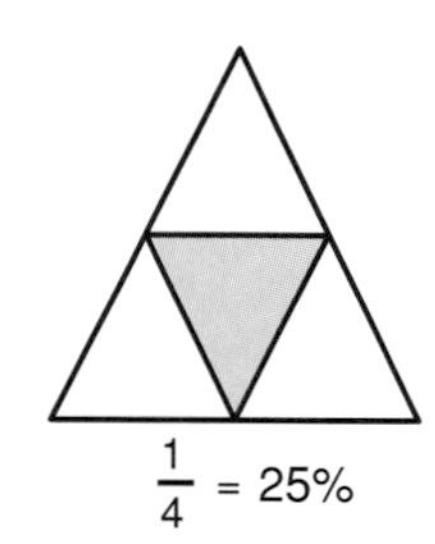

$\frac{1}{4}$ = 25%

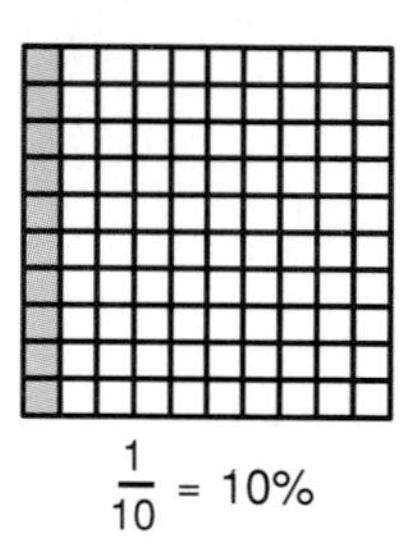

$\frac{1}{10}$ = 10%

Whenever you split the cost of a CD with your sister, or share a pizza, you usually work with fractions. But how often do you cut a pizza into a hundred parts? How can you express eating three pieces of pizza out of a total of eight pieces as a percent?

By looking at the model at the right or the number line below, you can see that $\frac{3}{8}$ is less than $\frac{1}{2}$ and greater than $\frac{1}{4}$. You can estimate that $\frac{3}{8}$ is between 50% and 25%.

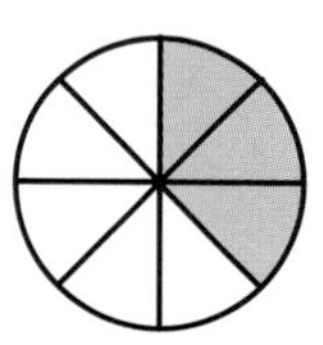

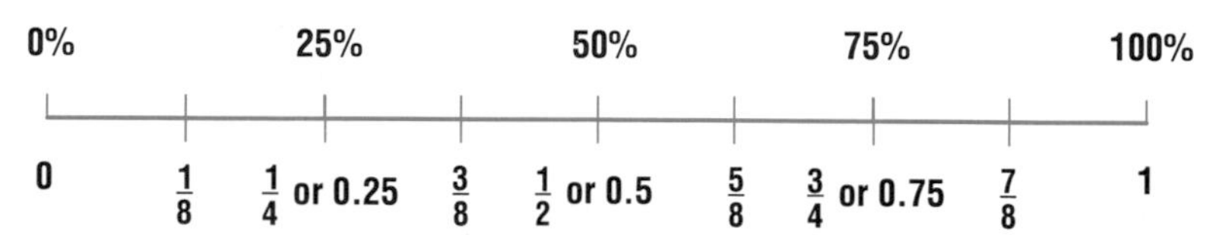

LOOKBACK

You can review ratios on page 338 and proportions on page 344.

To find the percent (x), you can solve this proportion.

$$\frac{3}{8} = \frac{x}{100}$$ *3 pieces out of 8 eaten*

$$3 \cdot 100 = 8x$$ *Write the cross products.*

$$300 = 8x$$ *Divide each side by 8.*

300 [÷] 8 [=] 37.5

$$x = 37.5$$

The solution is 37.5. So, $\frac{3}{8}$ or 37.5% of the pizza was eaten.

In the pizza proportion, 3 is called the **percentage** (P). The number 8 is called the **base** (B). The ratio $\frac{37.5}{100}$ is called the **rate.**

$$\frac{3}{8} = \frac{37.5}{100} \rightarrow \frac{\text{Percentage}}{\text{Base}} = \text{Rate}$$

If r represents the percent, the proportion can be written as $\frac{P}{B} = \frac{r}{100}$. This proportion is called the **percent proportion.** You can think of the percent proportion as a comparison. In the pizza proportion, the part eaten is compared to the whole pizza.

slices of pizza eaten →	$\frac{3}{8} = \frac{37.5}{100}$	← *percent eaten*
slices in whole pizza →		← *percent in whole*

You can use the percent proportion to solve percent problems.

DID YOU KNOW

In 1991, 25% of Americans owned compact disc players.

Example 1 ***Problem Solving***

Consumer Awareness The purchase price of a CD player is \$127. The state tax rate is 6% of the purchase price. Find the total price.

$$\frac{P}{B} = \frac{r}{100}$$

$$\frac{P}{127} = \frac{6}{100}$$ *Replace B with 127 and r with 6.*

$$P \cdot 100 = 127 \cdot 6$$ *Find the cross products.*

127 [×] 6 [÷] 100 [=] 7.62 *Find 127 · 6. Then divide by 100.*

$$P = 7.62$$ *The tax is \$7.62.*

The total cost is \$127 + \$7.62 or \$134.62.

Example 2 ***Problem Solving***

Smart Shopping A bicycle is on sale for \$115. This is 80% of the regular price. What is the regular price?

$$\frac{P}{B} = \frac{r}{100}$$

$$\frac{115}{B} = \frac{80}{100}$$ *Replace P with 115 and r with 80.*

$$115 \cdot 100 = B \cdot 80$$ *Find the cross products.*

$$11{,}500 = 80B$$ *Divide each side by 80.*

11500 [÷] 80 [=] 143.75

$$143.75 = B$$

The regular price is \$143.75.

Example 3

Express $\frac{3}{5}$ as a percent.

$\frac{P}{B} = \frac{r}{100}$

$\frac{3}{5} = \frac{r}{100}$ *Replace P with 3 and B with 5.*

$3 \cdot 100 = 5 \cdot r$ *Find the cross products.*

$300 = 5r$ *Divide each side by 5.*

300 [÷] 5 [=] 60

$60 = r$ $\frac{3}{5}$ is equivalent to 60%.

Checking for Understanding

Communicating Mathematics

Read and study the lesson to answer each question.

1. **Tell** what is meant by percent.
2. **Write** a proportion you would use to find 53% of 215.
3. **Tell** why 37.5% of the figure at the right is shaded.

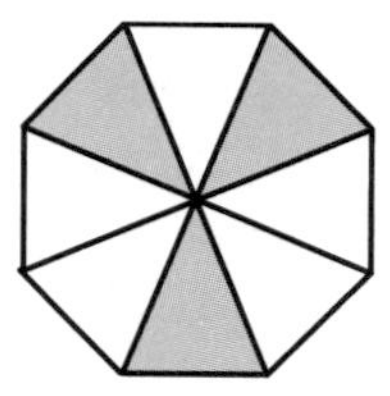

Guided Practice

Express each fraction as a percent.

4. $\frac{17}{25}$
5. $\frac{35}{100}$
6. $\frac{7}{8}$
7. $\frac{9}{10}$
8. $\frac{1}{20}$

Match each question with its corresponding proportion.

9. 24 is 25% of what number? a. $\frac{24}{25} = \frac{r}{100}$
10. 24 is what percent of 25? b. $\frac{24}{B} = \frac{25}{100}$
11. What number is 25% of 24? c. $\frac{P}{24} = \frac{25}{100}$

Write a percent proportion to solve each problem. Round answers to the nearest tenth.

12. Find 19% of 532.
13. Find 12.5% of 88.
14. If a man weighs 185 pounds, his body contains about 111 pounds of water. The water weight is what percent of his body weight?

Exercises

Independent Practice

Express each fraction as a percent.

15. $\frac{9}{100}$
16. $\frac{1}{8}$
17. $\frac{7}{10}$
18. $\frac{3}{4}$
19. $\frac{23}{50}$
20. $\frac{5}{8}$
21. $\frac{2}{3}$
22. $\frac{27}{30}$
23. $\frac{4}{5}$
24. $\frac{15}{40}$

Write a percent proportion to solve each problem. Round answers to the nearest tenth.

25. What is 35% of 230?
26. Find 37.5% of 104.
27. 50 is what percent of 400?
28. 17.8 is what percent of 178?
29. 57 is 30% of what number?
30. 80 is 45% of what number?
31. Twenty-eight is 25% of what number?
32. Fifteen is 40% of what number?
33. Jim has autographs from 16 of the 25 members of the Pittsburgh Pirates. The number of autographs is what percent of the team?

Mixed Review

34. Mr. and Mrs. Dixon are having their living room ceiling painted. The living room is 14 feet wide and 20 feet long. A painter charges 24¢ for each square foot to be painted. How much will it cost the Dixons to have their ceiling painted? *(Lesson 1-1)*
35. **Statistics** Refer to the line plot in the Example on page 137. Find the median and upper and lower quartiles of the data in the line plot. *(Lesson 4-6)*
36. **Geometry** Refer to the figure in Example 2 on page 178. Find $m\angle 8$. *(Lesson 5-1)*
37. Write $\frac{40}{72}$ in simplest form. *(Lesson 6-5)*
38. Find $\sqrt{220}$ to the nearest tenth. *(Lesson 8-3)*
39. In the figure at the right, find the value of x to the nearest degree. *(Lesson 9-10)*

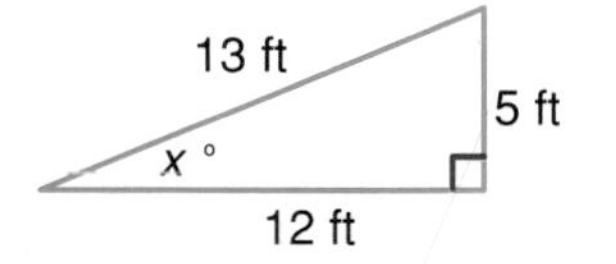

Problem Solving and Applications

40. **Sports** Eleven out of 48 members of the football team are on the field. What percent of the team members are playing?
41. **Critical Thinking** Katrina made 56% of her free throws in the first half of the basketball season. If she makes 7 shots out of the next 13 attempts, will it help or hurt her average? Explain.
42. **Recycling** The graph at the right shows what products in the United States were recycled most often in 1990.
    a. Americans recycled a record 63.6% of the aluminum cans produced. If 55 billion cans were recycled, how many were produced?
    b. Lake Charleston estimates it recycled 2,500 of the 6,250 glass bottles purchased in 1991. How does this compare with the national average in 1990?

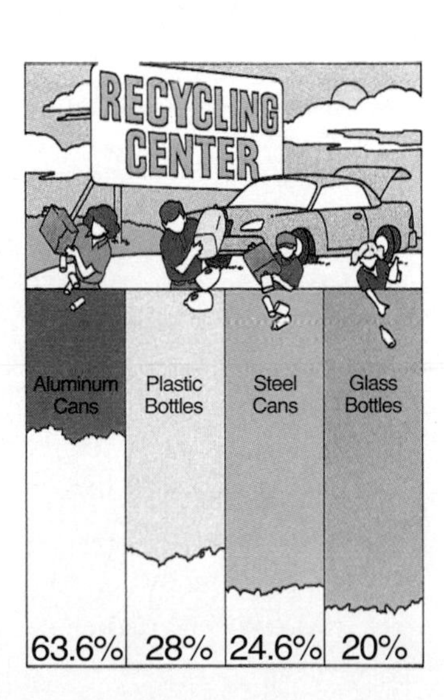

10-2 Fractions, Decimals, and Percents

Objectives

Express decimals and fractions as percents.
Express percents as fractions and decimals.

Jackie wants to buy a new skateboard. She finds two stores that have the same model on sale. If the original price is the same at both stores, which store offers the better sale price?

To find the better price, you need to compare $\frac{2}{3}$ and 70%.

One way to compare them is to express $\frac{2}{3}$ and 70% as decimals and compare the decimals.

$\frac{2}{3} = 2 \div 3$

2 [÷] 3 [=] 0.6666667

$\frac{2}{3} \approx 0.67$

$70\% = \frac{70}{100}$ *Definition of percent*

$= 0.70$ or 0.7

Since 0.70 is greater than 0.67, Henry's offers the better sale price.

LOOKBACK

You can review expressing fractions as decimals on page 227.

You learned how to express fractions as decimals in Chapter 6, and in Lesson 10-1 you learned how to express fractions as percents. Fractions, decimals, and percents are different ways to name the same number.

To express a decimal as a percent, first write the decimal as a fraction with a denominator of 100. Then, express the fraction as a percent.

Mental Math Hint

In Examples 1–3, notice the percent has the same number as the result of multiplying the decimal by 100.

Examples

Express each decimal as a percent.

1 **0.6**

$0.6 = 0.60$

$= \frac{60}{100}$ or 60%

2 **0.78**

$0.78 = \frac{78}{100}$ or 78%

3 **0.354**

$0.354 = \frac{354}{1{,}000}$

$= \frac{35.4}{100}$ or 35.4% *Divide the numerator and the denominator by 10, so that the denominator is 100.*

To express a fraction as a percent, you can solve a percent proportion or you can use a calculator to express the fraction as a decimal. Then express the decimal as a percent like you did in Examples 1–3.

Examples

Express each fraction as a percent.

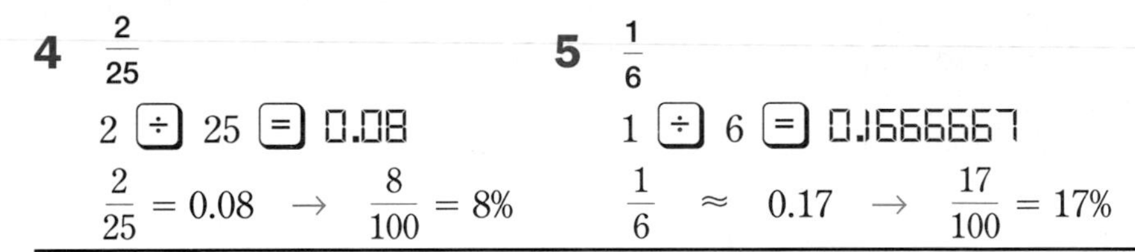

The following chart provides commonly-used fractions and their percent equivalents.

Mental Math Hint

It is helpful to know fraction-percent equivalencies by memory.

Fraction-Percent Equivalencies					
Fraction	Percent	Fraction	Percent	Fraction	Percent
$\frac{1}{2}$	50%	$\frac{1}{5}$	20%	$\frac{5}{6}$	$83\frac{1}{3}\%$
$\frac{1}{3}$	$33\frac{1}{3}\%$	$\frac{2}{5}$	40%	$\frac{1}{8}$	$12\frac{1}{2}\%$
$\frac{2}{3}$	$66\frac{2}{3}\%$	$\frac{3}{5}$	60%	$\frac{3}{8}$	$37\frac{1}{2}\%$
$\frac{1}{4}$	25%	$\frac{4}{5}$	80%	$\frac{5}{8}$	$62\frac{1}{2}\%$
$\frac{3}{4}$	75%	$\frac{1}{6}$	$16\frac{2}{3}\%$	$\frac{7}{8}$	$87\frac{1}{2}\%$

Mini-Lab

Work with a partner.

Materials: 10 × 10 grid, colored pencils

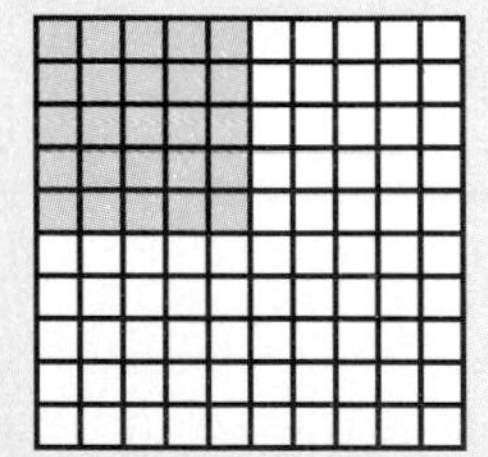

- Shade a 5 × 5 section in one corner of the grid.
- Use a different colored pencil to shade each of the other three 5 × 5 corner sections.

Talk About It

a. How many small squares are in the grid?
b. How many small squares are in each colored section?
c. What percent is represented by each colored section?
d. How many 5 × 5 sections are there?
e. What fraction of the grid is represented by each colored section?

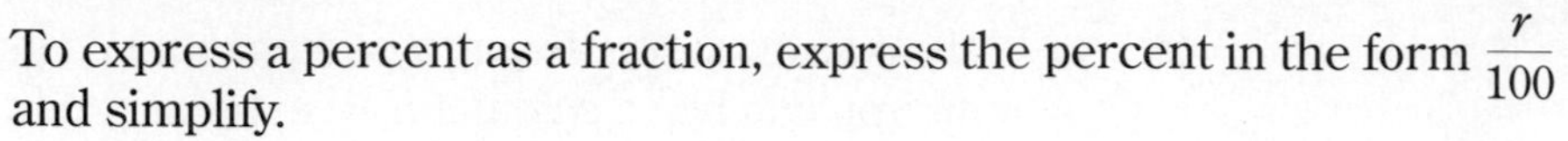

To express a percent as a fraction, express the percent in the form $\frac{r}{100}$ and simplify.

Examples

Express each percent as a fraction in simplest form.

6 6%

$$6\% = \frac{6}{100} = \frac{3}{50}$$

7 45%

$$45\% = \frac{45}{100} = \frac{9}{20}$$

8 $87\frac{1}{2}\%$

$$87\frac{1}{2}\% = \frac{87\frac{1}{2}}{100} = 87\frac{1}{2} \div 100 = \frac{\overset{7}{\cancel{175}}}{2} \times \frac{1}{\underset{4}{\cancel{100}}} = \frac{7}{8}$$

To express a percent as a decimal, express the percent in the form $\frac{r}{100}$ and then express the fraction as a decimal.

Examples

Express each percent as a decimal.

9 75%

$$75\% = \frac{75}{100} \text{ or } 0.75$$

10 20%

$$20\% = \frac{20}{100} \text{ or } 0.20 \text{ or } 0.2$$

11 5.5%

$$5.5\% = \frac{5.5}{100}$$

$$= \frac{55}{1{,}000}$$ *Multiply the numerator and the denominator by 10, so that the numerator is a whole number.*

$$= \frac{55}{1{,}000} \text{ or } 0.055$$

Checking for Understanding

Communicating Mathematics

Read and study the lesson to answer each question.

1. **Tell** how to express a decimal as a percent.
2. **Tell** how you would express a percent as a fraction.
3. **Write** a shortcut that you might use to express a decimal as a percent.

Guided Practice

Express each decimal as a percent.

4. 0.7 5. 0.605 6. 0.26 7. 0.02

Express each percent as a fraction in simplest form.

8. 65% 9. 6.5% 10. 12.5% 11. 96%

Express each percent as a decimal.

12. 78% 13. 9% 14. 12.3% 15. 8.4%

Exercises

Independent Practice

Determine which is greater.

16. 15% or $\frac{1}{8}$ 17. 0.3 or 3.2% 18. $\frac{1}{4}$ or 28%

Express each decimal as a percent.

19. 0.18 20. 0.08 21. 0.704
22. 0.039 23. 0.553 24. 0.6306

Express each percent as a fraction in simplest form.

25. 58% 26. 37% 27. 44.5%
28. $83\frac{1}{3}\%$ 29. 18.25% 30. $12\frac{3}{4}\%$

Express each percent as a decimal.

31. 28% 32. 74% 33. 84.25%
34. 81.5% 35. 38.4% 36. 9.01%

37. Express the ratio 11:55 as a decimal, as a percent, and as a fraction in simplest form.

Mixed Review

38. Solve $r = 9 + (-14) + 5 + (-21)$. Check by solving another way. *(Lesson 3-4)*
39. Factor -162 completely. *(Lesson 6-2)*
40. Solve $\frac{z}{3} = -2.4$. *(Lesson 7-10)*
41. **Geometry** How can you tell if two polygons are similar? *(Lesson 9-5)*
42. Find 28% of 250. *(Lesson 10-1)*

Problem Solving and Applications

43. **Critical Thinking** A pizza with a diameter of 8 inches costs $6. A pizza with a diameter of 16 inches costs $12.
    a. The area of the 8-inch pizza is what percent of the area of the 16-inch pizza?
    b. Which is the better buy? Explain.
44. **Geometry** What percent of the area of the square at the right is shaded? Express the percent as a decimal.

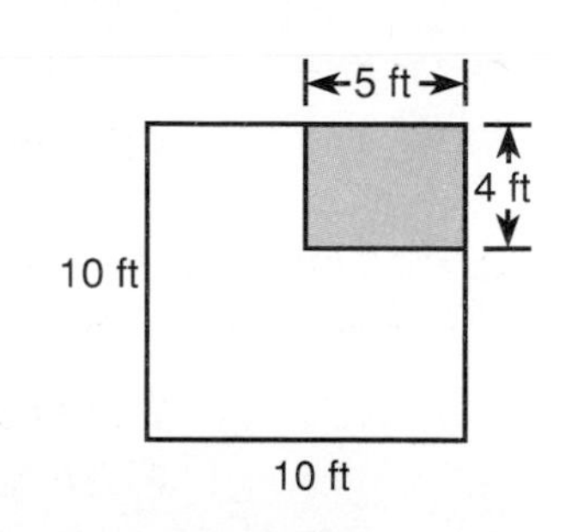

10-3 Large and Small Percents

Objective

Express percents greater than 100 or less than 1 as decimals and fractions.

Jason organized a neighborhood carnival to raise money for the Muscular Dystrophy Association. In order to make his goal, he needed to collect \$30 an hour. The first hour he collected \$35. What percent of his hourly goal did he collect the first hour?

100% of his hourly goal is \$30, so Jason collected more than 100% of his hourly goal.

You can use the percent proportion to find the exact percent.

$$\frac{P}{B} = \frac{r}{100}$$

$$\frac{35}{30} = \frac{r}{100}$$

$$35 \cdot 100 = 30 \cdot r$$

$$3{,}500 = 30r$$

3500 ÷ 30 = 116.66667

$$116.67 \approx r$$

During the first hour, Jason collected about 117% of his hourly goal.

Estimation Hint

THINK:
30 is 100% of 30.
3 is 10% of 30.
6 is 20% of 30.
35 is almost 120% of 30.

Mini-Lab

Work with a partner.

Materials: graph paper, colored pencils

- Draw a 10×10 square on a piece of graph paper. Separate the square into 10 equal parts as shown at the right.
- If the 10×10 square represents 100%, shade a region that represents 50%.
- Shade a region that represents 1%.

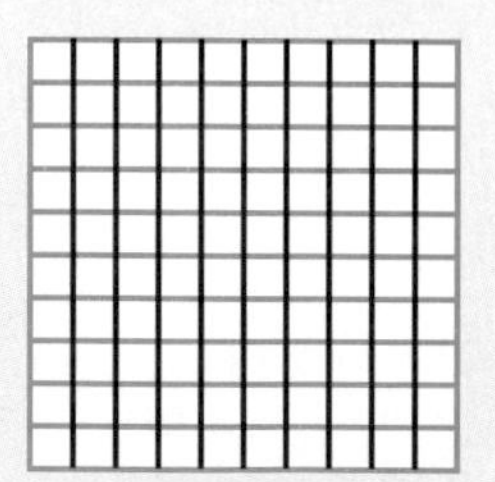

Talk About It

a. How can you shade a region to represent $\frac{1}{2}$%?

b. Would a $\frac{1}{2}$% region be larger or smaller than the 1% region?

c. How would a $\frac{1}{2}$% region compare to the 50% region?

Professional fund raisers usually work at non-profit organizations. Their main task is to plan, organize, and execute events such as telethons or sales to raise money for a specific cause.

Fund raisers must have a college degree and an ability to estimate, budget, and organize information.

For additional information, contact:
National Management Association
2210 Arbor Blvd.
Dayton, OH 45439

To express a percent greater than 100 or less than 1 as a fraction or a decimal, you can use the same procedure you used for percents from 1 to 100.

Examples

Express each percent as a fraction or mixed number in simplest form.

1 **0.2%**

$$0.2\% = \frac{0.2}{100} = \frac{2}{1{,}000} = \frac{1}{500}$$

2 **135%**

$$135\% = \frac{135}{100} = 1\frac{35}{100} = 1\frac{7}{20}$$

3 $\frac{1}{8}$**%**

$$\frac{1}{8}\% = \frac{\frac{1}{8}}{100}$$ *The fraction bar indicates division.*

$$= \frac{1}{8} \div 100$$

$$= \frac{1}{8} \times \frac{1}{100}$$ *To divide by 100, multiply by its multiplicative inverse, $\frac{1}{100}$.*

$$= \frac{1}{800}$$

Express each percent as a decimal.

LOOKBACK You can review division of fractions on page 288.

4 **123%**

$$123\% = \frac{123}{100} \text{ or } 1.23$$

5 **0.9%**

$$0.9\% = \frac{0.9}{100} = \frac{9}{1{,}000} \text{ or } 0.009$$

Checking for Understanding

Communicating Mathematics

Read and study the lesson to answer each question.

1. **Write** a shortcut you might use to express a percent greater than 100 as a decimal.
2. **Draw** a model of 150%.
3. **Tell** how you would draw a model of $\frac{1}{4}$%.
4. **Tell** how you know that 50% is not equal to $\frac{1}{2}$%.

Guided Practice Express each percent as a fraction in simplest form.

5. 0.02% 6. 0.1% 7. 175% 8. $\frac{1}{5}\%$

Express each percent as a decimal.

9. 178% 10. 201.2% 11. 0.6% 12. 0.05%

Replace each ■ with $<$ or $>$ to make a true statement.

13. A percent greater than 100 is a number ■ 1.
14. A percent less than 1 is a number ■ $\frac{1}{100}$.

Exercises

Independent Practice Express each percent as a fraction or mixed number in simplest form.

15. 0.3% 16. 0.04% 17. 760% 18. 12,309.2%
19. 243% 20. $\frac{3}{4}\%$ 21. $\frac{4}{25}\%$ 22. $33\frac{1}{3}\%$

Express each percent as a decimal.

23. 212% 24. 1,819% 25. 0.03% 26. 10.088%
27. $\frac{7}{10}\%$ 28. $\frac{13}{20}\%$ 29. 0.008% 30. $16\frac{2}{5}\%$
31. Order 0.8, 67%, 7 and $\frac{3}{8}\%$ from least to greatest.

Mixed Review

32. Solve $2.4m = 14.4$ *(Lesson 2-4)*
33. Express 0.000084 in scientific notation. *(Lesson 6-11)*
34. **Geometry** Refer to the illustration in Exercise 5 on page 321. How high will the ladder reach if the bottom of the ladder is 4 feet from the wall? *(Lesson 8-6)*
35. Name the eighth term in the sequence 81, 27, 9, ... *(Lesson 7-6)*
36. Express 38.5% as a decimal and as a fraction in simplest form. *(Lesson 10-2)*

Problem Solving and Applications

37. **Economics** Between 1980 and 1989, the number of people in the United States reporting annual incomes of more than $500,000 increased by 9.85%. Express the percent of increase as a decimal.
38. **Critical Thinking** The price of gasoline has doubled in the last decade. Explain whether or not this means that the price has increased by 200%.
39. **Consumer Awareness** The purchase price of a camera is $84. The state tax rate is 5.5% of the purchase price. Explain whether or not this means that the total cost is 105.5% of the purchase price.

40. **Journal Entry** Write a sentence explaining how you can tell when a decimal represents more than 100% or less than 1%.

Problem-Solving Strategy

10-4 Solve a Simpler Problem

Objective

Solve problems by first solving a simpler problem.

An understudy is an actor who learns the part of another actor in case a substitute is needed. There are 32 students trying out for the spring drama production. Twenty-five% of them will serve as understudies. How many students will be understudies?

Explore *What do you know?*
There are 32 students trying out and 25% of them will be understudies.
What are you trying to find out?
How many students will be understudies.

Plan Instead of multiplying 25% and 32, look for a simpler way to solve the problem. Use mental math and the fraction equivalent to 25%.

Solve The product of 25% and 32 is the same as the product of $\frac{1}{4}$ and 32, because $25\% = \frac{1}{4}$.

$32 \times \frac{1}{4} = 32 \div 4$ or 8

Multiplying by $\frac{1}{4}$ is the same as dividing by 4.

Examine Since the number of students trying out (32) is a multiple of 4, it is simpler to find $\frac{1}{4}$ of 32 mentally than to use 25%.

Solving a simpler problem is a strategy that also involves setting aside the original problem and solving one or more simpler, similar problems.

Example

What is the total number of squares of any size in the 6 × 6 square at the right?

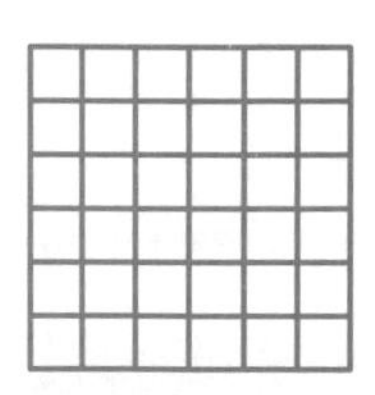

Find the number of squares in a 1 × 1 square, a 2 × 2 square, and a 3 × 3 square. Look for a pattern that you can extend to a 6 × 6 square.

1 × 1 square

1 × 1's: 1 *$1 = 1^2$*

2 × 2 square

1 × 1's: 4 *$4 = 2^2$*
+ 2 × 2's: 1 *$1 = 1^2$*
5

3 × 3 square

1 × 1's: 9 *$9 = 3^2$*
2 × 2's: 4 *$4 = 2^2$*
+ 3 × 3's: 1 *$1 = 1^2$*
14

Notice that the total number of squares is the sum of the numbers of squares up to that point. For a 6 × 6 square, there would be:

$$1^2 + 2^2 + 3^2 + 4^2 + 5^2 + 6^2 = 1 + 4 + 9 + 16 + 25 + 36$$
$$= 91 \text{ squares}$$

Checking for Understanding

Communicating Mathematics

Read and study the lesson to answer each question.

1. **Tell** how you can use fraction-decimal-percent equivalencies to solve a problem.
2. **Tell** how solving a simpler problem can save time.

Guided Practice

Solve by solving a simpler problem.

3. Find 20% of 525.
4. Find the sum of the whole numbers from 1 to 200.
5. There are 210 people visiting the Exhibit Hall. Thirty percent of them are students. How many visitors are students?
6. How many cuts are needed to separate a long board into 19 smaller parts?

Problem Solving

Practice

Solve using any strategy.

Strategies

Look for a pattern.
Solve a simpler problem.
Act it out.
Guess and check.
Draw a diagram.
Make a chart.
Work backward.

7. Tracie bought a leather jacket for $220. She paid $14.30 in sales tax. What was the sales tax rate?
8. A number is tripled and then -16 is added to it. The result is 2. What is the number?
9. A frozen lean dinner entree contains 200 calories. About 25% of its calories are from fat. How many calories are from fat?
10. Benjamin bought some cans of peaches for $1.29 each and some cartons of cream for $2.16 each. He spent a total of $12.51. How many of each item did he buy?
11. Four-fifths of the state tax returns filed were residents of the state for the entire year. If 160,000 returns were filed, how many people were residents for the entire year?
12. Which is the least number of clothespins you need to hang 8 towels on a clothesline to dry? Assume it takes 2 clothespins to hang one towel.

13. **Data Search** Refer to page 666. What fraction of home buyers had incomes of $40,000 or less?

14. Maryann just got a job at the Too Good Yogurt Stand. The manager gave her 3 shirts and 3 pairs of pants to wear as a uniform. She has a red shirt, a gray shirt, and a white shirt and a black pair of pants, a navy blue pair of pants, and a white pair of pants. How many different combinations of uniforms can she choose from?

10-5 Percent and Estimation

Objective

Estimate by using fractions, decimals, and percents interchangeably.

In 1990, *Putt-Putt Golf of America®* surveyed 3,400 customers. One of the items on the survey asked for the players' age. About 11% of those surveyed said they were between the ages of 13 and 15. About how many of those surveyed were in this age group?

You can estimate the number of 13–15 year olds that were surveyed by using the fraction method.

11% is slightly more than 10% or $\frac{1}{10}$.

$\frac{1}{10}$ of 3,400 is 340.

About 340 players surveyed were 13–15 years old.

Examples

1 Estimate 13% of 48.

13% is about 12.5% or $\frac{1}{8}$.

$\frac{1}{8}$ of 48 is 6.

13% of 48 is about 6.

2 Estimate 0.6% of 205.

0.6% is about half of 1%,
205 is slightly more than 200.

1% of 200 is 2 and $\frac{1}{2}$ of 2 is 1.

0.6% of 205 is about 1.

LOOKBACK

You can review compatible numbers on page 12.

3 Estimate 25% of 78.

$25\% = \frac{1}{4}$ and 78 is about 80, which is compatible with $\frac{1}{4}$.

$\frac{1}{4}$ of 80 is 20.

25% of 78 is about 20.

4 Estimate 30% of $118.

30% is about $33\frac{1}{3}\%$ or $\frac{1}{3}$ and $118 is about $120, which is compatible with $\frac{1}{3}$. $\frac{1}{3}$ of $120 is $40.

30% of $118 is about $40.

Problem Solving Hint

Use the *solve a simpler problem* strategy.

Example 5 *Problem Solving*

Consumer Math Rick took his mother out to dinner for her birthday. When the bill came, Rick's mother reminded him that it is customary to tip the server about 15% of the bill. If the bill was for $19.60, what should Rick leave for the tip?

A 10% tip is easy to figure. A 15% tip is equal to a 10% tip plus half of a 10% tip.

Estimation Hint

To find the area of an irregular region, determine how many squares contain any part of the region and the number of full squares within the region. Then find the mean of the two measures.

$\frac{15 + 35}{2} = 25$

$19.60 is almost $20.
10% of 20 is 2.
So, 15% of $20 is $\$2 + \frac{1}{2} \cdot \$2$ or $3.
Rick should leave a $3 tip.

Example 6 *Connection*

Geometry Estimate what percent of the area of the large square is shaded.

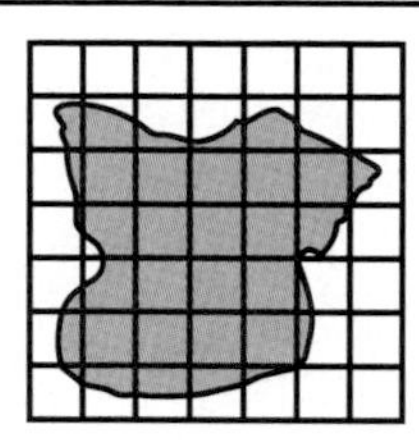

About 25 squares are shaded out of 49.

$\frac{25}{49}$ is about $\frac{25}{50}$ or $\frac{1}{2}$.

$\frac{1}{2} = 50\%$.

About 50% is shaded.

Checking for Understanding

Communicating Mathematics

Read and study the lesson to answer each question.

1. **Tell** how you can estimate 23% of $98.95 using fractions and compatible numbers.
2. **Draw** a triangle on graph paper. Shade about $\frac{1}{3}$ of the area.
3. **Tell** how you can estimate a 15% tip on a $24.90 dinner.

Guided Practice

Determine which is the better estimate.

4. 46% of 80 — **a.** less than 40 — **b.** greater than 40
5. 22% of 300 — **a.** less than 60 — **b.** greater than 60
6. 107% of $42 — **a.** less than $42 — **b.** greater than $42
7. $\frac{1}{4}$% of 4,000 — **a.** less than 40 — **b.** greater than 40

Estimate.

8. 32% of 89
9. 14% of 78
10. 88% of 61

Estimate the percent.

11. 7 out of 16
12. 8 out of 13
13. $\frac{15}{11}$

Estimate the percent of the area shaded.

14. 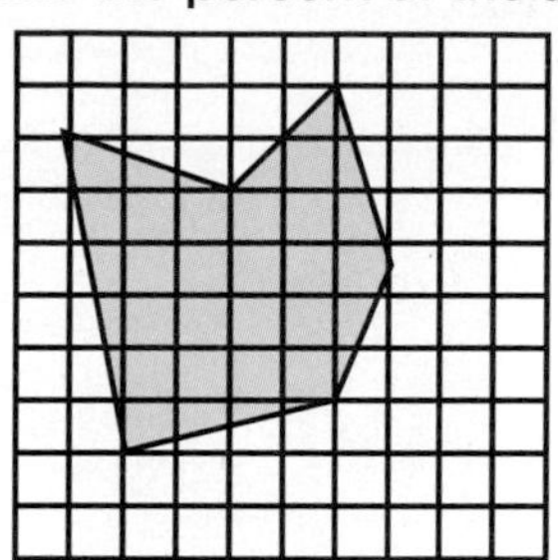

15. 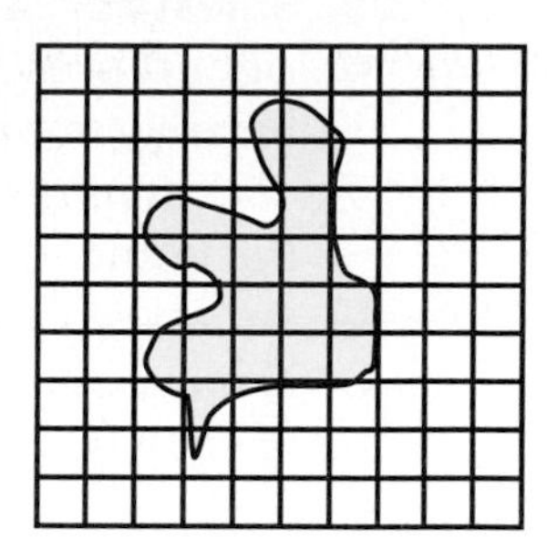

Exercises

Independent Practice

Estimate.

16. 73% of 65
17. 16% of 55
18. 19% of 72
19. 68% of 33
20. 29% of 50
21. 12.4% of 39
22. How would you determine if 8% of 400 is greater than 4% of 180?
23. What compatible numbers could you use to estimate 32% of 154?

Estimate the percent of the area shaded.

24. 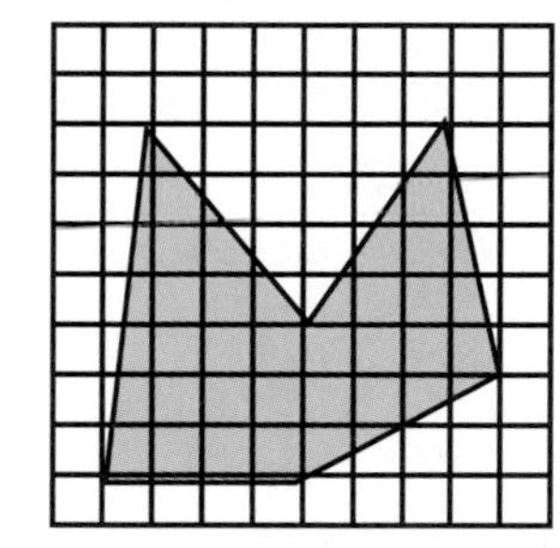

25. 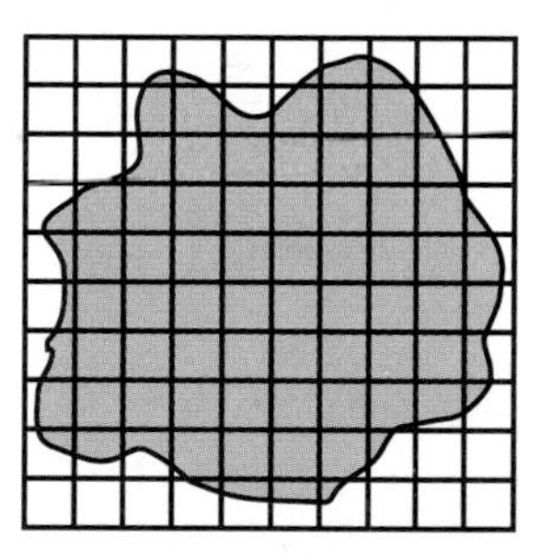

Estimate the percent.

26. 4 out of 25
27. 24 out of 62
28. 61 out of 88
29. 12 out of 60
30. 7 out of 56
31. 38 out of 159
32. How would you estimate Janice's foul shooting percent if she made 13 foul shots in 22 attempts?

Mixed Review

33. **Geometry** Find the area of a parallelogram whose height is 6.5 feet and whose base is 9 feet. *(Lesson 2-9)*
34. **Probability** Two dice are rolled. Find the probability that the sum of the numbers showing is seven. *(Lesson 6-8)*
35. Express the ratio $\frac{\$2.88}{9 \text{ ounces}}$ as a unit rate. *(Lesson 9-1)*
36. Express $\frac{2}{5}$% as a fraction in simplest form. *(Lesson 10-3)*

Problem Solving and Applications

37. **Critical Thinking** If x is 285% of y, then y is about what percent of x?
38. **Ecology** In a survey of 1,413 shoppers, 6% said they would be willing to pay more for environmentally safe products. About how many shoppers said they would pay more?

39. **Statistics** The graph at the right represents the six beverages with the highest consumption rates outside the home. If each household surveyed consumed four drinks, estimate the number of each type consumed.

Beverage	Estimate
Soft drinks	
Coffee	
Tea	
Other	
Milk	
Bottled water	

40. **Journal Entry** Write a few sentences explaining a method of estimating with percents.

10 Assessment: Mid-Chapter Review

Write a proportion to solve each problem. *(Lesson 10-1)*

1. What number is 40% of 60?
2. Seventy-five is 30% of what number?

Express each fraction as a percent. *(Lesson 10-1)*

3. $\frac{44}{100}$
4. $\frac{2}{3}$
5. $\frac{5}{8}$
6. $\frac{2}{5}$

Express each percent as a decimal and as a fraction in simplest form. *(Lessons 10-2, 10-3)*

7. 80%
8. 9%
9. $37\frac{1}{2}\%$
10. 7.09%
11. 0.4%
12. 118%
13. $\frac{7}{8}\%$
14. $1\frac{3}{50}\%$
15. There are 24 girls on the cheerleading squad. If 25% of them are freshmen, how many cheerleaders are freshmen? *(Lesson 10-4)*

Determine which is the best estimate. *(Lesson 10-5)*

		a.	b.	c.
16.	47.5% of 600	a. 3	b. 30	c. 300
17.	17% of 42	a. 0.8	b. 8	c. 80
18.	7 out of 41	a. 1.7%	b. 17%	c. 170%
19.	$\frac{9}{16}\%$ of 415	a. 200	b. 20	c. 2
20.	108% of 1,988	a. 22	b. 220	c. 2,200

Cooperative Learning

10-5B Percent Scavenger Hunt

A Follow-Up of Lesson 10-5

Objective

Estimate area using nonstandard units of measure.

Materials

measuring tape

Look at the door of your classroom. Can you imagine using the door as a tool for measuring area? In this lab you will do that as you go on a scavenger hunt for the area of other objects.

Try this!

Work with a partner.

- Copy the table below. Look at the door. Now look for an object with a surface whose area appears to be about 50% of the area of the door. Write this object in the first row of the second column of your table. Continue this procedure until you complete the second column of the table.
- Using a measuring tape, find the dimensions of the door and the other objects you identified in the second column. Complete the two middle columns of your table.
- Express the ratio of the area of each object in the second column to the area of the door. Express each result as a percent in the last column.

Estimated Area of the Door	Object	Area of Object	Area of Door	Ratio of Areas (%)
50%				
25%				
$133\frac{1}{3}$%				
$\frac{5}{8}$				
0.9				
0.5%				

What do you think?

1. Select one of the objects you chose. Explain how you used fractions or percents for your estimation.
2. Discuss with another pair of classmates how you can estimate area by comparing areas. Present an estimation problem to the class that could be solved using the same procedure.

Planning a Fund-Raising Event

Situation

The Pep Club is planning a fund-raising activity scheduled to take place at the All-State Baseball Championship being held at your school. The club has decided to sell hot dogs. You have been assigned the task of organizing the toppings you will use to offer four kinds of hot dogs. How much should you charge for each kind of hot dog to make a profit?

Hidden Data

Getting the supplies: Who will do the shopping before the game?
Preparation of the food: Where will the food be stored and cooked?
Attendance: How many people are expected to attend each game?
Complimentary condiments: Will you offer some ingredients without charge? What would they be?

Pricing Your Ingredients

Ingredient	Amount per package	Price per package
hot dogs	8 hot dogs	$2.96
hot dog buns	10 buns	$1.20
hot dog chili	2 cups	$3.09
cheese	15 slices per pound	$2.78 (pound)
sauerkraut	1.3 cups per can	$0.89
onions	5 medium per pound	$0.30 (pound)

HOT DOG MENU

#1 Hot dog with chili, cheese, and onions
#2 Hot dog with onions
#3 Hot dog with sauerkraut
#4 Hot dog with cheese

Analyzing the Data

1. About how much does one plain hot dog with bun cost?
2. About how much does one slice of cheese cost?
3. How much does hot dog #1 cost to prepare?

Making a Decision

4. **About how much** profit do you wish to make on each sale?
5. **Will you** encourage people to buy the hot dog with the most toppings to make a larger profit? Explain.
6. **How will you** know if the prices you select are reasonable?
7. **What special arrangements** will you have to make to prepare the food and keep it warm?

Instructions

To concession stand workers

#1 CHILI-CHEESE-ONIONS
Use:
1 hot dog/ 1 bun
$\frac{1}{4}$ cup chili
1 slice cheese
2 tablespoons onion

#2 ONIONS ONLY
Use:
1 hot dog/ 1 bun
3 tablespoons onion

#3 SAUERKRAUT ONLY
Use:
1 hot dog/ 1 bun
$\frac{1}{3}$ cup sauerkraut

#4 CHEESE ONLY
Use:
1 hot dog/ 1 bun
2 slices cheese

Making Decisions in the Real World

8. **Gather information** about the cost of a hot dog with similar toppings at fast food restaurants, stadiums, and the school cafeteria. Where were the hot dogs more expensive? How did this price compare with your prices?

10-6 The Percent Equation

Objective

Solve problems using the percent equation.

In 1991, nearly 400,000 foreign students were studying at United States colleges. How many of the students were from Asia?

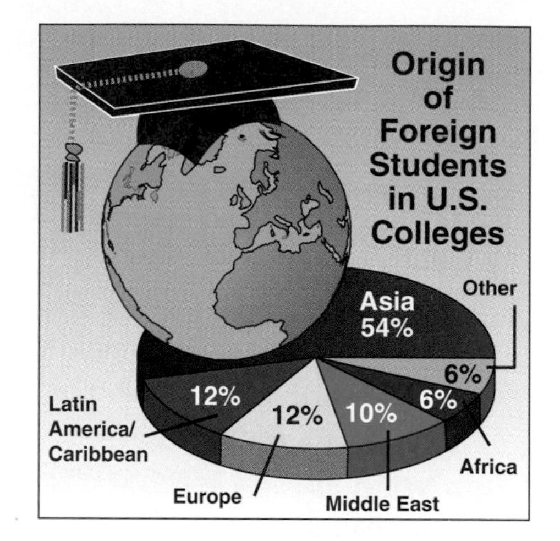

Estimation Hint

THINK:
54% is slightly more than 50%.
$50\% = \frac{1}{2}$
$\frac{1}{2}$ of 400,000 is 200,000.

You can use an equation to solve this problem.

Let s = the number of students from Asia. Write an equation. Write the percent as a decimal.

$s = 0.54 \cdot 400{,}000$

0.54 [×] 400000 [=] 216000

$s = 216{,}000$

There were 216,000 students from Asia studying in the United States. *Compared to the estimate, the answer is reasonable.*

TEEN SCENE

About 5.5% of incoming college freshmen chose business administration as their major in 1990, up one percent from 1989.

You can express the percent proportion you studied in Lesson 10-1 in a form that is easier to use when the rate and base are known.

$\frac{P}{B} = \frac{r}{100}$	*Write the percent proportion.*
$\frac{P}{B} \cdot B = \frac{r}{100} \cdot B$	*Multiply each side by B.*
$P = \frac{r}{100} \cdot B$	*Remember, the ratio $\frac{r}{100}$ is called the rate.*
$P = R \cdot B$	*Let R represent $\frac{r}{100}$. Replace $\frac{r}{100}$ with R.*

The equation is now in the form $P = R \cdot B$. You can use this equation to solve percent problems like the one at the beginning of the lesson.

Example 1

Find 6% of $725. *Estimate: 1% of 700 is 7; 6 · 7 = 42*

What number is 6% of 725?

$P = 0.06 \cdot 725$ *Write in P = R · B form. P represents the percentage.*

0.06 [×] 725 [=] 43.5

$P = 43.5$

6% of $725 is $43.50. *Compare with the estimate.*

Example 2

12 is 40% of what number? *Estimate: 40% is about $\frac{1}{2}$ · 12 is $\frac{1}{2}$ of 24.*

$P = R \cdot B$

$12 = 0.4 \cdot B$ *Replace P with 12 and R with 40% or 0.4*

$12 = 0.4B$ *Solve for B.*

12 [÷] 0.4 [=] 30 *Divide each side by 0.4.*

$30 = B$

12 is 40% of 30. *Check your answer by finding 40% of 30.*

Example 3 ***Problem Solving***

Entertainment The price of a home video game system is $198. If the sales tax on the video game is $8.91, what is the sales tax rate?

$P = R \cdot B$

$8.91 = R \cdot 198$ *Replace P with 8.91 and B with 198.*

8.91 [÷] 198 [=] 0.045 *Divide each side by 198.*

$R = 0.045$ or 4.5%

The sales tax rate is 4.5%.

Checking for Understanding

Communicating Mathematics

Read and study the lesson to answer each question.

1. **Write** an equation in the form $P = R \cdot B$ that you would use to determine the percent score of correctly answering 28 out of 34 questions on a history test.
2. **Tell** what number you would use for R in the equation $P = R \cdot B$ if the rate is 35%.
3. **Tell** whether the percentage is greater than or less than the base, when the rate is less than 100%.

Guided Practice

Write an equation in the form $P = R \cdot B$ for each problem. Then solve.

4. What number is 47% of 52?
5. What number is 56% of 80?
6. What percent of 52 is 25?
7. What percent of 3,600 is 2?
8. $48 is 30% of what amount?
9. $64 is what percent of $78?
10. What percent of 4.8 is 1.12?
11. Find 16.5% of 60.
12. Thirty percent of what amount is $5,000?

Exercises

Independent Practice

Solve.

13. What is 81% of 11.2?
14. What percent of 90 is 36?
15. Fifty is 10% of what number?
16. Find 0.5% of 3,200.
17. What is 7.4% of 40?
18. What percent of 21 is 96?
19. 20% of what number is 65?
20. Find 28% of $231.90.
21. Sixty-six is what percent of 55?
22. $17 is what percent of $51?
23. $54 is 108% of what amount?
24. What is 124% of 72?
25. Sixteen is $66\frac{2}{3}$% of what number?
26. $6 is what percent of $50?
27. $6,899.70 is 105.5% of how many dollars?
28. There are 35 students. Eight of them are wearing orange. What percent are wearing orange?

Mixed Review

29. **Algebra** In trapezoid *STUV*, $m\angle S = 90°$, $m\angle T = 3x°$, $m\angle U = 60°$, and $m\angle V = 90°$. Find the value of x. *(Lesson 5-4)*

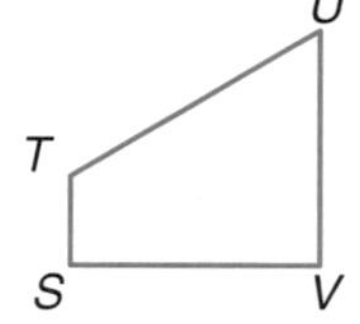

30. Find the multiplicative inverse of $-\frac{1}{a}$. *(Lesson 7-4)*
31. If $x^2 = 121$, what is the value of x? *(Lesson 8-3)*
32. Estimate 20.5% of 89. *(Lesson 10-5)*

Problem Solving and Applications

33. **Royalties** Marcela writes songs and receives a royalty of 8% of the CD sales. Last month she received a check for $3,896. How much money was spent on the CDs with her songs?
34. **Taxes** Mrs. Harper stayed overnight at a hotel during a mathematics teachers' convention. The cost of the room was $84 and the room tax was $10.92. What was the tax rate?

35. **Portfolio Suggestion** Select an item from this chapter that shows how you used a calculator or computer. Place it in your portfolio.
36. **Critical Thinking** Akiko adds 10% of a number to the number. Then she subtracts 10% of the total. Is the result equal to her original number? Explain.
37. **Sales** Jeremiah Morrison bought a new electric guitar and amplifier for $229.99. In addition, he paid a $5\frac{1}{2}$% state sales tax. What is the total amount he paid?
38. **Education** Bryan scored 84% on his language arts test. If he answered 21 questions correctly, how many questions were on the test?
39. **Sales** The sales tax on a $21 purchase was $1.47. What was the rate of sales tax?
40. **Make Up a Problem** Find a graph or newspaper article that involves the use of percents. Write a problem that can be solved using the equation $P = R \cdot B$.

Statistics Connection

10-7 Circle Graphs

Objective

Construct circle graphs.

Words to Learn

circle graph

The table below shows the average attendance at four major attractions.

Average Visitors Per Day in 1991	
Dollywood—Pigeon Forge, Tennessee	4,408
Opryland, USA—Nashville, Tennessee	5,303
Disneyland—Anaheim, California	31,983
Disney World/Epcot Center—Orlando, Florida	77,135

In Chapter 4, you learned how to make stem-and-leaf plots to display sets of data, and you learned how to summarize data in a box-and-whisker plot. How do you display a comparison of parts of a set of data to the whole set? For example, how would you compare Dollywood visitors to the total visitors at all four attractions?

A good way to display this information is to use a *circle graph*. A **circle graph** is used to compare parts of a whole. Study the example to find out how to make a circle graph.

Example

Construct a circle graph using the data in the table above.

- Find the total number of visitors.

  4408 [+] 5303 [+] 31983 [+] 77135 [=] 118829

- Find the ratio that compares the visitors at each park to the total number of visitors. Round each ratio the the nearest thousandth.

Since you are comparing the number of visitors at each park to the total, each ratio represents a percent of the whole circle.

Dollywood: $\frac{4,408}{118,829}$

4408 [÷] 118829 [=] 0.0370953 *0.0370953 ≈ 0.037 or 3.7%*

Opryland, USA: $\frac{5,303}{118,829}$

5303 [÷] 118829 [=] 0.0446272 *0.0446272 ≈ 0.045 or 4.5%*

Disneyland: $\frac{31,983}{118,829}$

31983 [÷] 118829 [=] 0.2691515 *0.2691515 ≈ 0.269 or 26.9%*

Disney World: $\frac{77,135}{118,829}$

77135 [÷] 118829 [=] 0.6491261 *0.6491261 ≈ 0.649 or 64.9%*

- Since there are 360° in a circle, multiply each ratio by 360 to find the number of degrees for each section of the graph. Round to the nearest degree.

  Dollywood: 0.037 [×] 360 [=] 13.32 *13.32 ≈ 13°*

  Opryland, USA: 0.045 [×] 360 [=] 16.2 *16.2 ≈ 16°*

  Disneyland: 0.269 [×] 360 [=] 96.84 *96.84 ≈ 97°*

  Disney World: 0.649 [×] 360 [=] 233.64 *233.64 ≈ 234°*

- Use a compass to draw a circle and a radius as shown at the right.

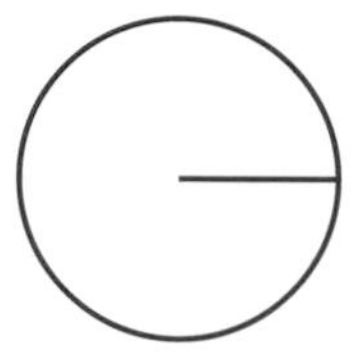

- Use a protractor to draw an angle of 13°. *Actually, you can start with any of the four angles.*

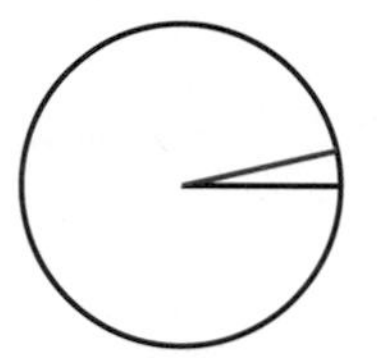

- From the new radius, draw the next angle. Repeat for the remaining angles. Label each section and give the graph a title.

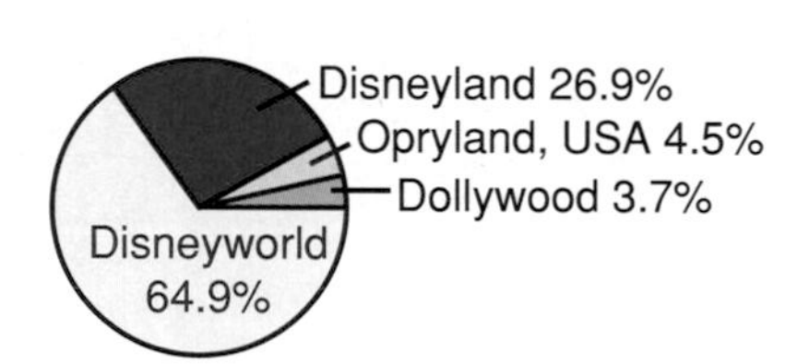

Visitors at Selected Parks

Checking for Understanding

Communicating Mathematics

Read and study the lesson to answer each question.

1. **Tell** why circle graphs are used to show data.
2. **Tell** how to find the number of degrees for a section of a circle graph that represents the number of girls in your math class.
3. **Write** what the sum of the percents in a circle graph should be.
4. **Draw** a circle graph to represent the percent of girls and the percent of boys in your class today.

Guided Practice

Use the table at the right to answer Exercises 5–8.

5. What is the expected total of the percent column?
6. What is the expected total of the degree column?
7. Copy and complete the table.
8. Make a circle graph of the data.

Monthly Budget

Category	Amount	% of total	Degrees in graph
Housing	$775		
Food	$375		
Insurance	$225		
Transportation	$475		
Other	$650		
Total			

Exercises

Independent Practice

9. **Statistics** The chart shows the number of patents issued to foreign inventors on an average day in the United States. Make a circle graph to display the data.

Country	Number
Canada	5
Great Britain	8
France	8
West Germany	22
Japan	47

10. **Probability** What is the probability of selecting a free prize in a classroom raffle if there are 2 winning tickets out of a total of 9 tickets? Make a circle graph to show the probability of winning and of losing.

Mixed Review

11. Find the absolute value of -28. *(Lesson 3-1)*
12. Find the LCM of 18, 34, and 6. *(Lesson 6-9)*
13. What percent of 32 is 12? *(Lesson 10-6)*

Problem Solving and Applications

14. **Critical Thinking** Pam was preparing a circle graph to display the data in the chart at the right. When she multiplied each ratio by 360 to find the degrees of each section, she noticed the total was 361°. Why is the total more than 360° and how can Pam get the correct total?

Areas (in square miles) of the Oceans of the World	
Pacific Ocean	64,186,300
Atlantic Ocean	33,420,000
Indian Ocean	28,350,500
Arctic Ocean	5,105,700

15. **Statistics** On an average day in the United States 2,749 people enroll in a foreign language course. Of these, the average enrollments are: Chinese, 46; Japanese, 64; Russian, 93; French, 754; and Spanish, 1,127. Make a circle graph to display the data.

16. **Research** Take a survey of the people in your class on the amount of time they spent watching television during the past 24 hours. Make a circle graph to show what percent of the class watched less than one hour, 1–2 hours, 2–3 hours, or more than 3 hours.

17. **Nutrition** A restaurant's survey asked each customer how often they eat salad. The results are listed in the chart at the right. Make a circle graph of the data.

Frequency of eating salad	number
once or twice a day	100
several times a week	270
once a week	60
1–3 times a month	50
less than once a month	20

18. **Data Search** Refer to pages 378 and 379.
    a. Construct a circle graph to display what happened to the waste generated per person per day in 1960.
    b. Construct a circle graph to display what happened to the waste generated per person per day in 1988.
    c. What conclusion can you make by comparing the two graphs?

10-8 Percent of Change

Objective

Find the percent of increase or decrease.

Ronald Reagan was the fortieth president of the United States. During his first term, the number of people who were unemployed was 10,678,000. By the end of his second term, 6,701,000 people were unemployed. You can express the decrease from about 11 million to about 7 million by using percents.

Mini-Lab

Work with a partner.

Materials: graph paper, pencil, scissors

- Draw a rectangle 11 units long and 1 unit wide on graph paper. Draw another rectangle 7 units long and 1 unit wide.

- Cut out the shorter rectangle and place it over the longer rectangle. Shade the squares that are not covered.

Talk About It

a. Describe the relationship between the area of the longer rectangle and the area of the shorter rectangle.

b. What do the shaded squares represent?

c. What is the ratio of the area of the shaded squares to the total area of the longer rectangle? Express the ratio as a percent.

The Mini-Lab shows how you can find the estimated percent decrease in the number of unemployed people. You can compute the actual percent of decrease by using the percent equation and a calculator.

Estimation Hint

THINK: $\frac{4{,}000{,}000}{10{,}000{,}000} = \frac{4}{10}$

$\frac{4}{10} = 40\%$

- Subtract to find the amount of decrease.

  10678000 [−] 6701000 [=] 3977000

- Use the form $P = R \cdot B$. Compare the amount of decrease to the original amount.

  3,977,000 is what percent of 10,678,000?

  $3{,}977{,}000 = R \cdot 10{,}678{,}000$

  3977000 [÷] 10678000 [=] 0.372448

  $R \approx 0.37$ or 37%

  The unemployment rate decreased about 37%.

You can also use the percent proportion to find the percent of change. The base represents the original amount and the amount of change is the percentage.

Estimation Hint

THINK: $\frac{20}{30} = \frac{2}{3}$

Since $\frac{18}{30} < \frac{20}{30}$, the percent of increase is less than $66\frac{2}{3}\%$.

Example 1 *Problem Solving*

Business John Clark sells the merchandise in his sporting goods store for more than what he pays for it. This increase in price is called the *markup.* It is used to cover expenses and make a profit. Mr. Clark's cost for a pair of roller blades is \$30. If he sells them for \$48, what is his markup rate?

$48 - 30 = 18$ *Find the amount of increase.*

$\frac{18}{30} = \frac{r}{100}$ *Write the percent proportion. The original amount is \$30.*

$18 \cdot 100 = 30 \cdot r$ *Find the cross products.*

$1{,}800 = 30r$ *Divide each side by 30.*

1800 ÷ 30 = 60

$60 = r$

The markup rate is 60%.

Example 2 *Problem Solving*

Earning Money At the beginning of the year, Shawna received an allowance of \$8 per week. Now that she has a job after school she receives \$3 per week. Find the percent of decrease in her allowance.

$8 - 3 = 5$ *Find the amount of decrease*

$\frac{5}{8} = \frac{r}{100}$ *Write the percent proportion.*

$5 \cdot 100 = 8 \cdot r$ *Find the cross products.*

$500 = 8r$ *Divide each side by 8.*

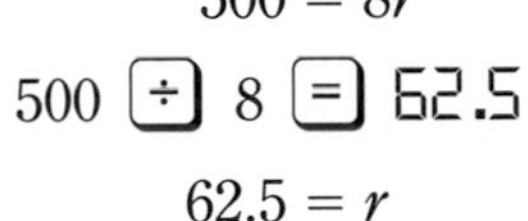

$62.5 = r$

The percent of decrease is 62.5%.

Checking for Understanding

Communicating Mathematics

Read and study the lesson to answer each question.

1. **Tell** what the first step is when finding the percent of change.
2. **Tell** how you know which number to use as the base when finding a percent of change.
3. **Draw** a picture to show a decrease of 20%.

Guided Practice Estimate the percent of change.

4. old: $4
   new: $6
5. old: $56.12
   new: $62
6. new: $88
   old: $76

Find the percent of change. Round to the nearest whole percent.

7. old: $3
   new: $2
8. old: $72
   new: $80
9. old: $533
   new: $600

Exercises

Independent Practice Estimate the percent of change.

10. old: $5
    new: $7
11. old: $48.88
    new: $36.99
12. old: $44
    new: $39

Find the percent of change. Round to the nearest whole percent.

13. old: $6
    new: $5
14. old: $48
    new: $64
15. old: $221
    new: $300
16. old: $21
    new: $30
17. old: $90
    new: $80
18. old: $315
    new: $400
19. Find the percent of change in taxes if the old taxes were $54.00 and the new taxes are $57.78.

Mixed Review

20. How many pints are in 9 cups? *(Lesson 1-7)*
21. **Geometry** The length of a leg of a 45°– 45° right triangle is 6.5 feet. Find the length of the hypotenuse to the nearest tenth. *(Lesson 8-8)*
22. **Statistics** On an average Saturday, Kelly spends 9 hours sleeping, 4 hours watching TV, 1.5 hours eating, 5 hours with her friends, and 4.5 hours with her family. Make a circle graph to display the data. *(Lesson 10-7)*

Problem Solving and Applications

23. **Critical Thinking** A store has a 50% markup on sweaters. But, the sweaters do not sell well at the listed price. So, the sweaters are put on sale at 50% of the listed price. All the sweaters are sold. Did the store break even, make a profit, or lose money? Explain.
24. **Traffic Safety** The graph at the right shows the percentage of drivers who wore seat belts each year from 1982 to 1991.
    a. What is the percent of increase from 1982 to 1991 in the percentages of drivers using seat belts?
    b. For which year was the rate of increase greatest?

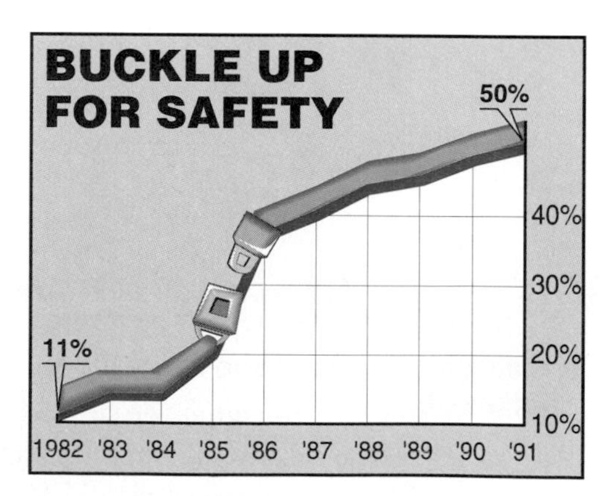

25. **Journal Entry** Write one or two sentences explaining how to find a percent of change.

10-9 Discount

Objective

Solve problems involving discounts.

Words to Learn

discount
sale price

The 18-speed mountain bike Diana wants to buy usually sells for $198. If she buys it on sale at 25% off, how much will she pay?

The amount by which the regular price is reduced is called the **discount.** The **sale price** is the price after the discount has been subtracted.

You can find the sale price in one of two ways.

Estimation Hint

THINK

$25\% = \frac{1}{4}$

$\frac{1}{4}$ of 200 is 50.

The discount is about $50. The sale price is about $200 − $50 or $150.

Method 1

Find the amount of the discount.
25% of $198 is the discount.

0.25 [×] 198 [=] 49.5

Diana will save $49.50.

Subtract the discount from the regular price.

198 [−] 49.5 [=] 148.5

The sale price is $148.50.

Method 2

Find the percent paid.
100% − 25% = 75%

Diana will pay 75% of the regular price.

Find the amount paid.
75% of $198 is the sale price.

0.75 [×] 198 [=] 148.5

Example *Problem Solving*

Smart Shopping Mitchell Li bought a pair of pants that had been reduced from $39 to $32. What was the percent of discount?

39 − 32 = 7 *Find the discount amount.*

7 is what percent of 39?

$7 = R \cdot 39$ *Write in $P = R \cdot B$ form.*

7 [÷] 39 [=] 0.1794872 *Divide each side by 39.*

$R \approx 0.1794872$ or about 0.18

The percent of discount is about 18%.

Technology Activity

You can learn how to use a spreadsheet to find the sale price of an item in Technology Activity 5 on page 662.

Checking for Understanding

Communicating Mathematics

Read and study the lesson to answer each question.

1. **Tell** how to find the price of a \$35 sweater on sale at 20% off.
2. **Write** the equation you would use to find the rate of discount on a video tape that is on sale for \$3.74 less than the regular price of \$19.98.

Guided Practice

Find the discount to the nearest cent.

3. \$49 athletic shoes, 20% off
4. \$300 stereo, $33\frac{1}{3}$% off
5. \$8.60 tie, 40% off
6. \$38.50 dress, 15% off

Find the sale price of each item.

7. \$899 sofa, 10% off
8. \$210 suit, 5% off
9. \$37.95 sweatsuit, 25% off
10. \$40 jacket, 30% off

Find the percent of discount.

11. regular price: \$65
discount: \$25
12. regular price: \$40
sale price: \$35

Exercises

Independent Practice

Find the amount of discount and the sale price of each item to the nearest cent.

13. \$29.95 jeans, 25% off
14. \$3 socks, 30% off
15. \$14.50 CD, 15% off
16. \$13 book, 20% off
17. \$119.50 lamp, $\frac{1}{3}$ off
18. \$3.59 tennis balls, $\frac{1}{4}$ off

Find the percent of discount.

19. regular price: \$89
discount: \$22.25
20. regular price: \$75
discount: \$30
21. regular price: \$72
sale price: \$36
22. regular price: \$108
sale price: \$72
23. What is the discount on a \$22.75 shirt on sale for 15% off?
24. What is the sale price of a \$169.95 camera on sale at 10% off?
25. Find the discount rate on a \$20 watch that regularly sells for \$25.
26. What is the regular price of a pair of jeans that sold for \$25.95 at a 30% off sale?

Mixed Review

27. Solve $n - 12 = 8$. *(Lesson 2-2)*
28. Solve $-3\frac{5}{8}(-5\frac{1}{4}) = s$. Write the solution in simplest form. *(Lesson 7-3)*
29. **Geometry** Find the value of x in the figure at the right. Round your answer to the nearest tenth. *(Lesson 9-9)*
30. **Consumer Math** Find the percent of change in gasoline prices if the old price per gallon was \$1.15 and the new price is \$1.19. *(Lesson 10-8)*

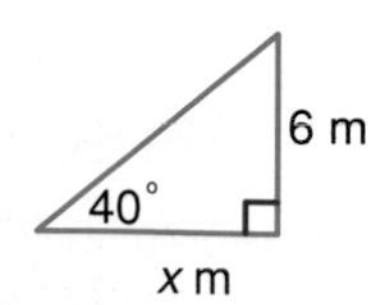

Problem Solving and Applications

31. **Telecommunication** When you call long distance within your calling zone at night or on weekends, you get discounts off the regular cost. The times and discount rates are shown in the chart at the right.

    a. The regular cost of a call is \$7.60. What is the discount if the call is made at 11:15 P.M.?

    b. The regular cost of a call is \$4.20. If the call is made Wednesday at 6:00 P.M., what is the cost?

    c. Bill made a call and is charged \$10.70. What percent of discount did he receive if the regular cost is \$26.75?

LONG DISTANCE CALLING

	M	T	W	T	F	S	S
8 A.M. to 5 P.M.							
5 P.M. to 11 P.M.							
11 P.M. to 8 A.M.							

□ Weekday (full rate)
□ Evening (30% discount)
□ Night and Weekend (60% discount)

32. **Critical Thinking** A jewelry store is having a 25% off sale. On Saturday only, all sale items are marked an additional 10% off. The Saturday sale price offers what rate of discount from the original price?

33. **Critical Thinking** During a one-day sale, a bedspread is discounted 25% to \$51.

    a. What was the original price?

    b. After the sale is over, what must the percent of increase be to return the price of the bedspread to the original price?

34. **Mathematics and Marketing** Read the following paragraphs.

> Marketing involves many activities that direct the flow of goods and services from producers to consumers. Consumer goods are sold through a variety of channels including retail stores, which often offer lower prices and a large selection of product.
>
> Careful studies of consumers and their buying habits enable marketers to determine the best times to have promotions or sales. Marketers often use spreadsheet software to calculate the percent of decrease for different items.

This spreadsheet shows data for one store's sale merchandise. The percent of decrease is found by entering a formula into a cell. The formula in cell E2 is C2/B2*100. Find the values in cells E3 and E4.

	A	B	C	D	E
1	ITEM	REGULAR PRICE	DISCOUNT	DISCOUNT PRICE	PERCENT DECREASE
2	REFRIGERATOR	\$900.00	\$102.60	\$797.40	11.4
3	MICROWAVE	\$450.00	\$90.00	\$360.00	
4	DRYER	\$389.50	\$54.53	\$334.97	

10-10 Simple Interest

Objective

Solve problems involving simple interest.

Words to Learn

interest
principal
rate
time

Stephen Sopher found a United States savings bond his uncle bought when he was born. The savings bond had matured and was worth \$50 eight years ago. Stephen is curious about how much the savings bond is worth now. If the savings bond earned $6\frac{1}{4}$% simple interest annually for the last 8 years, what is its value now?

Simple **interest** (I) is calculated on the **principal** (p), which is the amount of money in an account. The **rate** (r), is a percent, and the **time** (t), is given in years. You can use the formula below to find out how much interest Stephen's savings bond earned over the past eight years.

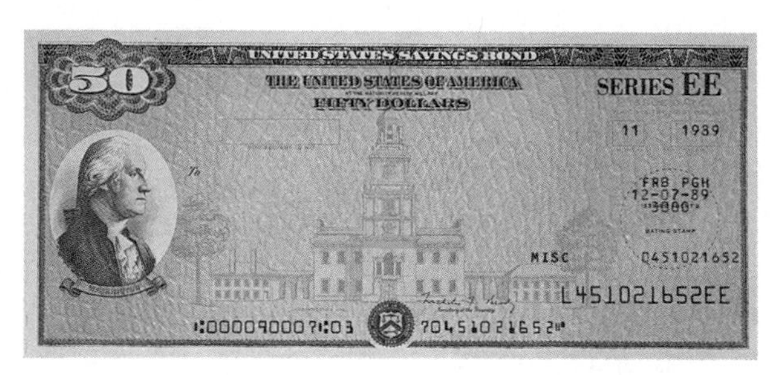

interest = principal · rate · time

$I = prt$

$I = 50 \cdot 0.0625 \cdot 8$

50 [×] 0.0625 [×] 8 [=] 25 So, $I = 25$.

The savings bond earned \$25 interest in the last eight years. Stephen's savings bond is now worth \$75.

Example 1 ***Problem Solving***

Earning Money Yolanda sold calendars to her paper route customers and opened a savings account with the \$300 she earned. Her account earns $5\frac{3}{4}$% interest annually. If she does not withdraw or deposit any money for 15 months, how much will be in her account?

First, find the amount of interest earned by using the simple interest formula.

$I = prt$ *Replace p with 300, r with 0.0575, and t with $\frac{15}{12}$ since 15 months is $\frac{15}{12}$ year.*

$I = 300 \cdot 0.0575 \cdot \frac{15}{12}$

300 [×] 0.0575 [×] 15 [÷] 12 [=] 21.5625

$I = 21.5625$

The interest earned after 15 months is \$21.56.

Then, add the interest to the savings: 300 [+] 21.56 [=] 321.56.

After 15 months, the account will contain \$321.56.

DID YOU KNOW

Outstanding consumer credit increased from \$131.6 billion in 1970 to \$728.9 billion in 1988.

Interest is charged to you when you borrow money. When this happens, the principal is the amount borrowed. If merchandise is bought using a credit card, a bank or the store is actually lending the money.

Example 2 *Problem Solving*

Consumer Math **Jessica's mother used her credit card to purchase a $380 stereo. The bank that issued the credit card charges an annual interest rate of 19.8%. Each month they charge interest on the unpaid balance of the account. If Jessica's mother does not make any more purchases and makes a $50 payment, how much interest will she be charged next month?**

By making a $50 payment, the balance, p, for the next month is $330.

$I = prt$

$I = 330 \cdot 0.198 \cdot \frac{1}{12}$ *Replace p with $330, r with 0.198, and t with* $\frac{1}{12}$.

330 [×] 0.198 [÷] 12 [=] 5.445

Jessica's mother will be charged $5.45 in interest next month.

Example 3 *Problem Solving*

Finance **Dan's father borrowed $7,800 to buy a new car. He paid the money back in equal payments of $243.75 for the next 48 months. Find the simple interest rate on his loan.**

First, find the amount he repaid. → 48 [×] 243.75 [=] 11700

Then find the amount of interest. → 11700 [−] 7800 [=] 3900

$I = prt$

$3,900 = 7,800 \cdot r \cdot 4$ *Replace I with 3,900, p with 7,800 and t with 4 since 48 months is 4 years.*

$3,900 = 31,200r$

3900 [÷] 31200 [=] 0.125 *Divide each side by 31,200.*

The simple interest rate on the loan was 12.5%.

Checking for Understanding

Communicating Mathematics

Read and study the lesson to answer each question.

1. **Write** how you would express nine months in the simple interest formula.
2. **Tell** the difference between the principal used to find interest on a loan and on a savings account.
3. **Tell** why it is a good idea to pay off the entire balance on credit card accounts each month.

Guided Practice **Find the simple interest to the nearest cent.**

4. \$345 at $6\frac{1}{4}$% for 6 months
5. \$400 at 15% for 18 months
6. \$1,088 at 18% for 1 year
7. \$62.25 at 5.5% for 9 months

Find the total amount in each account.

8. \$615 at 7% for 8 months
9. \$120 at $12\frac{3}{4}$% for $1\frac{1}{2}$ years.
10. \$118 at 5.5% for 19 months
11. \$217.75 at 6% for 36 months

Find the annual rate of simple interest.

12. interest: \$31.50; principal: \$700; time: 9 months
13. interest: \$142.80; principal: \$560; time: 2 years

Exercises

Independent Practice **Find the simple interest to the nearest cent.**

14. \$205 at $6\frac{1}{4}$% for 9 months
15. \$500 for 15 months at 12%
16. \$78.75 at 6.5% for 8 months
17. \$2,108 for 2 years at 16%
18. \$100 at 5% for 18 months
19. \$4,000 at 13.5% for 21 months

Find the total amount in each account.

20. \$708 at 8% after 6 months
21. \$200 at 6.75% after 9 months
22. \$235 at 5.25% after 14 months
23. \$176.77 at 6% after 6 years
24. \$1,860 at $7\frac{1}{2}$% after 5 years
25. \$10,000 at 13% after 6 months

Find the annual rate of simple interest.

26. principal: \$4,000; total amount: \$5,920; time: 3 years
27. principal: \$3,200; total amount: \$3,632; time: 18 months
28. Midori borrowed \$9,800 and paid back \$257.25 a month for five years. What was the annual rate of simple interest?

Mixed Review

29. **Statistics** Refer to the graph in Example 1 on page 133. How does the number of people who watch 0–5 hours compare to the number of people who watch 15–17 hours? *(Lesson 4-2)*
30. Solve $6p = \frac{5}{9}$. *(Lesson 7-10)*
31. On a map, the distance between two cities is $2\frac{3}{4}$ inches. Find the actual distance between the cities if the scale is 1 in.:20 mi. *(Lesson 9-7)*
32. **Smart Shopping** Find the amount of discount and the sale price of a pair of \$89 shoes on sale for 30% off. *(Lesson 10-9)*

Problem Solving and Applications

33. **Finance** Doug's savings account earned $27.93 interest in 6 months. The interest rate is $5\frac{1}{4}$%. How much was in his account to earn that amount of simple interest?

34. **Critical Thinking** Maura can use her credit card to get a cash advance from an automatic teller machine or she can get a loan from her credit union. The interest rate on cash advances is 1.5% per month and the credit union rate is 16% per year. Which is better for Maura? Explain.

35. **Critical Thinking** Heather paid $50 for a United States savings bond that will double in value in 8 years. What is the simple interest rate on this savings bond?

36. **Finance** Tim opened a savings account several years ago and deposited $100. He never made any more deposits or withdrawals. His last statement from the bank listed the balance at $191. If the account earned a simple interest rate of 7%, how long ago did Tim open his savings account?

37. **Computer Connection** A spreadsheet like the one shown below can be used to generate a simple interest table for various account balances.

	A	B	C	D	E
1	Principal	Rate	Time	Interest	New Balance
2					
3	500	=B2/100	=C2	=A3*B3*C3	=A3+D3
4	1000	=B2/100	=C2	=A4*B4*C4	=A4+D4
5	1500	=B2/100	=C2	=A5*B5*C5	=A5+D5
6	2000	=B2/100	=C2	=A6*B6*C6	=A6+D6
7	2500	=B2/100	=C2	=A7*B7*C7	=A7+D7
8	3000	=B2/100	=C2	=A8*B8*C8	=A8+D8
9	3500	=B2/100	=C2	=A9*B9*C9	=A9+D9

a. Why is the rate in column B divided by 100?

b. Suppose you want to make a table to find the balance if the interest rate is $5\frac{1}{2}$% and the time period is 3 years. What value will be in cell E6?

c. Your teacher asks you to add a new row to the spreadsheet to represent a principal of $4,000. List each of the cell entries (A10, B10, C10, D10, E10) you would enter.

d. How would you modify the spreadsheet if you wanted to calculate the simple interest on a principal of $900 at a rate of 6% for a 15-month period?

Chapter 10

Study Guide and Review

Study Guide and Review

Communicating Mathematics

State whether each sentence is *true* or *false*. If false, replace the underlined word or number to make a true sentence.

1. A percent is a ratio that compares a number to 10.
2. The decimal 0.25 is equivalent to 2.5%.
3. 12 is 10% of 120.
4. The amount by which a regular price is reduced to find the sale price is called the discount.
5. The interest is the amount of money in an account.
6. To express a percent as a fraction, express the percent in the form $\frac{r}{100}$ and simplify.
7. A circle graph is used to compare parts of a set of data to the whole set.

8. Explain how to express a fraction as a percent.
9. Tell how to express 160% as a mixed number.
10. Write how you would find the simple interest on \$400 invested at $5\frac{1}{2}$% for 1 year.

Self Assessment

Objectives and Examples	***Review Exercises***
Upon completing this chapter, you should be able to:	*Use these exercises to review and prepare for the chapter test.*

- solve problems using the percent proportion *(Lesson 10-1)*

96 is what percent of 240?

$$\frac{96}{240} = \frac{r}{100}$$
$$96(100) = 240(r)$$
$$r = 40$$

So, 96 is 40% of 240.

Use a proportion to solve each problem.

11. What is 28% of 400?
12. 54 is what percent of 180?
13. 61.5 is 75% of what number?
14. 220 is what percent of 800?
15. Find 62.5% of 128.

- express percents as fractions and decimals *(Lesson 10-2)*

Express 40% as a fraction.

$$40\% = \frac{40}{100} = \frac{2}{5}$$

Express each percent as a fraction in simplest form.

16. 65%
17. 8%
18. 30%
19. 17.5%

Objectives and Examples

- express percents greater than 100 or less than 1 as decimals and fractions *(Lesson 10-3)*

Express 120% as a mixed number.

$120\% = \frac{120}{100} = 1\frac{20}{100} = 1\frac{1}{5}$

Review Exercises

Express each percent as a fraction or mixed number in simplest form.

20. 145%
21. $\frac{4}{5}\%$
22. 428%
23. $\frac{3}{7}\%$

- estimate by using fractions, decimals, and percents interchangeably *(Lesson 10-5)*

Estimate 20% of 32.

$20\% = \frac{1}{5}$ *Round 32 to 30.*

$\frac{1}{5}$ of 30 is 6. So, 20% of 32 $\approx$ 6.

Estimate.

24. 8% of 64
25. 66% of 31
26. 59% of 80
27. 26% of 37
28. 98% of 20
29. 49% of 56

- solve problems using the percent equation *(Lesson 10-6)*

Find 15% of 600.

$P = R \cdot B$

$P = 0.15 \cdot 600$ *R = 15% or 0.15, B = 600*

$P = 90$

Solve.

30. What is 36% of 75?
31. What percent of 80 is 15?
32. Find 30% of $128.50.
33. Forty-two is 28% of what number?

- construct circle graphs *(Lesson 10-7)*

A circle graph is used to compare parts of a set of data to the whole set.

34. Make a circle graph of the number of hours each week that you study, talk on the telephone, watch television, and sleep.

- find the percent of increase or decrease *(Lesson 10-8)*

Find the percent of increase of a sweater that had cost $20 and now costs $25.

$25 − $20 = $5

$\frac{P}{B} = \frac{r}{100}$

$\frac{5}{20} = \frac{r}{100}$ *P = 5, B = 20*

$r = 25$

The percent of increase is 25%.

Find the percent of change. Round to the nearest whole percent.

35. old: $30
new: $36

36. old: $250
new: $210

37. old: $48
new: $60

- solve problems involving discounts *(Lesson 10-9)*

What is the sale price of a $15 CD on sale at 10% off?

$0.10 \cdot 15 = 1.50$

$15 - 1.50 = 13.50$

The sale price is $13.50.

Find the amount of discount and the sale price of each item.

38. $54.95 video game, 20% off
39. $60 watch, 15% off
40. $120 bicycle, 30% off
41. $84 clock, $\frac{1}{4}$ off

Objectives and Examples	Review Exercises
• solve problems involving simple interest *(Lesson 10-10)* Find the simple interest of \$250 at 6% for 8 months. $I = 250 \cdot 0.06 \cdot \frac{8}{12}$ $= 10$ The interest is \$10.	**Find the simple interest to the nearest cent.** **42.** \$820 at $7\frac{1}{2}$% for 1 year **43.** \$1,728 at 5.75% for 9 months **44.** \$96 at 6% for 15 months **45.** \$240 at $6\frac{1}{4}$% for 18 months

Applications and Problem Solving

46. Randy hired a rock band to play at the Snowflake Dance. They charge \$2,000 per performance. The manager required a 15% deposit. How much did Randy pay on deposit? *(Lesson 10-4)*

47. Smart Shopping Emilio wants to buy a CD player that costs \$149.95. If he waits until the CD player is on sale at 20% off to buy it, how much will he save? *(Lesson 10-9)*

48. Statistics 55% of Mrs. Kackley's math class are boys. What fraction of her class are boys? *(Lesson 10-2)*

49. Taxes Find the percent of change if the sales tax increased from 5% to $5\frac{1}{2}$%. *(Lesson 10-8)*

Curriculum Connection Projects

- **Home Economics** Survey your class to find how many like each flavor of potato chips. Find the percent of the class that likes each flavor. Draw a circle graph of your results.
- **Health** Have a friend measure your pulse while you are resting and again after you have run in place for 2 minutes. Find the percent of increase in your heart rate.
- **Consumer Awareness** Find how much interest you would pay in a year if you used a credit card to pay for a CD player.

Read More About It

Corcoran, Barbara. *The Strike.*
Cribb, Joe. *Money.*
Dickens, Charles. *The Christmas Carol.*
Scott, Elain. *The Banking Book.*

Chapter

10 Test

Solve.

1. Seventeen is $33\frac{1}{3}\%$ of what number?
2. What is $\frac{3}{4}\%$ of 2,400?
3. Fifty-eight is 80% of what number?
4. What percent of 45 is 18?
5. What percent of 28 is 42?
6. Express $\frac{8}{5}$ as a percent.

Express each percent as a decimal and as a fraction.

7. 12.8%
8. 285%
9. 35%
10. $\frac{1}{4}\%$

Refer to the table at the right to answer the following questions.

Number of marbles in a bag	
red	6
blue	4
green	10

11. What is the percent of blue marbles to the total number of marbles?
12. If the data in the table were used to make a circle graph, what would be the number of degrees for each section of the graph?
13. Rachel received a score of 70 out of 100 on an English test. On the next test she received a score of 84 out of 100. Find the percent of increase.

Find the amount of discount and the sale price of each item.

14. \$28 shirt, 15% off
15. \$299 recliner, 20% off
16. \$16 sweatshirt, $\frac{1}{4}$ off
17. \$59.90 lamp, 10% off
18. Find the discount rate if the regular price is \$60 and the sale price is \$42.

Determine which is the best estimate.

19. $\frac{7}{16}\%$ of 595 — a. 3 — b. 30 — c. 300
20. 24% of 60 — a. 1.5 — b. 15 — c. 150
21. 12.3% of 74 — a. 90 — b. 9 — c. 0.9
22. 204% of 1,309 — a. 26 — b. 260 — c. 2,600
23. Find the annual rate of simple interest for a loan if the principal borrowed is \$3,000, the interest paid is \$540, and the length of the loan is 18 months.
24. Find the total amount in a savings account if the principal is \$600, the interest rate is $5\frac{1}{2}\%$, and the time is 9 months.
25. The lockers in the hall outside Sam's homeroom are numbered consecutively and start with number 265 and end with number 357. How many lockers are in the hall?

Bonus Express 100% as a fraction in simplest form.

Chapter Test

Chapter

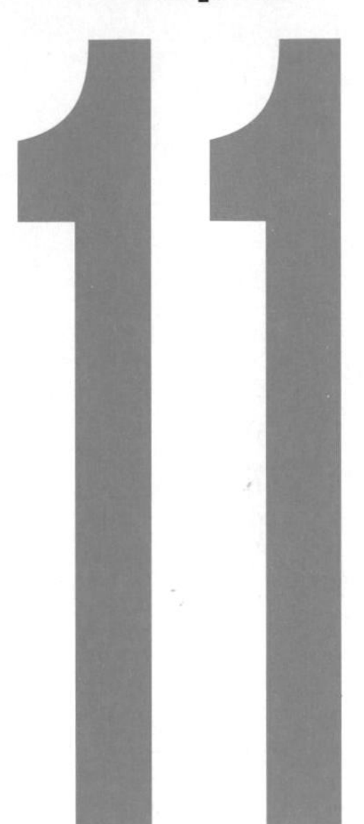

Algebra: Functions and Graphs

Spotlight on Oceans

Have You Ever Wondered...

- What happens to the temperature of the ocean water as depth increases?
- What kinds of life live at different depths in the ocean?

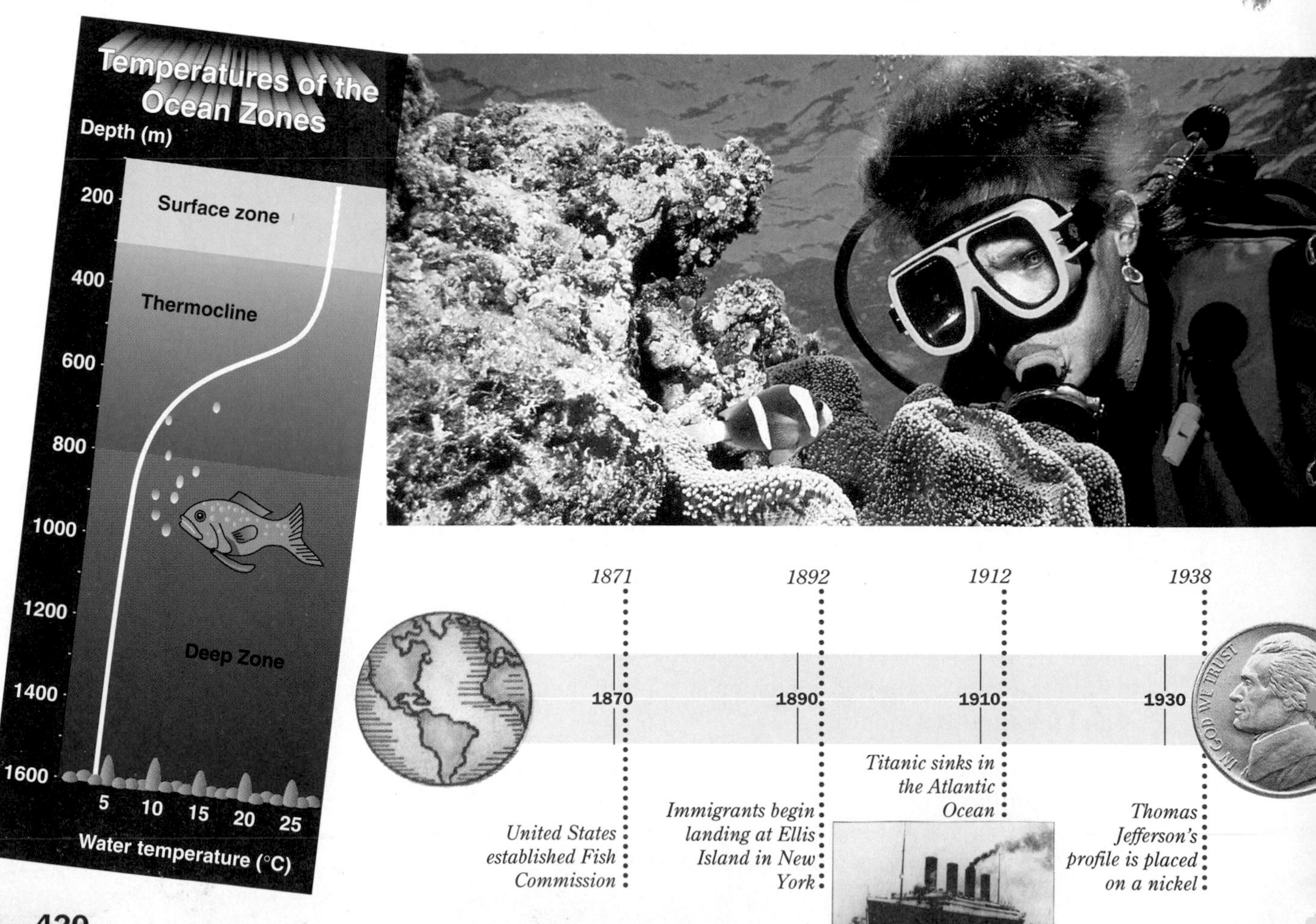

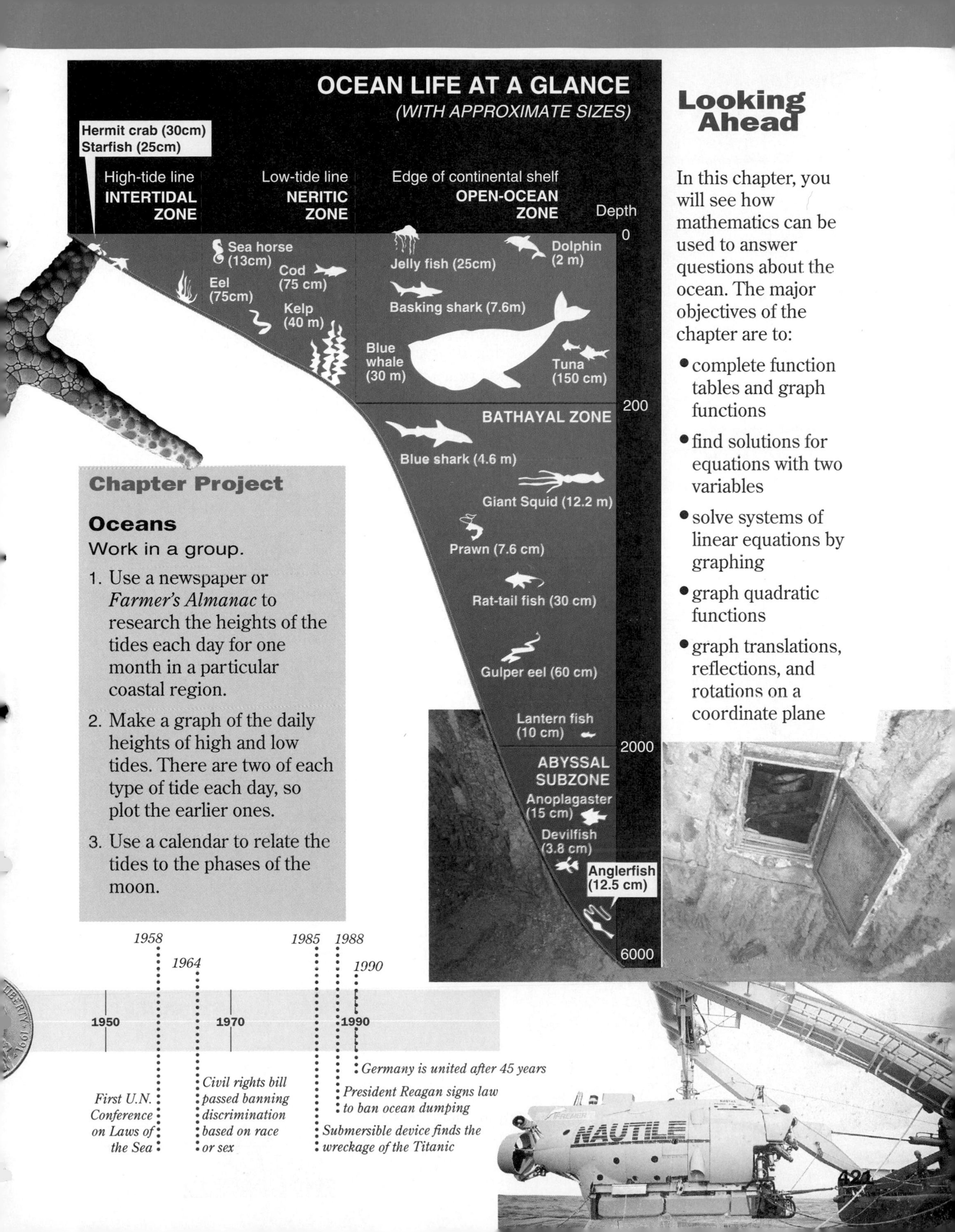

Looking Ahead

In this chapter, you will see how mathematics can be used to answer questions about the ocean. The major objectives of the chapter are to:

- complete function tables and graph functions
- find solutions for equations with two variables
- solve systems of linear equations by graphing
- graph quadratic functions
- graph translations, reflections, and rotations on a coordinate plane

Chapter Project

Oceans

Work in a group.

1. Use a newspaper or *Farmer's Almanac* to research the heights of the tides each day for one month in a particular coastal region.
2. Make a graph of the daily heights of high and low tides. There are two of each type of tide each day, so plot the earlier ones.
3. Use a calendar to relate the tides to the phases of the moon.

1950 — 1970 — 1990

1958 First U.N. Conference on Laws of the Sea

1964 Civil rights bill passed banning discrimination based on race or sex

1985 Submersible device finds the wreckage of the Titanic

1988 President Reagan signs law to ban ocean dumping

1990 Germany is united after 45 years

11-1 Functions

Objective
Complete function tables.

Words to Learn
function
domain
range

Do you have a pet? Nearly one-third of all American households have some sort of pet. The most popular pet among households with children is a dog. To help humans relate to how old their dog is, veterinarians have used the rule that one year of a dog's life is equivalent to seven years of human life. So a dog 10 years old is 70 years old in human years.

The heaviest domestic dog is the St. Bernard, which weighs up to 100 kilograms and stands 70 centimeters high at the shoulders.

For every dog age, there is a corresponding human age. This relationship is an example of a **function.** In a function, one or more operations are performed on one number to get another. Thus, the second number is a function of the first.

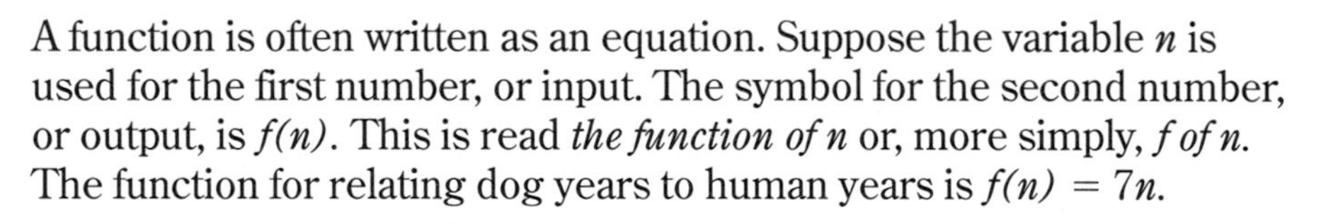

A function is often written as an equation. Suppose the variable n is used for the first number, or input. The symbol for the second number, or output, is $f(n)$. This is read *the function of n* or, more simply, *f of n*. The function for relating dog years to human years is $f(n) = 7n$.

Let's use a table to evaluate the function for equating dog years and human years.

Example 1

Copy and complete the table at the right to find out the human ages of dogs ages 3 through 6.

INPUT dog years	RULE	OUTPUT human years
n	$7n$	$f(n)$
3		
4		
5		
6		

Tables like these are called function tables.

To find each output, substitute each dog's age for n and multiply by 7.

$f(n) = 7n$
$f(3) = 7 \cdot 3$ or 21
$f(4) = 7 \cdot 4$ or 28
$f(5) = 7 \cdot 5$ or 35
$f(6) = 7 \cdot 6$ or 42

INPUT	RULE	OUTPUT
n	$7n$	$f(n)$
3	7(3)	21
4	7(4)	28
5	7(5)	35
6	7(6)	42

The dog ages in human years are 21, 28, 35, and 42.

The set of input values in a function is called the **domain.** The set of output values is called the **range.** So the domain contains all values of n, and the range contains all values of $f(n)$.

Example 2

Make a function table to find the range of $f(n) = 3n + 5$ if the domain is {-2, -1, 0, 3, 5}.

First make the function table. List the values in the domain, the input values.

n	$3n + 5$	$f(n)$
-2		
-1		
0		
3		
5		

Substitute each member of the domain into the function to find the values of $f(n)$, the output values.

n	$3n + 5$	$f(n)$
-2	3(-2) + 5	-1
-1	3(-1) + 5	2
0	3(0) + 5	5
3	3(3) + 5	14
5	3(5) + 5	20

The range is {-1, 2, 5, 14, 20}.

Checking for Understanding

Communicating Mathematics

Read and study the lesson to answer each question.

1. **Tell** another name for the set of input numbers.
2. **Tell** another name for the set of output numbers.
3. **Write** in your own words a definition for function.
4. **Tell** how to find $f(6)$ if $f(n) = 2n^2 - 18$.

Guided Practice

Complete each function table.

5. $f(n) = -5n$

n	$-5n$	$f(n)$
-4	-5(-4)	
-2	-5(-2)	
0		
2.5		

6. $f(n) = 2n + (-6)$

n	$2n + (-6)$	$f(n)$
-2		
-1		
0		
$\frac{1}{2}$		

7. Find $f(-3)$ if $f(n) = -2n - 4$.
8. Find $f\left(\frac{1}{3}\right)$ if $f(n) = 15 - 3n$.
9. Find $f(0.25)$ if $f(n) = 100n$.

Exercises

Complete each function table.

Independent Practice

10. $f(n) = n + 5$

n	$n + 5$	$f(n)$
-2		
-1		
0		
1		
2		

11. $f(n) = 3n$

n	$3n$	$f(n)$
-1		
0		
$\frac{2}{3}$		
1		

12. $f(n) = 2n + 3$

n	$2n + 3$	$f(n)$
-2.5		
-1.5		
0.5		
1		
2		

13. $f(n) = -0.5n + 1$

n	$-0.5n + 1$	$f(n)$
-4		
-2		
0		
2.5		
8		

14. Find $f(-8)$ if $f(n) = 3n + 24$.

15. Find $f\left(\frac{4}{5}\right)$ if $f(n) = -5n - 4$.

16. Find $f(1.5)$ if $f(n) = n^2 + 1$.

17. Find $f(-3.4)$ if $f(n) = 2n^2 - 5$.

Mixed Review

18. **Geometry** Draw two similar right triangles. *(Lesson 5-6)*

19. **Geometry** If the length of the hypotenuse of a 30°–60° right triangle is 12 centimeters, find the length of the side opposite the 30° angle. *(Lesson 8-8)*

20. **Finance** Mrs. Sanchez borrowed $3,200 to build a garage. She made 12 monthly payments of $292 to pay back the simple interest loan. Find the interest on her loan. *(Lesson 10-10)*

Problem Solving and Applications

21. **Critical Thinking** Use $f(n)$ to write the equation that describes the function represented by the table at the right.

n	$f(n)$
-3	2
-1	4
1	6
3	8

22. **Computer Connection** The number of hours each employee works is entered in column A, the hourly wage is entered in column B, and the fee for cleaning uniforms is in column C. Column D records the pay before taxes. The formula in cell D1 is A1 * B1 − C1. The formula acts like the rule of a function. Find the pay for each employee.

	A	B	C	D
1	40	$4.30	$5	
2	32	$5.75	$4	
3	30	$6.00	$6	
4	27	$4.80	$3	

11-2 Graphing Functions

Objective

Graph functions by using function tables.

Sometimes a function table, like the one at the right, shows only *n* and *f(n)* values. This table shows the total cost of 92-octane gas at a station in 1992. Another way to present the input and output values of a function is by using ordered pairs. Then the function can be graphed.

Gallons	Total Cost ($)
n	*f(n)*
1	1.16
2	2.32
3	3.48
4	4.64

To form ordered pairs, let the *x*-coordinate be the value of *n*. The *y*-coordinate is the value of *f(n)*. For the table, the ordered pairs are as follows.

(1, 1.16), (2, 2.32), (3, 3.48), (4, 4.64)

The graph of these ordered pairs is shown at the right. Notice that the *x*-axis is labeled *n* and the *y*-axis is labeled *f(n)*. *What pattern do the points suggest?*

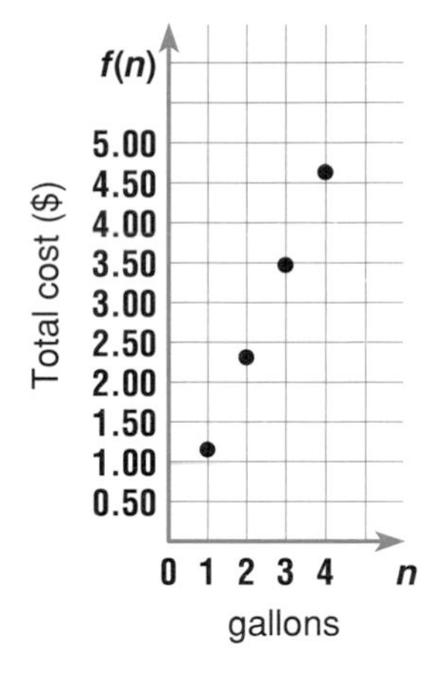

TEEN SCENE

With gasoline prices changing frequently, it is a good idea to estimate the cost and the amount of gasoline you can purchase. It would be embarrassing not to have enough money to pay for a fill up or to run out of gas.

Example 1 ***Problem Solving***

Meteorology In December 1991, a rain system hit central Texas dumping several inches of rain in a few days. The Colorado River crested at over 46 feet, which was 7 feet, or 84 inches, above flood stage. After cresting, the river fell at a rate of 4 inches per hour. If *n* represents the number of hours, $f(n) = 84 - 4n$ represents the height of the river above flood stage after each hour. Draw a graph of this function.

First, make a function table and find the values of *f(n)* for each value of *n* in the table. Then write the ordered pairs.

Graph the ordered pairs. Draw the line the points suggest.

n	84 − 4*n*	*f(n)*	(*n*, *f(n)*)
1	84 − 4(1)	80	(1, 80)
3	84 − 4(3)	72	(3, 72)
5	84 − 4(5)	64	(5, 64)
7	84 − 4(7)	56	(7, 56)

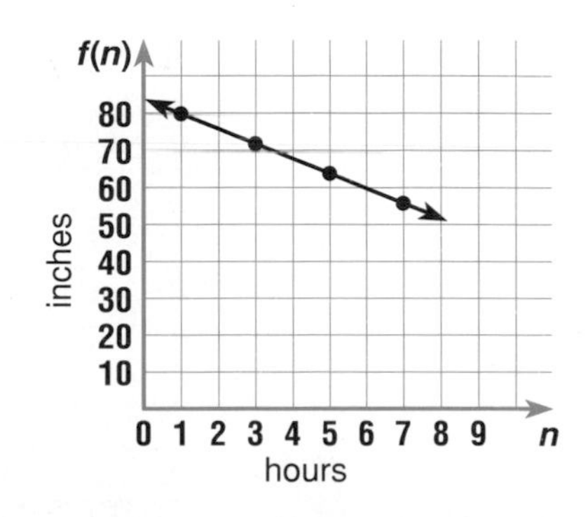

LOOKBACK

You can review graphing points on page 117.

Not all graphs of functions form a line, and the graphs of many functions lie in more than the first quadrant.

Example 2

Problem Solving Hint

Look for a pattern in the positions of the points. Use the pattern to draw the complete graph.

Graph the function $f(n) = n^2$.

Make a table of values for n and find $f(n)$.

n	f(n)	(n, f(n))
-4	16	(-4, 16)
-3	9	(-3, 9)
-1	1	(-1, 1)
0	0	(0, 0)
1	1	(1, 1)
2	4	(2, 4)
4	16	(4, 16)
5	25	(5, 25)

Then graph each ordered pair.

There are infinitely many values possible for n. If you graphed all these values, the points would suggest a smooth curve. Draw the curve suggested by our points.

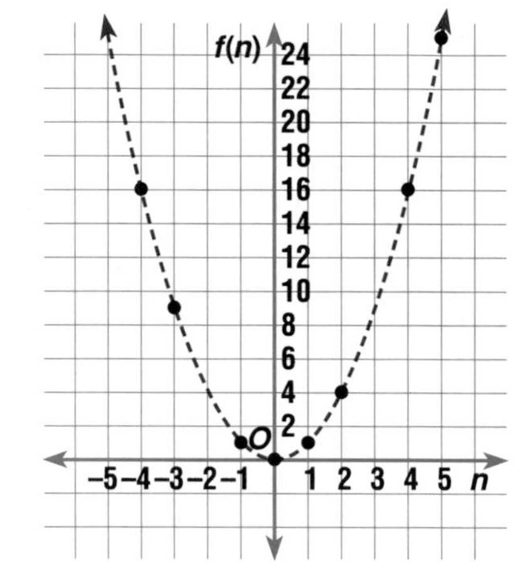

Checking for Understanding

Communicating Mathematics **Read and study the lesson to answer each question.**

1. **Write** how you can use a function table to graph a function.
2. **Tell** what $f(0)$ means in terms of the flood described in Example 1.
3. **Tell** how long it would take the Colorado River to recede at the rate of 4 inches per hour. What type of weather might affect this?
4. **Show** which axes are used for the input numbers and output numbers when graphing a function.

Guided Practice **Copy and complete each function table. Then graph the function.**

5. $f(n) = n + 4$

n	f(n)	(n, f(n))
-1		
1		
2		
4		
5		

6. $f(n) = \frac{8}{n}$

n	f(n)	(n, f(n))
$\frac{1}{2}$		
2		
4		
8		
16		

Exercises

Independent Practice

Complete each function table. Then graph the function.

7. $f(n) = 3n + 1$

n	$f(n)$	$(n, f(n))$
0		
1		
2		
3		

8. $f(n) = 6n$

n	$f(n)$	$(n, f(n))$
-1		
1		
2.5		
3.5		
4		

9. $f(n) = \frac{16}{n}$

n	$f(n)$	$(n, f(n))$
8		
2		
4		
$\frac{8}{3}$		

10. Choose values for n and graph $f(n) = 0.25n + 3$.
11. Choose values for n and graph $f(n) = n^2 + (-2)$.

Mixed Review

12. **Geometry** Find the circumference of a circle with a radius of 2.7 millimeters. *(Lesson 7-8)*
13. Find $f\left(-\frac{3}{4}\right)$ if $f(n) = 8n - 1$. *(Lesson 11-1)*

Problem Solving and Applications

14. **Critical Thinking** Graph the absolute value function $f(n) = |n + 2|$. Describe the shape and range of this graph.
15. **Domestic Engineering** Suppose a coffee pot heats up at a rate of 8° C per minute. The water put into the pot is at 20° C.
    a. Graph the function $f(n) = 8n + 20$ to see how the temperature of the water in the pot rises.
    b. In a coffee pot, does the temperature continue to rise forever? Explain your answer. Does this affect the domain?

16. **Critical Thinking** The Post Office rounds the weight of a letter up to the next ounce when figuring the cost of mailing a letter. The first table below shows the postal rates for first-class mail as of March, 1992.

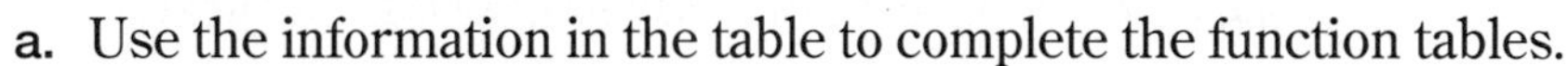

    a. Use the information in the table to complete the function tables.

Oz	Cost
1	29¢
2	52¢
3	75¢
4	98¢

n oz	$f(n)$ cents
0.1	
0.3	
0.6	
1.0	

n oz	$f(n)$ cents
1.1	
1.5	
1.9	
2.0	

n oz	$f(n)$ cents
2.1	
2.4	
2.8	
3.0	

n oz	$f(n)$ cents
3.1	
3.2	
3.7	
4.0	

    b. Graph the ordered pairs that can be formed from these function tables. What pattern do you notice?
    c. This type of function is called a *step function*. Why do you think this name was chosen?

17. **Journal Entry** Does graphing a function help you better understand the real-life situation that the equation describes? Explain.

11-3 Equations with Two Variables

Objective
Find solutions for equations with two variables.

All of the functions we have used in Lessons 11-1 and 11-2 have been written using $f(n)$. Many functions do not use this notation. Instead they use two variables—one represents the input and the other represents the output. You are already familiar with many of these functions.

Changing Celsius to Fahrenheit	$F = \frac{9}{5}C + 32$
Rectangle with area of 64 square units	$64 = \ell w$
Parallelogram with perimeter of 36 units	$36 = 2a + 2b$
The distance you can run at 5 miles per hour	$d = 5t$

When you find values for the two variables that make the equation true, these values form an ordered pair that is a solution of the equation. Most equations with two variables have an infinite number of solutions.

In equations with two variables, one variable represents the input and the other represents the output of the function. When x and y are used, x usually represents the input.

Example 1

LOOKBACK You can review solving equations on page 67.

Find four solutions for $3x - 6 = y$. Write the solutions as a set of ordered pairs.

First make a function table to find the ordered pairs. Use x for n and y for $f(n)$.

Choose any four values for x. Then complete the table.

x	$3x - 6$	y	(x, y)
-2	3(-2) − 6	-12	(-2, -12)
0	3(0) − 6	-6	(0, -6)
1	3(1) − 6	-3	(1, -3)
5	3(5) − 6	9	(5, 9)

The set of ordered pairs that are solutions for an equation is called a solution set for the equation.

Four solutions for the equation $3x - 6 = y$ are {(-2, -12), (0, -6), (1, -3), (5, 9)}.

In Example 1, we found four ordered pairs that were solutions for $3x - 6 = y$. There are many more. *Can you name another?*

Exercises

Independent Practice

Copy and complete the table for each equation.

8. $y = 0.2x + 7$

x	y
-5	
0	
5	
10	

9. $y = \frac{x}{3} + 4$

x	y
-3	
3	
6	
8	

10. $y = -5x - 1$

x	y
-2	
-1	
0	
2	
4	

Find four solutions of each equation.

11. $y = 2x + 1$
12. $y = -x + 1$
13. $y = \frac{x}{2} + 3$
14. $y = 2.5x + 1$
15. $y = -2x + 3$
16. $y = 15x - 57$

17. **Number Sense** *Twice the sum of two numbers is 16.* Determine which ordered pairs from the set {(12, 4), (6, 2), (3, 5), (-45, 53)} are solutions for the verbal sentence.

Mixed Review

18. Use mental math to find 8(35). *(Lesson 1-2)*
19. Express $2.\overline{44}$ as a mixed number in simplest form. *(Lesson 6-7)*
20. Choose values for n and graph $f(n) = \frac{n}{2} - 1$. *(Lesson 11-2)*

Problem Solving and Applications

21. **Number Sense** Write an equation for each verbal sentence. Find three solutions for each equation.
    a. The second number is nine more than the first.
    b. The second number is three more than twice the first.
    c. The sum of two numbers is 0.

22. **Critical Thinking** The equations $y = 3x - 1$ and $6x - 2y = 2$ have the same solution set.
    a. Find four ordered pairs to show this statement is true.
    b. Why does $y = 3x - 1$ seem easier to solve than $6x - 2y = 2$?

23. **Biology** You can tell the approximate temperature in degrees Fahrenheit by counting the chirps a cricket makes in 15 seconds. The rule is $t = c + 40$. Complete the table to find the temperature for each number of chirps.

chirps in 15 seconds (*c*)	12	23	31	47	50
temperature in °F (*t*)					

24. **Data Search** Refer to page 666.
Buyers can usually borrow about $2\frac{1}{2}$ times their annual income for a home mortgage. What amounts could be borrowed at each income level on the graph?

Example 2 *Problem Solving*

Number Sense One number is one less than three times another number. Determine which ordered pairs in the set {(-1, -4), (0, 2), $\left(\frac{1}{3}, 1\right)$, (1.5, 3), (2, 5)} are solutions for the two numbers.

First, translate the verbal sentence to an equation. Let x = one number and y = the other number. The equation that describes the problem is $y = 3x - 1$.

Let's test each ordered pair in the equation.

Test (-1, -4)

$y = 3x - 1$

$-4 \stackrel{?}{=} 3(-1) - 1$ *Replace y with -4 and x with -1.*

$-4 \stackrel{?}{=} -3 - 1$

$-4 = -4$ ✓ (-1, -4) is a solution.

Test (0, 2)

$2 \stackrel{?}{=} 3(0) - 1$

$2 \neq -1$

(0, 2) is not a solution.

Test $\left(\frac{1}{3}, 1\right)$

$1 \stackrel{?}{=} 3\left(\frac{1}{3}\right) - 1$

$1 \neq 0$

$\left(\frac{1}{3}, 1\right)$ is not a solution.

Test (1.5, 3)

$3 \stackrel{?}{=} 3(1.5) - 1$

$3 \neq 3.5$

(1.5, 3) is not a solution.

Test (2, 5)

$5 \stackrel{?}{=} 3(2) - 1$

$5 = 5$ ✓

(2, 5) is a solution.

The ordered pairs (-1, -4) and (2, 5) are solutions.

Checking for Understanding

Communicating Mathematics

Read and study the lesson to answer each question.

1. **Write** a sentence to tell how finding solutions to an equation with two variables is like finding ordered pairs of a function.
2. **Show** how you could use the method in Example 2 to check the solutions in Example 1.

Guided Practice

3. Determine which ordered pairs in the set {(-2, -1), $\left(\frac{1}{2}, 2\right)$, (0, 1), (3, -3)} are solutions for the equation $y = -2x + 3$.

Copy and complete the table for each equation.

4. $y = x + 1$

x	y
-3	
-1	
1	
2	

5. $y = -0.5x$

x	y
-4	
2	
0	
6	

6. $y = 5x - 2$

x	y
$-\frac{4}{5}$	
2	
$\frac{7}{15}$	

7. Find four solutions to $y = x + 3$.

Cooperative Learning

11-4A Graphing Linear Functions

A Preview of Lesson 11-4

Objective

Graph a relationship that can be described by a linear function.

Materials

large rubber band
2 paper clips
ruler
paper cup
marbles

When scientists perform experiments, they often graph the relationships they find and look for a pattern in the points they graph.

Try this

Work in groups of four.

- Punch a small hole in the bottom of the cup. Place one paper clip onto the rubber band. Push the other end of the rubber band through the hole in the bottom of the cup. Attach the other paper clip to act as a hook.

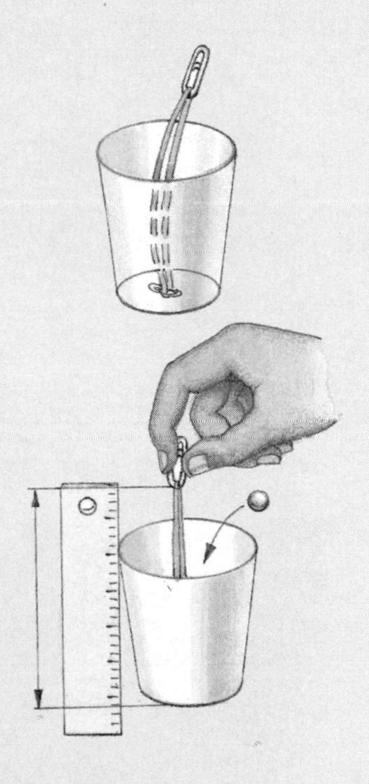

- Have one person hold the suspended cup. Drop one marble into the cup. Record the distance from the base of the paper clip to the bottom of the cup.
- Drop another marble in. Record the distance again.
- Keep dropping marbles in and recording distances until you have 10 sets of distances.

What do you think?

1. Make a table of ordered pairs. Let x represent the number of marbles and y represent the distance from the paper clip to the bottom of the cup.
2. Suppose you were to graph these ordered pairs. Make a guess about the pattern they might show. Then graph the ordered pairs.
3. Was your guess correct? What pattern do the points seem to suggest?

LOOKBACK

You can review scatter plots on page 157.

Extension

4. **Statistics** Make a scatter plot by graphing the data from all groups in your class. See if you can find a linear pattern. In statistics, this line is called the *best-fit line.* The line does not go through all the points. It is used to write an equation to generalize a set of data.

11-4 Graphing Linear Functions

Objective

Graph linear functions by plotting points.

Words to Learn

linear function

How long will the batteries last in this toy? Battery companies spend millions of dollars each year to convince you that their batteries last the longest.

The formula, $d = rt$, where d is the distance, r is the rate, and t is the time, can be used to find how far this toy will go. Suppose this toy travels in a straight line at a rate of 0.5 feet per second. By substituting 0.5 for r, the formula becomes $d = 0.5t$.

Technology Activity

You can use a graphing calculator to explore graphs of linear equations in Technology Activity 6 on page 663.

Mini-Lab

Draw a graph of all ordered pairs that are solutions of the equation $d = 0.5t$.

Materials: graph paper

- Make a function table to find ten solutions for the equation.

t	d	(t, d)
3	1.5	(3, 1.5)
5	2.5	(5, 2.5)
6	3	(6, 3)
10	5	(10, 5)

- The solutions include the ordered pairs, (3, 1.5), (5, 2.5), (6, 3), (10, 5), and so on. *(6, 3) means the toy travels 3 feet in 6 seconds.*
- Graph all the ordered pairs and look for a pattern in the points. *Since all the coordinates are positive, you need only show the first quadrant of the coordinate plane.*

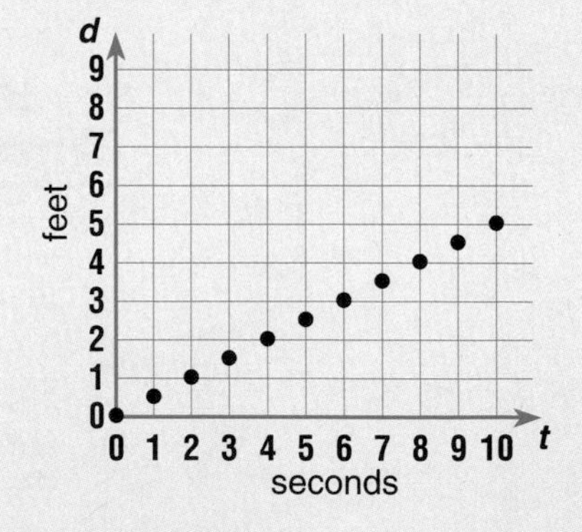

Talk About It

a. What figure is suggested by these ten points?

b. Sketch this figure.

c. Find four more solutions. Do they lie on the figure you sketched?

d. What conclusion could you make about other solutions of $d = 0.5t$?

In the Mini-Lab, you saw that all solutions of the equation suggested a straight line. An equation in which the graphs of the solutions form a line is called a **linear function.**

The graphs of linear functions that interpret real-life activities often lie in the first quadrant. However, most linear functions lie in at least two quadrants of the coordinate plane.

Estimation Hint

Before graphing a function, look at your ordered pairs. Estimate how long each axis should be. The graph may be more manageable if each unit on the coordinate plane represents 2, 3, 5, or some greater number, instead of 1.

Example

Graph $y = -3x + 10$.

First make a function table. List at least three values for x.

x	y	(x, y)
-3	19	(-3, 19)
0	10	(0, 10)
3	1	(3, 1)

Graph the ordered pairs. Connect them with a line. Put arrows on the ends of the line to show that the line continues indefinitely.

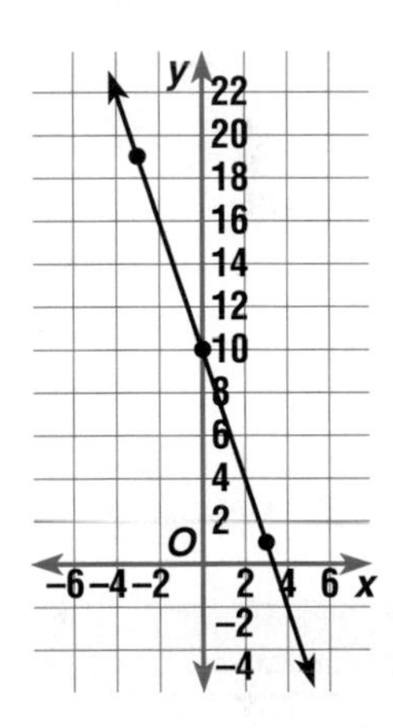

Checking for Understanding

Communicating Mathematics

Read and study the lesson to answer each question.

1. **Write** a sentence explaining why the graph of a linear function is a line.
2. **Tell** why you should graph at least three ordered pairs when only two points are needed to draw the line.
3. **Show** how you would label the axes to graph the ordered pairs (10, 50), (20, 100), (30, 150), (40, 200).

Guided Practice

Copy and complete each function table. Then graph the function.

4. $y = 5x$

x	y	(x, y)
-4		
-1		
0		
2		

5. $y = 25 - x$

x	y	(x, y)
-5		
0		
5		
10		

6. Graph $y = 3x + 1$.

7. Graph $y = 15 - 3x$.

Exercises

Independent Practice

Graph each function.

8. $y = 8x$
9. $y = -4x$
10. $y = 5x + 3$
11. $y = 6x + 5$
12. $y = 5x - 10$
13. $y = 1.5x + 2.5$
14. $y = \frac{x}{3} + 5$
15. $y = 6 - \frac{x}{2}$
16. $y = 75 - 3x$

Mixed Review

17. **Algebra** Write *six more dollars than Atepa has* as an algebraic expression. *(Lesson 2-6)*

18. **Geometry** Make an Escher-like drawing for the pattern shown at the right. Use two rows of three squares as your base. *(Lesson 5-7)*

19. **Geometry** Find the area of a triangle whose base is 2.5 inches and whose height is 3.5 inches. *(Lesson 7-7)*

20. Find four solutions for $y = \frac{x}{3} + 8$. Write the solution set. *(Lesson 11-3)*

Problem Solving and Applications

21. **Weather** Suppose you didn't know the formula for changing temperatures in degrees Celsius to degrees Fahrenheit, but you could remember that the freezing points for water are 0° C and 32° F and the boiling points are 100° C and 212° F.
    a. Write two ordered pairs that relate Celsius and Fahrenheit.
    b. Graph the linear function that contains these two points.
    c. How could you use this graph to find other equivalents of other Celsius and Fahrenheit temperatures?

22. **Physics** When there is a storm, you see the lightning before you hear the thunder. If you see a flash of lightning and hear the thunder 5 seconds later, the lightning is about 1 mile away. If t is the time in seconds and d is the distance in miles, the function that describes this relationship is $t = 5d$. Graph this function.

23. **Critical Thinking** Graph the functions $y = 5x$ and $y = 5x + 5$ on the same coordinate plane. Describe what type of lines these are.

24. **Journal Entry** Write a few sentences that tell an advantage to looking at the graph of a function rather than just looking at its table of values.

11-5 Graphing Systems of Equations

Objective

Solve systems of linear equations by graphing.

Words to Learn

system of equations

Miss Taggers is the eighth grade girls' basketball coach at Union Middle School. She asked the video club to video tape all the home games. The club needs to rent a camcorder while theirs is being repaired.

Video Town charges a \$30 rental fee plus \$35 per day. All-Pro Rental Services charges a \$45 rental fee and \$30 per day for the same camcorder. Which company should they rent from?

You can make a graph of the information to help you answer this qucstion. Let x represent the number of days for each rental. Let y represent the total cost for the rental. An equation can be written for each store.

Video Town: $y = 30 + 35x$ **All-Pro Rental:** $y = 45 + 30x$

Example 1 *Problem Solving*

Smart Shopping **Use a graph to determine which store offers the better deal in renting a video camcorder.**

Graph both equations on the same coordinate plane. You can use a calculator to quickly find the y-values for each x.

For $x = 4$, in $y = 30 + 35x$: 30 [+] 35 [×] 4 [=] 170

Video Town $y = 30 + 35x$	
x	y
1	65
2	100
3	135
4	170

All-Pro $y = 45 + 30x$	
x	y
1	75
2	105
3	135
4	165

Now graph each function. Since the y values are great, let each unit on the y-axis equal \$10.

The graphs intersect at (3, 135).

Up to a 3-day rental, Video Town is cheaper.

For a 3-day rental, they cost the same.

For more than a 3-day rental, All-Pro is cheaper.

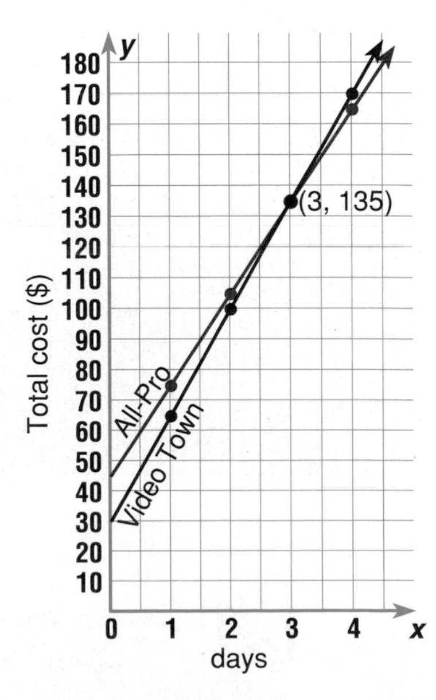

Calculator Hint

It is sometimes helpful to use the memory key [STO], or [M+], when evaluating equations with great numbers or decimals. In the equation $4{,}235x + 26 = y$, you can store 4,235 in the memory and use the [RCL] key to enter it instead of keying in all four digits each time.

In Example 1, the graph of the ordered pair (3, 135) is a point on both lines. This means (3, 135) is a solution to both equations. When you find a common solution for two or more equations, you have solved a **system of equations.** The ordered pair for the point where the graphs of the equations meet is the solution for the system of equations. To check a solution for a system, you must check that solution in *each* of the equations of the system.

Example 2

Graph the system of equations $y = 2x + 1$ and $y = -x + 7$. Then find the solution to the system.

Make a table for each equation.

$y = 2x + 1$	
x	y
-1	-1
0	1
3	7

$y = -x + 7$	
x	y
-2	9
0	7
2	5

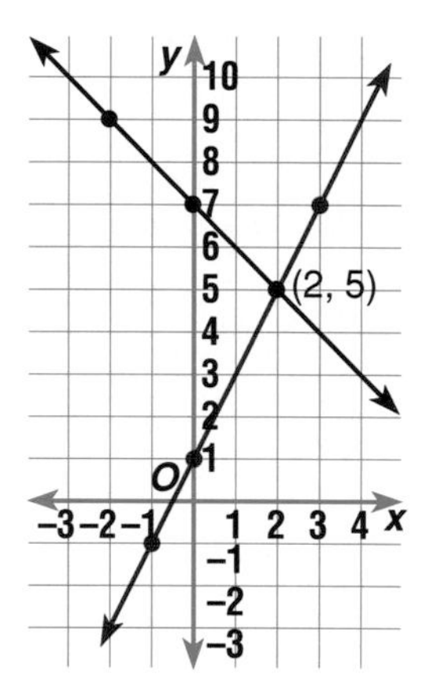

Graph each equation. Locate the point where they intersect.

The lines intersect at the point whose coordinates are (2, 5). So, the solution for the system of equations is (2, 5).

Check your solution in *both* equations.

$y = 2x + 1$
$5 \stackrel{?}{=} 2(2) + 1$ $x = 2, y = 5$
$5 \stackrel{?}{=} 4 + 1$
$5 = 5$ ✓

$y = -x + 7$
$5 \stackrel{?}{=} -(2) + 7$ $x = 2, y = 5$
$5 \stackrel{?}{=} -2 + 7$
$5 = 5$ ✓

The solution checks in both equations.

Mini-Lab

Work with a partner. Graph the system of equations $y = x + 3$ and $y = x - 2$ on the same coordinate plane.
Materials: graph paper

- Make a table for each equation.
- Graph the ordered pairs and draw the lines.

Talk About It

a. What type of lines are these?
b. Where do these lines meet?
c. What can you conclude about the solution to this system of equations?

Checking for Understanding

Communicating Mathematics

Read and study the lesson to answer each question.

1. **Tell** what is meant by a system of equations.
2. **Write** a sentence to describe the solutions for a system of equations.
3. **Show** why (1, 3) is the solution for the system $x + y = 4$ and $2x - y = -1$.
4. **Draw** the graphs of a system of equations that has no solution.
5. **Tell** the solution for the system of equations at the right.

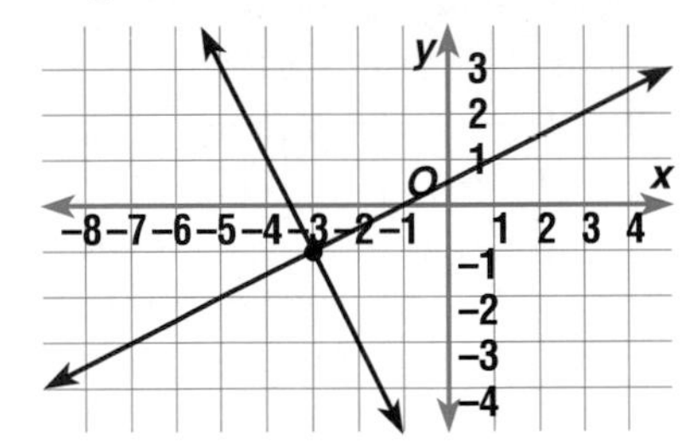

Guided Practice

Lines *a*, *b*, *c*, and *d* are graphs of four equations. Use the graphs to find the solution for each system of equations.

6. equations a and b
7. equations a and c
8. equations a and d
9. equations b and c
10. equations b and d
11. equations c and d

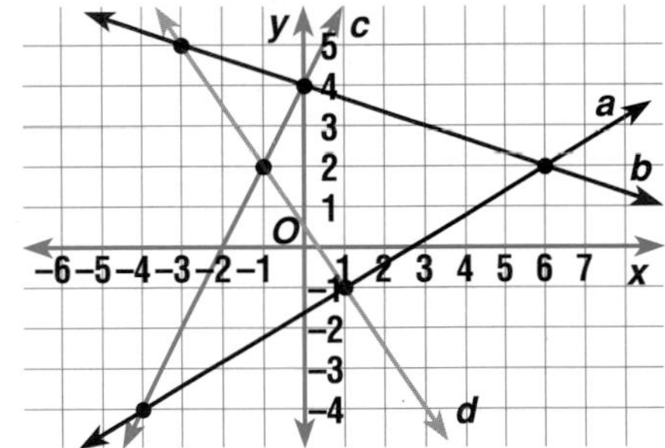

12. Solve the system $y = 3 + x$ and $y = 2x + 5$ by graphing.
13. Solve the system $y = 3x + 1$ and $y = 3x - 1$ by graphing.

Exercises

Independent Practice

Solve each system of equations by graphing.

14. $y = x - 4$
$y = -4x + 16$
15. $y = x$
$y = -x + 4$
16. $y = 2x$
$y = -2x + 4$
17. $y = 4x - 15$
$y = x + 3$
18. $y = -2x - 6$
$y = -2x - 3$
19. $y = -x - 1$
$y = -2x + 4$
20. Find the solution for the system $y = 4x - 13$ and $y = -\frac{1}{2}x + 5$.
21. Find the solution for the system $2x + y = 5$ and $x - y = 1$.

Mixed Review

22. **Statistics** Determine whether a scatter plot of the speed of a car and the stopping distance would show a positive, negative, or no relationship. *(Lesson 4-8)*
23. Replace ● with >, <, or = in $\frac{5}{6}$ ● $\frac{7}{9}$. *(Lesson 6-10)*
24. Find the best integer estimate for $\sqrt{60}$. Then check your estimate by using a calculator. *(Lesson 8-2)*
25. Graph $y = 3x + 1$. *(Lesson 11-4)*

Problem Solving and Applications

26. **Home Maintenance** Mr. Gill needs some rewiring done in his garage. One electrician he called charges a $35 house call fee plus $25 per hour for labor. A second electrician charges a $20 house call fee plus $30 per hour for labor.
    a. Graph the equations $C = 25t + 35$ and $C = 30t + 20$ to represent the fees charged by each electrician.

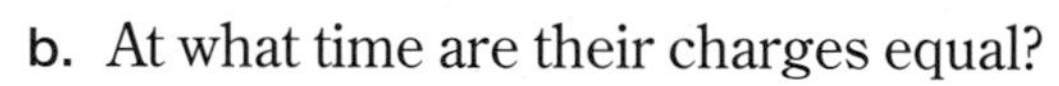

    b. At what time are their charges equal?
    c. Mr. Gill estimates the job will take 16 hours to complete. Which electrician should he use?

27. **Geometry**
    a. Graph the equations $y = 2x + 5$ and $y = -\frac{1}{2}x + 10$.
    b. What is the solution of this system of equations?
    c. Measure the angles formed by these two lines. What type of lines are they?

28. **Critical Thinking**
    a. Graph $2x + y = 6$ and $4x + 2y = 12$.
    b. What do you notice about these two lines?
    c. What do you think is the solution of this system?

11 Assessment: Mid-Chapter Review

1. Make a function table for $f(n) = -2n + 5$. *(Lesson 11-1)*
2. Use the table in Exercise 1 to graph $f(n) = -2n + 5$. *(Lesson 11-2)*
3. Find four solutions of $y = \frac{x}{3} + 4$. *(Lesson 11-3)*
4. Graph $y = 10 - 2x$. *(Lesson 11-4)*
5. What is the solution of the system of equations graphed at the right? *(Lesson 11-5)*

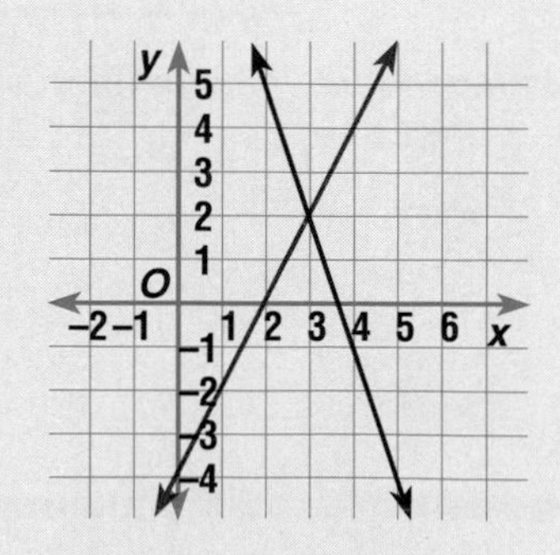

Problem-Solving Strategy

11-6 Use a Graph

Objective

Solve problems by using a graph.

Mrs. Degas works as a consultant for mass media corporations. She makes recommendations to companies about new products and trends in television technology. During a recent presentation, she used the graph below to show the amount of change in TV gadgets over a 45-year period. How long did it take for color televisions to become a part of almost all the homes in America?

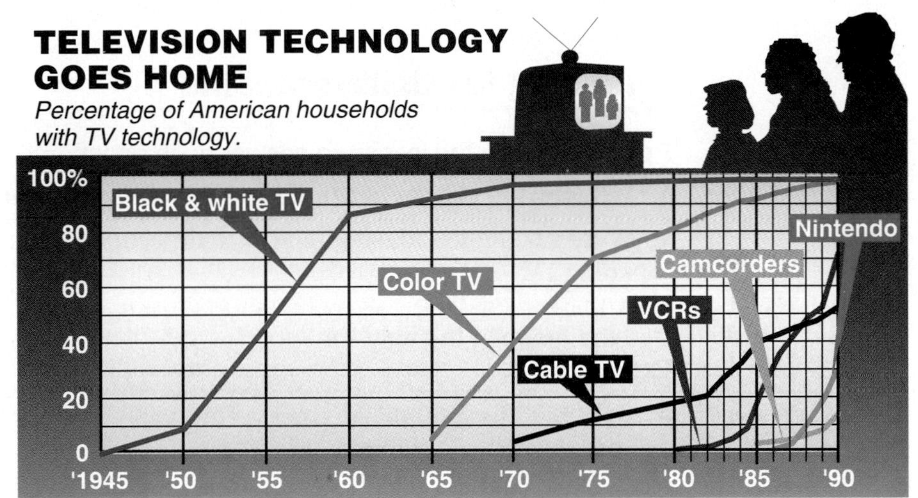

Graphs are often used to do comparison shopping. You may use a graph to decide which bank offers the best deal on a credit card or short term loan.

Explore

What do you know?

The graph shows the percentage of American households with TV technology.

The graph also shows the beginning years for each type of TV technology.

What do you need to find out?

How long did it take for color television to become a part of almost all American households?

Plan

Find when color television first appeared.

Follow the line from that point to the year where the line nears 100%. The number of years will be the difference of the dates for the two points.

Solve

The line for color television begins at 1965. It continues upward until it levels off near 100% at 1989.

Subtract these two years: $1989 - 1965 = 24$

It took 24 years for color TV to become a part of almost all American households.

Example

A client asks Ashley how many years it took for VCRs to become as popular as cable television. How do you think she should respond?

The point where VCRs became as popular as cable television is the point where their graphs intersect, which is about 1988. Even though cable television existed 10 years before VCRs, use the first year they both existed, 1980. To find how long this took, subtract 1980 from 1988.

$$1988 - 1980 = 8$$

It took about 8 years for VCRs to become as popular as cable television.

Checking for Understanding

Communicating Mathematics

Read and study the lesson to answer each question.

1. **Tell** why the graphs of the different technologies are not straight lines.
2. **Write** a sentence to explain how to determine when half of the households had color TV.

Guided Practice

Use the graph at the right for Exercises 3-5.

3. Estimate the amount of money banks lent businesses in April 1991.
4. In which month did business loans see its biggest increase?
5. Describe the overall trend in bank lending.

BANKING, 1991

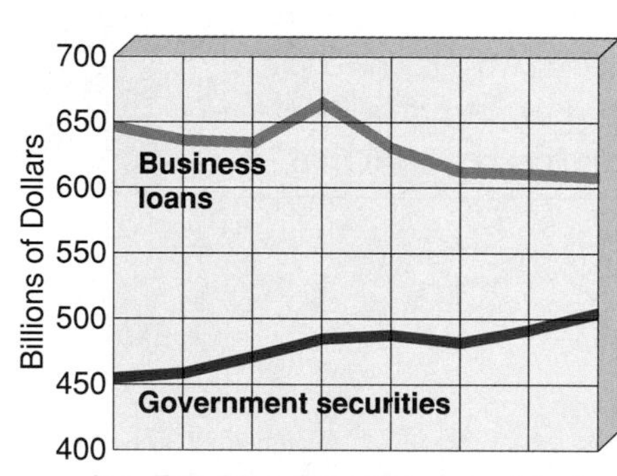

Problem Solving

Practice

Solve using any strategy.

6. Which graph below shows a consistent growth in profits made from carnation sales over the last 14 Valentine's Days? Explain your answer.

a. 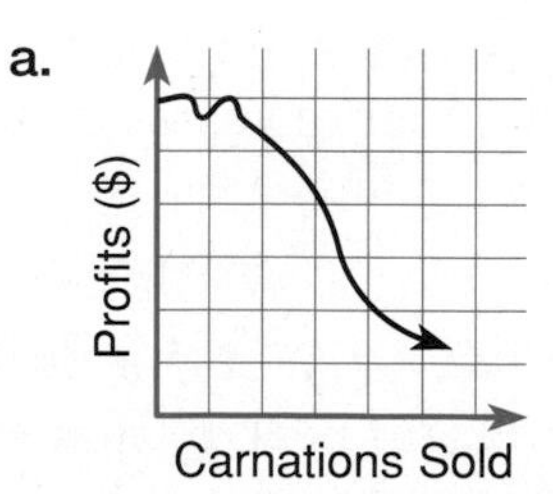

b. 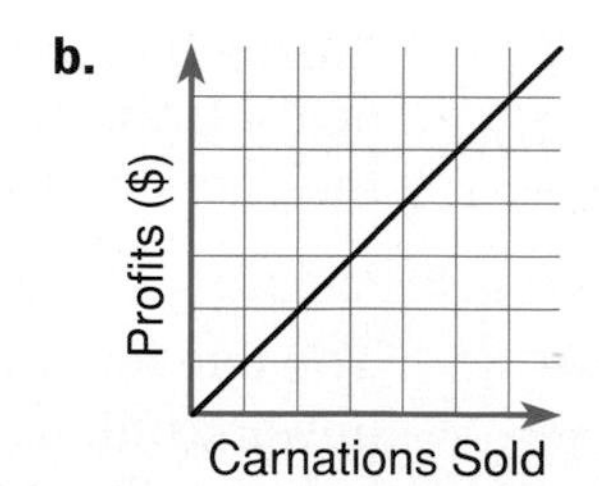

c. 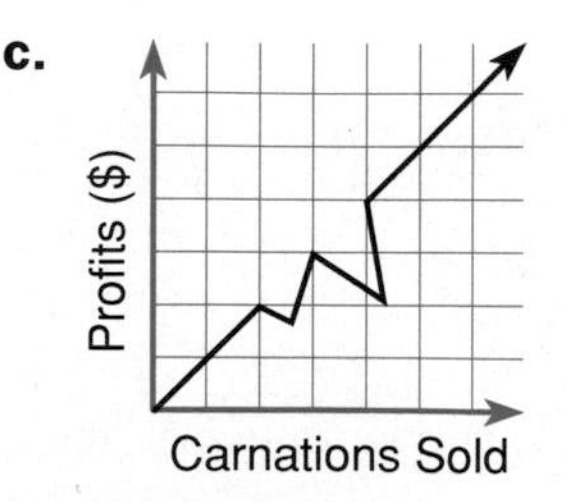

Strategies

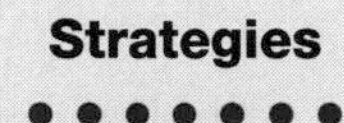

Look for a pattern.
Solve a simpler problem.
Act it out.
Guess and check.
Draw a diagram.
Make a chart.
Work backward.

7. A number is doubled and then -19 is added to it. The result is 11. What is the number?

8. Risa, James, and Corrine each have different stones in their class rings. Their stones are onyx, sapphire, and amethyst. James's stone is not amethyst. Risa wishes hers was onyx. Corrine's stone is blue. What kind of stones does each student have?

Use the graph below for Exercises 9-11.

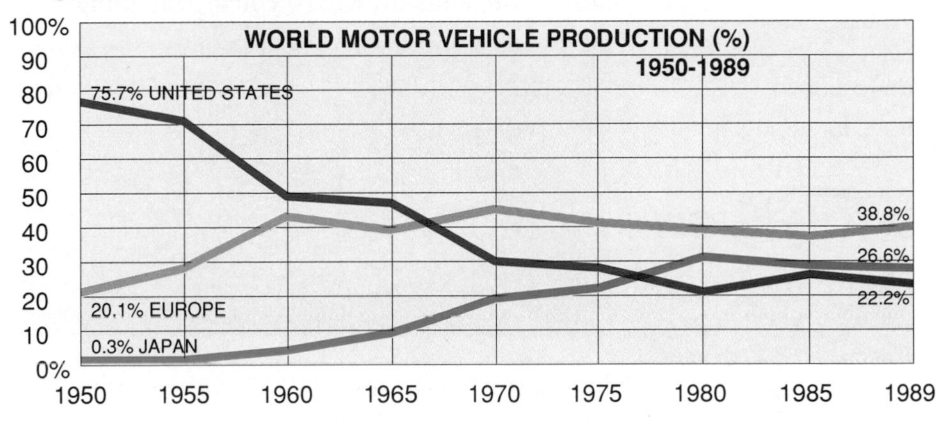

9. During which 5-year period did the United States' percent of motor vehicle production drop most significantly?
10. Estimate in what year Japanese production of motor vehicles first equaled production in the United States.
11. Describe the trend in motor vehicle production suggested by this graph.

12. **Data Search** Refer to pages 420 and 421.
    Make a scatter plot of the average size of marine organisms versus the depth of the water in which they are found.

SAVE THE PLANET • SAVE THE PLANET •

Save Planet Earth

Appliances in the Home Energy specialists agree that we can have a significant impact on the environment if we maintain major appliances like refrigerators, air conditioners, and washing machines to make sure they are kept in peak working condition.

How you can help

- Use the microwave or toaster oven instead of heating up the oven in the stove. Both are more energy-efficient than a conventional oven.
- Don't switch your air conditioner to a colder setting when you turn it on. It won't cool the room any faster, but it will waste energy.
- Since washers use 32 to 59 gallons of water per cycle, you'll save water if you wait until you have a full load of wash to do. If you must do a smaller load, choose the appropriate water level.

11-7 Graphing Quadratic Functions

Objective
Graph quadratic functions.

Words to Learn
quadratic function

Eyeglasses are worn by 431 of every 1,000 boys and men and 509 of every 1,000 girls and women. By the time they are 45 years old, 883 of every 1,000 Americans wear glasses or contact lenses.

If an oil tanker is 4 miles from shore, will you be able to see it? The answer depends on where you are standing.

In order to see an object *d* miles from shore, the eyes of a person with 20/20 vision must be at a certain height, $f(d)$. This can be described by the function $f(d) = \frac{2}{3}d^2$. Since the greatest power in this function is 2, it is called a **quadratic function.**

If you wanted to know how high your eye level needs to be for every mile you wanted to see from shore, you could evaluate $f(d)$ for several values of *d*. A better way is to graph the function. To graph a quadratic function, follow the same steps you used to graph a linear function.

Example 1 ***Problem Solving***

Physics **Graph $f(d) = \frac{2}{3}d^2$ to find the height your eyes need to be to see an oil tanker 4 miles from shore.**

Make a function table.

d	f(d)	(d, f(d))
1	$\frac{2}{3}(1) \approx 0.7$	(1, 0.7)
2	$\frac{2}{3}(2)^2 \approx 2.7$	(2, 2.7)
3	$\frac{2}{3}(3)^2 = 6$	(3, 6)
4	$\frac{2}{3}(4)^2 \approx 10.7$	(4, 10.7)

Graph the ordered pairs.

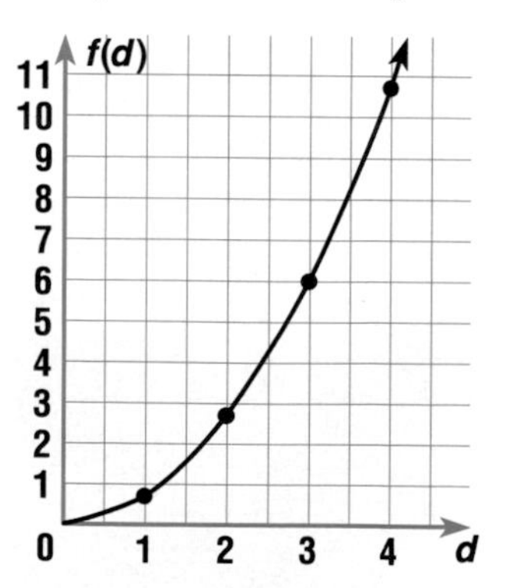

Notice that the points suggest a curve. Sketch a smooth curve to connect the points. The height needed to see an oil tanker 4 miles from shore is about 11 feet.

Calculator Hint

Use the x^2 key on your calculator to quickly find values of A. For $s = 1.5$, enter 1.5 x^2.

Example 2 *Connection*

Geometry **The area of a square is found by using the formula $A = s^2$. Graph this quadratic function to estimate the area of a square whose side is 3.5 units long.**

Make a table. Then graph the ordered pairs.

s	*A*	*(s, A)*
1	1	(1, 1)
1.5	2.25	(1.5, 2.25)
2	4	(2, 4)
2.5	6.25	(2.5, 6.25)
3	9	(3, 9)
4	16	(4, 16)

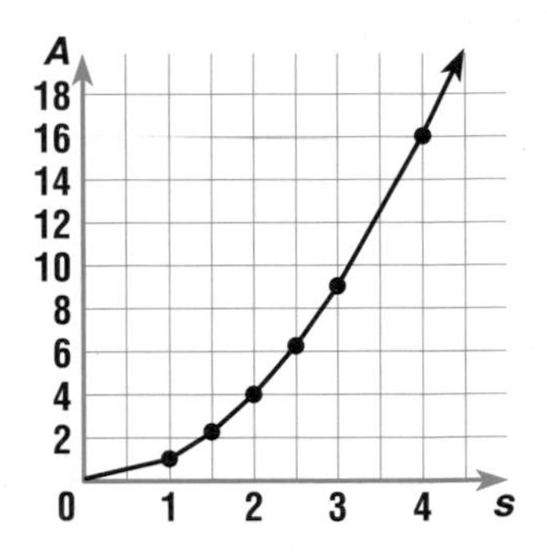

Graph the ordered pairs. Draw a smooth curve to connect the points.

The area of a square with sides 3.5 units long is about 12 square units.

The graphs of many quadratic functions used in real-life applications are only in the first quadrant. Actually, the graph of a quadratic function or equation can be in any quadrant.

Example 3

Graph $y = -2x^2 + 10$.

Make a table. Graph the ordered pairs.

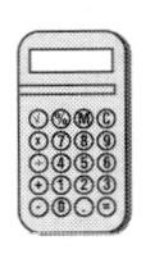

x	$-2x^2 + 10$	*y*	*(x, y)*
-2	$-2(-2)^2 + 10$	2	(-2, 2)
-1.5	$-2(-1.5)^2 + 10$	5.5	(-1.5, 5.5)
-1	$-2(-1)^2 + 10$	8	(-1, 8)
0	$-2(0)^2 + 10$	10	(0, 10)
1	$-2(1)^2 + 10$	8	(1, 8)
2	$-2(2)^2 + 10$	2	(2, 2)
2.5	$-2(2.5)^2 + 10$	-2.5	(2.5, -2.5)

The graphs of the ordered pairs suggest a downward curve. Connect the points with a smooth curve.

A graph that has this shape is called a parabola. A parabola can also curve upward, to the left, or to the right.

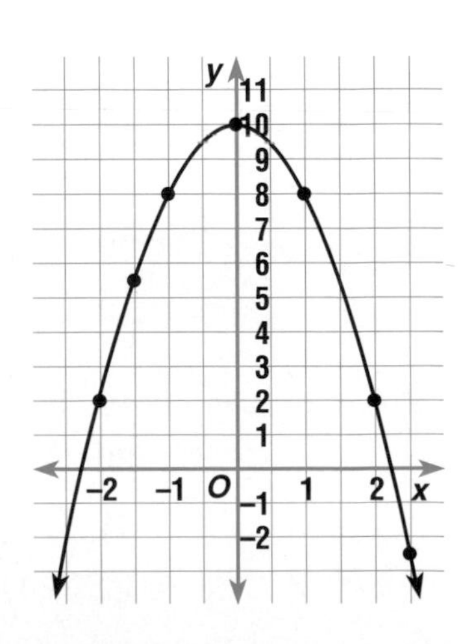

Checking for Understanding

Communicating Mathematics

Read and study the lesson to answer each question.

1. **Tell** what the greatest power is in a quadratic function.
2. **Draw** the general shape of the graph of a quadratic function.
3. **Write** a sentence to explain why the graphs in Examples 1 and 2 only used the first quadrant of the coordinate plane.

Guided Practice

Complete each function table. Then graph the function.

4. $f(x) = x^2$

x	f(x)	(x, f(x))
-2		
-1		
0		
1		
2		

5. $y = x^2 - 1$

x	y	(x, y)
-2		
-1		
0		
1		
2		

6. $y = 2x^2 + 1$

x	y	(x, y)
-2		
-1.5		
0		
3		
4		

7. $f(x) = -x^2$

x	f(x)	(x, f(x))
-3		
-1		
0		
1.5		
2		

Exercises

Independent Practice

Graph each quadratic function.

8. $f(n) = 3n^2$
9. $f(n) = 5n^2 + 1$
10. $f(n) = -n^2$
11. $y = -2x^2$
12. $y = \frac{1}{2}x^2 + 2$
13. $y = 1.5x^2 - 1$
14. $y = -1.5x^2 - 1$
15. $f(n) = 10 - n^2$
16. $y = x^2 + x$
17. Determine which ordered pairs from the set {(-2, 8), (-1, -7), (0, -4), (1, -1), (2, -8)} are solutions for $y = 3x^2 - 4$.
18. Determine which ordered pairs from the set {(-2, 2), (-1.5, -3.75), (2, 0), (3.5, -8.75), (0, 0)} are solutions for $f(x) = -x^2 + x$.

Mixed Review

19. Evaluate $6a^2 + b$ if $a = 4$ and $b = 1$. *(Lesson 2-1)*
20. Solve $j = -496 \div 16$. *(Lesson 3-7)*
21. Evaluate $a + b - c$ if $a = \frac{1}{2}$, $b = 4\frac{3}{5}$, and $c = 1\frac{3}{4}$. *(Lesson 7-2)*
22. Solve the system $y = -x + 5$ and $y = 2x - 4$ by graphing. *(Lesson 11-5)*

Problem Solving and Applications

23. **Critical Thinking** Study your graphs in Exercises 8–16.
    a. Which graphs turn upward and which turn downward?
    b. Study the quadratic equations. Describe how you can look at the equation and tell if the graph turns upward or downward without graphing it.

24. **Physics** If a ball is thrown into the air, its height (in feet) at time t (in seconds) is given by the function $h(t) = 96t - 16t^2$. Make a graph of this function for the first six seconds a ball is in the air.

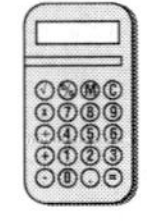

25. **Business** The Solar Heating Company makes circular thermal blankets for swimming pools. The area of the material in each blanket can be described by the function $f(r) = \pi(r + 0.5)^2$, where r is the radius in feet. Graph this function.

26. **Geometry** The area of a square can be described by the function $f(s) = s^2$. The perimeter of a square can be described by the function $f(s) = 4s$.
    a. Graph these two functions on the same coordinate plane.
    b. When do the area and the perimeter have the same measure?

27. **Critical Thinking**
    a. Graph $y = x^2$, $y = \frac{1}{2}x^2$, and $y = 2x^2$ on the same coordinate plane.
    b. How do the graphs of $y = \frac{1}{2}x^2$ and $y = 2x^2$ differ from the graph of $y = x^2$?
    c. Graph $y = x^2 + 2$. How does it differ from the graph of $y = x^2$?
    d. Make a statement about how numbers multiplied by x^2 or added to x^2 affect the position of the graph.

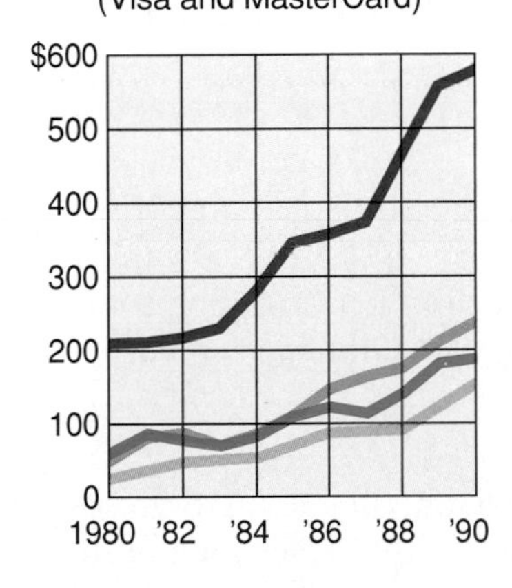

28. **Mathematics and Consumer Credit** Read the following paragraphs.

> One type of consumer loan is the installment loan. These loans are repaid in two or more payments and usually include car loans, home equity loans, personal loans, and credit card purchases.
>
> Although credit card spending makes shopping easier, it is also a very expensive way to shop if you pay installments. Finance charges on credit cards generally run higher than those of personal and business loans, and their annual percentage rate can go as high as 23%.

Credit card spending has increased dramatically since 1980.
   a. Use the graph at the left to determine the increase of total credit card spending from 1980 to 1990.
   b. During what years did bank card spending begin to increase?

Geometry Connection

11-8 Translations

Objective

Graph translations on a coordinate plane.

Words to Learn

translation

Most adventure movies today involve the use of special effects. These special effects are an application of *motion geometry.* Filmmakers use computers to move objects to create different illusions. In Chapter 5, you learned that a sliding motion in a certain direction is called a **translation.**

Mini-Lab

Work with a partner.
Materials: graph paper, straightedge, colored pencils

LOOKBACK

You can review translations on page 202.

- Graph $\triangle ABC$ with vertices $A(-7, 3)$, $B(-2, 2)$, and $C(-4, 5)$ on a coordinate plane.

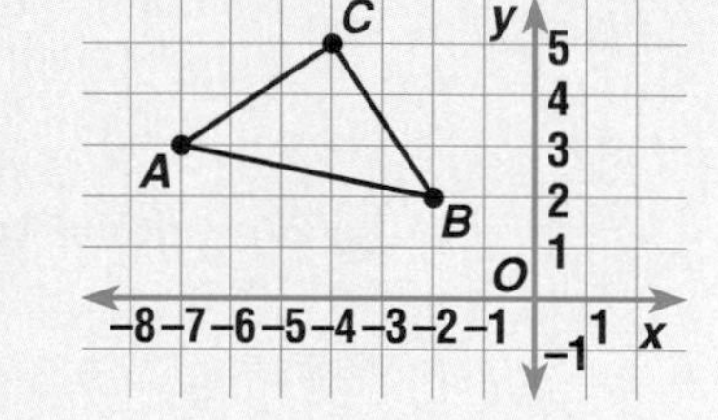

- Trace $\triangle ABC$ and cut it out.
- Place the cutout over $\triangle ABC$. Then translate the cutout 4 units left. Then translate it 3 units down. Trace the cutout with a colored pencil.

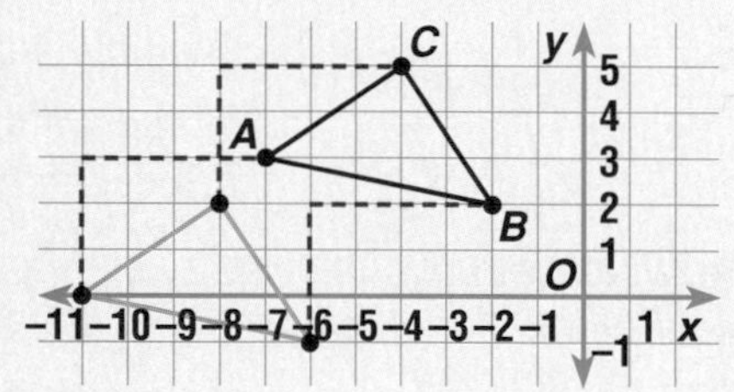

Talk About It

a. Name the three ordered pairs that describe the location of the three vertices after the translation.

b. Compare the coordinates of the vertices of the translation to the coordinates of the original vertices. What do you notice?

If you think of movements in terms of positive and negative, to the left or down is negative, and to the right or up is positive. If we look at the results from the Mini-Lab, the translation can be written as $(-4, -3)$. If we add -4 to the x-coordinate and -3 to the y-coordinate of each vertex of the original triangle, the results are the coordinates of the vertices of the translated triangle.

Translation	**In words:** To translate a point as described by an ordered pair, add the coordinates of the ordered pair to the coordinates of the point.
	Arithmetic (2, 3) moved (1, 1) becomes (3, 4) **Algebra** (x, y) moved (a, b) becomes $(x + a, y + b)$

Example 1

The vertices of $\triangle RST$ are $R(-2, 1)$, $S(1, 3)$, and $T(3, 0)$. Graph $\triangle RST$. Then graph the triangle after a translation 5 units right and 2 units up.

The location of a point after being moved is often written using a prime. So the new coordinates of R are written as R', S as S', and T as T'. *R' is read R prime.*

vertex		*5 right, 2 up*		*translation*
$R(-2, 1)$	+	(5, 2)	→	$R'(3, 3)$
$S(1, 3)$	+	(5, 2)	→	$S'(6, 5)$
$T(3, 0)$	+	(5, 2)	→	$T'(8, 2)$

The coordinates of the vertices are $R'(3, 3)$, $S'(6, 5)$, and $T'(8, 2)$.

Graph R', S', and T' and draw the triangle.

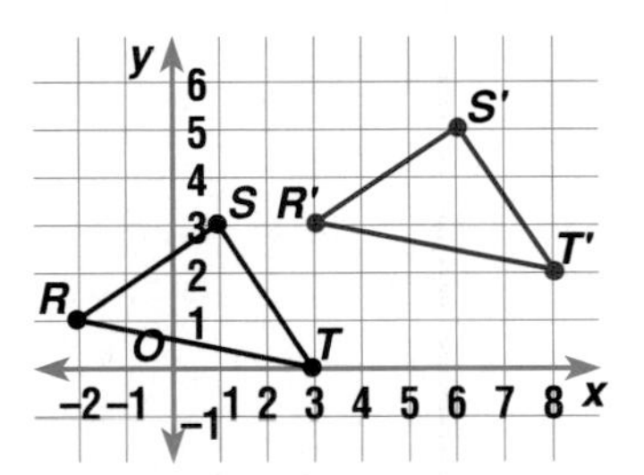

Sometimes you may need to know how a figure was moved. You can use your knowledge of translations and algebra to find the ordered pair that describes the translation.

Example 2 *Connection*

Algebra Rectangle *RICK* has vertices $R(1, 3)$, $I(4, 0)$, $C(3, -1)$, and $K(0, 2)$. Use an equation to find the ordered pair that describes the translation if R' has coordinates (4, -1). Then graph rectangle $R'I'C'K'$.

Let (a, b) represent the ordered pair for the translation.

We know that $R(1, 3) + (a, b) \rightarrow R'(4, -1)$. Let's write an equation for each coordinate.

x-coordinate	*y-coordinate*
$1 + a = 4$	$3 + b = -1$
$a = 3$	$b = -4$

A positive value for a means move 3 units right. A negative value for b means move 4 units down. The ordered pair for the translation is (3, -4).

Now use (3, -4) to find the coordinates of I', C', and K'.

$I(4, 0) + (3, -4) \rightarrow I'(7, -4)$
$C(3, -1) + (3, -4) \rightarrow C'(6, -5)$
$K(0, 2) + (3, -4) \rightarrow K'(3, -2)$

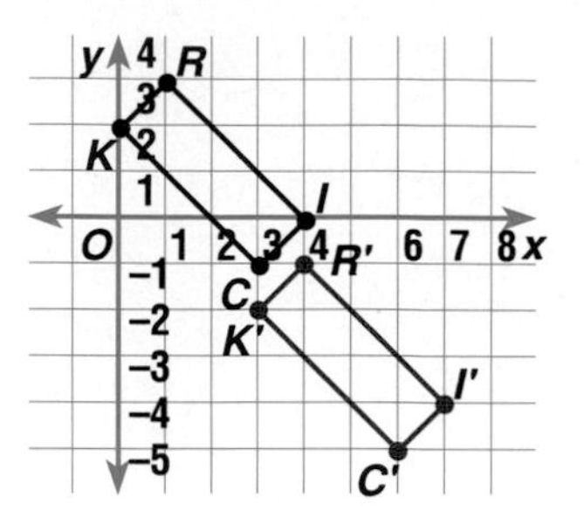

To check your work, graph rectangle *RICK* and rectangle *R'I'C'K'*. Is the translation correct?

Checking for Understanding

Communicating Mathematics

Read and study the lesson to answer each question.

1. **Tell** what type of movement results from a translation.
2. **Write** a sentence to describe the translation named by (-2, 3).
3. **Draw** a triangle on a coordinate plane. Show the position of the triangle after a translation named by (-2, 3).

Guided Practice

Name the coordinates of the ordered pair needed to translate each point *A* to point *B* to complete each scene.

4. 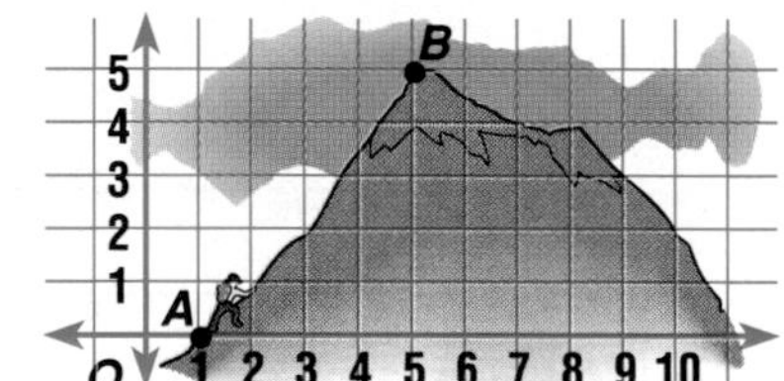

5. 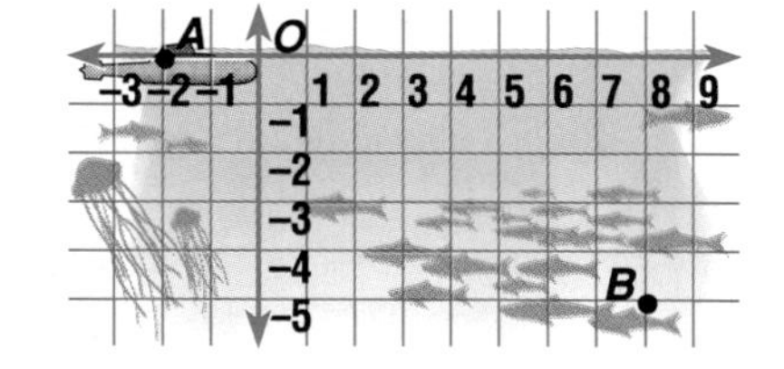

6. Translate $\triangle PQR$ with vertices $P(0, 0)$, $Q(-3, -4)$, and $R(1, 3)$ 6 units left and 3 units up. Then graph $\triangle P'Q'R'$.
7. Square $WXYZ$ has vertices $W(2, 1)$, $X(4, 3)$, $Y(2, 5)$, and $Z(0, 3)$. Find the coordinates of the vertices of $W'X'Y'Z'$ after a translation described by (-1, 3).

Exercises

Independent Practice

Find the coordinates of the vertices of each figure after the translation described. Then graph the figure and its translation.

8. $\triangle ABC$ with vertices $A(-5, -2)$, $B(-2, 3)$, and $C(2, -3)$, translated by (6, 3)
9. rectangle $PQRS$ with vertices $P(-4, 1)$, $Q(2, 4)$, $R(3, 2)$, and $S(-3, -1)$, translated by (-1, 4)

10. Three vertices of rectangle $RSTU$ are $R(-5, 5)$, $S(-1, 5)$, and $T(-1, 1)$.
   a. Graph the three vertices and find the fourth vertex of the rectangle.
   b. Find the coordinates of the vertices of $R'S'T'U'$ after a translation described by $(8, -5)$. Graph $R'S'T'U'$.

11. Triangle RST has vertex $R(-2, 3)$. When translated, R' has coordinates $(3, 5)$. Describe the translation using an ordered pair.

12. Pentagon $ABCDE$ has vertices $A(-2, -1)$, $B(0, -1)$, $C(1, 1)$, $D(-1, 3)$, and $E(-3, 1)$. After a translation, the coordinates of C' are $(-1, 2)$.
   a. Describe the translation using an ordered pair.
   b. Graph pentagon $ABCDE$ and pentagon $A'B'C'D'E'$.

Mixed Review

13. **Geometry** The lengths of the sides of a triangle are 24 feet, 10 feet, and 26 feet. Determine whether the triangle is a right triangle. *(Lesson 8-5)*

14. Graph the quadratic function $y = 2x^2 - 8$. *(Lesson 11-7)*

Problem Solving and Applications

15. **Make a Model** Marc misread the row and seat assignment on his theater ticket. Instead of seat (C, 3), he had seat (G, 13).
   a. Use a graph to model this situation.
   b. Tell the number of seats over and back he must move.

16. **Video Games** Nintendo® estimates that in 1991, 35% of the homes in the United States will have a Nintendo® game system. Study the graph below.
   a. Name each translation that moves Mario from each year to the next year.
   b. Approximately how many households had Nintendo® in 1991?
   c. Study the pattern of the increases. Describe how you think Nintendo® could analyze their success.

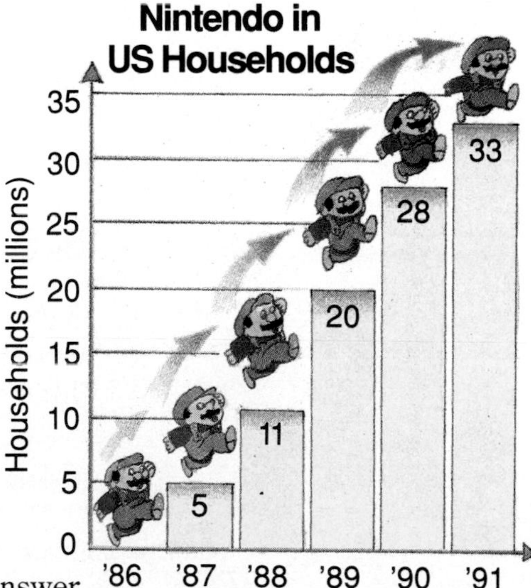

17. **Critical Thinking** A triangle is translated by $(3, -5)$. Then the result is translated by $(-3, 5)$. Without graphing, what is the final position of the triangle? Write an argument to defend your answer.

18. **Portfolio Suggestion** Review the items in your portfolio. Make a table of contents of the items, noting why each item was chosen. Replace any items that are no longer appropriate.

11-8B Slope

A Follow-Up of Lesson 11-8

Objective

Use translations and slope to find other solutions of a linear function.

Materials

graph paper
straightedge

Remember that the graphs of linear functions are lines. The steepness of a line is called its **slope.** The slope is also related to the translation that relates any two points on the line.

Try this

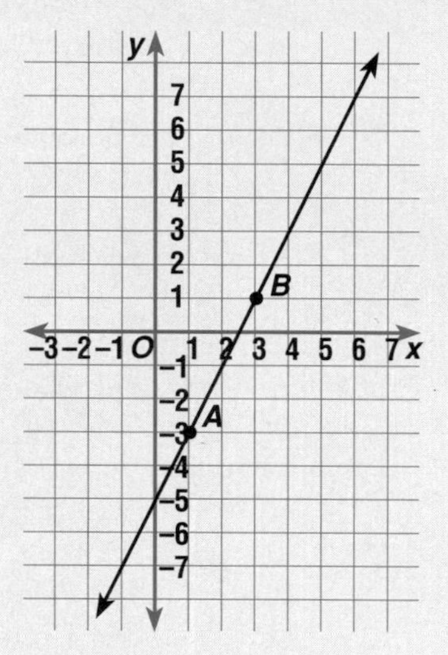

- Copy the graph shown at the right.
- Name the ordered pair that describes the translation from A to B.
- The slope of the line is defined as $\frac{y \text{ move from } A \text{ to } B}{x \text{ move from } A \text{ to } B}$. Use the translation to define the slope of this line.
- Use the slope to translate point B to a new location. Label this point C.
- Use the slope to translate point C to a new location. Label this point D.

What do you think?

1. What do you notice about points A, B, C, and D?
2. Find the ordered pair that describes the translation from B to A.
   a. Write this as the slope.
   b. How does this compare to the slope from A to B?
3. The line is the graph of the function $y = 2x - 5$.
   a. Use a table to find two other solutions for the function. Write them as ordered pairs.
   b. Find a translation that relates one ordered pair to the other.
   c. Write the translation as the slope. How does this slope compare with the slope from A to B?

Extension

One solution and the slope of the graph of a function are given.

a. Use the ordered pair and the slope to graph the function.

b. Name two other solutions.

4. (3, 2), $\frac{1}{2}$
5. (-2, -4), $\frac{-4}{3}$
6. (0, 0), $\frac{2}{1}$

Geometry Connection

11-9 Reflections

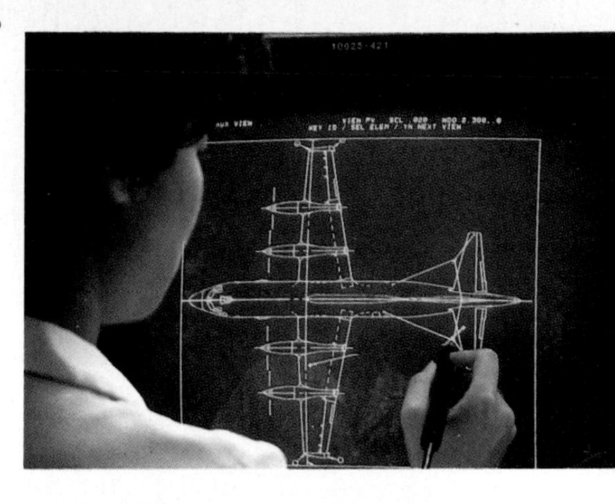

Objective

Graph reflections on a coordinate plane.

Words to Learn

symmetric
line of symmetry
reflection

Advertisers are always trying to catch your eye with flashy new designs and logos. Most of these new creations are generated using a computer. Graphic designers develop many designs using figures that can be folded into two identical parts. Each part of these designs is **symmetric** to the other part. In Chapter 5, you learned that this fold line is called the **line of symmetry.** One part of the figure is a **reflection** of the other part.

You can review reflections and symmetry on pages 192 and 193.

When a figure is reflected on a coordinate plane, every point of the original figure has a corresponding point on the other side of the line of symmetry.

Mini-Lab

Work with a partner.

Materials: graph paper, straightedge, scissors

- Graph $\triangle ABC$ with vertices $A(2, 3)$, $B(7, 1)$, and $C(5, 5)$.
- Count how many units point A is from the x-axis. Then count that many units on the opposite side of the axis. Label this point, A'.
- Count how many units B and C are from the x-axis. Use the same method to graph points B' and C'.
- Draw $\triangle A'B'C'$. Cut out the coordinate plane and fold it along the x-axis.

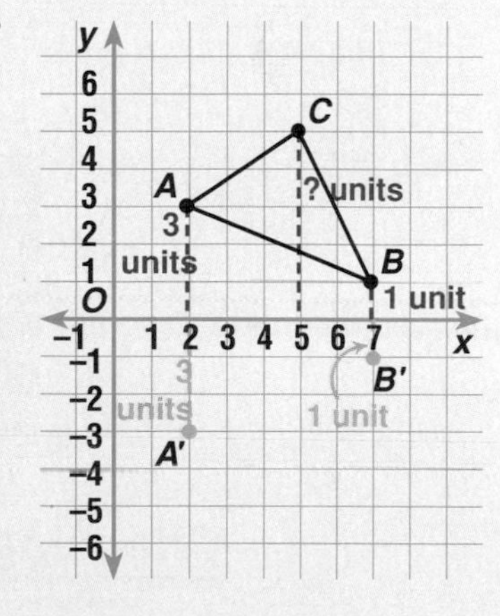

Talk About It

a. What do you notice about the two triangles when you fold the paper?

b. Compare the coordinates of A with A', B with B', and C with C'. What pattern do you notice?

The reflection in the Mini-Lab is a reflection over the x-axis. The x-axis is the line of symmetry. Each point of $\triangle A'B'C'$ corresponds to a point of $\triangle ABC$. The coordinates of the points also correspond in a special way.

Reflection over the *x*-axis	**In words:** To reflect a point over the x-axis, use the same x-coordinate and multiply the y-coordinate by -1.	
	Arithmetic	**Algebra**
	(2, 3) becomes (2, -3)	(x, y) becomes $(x, -y)$

What do you suppose happens if you reflect the figure over the y-axis?

Example

Reflect trapezoid *PQRS* over the *y*-axis if the vertices are *P*(4, -2), *Q*(8, -2), *R*(8, 4), and *S*(2, 4).

Count how many units each vertex is from the y-axis and graph the corresponding point on the opposite side of the axis.

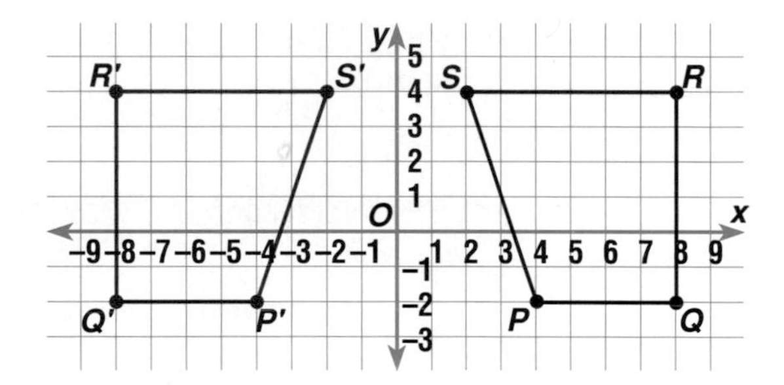

Compare the original coordinates to those of the reflected points. What do you notice?

Reflection over the *y*-axis	**In words:** To reflect a point over the y-axis, multiply the x-coordinate by -1 and use the same y-coordinate.	
	Arithmetic	**Algebra**
	(2, 3) becomes (-2, 3)	(x, y) becomes $(-x, y)$

Checking for Understanding

Communicating Mathematics

Read and study the lesson to answer each question.

1. **Tell** how the meaning of a mirror reflection relates to a geometric reflection.
2. **Write** a quick way to remember each rule for reflecting a figure over one of the axes.
3. **Draw** the figure at the right. Then draw all of the lines of symmetry.

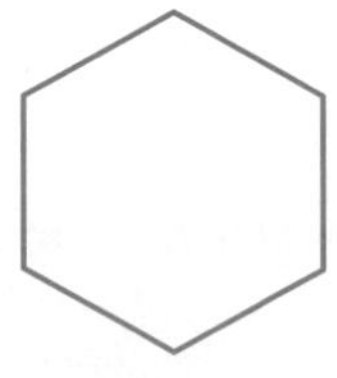

Guided Practice Name the line of symmetry for each pair of figures.

4. 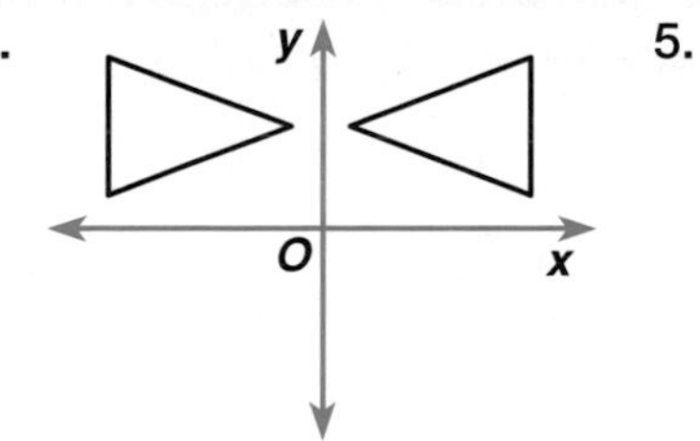

5. 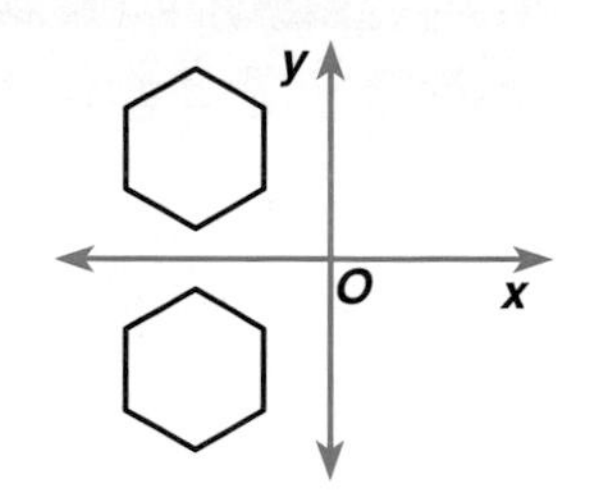

6. 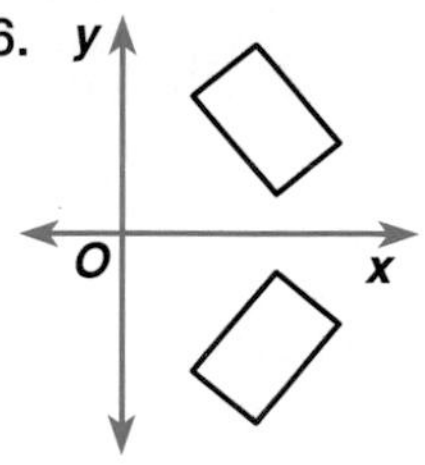

7. Graph $\triangle COW$ with vertices $C(3, 3)$, $O(0, 0)$, and $W(6, -1)$.
   a. Reflect $\triangle COW$ over the x-axis.
   b. Reflect $\triangle COW$ over the y-axis.

Exercises

Independent Practice

8. Graph $\triangle RAP$ with vertices $R(-8, 4)$, $A(-3, 8)$, and $P(-2, 2)$.
   a. Find the coordinates of the vertices after a reflection over the y-axis.
   b. Graph $\triangle R'A'P'$.
9. Graph parallelogram $MONY$ with vertices $M(1, 2)$, $O(0, 0)$, $N(-5, 0)$, and $Y(-4, 2)$.
   a. Find the coordinates of the vertices after a reflection over the x-axis.
   b. Graph parallelogram $M'O'N'Y'$.
10. Graph rectangle $EASY$ with vertices $E(-3, 3)$, $A(3, 3)$, $S(3, -3)$, and $Y(-3, -3)$.
    a. Reflect rectangle $EASY$ over the x-axis.
    b. On the same coordinate plane, reflect rectangle $EASY$ over the y-axis.
    c. Write a statement comparing your three graphs.

Mixed Review

11. Solve $\frac{2}{n} = \frac{7}{98}$. *(Lesson 9-2)*
12. Translate $\triangle ABC$ 2 units right and 3 units up if its vertices are $A(-3, -2)$, $B(-1, 1)$, and $C(2, -1)$. Graph $\triangle ABC$ and $\triangle A'B'C'$. *(Lesson 11-8)*

Problem Solving and Applications

13. **Critical Thinking** An isosceles triangle has one vertex at (0, 6) and another at (-3, 0). Use reflections to graph two different triangles that meet these requirements.
14. **Algebra** Quadratic functions have vertical lines of symmetry.
    a. Complete the table at the right for the function $f(n) = n^2$.
    b. Graph these ordered pairs. Then use symmetry to complete the other half of the graph.

n	*f(n)*
0	
1	
2	
3	
4	

11-10 Rotations

Geometry Connection

Objective

Graph rotations on a coordinate plane.

Words to Learn

rotation

The first windmills were found in A.D. 644 in Persia (now known as Iran) and used to grind grain. In A.D. 1220, Ghenghis Kahn captured Persian mill builders to take them to China to build windmills for irrigating fields. The most familiar windmills are those in the Netherlands. These mills were used to pump water from the soggy lands uncovered after the dikes were built.

The air flowing through the vanes of the windmills creates a movement called **rotation.** In Chapter 5, you learned that rotations move a figure about a central point.

LOOKBACK

You can review rotations on page 203.

Mini-Lab

Work in groups of three.

Materials: protractor

- The graph models the vanes of a Dutch windmill. Record the coordinates of each lettered point.
- Measure $\angle COG$, $\angle GOK$, $\angle KOP$, and $\angle POC$. Record your measurements.

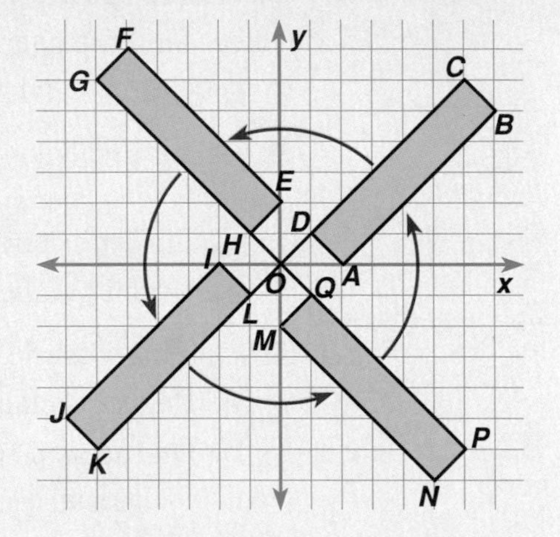

- As the windmill turns, each point of one vane will occupy the previous location of the corresponding point on another vane. Make a list of the corresponding vertices of the four vanes.

Talk About It

a. What shape is each vane of the windmill?
b. Which direction is this windmill turning?
c. How do the coordinates of the vertices of vane *ABCD* compare with those of vane *EFGH?*
d. How many degrees did the vane rotate to move from point *C* to point *G?*
e. Compare the coordinates of the vertices of vane *ABCD* with those of vane *IJKL.*
f. How many degrees did the vane rotate to move from point *C* to point *K?*

As you discovered in the Mini-Lab, the coordinates of rotated points are related for every 90° they turn.

Rotation of 90° counter-clockwise	**In words:** To rotate a figure 90° counterclockwise, switch the coordinates of each point and then multiply the first coordinate by -1. **Arithmetic** $A(2, 3) \rightarrow A'(-3, 2)$ **Algebra** $A(x, y) \rightarrow A'(-y, x)$
Rotation of 180°	**In words:** To rotate a figure 180°, multiply both coordinates of each point by -1. **Arithmetic** $A(2, 3) \rightarrow A'(-2, -3)$ **Algebra** $A(x, y) \rightarrow A'(-x, -y)$

Examples

1 **Triangle ABC has vertices $A(1, 3)$, $B(6, 7)$, and $C(9, 1)$. Graph $\triangle ABC$ and rotate it 180°. Then graph $\triangle A'B'C'$.**

To rotate $\triangle ABC$ 180°, multiply each coordinate by -1.

$A(1, 3) \rightarrow A'(-1, -3)$
$B(6, 7) \rightarrow B'(-6, -7)$
$C(9, 1) \rightarrow C'(-9, -1)$

Now graph the ordered pairs for points A', B', and C' and draw $\triangle A'B'C'$.

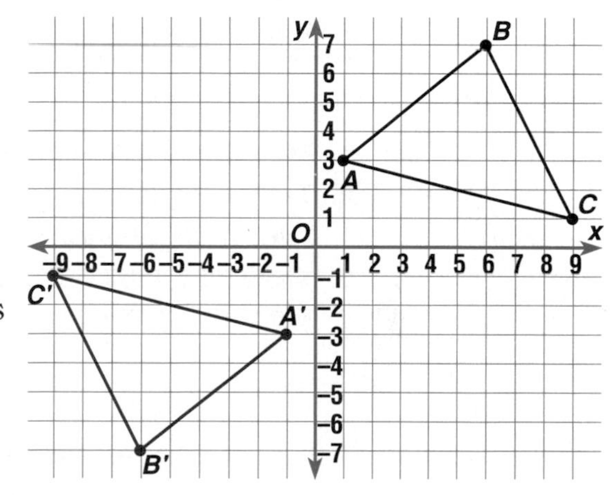

Estimation Hint

You can mentally check your figure by laying the corner of another piece of paper at the origin. Move it until A and A' lie on the edges of the paper. If you can't do this, you've probably made a mistake.

2 **Use $\triangle ABC$ from Example 1. Rotate it 90° counterclockwise. Then graph $\triangle A'B'C'$.**

To rotate $\triangle ABC$ 90° counterclockwise, switch the coordinates and multiply the first by -1.

$A(1, 3) \rightarrow (3, 1) \rightarrow A'(-3, 1)$
$B(6, 7) \rightarrow (7, 6) \rightarrow B'(-7, 6)$
$C(9, 1) \rightarrow (1, 9) \rightarrow C'(-1, 9)$

Now graph the ordered pairs for points A', B', and C' and draw $\triangle A'B'C'$.

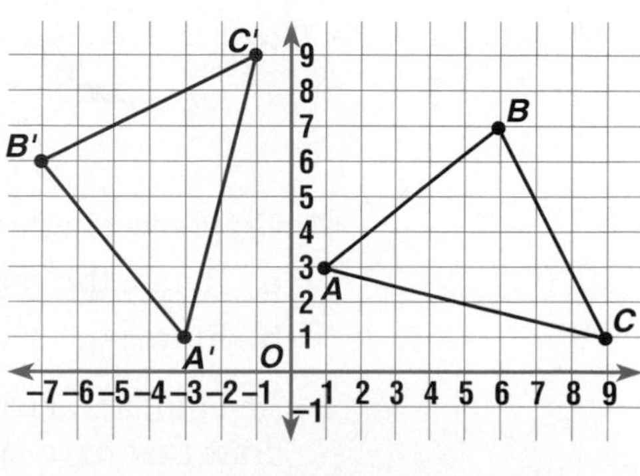

You can review rotational symmetry on page 193.

In Chapter 5, you learned that some figures have line symmetry. Other figures have rotational symmetry. That is, if you turn them around their center point, there is at least one other position in which the figure looks the same as it did originally.

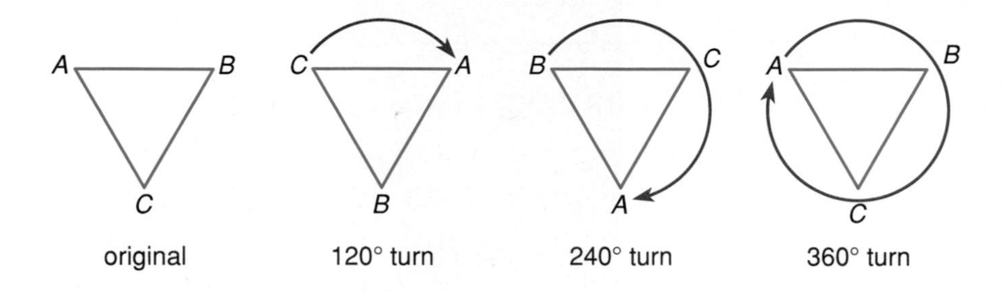

While the vertices are in different locations, the figure itself looks the same after each turn. An equilateral triangle has rotational symmetry at clockwise turns of 120°, 240°, and 360°.

Checking for Understanding

Communicating Mathematics

Read and study the lesson to answer each question.

1. **Tell** three examples of rotating objects you see every day.
2. **Tell** what quadrant a triangle will be in if it is rotated 180° from its location in the second quadrant.
3. **Show** how you could use the 90° rotation to find a 180° rotation if you forgot the rule for 180°.

Guided Practice

Determine whether each pair of figures represents a rotation. Write *yes* or *no*.

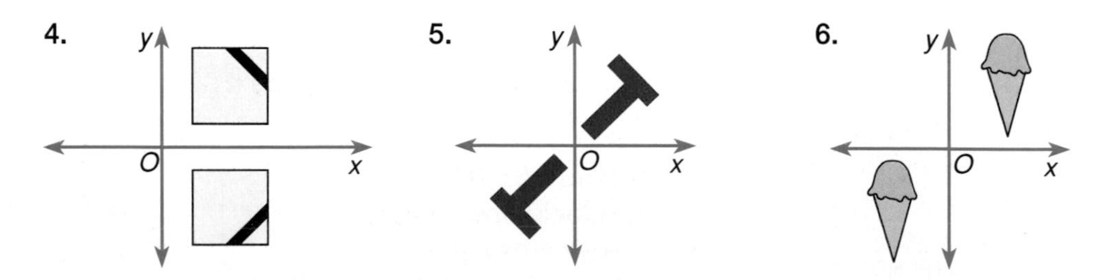

7. Graph rectangle $HAIR$ with vertices $H(-3, 4)$, $A(-5, 4)$, $I(-5, -2)$, and $R(-3, -2)$.
   a. Rotate the rectangle 90° counterclockwise, and graph $H'A'I'R'$.
   b. Rotate the rectangle 180°, and graph $H'A'I'R'$.
8. An equilateral triangle has rotational symmetry. What other figures can you draw that have sides with equal length and rotational symmetry?

Exercises

Independent Practice

9. Triangle RST has vertices $R(-1, -3)$, $S(-6, -9)$, and $T(-8, -5)$.
   a. Graph $\triangle RST$.
   b. Find the coordinates of the vertices of $\triangle R'S'T'$ after a 90° counterclockwise rotation.
   c. Graph $\triangle R'S'T'$.

10. Trapezoid $ABCD$ has vertices $A(3, -2)$, $B(7, -2)$, $C(9, -7)$, and $D(1, -7)$.
    a. Graph trapezoid $ABCD$ and its 180° rotation.
    b. Rotate trapezoid $A'B'C'D'$ 180°. What is the result?

11. A triangle is rotated 180°. The coordinates of the vertices of the rotated triangle are (4, -1), (1, -4), and (5, 8). What are the coordinates of the original triangle?

12. Copy the figure at the right.
    a. Does this figure have rotational symmetry?
    b. If so, find the degree turns that show this symmetry.

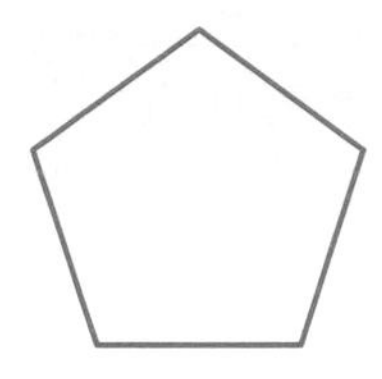

Mixed Review

13. Solve $s - 34 = 71$. Check your solution. *(Lesson 2-3)*

14. Find the GCF of 84 and 36. *(Lesson 6-4)*

15. Graph square $BART$ with vertices $T(1, -1)$, $R(3, -1)$, $A(3, -3)$, and $B(1, -3)$. *(Lesson 11-9)*
    a. Reflect square $BART$ over the x-axis.
    b. Reflect square $BART$ over the y-axis.

Problem Solving and Applications

16. **Geometry** Graph pentagon $EIGHT$ with vertices $E(-6, 5)$, $I(-2, 5)$, $G(-2, 2)$, $H(-4, 1)$, and $T(-6, 2)$.
    a. Reflect $EIGHT$ over the x-axis. Then reflect $E'I'G'H'T'$ over the y-axis.
    b. Graph $EIGHT$ on another coordinate plane. Rotate it 180°.
    c. Write a sentence comparing your two graphs.

17. **Entertainment** In a standard deck of playing cards, the 10 of hearts has rotational symmetry. What other cards have rotational symmetry?

18. **Critical Thinking** Find a rule for rotating a figure 90° in a clockwise direction.

19. **Journal Entry** Write a few sentences telling how you think rotations and reflections might be used in computer design. Include specific examples.

Study Guide and Review

Communicating Mathematics

Choose a word from the list at the right to correctly complete each sentence.

linear
quadratic
solution set
reflection
translation
domain
range

1. The __?__ is the set of input values of a function.
2. The __?__ is the set of output values of a function.
3. The set of ordered pairs that are solutions for an equation is called a __?__ for the equation.
4. A function in which the graphs of the solutions form a line is called a __?__ function.
5. A function in which the greatest power is two is called a __?__ function.
6. The movement of a figure 2 units right and 4 units down is a __?__ .
7. $A(2, 1) \rightarrow A'(-2, 1)$ describes a __?__ over the y-axis.
8. Tell how to determine if an ordered pair is a solution of a function.
9. Describe how to rotate a triangle 90° counterclockwise.

Self Assessment

Objectives and Examples

Upon completing this chapter, you should be able to:

Review Exercises

Use these exercises to review and prepare for the chapter test.

- complete function tables *(Lesson 11-1)*

Find $f(4)$ if $f(n) = 3n - 1$.

$f(4) = 3(4) - 1$
$f(4) = 12 - 1$
$f(4) = 11$

Copy and complete the function table.

10. $f(n) = 2 - 4n$

n	$2 - 4n$	$f(n)$
-1		
0		
2		

- graph functions using function tables *(Lesson 11-2)*

Graph $f(n) = n + 1$.

n	$f(n)$	$(n, f(n))$
-2	-1	(-2, -1)
0	1	(0, 1)
2	3	(2, 3)

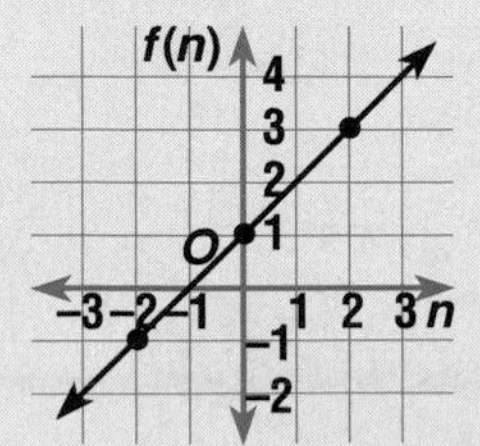

Make a function table for each function. Then graph the function.

11. $f(n) = 4n + 1$
12. $f(n) = \frac{1}{2}n - 2$
13. $f(n) = -3n$

Objectives and Examples	**Review Exercises**
• find solutions of equations with two variables *(Lesson 11-3)* Find a solution of $y = 6x + 8$. Let $x = 2$. Then $y = 6(2) + 8 = 20$. A solution of $y = 6x + 8$ is (2, 20).	**Find four solutions of each equation.** 14. $y = -1.5x - 1$ 15. $y = x + 4$ 16. $y = -5x + 7$
• graph linear functions by plotting points *(Lesson 11-4)* Make a function table and choose at least three values for x. Then graph the ordered pairs and connect them with a line.	**Graph each linear function.** 17. $y = -6x$ 18. $y = 2x + 7$ 19. $y = -3.5x + 1.5$ 20. $y = \frac{x}{2} - 1$
• solve systems of linear equations by graphing *(Lesson 11-5)* The coordinates of the point where the graphs of the equations intersect is the solution to the system of equations.	**Solve each system of equations by graphing.** 21. $y = 6x$, $y = x + 5$ 22. $y = 4x - 6$, $y = x + 3$
• graph quadratic functions by plotting points *(Lesson 11-7)* Make a function table. Then graph the ordered pairs and draw a smooth curve connecting the points.	**Graph each quadratic function.** 23. $y = \frac{1}{2}x^2 + 3$ 24. $y = x^2 - 1$ 25. $f(n) = 4 - n^2$ 26. $y = -1.25x^2 - 1.5$
• graph translations on a coordinate plane *(Lesson 11-8)* To translate a point as described by the ordered pair (a, b), add a to the x-coordinate and add b to the y-coordinate.	**Graph each figure and its translation.** 27. rectangle $RSTU$ with vertices $R(-4, 1)$, $S(-2, 1)$, $T(-2, -1)$, and $U(-4, -1)$, translated by (3, 4) 28. $\triangle DEF$ with vertices $D(1, 1)$, $E(2, 4)$, and $F(4, 2)$, translated by (-5, -3)
• graph reflections on a coordinate plane *(Lesson 11-9)* To reflect a point over the x-axis, use the same x-coordinate and multiply the y-coordinate by -1. $A(1, 2) \rightarrow A'(1, -2)$ To reflect a point over the y-axis, multiply the x-coordinate by -1 and use the same y-coordinate. $B(-2, 3) \rightarrow B'(2, 3)$	29. Graph rectangle $ABCD$ with vertices $A(2, 5)$, $B(6, 5)$, $C(6, 3)$, and $D(2, 3)$ and its reflection over the x-axis. 30. Graph $\triangle CAR$ with vertices $C(-4, -5)$, $A(-3, -2)$, and $R(-5, -3)$ and its reflection over the y-axis.

Objectives and Examples	***Review Exercises***
• graph rotations on a coordinate plane *(Lesson 11-10)* To rotate a point 90° counterclockwise, switch the coordinates and multiply the first coordinate by -1. $C(3, 4) \rightarrow C'(-4, 3)$ To rotate a point 180°, multiply both coordinates by -1. $D(-2, 4) \rightarrow D'(2, -4)$	**31.** Graph square $LATE$ with vertices $L(1, 4)$, $A(4, 7)$, $T(7, 4)$, and $E(4, 1)$ and its rotation of 90° counterclockwise. **32.** Graph $\triangle XYZ$ with vertices $X(3, 1)$, $Y(5, -2)$, and $Z(2, -4)$ and its 180° rotation.

Applications and Problem Solving

33. Business One company's profits can be described by the equation $y = 300x - 400$. Another company's profits can be described by the equation $y = 300x + 100$. They plan to merge when their profits are the same amount. At what point will that occur? Explain your answer. *(Lessons 11-5 and 11-6)*

34. Geometry Find the base and height of three different triangles whose area is 12 square inches. Write the solutions as ordered pairs. *(Lesson 11-3)*

Curriculum Connection Projects

- **Drafting** Graph large block versions of your initials, labeling the x- and y-coordinates of key points on each letter. Then graph a translation, a reflection, and a rotation of your initials.
- **Physical Education** As muscles are used continuously, they tire. Have a friend hold a math book in his/her hand with the arm outstretched. Measure the distance from the floor to your friend's hand every 30 seconds for 5 minutes. Graph your results.

Read More About It

Harman, Carter. *A Skyscraper Goes Up.*
Dobbler, Lavina. *I Didn't Know That.*
Dickenson, Peter. *Eva.*
Jonas, Ann. *Round Trip.*

Chapter 11 Test

Make a function table for each function. Then graph the function.

1. $f(n) = -4n$
2. $f(n) = 2n - 2$
3. Find $f\left(-\frac{1}{4}\right)$ if $f(n) = \frac{16}{n}$.

Find four solutions for each equation.

4. $y = \frac{x}{4} + 1$
5. $y = -2.5x - 5$
6. $y = 3 - x$
7. $y = 9x - 15$

Graph each function.

8. $y = \frac{x}{3} - 2$
9. $y = -\frac{1}{2}x^2 + 6$
10. $f(n) = 3n^2 - 1$
11. $y = 35 - 4x$
12. Find the solution for the system $5x + 2 = y$ and $2x - 1 = y$ by graphing.
13. Without graphing, determine if (3, 10) is the solution to the system of equations $y = \frac{1}{3}x + 7$ and $y = 4x - 5$. Why or why not?

Rectangle *PQRS* has vertices *P*(-5, -2), *Q*(-3, -2), *R*(-3, -5), and *S*(-5, -5). After a translation, the coordinates of *P′* are (2, 7).

14. Describe the translation using an ordered pair.
15. Find the coordinates of *Q′*, *R′*, and *S′*.

Triangle *CAT* has vertices *C*(-5, 2), *A*(-2, 3), and *T*(-3, 6).

16. Find the coordinates of the vertices after a reflection over the *y*-axis.
17. Find the coordinates of the vertices after a reflection over the *x*-axis.

Triangle *ABC* is rotated 180°. The vertices of the rotated triangle are *A′*(4, 4), *B′*(1, 2), and *C′*(3, 1).

18. What are the coordinates of $\triangle ABC$?
19. List the coordinates of triangle *A′B′C′* after a rotation of 90° counterclockwise.
20. **Business** About how much higher was the daily circulation of the *Denver Post* in 1991 than in 1989?

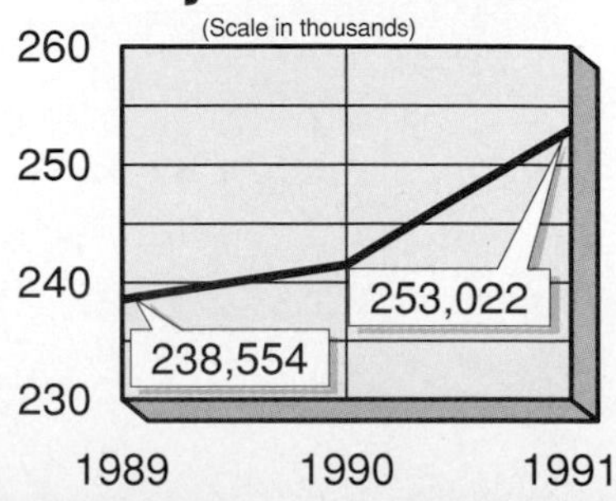

Bonus Find a rule involving ordered pairs for reflecting a point over the line $y = x$.

Chapter

Area and Volume

Spotlight on Sports

Have You Ever Wondered...

- Which stadium in the United States can seat the most people?
- Which holds more air, a basketball or a volleyball?

At the athletic shoe factory

TOO SPRINGY.

12·25

10 largest stadiums in the United States

Name and location	Seating capacity*
Rose Bowl, Pasadena, Calif.	104,000
Michigan Stadium (University of Michigan), Ann Arbor	101,701
Neyland Stadium (University of Tennessee), Knoxville	93,515
Memorial Coliseum, Los Angeles	92,516
John F. Kennedy Stadium, Philadelphia, Pa.	90,000
Ohio Stadium (Ohio State University), Columbus	90,000
Beaver Stadium (Pennsylvania State University), University Park	85,210
Stanford Stadium (Stanford University), Stanford, Calif.	84,892
Sanford Stadium (University of Georgia), Athens	82,122
Gator Bowl, Jacksonville, Fla.	82,000

*Capacities are listed for football games and may include temporary seats

Chapter Project

Sports

Work in a group.

1. Keep a record of any objects you see or use in a week that are in the shape of a circle, a sphere, or a cylinder. Include as many sports objects as you can.
2. Measure the circumference of the curved edge or surface of each of these objects.
3. Prepare a chart or graph to compare the circumferences of these objects.

Offical Sizes of the Balls Used in Various Sports

Basketball
Circumference 30 inches....weight 20-22 ounces

Baseball
Circumference 9-9.25 inches....weight 5-5.25 ounces

Soccer
Circumference 27-28 inches....weight 14-16 ounces

Volleyball
Circumference 26 inches....weight 9.25 ounces

Tennis
Circumference 7.75-8.25 inches....weight 2-2.062 ounces

Polo
Circumference 9.5-11 inches....weight 3.5-4.5 ounces

Ping-Pong
Circumference 4.4-4.7 inches....weight 0.083-0.091 ounces

Football
11 inches long, 7 inch diameter in center....weight 14-15 ounces

Looking Ahead

In this chapter, you will see how mathematics can be used to answer questions about sports. The major objectives of the chapter are to:

- find the area of circles
- solve problems by making a model
- find the surface area of prisms and cylinders
- sketch three-dimensional figures and find the volume of the figures
- describe a measurement by using precision and significant digits

12-1 Area of Circles

Objective

Find the area of circles.

Georgio shows the actual pans he uses in his pizzeria so customers can see the size of the pizza they are ordering. If you can get one large pizza for \$8.99, or two medium pizzas for \$9.99, which offer is the better buy?

TEEN SCENE

The first pizza made with tomatoes and cheese was created for Queen Marghenta of Italy in 1889. The pizza maker, Raffaele Esposito, used ingredients that matched the colors of the Italian flag: red (tomatoes), white (mozzarella cheese), and green (basil).

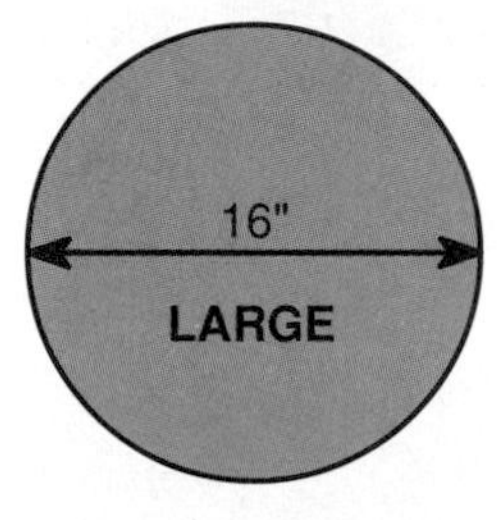

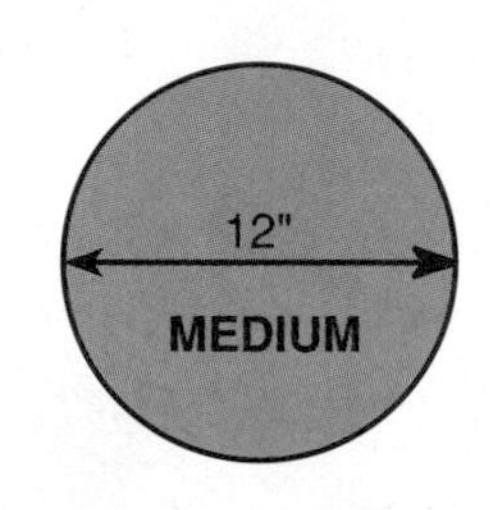

In order to solve this problem, we need to find the area of each size of pizza. The formula for the area of a circle is related to the formula for the area of a parallelogram.

Suppose you draw several radii of a circle equally-spaced. Then cut the circle along the radii to form wedge-like pieces. Rearrange the pieces to form a parallelogram-shaped figure.

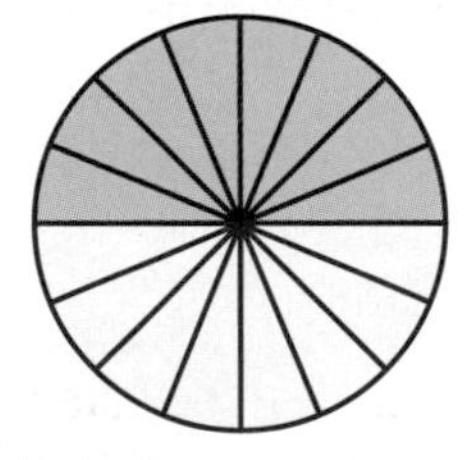

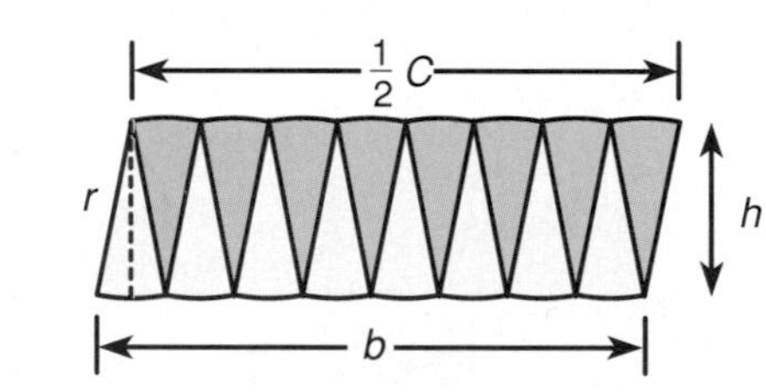

The length of each wedge from the point to its edge is the same as the radius of the circle. So the height of the parallelogram is r. The curved sides of the wedges form the circumference of the circle. The base of the parallelogram is made of half of these curves, or half the circumference of the circle.

$A = bh$ — *area of parallelogram*

$A = \frac{1}{2}C \cdot r$ — $b = \frac{1}{2}C,\ h = r$

$A = \frac{1}{2} \cdot 2\pi r \cdot r$ — $C = 2\pi r$

$A = \pi r^2$

LOOKBACK

You can review circles on page 284.

So, the formula for the area of a circle is $A = \pi r^2$.

Area of a Circle	**In words:** The area (A) of a circle equals π times the radius (r) squared. **In symbols:** $A = \pi r^2$

Example 1

Find the area of a circle with a radius of 8 inches to the nearest square inch.

Using paper and pencil

$A = \pi r^2$

$A \approx 3.14(8^2)$ *Use 3.14 for* π.

$A \approx 3.14(64)$

$A \approx 200.96$

Using a calculator

Use the π key.

[π] [×] 8 [x²] [=] 201.06193

The area of the circle is about 201 square inches.

You can use the area formula for circles to find areas of circular-shaped objects. Since most calculators have a π key, we will use a calculator to compute the measurements in this chapter.

Example 2 *Problem Solving*

> **Estimation Hint**
>
> Since $\pi \approx 3$, you can estimate the area of any circle by squaring the radius and multiplying by 3.

Road Construction **A Kentucky visitors' center along I-75 is surrounded by a circular driveway. Find the cost of repaving this driveway if repaving costs $0.89 per square foot.**

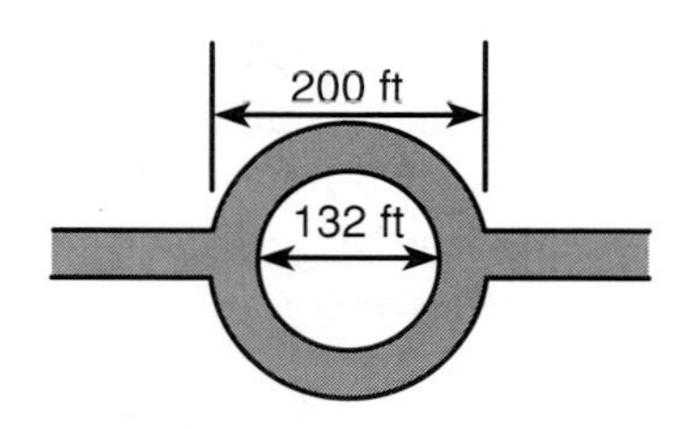

Notice that the edges of the driveway are formed by two circles. The area of the driveway would be the area of the larger circle minus the area of the smaller circle.

Remember that the radius of a circle is half its diameter.

area of large circle

$d = 200$, so $r = \frac{200}{2}$ or 100

$A = \pi r^2$

$A = \pi \cdot 100^2$

$A \approx 31{,}415.927$

area of small circle

$d = 132$, so $r = \frac{132}{2}$ or 66

$A = \pi r^2$

$A = \pi \cdot 66^2$

$A \approx 13{,}684.778$

area of driveway = area of large circle − area of small circle

$A \approx 31{,}415.927 - 13{,}684.778$ or $17{,}731.149$

The area of the driveway is about 17,731 square feet. The cost of paving would be $0.89 · 17,731 or $15,780.59.

Remember that the probability of an event is defined as the ratio of the number of ways something can happen to the total possible outcomes. Probability can also be related to the area of a figure.

Example 3 *Connection*

Probability Suppose you throw a dart at random at the dart board at the right and you hit the board. What is the probability that the dart lands in the red section?

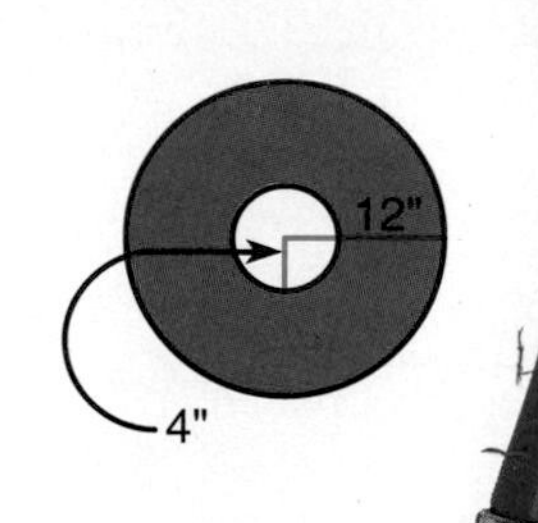

To find the probability of landing in the red section, you need to know the area of the red section and the area of the entire dart board.

Mental Math Hint

Use the distributive property to combine numbers containing π.

$144\pi - 16\pi =$

$(144 - 16)\pi =$

128π.

$$\begin{aligned}\text{area of red section} &= \text{area of dart board} - \text{area of inner circle}\\ &= \pi \cdot (12)^2 - \pi \cdot (4)^2\\ &= 144\pi - 16\pi \text{ or } 128\pi\end{aligned}$$

$$\begin{aligned}P(\text{landing in red section}) &= \frac{\text{area of red section}}{\text{total area}}\\ &= \frac{128\pi}{144\pi} \quad \textit{The GCF of the numerator and denominator is } 16\pi.\\ &= \frac{8}{9}\end{aligned}$$

The probability of landing in the red section of the dart board is $\frac{8}{9}$.

Checking for Understanding

Communicating Mathematics

Read and study the lesson to answer each question.

1. **Tell** why the area of a circle is always given in square units.
2. **Tell** how you can estimate the area of a circle whose radius is 5 meters.
3. **Tell** how you would find the area of a circle if you only know the diameter of the circle.
4. **Show** how you would find the area of the semicircle shown at the right.

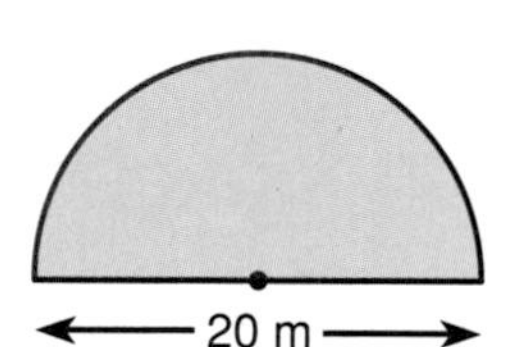

Guided Practice

Find the area of each circle to the nearest tenth.

5. 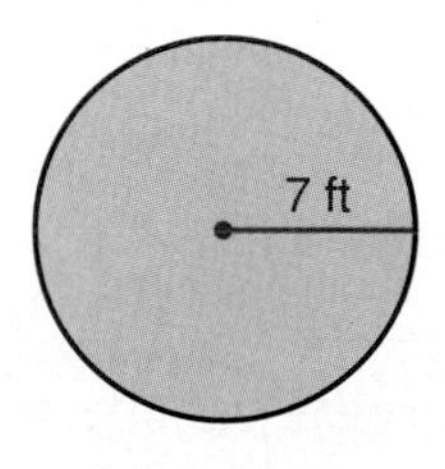

6. 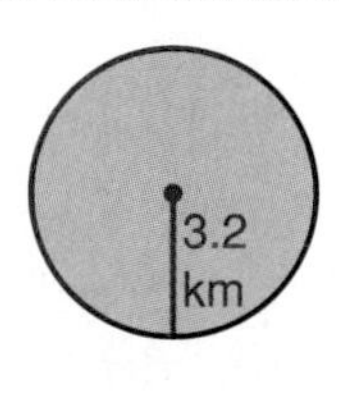

7. 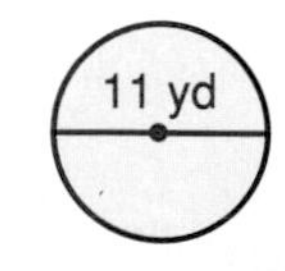

8. Jolie dropped a dime down a wishing well without looking. There's a bucket at the bottom of the well. What is the probability the dime will land in the bucket?

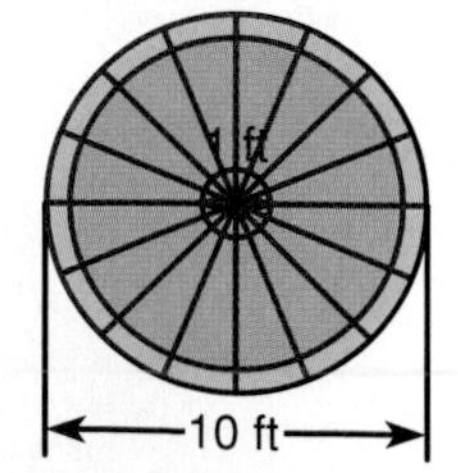

Exercises

Independent Practice

Find the area of each circle to the nearest tenth.

9. 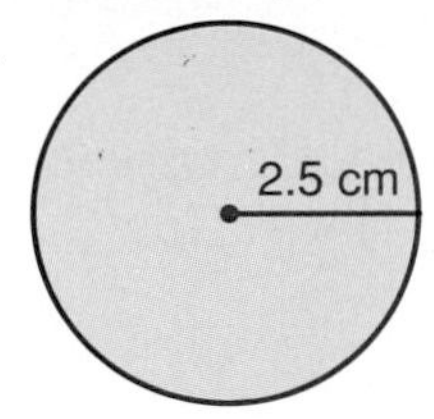

10. 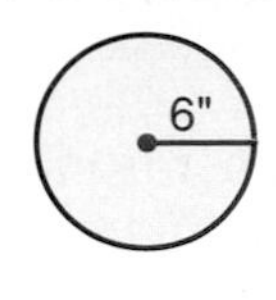

11. 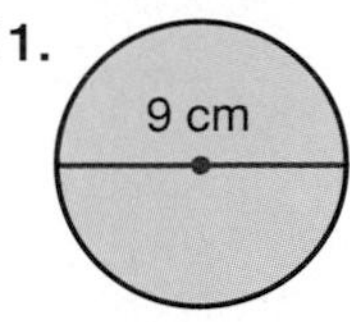

12. Find the area of a circle whose radius is 13 feet.

13. Find the area of a circle whose diameter is 10 meters.

14. Find the radius of a circle whose area is 49π square inches.

Find the area of each shaded region.

15. 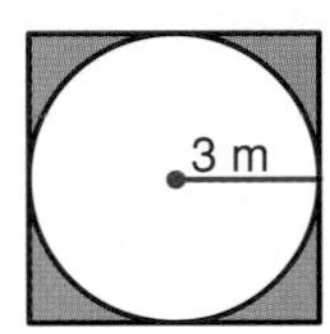

16. 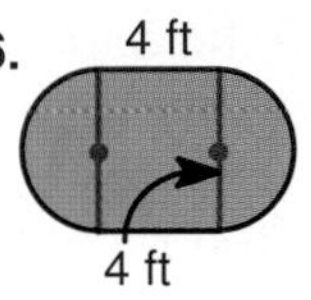

17. 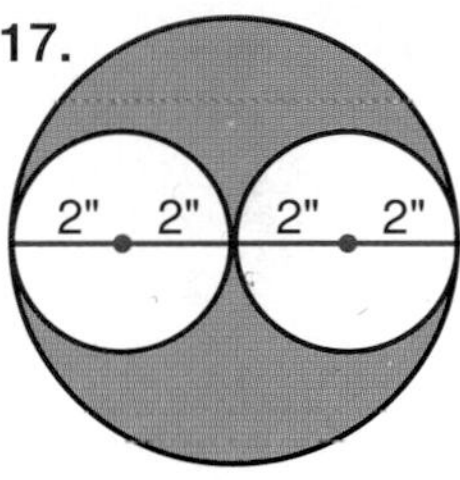

Mixed Review

18. **Algebra** Find the value of x in $\triangle RST$ if $m\angle R = 54°$, $m\angle S = 90°$, and $m\angle T = 3x°$. *(Lesson 5-3)*

19. Find the distance between points $A(-5, -4)$ and $B(3, -2)$. Express your answer to the nearest tenth. *(Lesson 8-7)*

20. **Geometry** Graph $\triangle HIJ$ with vertices $H(2, 2)$, $I(4, -1)$, and $J(1, -2)$. Then graph $\triangle H'I'J'$ after a rotation of 90° counterclockwise. *(Lesson 11-10)*

Problem Solving and Applications

21. **Probability** Louie is blindfolded and throws darts at the dart board shown at the right.
    a. What is the area of each ring of the board?
    b. Suppose Louie hits the board each time he throws a dart. What is the probability of landing in the white ring?

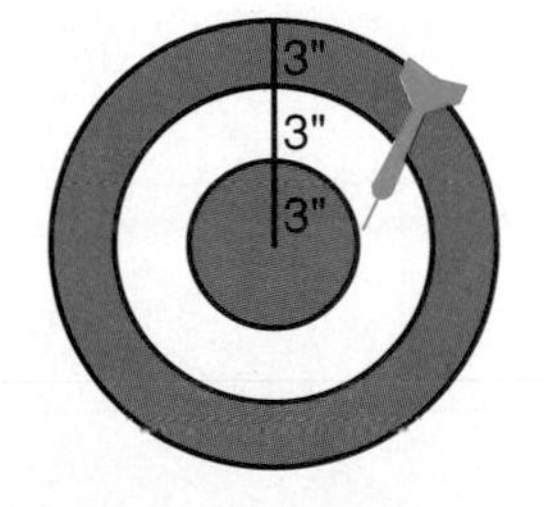

22. **Critical Thinking** What happens to the circumference and area of a circle when the length of the radius is doubled? Make a drawing.

23. **Smart Shopping** Refer to the pizza problem at the beginning of the lesson.
    a. Find the area of the large pizza and two medium pizzas.
    b. Use a calculator to find the cost per square inch of each pizza deal. Use your calculator displays to find which costs less per square inch.
    c. Now use your results rounded to the nearest cent. Which is the better deal?

Problem-Solving Strategy

12-2 Make a Model

Objective
Solve problems by making a model.

Materials
cubes

A set designer is creating a mirrored staircase for a dance number. The plan shows the top, front, and side views. How many cubes are needed to build the staircase?

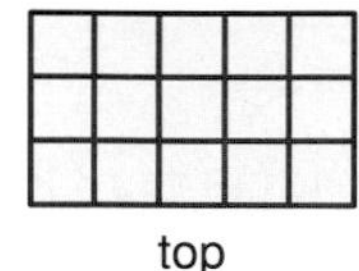
top

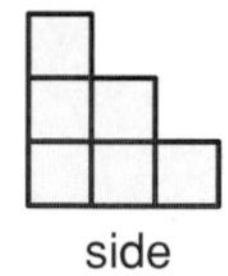
side

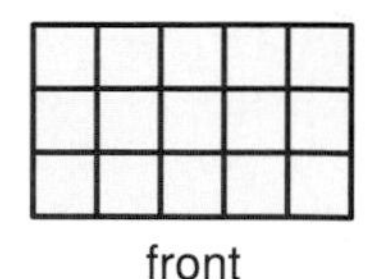
front

Example

Make a model to determine how many cubes are needed to build the staircase for the dance number.

Explore *What do you know?*
You know the shape and dimensions of the staircase from the squares shown on the plan.

What do you need to find out?
You need to find the number of cubes the designer needs to build the staircase.

Plan Use each of the views of the staircase to determine the dimensions. Then use cubes to build a model.

Solve The view from the top shows 15 cubes arranged in a rectangle, 5 cubes by 3 cubes.

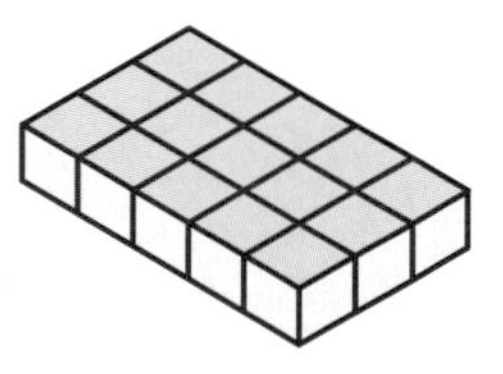

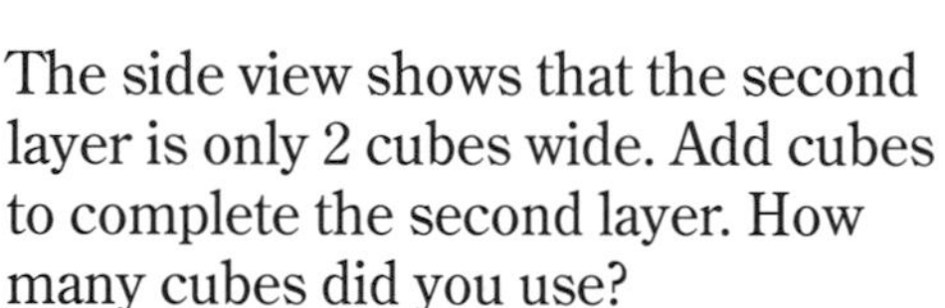

The side view shows that the second layer is only 2 cubes wide. Add cubes to complete the second layer. How many cubes did you use?

The side view also shows that the top layer is only 1 cube wide. Add cubes to complete the top layer.

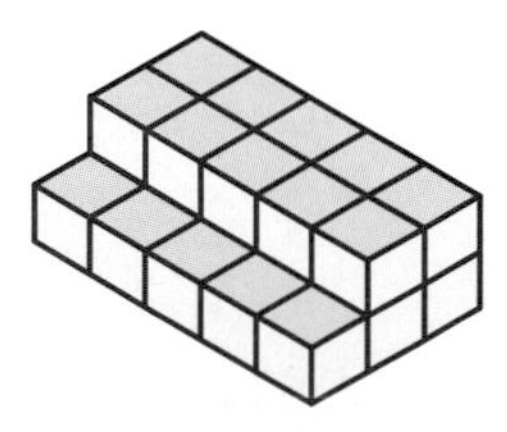

The front view shows 3 layers of 5 cubes each. Look at the front of your model to make sure you have this number of layers and cubes.

Total the number of cubes you used in each layer.

$15 + 10 + 5 = 30$ cubes

Examine Turn your model to make sure that each view is correct. If your model matches each view, the model is correct.

Checking for Understanding

Communicating Mathematics

Read and study the lesson to answer each question.

1. **Tell** how you determined how many cubes were in each layer.
2. **Show** another way to build the model by starting with the side view first.

Guided Practice

Solve. Use the make-a-model strategy.

3. Cindy is placing glass decorations near the ceiling at a door entrance she is designing. Find how many glass cubes she needs for two decorations with the plans shown below.

top

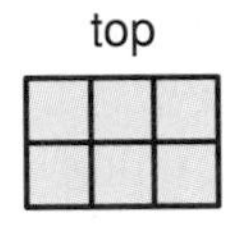

side

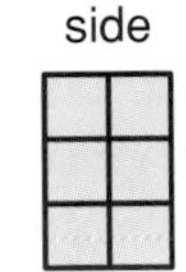

front

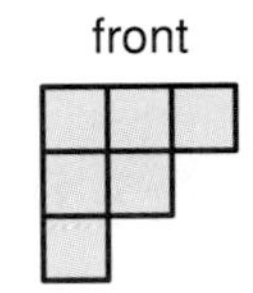

Problem Solving

Practice

Strategies

- Look for a pattern.
- Solve a simpler problem.
- Act it out.
- Guess and check.
- Draw a diagram.
- Make a chart.
- Work backward.

Solve using any strategy.

4. Use the model in the Example to determine how many mirrored tiles would be needed to cover the staircase if each tile is the same size as a face of one of the cubes.
5. Natasha paid $45 for a perm that was on sale at 40% off. What was the original price of the perm?
6. Toni, Matsue, and Juana all like pizza. One likes her pizza plain. One likes her pizza with mushrooms. One likes her pizza with anchovies. Use the following clues to find out which kind of pizza each girl likes.
   - Toni doesn't know the girl who likes her pizza plain.
   - Matsue's favorite kind of pizza is cheaper than pizza with mushrooms.
   - The girl who likes mushrooms is Toni's cousin.

7. Edu-Toys is designing a new package to hold a set of 30 alphabet blocks. Each block is a cube with each edge of the cube being 2 inches long. A box of which dimensions will hold the set of blocks without the need for some type of filler?

   a. 8 inches by 4 inches by 12 inches
   b. 3 inches by 6 inches by 13 inches
   c. 6 inches by 4 inches by 10 inches

12-3 Three-Dimensional Figures

Objective

Sketch three-dimensional figures from different perspectives.

Words to Learn

solid
face
edge
vertex
base

Edmonia Lewis was born in upstate New York. Her mother was a Chippewa Indian and her father was a freedperson. One of her works is a bust of Henry Wadsworth Longfellow, done for the Harvard College Library.

Edmonia Lewis (1845–1890) was America's first African-American woman to be recognized as a talented sculptor. She strove to capture the emotions of the people in her sculptures, unlike many of her contemporaries.

Sculpture is a three-dimensional art. The sculptor must look at the sculpture from many perspectives to make sure each detail has been captured. In geometry, we study three-dimensional figures called **solids** that have certain details that make them unique. Some common solids are shown below.

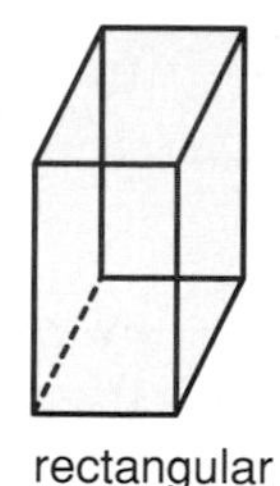
rectangular prism

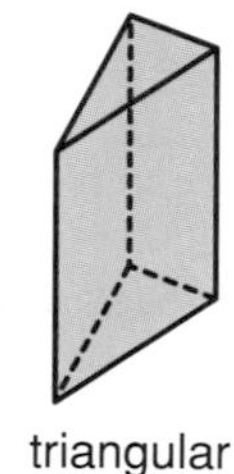
triangular prism

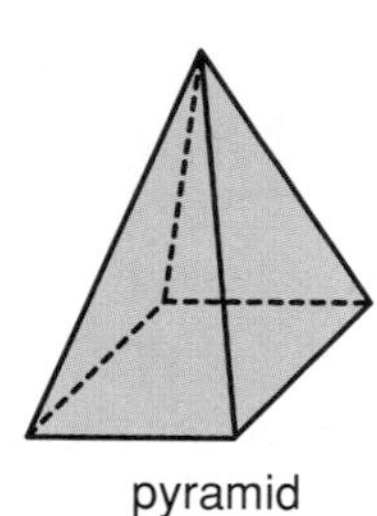
pyramid

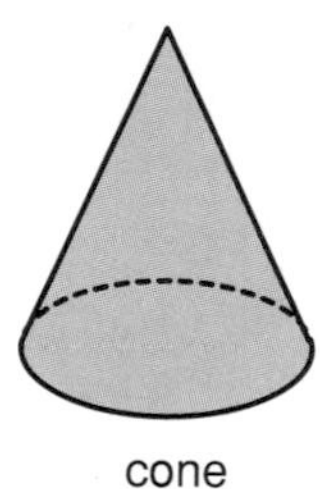
cone

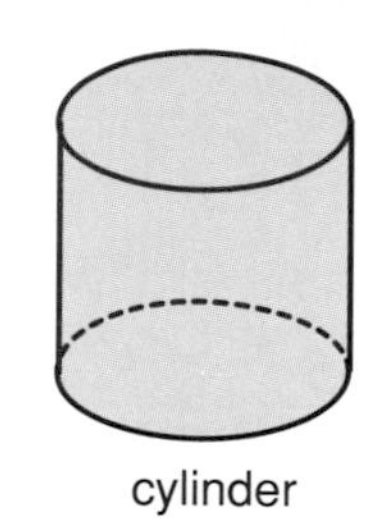
cylinder

Before sculpting, many artists sketch different perspectives of their figures. In geometry, it is often helpful to sketch figures before solving a problem related to those figures.

Example 1

Use dot paper to sketch a rectangular prism that is 4 units long, 3 units high, and 5 units deep.

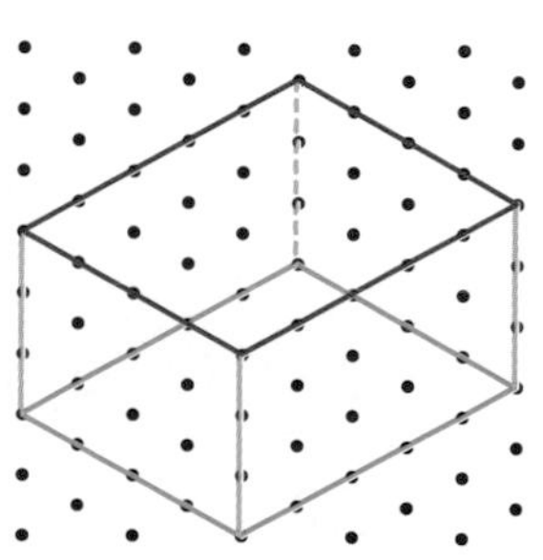

Step 1 Lightly draw the edges of the bottom of the prism that are 4 units by 5 units. Complete the other two bottom edges.

Step 2 Lightly draw vertical segments at the vertices of the base. Each segment is 3 units high.

Step 3 Complete the top of the prism.

Step 4 Go over your lines. Use dashed lines for the edges of the prism you can't see from your perspective and solid lines for the edges you can see.

You could draw several other perspectives of the same prism. Suppose you sat it up on end or rolled it over on its side.

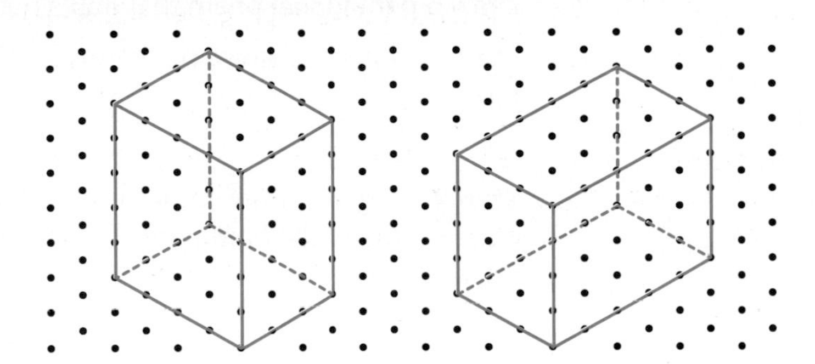

Prisms, like the ones above, have flat surfaces. The surfaces of a prism are called **faces.** The faces meet to form the **edges** of the prism. The edges meet at corners called **vertices.**

All prisms have at least one pair of faces that are parallel and congruent. These are called **bases** and are used to name the prism. The prisms shown above are *rectangular prisms.* As you can see, the bases of a rectangular prism change depending on how the prism is positioned.

Examples

Tell the dimensions of the base and height of each rectangular prism.

2

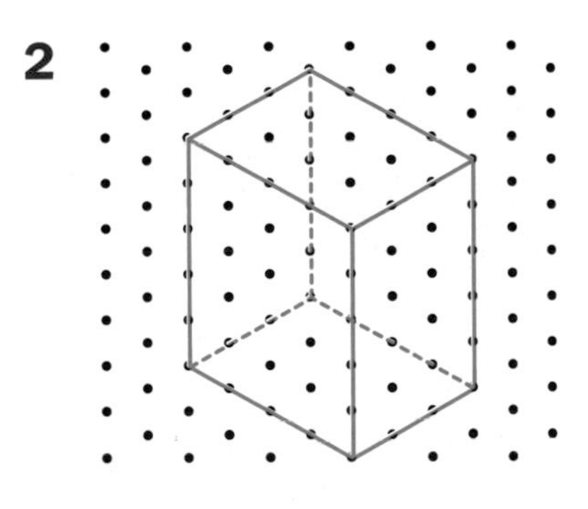

The base is 3 units by 4 units.
The height is 5 units.

3

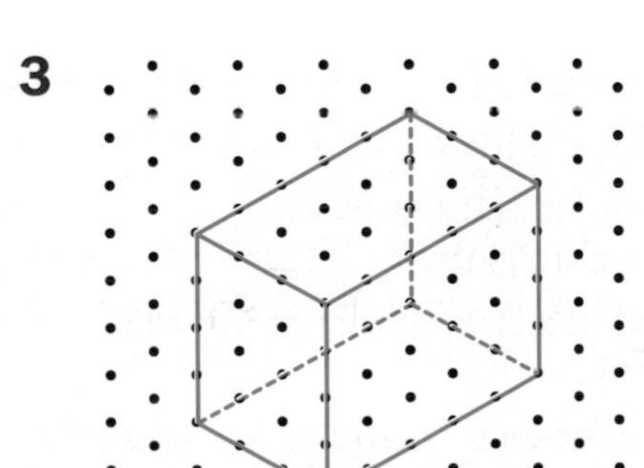

The base is 5 units by 3 units.
The height is 4 units.

Other geometric shapes can be used for the bases of a prism.

Example 4

Draw a pentagonal prism that has a height of 4 units.

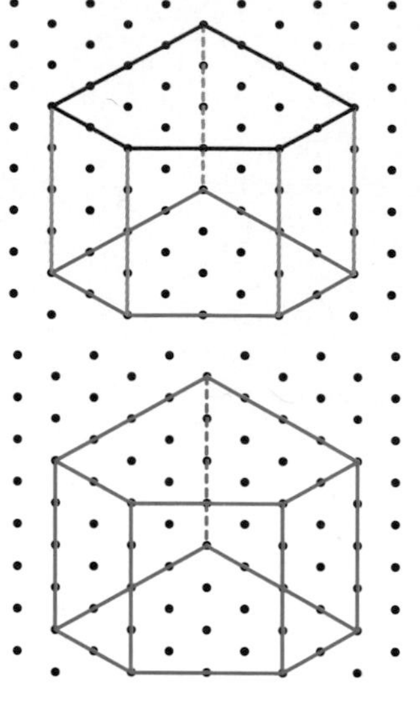

Step 1 Lightly draw a pentagon for the base.

Step 2 Lightly draw lines 4 units high from each vertex of the pentagon. Then draw the top pentagon.

Step 3 Go over the lines you have drawn. Make the edges you can see solid and use dashed lines for the edges you can't see.

Checking for Understanding

Communicating Mathematics

Read and study the lesson to answer each question.

1. **Tell** which solids pictured in this lesson have one or more surfaces that can be rectangles.
2. **Write** the steps you would use to draw a prism that has a triangle for a base.
3. **Show** how to determine which lines are solid and which lines are dashed in a geometric drawing.

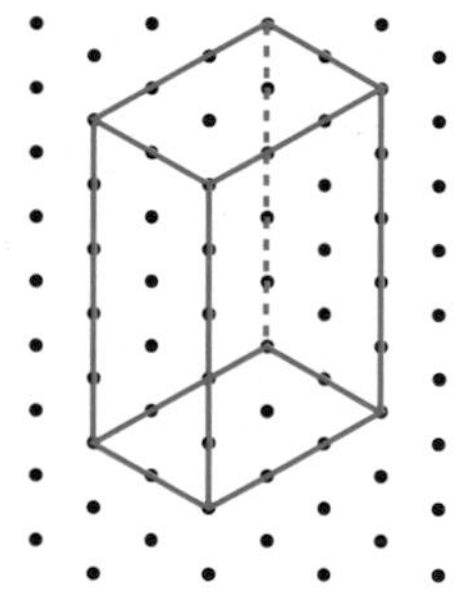

Guided Practice

4. Tell the dimensions of the prism shown at the right.
5. Draw another view of the pentagonal prism in Example 4.

Exercises

Independent Practice

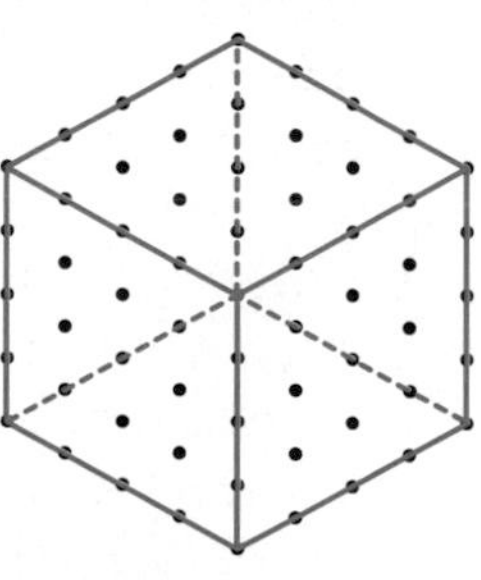

6. Tell the dimensions of the prism shown at the right.
   a. What is the height of the prism?
   b. What type of prism is this?
7. Draw three views of a rectangular prism that is 3 units by 2 units by 1 unit.
8. Draw a prism that has a hexagon as a base and is 3 units tall.
9. Explain how you would draw a pyramid that has a parallelogram as a base.
10. Write a few sentences to tell how you would draw a cylinder. Then draw one.

11. **Make a Model** of the figure described at the right. Then make a drawing of your model.

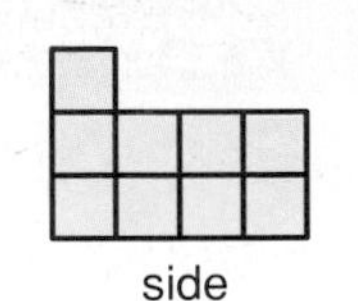
side

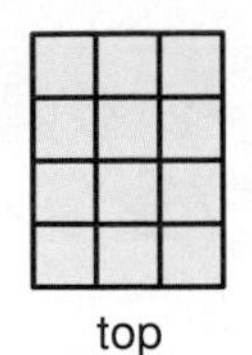
top

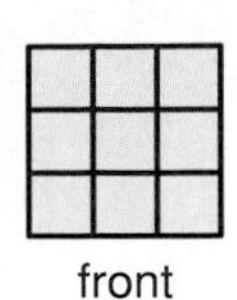
front

Mixed Review

12. **Statistics** Find the interquartile range of the set {73, 70, 71, 64, 74, 71, 78, 68, 79}. *(Lesson 4-6)*
13. Solve $\frac{3}{17} = \frac{n}{68}$. *(Lesson 9-2)*
14. **Geometry** Find the area of a circle whose diameter is 12.4 meters. *(Lesson 12-1)*

Problem Solving and Applications

15. **Home Decorating** Miki has eight plastic milk crates that he can arrange in his room to store his books, Nintendo® games, and music. His mom has said he can arrange them in his room any way he wishes. Draw at least three ways he can arrange the crates so that he can store items in each of them.
16. **Critical Thinking** Copy the figure at the right, which is the outline of two prisms laying side by side. One prism is 3 units by 1 unit by 1 unit. The other prism is 3 units by 3 units by 1 unit. Complete the drawing to show each prism.

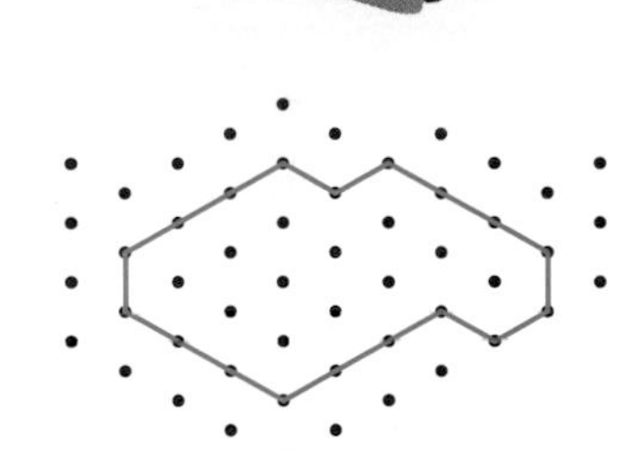

17. **Journal Entry** Write a list of several careers that would involve drawing three-dimensional figures. Tell how they would use them.

CULTURAL KALEIDOSCOPE

Federico Peña

In 1983, Federico Peña became the first Hispanic to be elected mayor of Denver, Colorado. He graduated from college with a degree in law and practiced law for 10 years before becoming mayor. During Mayor Peña's two terms in office, Colorado was devastated by one of the worst recessions in the state's history. Despite the economic conditions, Mayor Peña's focus was to change the economy and create new jobs and industry for the people of Denver.

Perhaps his most significant contribution was winning voter approval for the construction of Denver International Airport, the only major airport project to be approved in the United States since 1970. More than 60 design and engineering firms submitted ideas for this 54 square mile airport site.

Other accomplishments of Mayor Peña include bringing the Colorado Rockies, a National League expansion baseball team, to Denver and winning voter approval for a 937,000-square foot convention center.

Cooperative Learning

12-3B Nets

A Follow-Up of Lesson 12-3

Objective

Recognize a solid from its net and sketch it.

Materials

scissors
graph paper
tape

Every solid with at least one flat surface can be formed from a pattern called a **net.**

Try this!

Work in groups of four.

- Draw the figure shown at the right on graph paper. *What geometric figures form the parts of this net?*

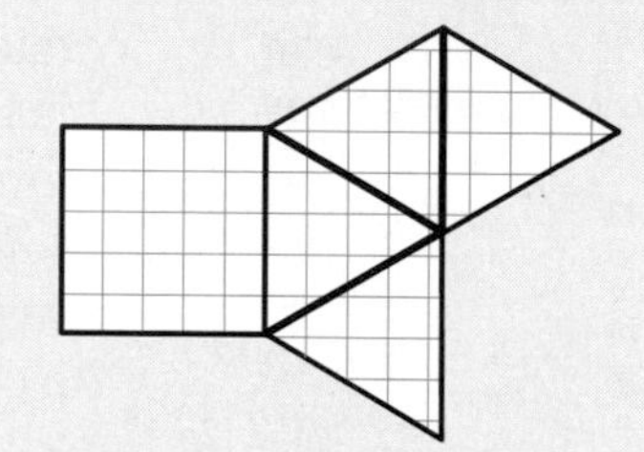

- Cut the figure out and fold along the heavy lines. Tape the edges together to form a solid.
- Have each person in the group sketch their perspective of the solid formed when the edges were taped together.

What do you think?

1. What solid did you form when the edges were taped together?
2. Compare your sketch of the solid with others in your group. What differences and similarities do you see?
3. Study your model. Is there another net that would produce this same solid? If so, draw it.
4. Describe the solid formed by each net. Then sketch the solid.

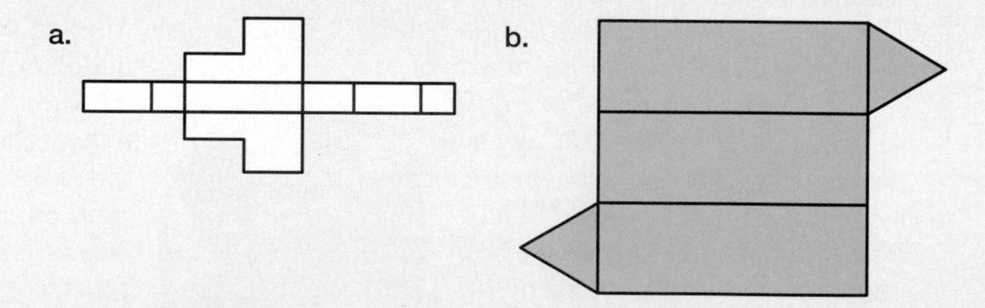

Application

5. **Manufacturing** The Blocko Company uses nets of plastic to form cubes used for support in cloth-covered alphabet blocks. Use graph paper to draw all possible nets they might consider to manufacture a cube.

12-4 Surface Area of Prisms

Objective

Find the surface area of rectangular and triangular prisms.

Words to Learn

surface area
triangular prism

Vegetables, such as carrots and celery, are often cut at a diagonal so that the pieces have a greater surface area. They cook faster when cut this way. Why?

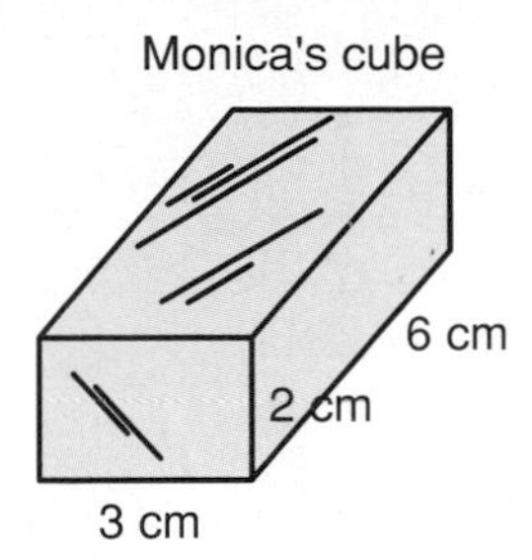

After measuring cooling rates in a science lab, Monica was trying to determine why her solution cooled at a different rate than Larry's solution. They each had the same amount of solution, and each block of ice placed in the solution had the same weight. The only noticeable difference was the shape of the two blocks of ice.

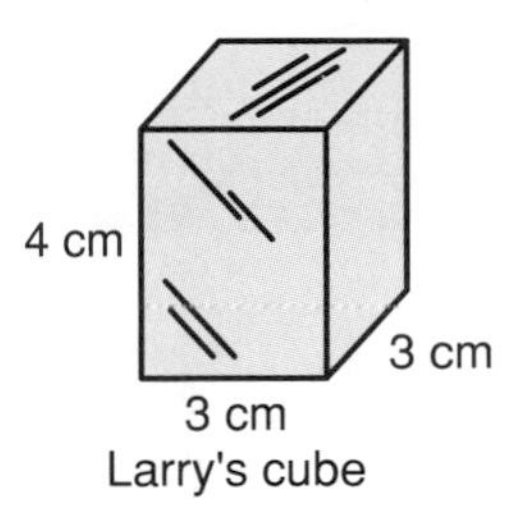

Before the experiment, Monica's block measured 3 centimeters wide, 6 centimeters long, and 2 centimeters thick. Larry's block measured 3 centimeters wide, 4 centimeters long, and 3 centimeters thick. Their teacher suggested that finding the surface area of each block of ice would explain the difference in cooling rates.

An ice cube is an example of a rectangular prism. To find the **surface area** of a prism, you need to find the area of each face and then add.

Mini-Lab

Work with a partner.

Materials: graph paper, scissors, tape

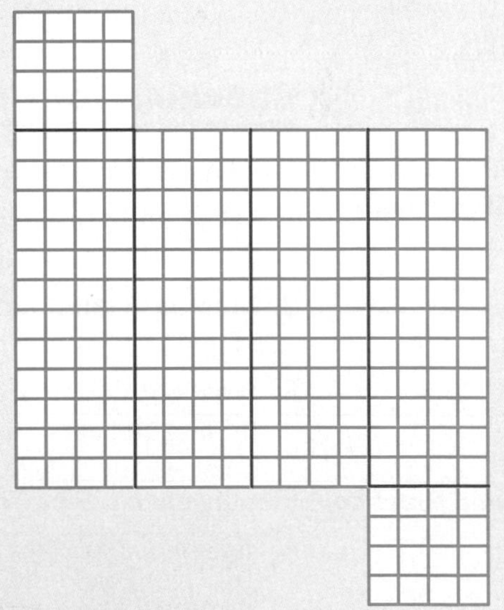

- Draw the pattern shown at the right on graph paper and cut it out. *This pattern is an example of a net.*
- Fold the pattern along the lines and tape the edges to form a rectangular prism.

Talk About It

a. What is the length, width, and height of the prism?

b. What is the area of each base?

c. What is the area of each of the other four faces?

d. What is the total surface area?

A **triangular prism** is a prism whose bases are triangles. In Lesson 12-3, we learned that geometric figures can be drawn from several perspectives. The prism at the right is drawn so that the bases are *not* on the top and bottom.

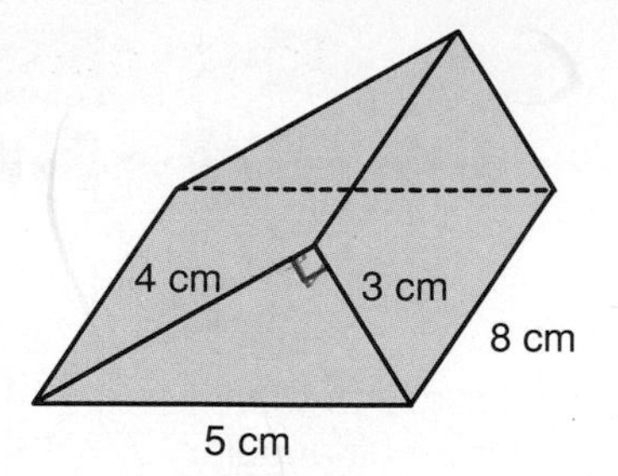

Example ***Problem Solving***

Physics **When light travels through a prism, it is refracted to form the colors of the rainbow. A special coating can be placed on a prism like the one shown above to enhance the brilliance of the rainbow. The amount of coating needed is based on the surface area of the prism. Find the surface area of the prism.**

To find the surface area of this prism, we need to find the area of each of the faces. Since the bases are triangles, use the formula $A = \frac{1}{2}bh$ to find the area of each triangle. The other three faces are rectangles.

face	dimensions	area	
one base	$b = 4, h = 3$	$A = \frac{1}{2}(4)(3)$ or	6
second base	$b = 4, h = 3$	$A = \frac{1}{2}(4)(3)$ or	6
face	$b = 8, h = 3$	$A = 8(3)$ or	24
face	$b = 8, h = 4$	$A = 8(4)$ or	32
face	$b = 8, h = 5$	$A = 8(5)$ or	40
		TOTAL	108

The surface area of the prism is 108 square inches.

Checking for Understanding

Communicating Mathematics

Read and study the lesson to answer each question.

1. **Tell** what figures you see when you view a triangular prism from different perspectives.
2. **Draw** a rectangular prism that has all six sides congruent. What do we call this shape?
3. **Write** a quick way to find the surface area of the prism in the Mini-Lab without finding the area of each face.

Guided Practice

Identify each prism. Then find its surface area.

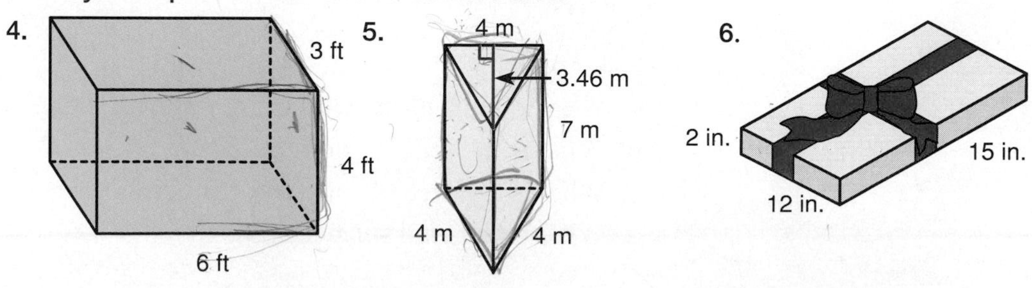

Exercises

Independent Practice

Find the surface area of each prism.

7.

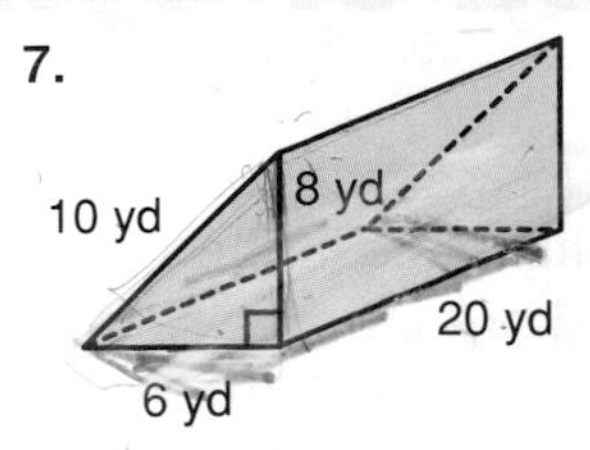

8.

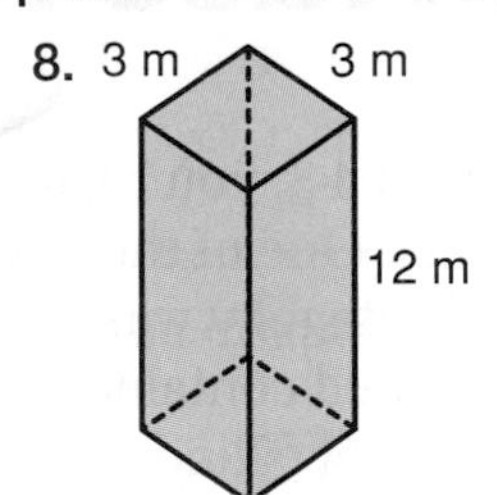

9.

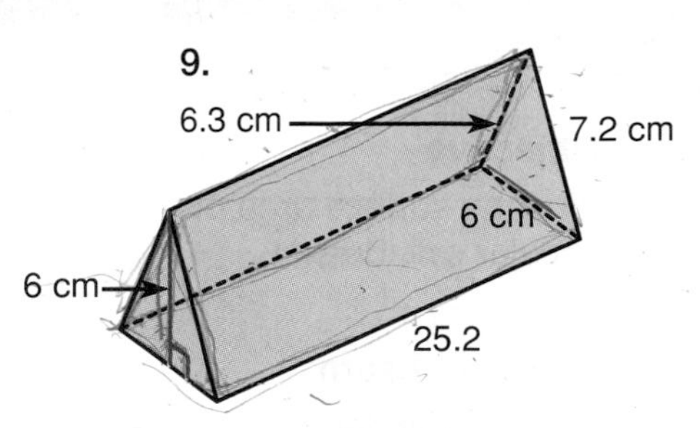

10.

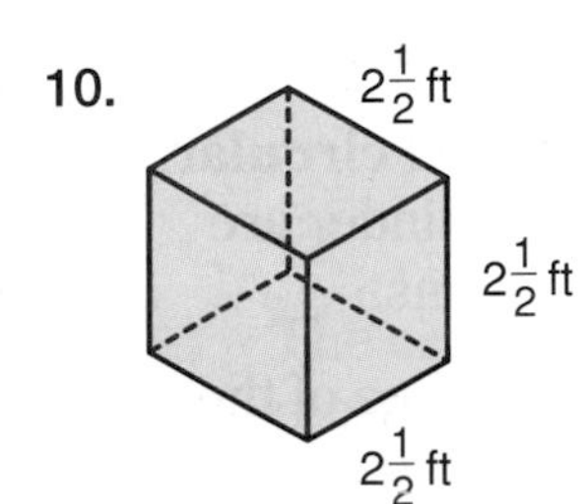

11.

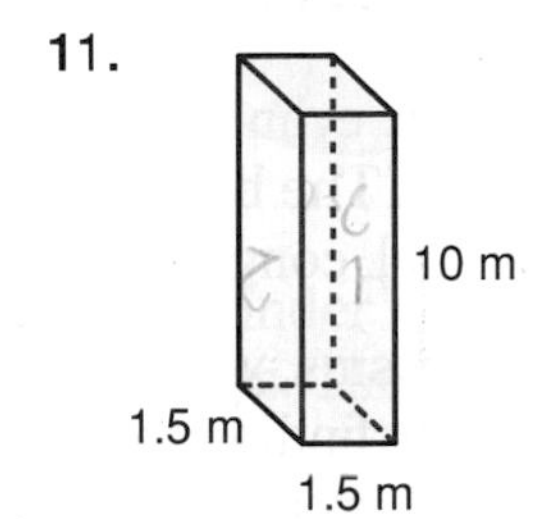

12.

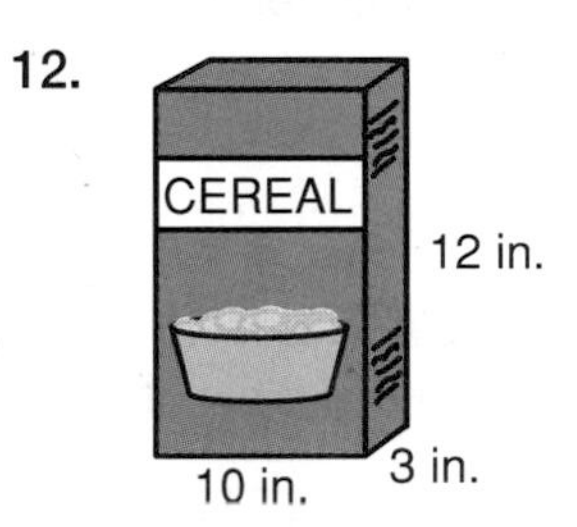

Draw a figure to show a prism for each description. Then find the surface area of the prism.

13. a rectangular prism with a length of 4 centimeters, a width of 7 centimeters, and a height of 3 centimeters

14. a triangular prism whose bases are right triangles with sides 5 meters, 12 meters, and 13 meters and whose height is 10 meters

15. a rectangular prism with all edges 8 feet long

Mixed Review

16. Evaluate the expression $2^4 \cdot 4^3$. *(Lesson 1-9)*

17. Use divisibility rules to determine if 726 is divisible by 4. *(Lesson 6-1)*

18. Draw a hexagonal prism that is 5 units tall. *(Lesson 12-3)*

Problem Solving and Applications

19. **Science** Refer to the science experiment on page 475. Which block of ice would cool the solution the quickest? Explain your answer.

20. **Critical Thinking** Suppose you had eight identical wooden blocks you are gluing together to form a prism. What is the shape of the prism that would require the least amount of paint to cover it?

21. **Design** Mr. Mayoree has been commissioned to build a glass display case in the shape of a trapezoidal prism. The case will be made of panes of glass held together by brass edging. Find how many square inches of glass he needs to build this display case.

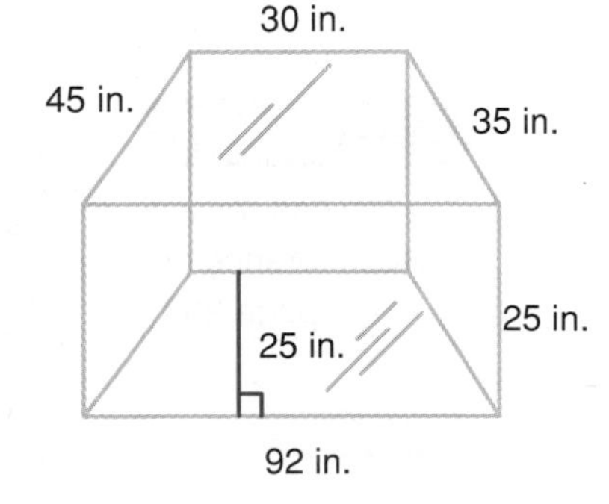

22. **Journal Entry** Find an example of a prism in your house and measure it. Write a sentence about how surface area can be used with this object.

Exercises

Independent Practice **Find the surface area of each cylinder.**

8. 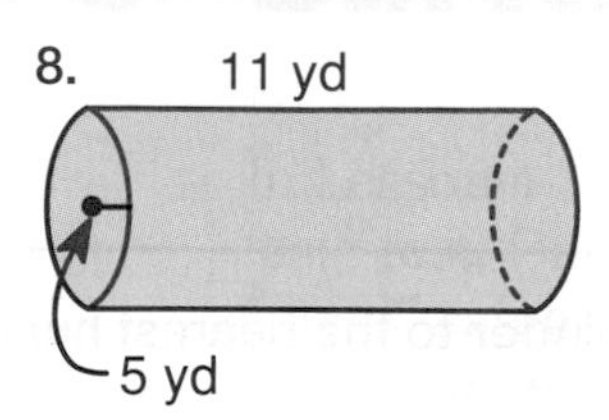

9. 

10. 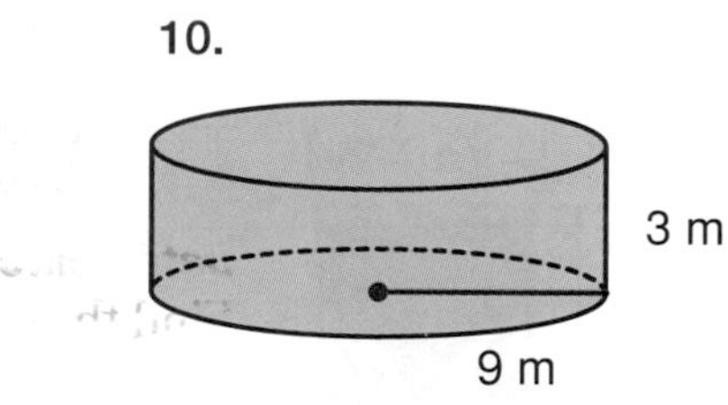

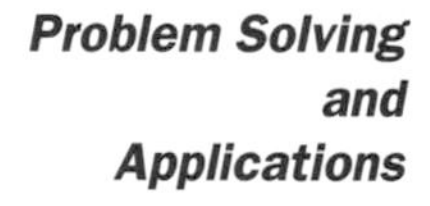

Draw each cylinder. Then find its surface area.

11. The radius of the base is 3 inches and its height is 5 inches.
12. The diameter of the base is 30 centimeters and its height is 12 centimeters.
13. The area of its base is 24.2 square meters and its height is 20 meters.

Mixed Review

14. Solve $6b - 5 < 7$. Show the solution on a number line. *(Lesson 2-10)*
15. Solve $h = 8\frac{3}{4} \div \frac{1}{2}$. *(Lesson 7-9)*
16. Express 35.2% as a decimal. *(Lesson 10-3)*
17. **Geometry** Draw a rectangular prism that is 5 inches long, 3 inches wide, and 2 inches tall. Then find the surface area of the prism. *(Lesson 12-4)*

Problem Solving and Applications

18. **Food** Wisconsin produces more cheese than any other state in the United States. When cheese is made, the curd is pressed into large hoops lined with cheesecloth. The cheese is then removed from the hoops and sliced into wheels. Some cheeses are sealed in wax or a thin plastic film to protect their moisture. Find the surface area of the plastic film on the small cheese wheel shown at the right.

19. **Critical Thinking** Will the surface area of a cylinder increase more if you double the height or double the radius of the base? Give examples to support your answer.

20. **Design** The model at the right is a preliminary sketch of a type of trash can for the school cafeteria. The school plans to paint them using the school's colors.
    a. Find the surface area of the container.
    b. The enamel paint they are going to use covers 200 square feet per gallon. Approximately how many trash cans can be covered with 1 gallon of paint?

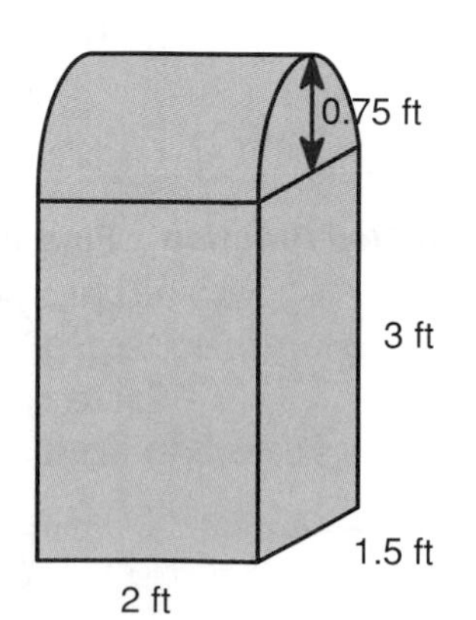

21. **Manufacturing** Morton® salt packages its table salt in cylinder-shaped containers whose base has a diameter of $3\frac{1}{2}$ inches and whose height is $5\frac{1}{2}$ inches. Morton places a label around the curved surface of the cylinder. If the label is placed $\frac{1}{4}$ inch from the top and bottom, what is the surface area of the label?

DATA SEARCH

22. **Data Search** Refer to pages 462 and 463.
Find the approximate areas of the circles formed by the edges of a basketball, a baseball, and a soccer ball when they are cut exactly through their centers.

12 Assessment: Mid-Chapter Review

1. Find the area of a circle whose radius is 6 cm. *(Lesson 12-1)*

2. **Probability** Shu Ping likes to play a game at the arcade where he rolls a ball up a ramp and it lands in one of four hoops. What is the probability that Shu Ping will score 50 points on the first ball? *(Lesson 12-1)*

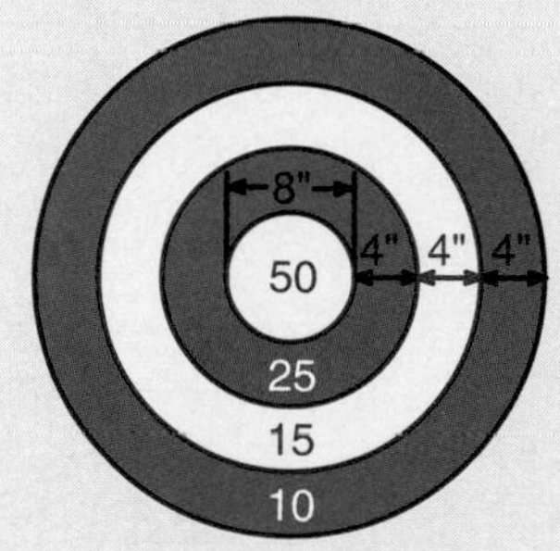

3. Use cubes to build a model of the solid from the perspectives shown. *(Lesson 12-2)*

side top front

4. Sketch two different views of a rectangular prism that is 4 feet wide by 3 feet tall by 1 foot long. *(Lesson 12-3)*

Find the surface area of each solid. *(Lessons 12-4, 12-5)*

5.

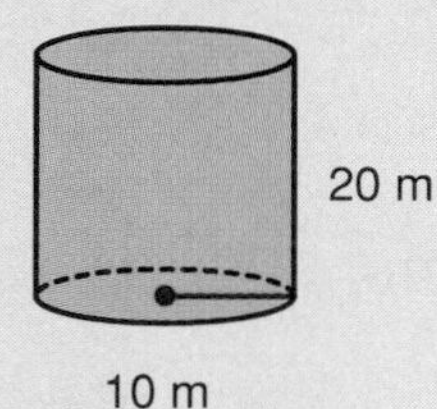

6.

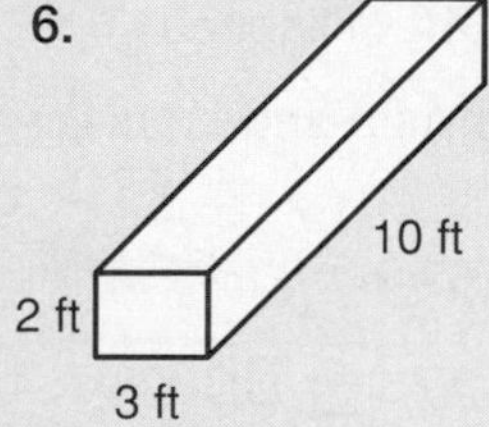

7.

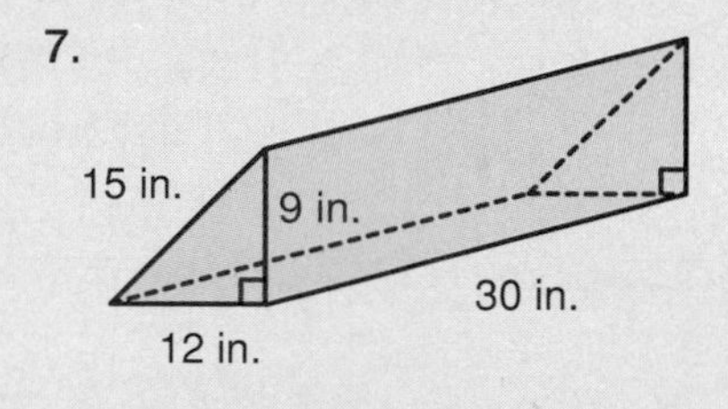

12-6 Volume of Prisms and Cylinders

Objective

Find the volume of prisms and circular cylinders.

Words to Learn

volume

In Lessons 12-4 and 12-5, you found the amount of material it takes to make prisms and cylinders. However, if you are very thirsty, you don't care how much material it takes to make a drink box or a can of juice. You want to know which one has more juice inside. In this lesson, you will learn how to find the amount of space occupied by a prism or cylinder.

In geometry, **volume** is the measure of the space occupied by a solid. It is measured in cubic units. Two common units of measure for volume are the cubic centimeter (cm^3) and the cubic inch (in^3).

In Lesson 12-2, you used cubes to make models of solid shapes. The prism model at the right was built with 24 cubes. If each cube is a centimeter cube, the volume of the prism is 24 cubic centimeters.

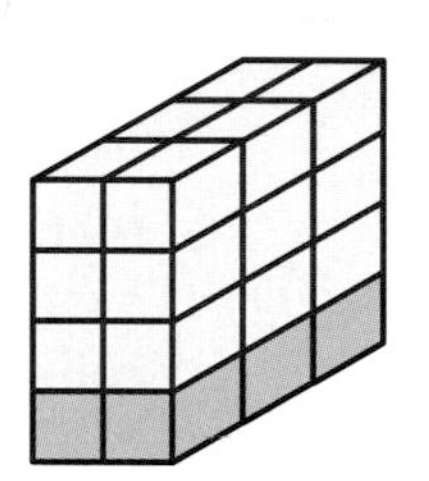

The dimensions of the prism are 2 centimeters by 3 centimeters by 4 centimeters. The model is made of 4 layers. Each layer contains 6 cubes. Notice that 6 is the area of the base, found by multiplying $3 \cdot 2$.

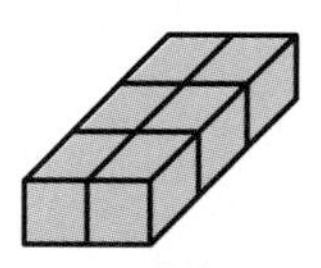

Volume of a Rectangular Prism	**In words:** The volume (V) of a rectangular prism is the area of the base (B) times the height (h). **In symbols:** $V = Bh$ or, since $B = \ell w$, $V = \ell wh$

Example 1

Find the volume of a rectangular prism with length 15 inches, width 13 inches, and height 17 inches.

First draw the prism and label the dimensions.

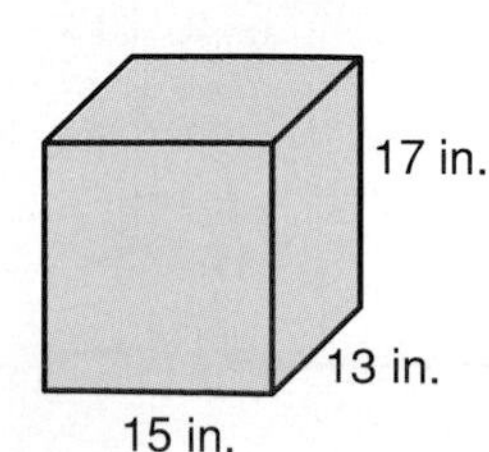

$V = \ell wh$

$V = 15(13)(17)$ *$\ell = 15, w = 13, h = 17$*

15 [×] 13 [×] 17 [=] 3315

The prism has a volume of 3,315 cubic inches.

To find the volume of any other type of prism, you can still use the $V = Bh$ formula.

Example 2

Find the volume of the triangular prism.

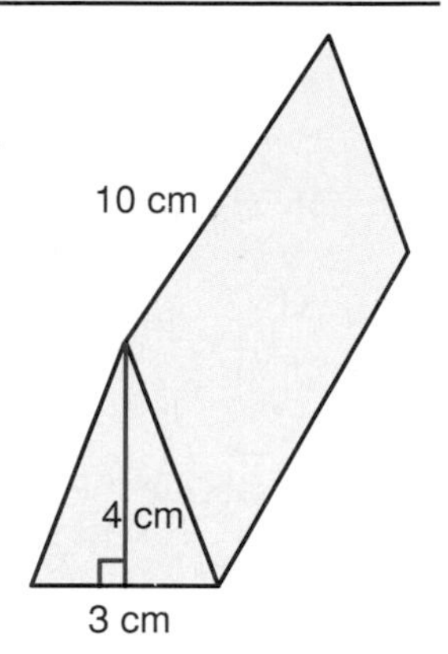

First find the area of the base, which is a triangle.

$A = \frac{1}{2}bh$ *Formula for area of a triangle.*

$A = \frac{1}{2} \cdot 3 \cdot 4$ *Replace b with 3 and h with 4.*

$A = 6$ *The area of the base is 6 cm².*

Now find the volume of the prism.

$V = Bh$ *Formula for the area of a prism*

$V = 6 \cdot 10$ *Replace B with 6 and h with 10.*

$V = 60$ The volume of the triangular prism is 60 cm^3.

To find the volume of a cylinder, you again use the formula, $V = Bh$. However, since the area of the base of a cylinder is the area of a circle (πr^2), the formula becomes $V = \pi r^2 h$.

Volume of a Cylinder	**In words:** The volume (V) of a cylinder is the area of the base (B) times the height (h). **In symbols:** $V = Bh$ or $V = \pi r^2 h$

Example 3 *Problem Solving*

Home Economics You bought a 10-pound sack of sugar. The sack is a rectangular prism 9 inches by 8 inches by 5 inches. You have two cylinder-shaped canisters into which the sugar can be poured. Which canister is better suited to hold the sugar?

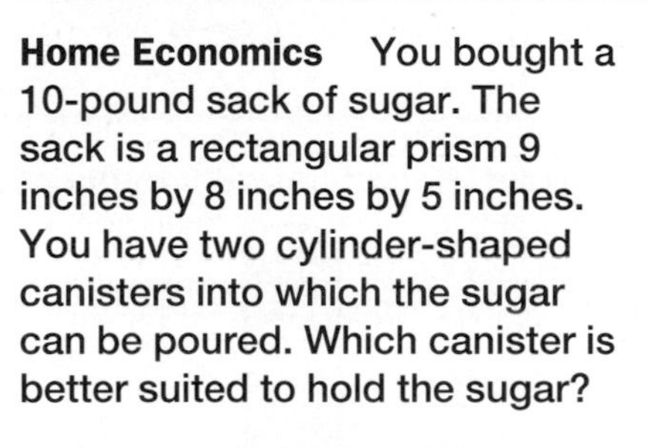

First find the volume of the sugar.

$V = \ell wh$

$V = 9 \cdot 8 \cdot 5$ or 360 in^3

Now find the volume of each cylinder.

$V = \pi r^2 h$ $V = \pi r^2 h$

$V = \pi(3)^2 \cdot 10 \approx 282.74$ in^3 $V = \pi(4)^2 \cdot 10 \approx 502.65$ in^3

The first canister is not large enough to contain the sugar. The second canister is the better choice.

Calculator Hint

Use your memory keys to save key strokes. The volume of each canister has factors of π and 10.

 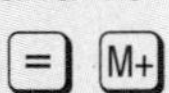 

π × 10 = M+ stores the product. The second volume can be found using MR × 4 x² = 502.65482.

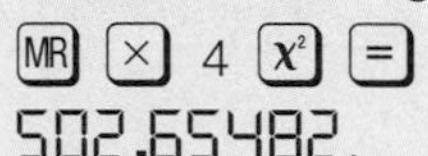

Some calculators may have memory keys named STO and RCL.

Checking for Understanding

Communicating Mathematics **Read and study the lesson to answer each question.**

1. **Write** a sentence to state how the volume formulas for the rectangular prism, triangular prism, and cylinder are alike.
2. **Write** a sentence to state how the volume formula for a rectangular prism is different from the formula for a cylinder.
3. **Tell** why it is impossible to find the volume of a square.
4. **Draw** a cube and find its volume. What is a short way to write the formula for the volume of a cube?
5. **Tell** why the volumes of the cylinders in Example 3 are approximations.

Guided Practice **Find the volume of each solid. Round answers to the nearest tenth.**

6.
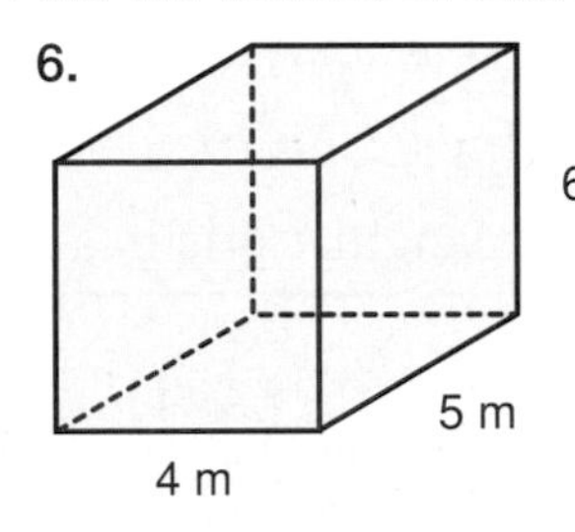

7.
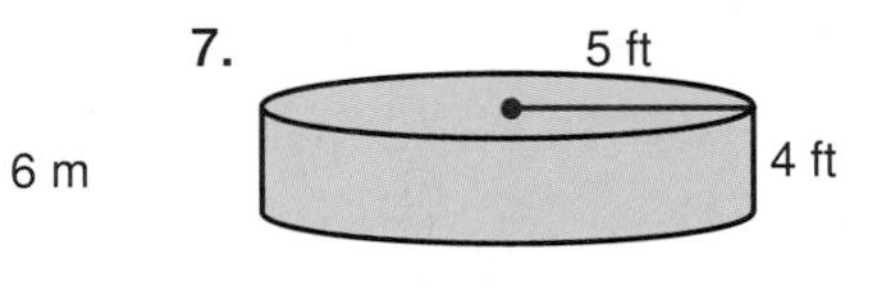

8.
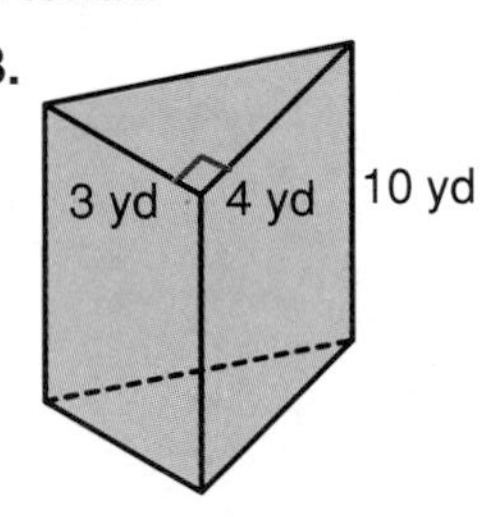

9. 6 cm, 5 cm

10.
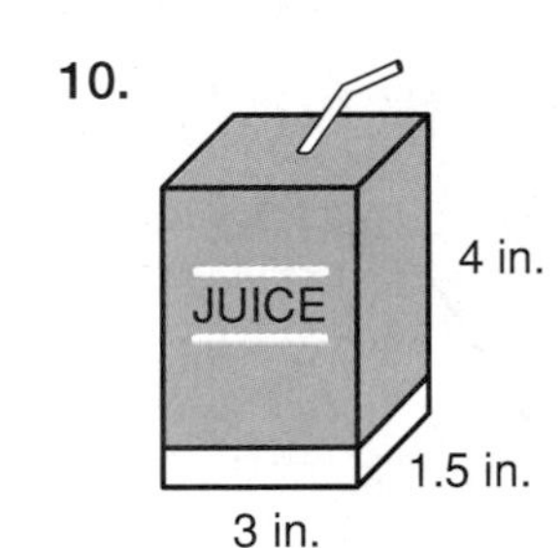

11.
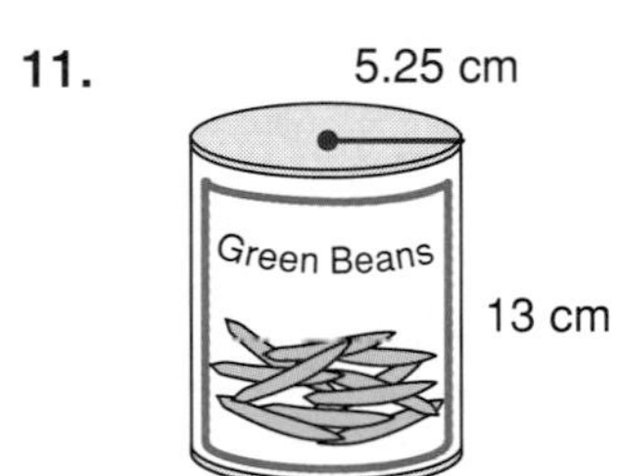

Exercises

Independent Practice **Find the volume of each solid. Round answers to the nearest tenth.**

12.
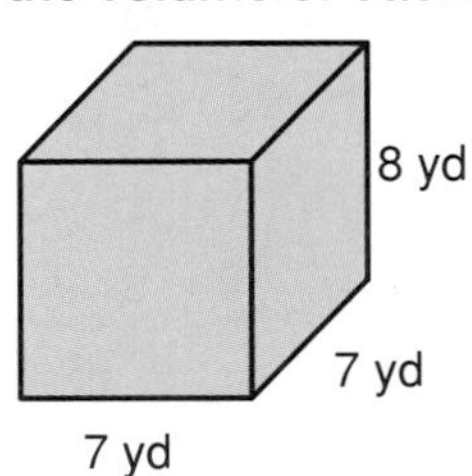

13.
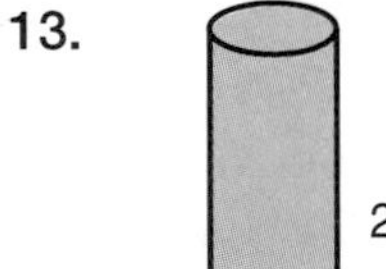
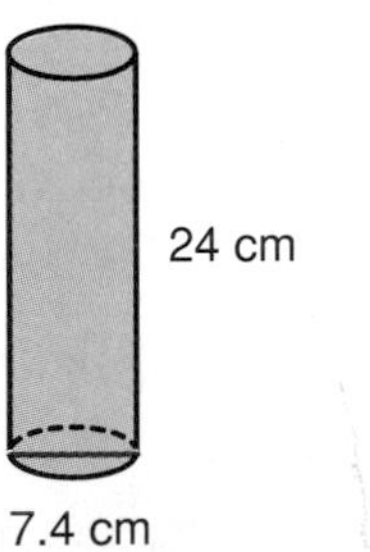

14.
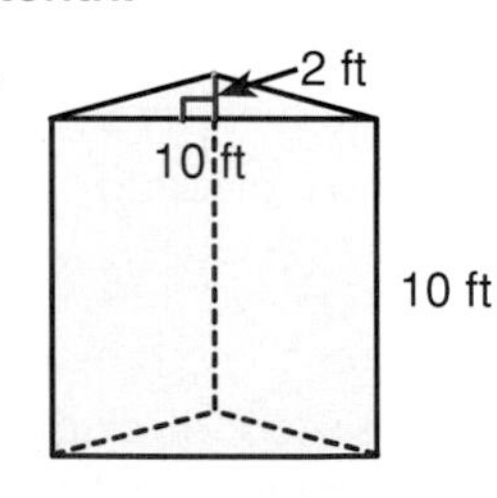

15.
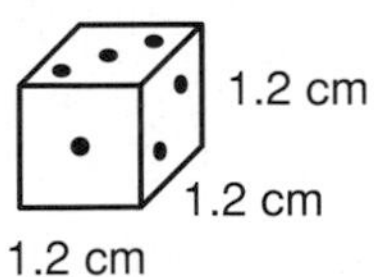

16.
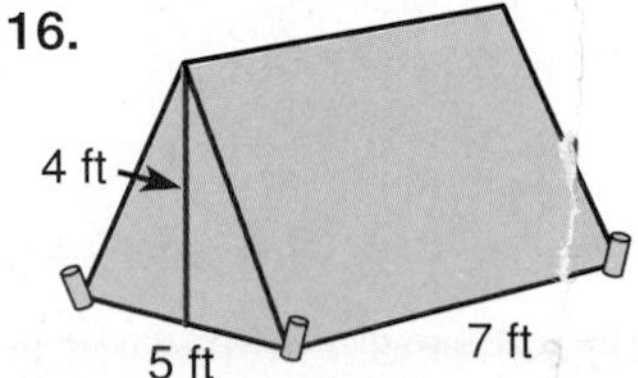

17.
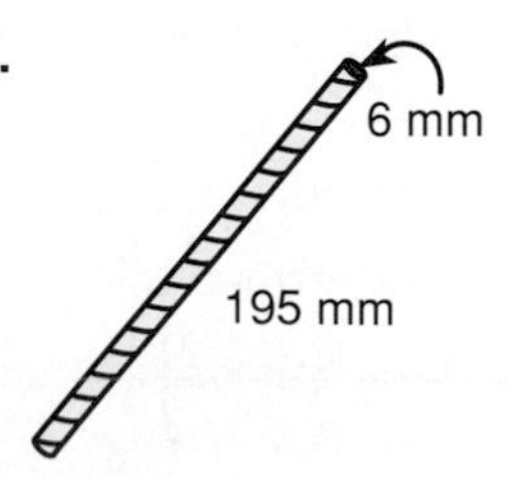

Draw each figure. Then find its volume.

18. Find the volume of a cylinder whose diameter is 12 meters and is 14 meters tall.

19. Find the volume of a triangular prism that is 10 feet tall and has a base that is a right triangle with legs 5 feet and 12 feet.

20. Find the volume of a pentagonal prism that is 5 centimeters tall and has a base that has an area of 25.4 square centimeters.

Mixed Review

21. **Geometry** Draw two congruent trapezoids. *(Lesson 5-6)*

22. Find the best integer estimate for $-\sqrt{76}$. Then check your estimate by using a calculator. *(Lesson 8-2)*

23. Find the surface area of a cylinder 9 centimeters tall with a radius of 4 centimeters. *(Lesson 12-5)*

Problem Solving and Applications

24. **Marketing** The graph uses cylinders to show the change in sales of music media from 1986 to 1992.

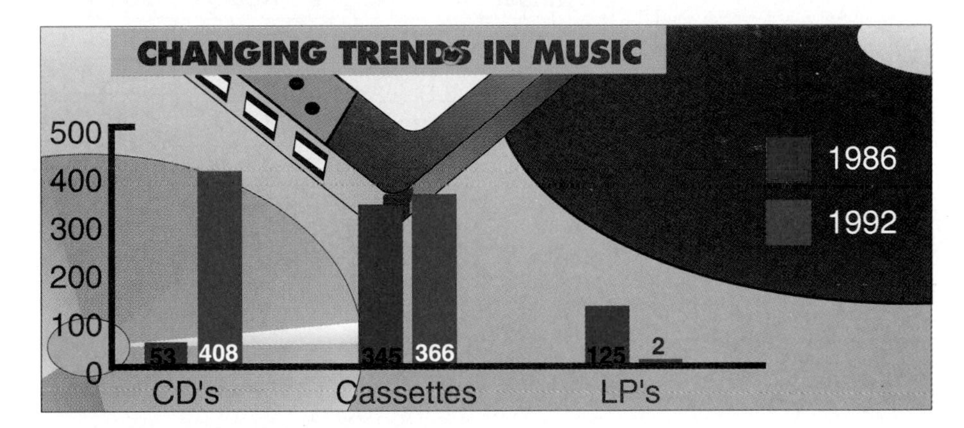

a. Which type of music medium had a decrease in sales from 1986 to 1992?

b. Which type of music medium had the greatest percentage increase in sales?

c. Write a sentence to explain why you think the amounts in each category increased or decreased.

d. Suppose you built a three-dimensional model of this graph and the diameter of each cylinder was 50 millimeters. What would be the volume of each cylinder, if each million sold was 1 millimeter high?

25. **Critical Thinking** A rectangular ditch is dug to hold a pipe being installed for a drainage system. What is the volume of dirt to be filled in around the pipe? *Assume the pipe is the same length as the ditch.*

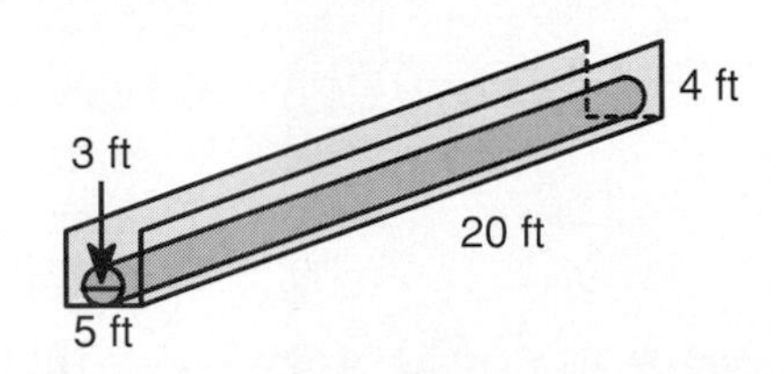

26. **Journal Entry** Write about some real objects that will help you remember what a cube and a cylinder look like. Do you think that any of these objects were designed to have a specific volume?

Cooperative Learning

12-6B Surface Area and Volume

A Follow-Up of Lesson 12-6

Objective

Investigate how surface area and volume are related.

Materials

soft drink can
calculator

Containers are often manufactured so that the least amount of materials is used for a given volume. This is an application of surface area and volume.

Try this!

Work in groups of three.

- Find the height and radius of the base of a soft drink can to the nearest millimeter. Find the volume and surface area. Record your findings in the first row of a chart like the one below.

Cylinder	Height	Radius	Volume	Surface Area
soft drink can	120	33	410,543	31,723
#1				
#2				

The first aluminum soft drink can appeared on grocery shelves in 1963.

- Alter the height and radius measurements to create a new cylinder that has approximately the same volume as the soft drink can. Then find the surface area of this cylinder. Record your findings in the second row of the chart.
- Use your calculator to create four more cylinders that have approximately the same volume as the soft drink can. Record your findings in the appropriate rows.

What do you think?

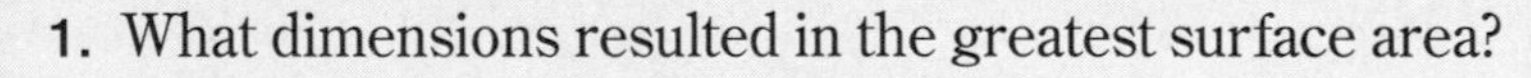

1. What dimensions resulted in the greatest surface area?
2. What dimensions resulted in the least surface area?
3. When the radius is greater than the radius of the original can, what happens to the height of the cylinder?

Application

4. **Manufacturing** If you were manufacturing a soft drink can, which dimensions would you probably want to use for your can? Explain your answer.
   a. Why do you think soft drink manufacturers have chosen the size that is used?
   b. If it costs 0.016¢ per square inch to manufacture a can, how much would a company save in producing 100,000 cans by changing to a smaller surface area for the same volume?

12-7 Volume of Pyramids and Cones

Objective

Find the volume of pyramids and circular cones.

Words to Learn

altitude
circular cone

The volume of the pyramid of Quetzacoatl in Cholula, Mexico is reported to be about 116 million cubic feet. If a rectangular prism having the same dimensions as the pyramid were built, it would have a volume of 348 million cubic feet.

What is the ratio of the volume of the pyramid to the volume of the prism?

$$\frac{\text{volume of pyramid}}{\text{volume of prism}} = \frac{116 \text{ million ft}^3}{348 \text{ million ft}^3} \text{ or } \frac{1}{3}$$

You can review ratios on page 338.

This ratio suggests that the volume of a pyramid is one-third the volume of the prism into which it will fit. So, $V = \frac{1}{3}Bh$.

The model shows how many pyramid-fulls of sand it takes to fill the prism.

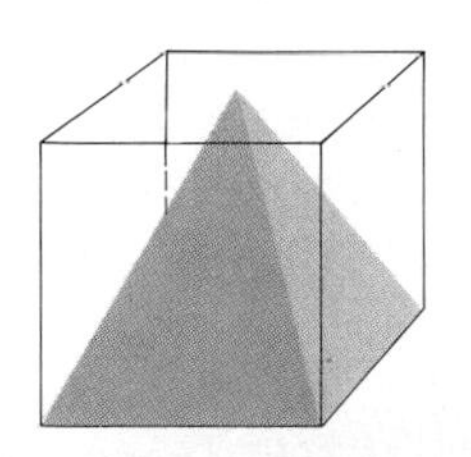

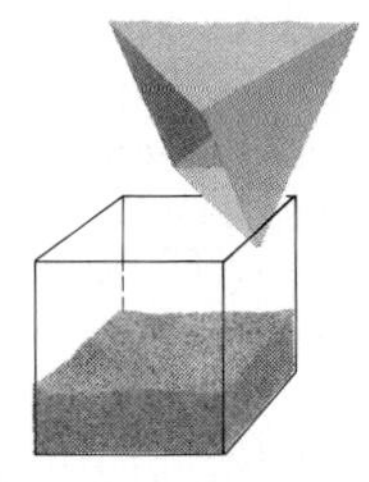

Volume of a Pyramid	**In words:** The volume of a pyramid equals one-third the area of the base (B) times the height (h). **In symbols:** $V = \frac{1}{3}Bh$

In June 1990, twelve students of Nanyang Technological Institute in Singapore, constructed a pyramid consisting of 263,810 bottle caps!

The segment that goes from the vertex of a pyramid to its base and is perpendicular to the base is called the **altitude.** The height of a pyramid is measured along the altitude. Like prisms, a pyramid can be named by the shape of its base.

Example 1

Find the volume of the square pyramid.

vertex
6 in.
4 in.
4 in.

$V = \frac{1}{3}Bh$

$V = \frac{1}{3}s^2h$ *Since the base is a square, $B = s^2$.*

$V = \frac{1}{3} \cdot (4)^2 \cdot 6$ *Replace s with 4 and h with 6.*

$V = \frac{96}{3}$ or 32

The volume of the square pyramid is 32 cubic inches.

Remember that the formula for volume of a cylinder was derived from the $V = Bh$ formula. A **circular cone** can fit inside a cylinder in the same way a pyramid fits inside a prism. The relationship of the volumes is also the same.

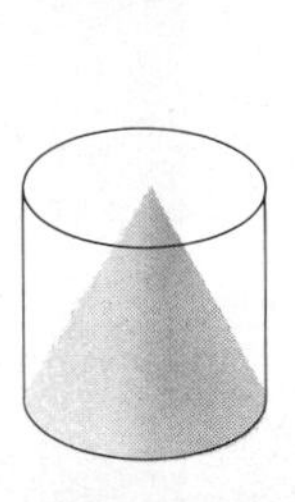 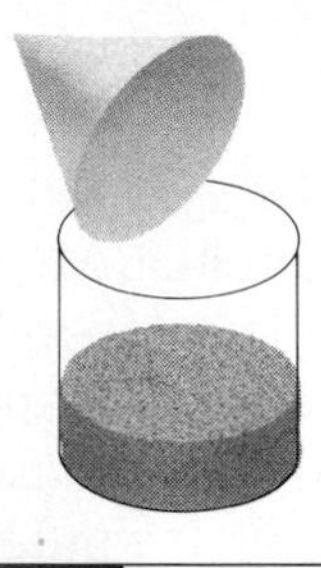  

Volume of a Cone	**In words:** The volume (V) of a cone equals one-third the area of the base (B) times the height (h). **In symbols:** $V = \frac{1}{3}Bh$ or $V = \frac{1}{3}\pi r^2 h$

Example 2 ***Problem Solving***

Food Drumstick® ice cream treats are ice cream cones filled to the top by combinations of ice cream, fudge, nuts, and other ingredients. The level cone is covered by paper to protect it. Find the volume of the Drumstick® treat shown at the right.

2.5 cm

15 cm

$V = \frac{1}{3}Bh$

$V = \frac{1}{3}(\pi r^2)h$ *Replace B with πr^2.*

$V = \frac{1}{3}(\pi \cdot 2.5^2) \cdot 15$ *Replace r with 2.5 and h with 15.*

$V = 1$ ÷ 3 × π × 2.5 x² × 15 = 

The volume of the Drumstick® is about 98.17 cubic centimeters.

Checking for Understanding

Communicating Mathematics

Read and study the lesson to answer each question.

1. **Tell** what shape forms the faces of any pyramid regardless of the shape of the base.
2. **Write** a sentence to tell how the formulas for the volume of a pyramid and the volume of a cone are alike.
3. **Show** which figure at the right has a greater volume.

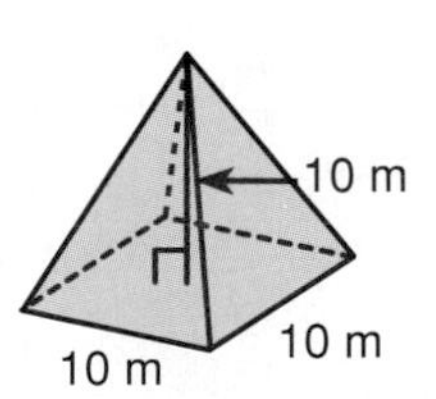

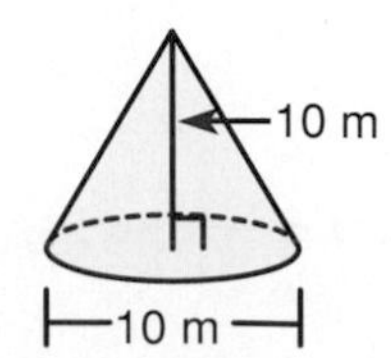

Guided Practice Find the volume of each solid. Round answers to nearest tenth.

4.
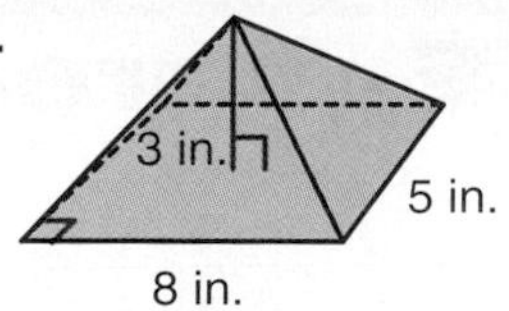

5.
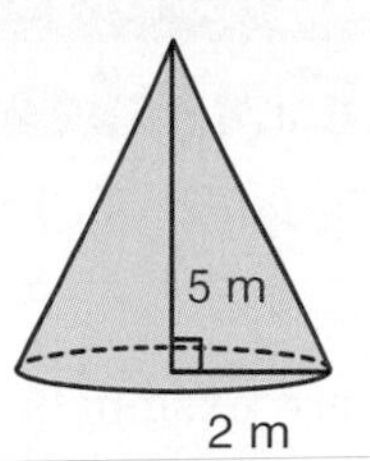

6.
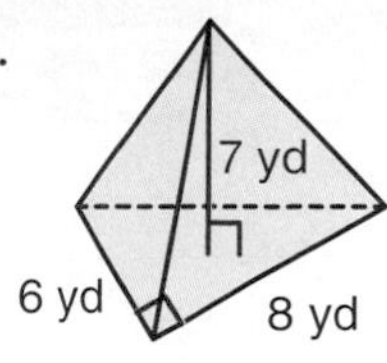

Exercises

Independent Practice Find the volume of each solid. Round answers to the nearest tenth.

7.
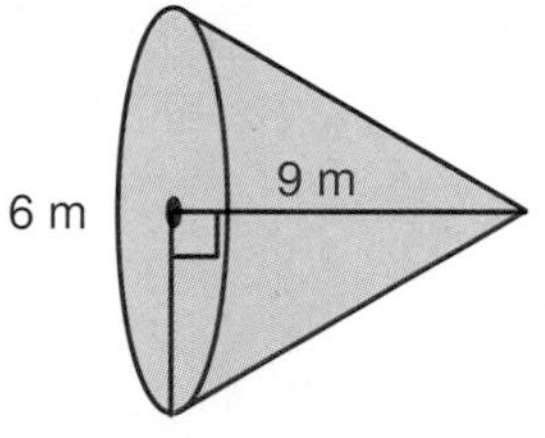

8.
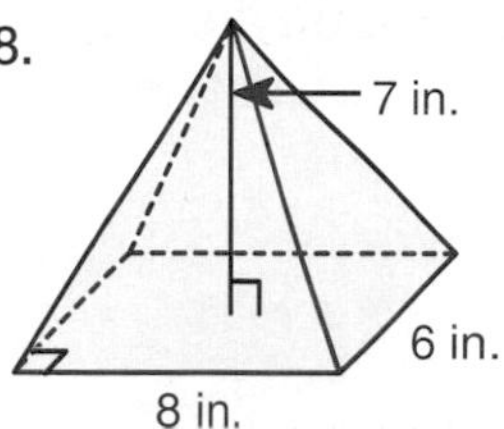

9.
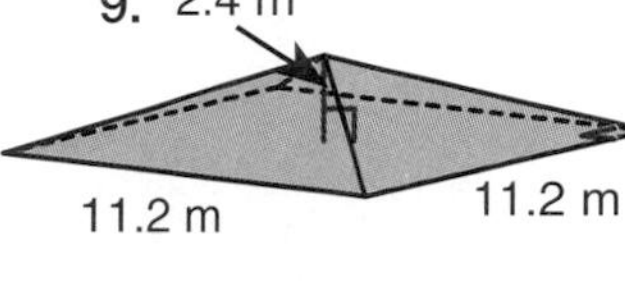

10.
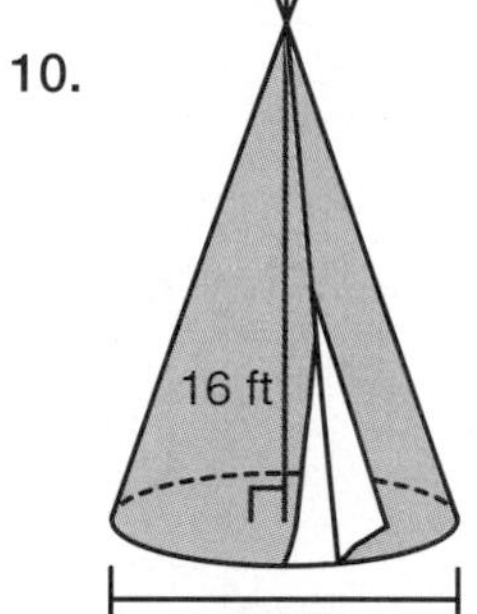

11.
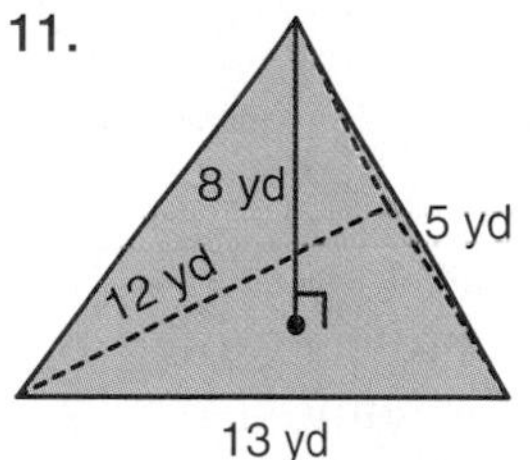

12.
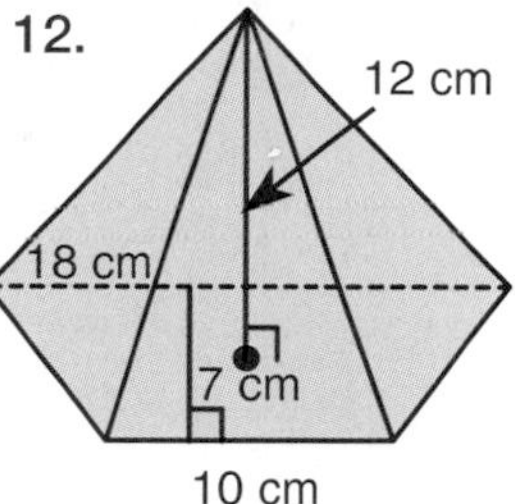

13. Find the volume of a pyramid 9 cm tall, with a base area of 15 cm^2.

14. Find the volume of a cone 12 feet tall, with a base area of 235.2 square feet.

Mixed Review

15. Write the next four terms of an arithmetic sequence that begins with 21 and has a common difference of 6. *(Lesson 7-5)*

16. Graph ΔDEF with vertices $D(3, 6)$, $E(10, 8)$, and $F(9, 3)$. Then graph with a scale factor of 2. *(Lesson 9-8)*

17. Find the volume of a cylinder whose diameter is 8 feet and height is 10 feet tall. *(Lesson 12-6)*

Problem Solving and Applications

18. **Critical Thinking** Suppose the dimensions of a prism are doubled. What changes would have to be made in the dimensions of a pyramid that fits inside the original prism so that the ratio of the volumes of the new pyramid and new prism would still be 1:3?

19. **Manufacturing** The funnel at the right holds the honey that goes into Nature's Best cookies. What is the maximum volume of the funnel? *Hint: What two solids make up this funnel?*

12 cm
6 cm
4 cm
1 cm

20. **Portfolio Suggestion** Select an item from this chapter that shows how you used a calculator or computer. Place it in your portfolio.

Cooperative Learning

12-7B Exploring Spheres

A Follow-Up of Lesson 12-7

Objective

Explore the surface area and volume of a sphere.

Materials

3 styrofoam balls
scissors
tape
straight pins

Balls are examples of spheres. Spheres, like circles, have a radius and circumference. Spheres also have surface area and volume.

Try this!

- Cut one ball in half. Trace around the edge to draw a circle. Cut out the circle.
- Fold the circle in half three times. Unfold and cut the 8 sections apart. Tape them in the pattern shown.

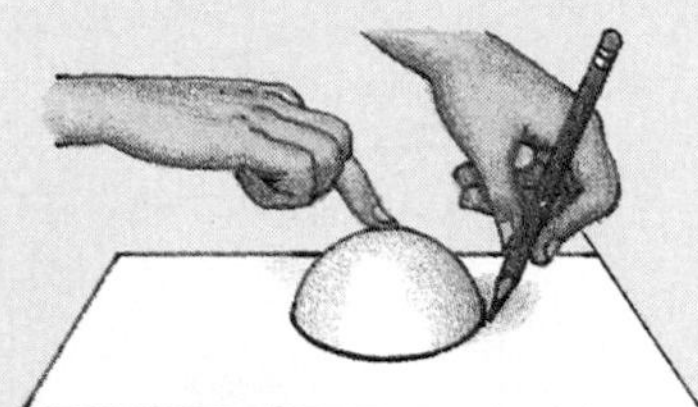

- Use pins to attach the pattern to another ball the same size as the first.

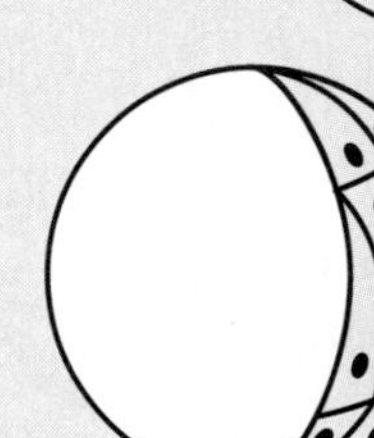
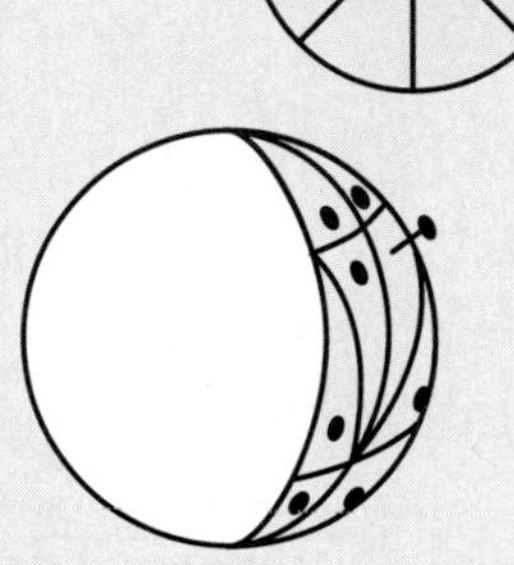

What do you think?

1. What part of the ball did the circle cover?
2. How many patterns would it take to cover the entire ball?
3. What is the area of each 8-section pattern?
4. Write an expression for the surface area of the ball.

Extension

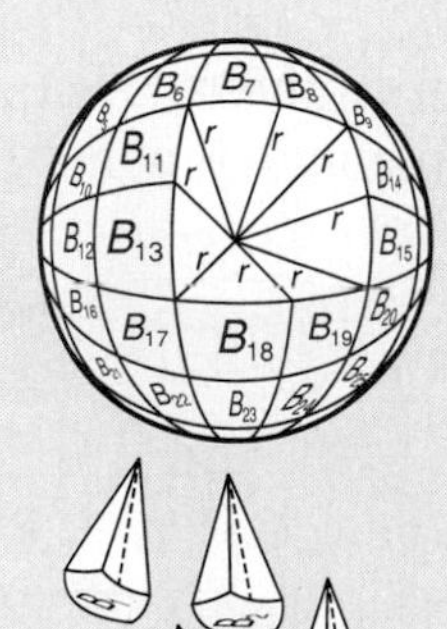

5. Suppose you cut a ball into small wedges so that each wedge has its vertex at the center.
   a. What shape does each wedge resemble?
   b. Describe the height and the base of that shape in relationship to the sphere?
   c. Write a sentence to describe each step for the following development of the formula for volume of a sphere.

   $V = \frac{1}{3}B_1h + \frac{1}{3}B_2h + \frac{1}{3}B_3h + \ldots$

   $V = \frac{1}{3}B_1r + \frac{1}{3}B_2r + \frac{1}{3}B_3r + \ldots$

   $V = \frac{1}{3}r(B_1 + B_2 + B_3 + \ldots)$

   $V = \frac{1}{3}r$ (surface area of the sphere)

   $V = \frac{1}{3}r(4\pi r^2)$ or $\frac{4}{3}\pi r^3$ *The formula for the volume of a sphere.*

12-8 Precision and Significant Digits

Objective

Describe a measurement using precision and significant digits.

Words to Learn

precision
significant digits

The winners in some high school track and field events are determined by the distances they achieve. Events involving shorter distances, such as high jump and shot put are measured to the nearest quarter inch. Events involving longer distances, such as the discus and javelin throws, are measured to the nearest half inch.

The **precision** of a measurement depends on the unit being used. The smaller the unit the more precise the measurement is.

Example 1 ***Problem Solving***

Health **The school nurse measures the heights of kindergarten students at the beginning of the year. How precise must the measurements be?**

- If she measures them to the nearest foot, some students would show no growth.
- If she measured them to the nearest inch, many would show growth.
- If she measured them to the nearest half-inch, most would show growth.
- Measuring them to the nearest tenth-inch would be more precise, but very difficult to do accurately.

So, the nearest half-inch would probably be precise enough.

The digits used in your measurements are **significant digits.** Significant digits are all the digits of a measurement that are known to be accurate plus one estimated digit.

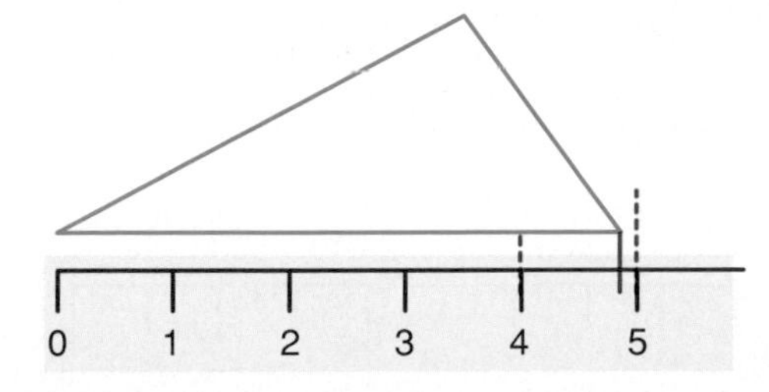

Suppose you wanted to measure the side of a triangle. The ruler you are using is marked in centimeters.

You can see that the side is longer than 4 centimeters, but shorter than 5. You estimate that the measurement is about 0.8 of the way from 4 to 5 centimeters. Your measurement is 4.8 centimeters. The measurement has two significant digits, the 4 that you know is accurate and the 8 that you estimate.

"When am I ever going to use this?"

An interior decorator strives to create an effect with color, furniture, fabric, wallpaper, and accessories. The decorator must be able to use proportions to create scale drawings. The measurements of windows and rooms must be precise to calculate the surface area needed of fabric, wallpaper, and paint.

For more information, contact your local chapter of the American Society of Interior Designers.

When determining significant digits, zeros sometimes pose a problem. If a zero does not fall between two significant digits and is only a placeholder for the placement of the decimal point, it is not a significant digit.

Number	20.3	4,200	0.00251	0.0580
Number of Significant Digits	3	2	3	3

Example 2

A pipe is measured as 12.73 meters long. Analyze this measurement.

To analyze a measurement, you use significant digits to determine how precise this measurement is.

- There are four significant digits. This means you know 12.7 to be accurate and the 3 in the hundredths place is an estimate. So, the measurement is exact to 0.1 meter.
- The estimated hundredths digits tells you the length is closer to 12.73 meters than to 12.7 or 12.8 meters.

When measurements are added, the sum can be no more precise than the least precise measurement. The least precise measurement has the fewest significant digits.

Example 3 ***Connection***

Geometry **The lengths of the sides of a triangle are measured as 18.64 cm, 7.092 cm, and 10.4 cm. Find the perimeter.**

$P = 18.64 + 7.092 + 10.4$ or 36.132 cm

The least precise measurement is 10.4. Its last significant digit is in the tenths place. Round the answer to tenths.

$36.132 \rightarrow 36.1$ The perimeter is about 36.1 cm.

Checking for Understanding

Communicating Mathematics

Read and study the lesson to answer each question.

1. **Tell** if it is truly possible to measure something exactly. Explain your answer.
2. **Tell** what unit of measure you would use to say how far you live from school. Tell why you chose this unit.
3. **Show** how precise your measurements can be using the ruler below.

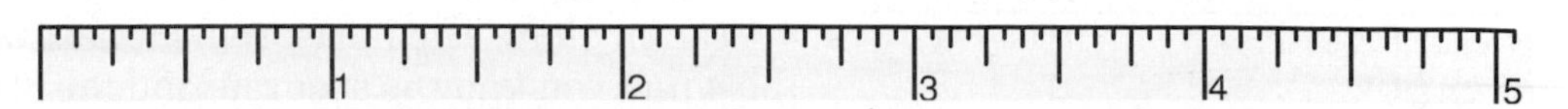

Guided Practice

4. A flag is 5.50 feet long.
   a. How many significant digits does this measurement have?
   b. How precise is this measurement?
5. Suppose each number in the table at the top of page 492 represents the measured lengths of boards in meters. How precise is each measurement?
6. Which would be the most precise measurement for a can of tomatoes: 2 pounds, 34 ounces, or 34.3 ounces? Explain your answer.
7. In 1990, the United States school systems spent $212,900,000,000 on education. Analyze this measurement.

Exercises

Independent Practice

Use significant digits to analyze each measurement.

8. 5.2 feet	9. 3.08 inches	10. 0.034 mm
11. 4.003 miles	12. 0.20020 cm	13. 4,300 km

14. Which is the more precise measurement for the length of a pencil, 18 cm or 18.0 cm? Explain your answer.
15. Which is the more precise measurement for the height of the Statue of Liberty, 100 yards or 305 feet?
16. A wire's length is measured as 0.02 meters.
    a. How many significant digits are there in 0.02?
    b. Suppose the measurement was actually 2.2 centimeters. How many significant digits are there in 2.2?
    c. Compare the preciseness of these two measurements.

Mixed Review

17. **Statistics** Refer to the data in Exercise 35 on page 158. Make a back-to-back stem-and-leaf plot of the data. *(Lesson 4-4)*
18. Express 9.4×10^{-5} in standard form. *(Lesson 6-11)*
19. **Geometry** Find the volume of a cone whose radius is 3.5 meters and whose height is 12 meters. *(Lesson 12-7)*

Problem Solving and Applications

20. **Critical Thinking** To be precise when multiplying measurements, your rounded answer should have the same number of significant digits as the least precise measurement. With that in mind, find the area of a circle whose radius is 4.2 centimeters.

21. **Data Search** Refer to page 668. How precise are the time measurements for the 1,500-meter race?
22. **Clothing** Men's dress shirts are sized according to neck size. The sizes may range from $14\frac{1}{2}$ to $17\frac{1}{2}$ in half-inch intervals. A size $16\frac{1}{2}$ means the shirt fits a man whose neck is $16\frac{1}{2}$ inches around. How do these shirt sizes relate to precision?

23. **Computer Connection** The following program finds the volume of a sphere. How precise is the output of the program?

```
10    INPUT R
20    PRINT 4/3 * 3.141592654 * R^3
```

Chapter 12

Study Guide and Review

Communicating Mathematics

Choose the letter of the correct word or words to complete each statement.

1. In geometry, three-dimensional figures are called __?__ .
2. __?__ is the measure of the space occupied by a solid.
3. The flat surfaces of a prism are called __?__ .
4. The area of a __?__ is π times the radius squared.
5. The __?__ of a measurement depends on the unit being used.
6. The bases of a __?__ are two parallel, congruent circular regions.
7. The volume of a __?__ is one-third the area of the base times the height.

a. pyramid
b. cylinder
c. rectangular prism
d. circle
e. precision
f. solids
g. faces
h. surface area
i. volume

8. Write in your own words how to find the surface area of a rectangular prism.
9. What geometric shape describes the curved surface of a cylinder?
10. Explain how to find the volume of a cylinder if you know the diameter and the height.

Self Assessment

Objectives and Examples	***Review Exercises***
Upon completing this chapter, you should be able to:	*Use these exercises to review and prepare for the chapter test.*

- find the area of circles *(Lesson 12-1)*

Find the area of a circle with a radius of 6 feet to the nearest foot.

$A = \pi r^2$

$A \approx 3.14(6^2)$

$A \approx 113.04$

The area of the circle is *about* 113 square feet.

Find the area of each circle to the nearest tenth.

11.

4 m

12.

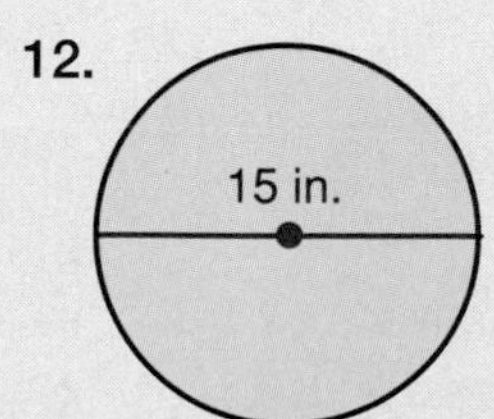

Objectives and Examples	*Review Exercises*

- sketch three-dimensional figures from different perspectives *(Lesson 12-3)*

Make the edges you can see solid and the edges you can't see dashed.

Sketch each solid.

13. a rectangular prism that is 3 units by 4 units by 6 units
14. a prism that has a hexagon as a base and is 6 units tall
15. a rectangular prism with all edges 4 units

- find the surface area of rectangular and triangular prisms *(Lesson 12-4)*

To find the surface area of a prism, find the area of each face. Then add all the areas.

Draw each prism. Then find its surface area.

16. a rectangular prism with a length of 2 feet, a width of 3 feet, and a height of 5 feet
17. a rectangular prism with all edges 3 centimeters long

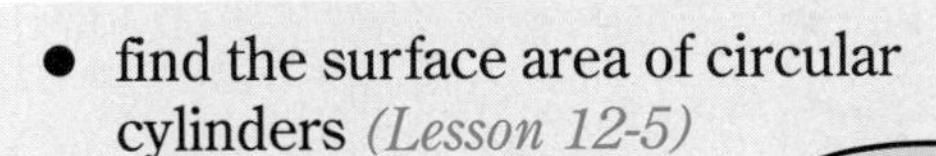

- find the surface area of circular cylinders *(Lesson 12-5)*

Surface area = areas of base + base + curved surface

$\pi r^2 + \pi r^2 + 2\pi rh$

Draw a cylinder. Then find its surface area.

18. The radius of the base is 5 meters and its height is 8 meters.
19. The diameter of the base is 40 yards and its height is 16 yards.

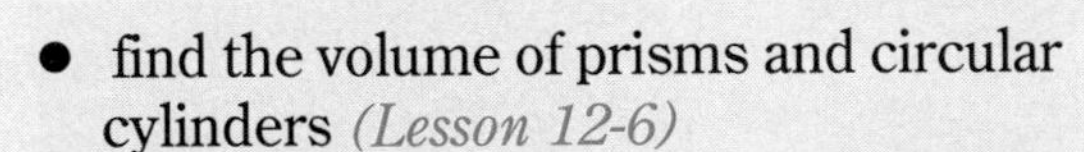

- find the volume of prisms and circular cylinders *(Lesson 12-6)*

Find the volume of the rectangular prism.

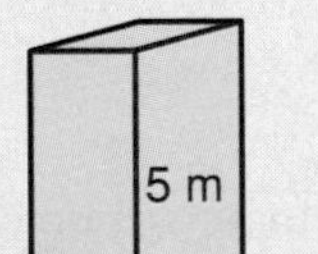

$V = \ell wh$

$V = 2(3)(5)$ or 30

The prism has a volume of 30 cubic meters.

Draw each figure. Then find its volume.

20. a cylinder with a radius of 8 inches and is 12 inches tall
21. a rectangular prism with a length of 6 meters, a width of 9 meters, and a height of 7 meters

- find the volume of pyramids and circular cones *(Lesson 12-7)*

Find the volume of the square pyramid.

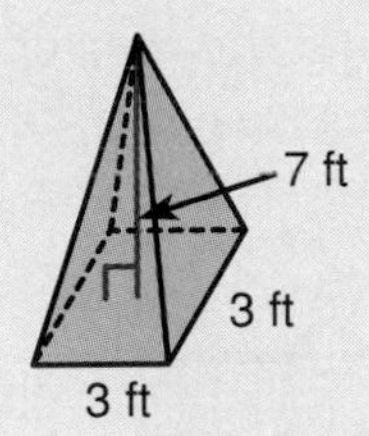

$V = \frac{1}{3}Bh$

$V = \frac{1}{3}(3 \cdot 3)7$ or 21

The volume of the square pyramid is 21 cubic feet.

Find the volume of each solid.

22. 23.

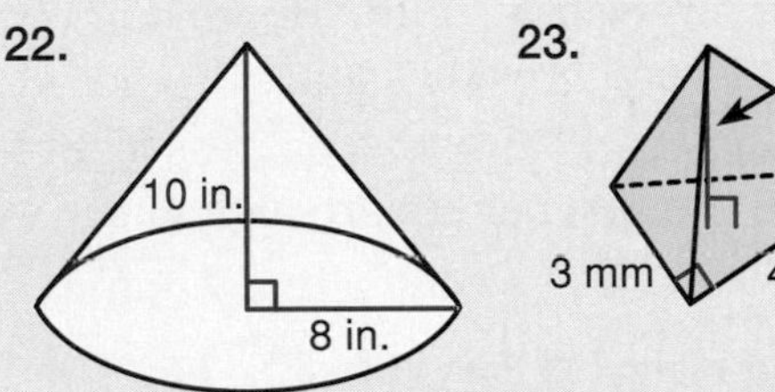

Objectives and Examples	*Review Exercises*
• describe a measurement using precision and significant digits *(Lesson 12-8)* Analyze 4.07 meters. There are three significant digits. 4.0 is accurate and the 7 in the hundredths place is an estimate. The measurement is closer to 4.07 meters than it is 4 meters or 4.1 meters.	Tell how many significant digits are in each measurement. **24.** 0.04 millimeters **25.** 12.8 feet **26.** 2,400 yards **27.** 28.20 meters **28.** 0.3009 centimeters **29.** Use significant digits to analyze 5.02 miles.

Applications and Problem Solving

30. Music Find the area of the top of a compact disc if its diameter is 12 centimeters and the diameter of the hole in its center is 1.5 centimeters. *(Lesson 12-1)*

31. Art An artist is creating a pyramid piece of art using differently-colored marble cubes. There is one cube on the top, two cubes in the next layer, three cubes in the next layer, and so on until 10 layers are completed. How many cubes did the artist use to create this piece of art? *(Lesson 12-2)*

32. Food A cardboard salt container is in the shape of a circular cylinder. The top has a diameter of 8.4 centimeters, and the container is 13.8 centimeters tall. Find the volume of this container. *(Lesson 12-6)*

Curriculum Connection Projects

- **Consumer Awareness** Find the surface area and the volume of two sizes of your favorite canned food and breakfast cereal.
- **Smart Shopping** Think of a pizza as a circle. Find the area of three different sizes of pizza. Then find the cost per square inch to see which size is the best buy for the amount of pizza you get.

Read More About It

Thompson, Julian. *The Taking of Mariasburg.*
Muller, Robert. *The Great Book of Math Teasers.*
Roth, Charlene Davis. *The Art of Making Puppets and Marionettes.*

Chapter

12 Test

Find the area of each shaded region to the nearest tenth.

1. 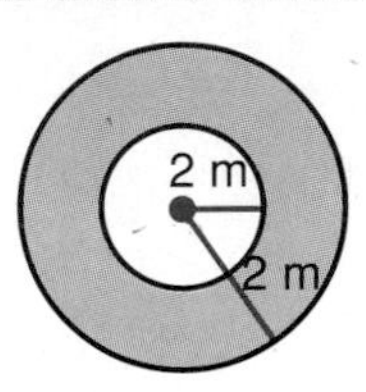

2. 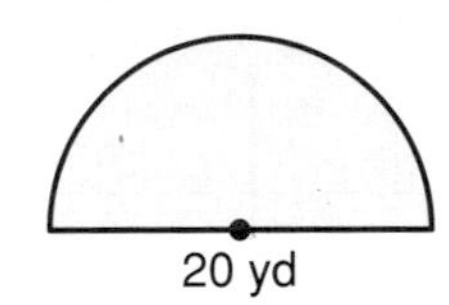

3. 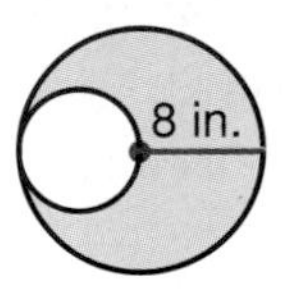

4. Draw two views of a rectangular prism whose dimensions are 4 units by 3 units by 5 units.

Find the surface area of each solid to the nearest tenth.

5. 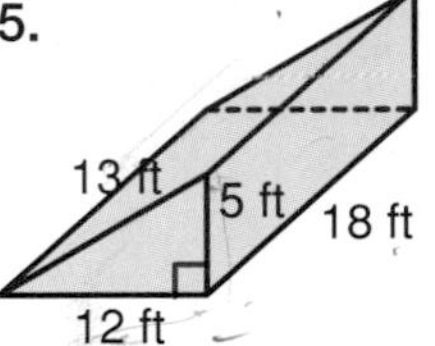

6. 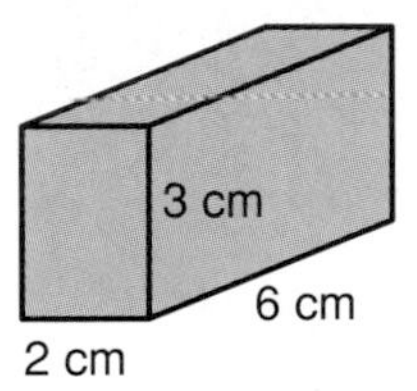

7. 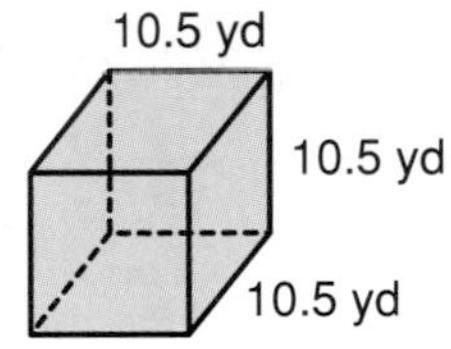

8. 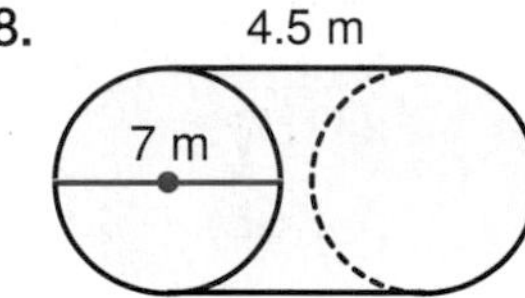

9. Draw a prism that has a pentagon as a base and is 3 units tall.

Find the volume of each solid to the nearest tenth.

10. 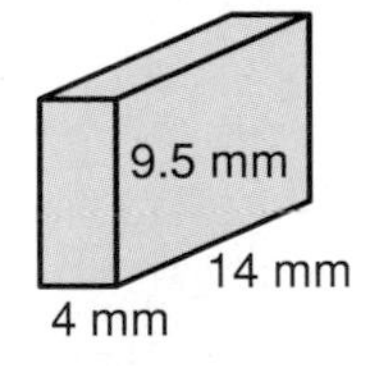

11. 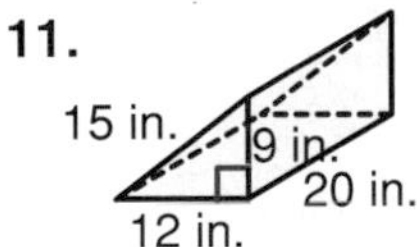

12. 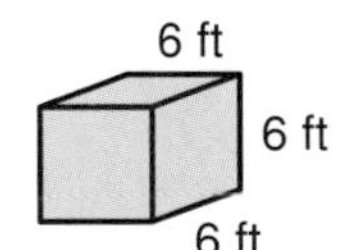

13. 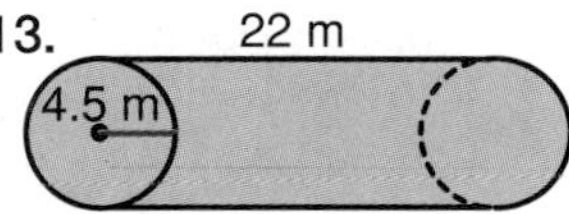

14. 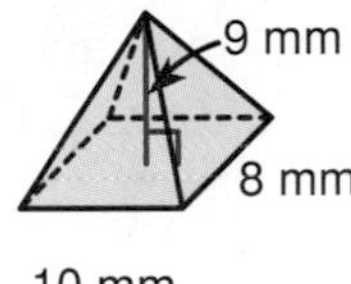

15. 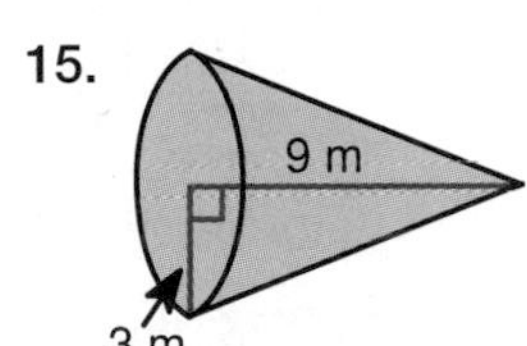

16. 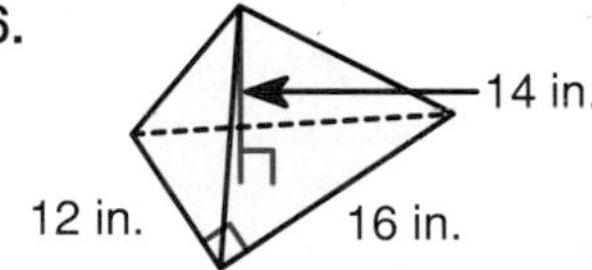

17. How many significant digits are in 0.004?
18. Use significant digits to analyze 8.20 feet.
19. A grain silo has a diameter of 20 feet and is 52 feet tall. Another grain silo has a diameter of 30 feet and is 38 feet tall. Which silo has a greater volume?
20. Phillipe counted six cyclists going past his house. His sister Rene counted 17 wheels. The cyclists rode bicycles and tricycles. Use a model to find how many of each Phillipe and Rene saw.

Bonus How could you find the surface area of a pyramid?

Standardized Format, Chapters 1–12

Academic Skills Test

Directions: Choose the best answer. Write A, B, C, or D.

1. Bonnie is reading a 186-page book. She needs to read twice as many pages as she has already read. How many pages has she read?

A 372 pages **B** 124 pages
C 93 pages **D** 62 pages

2. Max is 4 years younger than three times Carl's age. Carl is 14. Which equation can you solve to find Max's age?

A $3(m) - 4 = 14$
B $m - 4 = 14$
C $m + 4 = 3(14)$
D $m - 4 = 3(14)$

3. Which figure has rotational symmetry?

A

B 

C 

D 

4. To the nearest tenth, what is the distance between points A and B?

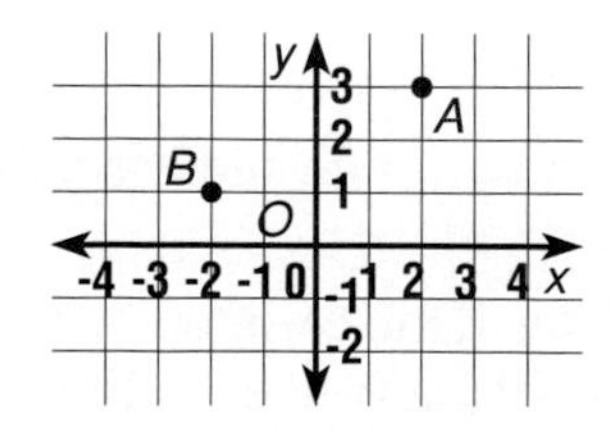

A 2.2 units **B** 2.3 units
C 3.7 units **D** 4.5 units

5. 0.3 =

A 3% **B** 0.3%
C 0.03% **D** none of these

6. Which is a solution set for the equation $y = 2x - 4$?

A {(-1, -6), (1, -2)}
B {(-6, -1), (-2, 1)}
C {(1, -2), (3, -2)}
D {(-2, 1), (2, 3)}

7. A circular clock has a diameter of 14 inches. Which expression shows the area of the face of the clock?

A $\pi \cdot 7 \cdot 7$ **B** $\pi \cdot 28$
C $\pi \cdot 14$ **D** $\pi \cdot 14 \cdot 14$

8. The bases of a triangular prism are right triangles with sides 3 cm, 4 cm, and 5 cm. The height is 10 cm. What is its surface area?

A 600 cm^2 **B** 60 cm^2
C 150 cm^2 **D** 132 cm^2

9. To the nearest 10 cm^2, what is the area of the label on the can shown below?

A 380 cm^2
B 530 cm^2
C 830 cm^2
D 940 cm^2

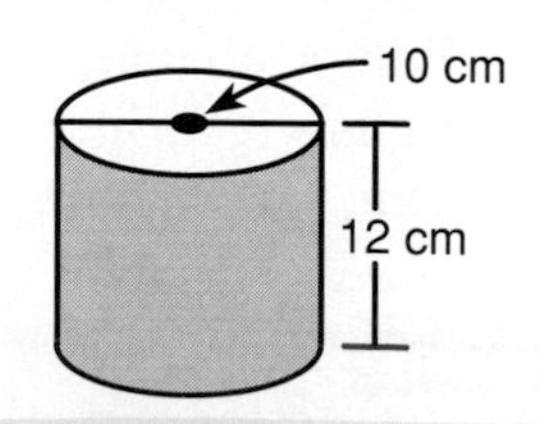

Directions: Write A if the quantity in Column A is greater. Write B if the quantity in Column B is greater. Write C if the quantities are equal. Write D if there is not enough information to decide.

	Column A	Column B
10.	$5 + \lvert -4 \rvert$	$\lvert 5 \rvert + \lvert -4 \rvert$
11.	the mean	the median
	Data: 12, 15, 19, 11, 20, 18, 16, 15, 14, 14	

12.

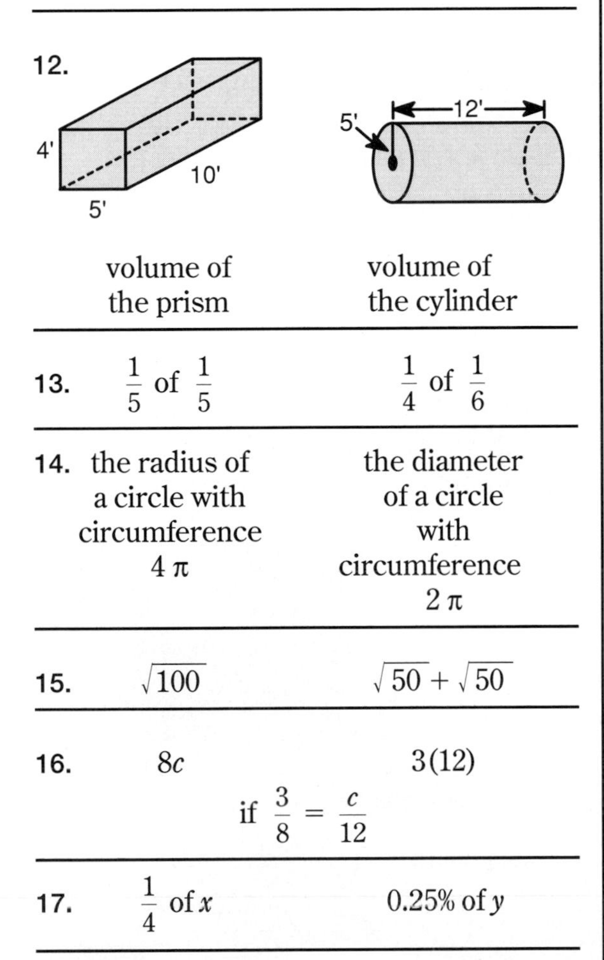

	Column A	Column B
12.	volume of the prism	volume of the cylinder
13.	$\frac{1}{5}$ of $\frac{1}{5}$	$\frac{1}{4}$ of $\frac{1}{6}$
14.	the radius of a circle with circumference 4π	the diameter of a circle with circumference 2π
15.	$\sqrt{100}$	$\sqrt{50} + \sqrt{50}$
16.	$8c$	$3(12)$
	if $\frac{3}{8} = \frac{c}{12}$	
17.	$\frac{1}{4}$ of x	0.25% of y

18.

8 cm, 4 cm, 7 cm, A, B, C

	Column A	Column B
18.	$\tan A$	$\tan B$

Test-Taking Tip

In test questions like Exercises 10–23 treat the two expressions given as two sides of an inequality. Substitute values for any variables to get a sense of the correct answer. Be sure to use many types and combination of numbers. Do not make assumptions.

If the use of As and Bs in both the column names and answer choices is confusing, rename the columns with another pair of letters or numbers, such as X and Y or I and II.

	Column A	Column B
19.	cost of \$50 jeans, 20% off	cost of \$50 jeans, \$10 off
20.	8% simple interest on \$550 for 1 year	4.5% simple interest on \$500 for 2 years
21.	$f(5)$	10
	if $f(n) = 2x + 3$	
22.	the LCM of 100 and 36	the LCM of 50 and 90
23.	$\frac{3}{25}$	0.15

Chapter

13 Discrete Math and Probability

Spotlight on Food

Have You Ever Wondered...

- What the shelf life of a bag of pretzels is?
- What kinds of juices are found in most refrigerators?

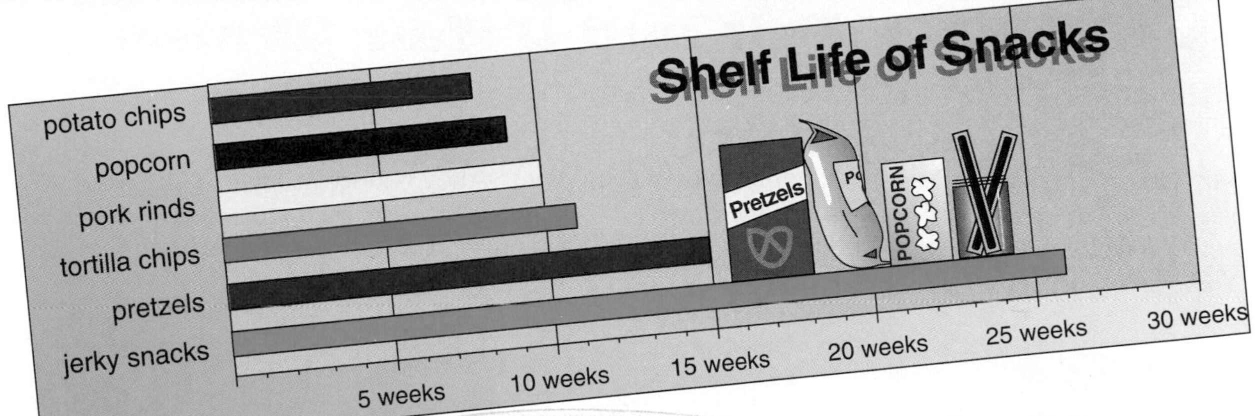

Looking Ahead

In this chapter, you will see how mathematics can be used to answer questions about food. The major objectives of the chapter are to:

- count outcomes using a tree diagram or the Fundamental Principal of counting
- find the number of permutations or combinations of a group of objects
- identify patterns in Pascal's Triangle
- solve problems by acting them out or using a simulation
- use samples to predict actions in larger groups of data

Chapter Project

Food

Work in a group.

1. Take a survey of your classmates to find out their favorite kinds of fruit juice.
2. Find what percent of those you surveyed like each type of juice.
3. Estimate what results you would expect if you surveyed 12 adults about their favorite juice. Then take a survey of adults.
4. Make circle graphs to demonstrate your two sets of data. Compare the two graphs.

Cooperative Learning

13-1A Fair and Unfair Games

A Preview of Lesson 13-1

Objective

Discover how to determine whether a game is fair or unfair.

Materials

3 counters
pencil

Many people enjoy playing games they believe are fair. A *fair game* is defined as one in which each player has an equal chance of winning. In an *unfair game,* players do *not* have an equal chance of winning.

Have you ever played *Scissors, Paper, Stone?* This ancient game, also known as *Hic, Haec, Hoc,* is played all over the world. On the count of three, two players simultaneously display one hand with either two fingers forming a V (scissors), an open hand (paper), or a fist (stone). The winner of the game is decided by the following rules.

a. scissors cut paper
b. paper wraps stone
c. stone breaks scissors

If both players pick the same object, the round is a draw.

Activity One

Work with a partner.

- Play 20 rounds of *Scissors, Paper, Stone* with your partner. Record the number of times each player wins in a table like the one below.

Player A	Player B	Winner

- Make a list of all possible outcomes.

What do you think?

1. How many different outcomes are possible?
2. How many ways can player A win?
3. How many ways can player B win?
4. How many outcomes are a draw?
5. Is each outcome equally likely?
6. Is *Scissors, Paper, Stone* a fair game? Explain.

Activity Two

- On one of three counters, write or tape an A on one side and a B on the other side. On the second counter, write or tape an A on one side and a C on the other. On the third counter, write or tape a B on one side and a C on the other. One player tosses all three counters. Player 1 wins if any two counters match. Player 2 wins if all three counters are different.
- Play 20 rounds of this game with your partner. Record the number of times each player wins.
- Determine all the possible outcomes for this game.

What do you think?

7. How many different outcomes are possible?
8. How many ways can Player 1 win?
9. How many ways can Player 2 win?
10. Is it possible to have a draw in this game?
11. Is each outcome equally likely?
12. Is this game fair or unfair? Explain.

Application

13. Four marbles are in a bag. Two are red and two are blue. Two marbles are drawn from the bag. Player X wins if the marbles are the same color. Player Y wins if the marbles are different colors. Is this game fair or unfair? Be prepared to defend your answer.

Extension

14. Design and play a fair game similar to any of those you have played in this lab.

13-1 Counting Outcomes

Objective

Count outcomes using a tree diagram or the Fundamental Principle of Counting.

Words to Learn

outcome
tree diagram
Fundamental Principle of Counting

The games you played in Mathematics Lab 13-1A involved counting outcomes by listing them. In this lesson you will learn two other ways to count outcomes.

The Hopi Indians invented a game of chance called Totolospi. This game was played with three cane dice, a counting board inscribed on stone, and a counter for each player. Each cane die can land round side up (R) or flat side up (F). In Totolospi for two players, each player places a counter on the nearest circle. The moves of the game are determined by tossing the three cane dice.

- Advance 2 lines with three round sides up (RRR).
- Advance 1 line with three flat sides up (FFF).
- Lose a turn with any other combination.

The player reaching the opposite side first wins.

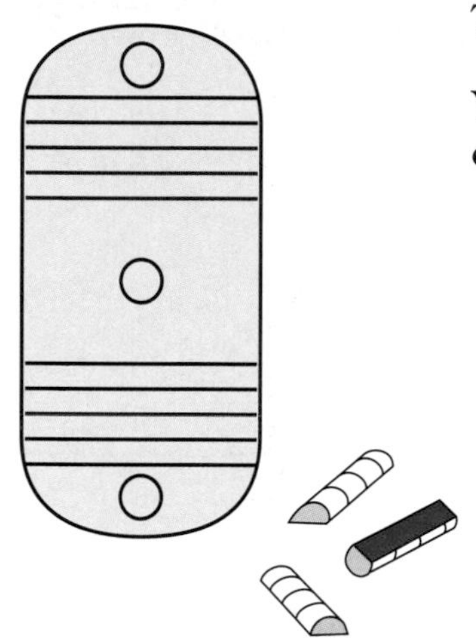

You can draw a diagram to find the number of possible combinations or **outcomes.**

First Die	Second Die	Third Die	Outcome
round R	R	R	→ RRR
		F	→ RRF
	F	R	→ RFR
		F	→ RFF
flat F	R	R	→ FRR
		F	→ FRF
	F	R	→ FFR
		F	→ FFF

There are eight possible outcomes.

The diagram above is called a **tree diagram.** You can also find the total number of outcomes by multiplying. This principle is known as the **Fundamental Principle of Counting.**

Fundamental Principle of Counting	If event M can occur in m ways and is followed by event N that can occur in n ways, then the event M followed by the event N can occur in $m \cdot n$ ways.

DID YOU KNOW

Without considering strategy in a game of chess, there are 400 possible ways of playing the first round of moves.

The number of possible outcomes for the Totolospi game can be determined by using this principle.

number of ways first die can land	×	*number of ways second die can land*	×	*number of ways third die can land*	=	*number of possible outcomes*
2	×	2	×	2	=	8

Examples

1 **Each spinner at the right is spun once. How many outcomes are possible?**

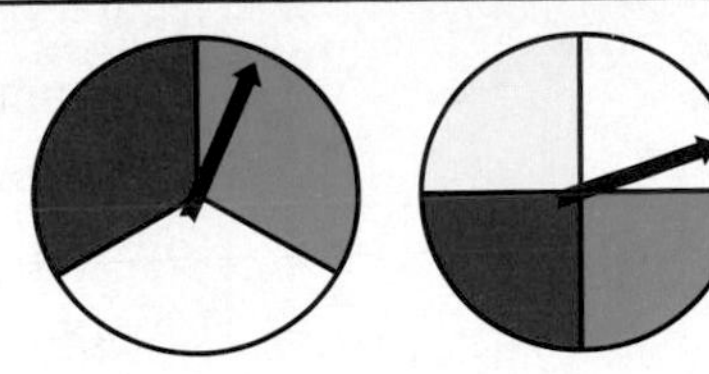

number of outcomes for first spinner	×	*number of outcomes for second spinner*	=	*number of possible outcomes*
3	×	4	=	12

There are 12 possible outcomes.

2 **The school store has small (S), medium (M), and large (L) sweatshirts in red (R), gray (G), or white (W). If the school store has one of every possible size and color sweatshirt, what is the probability of selecting a large red sweatshirt at random?**

LOOKBACK

You can review probability on page 233.

Use a tree diagram to find all the possible outcomes.

Size	Color	Outcome
S	R	SR
	G	SG
	W	SW
M	R	MR
	G	MG
	W	MW
L	R	LR
	G	LG
	W	LW

There are 9 possible outcomes.

$$P(\textit{large red}) = \frac{1}{9}$$

The probability of selecting a large red sweatshirt is $\frac{1}{9}$.

Checking for Understanding

Communicating Mathematics

Read and study the lesson to answer each question.

1. **Draw** a tree diagram to list all the outcomes for Example 1.
2. **Write** a problem that corresponds to the tree diagram at the right.

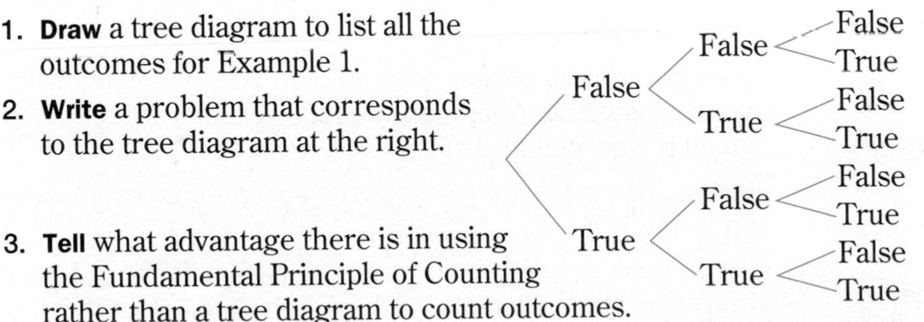

3. **Tell** what advantage there is in using the Fundamental Principle of Counting rather than a tree diagram to count outcomes.
4. **Tell** if you think the possible outcomes in Totolospi, the Hopi Indian game, are equally likely. Explain.

Guided Practice

5. Two coins are tossed and a die is rolled.
   a. Draw a tree diagram that represents the situation.
   b. How many outcomes are possible?
   c. How many outcomes show two heads?
   d. How many outcomes show a 6 on the die?
   e. How many outcomes show two tails and a 3 on the die?
   f. What is the probability of one head and an even number on the die?
6. A quiz has five true-false questions. How many outcomes for giving answers to the five questions are possible?

Exercises

Independent Practice

Draw a tree diagram to find the number of possible outcomes for each situation.

7. A die is rolled. Then a coin is tossed.
8. The spinner at the right is spun three times.
9. A restaurant offers a choice of orange, tomato, or grape juice with a choice of bacon or ham.
10. Phones come in wall or desk models with straight or coiled cords. They come in three colors, black, almond, and green.

State the number of possible outcomes for each event.

11. Four coins are tossed.
12. Two dice are rolled.
13. Mark has three pairs of shorts, four shirts, and two pairs of athletic shoes. How many three-piece outfits are possible?
14. Suppose there are only two colors of sweatshirts in Example 2. How many outcomes are possible?
15. A restaurant offers three types of pasta, three types of sauce, and three types of meat. How many combinations of one pasta, one sauce, and one meat are possible?

Mixed Review

16. Solve $\frac{f}{2} + 10 = 14$. Check your solution. *(Lesson 2-7)*
17. Solve $b = -3\frac{1}{3}(-6\frac{3}{5})$. Write your solution in simplest form. *(Lesson 7-3)*
18. How many significant digits are in the measurement, 14.4 centimeters? *(Lesson 12-8)*

Problem Solving and Applications

19. **Business** The Appliance Store found that customers preferred washers, dryers, and refrigerators in almond, black, and white. Use a tree diagram to find all the possibilities of appliances that are available.
20. **Algebra** If x coins are tossed, write an algebraic expression for the number of possible outcomes.
21. **Critical Thinking** When constructing a tree diagram, does it make a difference which event is listed first? Explain.

22. **Journal Entry** Write one or two sentences explaining the Fundamental Principle of Counting.

13-2 Permutations

Objective

Find the number of permutations of objects.

Words to Learn

permutation
factorial

The track coach at Jay Neff Middle School is preparing the lineup for the 400-meter relay team. She must choose four runners from the six that have been practicing together. As she looks over the six names, she wonders how many possible arrangements there are.

She reasons that any of the 6 runners can start the race. Once that runner is chosen, there are 5 runners left who can run second. After the second runner is chosen, there are 4 runners left who can run third. Finally, there are 3 runners left who can run fourth. Using the Fundamental Principle of Counting, she finds there are $6 \times 5 \times 4 \times 3$, or 360, possible arrangements.

An arrangement or listing in which order is important is called a **permutation.** In the above example, the symbol $P(6, 4)$ represents the number of permutations of 6 runners taken 4 at a time.

Definition of *P(n, r)*	**In words**: $P(n, r)$ means the number of permutations of n things taken r at a time. **Arithmetic** $P(6, 4) = 6 \cdot 5 \cdot 4 \cdot 3$ **Algebra** $P(n, r) = n \cdot (n - 1) \cdot (n - 2) \cdot \ldots \cdot (n - r + 1)$

Mini-Lab

Work in groups of three or four.
Materials: 4 different colored pencils or markers

- On your notebook paper, draw four columns. Use the lines of the paper to complete a grid so that each row has four units.
- On the first row, color each unit a different color.
- On the next row, use the same four colors to color the units, but do not repeat the pattern you used in the first row.
- Continue coloring the units in each row until you have created all possible arrangements of the four colors.

Talk About It

a. How many rows did you color?

b. Is this an example of a permutation? Why?

Calculator Hint

The factorial key [x!] on your calculator provides a fast way to compute factorials. Find 7!.

7 [x!] 5040

The number of permutations in the Mini-Lab can be expressed as $P(4, 4)$. Notice that $P(4, 4)$ means the number of permutations of 4 things taken 4 at a time.

$$P(4, 4) = 4 \cdot 3 \cdot 2 \cdot 1$$

The mathematical notation 4! also means $4 \cdot 3 \cdot 2 \cdot 1$. The symbol 4! is read *four* ***factorial.*** $n!$ means the product of all counting numbers beginning with n and counting backward to 1. We define 0! as 1.

Example 1 ***Problem Solving***

Elections Min, Karen, Linda, Juan, and Michael are running for student council governor. The student who receives the second highest number of votes will be lieutenant governor. How many ways can the students be elected to these 2 positions? Assume there are no ties.

There are 5 students running, but only 2 will be elected. The order of finish is important. So, you must find the number of permutations of 5 students taken 2 at a time.

$$P(5, 2) = 5 \cdot 4 \text{ or } 20$$

There are 20 different ways for the students to be elected.
You can draw a tree diagram to check the answer.

Example 2 ***Problem Solving***

Sports There are five members on the golf team. In how many different orders can they tee off?

You must find the number of permutations of 5 students taken 5 at a time.

$$\begin{aligned} P(5, 5) &= 5! \\ &= 5 \cdot 4 \cdot 3 \cdot 2 \cdot 1 \\ &= 120 \end{aligned}$$

They can tee off in 120 different orders.

Checking for Understanding

Communicating Mathematics

Read and study the lesson to answer each question.

1. **Draw** a tree diagram to show the number of different ways 3 books can be stacked 2 books at a time.
2. **Tell** what 6! means.

Guided Practice

Find each value.

3. $P(4, 3)$
4. $P(12, 4)$
5. 3!
6. 7!

How many different ways can the letters of each word be arranged?

7. STUDY
8. MATH
9. EQUALS
10. FUN

Exercises

Independent Practice

Find each value.

11. $P(6, 3)$ **12.** $0!$ **13.** $P(8, 4)$ **14.** $9!$

15. $6!$ **16.** $P(10, 5)$ **17.** $5!$ **18.** $P(8, 8)$

19. In how many different ways can you arrange the letters in the word *cards* if you take the letters 4 at a time?

20. How many 4-digit whole numbers can you write using the digits 2, 3, 5, and 8? In each number you write, use each digit only once.

21. How many ways can 5 members of a family be seated in a theater if the father is seated on the aisle?

Mixed Review

22. Solve $r = \frac{352}{-11}$. *(Lesson 3-7)*

23. Find the GCF of 28, 126, and 56. *(Lesson 6-4)*

24. What percent of 70 is 42? *(Lesson 10-6)*

25. A test has 5 multiple choice questions. Each question has three choices. How many outcomes for giving answers to the 5 questions are possible? *(Lesson 13-1)*

Problem Solving and Applications

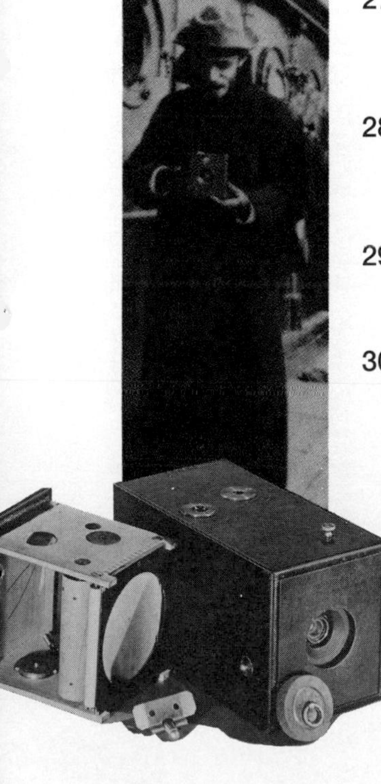

26. Sports There are six runners in a race. Medals will be given to the first three runners who finish the race. How many ways can the medals be awarded?

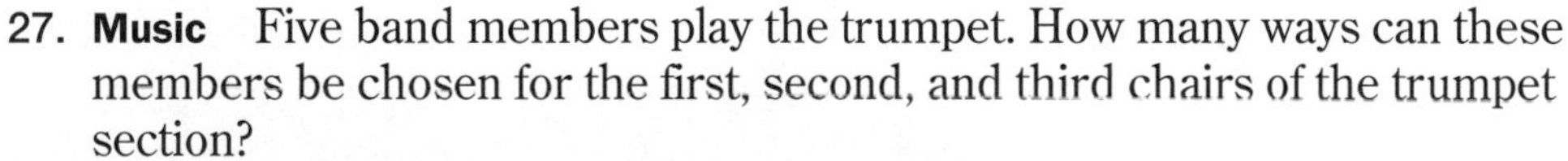

27. Music Five band members play the trumpet. How many ways can these members be chosen for the first, second, and third chairs of the trumpet section?

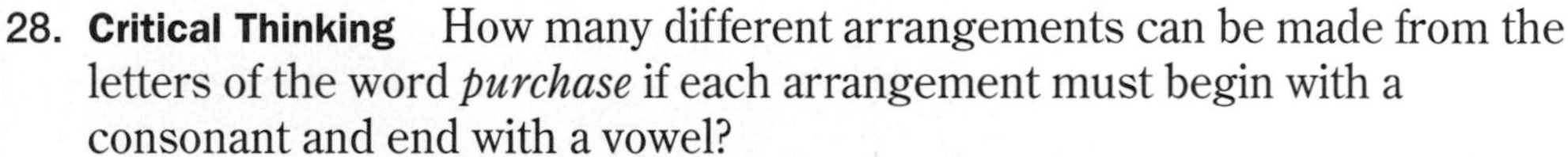

28. Critical Thinking How many different arrangements can be made from the letters of the word *purchase* if each arrangement must begin with a consonant and end with a vowel?

29. Make Up a Problem Write a problem where you would need to find the number of permutations of ten things taken three at a time.

30. Mathematics and Photography Read the following paragraph.

The camera you use today is a great improvement over the first Kodak camera invented by George Eastman in 1888. To take a photo, you press the shutter of the camera, which lets light through the lens for a fraction of a second. The light forms a picture on the film inside the camera. The film is coated with chemicals that change slightly when light falls on them. To see the picture, the film has to be placed in other chemicals so that the picture develops. Edwin H. Land developed the Polaroid camera in 1947, which developed the black-and-white photos in the camera while you waited.

In how many orders can a photographer arrange nine cheerleaders in a line for a photo in a yearbook?

Decision Making

Ordering Inventory

Situation

Every spring your school sponsors a community-wide flea market. The school allows various school groups to have booths at the flea market to raise money for special projects. Your committee is in charge of the School Pride booth, which last year brought in $1,700, the best of any of the groups. You know how much inventory you had for last year's booth. What and how much of each item do you buy this year to guarantee another successful sale?

Hidden Data

Cost of returns: Are you allowed to return unsold items for credit?
Cost of credit: If you must buy on credit, will it be for more than 30 days? What percent of interest is charged after 30 days?
Taxes: By being a school group, are you exempt from paying taxes?
Cost of shipping: Do you know the cost of shipping your entire order?

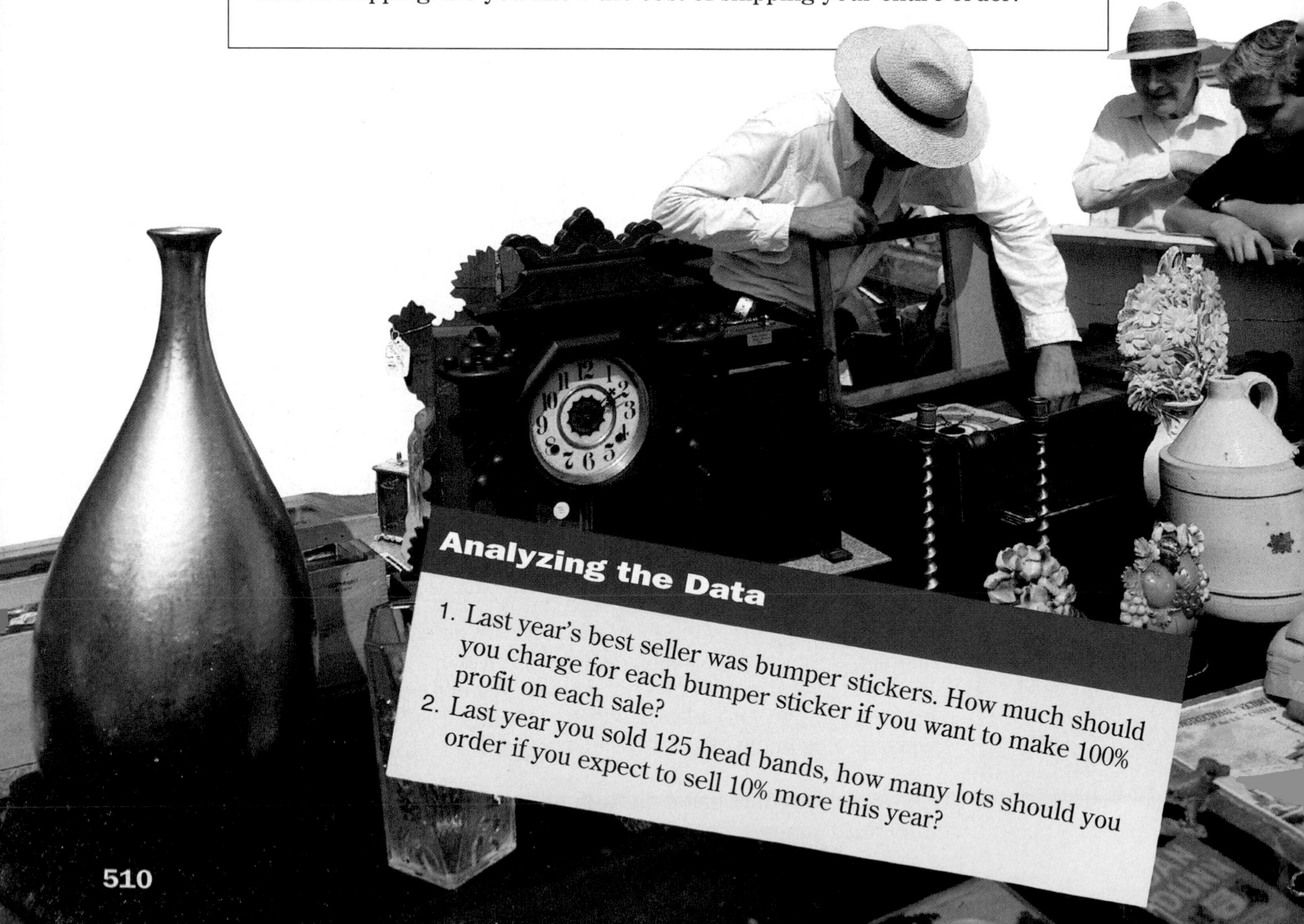

Analyzing the Data

1. Last year's best seller was bumper stickers. How much should you charge for each bumper sticker if you want to make 100% profit on each sale?
2. Last year you sold 125 head bands, how many lots should you order if you expect to sell 10% more this year?

Making a Decision

3. **Which colors** should you choose for your items with logos?
4. **Will you** accept checks for items purchased?
5. **Would it** be wise to take a survey of students to see if banners would be a good item to add to the inventory for the booth this year?

Making Decisions in the Real World

6. **Interview** some local merchants to see what factors they consider when ordering inventory.
7. **Investigate** other catalogs for ordering school pride items. Are there others that might offer better opportunities for your committee?

13-3 Combinations

Objective

Find the number of combinations of objects.

Words to Learn

combination

The Yogurt Oasis has a choice of 10 different toppings for their sundaes. How many different sundaes with 3 toppings can they serve?

In this problem, order is *not* important. That is, *raisins, peanuts, chocolate chips* is the same as *chocolate chips, raisins, peanuts. You will solve this problem in Example 1.*

Arrangements or listings, like these, where order is not important are called **combinations.**

TEEN SCENE

Ice cream cones were invented at the St. Louis World's Fair in 1904. A vendor selling ice cream ran out of dishes. A pastry-seller rolled up one of his wafers and asked the vendor to put ice cream in it. Together, they sold the first ice cream cones.

Mini-Lab

Work with a partner to find the number of ways 5 cards can be selected 3 at a time.

Materials: 5 small index cards

- Mark five index cards with an A, B, C, D, and E. Select any three cards. Record your selections.
- Select another three cards, but not the same group you chose before. Record your selections.
- Continue selecting groups of three cards until all possible groups are recorded.

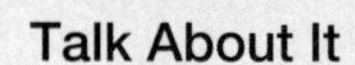

Talk About It

a. How many groups did you record?

b. In how many different orders can three letters be arranged?

c. Find $P(5, 3)$.

d. What is the relationship between the number of groups, the number of arrangements of three cards, and $P(5, 3)$?

A quick way to find the number of groupings, or combinations, of 5 cards taken 3 at a time, $C(5, 3)$, is to divide the number of permutations, $P(5, 3)$, by the number of orders 3 cards can be arranged, which is 3!

$$C(5, 3) = \frac{P(5, 3)}{3!}$$ *Dividing by 3! eliminates the combinations that are the same.*

$$= \frac{5 \cdot 4 \cdot 3}{3 \cdot 2 \cdot 1}$$

$$= \frac{60}{6} \text{ or } 10$$ There are 10 groups.

Definition of $C(n, r)$	**In words:** $C(n, r)$ means the number of combinations of n things taken r at a time.
	Arithmetic $C(9, 3) = \frac{9 \cdot 8 \cdot 7}{3!}$ or 84 — **Algebra** $C(n, r) = \frac{P(n, r)}{r!}$

Example 1

Determine how many different sundaes with three toppings Yogurt Oasis can serve.

You need to find $C(10, 3)$.

$$C(10, 3) = \frac{P(10, 3)}{3!} = \frac{10 \cdot 9 \cdot 8}{3 \cdot 2 \cdot 1} = \frac{720}{6} \text{ or } 120$$

Yogurt Oasis can serve 120 different sundaes with three toppings.

Example 2 ***Problem Solving***

Travel **Eight students are eligible to compete in the Mansfield University Writing Contest. Their sponsor, Miss Valentine, will take four of them in her car. Another parent will take the others. How many different groups of these students could she take?**

Order does not matter, so you need to find the number of combinations of 8 things taken 4 at a time.

$$C(8, 4) = \frac{P(8, 4)}{4!} \text{ or } \frac{8 \cdot 7 \cdot 6 \cdot 5}{4!}$$

(8 × 7 × 6 × 5) ÷ 4 x! = 70

She can select from 70 different groups of 4 students.

Checking for Understanding

Communicating Mathematics

Read and study the lesson to answer each question.

1. **Tell** how you would find how many combinations of 5 cards there are in a standard deck of 52 cards.
2. **Write** an expression to represent the number of five-person committees that could be formed in your class.
3. **Tell,** without computing, which is greater, $P(3, 2)$ or $C(3, 2)$. Explain why.
4. **Draw** a model to show the number of combinations of 3 geometric shapes taken 2 at a time.

Guided Practice Find each value.

5. $C(5, 2)$
6. $C(4, 4)$
7. $C(3, 2)$
8. $C(12, 5)$

Determine whether each situation is a permutation or a combination.

9. three books in a row
10. a team of 5 players from 11
11. six CDs from a group of ten
12. arranging 9 model cars in a line
13. How many 3-letter combinations can be made from the letters in the word CARD?

Exercises

Independent Practice Find each value.

14. $C(7, 2)$
15. $C(6, 5)$
16. $C(9, 4)$
17. $C(25, 9)$

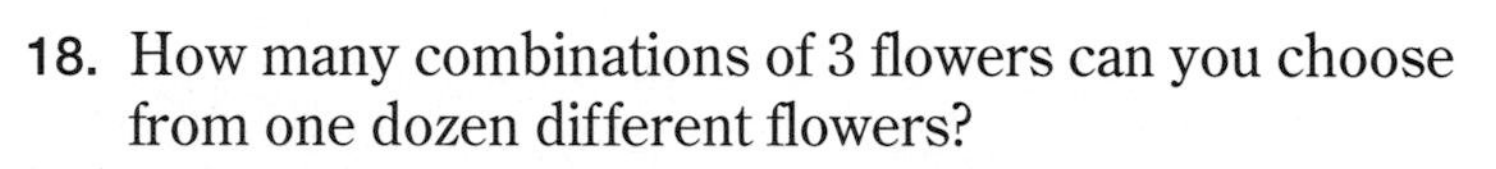

18. How many combinations of 3 flowers can you choose from one dozen different flowers?
19. How many different 5-card hands is it possible to deal from a standard deck of 52 cards?
20. How many different 3-digit numbers can you write using the digits 1, 2, 3, 5, and 8 only once in each number?
21. How many ways can you choose 3 shirts from 8 shirts in a closet?
22. How many different volleyball teams of 6 players can be formed from a squad of 12 players?

Mixed Review

23. How many liters are in 78 milliliters? *(Lesson 1-6)*
24. **Geometry** Refer to the figure in Example 2 on page 178. Find $m\angle 4$. *(Lesson 5-1)*
25. The distance between two locations on a map is $2\frac{1}{4}$ inches. Find the actual distance, if the scale on the map is 1 inch:100 miles. *(Lesson 9-7)*
26. Find the value of $P(8, 3)$. *(Lesson 13-2)*

Problem Solving and Applications

27. **Critical Thinking** How many ways can a study group of 2 males and 3 females be formed from a class of 18 males and 12 females?
28. **Business** A pizza shop has 12 different toppings from which to choose. This week, if you buy a 2-topping pizza, you get 2 more toppings free. How many different ways can the special pizza be ordered?
29. **Geometry** Eight points are marked on a circle. How many different line segments can be drawn between pairs of points?

30. **Critical Thinking** Write a problem for which 792 combinations is the answer.
31. **Journal Entry** Write one or two sentences explaining the difference between a permutation and a combination.

13-4 Pascal's Triangle

Objective

Identify patterns in Pascal's Triangle.

Words to Learn

Pascal's Triangle

1

1 1

1 2 1

1 3 3 1

■ ■ ■ ■ ■

Pascal's Triangle

The student council at Abington Junior High is having a pizza sale to raise money. They need to know how many toppings will be offered so they can advertise how many types of pizza will be available.

Their advisor prepared a table so they could begin working on their posters as soon as the number of toppings is determined.

The columns under *Number of Toppings Taken at a Time* lists how many combinations are possible for the number of toppings taken 0 at a time, 1 at a time, 2 at a time, 3 at a time, and 4 at a time. For example, the number of different 2-topping pizzas that can be made when 4 toppings are offered is 6. $C(4, 2) = 6$

Number of Toppings Offered	Number of Toppings Taken at a Time					Types of Pizza
	0	1	2	3	4	
0	1					1
1	1	1				2
2	1	2	1			4
3	1	3	3	1		8
4	1	4	6	4	1	16

4 toppings taken 2 at a time ⟶

Do you see a triangular pattern formed by the ones in the table? This pattern is known as a Chinese Triangle. When all the rows are centered over the bottom row, they form **Pascal's Triangle,** named after French mathematician Blaise Pascal who studied its patterns and applied it to the study of probability.

Mini-Lab

Work with a partner.

- Copy Pascal's Triangle and add another two rows on your copy. The top row is row 0.
- Then draw four tree diagrams to represent the possible outcomes for tossing 2, 3, 4, and 5 coins.

Talk About It

a. Which row of Pascal's Triangle matches the number of outcomes for tossing 2 coins? 3 coins? 4 coins? 5 coins?

b. When tossing 3 coins, how many ways can the outcome of 2 heads and 1 tail result?

c. Where is the answer to part b. located in Pascal's Triangle?

Examples

Use Pascal's Triangle to answer each question.

1 **How many combinations are possible for 5 things taken 3 at a time?**

Since there are 5 things, you would use row 5.

Row 5	1	5	10	10	5	1
Number Taken at a Time	0	1	2	3	4	5

The fourth number in the fifth row of Pascal's Triangle is 10.

There are 10 combinations of 5 things taken 3 at a time.

Check: $C(5, 3) = \frac{P(5, 3)}{3!}$

$= \frac{5 \cdot 4 \cdot 3}{3 \cdot 2 \cdot 1}$ or 10 ✓

2 **How many different committees of 4 students can be taken from a group of 6 students?**

Since there are 6 students, you would use row 6.

Row 6	1	6	15	20	15	6	1
Number Taken at a Time	0	1	2	3	4	5	6

The fifth number in the sixth row of Pascal's Triangle is 15.

There are 15 different committees of 4 students that can be formed from a group of 6 students.

Check: $C(6, 4) = \frac{P(6, 4)}{4!}$

$= \frac{6 \cdot 5 \cdot 4 \cdot 3}{4!}$

(6 × 5  4  3  ÷ 4   15 ✓

Checking for Understanding

Communicating Mathematics Read and study the lesson to answer each question.

1. **Tell** how the sum of the numbers in row 4 in Pascal's Triangle compares to the total number of outcomes possible for tossing 4 coins.
2. **Show** how to find the third number in row 7 of Pascal's Triangle.
3. **Write** the seventh and eighth rows of Pascal's Triangle.

Guided Practice Use Pascal's Triangle to find each value.

4. $C(5, 2)$ 5. $C(7, 6)$ 6. $C(8, 3)$ 7. $C(4, 4)$

Use Pascal's Triangle to answer each question.

8. How many different teams of 5 players can be taken from a squad of 8?
9. How many different 4-question quizzes can be formed from a test bank of 7 questions?
10. What is the second number in row 8?
11. What is the second number in row 28?

Exercises

Independent Practice

Use Pascal's Triangle to find each value.

12. $C(5, 3)$ 13. $C(4, 1)$ 14. $C(9, 5)$ 15. $C(10, 4)$

Use Pascal's Triangle to answer each question.

16. How many varieties of pizza are possible if 4 toppings are offered?
17. How many branches will there be in the last column of a tree diagram showing the outcomes of tossing 3 coins?
18. Six coins are tossed. What is the probability that all 6 will be heads?
19. How many combinations are possible when 6 things are taken 5 at a time?
20. In which row will you find 20 other than the twentieth row?
21. What does the 6 in row 4 mean?

Mixed Review

22. **Statistics** Find the mean and median for the set {155, 154, 160, 152, 164, 158, 161}. Round to the nearest tenth. *(Lesson 4-5)*
23. Solve $t = 5\frac{1}{6} + 4\frac{2}{9}$. Write your solution in simplest form. *(Lesson 7-1)*
24. Find four solutions for $y = 3x + 10$. Write the solution set. *(Lesson 11-3)*
25. Find the value of $C(10, 3)$. *(Lesson 13-3)*

Problem Solving and Applications

26. **Algebra** Let n be the number of a row in Pascal's Triangle. Write an expression for the sum of the numbers in that row.
27. **Critical Thinking** Ball bearings fall down a chute toward a tray. As they fall, they hit pegs. At each peg, there is an equal chance to go either left or right.
    a. If 16 ball bearings go through the chute, how many will be in the tray for each branch?
    b. If 64 ball bearings go through the chute, how many will be in the tray for each branch?

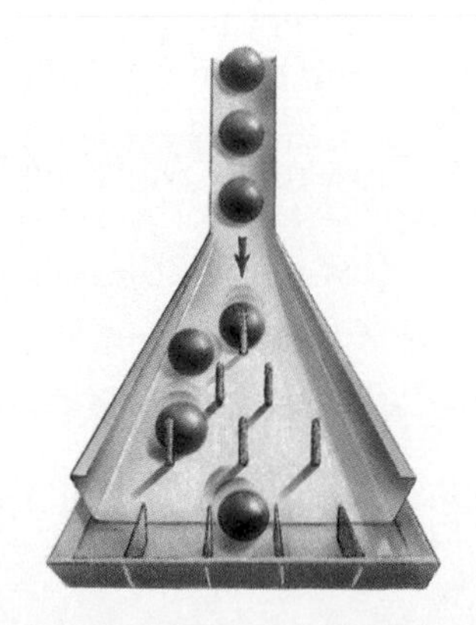

28. **Music** The Selingsgrove Middle School bell choir has 5 members who are each responsible for 2 bells. How many combinations are possible if 4 bells are rung at the same time?

29. **Business** A shoe salesperson has 10 different styles of boots. Only five styles can be displayed in the counter window at any one time. How many different groupings of boots can be displayed?

30. **Sports** The fastest two of eight runners will qualify for the final race. How many different combinations of two qualifiers are possible?

DATA SEARCH

31. **Data Search** Refer to pages 500 and 501. How many different combinations of juices are possible if there are three juices in the refrigerator?

13 Assessment: Mid-Chapter Review

1. A new style of jeans is available in 3 colors (black, white, and gray), and in 4 sizes (small, medium, large, and extra large). Draw a tree diagram that illustrates the outcomes. *(Lesson 13-1)*
2. A salad bar offers 2 different kinds of lettuce, 4 different dressings, and 3 different toppings. How many different salads of 1 lettuce, 1 dressing, and 1 topping can be made? *(Lesson 13-1)*
3. How many ways can 6 people line up to buy concert tickets? *(Lesson 13-2)*
4. How many different ways can 2 student council members be elected from 5 candidates? *(Lesson 13-3)*
5. A committee of 3 students is to be chosen from a group of 6. Use Pascal's Triangle to determine how many different committees are possible. *(Lesson 13-4)*

13-4B Patterns in Pascal's Triangle

A Follow-Up of Lesson 13-4

Objective

Discover numerical and visual patterns in Pascal's Triangle.

Materials

hexagonal grid
highlighter
calculator

LOOKBACK

You can review Fibonacci numbers on page 277.

A famous sequence of numbers is the Fibonacci sequence, 1, 1, 2, 3, 5, 8, ... The sequence begins with 1 and each number that follows is the sum of the previous two numbers.

The Fibonacci numbers and numerous other number patterns can be found in Pascal's Triangle.

Activity One

- Copy Pascal's Triangle and add another two rows to your copy.
- Draw diagonals beginning at each 1 along the left-hand side (passing under the 1 just above) and extend it to the far side of the triangle.
- Find the sum of the numbers each diagonal passes through.

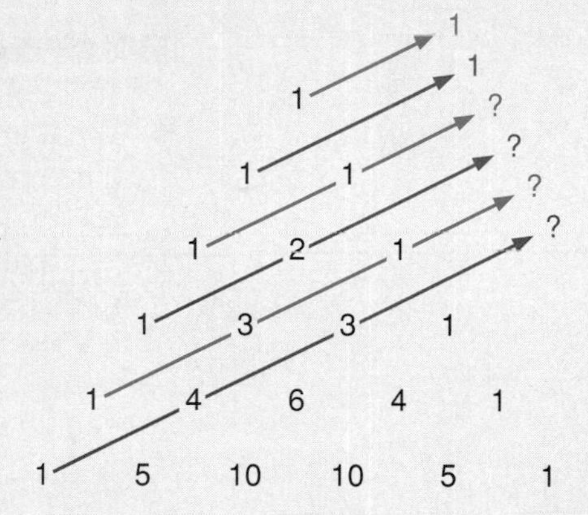

What do you think?

1. Do the sums form the Fibonacci sequence?
2. What should the sum along the next diagonal be?

Activity Two

- Find the product of the shaded ring of numbers in the Pascal's Triangle at the right.
- Find the product of any two other rings of the same shape and size in Pascal's Triangle.

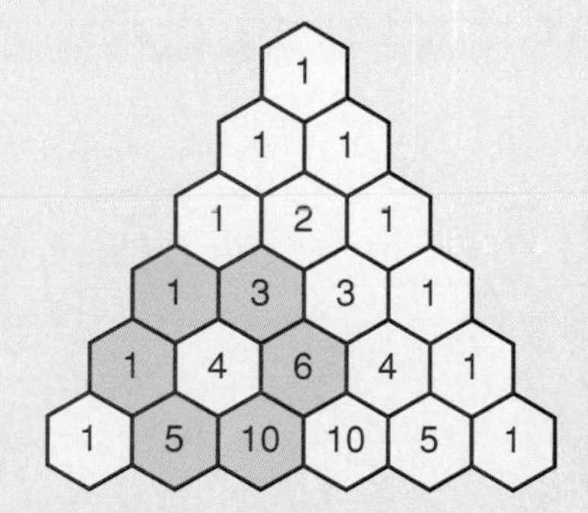

What do you think?

3. Determine the square root of the products of rings.
4. What conclusion can you make about the product of rings in Pascal's Triangle?

Activity Three

Work in groups of four.

- Using a hexagonal grid triangle like the one shown below, one person in the group highlights all the numbers that are multiples of 2.
- On another triangle, another person in the group highlights all the numbers that are multiples of 3.
- On a third triangle, another person highlights all the numbers that are multiples of 4.
- On a fourth triangle, another person highlights all the numbers that are multiples of 7.
- Each person should highlight as many rows as necessary to determine a geometrical pattern.

What do you think?

5. What general visual pattern is common in all four triangles? Is this true for multiples of all numbers? Why?
6. Which multiples of the numbers 2, 3, 4, and 7 have patterns that are symmetrical at *each* of the three vertices of the triangle?

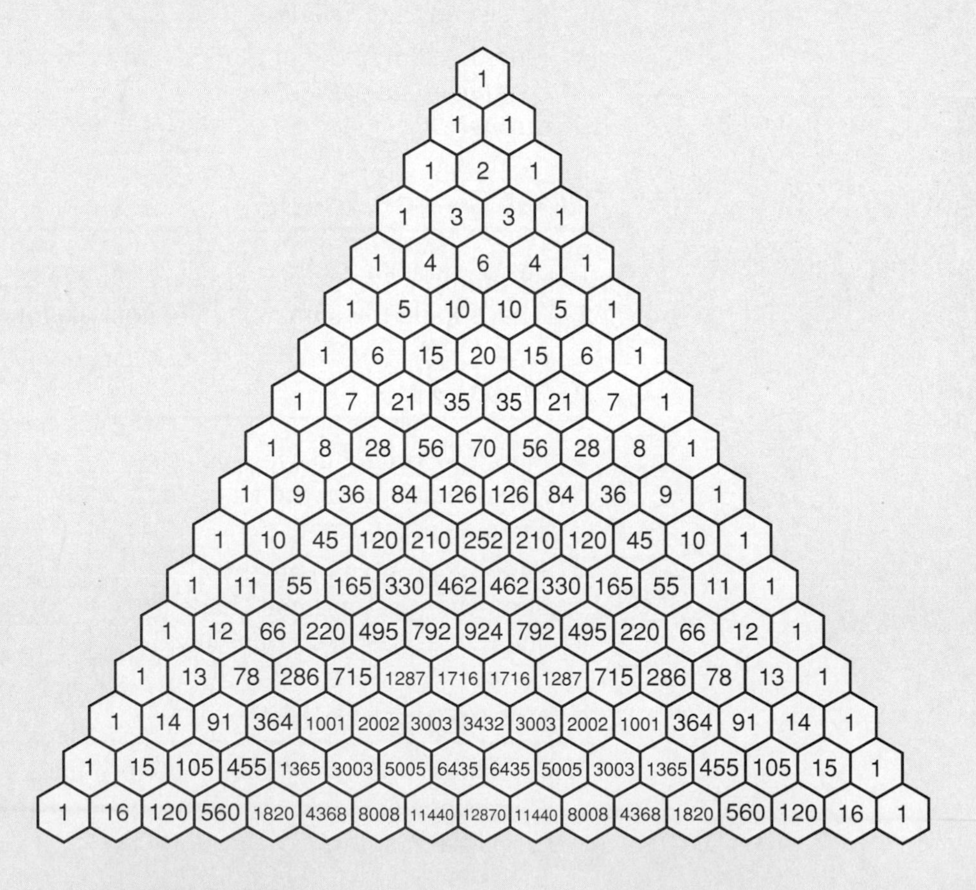

13-5 Probability of Compound Events

Objective

Find the probability of independent and dependent events.

Words to Learn

independent event
dependent event

You can review probability of simple events on page 233.

Do you have a pet? Human beings have always had pets. In Japan, children tame mice. In Australia, some children have pet kangaroos. In India, people make pets of mongooses.

Suppose you wanted to find if a household in the United States, chosen at random, had both pets and children. In the United States, $\frac{6}{10}$ of all households have some kind of pet. Also, approximately $\frac{1}{3}$ of all households have at least one child.

You can find the probability of pets and children in the same household by multiplying the probability of a household having a pet by the probability of a household with at least one child.

$$\frac{\textit{number of ways it can occur}}{\textit{total number of possible outcomes}} \begin{matrix} \rightarrow \\ \rightarrow \end{matrix} \frac{6}{10} \cdot \frac{1}{3} = \frac{6}{30} \text{ or } \frac{1}{5}$$

The probability of a household having both pets and children is $\frac{1}{5}$.

Selecting a household with a pet does not depend on the selection of a household having a child. We call these **independent events.** The outcome of one event does not affect the outcome of the other event.

Probability of Two Independent Events	
	In words: The probability of two independent events can be found by multiplying the probability of the first event by the probability of the second event.
	In symbols: $P(A \text{ and } B) = P(A) \cdot P(B)$

Example 1

Two dice are rolled. Find the probability that an odd number is rolled on one die and a composite number is rolled on the other.

$P(\text{odd number}) = \frac{1}{2}$ $\quad$ $P(\text{composite number}) = \frac{2}{6} \text{ or } \frac{1}{3}$

$P(\text{odd number and composite number}) = \frac{1}{2} \cdot \frac{1}{3} \text{ or } \frac{1}{6}$

The probability that the two events will occur is $\frac{1}{6}$.

Example 2 *Problem Solving*

Food The chart below lists the number and types of muffins found in a bakery. An oat muffin is chosen at random. Then a bran muffin is chosen at random. Find the probability that the bran muffin has cranberries and the oat muffin has nuts.

Type of Muffin	Number of Muffins		
	with Nuts	with Cranberries	Plain
BRAN	4	6	5
OAT	4	3	2

$P(\text{bran with cranberries}) = \frac{6}{15} \text{ or } \frac{2}{5}$ $P(\text{oat with nuts}) = \frac{4}{9}$

$P(\text{bran with cranberries and oat with nuts}) = \frac{2}{5} \cdot \frac{4}{9} \text{ or } \frac{8}{45}$

The probability that the two events will occur is $\frac{8}{45}$.

Sometimes the outcome of one event affects the outcome of another.

Mini-Lab

Work with a partner.

Materials: 3 red counters, 2 blue counters, 1 cup

- Place all the counters in the cup. Without looking, draw one counter from the cup. Do *not* put the counter back into the cup. Without looking, draw another counter from the cup.
- Copy and complete the tree diagram for this activity.

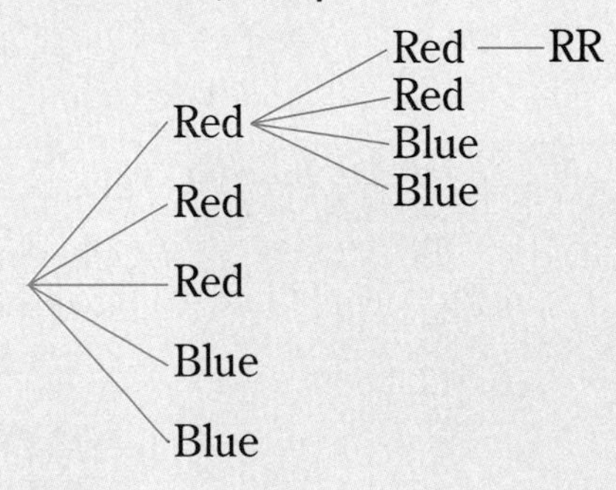

Talk About It

a. What is the probability of a red counter on the first draw?

b. If you drew a red counter on the first draw, what is the probability of drawing a red counter on the second draw?

c. How many equally likely outcomes are there?

d. How many outcomes show 2 red counters?

e. Does not replacing the first counter affect the number of outcomes? If so, how?

If the outcome of one event affects the outcome of another event, the events are called **dependent events.** Like independent events, the probability of two dependent events is also found by multiplying the probability of the first event by the probability of the second event.

Example 3

A paper bag contains 4 chocolate chip cookies, 5 oatmeal cookies, and 1 brownie. What is the probability that Julia pulls out an oatmeal cookie first, and then Joanne pulls out an oatmeal cookie too?

This is an example of dependent events because what Julia draws affects what Joanne draws.

First selection: $P(\text{oatmeal cookie}) = \frac{5}{10} \text{ or } \frac{1}{2}$

Second selection: $P(\text{oatmeal cookie}) = \frac{\text{number of oatmeal cookies left}}{\text{number of cookies left}}$

$= \frac{4}{9}$

$P(\text{2 oatmeal cookies}) = \frac{1}{2} \cdot \frac{4}{9}$

$= \frac{4}{18} \text{ or } \frac{2}{9}$

The probability that Joanne pulls out an oatmeal cookie after Julia pulls out an oatmeal cookie is $\frac{2}{9}$.

Checking for Understanding

Communicating Mathematics

Read and study the lesson to answer each question.

1. **Tell** what is meant by independent events.
2. **Write** an example of two dependent events.
3. **Tell** how to find the probability of two dependent events.

Guided Practice

A die is rolled and the spinner is spun. Find each probability.

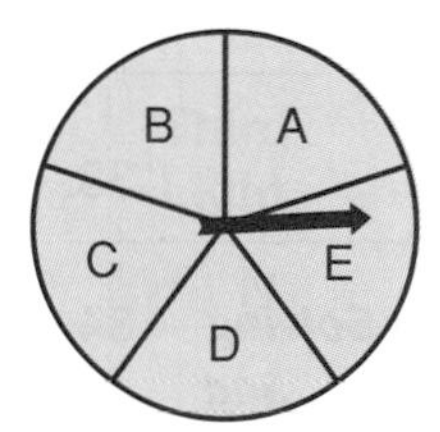

4. P(1 and B)
5. P(2 and C)
6. P(prime and D)
7. P(7 and E)
8. P(an even number and a vowel)
9. Do the probabilities in Exercises 4-8 represent dependent or independent events? Explain.

In a bag there are 5 red marbles, 2 yellow marbles, and 1 blue marble. Once a marble is selected, it is not replaced. Find the probability of each outcome.

10. a red marble and then a yellow marble
11. a blue marble and then a yellow marble
12. a red marble and then a blue marble
13. any color marble except yellow and then a yellow marble
14. a red marble three times in a row
15. Do the probabilities in Exercises 10-14 represent dependent or independent events? Explain.

Examine You can determine if your estimate is reasonable by finding the theoretical probability. Since the traffic lights operate independently, you can multiply the probability that each light is green.

0.4 [×] 0.4 [×] 0.4 [=] 0.064

The estimate is reasonable.

Checking for Understanding

Communicating Mathematics

Read and study the lesson to answer each question.

1. **Tell** what a simulation is.
2. **Write** another way to simulate the traffic light situation on page 525, using green and non-green slips of paper in a box.
3. **Tell** whether or not the results of a simulation will be exactly the same as computing the result. Explain your reasoning.

Guided Practice

Solve by acting it out.

4. Carmen has a record of making one free throw out of every two tries, and she averages 6 free throw attempts per game. Act it out to find the number of free throws she will make in the next game.
5. Ed has 6 ties. On Thursday through Sunday, Ed works at the mall. He chooses a tie at random to wear at his job. Act it out to find the probability that Ed wears the same tie more than once in his four-day week.
6. David has $1.15 made up of six United States coins. With the coins he has, he cannot make change for a dollar, a half dollar, a quarter, a dime, or a nickel. What are the six coins that David has?

Problem Solving

Practice

Strategies

Look for a pattern.
Solve a simpler problem.
Act it out.
Guess and check.
Draw a diagram.
Make a chart.
Work backward.

Solve using any strategy.

7. Marlene is reading a 216-page book. She needs to read twice as many pages as she has already read to finish the book. How many pages has she read so far?
8. There are 16 tennis players in a single elimination tournament. How many tennis matches will be played during the tournament?
9. Dawn is running late for an appointment. Without turning on the lights, she reaches into her jewelry box and pulls out two earrings one at a time. The jewelry box contains pairs of blue, silver, red, pink, and gold earrings. The earrings are identical except for color. Act it out to find the probability that Dawn will select a pair of red earrings.

10. Sam bought a jacket and a shirt. The total cost, not including tax, was \$62.50. The jacket cost four times as much as the shirt. How much did Sam spend on these additions to his wardrobe?

11. Mrs. Gossell, Ms. Alvarez, and Mrs. Yamaguchi teach at Watkins Middle School. One of the women is a mathematics teacher, one is a music teacher, and one is a social studies teacher. The music teacher, an only child, has taught the least number of years. Mrs. Yamaguchi who married Mrs. Gossell's brother, has taught more years than the math teacher. Name the subject each woman teaches.

12. Antonio places 10 pennies in a horizontal line on his desk. He adds enough pennies so that there are 5 pennies in a vertical line, 8 pennies in another vertical line, and 5 pennies in a diagonal line. What is the least number of pennies Antonio can use?

13. A number is added to 8, and the result is multiplied by 20. The final answer is 100. Find the number.

14. **Data Search** Refer to pagc 666. What was the median U.S. home price in 1992?

15. In a recent survey of 120 students, 50 students said they play baseball and 60 students said they play soccer. If 20 students play both sports, how many students do not play either baseball or soccer?

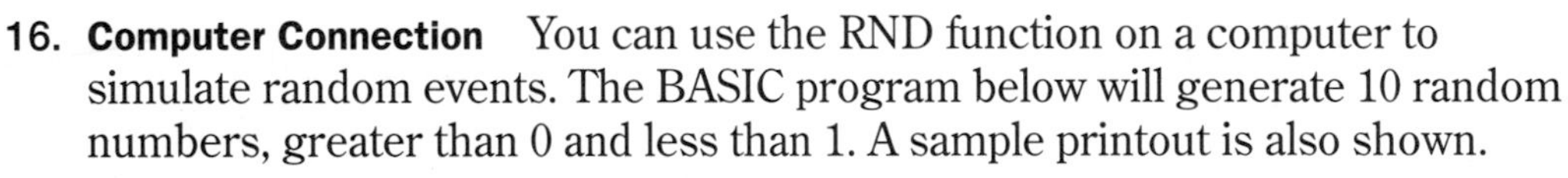

16. **Computer Connection** You can use the RND function on a computer to simulate random events. The BASIC program below will generate 10 random numbers, greater than 0 and less than 1. A sample printout is also shown.

```
10  FOR X = 1 TO 10
20  PRINT RND(X)
30  NEXT X
```

```
0.08067539
0.17157252
0.96578311
0.71980542
0.61702572
0.19884118
0.24290394
0.43047297
0.91472936
0.86625558
```

Run each program below on a computer. Write a sentence that describes the output. Describe a simulation for which you might use each program.

a.
```
10  FOR X = 1 TO 20
20  PRINT INT(2 * RND(X))
30  NEXT X
```

b.
```
10  FOR X = 1 TO 20
20  PRINT INT(6*RND(X) + 1)
30  NEXT X
```

13-7 Experimental Probability

Objective

Find experimental probability.

Words to Learn

experimental probability
theoretical probability

Sometimes you can't tell what the probability of an event is until you conduct an experiment. For example, suppose you toss a thumbtack 50 times and count the number of times it lands point up. The results of this experiment allow you to estimate the probability of that thumbtack landing point up.

Probabilities that are based on frequencies obtained by conducting an experiment, or doing a simulation as you did in the previous lesson, are called **experimental probabilities.** Experimental probabilities may vary when an experiment is repeated.

Probabilities based on physical characteristics like 4 sections of a spinner or 6 sides of a die, are called **theoretical probabilities.** Theoretical probability tells you *approximately* what should happen in an experiment.

Technology Activity

You can use a graphing calculator to explore probability in Technology Activity 7 on page 664.

Mini-Lab

Work with a partner.

Materials: paper bag containing 10 colored marbles

- Draw one marble from the bag, record its color, and replace it in the bag. Repeat this 10 times.
- Find the experimental probability for each color of marble.

$$\text{experimental probability} = \frac{\text{number of times color was drawn}}{\text{total number of draws}}$$

- Repeat both steps described above for 20, 30, 40, and 50 draws.

Talk About It

a. Is it possible to have a certain color marble in the bag and never draw that color?

b. Open the bag and compute the theoretical probability of drawing each color of marble.

c. Compare the experimental and theoretical probabilities.

Example

Three pennies are tossed 40 times. The results are displayed in the circle graph.

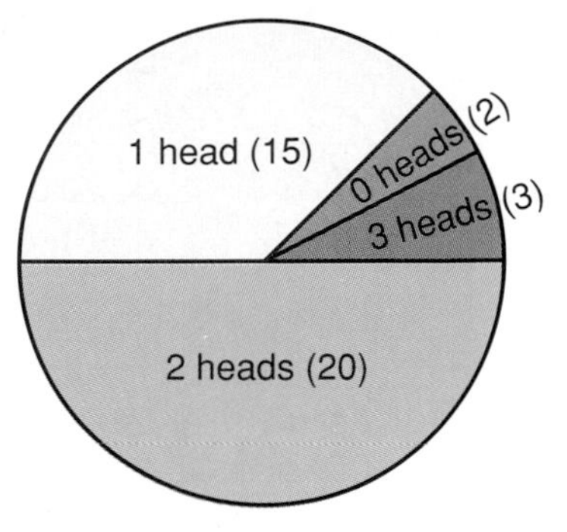

a. In this trial, what was the experimental probability of no heads?

The outcome of *no heads* happened twice. The experimental probability was $\frac{2}{40}$ or 0.05.

b. List the possible outcomes for tossing three coins.

HHH, HHT, HTH, HTT, THH, THT, TTH, TTT

Check your list by drawing a tree diagram.

c. What is the theoretical probability of no heads?

Since there are 8 outcomes when tossing three coins and only 1 outcome is no heads, the theoretical probability is $\frac{1}{8}$ or 0.125.

Checking for Understanding

Communicating Mathematics

Read and study the lesson to answer each question.

1. **Tell** why the experimental probability and theoretical probability of an event are not always the same.
2. **Draw** a bar graph that shows the results of the trial in the Example.

Guided Practice

Use the circle graph in the Example to answer Exercises 3 and 4.

3. Find the experimental probability of each outcome.
   a. P(1 head) b. P(2 heads) c. P(3 heads)
4. Find the theoretical probability of each outcome.
   a. P(1 head) b. P(2 heads) c. P(3 heads)

Exercises

Independent Practice

5. Ten students in Mr. Sopher's class believed that a coin is not fair. In other words, it appeared that the probability of getting heads was not $\frac{1}{2}$. To test this assumption, each student tossed the coin 40 times and recorded the results in the spreadsheet below.

STUDENT	1	2	3	4	5	6	7	8	9	10
NUMBER OF HEADS	21	22	18	26	21	21	19	22	18	29
NUMBER OF TAILS	19	18	22	14	19	19	21	18	22	11

   a. Find the P(heads) for each of the ten students. Then find P(heads) for all the students as a whole.
   b. Based on this data, does the coin appear to be fair?
   c. How could you get a better estimate of P(heads) for this coin?

6. Roll a die 60 times.
   a. Record the results.
   b. Based on your record, what is the probability of a 1?
   c. What is the probability of an even number? Explain.
   d. Are the probabilities in parts b and c examples of experimental or theoretical probabilities?

Mixed Review

7. Solve $p = -68 + 29$. *(Lesson 3-3)*
8. Find the GCF of 12 and 63. *(Lesson 6-4)*
9. Two die are rolled. Find the probability that a prime number is rolled on one die and an odd number is rolled on the other die. *(Lesson 13-5)*

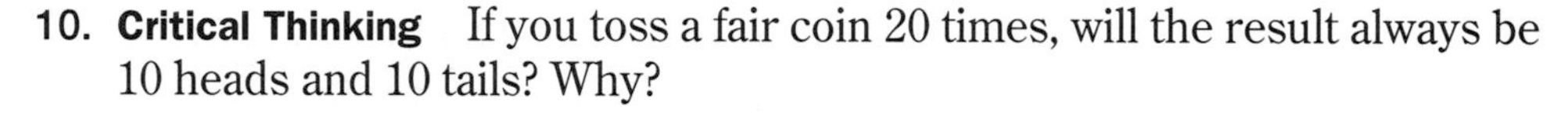

Problem Solving and Applications

10. **Critical Thinking** If you toss a fair coin 20 times, will the result always be 10 heads and 10 tails? Why?
11. **Portfolio Suggestion** Select one of the assignments from this chapter that you found especially challenging. Place it in your portfolio.
12. **Entertainment** The wheel on *Wheel of Fortune* is divided into 24 sections. If three of the sections have amounts higher than $1,000, what is the theoretical probability of the wheel stopping at one of these sections?

COMPUTER CONNECTION

13. **Computer Connection** In the "bonus situation" in basketball, the shooter is awarded one free throw. If he or she makes the basket, a bonus of one more free throw is awarded. Therefore, there are three possibilities:
    A. Shooter misses first free throw.
    B. Shooter makes first free throw, misses the second.
    C. Shooter makes both free throws.

    Suppose the probability that a certain shooter makes any given free throw is 67%. The following computer program simulates the results of 10,000 trials and prints the experimental probability of each situation.

```
10  FOR X = 1 TO 10000
20  LET Y = RND(X)
30  IF Y > 0.67 THEN CA = CA + 1: GOTO 70
40  LET Y = RND(X)
50  IF Y > 0.67 THEN CB = CB + 1: GOTO 70
60  CC = CC + 1
70  NEXT X
80  PRINT "P(A) = ";CA/10000: PRINT "P(B) = ";
    CB/10000: PRINT "P(C) = ";CC/10000
```

    a. Run the program and list the probability of each outcome.
    b. Modify the program if the probability that the shooter makes any given basket is 50%.

Cooperative Learning

13-8A Punnett Squares

A Preview of Lesson 13-8

Objective

Discover how experimental probability is used in biology.

Materials

40 yellow counters
40 red counters
2 paper bags

In the early 1900s, Reginald Punnett developed a model to show the possible ways genes can combine at fertilization. In a **Punnett square,** *dominant* genes are shown with capital letters. *Recessive* genes are shown with the lowercase of the same letter. Letters representing the parent's genes are placed on the outer sides of the Punnett square. Letters inside the boxes of the square show the possible gene combinations for their offspring.

Let **T** represent the dominant gene for tallness. Let **t** represent the recessive gene for shortness. A pea plant with **TT** genes is pure dominant and is tall. A pea plant with **tt** genes is recessive and is short. A pea plant with **Tt** genes is hybrid and is tall because it has a dominant gene. Notice that the capital letter goes first in hybrid genes.

Around 1865, an Austrian monk named Gregor Mendel explained what dominant and recessive traits were.

The Punnett square at the right represents a cross between two hybrid tall pea plants.

You can use a Punnett square to do a simulation that demonstrates how random fertilization works.

	T	t
T	TT	Tt
t	Tt	tt

Try this!

Work in groups of four to simulate the birth of 40 offspring.

- Place 20 yellow counters and 20 red counters into a paper bag. Label the bag *female parent.* This represents a parent with one dominant gene and one recessive gene.
- Place 20 yellow counters and 20 red counters into a second paper bag. Label this bag *male parent.* This represents a parent with one dominant gene and one recessive gene.
- Make a Punnett square to show the expected offspring of these parents. Use **R** for the dominant gene, red. Use **r** for the recessive gene, yellow.

Trial	Male Parent		Female Parent		Gene Pairs
	Red Counters	Yellow Counters	Red Counters	Yellow Counters	
1					
2					
3					
4					
40					

- Copy the table at the left.
- Shake the bags. Reach into each bag and, without looking, remove one counter. Record the colors of the counters in your table next to trial 1. Put the counters back into the bags from which they came.
- Repeat the previous steps 39 more times.

What do you think?

1. Out of 40 offspring, how many did you expect to be pure dominant (**RR**)? hybrid (**Rr**)? pure recessive (**rr**)?
2. Which of the three gene combinations did you expect to occur most often?
3. What is the theoretical probability of an offspring having pure dominant genes? hybrid genes? pure recessive genes?
4. Find the experimental probability. Describe how it compares to the theoretical probability.
5. What determines how offspring will look?

Application

In humans, free earlobes is a dominant trait over attached earlobes. Let F represent free earlobes and f represent attached earlobes. Draw a Punnett square for each parent combination below. Find the theoretical probability for the offspring.

6. FF, Ff
7. FF, FF
8. ff, ff
9. Ff, ff

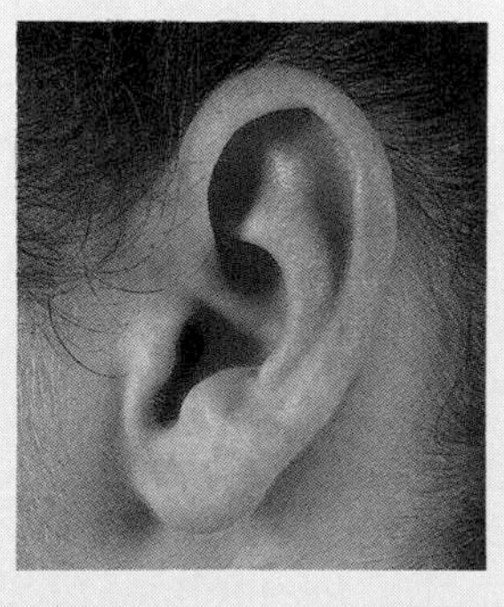

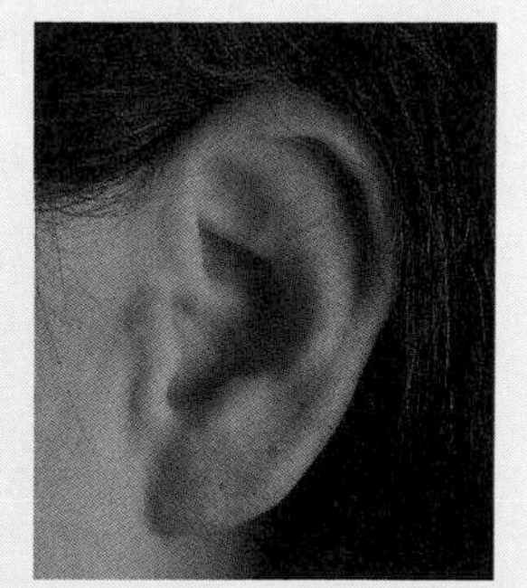

Extension

10. How does the prediction of 1 pure dominant, 2 hybrid, and 1 recessive offspring relate to Pascal's Triangle?

Statistics Connection

13-8 Using Experiments to Predict

Objective

Predict actions of a larger group using a sample.

Words to Learn

sample
population
random

In the 1990 census, it was estimated that more than 1.6% of the people living in the United States were not counted. This may sound like a small error, but it means over 4 million people were missed!

One of the main reasons for the inaccurate count was that many Americans did not return their census forms. Since it would take too much time and require too many workers to contact each of these people individually, the Census Bureau decided to have field-workers survey a smaller group of people called a **sample.** A sample is representative of a larger group called the **population.**

In the 1990 census, over 300,000 field workers were sent to one hundred geographical areas that were selected at **random.** In a random sample everything in the population has an equal chance of being selected. The results of the field workers were to be used to correct the undercount.

Example *Problem Solving*

Retail Sales The student council at Euclid Middle School plans to sell school jackets. They need to know how many of each size to order. They surveyed a sample of the school population. The sample consisted of one homeroom from each grade level. The results of the survey are shown in the frequency table at the right.

Size	Tally	Frequency
Small	𝍸 𝍸 𝍸 IIII	19
Medium	𝍸 𝍸 II	12
Large	𝍸 𝍸 𝍸 𝍸 𝍸 II	27
Extra Large	𝍸 𝍸 𝍸 II	17

Problem-Solving Hint

Notice that this is an example of the problem-solving strategy, *solve a simpler problem.*

a. How many students were surveyed for this sample?

$19 + 12 + 27 + 17 = 75$ *The sum of the frequencies is the sample size.*

75 students were surveyed.

b. What percent of the sample wear a large or medium jacket?

$12 + 27$ or 39 students wear a large or medium jacket.
Compare 39 to the total (75).

39 [÷] 75 [=] 0.52

52% of the sample wear a large or medium jacket.

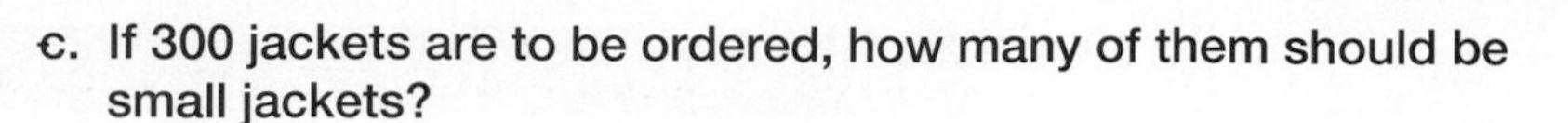

c. If 300 jackets are to be ordered, how many of them should be small jackets?

You can review proportions on page 344.

Set up a proportion and solve.

small jackets in sample → $\frac{19}{75} = \frac{x}{300}$ ← *small jackets to order*
total jackets in sample → ← *total jackets to order*

19 [×] 300 [÷] 75 [=] 76

76 small jackets should be ordered.

d. Would the student council have a representative sample if all seventy-five students were in the eighth grade?

No, because eighth graders will wear larger sizes, on average.

Checking for Understanding

Communicating Mathematics

Read and study the lesson to answer each question.

1. **Tell** how a sample group survey can be used to predict the actions of a whole population.
2. **Tell** what is the most important thing a surveyor needs to consider when a sample is used.

Guided Practice

Use the survey on favorite radio stations to answer each question.

3. What is the size of the sample?
4. What fraction chose WXGT?
5. For the 158,800 people in the listening range of these stations, how many would you expect to listen to WFRM?

Favorite Radio Station	
WFRM	40
WKIK	110
WXGT	180
WMOO	70

The Sunshine Orange Juice Company conducts a taste test. Out of 600 people that compare Sunshine Juice to Brand X, 327 prefer Sunshine Juice. Which statements below are true? Write *true* or *false*. Explain your answer.

6. "Consumers prefer Sunshine Juice 2 to 1."
7. "Over 50% of the people surveyed prefer Sunshine Juice."
8. "More people always choose Sunshine Juice over Brand X."

Use the survey on favorite soft drinks to answer each question.

9. What is the size of the sample?
10. What is the mode?
11. What fraction chose fruit juice?
12. If 7,200 people are expected to order drinks at the football playoff game, how many colas should the Band Boosters order?

Favorite Soft Drink	
Lemon-Lime	17
Cola	25
Root Beer	10
Fruit Juice	12
Ginger Ale	8

Exercises

Independent Practice

The bookstore sells 3-ring binders. All incoming sixth-grade students will need a binder. The binders come in 4 colors: red, green, blue, or yellow. The students who run the store decide to survey 50 sixth graders to find out their favorite color. Using this information, they will order binders to sell to the 450 students who will start the sixth grade in the fall.

13. From an alphabetical list, the students survey every fifth student. Is this a good sample? Why or why not?

14. Of the students surveyed, 25 chose red, 10 chose green, and 2 chose yellow. How many chose blue?

15. For the 450 students, how many of each color should be ordered?

Use the sample data on school jacket price ranges to answer each question.

16. What is the size of the sample?

17. If the price was less than \$29, about how many would buy a school jacket?

18. If there are 400 students in the eighth grade, about how many would buy a school jacket that cost \$26?

Amount Willing to Pay for School Jacket	
no more than \$25	8
no more than \$27	80
no more than \$29	82
no more than \$31	10

Mixed Review

19. Express $8\frac{9}{25}$ as a decimal. *(Lesson 6-6)*

20. **Algebra** Find the solution for the system $y = 2x - 7$ and $y = -2x + 9$. *(Lesson 11-5)*

21. Refer to the graph in the Example on page 529. Find $P(2 \text{ heads})$. *(Lesson 13-7)*

Problem Solving and Applications

22. **Military** The graph at the right shows the sizes of camouflage tops worn by one platoon of soldiers.

a. How many soldiers are in the platoon?

b. If the supply sergeant needed to order a new pattern of camouflage tops for 720 soldiers, how many of each size should be ordered?

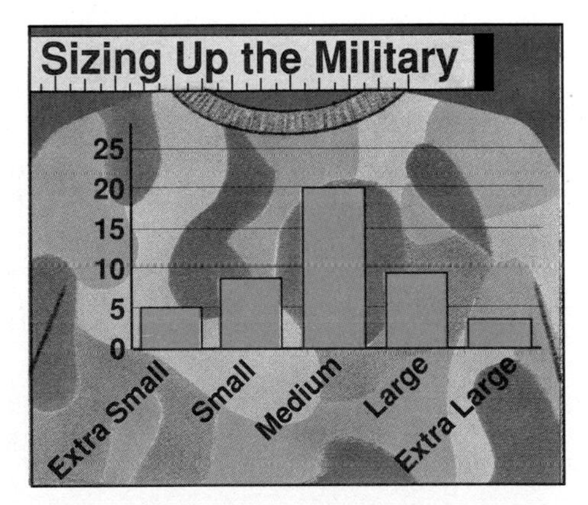

23. **Critical Thinking** A bag contains 100 bingo chips. Some are red. Some are green. Suppose you drew a chip out of the bag, recorded its color, and put it back. Then you kept doing this until your records showed 6 green chips and 14 red chips. How could you use these results to predict how many red chips and green chips are in the bag?

Chapter 13

Study Guide and Review

Communicating Mathematics

Choose the letter that best matches each phrase.

1. an arrangement or listing in which order is important
2. an arrangement or listing in which order is not important
3. the product of all counting numbers from n to 1
4. a diagram used to find the total number of outcomes of an event
5. a smaller group representative of a larger group
6. probabilities based on frequencies obtained in an experiment
7. probabilities based on physical characteristics

a. experimental probabilities
b. theoretical probabilities
c. population
d. sample
e. combination
f. permutation
g. tree diagram
h. $n!$
i. $P(n, r)$

8. Write the definition of the Fundamental Principle of Counting.
9. Explain how to find row 6 of Pascal's Triangle.
10. Tell what $C(8, 2)$ means.

Self Assessment

Objectives and Examples
Upon completing this chapter, you should be able to:

Review Exercises
Use these exercises to review and prepare for the chapter test.

- count outcomes using a tree diagram or the Fundamental Principle of Counting *(Lesson 13-1)*

Two coins are tossed. How many outcomes are possible?

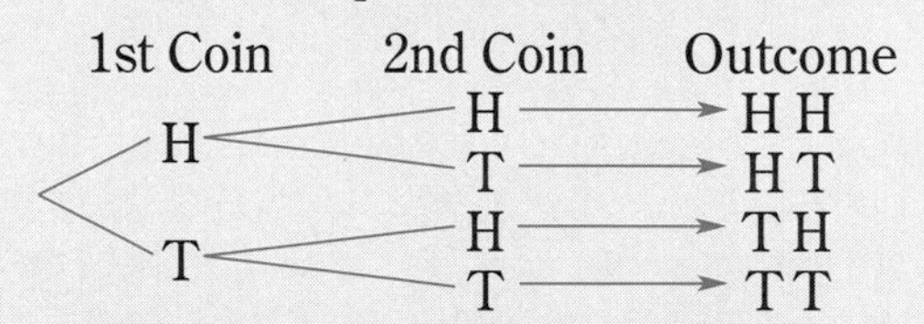

There are 4 outcomes.

State the number of possible outcomes for each event.

11. A diner offers three types of soft drinks, with or without ice.
12. A car comes in 2 models and 4 colors. There is a choice of a standard or automatic transmission.

- find the number of permutations of objects *(Lesson 13-2)*

The symbol $P(n, r)$ represents the number of permutations of n things taken r at a time.

$P(n, r) =$
$n \cdot (n-1) \cdot (n-2) \cdot \ldots \cdot (n-r+1)$

Solve.

13. How many different ways can 6 different books be arranged on a shelf?
14. How many different ways can you arrange the letters in the word OBJECTS if you take the letters 3 at a time?

Objectives and Examples	Review Exercises

Objectives and Examples

- find the number of combinations of objects *(Lesson 13-3)*

The symbol $C(n, r)$ represents the number of combinations of n things taken r at a time.

$$C(n, r) = \frac{P(n, r)}{r!}$$

Review Exercises

Solve.

15. How many different pairs of kittens can be selected from a litter of six?
16. How many different groups of 4 marbles can be chosen from a box containing 20 marbles?

- identify patterns in Pascal's Triangle *(Lesson 13-4)*

How many different groups of 3 people can be taken from a group of 5 people?

Row 5	1	5	10	10	5	1
number taken at a time	0	1	2	3 ↑	4	5

There are 10 different groups.

Use Pascal's Triangle to answer each question.

17. How many combinations are possible when 7 objects are chosen 4 at a time?
18. How many different kinds of 3-topping pizzas can be made when 6 toppings are offered?
19. Find the value of $C(9, 3)$.

- find the probability of independent and dependent events *(Lesson 13-5)*

The probability of two events can be found by multiplying the probability of the first event by the probability of the second event.

Two dice are rolled. Find each probability.

20. an even number and an odd number
21. a prime number and a composite number
22. 2 and a number divisible by 3

- find experimental probability *(Lesson 13-7)*

Experimental probabilities are probabilities based on frequencies obtained by conducting an experiment.

23. Toss two coins 40 times.
    a. Record the results.
    b. What is the experimental probability of no heads? Of one head and one tail?
    c. What are the theoretical probabilities for part b?

- predict actions of a larger group using a sample *(Lesson 13-8)*

If 18 out of a sample of 54 students prefer plain milk over chocolate milk, how many cartons of plain milk does the cafeteria need for 750 students?

plain milk sample → $\frac{18}{54} = \frac{p}{750}$ ← *plain milk order*
total sample → ← *total order*

18 [×] 750 [÷] 54 [=] 250

The cafeteria needs to order 250 cartons of plain milk.

The Goody Hot Dog Company conducted a survey. Participants in the survey were asked to choose one brand of hot dog they would buy. Of those surveyed, 96 said they would buy Goody brand, 68 said they would buy brand X, and 36 said they would buy brand Y.

24. What is the size of the sample?
25. What fraction chose brand Y?
26. For 8,000 people, how many would buy a Goody brand hot dog?

Applications and Problem Solving

27. **Sports** There are nine players on a baseball team that bat. How many different batting orders are possible? *(Lesson 13-2)*

28. **Business** Orange Crate Records is having a sale on CDs. They are offering 3 CDs for $32. If there are 40 CDs to choose from, how many combinations of 3 CDs can be chosen? *(Lesson 13-3)*

29. **Food** Frank's Restaurant has a Tuesday dinner special that offers a choice of soup or salad for an appetizer, a choice of steak, fish, or chicken for the main dish, and a choice of pudding or ice cream for dessert. Draw a tree diagram to find the number of possible outcomes for a complete meal. *(Lesson 13-1)*

30. Jay jogs past four benches in the park. Based on his observation, he estimates that any one bench is occupied 0.3 of the time. Act it out to find the probability that all four benches will be occupied as Jay jogs past. *(Lesson 13-7)*

Curriculum Connection Projects

- **Zoology** Survey your classmates to learn how many prefer dogs or cats as pets. Predict how many students in the school prefer each. Display the results in a creative way.

- **Health** List at least three of your favorite breakfast cereals and at least two of your favorite fruits to have on cereal. Find how many mornings you could have a different cereal-fruit combination.
- **Food** Find how many different three-item lunches you can assemble from the selections on today's lunch menu.

Read More About It

Asher, Herbert. *Polling and the Public: What Every Citizen Should Know.*
Blackwood, Gary. *The Dying Sun.*
Burns, Marilyn. *The I Hate Mathematics Book.*
McCauley, David. *Pyramid.*

Chapter

13 Test

Televisions come in three sizes, a 13-, 19-, or 27-inch screen, and with one or two speakers. They come with or without remote control.

1. Draw a tree diagram to find the number of possible outcomes.
2. If a television is chosen at random, what is the probability of selecting a 19-inch, two speaker television with remote control?
3. How many outcomes show a one-speaker television?

Find each value.

4. $P(8, 3)$
5. $C(15, 5)$
6. Michelle has 7 tennis trophies. How many ways can she arrange 4 of them in a row?
7. How many teams of 6 players can be chosen from 18 players?
8. In how many ways can 5 students stand in a line?

Use Pascal's Triangle to answer each question.

9. How many combinations of 3 baseball cards can be chosen from 5 baseball cards?
10. What does the 20 in row 6 mean?

In a bag there are 5 blue marbles, 3 red marbles, and 2 green marbles. Once a marble is selected, it is not replaced. Find the probability of each outcome.

11. a red marble and then a blue marble
12. two green marbles
13. Does the probability in Problem 11 represent dependent or independent events?

Two coins are tossed 20 times. No heads were tossed five times, one head was tossed eleven times, and two heads were tossed four times.

14. What is the experimental probability of two heads?
15. What is the theoretical probability of two heads?
16. Why is the experimental probability and theoretical probability of two heads different?

Use the survey on favorite chocolate bars to answer each question.

Favorite Chocolate Bar	
milk chocolate	13
with almonds	18
with crisps	11
with caramel	8

17. What is the size of this sample?
18. What percent of the sample chose a chocolate bar with almonds?
19. The school choir plans on selling chocolate bars to raise money for new uniforms. How many chocolate bars with almonds should they order if they plan on selling 800 chocolate bars?
20. Olivia reaches into a bag containing 3 apples, 2 pears, and 2 tangerines. Act it out to find the probability that she picks out 2 tangerines.

Bonus In how many ways can 2 boys and 2 girls stand in a line, if boys and girls are to alternate in line?

Chapter

14 Algebra: Investigations with Polynomials

Spotlight on Transportation

Have You Ever Wondered...

- How much subway fares have increased over the years?
- If the price of a rental car depends on the city in which it is rented?

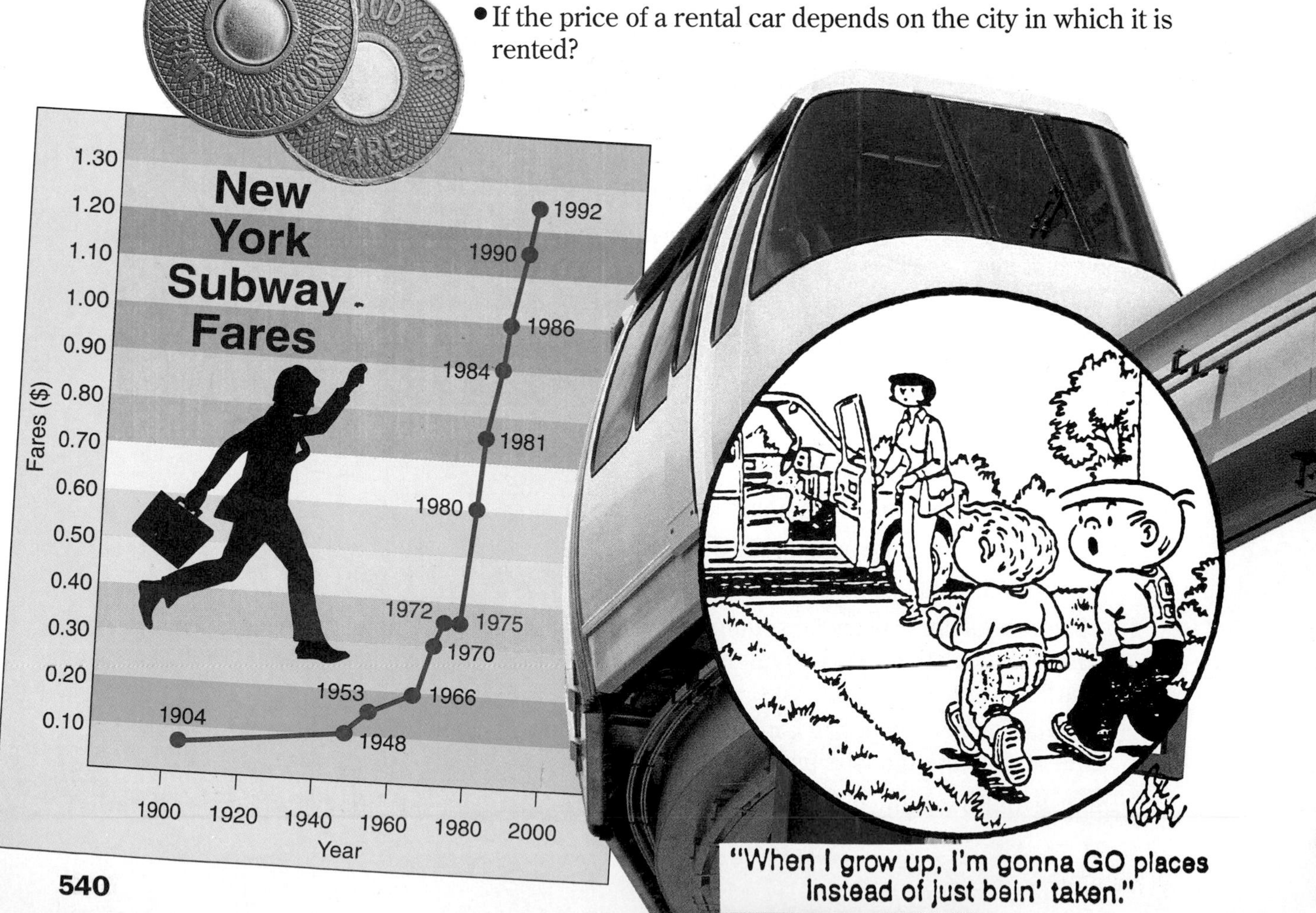

Chapter Project

Transportation

Work in a group.

1. Choose a place you would like to visit.
2. Determine how much it would cost to get there by different means of transportation, such as by car, by plane, by train, by bus, or by boat.
3. Make a poster of your information. Include the advantages and disadvantages of each type of transportation and which one you would most likely choose.

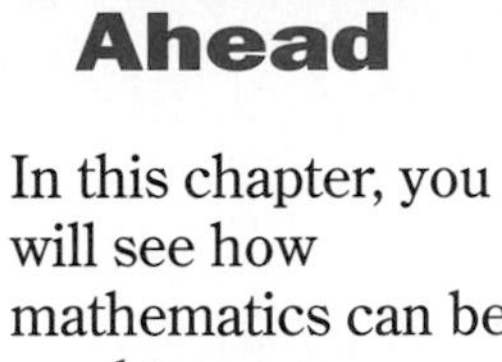

Looking Ahead

In this chapter, you will see how mathematics can be used to answer questions about transportation. The major objectives of the chapter are to:

- represent polynomials with area models
- use area models to simplify algebraic expression, and add, subtract, multiply, and divide polynomials.
- solve problems using logical reasoning

RENTING A CAR

City	Compact	Midsize	Full-size
Atlanta			
Avis	$41.00	$45.00	$46.00
Budget	$34.99	$36.99	$38.99
Hertz	$45.99	$49.99	$53.99
National	$26.90	$29.90	$33.90
Chicago			
Avis	$58.00	$60.00	$65.00
Budget	$53.00	$55.50	$61.00
Hertz	$56.99	$58.99	$63.99
National	$22.90	$22.90	$31.90
Dallas			
Avis	$43.00	$46.00	$47.00
Budget	$48.99	$52.99	$53.99
Hertz	$22.99	$25.99	$27.99
National	$35.90	$40.90	$41.90
Los Angeles			
Avis	$35.00	$39.00	$42.00
Budget	$22.89	$23.89	$28.99
Hertz	$46.99	$48.99	$51.99
National	$33.90	$34.90	$40.90
New York			
Avis	$56.98	$64.98	$67.98
Budget	$53.00	$59.00	$67.50
Hertz	$54.99	$57.99	$63.99
National	$45.90	$47.90	$55.90
St. Louis			
Avis	$37.00	$41.00	$43.00
Budget	$36.99	$37.99	$43.99
Hertz	$39.99	$45.99	$47.99
National	$25.00	$27.00	$30.00

Cooperative Learning

14-1A Area Models

A Preview of Lesson 14-1

Objective

Make area models for algebraic expressions.

Materials

yellow and red construction paper
scissors

Throughout this text, you have used rectangles to show multiplication problems. For example, the figure at the right shows the multiplication problem 4×5 as a rectangle that is 4 units wide and 5 units long. Its area is 20 square units.

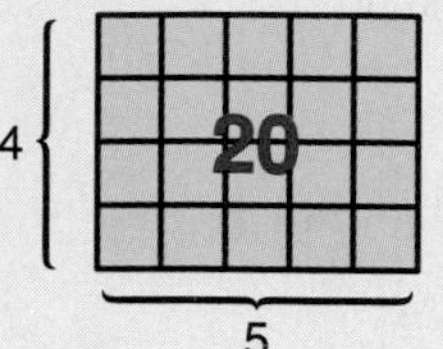

In this lab, you will show algebraic expressions using models called area tiles.

Try this!

Work with a partner.

- From yellow construction paper, cut out a square that is 1 unit long and 1 unit wide.

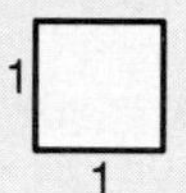

- Draw a line segment that is longer than 1 unit. Since the line segment can be any length, we will say that it is x units long. From yellow construction paper, cut out a rectangle that is 1 unit wide and x units long.

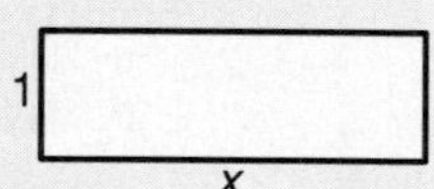

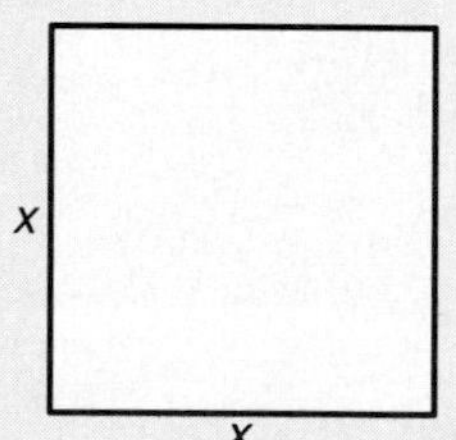

- Using the same measure for x, cut out a square that is x units long and x units wide from yellow construction paper.

What do you think?

1. Find the area of each shape and write the area on each tile.
2. For each tile, write a sentence that tells the relationship between the length, width, and area.

Extension

3. From yellow construction paper, cut out four more sets of tiles. Label each with its area.
4. From red construction paper, cut out five sets of tiles with the same dimensions as the yellow tiles. Label the areas -1, $-x$, and $-x^2$.

14-1 Area Models of Polynomials

Objective

Represent polynomials with area models.

Words to Learn

monomial
polynomial

When an engineer designs a new sports car, he or she often makes a model of the car. The model helps the engineer visualize the car and determine characteristics like structural strength and fuel efficiency. In mathematics we also use models to help us visualize concepts.

In the previous mathematics lab, you made area tiles.

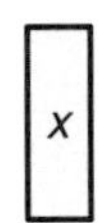

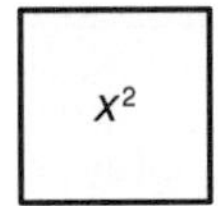

The expressions 1, x, and x^2 are called **monomials.** A monomial is a number, a variable, or a product of a number and one or more variables. You can model any monomial using area tiles.

Examples

Model each monomial using area tiles or drawings.

1 $2x^2$

To model this expression, you need 2 yellow x^2-tiles.

x^2 x^2

2 $-3x$

Use red tiles to model negative values. To model this expression, you need 3 red x-tiles.

$-x$ $-x$ $-x$

3 4

To model this expression, you need 4 yellow 1-tiles.

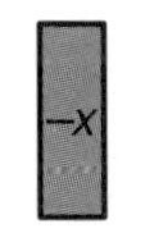

Problem Solving Hint

This text uses red and yellow tiles. You may want to use shaded and unshaded tiles or some other variation to show positive and negative values.

Sometimes, algebraic expressions contain more than one monomial. These expressions are called **polynomials.** A polynomial is the sum or difference of two or more monomials. You can also model polynomials with area tiles.

Examples

Model each polynomial using area tiles or drawings.

4 $2x^2 + 4x + 9$

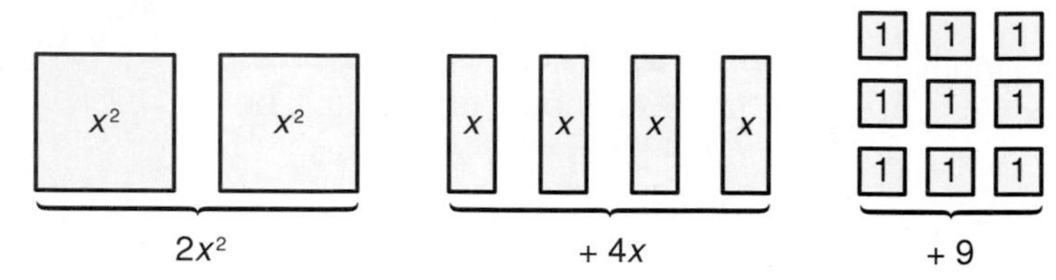

5 $x^2 - 2x - 3$

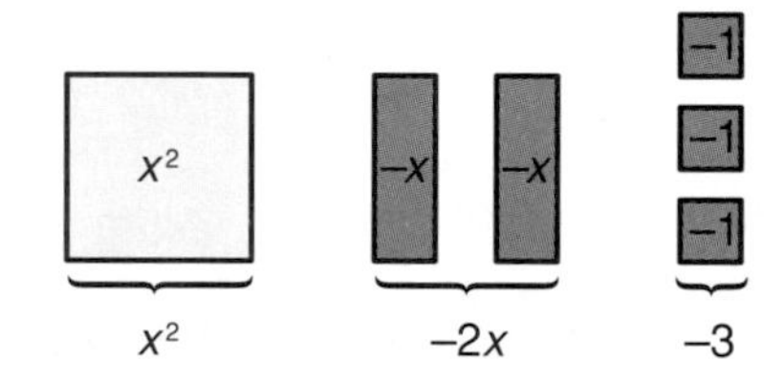

Polynomial expressions can be evaluated by replacing variables with numbers and then finding the value of the numerical expression.

Example 6 *Connection*

LOOKBACK

You can review order of operations on page 44.

Algebra Evaluate $3x^2 - 5x + 2$ if $x = -2$.

$$\begin{aligned} 3x^2 - 5x + 2 &= 3(-2)^2 - 5(-2) + 2 && \textit{Replace x with -2.} \\ &= 3(4) - 5(-2) + 2 && \textit{Use the order of operations.} \\ &= 12 - (-10) + 2 \\ &= 24 \end{aligned}$$

Checking for Understanding

Communicating Mathematics

Read and study the lesson to answer each question.

1. **Write** two expressions that are monomials and two that are polynomials.
2. **Tell** how to model a monomial like $-2x$.
3. **Write** the polynomial represented by the model at the right.
4. **Draw** a model of $-3x^2 + 2x - 5$.

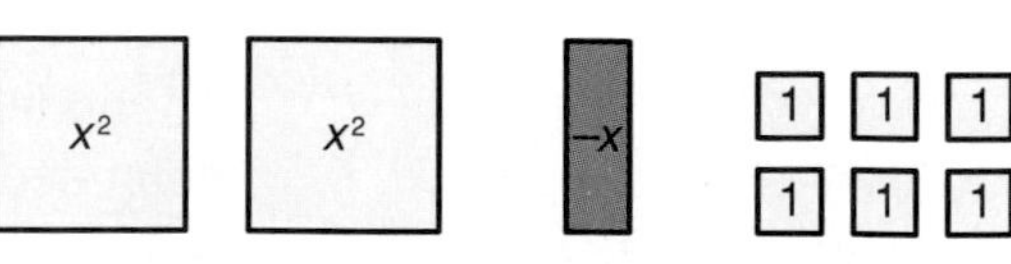

Guided Practice

Write a monomial or polynomial for each model.

5. 6.

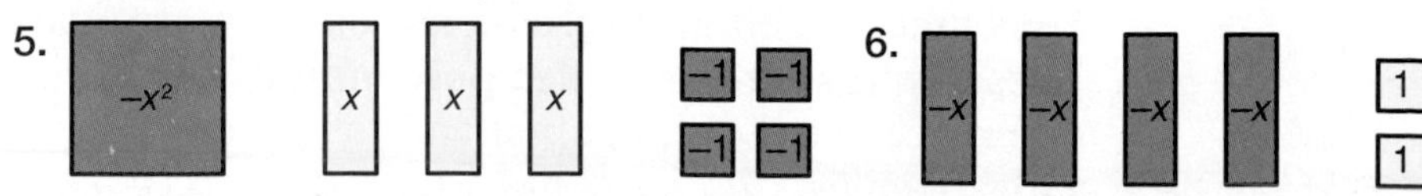

Model each monomial or polynomial using area tiles or drawings.

7. $3x^2$
8. $2x - 1$
9. $-x^2 + 5x - 6$
10. Evaluate $x^2 - 3x - 4$ if $x = 5$.

Exercises

Write a monomial or polynomial for each model.

11. 

12. 

13. 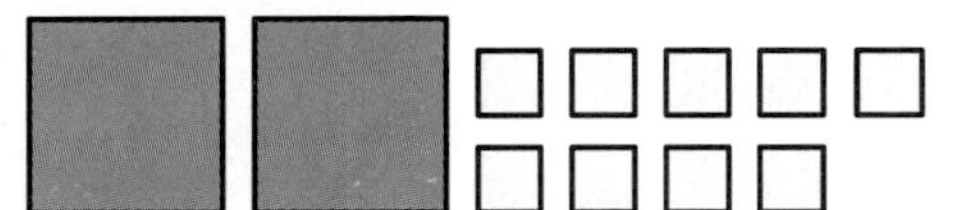

14. 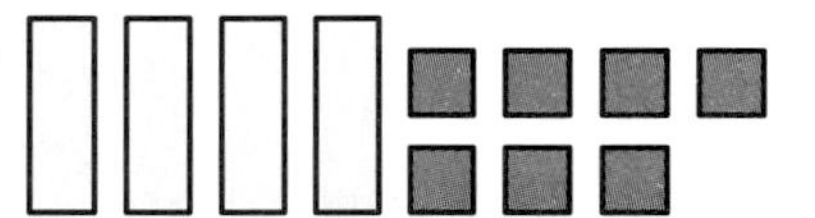

Model each monomial or polynomial using area tiles or drawings.

15. $-2x^2$
16. $5x + 3$
17. $3x^2 + 2x + 6$
18. $4x^2 - 3$
19. $-x^2 - 2x - 1$
20. $x^2 - 8$

Evaluate each expression.

21. $2x - 5$, if $x = 8$
22. $x^2 + 3x$, if $x = 2$
23. $x^2 - 10x + 25$, if $x = 5$
24. $-x^2 + 4x$, if $x = -1$

Mixed Review

25. **Geometry** Find the circumference of a circle with a radius of 8.2 millimeters. *(Lesson 7-8)*

26. Thirty is 60% of what number? *(Lesson 10-1)*

27. **Statistics** Refer to the favorite soft drink survey at the right. For 6,300 people, how much ginger ale should the Band Boosters order? *(Lesson 13–8)*

Favorite Soft Drink	
Flavor	**Number of Responses**
Lemon-Lime	17
Cola	25
Root Beer	10
Fruit	12
Ginger Ale	8

Problem Solving and Applications

28. **Business** Carla volunteers to get lunch for members of the Ecology Club. She represents their lunch order of 13 burgers, 7 shakes, and 10 orders of french fries as the polynomial expression $13b + 7s + 10f$. Determine the cost of the order if burgers are \$1.99 each, shakes are \$1.49 each, and fries are \$0.99 each.

29. **Critical Thinking** The model  represents an area of x. Suppose 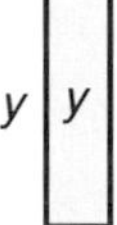 represents an area of y. Draw a model that represents an area of xy.

30. **Journal Entry** Make up a polynomial and use a drawing to represent it.

14-2 Simplifying Polynomials

Objective

Simplify polynomials using area models.

Words to Learn

term
like term
simplest form

Scouting is an organization that teaches young people to be good citizens and to develop their interests. Nearly 7 million young men and women belong to either the Boy Scouts or the Girl Scouts.

The members of Boy Scout Troop 92 were buying supplies for their spring camping trip. The Hawk Patrol bought one bag of apples, six boxes of granola bars, and two packages of recyclable paper plates. The Falcon Patrol bought two bags of apples, eight boxes of granola bars, and twelve packages of pre-sweetened drink mix.

The quantity bought can be described using polynomials. If x represents the number of granola bars in one box, $6x + 8x$ represents the total number of granola bars bought. We can model this situation with area tiles.

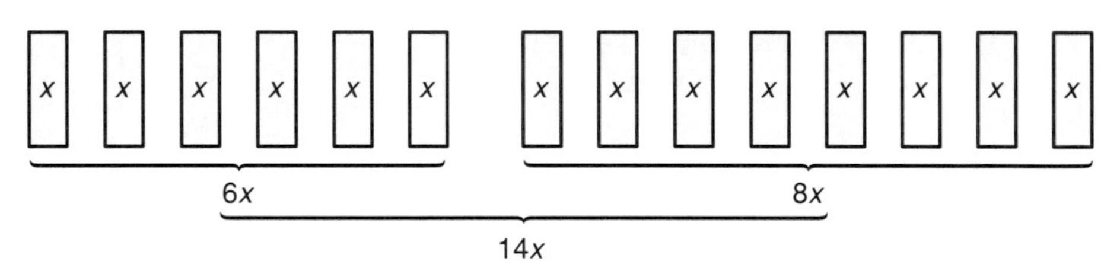

The polynomial $6x + 8x$ contains two monomials. Each monomial in the polynomial is called a **term.** The monomials $6x$ and $8x$ are called **like terms** because they have the same variable to the same power. When you use area tiles, you can recognize like terms because they have the same size and shape.

The model shown above suggests that you can simplify polynomials that have like terms. An expression that has no like terms is in **simplest form.** In simplest form, $6x + 8x$ is $14x$.

Example 1

Simplify $2x^2 + 3x^2 + 2x$.

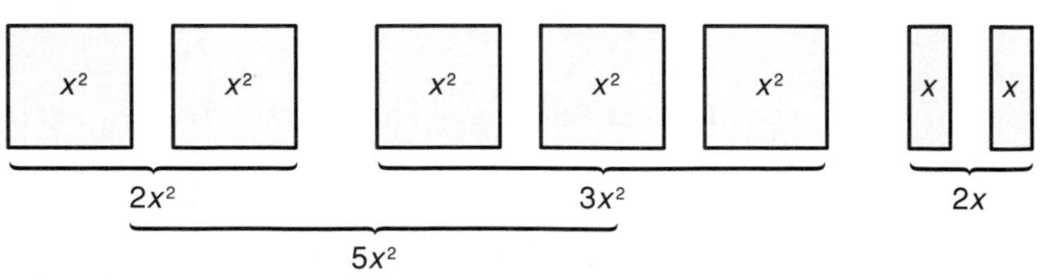

$2x^2 + 3x^2 + 2x = 5x^2 + 2x$

What do you suppose happens when you have both positive and negative terms in a polynomial? Let's use area tiles to find out.

Mini-Lab

Work with a partner to simplify $3x + 2 - 5x + 1$.

Materials: area tiles

- Model the polynomial.

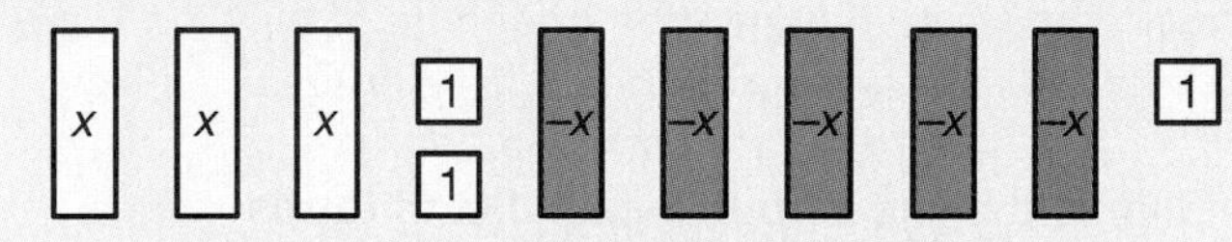

- Rearrange the tiles so that like terms are next to each other.

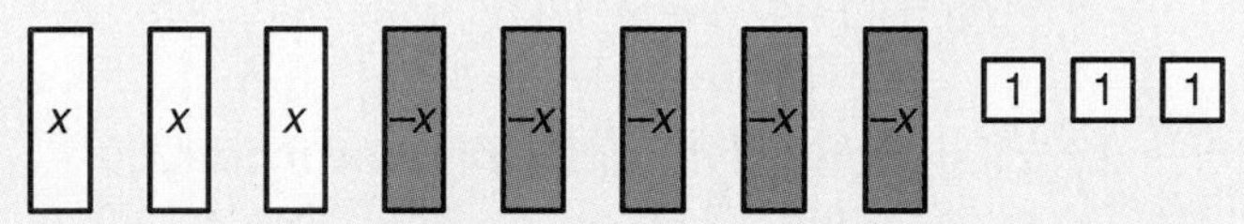

- When a positive tile is paired with a negative tile of the same size and shape, the result is called a *zero pair.* You can add or remove a zero pair without changing the value of the set. Remove all zero pairs.
- Write the polynomial for the tiles that remain.

LOOK BACK

You can review zero pairs on page 92.

Talk About It

a. How many zero pairs did you remove?

b. What kinds of tiles remained at the end?

c. What is the simplest form of the polynomial?

So far, we have used the variable x when dealing with polynomials. Other variables are possible.

Examples

Simplify each polynomial.

2 $y^2 + 2y^2$

These are like terms because they have the same variable to the same power.

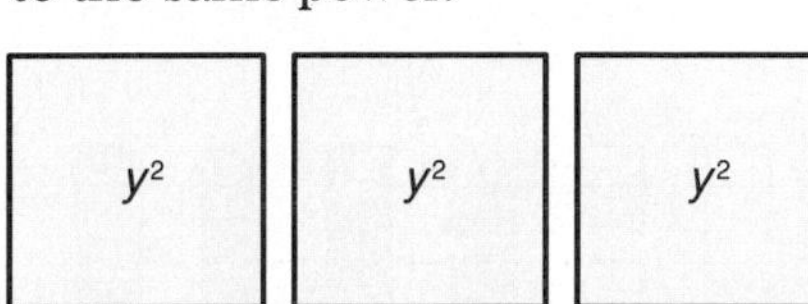

$y^2 + 2y^2 = 3y^2$

3 $3x + 2y$

These are not like terms because the variables are different.

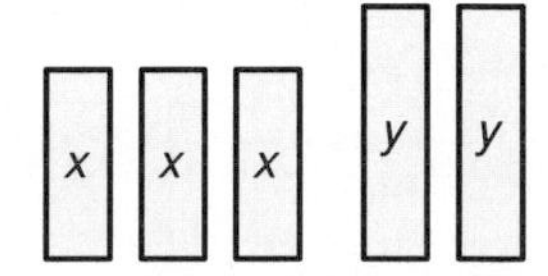

$3x + 2y$ is in simplest form.

Checking for Understanding

Communicating Mathematics

Read and study the lesson to answer each question.

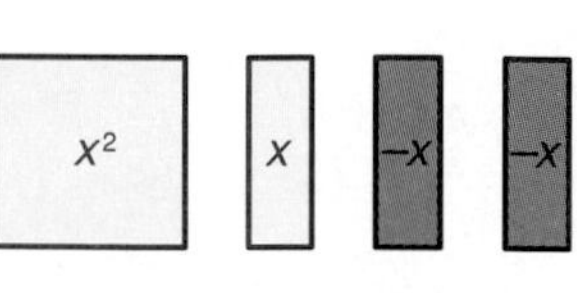

1. **Tell** how you can recognize like terms in a polynomial.
2. **Write** a polynomial containing three or more monomials to represent the model at the right.
3. **Draw** a model of $2x^2 + 3x - x^2 - 5x + 3$.
4. **Tell** whether $3a$ and $3b$ are like terms. Explain why or why not.

Guided Practice

Name the like terms in each list of terms.

5. $3x^2, 4x, 10x, -2x^2$
6. $4y, 8, 9, -2y, -3y$
7. $10x, 13y, 15z$
8. $-a^2, 4a^2, -2x^2$

Simplify each polynomial using the model.

9. $2x + 3 - x + x^2 - 4$

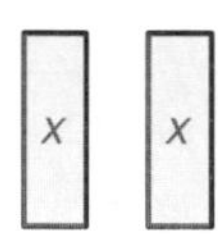

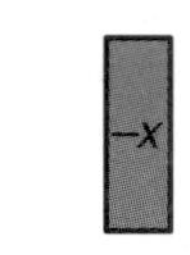

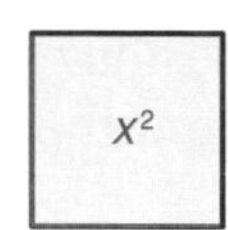

10. $3a^2 - 2a^2 + 3a$

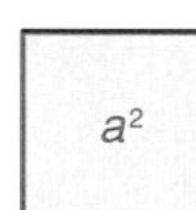

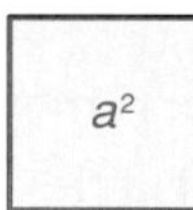

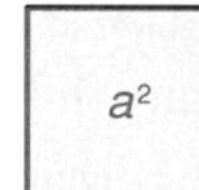

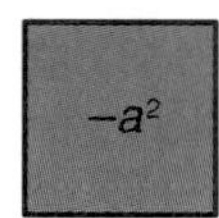

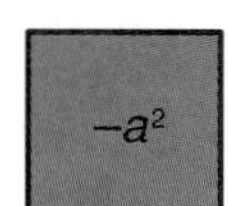

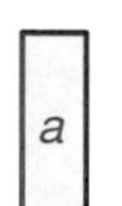

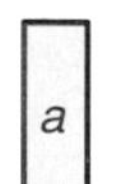

Exercises

Independent Practice

Name the like terms in each list of terms.

11. $6y, -4y^2, -11y$
12. $4, 3m^2, -5, 2m$
13. $15y^2, 2y, 8$
14. $7a, 6b, 10a, 14b$

Simplify each polynomial using the model.

15. $-x^2 + 4x + 2x + x^2$

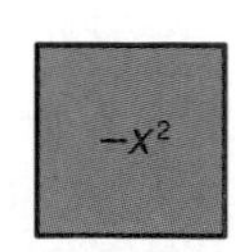

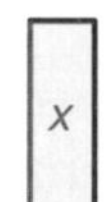

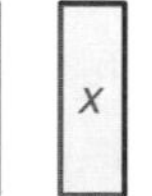

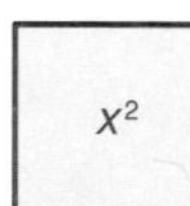

16. $2x + 3y - 4x - y$

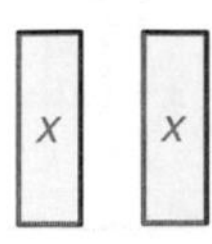

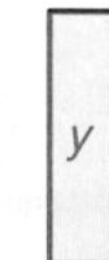

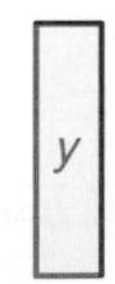

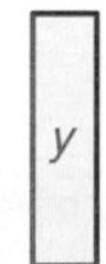

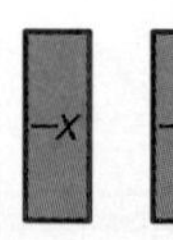

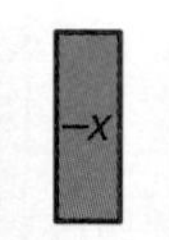

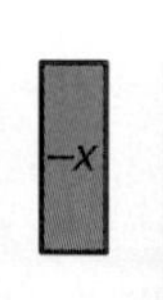

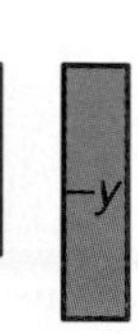

Simplify each polynomial. Use area tiles or drawings, if necessary.

17. $3x + 1 + 2x + 4$

18. $5y + 2 - 3y$

19. $2x^2 + 3 + 4x - 7$

20. $-3y^2 - 2y^2 - 4y + 3$

21. $a^2 - 5a - a^2 - 2a$

22. $10x + 3y - 8x + 5y$

Simplify each expression. Then evaluate if $a = -2$ and $b = 7$.

23. $2a + 5b + 7a + 9b$

24. $3a + 9b + 14a + 2b$

25. $4b + 3a - 2a + 5b$

26. $7a + 10b - 7b - 4a$

Mixed Review

27. Statistics Explain how to find the interquartile range from a box-and-whisker plot. *(Lesson 4-7)*

28. Express *\$9.60 for 8 feet* as a unit rate. *(Lesson 9-1)*

29. Graph $y = -2.5x + 3.5$ *(Lesson 11-4)*

30. Evaluate $2x^2 - 3x + 5$ if $x = -2$. *(Lesson 14-1)*

Problem Solving and Applications

31. Money On her way home from school, Karen stops in a convenience store to buy a large drink. In her backpack she finds three quarters, five dimes, and two nickels. In her pocket she had one quarter, three dimes and three nickels. Using q for quarters, d for dimes, and n for nickels, represent all the coins Karen had as a polynomial expression in simplest form. Then evaluate the simplified expression to determine how much money Karen has.

32. Critical Thinking Simplify $4x^3 + 2x^2 - x^3 + 4x^2 - 5x + 3$.

33. Data Search Refer to pages 540 and 541.
Determine the lowest total cost for renting a compact car in Atlanta for three days, and then a midsize car in Los Angeles for four days.

Save Planet Earth

Conservation of Trees About 10,000 years ago, more than 15 billion acres worldwide were covered with forest. Today, barely 10 billion acres of the world are forested. Why should you care about the world's forests? The relationship between trees and life on Earth is simple: we need oxygen and we produce carbon dioxide; trees require carbon dioxide and they produce oxygen. The loss of a tree not only reduces carbon dioxide consumption, it also releases the carbon dioxide stored in the tree.

How You Can Help

Plant a tree. Each tree that you plant will provide benefits for years to come.

14-3 Adding Polynomials

Objective

Add polynomials using area models.

People in the United States use the English system of measurement for most everyday weight measurement. It is customary to express many weights in combinations of pounds and ounces.

Twins occur about once in every 89 births. Triplets occur about once in 7,900 births.

Suppose twin baby boys were born today. One weighed 6 pounds 5 ounces and the other weighed 5 pounds 10 ounces. What was their combined weight?

$$\begin{array}{r} 6 \text{ lb} + 5 \text{ oz} \\ +\ 5 \text{ lb} + 10 \text{ oz} \\ \hline 11 \text{ lb} + 15 \text{ oz} \end{array}$$

Their combined weight was 11 pounds 15 ounces.

To find the combined weight, "like terms" are added. In this case, the like terms are pounds and ounces. Two or more polynomials can be added in a similar way.

Mini-Lab

Problem Solving Hint

It may be convenient to arrange like terms in columns before removing zero pairs.

Work with a partner to find $(x^2 - 2x + 4) + (2x^2 + 5x - 5)$.

Materials: area tiles

- Model each polynomial.

$x^2 - 2x + 4$ ➡

$2x^2 + 5x - 5$ ➡

- Combine like terms and remove all zero pairs.
- Write the polynomial for the tiles that remain.

Talk About It

a. How is this method like the method you used when simplifying polynomials?

b. What is $(x^2 - 2x + 4) + (2x^2 + 5x - 5)$?

Example 1

Write the two polynomials represented below. Then find their sum.

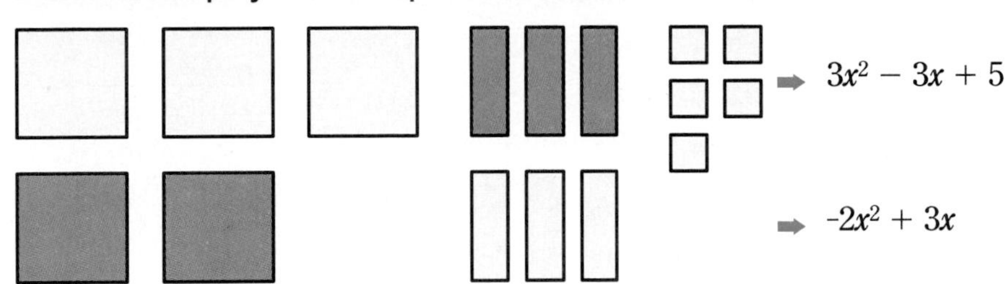

$(3x^2 - 3x + 5) + (-2x^2 + 3x) = x^2 + 5$

Example 2 *Connection*

Algebra **Find $(x^2 + 2x + 1) + (x^2 - 3x - 2)$. Then evaluate the sum for $x = -2$.**

$$\begin{array}{r} x^2 + 2x + 1 \\ + x^2 - 3x - 2 \\ \hline 2x^2 - 1x - 1 \end{array}$$

Arrange like terms in columns. Then add.

Now evaluate for $x = -2$.

$$\begin{aligned} 2x^2 - 1x - 1 &= 2(-2)^2 - 1(-2) - 1 \quad \textit{Replace x with -2.} \\ &= 2(4) - (-2) - 1 \\ &= 8 + 2 - 1 \text{ or } 9 \end{aligned}$$

Checking for Understanding

Communicating Mathematics

Read and study the lesson to answer each question.

1. **Write** the two polynomials represented below. Then find their sum.

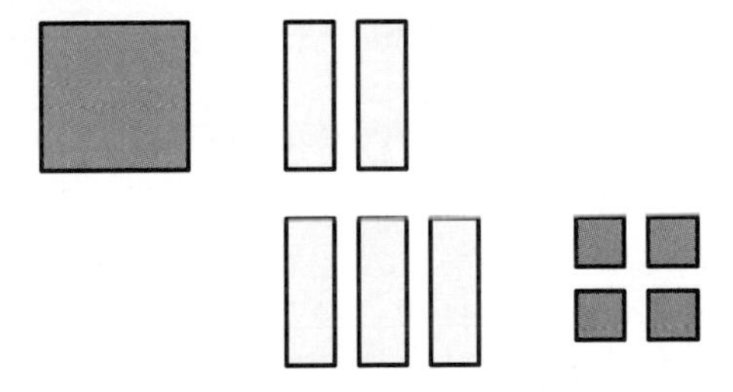

2. **Show** how to find the sum of $3x^2 - 3x + 5$ and $x^2 + 2x - 7$ using area tiles or drawings.

Guided Practice

Find each sum using area tiles or drawings.

3. $(2x + 5) + (-3x + 1)$
4. $(a^2 - 4a) + (2a^2 - 6a)$
5. $(-2x^2 + 4x - 6) + (-5x^2 - 3x + 7)$
6. Find the sum of $-3y^2 + 4y$ and $-1y^2 - 3y$. Then evaluate the sum for $y = 1$.

Exercises

Independent Practice

Find each sum. Use area tiles or drawings if necessary.

7. $\begin{array}{r} 5x^2 - 2x + 5 \\ + \ x^2 + 4x - 3 \\ \hline \end{array}$

8. $\begin{array}{r} -4r^2 + 3r - 2 \\ + 6r^2 - 5r - 7 \\ \hline \end{array}$

9. $\begin{array}{r} 2y^2 + \ y + 4 \\ + 3y^2 + 2y + 1 \\ \hline \end{array}$

10. $\begin{array}{r} 4a - 7b - 6c \\ + 3a + 5b + 2c \\ \hline \end{array}$

11. $(3x + 7y) + (9x + 5y)$
12. $(5m + 3n) + (4m + 2n)$
13. $(8s - 3t) + (s + 5t)$
14. $(5r - 7s) + (3r + 8s)$
15. $(3a^2 + 2a) + (7a^2 - 3)$
16. $(3x^2 - 5x + 7) + (-x^2 + 2x - 3)$
17. Find the sum of $-3x + 4$ and $5x^2$.

Find each sum. Then evaluate for $c = 8$ and $d = 5$.

18. $(2c + 5d) + (6c - 3d)$
19. $(4c + 3d + 2) + (3c - 4d - 1)$
20. $(-2c + 7d) + (8c - 8d)$
21. $(15c + 2d - 1) + (c - 3d + 2)$

Mixed Review

22. Evaluate $[3(18 - 2)] - 4^2$. *(Lesson 2-1)*
23. Solve $b^2 = 1.69$. *(Lesson 8-3)*
24. **Geometry** Find the volume of a cone with a radius of 2 meters and a height of 12 meters. *(Lesson 12-7)*
25. Simplify $6x - 2x + 3x^2 + x^2$. *(Lesson 14-2)*

Problem Solving and Applications

26. **Construction** A standard measurement for a window is *united inch.* You can find the united inches of a window by adding the length of the window to the width. If the length of a window is represented by the polynomial $x^2 + 2x - 3$, and the width is represented by $x + 3$, what is the size of the window in united inches?

27. **Geometry** Write and simplify an expression for the perimeter of each figure.

a.

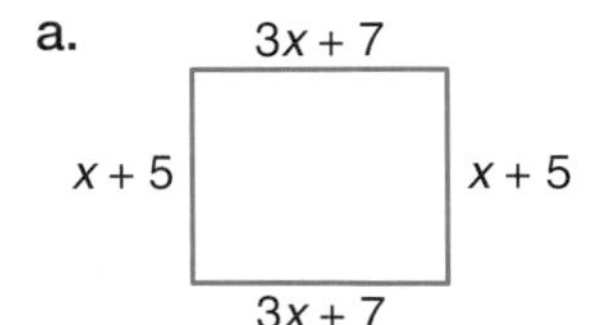

b.

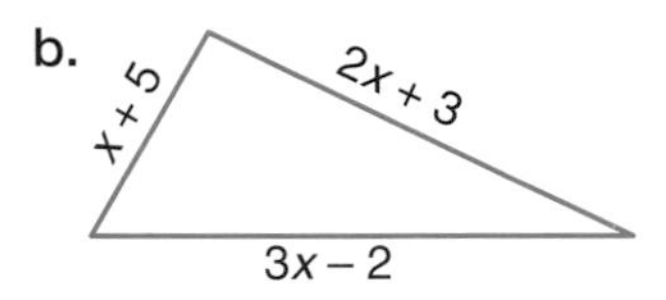

c.

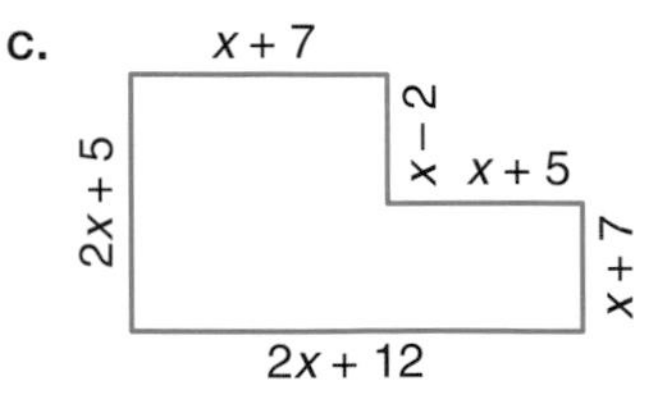

DATA SEARCH

28. **Data Search** Refer to page 668. What is the theoretical probability that two people in your math class share the same birthday?
29. **Critical Thinking** If $(5x - 13y) + (8x + 4y) = 13x - 9y$, what is $(13x - 9y) - (5x - 13y)$?

14-4 Subtracting Polynomials

Objective

Subtract polynomials using area models.

When you learned to subtract integers, you may have used counters. The figure at the right shows -8 − (-2). You start with 8 negative counters on the mat and remove 2 negative counters. Six negative counters remain. Therefore, -8 − (-2) = -6.

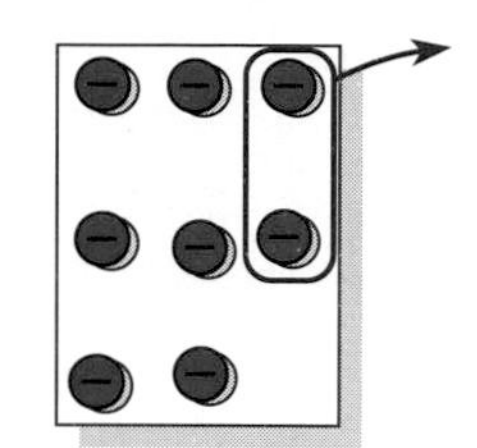

In a similar manner, you can subtract polynomials using area models. Consider the problem $(5x^2 + 6x + 7) - (3x^2 + 2x + 1)$. Model $(5x^2 + 6x + 7)$ using area tiles. Then remove 3 x^2-tiles, 2 x-tiles, and 1 1-tile.

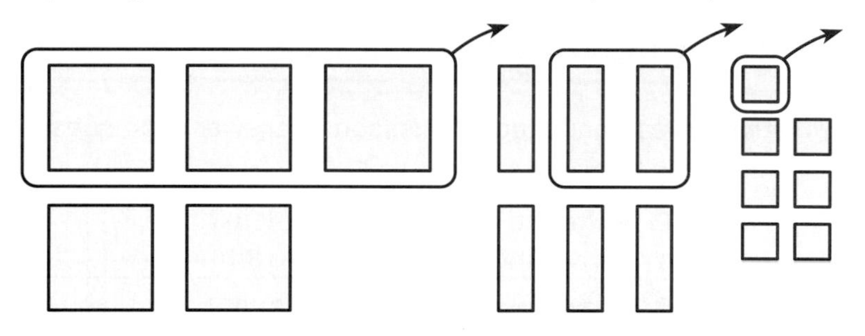

There are 2 x^2-tiles, 4 x-tiles, and 6 1-tiles remaining: $(5x^2 + 6x + 7) - (3x^2 + 2x + 1) = (2x^2 + 4x + 6)$.

Mini-Lab

Work with a partner to find $(2x + 5) - (-1x + 2)$.

Materials: area tiles

- Model the polynomial $2x + 5$.
- Now remove 1 negative x-tile and 2 1-tiles. You can remove the 1-tiles, but there are no negative x-tiles, so you can't remove $-1x$. Add a zero pair, then remove the negative x-tile.

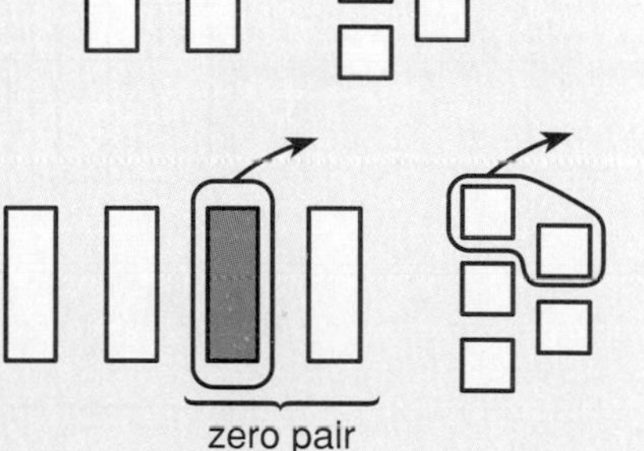

Talk About It

a. What kinds of tiles remain?

b. What is $(2x + 5) - (-1x + 2)$?

c. Why can you add or remove a zero pair?

You can review subtraction of integers on page 98.

To subtract an integer, you add its opposite. For example, $5 - 7 = 5 + (-7)$. In a similar manner, to subtract a polynomial, add the opposite of each term of the polynomial.

Examples

Find each difference.

1 $\begin{array}{r} 5x + 3 \\ -\ (2x + 1) \\ \hline \end{array} \rightarrow \left\{ \begin{array}{l} \textit{The opposite of } 2x \textit{ is } -2x. \\ \textit{The opposite of } 1 \textit{ is } -1. \end{array} \right\} \rightarrow \begin{array}{r} 5x + 3 \\ +\ (-2x - 1) \\ \hline 3x + 2 \end{array}$

2 $(2x^2 - 5x + 3) - (4x^2 + 2x - 1)$

$\begin{array}{r} 2x^2 - 5x + 3) \\ -\ (4x^2 + 2x - 1) \\ \hline \end{array} \rightarrow \begin{array}{r} 2x^2 - 5x + 3 \\ +\ (-4x^2 - 2x + 1) \\ \hline -2x^2 - 7x + 4 \end{array}$

Checking for Understanding

Communicating Mathematics

Read and study the lesson to answer each question.

1. **Tell** the opposite of $5x$.
2. **Write** the subtraction problem shown in the figure at the right.
3. **Show** how to find the difference of $2x^2 + 5x$ and $3x^2 - 2x$ using area tiles or drawings.

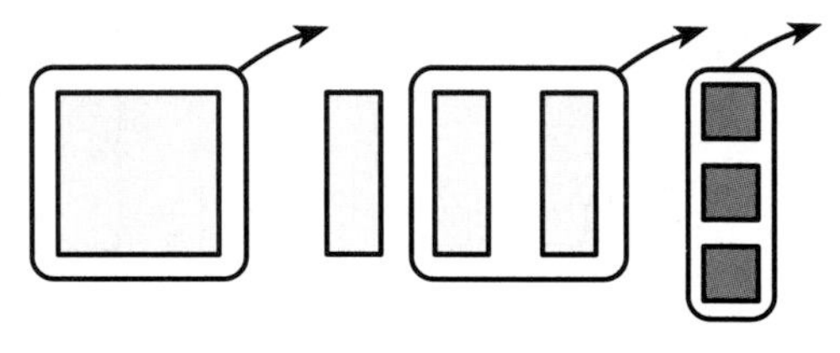

Guided Practice

State the opposite of each term of the polynomial.

4. x
5. $-5x^2$
6. 4
7. $10x^2 + 3x$

Find each difference using area tiles or drawings.

8. $(5x + 3) - (2x + 1)$

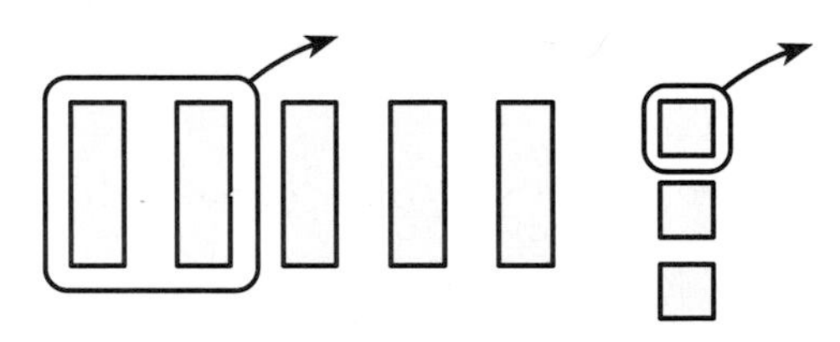

9. $(x^2 + 2x) - (2x^2 + x)$

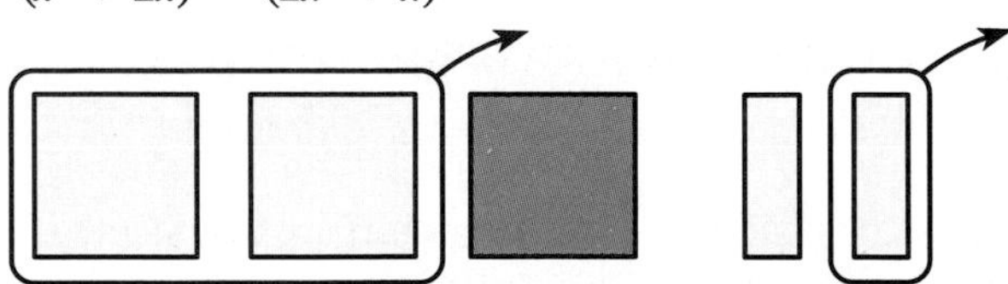

10. $(7x + 5) - (3x + 4)$
11. $(5x^2 - 3x + 2) - (3x^2 - 3)$
12. $(2x^2 - 5x - 1) - (x^2 - x - 1)$
13. $(-3x^2 + 2x + 1) - (x^2 + 3x - 1)$

Exercises

Independent Practice

Find each difference. Use area tiles or drawings if necessary.

14. $\begin{array}{r} 3x + 7 \\ -\ (2x + 5) \\ \hline \end{array}$

15. $\begin{array}{r} 4a^2 - 3a - 2 \\ -\ (2a^2 + 2a + 7) \\ \hline \end{array}$

16. $(9s - 1) - (7s + 2)$

17. $(-4a + 5) - (a - 1)$

18. $(5x^2 + 9) - (4x^2 + 9)$

19. $(10m - 2n) - (6m + 3n)$

20. $(6x^2 + 2x + 9) - (3x^2 + 5x + 9)$

21. $(4p^2 - 3p + 1) - (2p^2 - 2p)$

22. Find $(3r^2 - 3rt + t^2)$ minus $(2r^2 + 5rt - 3t^2)$.

23. What is $(7a^2 + ab - 2b^2)$ decreased by $(-a^2 - ab + b^2)$?

Mixed Review

24. Solve $j = -7(15)(-10)$ *(Lesson 3-6)*

25. Find the prime factorization of 48. *(Lesson 6-2)*

26. Find $P(8, 3)$. *(Lesson 13-2)*

27. Find $(-3m^2 + 6m - 2) + (-m^2 + 4m + 11)$. *(Lesson 14-3)*

Problem Solving and Applications

28. Geometry Find the length of the third side of the triangle shown at the right. The perimeter of the triangle is $7x + 2y$ units.

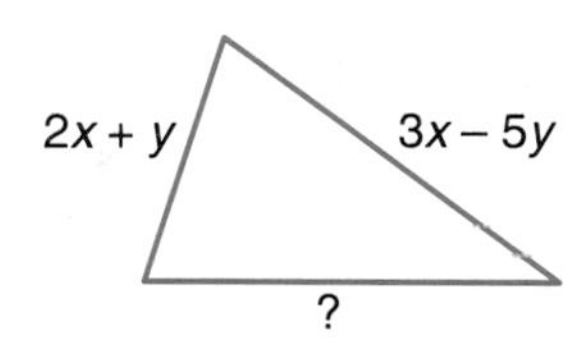

29. Critical Thinking The sum of two polynomials is $x^2 - 2x - 3$. The difference is $3x^2 - 4x + 5$. Find the polynomials.

30. Journal Entry Does using area tiles help you subtract polynomials? How do you remember which terms to combine without using the tiles?

14 Assessment: Mid-Chapter Review

Model each polynomial using area tiles or drawings. *(Lesson 14-1)*

1. $3x^2 - 2x + 7$

2. $-x^2 + 5x - 2$

Simplify each polynomial. Use area tiles or drawings, if necessary. *(Lesson 14-2)*

3. $3x^2 - 5 + 5x^2 - 2$

4. $7a + 2b + 3c$

Find each sum or difference. Use area tiles or drawings, if necessary. *(Lessons 14–3, 14-4)*

5. $(4x^2 - 5x) + (3x^2 + x)$

6. $(a^2 - 6) - (3a^2 + 1)$

7. $(3n + 6) - (n - 1)$

8. $(4x + 3y) + (-7x + 3y)$

14-5A Modeling Products

A Preview of Lesson 14–5

Objective

Model products with area tiles.

Materials

area tiles
product mat

The area tiles that you have been using are based on the fact that the area of a rectangle is the product of the width and length.

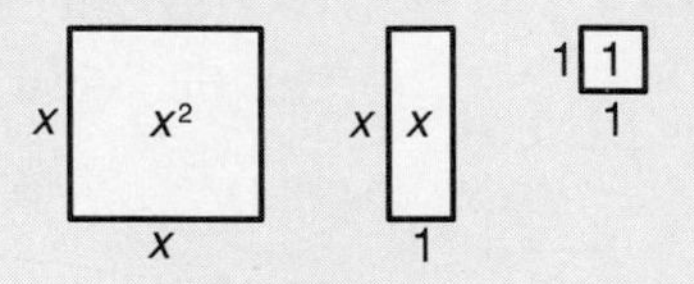

In this lab, you will use these area tiles to build more complex rectangles. These rectangles will help you understand how to find the product of simple polynomials. The width and length each represent a polynomial being multiplied; the area of the rectangle represents their product.

Try this!

Work with a partner to find $2(x + 1)$.

- You will make a rectangle with a width of 2 units and a length of $x + 1$ units. Use your area tiles to mark off the dimensions on a product mat.

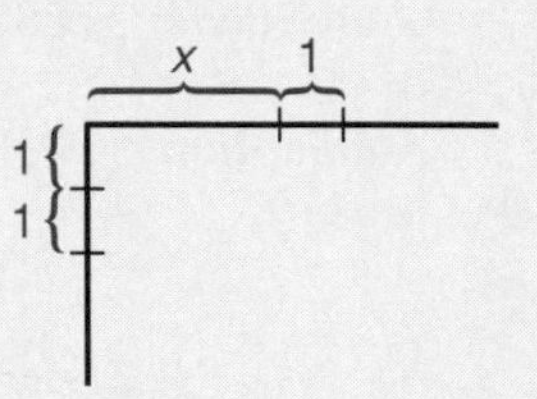

- Using the marks as a guide, fill in the rectangle with area tiles.

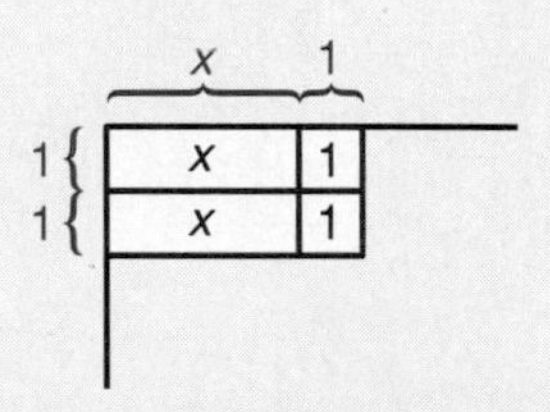

- The area of the rectangle is $x + x + 1 + 1$. In simplest form, the area is $2x + 2$. Therefore, $2(x + 1) = 2x + 2$.

Application

Find each product using area tiles.

1. $3(x + 3)$ **2.** $x(x + 2)$ **3.** $2(2x + 1)$

Extension

Make possible rectangles for each area. Then find each length and width.

4. $x^2 + 3x$ **5.** $4x + 8$ **6.** $6x + 6$

14-5 Multiplying a Polynomial by a Monomial

Objectives

Multiply a polynomial by a monomial and factor polynomials using area models.

Words to Learn

factoring

A carpenter was working with a piece of plywood that was 3 feet longer than it was wide.

To find the area of the plywood, you multiply the length by the width. Let x represent the width. Then $x + 3$ represents the length.

This diagram of the plywood shows that the area is $x(x + 3)$ square feet.

This diagram of the plywood shows that the area is $x^2 + 3x$ square feet.

x | $A = x(x + 3)$ | $x + 3$ ➡

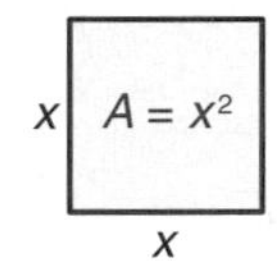

x | $A = 3x$ | 3

LOOKBACK

You can review factors on page 35 and the distributive property on page 9.

Since the areas are equal, $x(x + 3) = x^2 + 3x$. The expressions x and $(x + 3)$ are *factors* of the product $x^2 + 3x$.

The situation above shows how the distributive property can be used to multiply a polynomial by a monomial. Recall that the distributive property allows you to multiply the factor *outside* the parentheses by each term *inside* the parentheses.

Examples

Find each product.

1 $3(n + 4)$

$$3(n + 4) = 3 \cdot n + 3 \cdot 4$$
$$= 3n + 12$$

$n + 4$

3					
	n	1	1	1	1
	n	1	1	1	1
	n	1	1	1	1

2 $x(x - 4)$

$$x(x - 4) = x \cdot x - x \cdot 4$$
$$= x^2 - 4x$$

Sometimes you know the product of a polynomial and monomial and are asked to find the factors. This is called **factoring.** Using the area tiles, this means that you know the area of a rectangle and are asked to find the length and width.

Mini-Lab

Work with a partner to factor $3x + 6$.

Materials: area tiles

- Model the polynomial $3x + 6$.
- Try to form a rectangle with the tiles.
- Write an expression for the length and width.
- Repeat the procedure for the polynomial $3x + 5$.

Talk About It

a. What are the factors of $3x + 6$?

b. What are the factors of $3x + 5$?

c. Explain how to factor $x^2 + 8x$.

d. Name a polynomial that cannot be factored.

Checking for Understanding

Communicating Mathematics

Read and study the lesson to answer each question.

1. **Tell** what property you use to find the product of a monomial and a polynomial.

2. **Tell** the product of $2x$ and $3x + 1$ using the rectangle at the right.

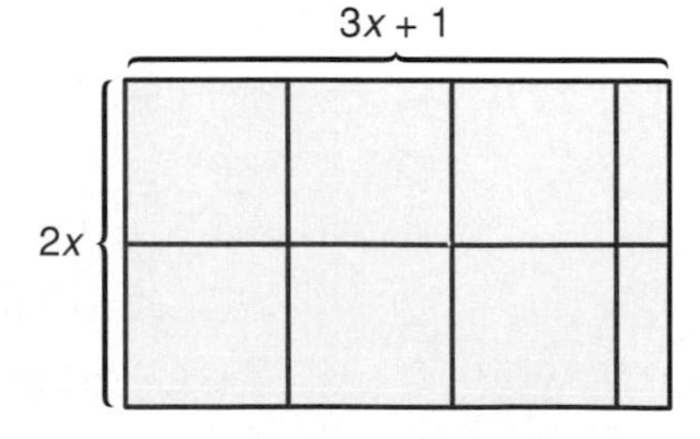

3. **Draw** a rectangle to model $x(x + 1)$ and another to model $(x + 1)x$. Explain how they are the same and how they are different.

4. **Show** how to factor the polynomial represented by the model shown below.

x^2 x x

Guided Practice **Find each product.**

5. $2(x + 5)$

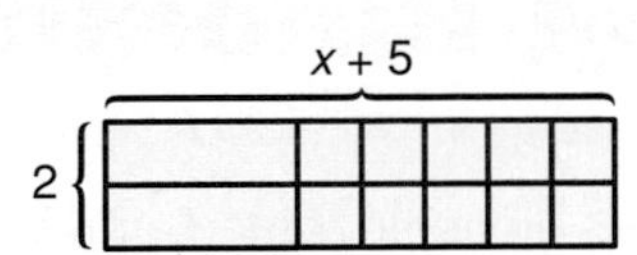

6. $x(x + 3)$

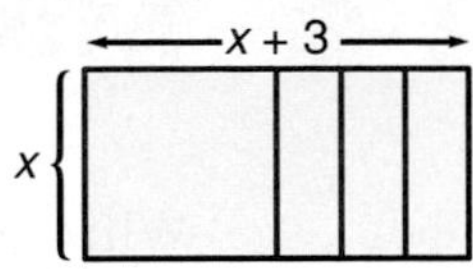

7. $5(y + 9)$
8. $y(y + 2)$
9. $a(a + 2)$

Factor.

10. $4z + 4$

z	1
z	1
z	1
z	1

11. $y^2 + 5y$

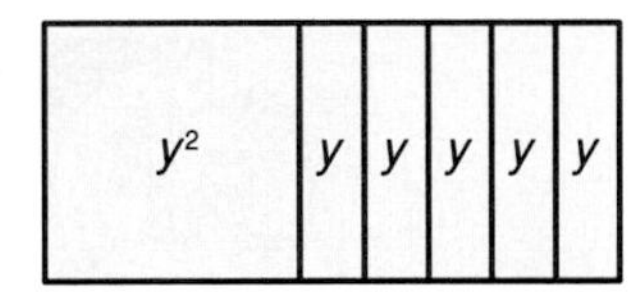

Exercises

Independent Practice **Find each product. Use area tiles or drawings if necessary.**

12. $6(n + 10)$
13. $4(b + 3)$
14. $3(2 + y)$
15. $d(d + 15)$
16. $3(2c + 4)$
17. $2x(2x + 1)$

Factor. Use area tiles or drawings if necessary.

18. $5x + 10$
19. $6x^2 + 3x$
20. $2x + 5$
21. $8a^2 + 8$
22. $12m + 6$
23. $4 + 20x$
24. Find the product of $2y$ and $4y + 1$.
25. Factor $2y + 8$.
26. Multiply $x + 2$ by $3x$.

Mixed Review

27. Solve $156 = y + 73$. *(Lesson 2-3)*
28. Find the percent of change if last month's electric bill was \$52.50 and this month's bill is \$67.20. *(Lesson 10-8)*
29. Graph $y = -3x^2 + 1$. *(Lesson 11-7)*
30. Find $(8x - 1) - (5x + 3)$. *(Lesson 14-4)*

Problem Solving and Applications

31. **Gardening** A square garden plot measures x feet on each side. Suppose you double the length of the plot and increase the width by 3 feet.
    a. Draw the new garden.

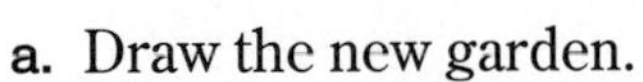

    b. Write two expressions for the area of the new plot.
    c. If the original plot was 10 feet on a side, what is the area of the new plot?

32. **Critical Thinking** A trapezoid has an area of 19.5 cm^2 and a height of 3 cm. Base 2 is 1 centimeter longer than twice the length of base 1. Find the length of base 1. Use $A = \frac{1}{2}h(b_1 + b_2)$.

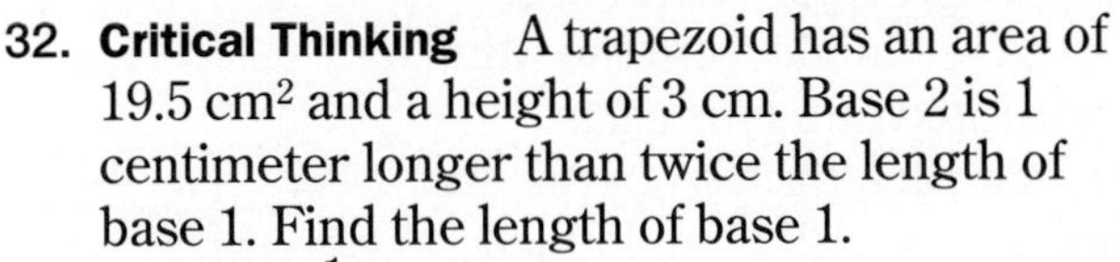

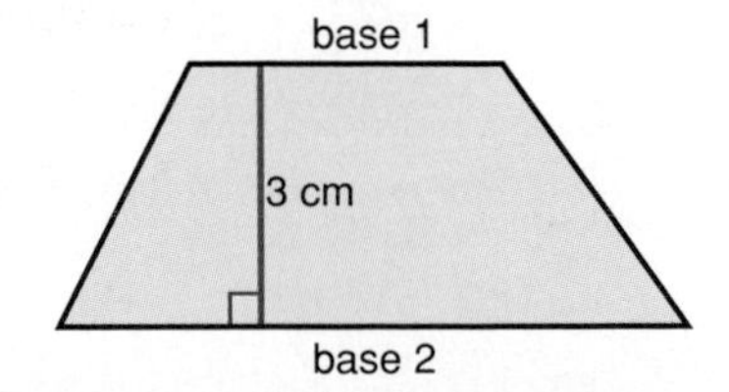

14-6 Multiplying Binomials

Objective

Multiply binomials using area models.

Words to Learn

binomial

What do the words bicycle, bicuspid, biped, and binomial have in common? They all begin with the prefix *bi-*, meaning two. A bicycle is a cycle with two wheels, a bicuspid is a tooth with two points, a biped is an animal with two feet, and a **binomial** is a polynomial with two terms. Some examples of binomials are $x + 3$, $2y - 1$, and $a + b$. You can find the product of simple binomials by using area tiles.

Mini-Lab

Work with a partner to find (x + 2)(x + 1).

Materials: area tiles, product mat

- Make a rectangle with a width of $x + 2$ and a length of $x + 1$. Use your area tiles to mark off the dimensions on a product mat.

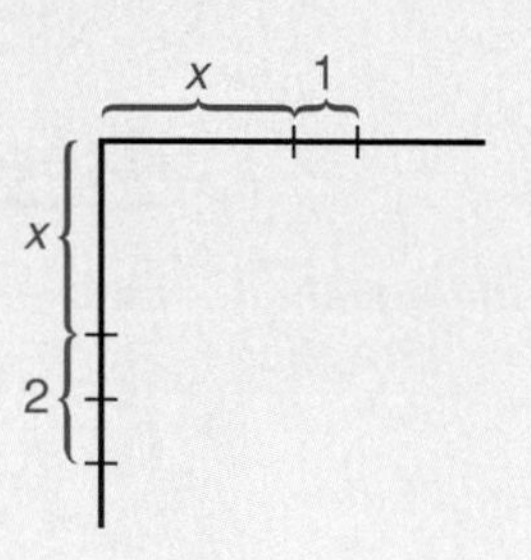

- Using the marks as a guide, fill in the rectangle with area tiles.

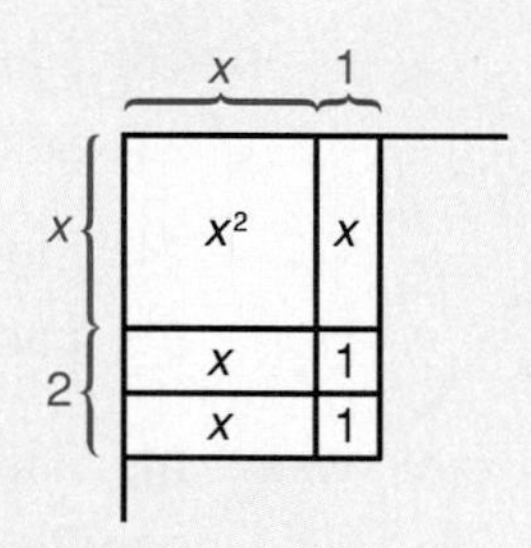

Talk About It

a. How is this method like the method you used when multiplying a monomial by a polynomial?

b. What is $(x + 2)(x + 1)$?

You can also use the distributive property to find the product of two binomials. The figure at the right shows the rectangle from the Mini-Lab, separated into four parts. Notice that each term from the first parentheses $(x + 2)$ is multiplied by each term from the second parentheses $(x + 1)$.

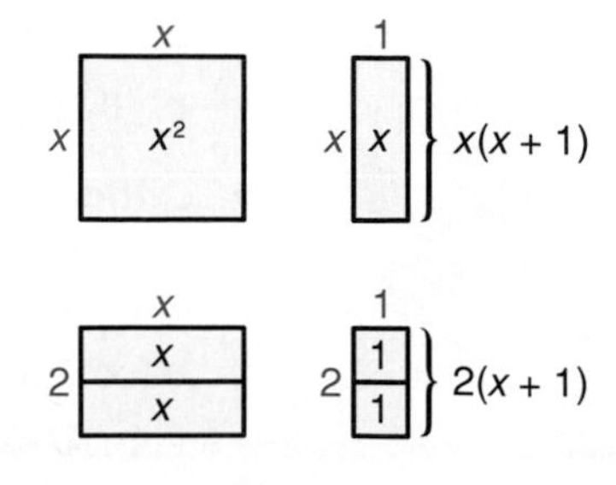

Examples

Find each product.

1 $(2x + 1)(x + 3)$

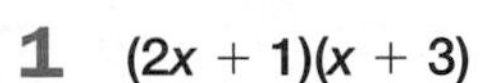

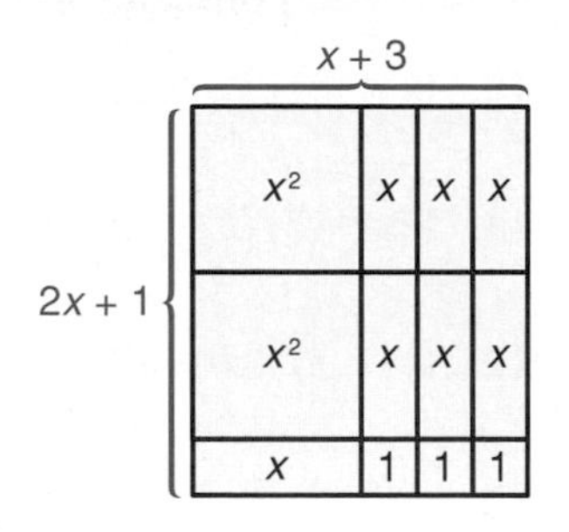

$$
\begin{aligned}
(2x + 1)(x + 3) &= 2x(x + 3) + 1(x + 3) \\
&= 2x^2 + 6x + 1x + 3 \\
&= 2x^2 + 7x + 3 \quad \textit{Simplify.}
\end{aligned}
$$

2 $(y + 4)(y + 1)$

$$
\begin{aligned}
(y + 4)(y + 1) &= y(y + 1) + 4(y + 1) \\
&= y^2 + 1y + 4y + 4 \\
&= y^2 + 5y + 4 \quad \textit{Simplify.}
\end{aligned}
$$

Checking for Understanding

Communicating Mathematics

Read and study the lesson to answer each question.

1. **Tell** the definition of *binomial.*
2. **Draw** a rectangle with a width of $(x + 3)$ and a length of $(2x + 2)$.
3. **Write** the product shown at the right.

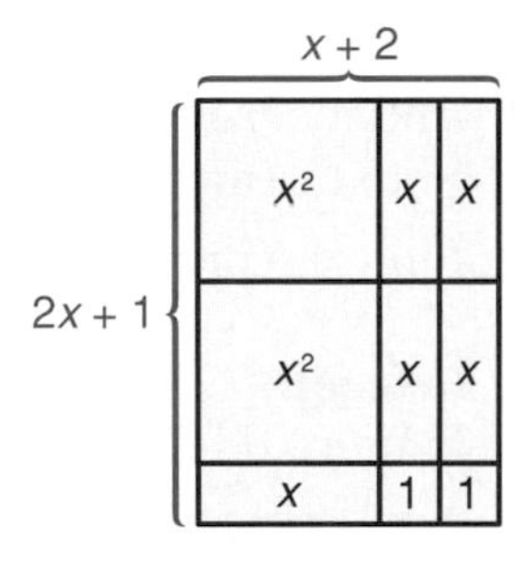

Guided Practice

Find each product.

4. $(x + 1)(x + 2)$

x + 2

x + 1

x^2	x	x
x	1	1

5. $(2x + 2)(2x + 3)$

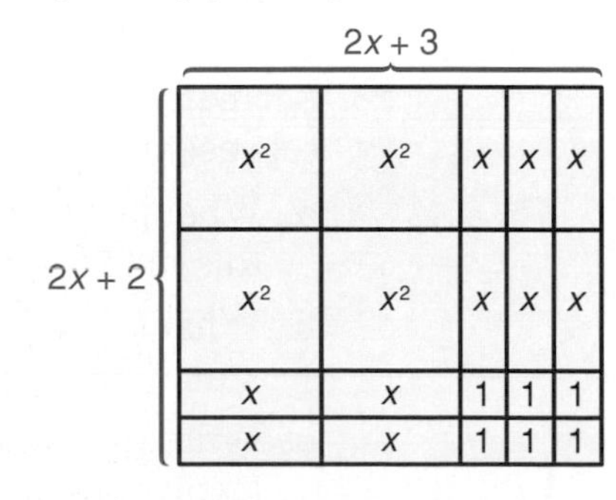

6. $(m + 4)(m + 3)$
7. $(3a + 1)(2a + 3)$
8. $(x + 1)(2x + 3)$
9. $(3z + 1)(4z + 5)$

Exercises

Independent Practice

Match each product with its corresponding model.

10. $(x + 5)(2x + 3)$

11. $(2x + 3)(x + 4)$

12. $(x + 3)(3x + 2)$

13. $(x + 3)(x + 3)$

a.

x^2	x	x	x
x	1	1	1
x	1	1	1
x	1	1	1

b.

x^2	x^2	x^2	x	x
x	x	x	1	1
x	x	x	1	1
x	x	x	1	1

c.

x^2	x^2	x	x	x
x	x	1	1	1
x	x	1	1	1
x	x	1	1	1
x	x	1	1	1
x	x	1	1	1

d.

x^2	x	x	x	x
x^2	x	x	x	x
x	1	1	1	1
x	1	1	1	1
x	1	1	1	1

Find each product. Use area tiles or drawings if necessary.

14. $(x + 3)(x + 1)$
15. $(x + 1)(x + 4)$
16. $(2x + 1)(x + 5)$
17. $(x + 2)(2x + 1)$
18. $(2x + 2)(2x + 3)$
19. $(x + 1)(x + 1)$
20. Find the product of $(2x + 5)$ and $(x + 1)$.

Mixed Review

21. Find $\sqrt{\frac{9}{16}}$. *(Lesson 8-1)*
22. Aiko borrowed \$2,400 and paid back \$116 a month for 24 months. Find the annual rate of simple interest. *(Lesson 10-10)*
23. Find the product of x and $(2x + 3)$. *(Lesson 14-5)*

Problem Solving and Applications

24. **Geometry** A square has dimensions of x feet $\times$ x feet. A rectangle is 4 feet longer and 3 feet wider than the square. Find the area of the rectangle.
25. **Critical Thinking** Write the multiplication problem shown in the figure at the right. Name the product.

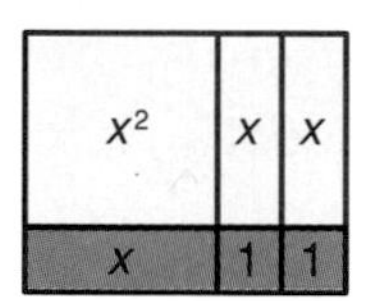

26. **Geometry** The model at the right represents the square of a binomial.
    a. What product of binomials does this model represent?
    b. What is the area of each small square and rectangle?
    c. Write the area of the square as a polynomial.

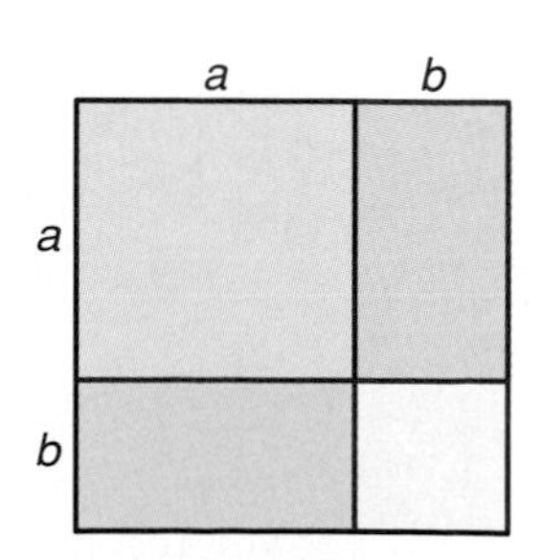

27. **Journal Entry** What concept in this chapter have you found most challenging? What do you think made it more difficult for you?

Cooperative Learning

14-6B Factoring Polynomials

A Follow-Up of Lesson 14-6

Objective

Factor polynomials using area models.

Materials

area tiles
product mat

From the previous lesson, you know that $(x + 1)(x + 2) = x^2 + 3x + 2$. The binomials $(x + 1)$ and $(x + 2)$ are the factors of $x^2 + 3x + 2$.

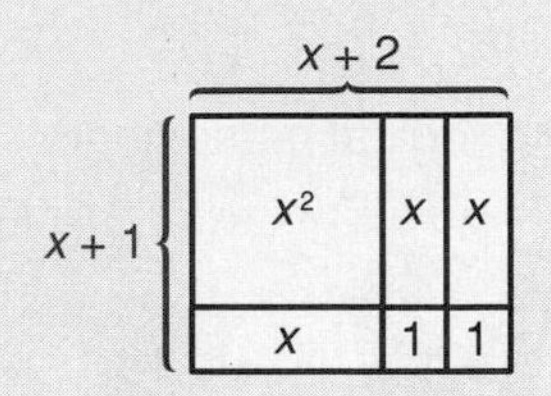

In this lab, you will use area tiles to find the factors of several polynomials. The polynomial can be factored if the tiles can be arranged into a rectangle.

Try this!

Work with a partner to factor $x^2 + 5x + 6$.

- Model the polynomial.

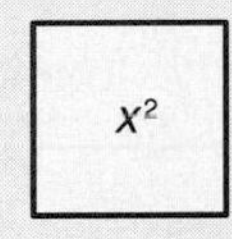

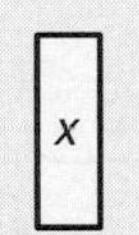
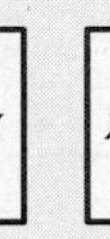
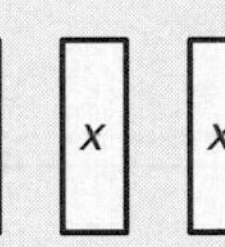
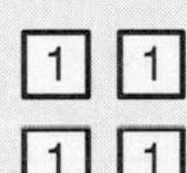

- Try to form a rectangle with the tiles. Use a product mat.

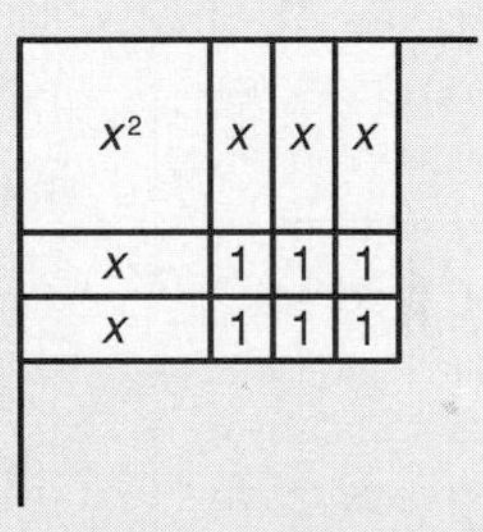

- Write an expression for the length and width of the rectangle.
- Repeat the procedure for the polynomial $x^2 + x + 1$.

What do you think?

1. What are the factors of $x^2 + 5x + 6$?
2. What are the factors of $x^2 + x + 1$?

Application

If possible, factor each polynomial using tiles or drawings.

3. $x^2 + 7x + 6$
4. $x^2 + 6x + 9$
5. $2x^2 + 7x + 6$
6. $x^2 + 4x + 5$

Problem-Solving Strategy

14-7 Use Logical Reasoning

Objective

Solve problems by using logical reasoning.

Throughout this text, you have learned many different problem-solving strategies. One of the most important strategies for problem solving is to use logical reasoning. You can apply this strategy in every situation, especially when you play games or solve puzzles.

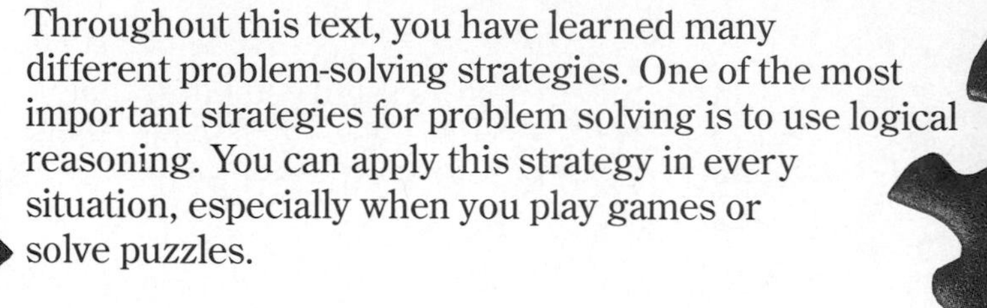

Example

Allan, Kevin, Eduardo, and Bob are friends. Each of them is on one of the following school teams: football, soccer, cross country, or golf. Allan is shorter than the boy who plays soccer. Eduardo only likes to play games with round balls. Bob has a problem with his knee and cannot run. Kevin practices kicking a ball as part of his training. Who plays each sport?

Explore There are four boys and four sports. You must match each boy with the sport he plays.

Plan Make a chart to organize the information. Use the clues to rule out possibilities.

Solve Put an X to show that Allan does not play soccer. Put two X's to show that Eduardo does not play football or run cross country. Put three X's to show that Bob does not play football, soccer, or cross country. Since only one student plays golf, put X's in the rest of the boxes in that row.

Now you can see that Eduardo plays soccer. Place an X to show that Kevin does not run cross country, which means he plays football. Therefore, Allan's sport is cross country.

	Allan	Kevin	Eduardo	Bob
football	✗	✓	✗	✗
soccer	✗	✗	✓	✗
cross country	✓	✗	✗	✗
golf	✗	✗	✗	✓

Examine Check the answer in the words of the problem. Allan runs cross country, Kevin plays football, Eduardo plays soccer, and Bob plays golf.

"When am I ever going to use this?"

A computer programmer writes a detailed plan or procedure for solving a problem in an ordered sequence of instructions.

Programming requires a degree in computer science with a working knowledge of computer languages and mathematics.

For more information, contact:
Association for Computing Machinery,
11 W. 42nd St.
3rd Floor
New York, NY 10036

Checking for Understanding

Communicating Mathematics

Read and study the lesson to answer each question.

1. **Tell** why you can put X's in the remaining boxes of a row after you have matched one of the boys with his sport.
2. **Tell** about a situation in your life in which you use logical reasoning.

Guided Practice

Solve using logical reasoning.

3. On average, it takes three minutes to saw through a log. How long will it take to saw a log into four pieces?
4. The Science and History Clubs are going on a research trip to the Museum of Natural History. There are 22 students in the Science Club and 26 students in the History Club. Five teachers will also make the trip. How many mini-buses will be needed if each bus holds 16 passengers?

Problem Solving

Practice

Strategies

Look for a pattern.
Solve a simpler problem.
Act it out.
Guess and check.
Draw a diagram.
Make a chart.
Work backward.

Solve using any strategy.

5. Without computing, choose the number that is the cube of 123. Explain your reasoning.

   a. 1,815,848 b. 1,860,867
   c. 1,953,125 d. 1,906,624

6. Brenda has four cats named Beanie, Tiger, Flower, and Snowball. One cat is white with brown markings, one is all white and one is all black. The fourth cat is calico. Snowball was all white as a kitten, but now has spots. Tiger and the black cat are brother and sister. Beanie does not get along with the black cat. Tiger has orange markings. What color is each cat?

7. **Mathematics and Architecture** Read the following paragraph.

Louis Henry Sullivan (1856–1924) was an American architect who, with his partner Dankmer Adler, designed over 100 buildings between 1881 and 1895. They are known for adapting modern methods to building design.

A section of a building is being redesigned into new offices. The current design has square-shaped offices on the perimeter of the floor. When the renovation is complete, each office will be 2 feet wider and 4 feet longer. Find the area of each new office in terms of the original length and width.

8. **Portfolio Suggestion** Review the items in your portfolio. Make a table of contents of the items, noting why each item was chosen. Replace any items that are no longer appropriate.

Chapter 14

Study Guide and Review

Communicating Mathematics

State whether each sentence is *true* or *false.* If false, replace the underlined word or number to make a true sentence.

1. The expression $b^2 - 3b$ is an example of a monomial.
2. A polynomial is the sum or difference of two or more monomials.
3. The additive inverse of $9y^2 - 5y + 2$ is $-9y^2 + 5y - 2$.
4. A polynomial with two unlike terms is called a *binomial.*
5. The product of $2m$ and $m^2 + 8m$ will have three terms.
6. Draw and label a rectangle with an area of $6x^2 + 2x$.
7. Write the definition of a monomial in your own words.
8. Explain how to factor $x^2 + 7x + 6$ using area tiles or drawings.

Self Assessment

Objectives and Examples

Upon completing this chapter, you should be able to:

- represent polynomials with area models *(Lesson 14-1)*

Model $x^2 + 2x + 5$.

$x^2 + 2x + 5$ ➡

Review Exercises

Use these exercises to review and prepare for the chapter test.

Model each polynomial using area tiles or drawings.

9. $4x - 6$
10. $2x^2 - 3$
11. $-3x^2 + 4x - 8$
12. $x^2 - 5$

- simplify polynomials using area models *(Lesson 14-2)*

Simplify $2b^2 + 7 + b^2$.

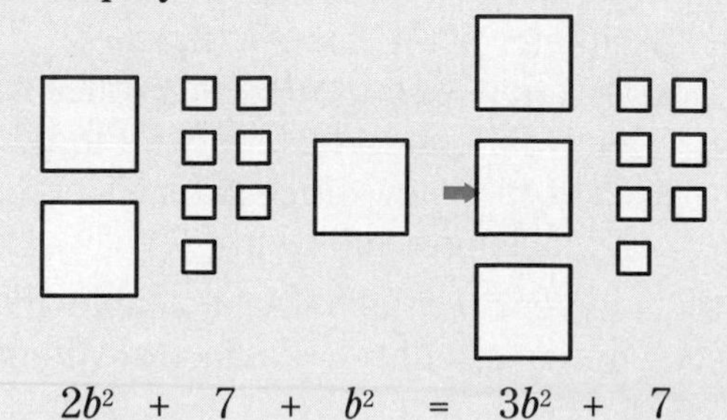

Simplify each polynomial. Use area tiles or drawings if necessary.

13. $4m^2 + 6m + 11m^2 + 2m$
14. $9p - 4p - 15$
15. $14x - 3x^2 - 8x + 5x^2$
16. $3a + 9b - a - 6b$
17. $17m^2 - 2m + 4m - 3m^2$

Objectives and Examples

- add polynomials using area models *(Lesson 14-3)*

Find $(2b^2 + 8b) + (4b^2 - 3b)$.

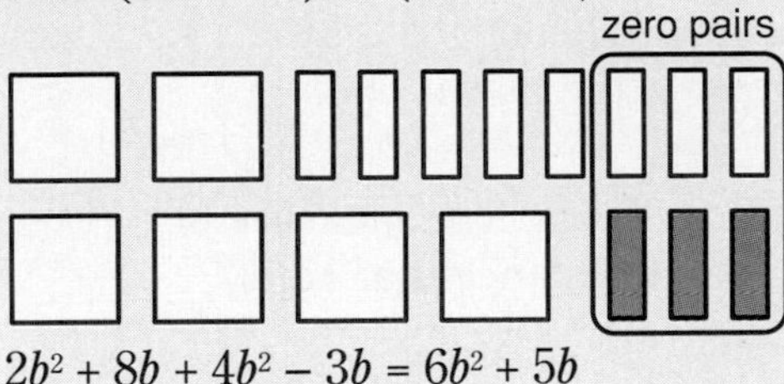

$2b^2 + 8b + 4b^2 - 3b = 6b^2 + 5b$

Review Exercises

Find each sum. Use area tiles or drawings if necessary.

18. $(11m^2 - 2m) + (4m^2 + 5m)$

19. $(5d + 1) + (9d + 7)$

20. $(2a^2 + 5a) + (2a^2 - 4a)$

21. $(b^2 - 4b + 2) + (3b^2 + b - 6)$

22. $(3x^2 - 8x) + (x^2 + 9x)$

- subtract polynomials using area models *(Lesson 14-4)*

Find $(8t^2 + 4) - (5t^2 - 2)$.

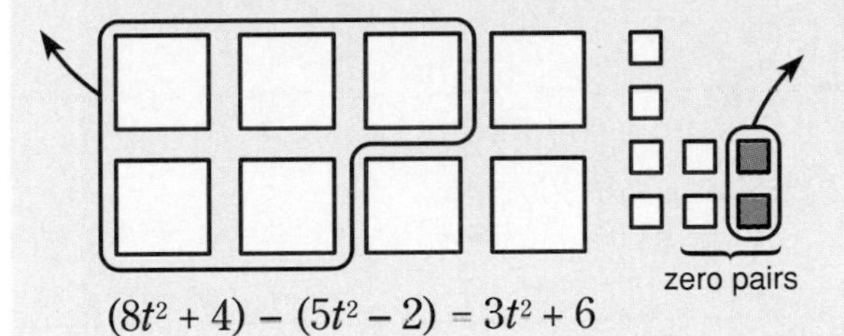

$(8t^2 + 4) - (5t^2 - 2) = 3t^2 + 6$

Find each difference. Use area tiles or drawings if necessary.

23. $(9g + 3) - (6g + 1)$

24. $(2m - 8) - (-2m + 3)$

25. $(4s^2 + 9) - (s^2 + 4)$

26. $(7k^2 - 2) - (2k^2 - 6k - 1)$

27. $(8p^2 + 4p - 7) - (6p^2 + 9p - 3)$

- multiply a polynomial by a monomial using area models *(Lesson 14-5)*

Find $7x(3x + 5)$

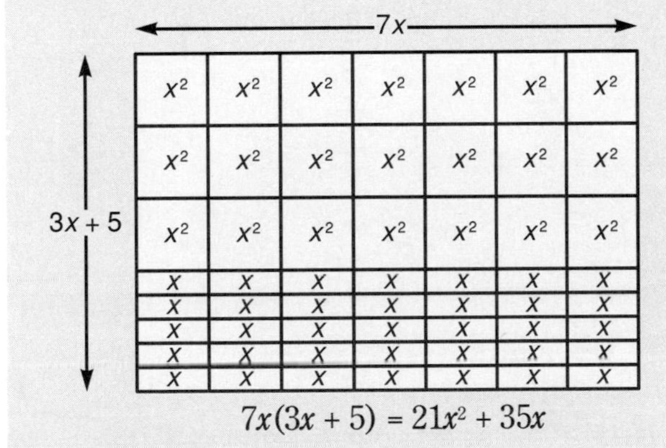

$7x(3x + 5) = 21x^2 + 35x$

Find each product. Use area tiles or drawings if necessary.

28. $5(2y + 4)$

29. $4z(z + 3)$

30. $c(3c + 1)$

31. $3t(t + 6)$

- multiply binomials using area models *(Lesson 14-6)*

Name the two binomials being multiplied and give their product.

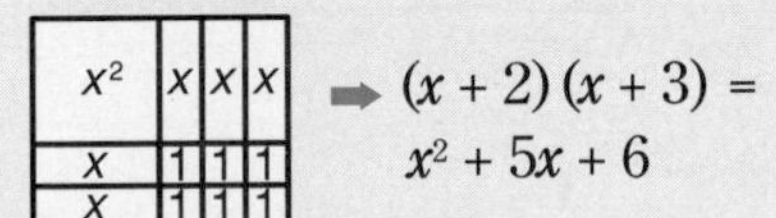

$\Rightarrow (x + 2)(x + 3) = x^2 + 5x + 6$

Find each product. Use area tiles or drawings if necessary.

32. $(x + 4)(x + 2)$

33. $(x + 3)(2x + 2)$

34. $(2x + 2)(2x + 5)$

35. $(x + 5)(x + 2)$

36. $(3x + 1)(x + 2)$

Applications and Problem Solving

37. **Interior Design** The amount of carpeting needed for the Glasers' living room is $2x^2 + x$ square meters. Find the dimensions of the living room. *(Lesson 14-5)*

38. **Gardening** Mrs. Keyser has a rectangular flower bed. The length is $3x - 1$ feet, and the width is $2x + 3$ feet. Find the perimeter of the flower bed. *(Lesson 14-3)*

39. **Exercise** Tami rode her bicycle around a square city block. The measure of each side of the block is $3x + 8$ yards. What is the total distance that Tami rode her bicycle? *(Lesson 14-5)*

40. **Geometry** The measures of two angles of a triangle are $-2x^2 - 3x + 9$ and $-3x^2 - x + 7$. Find the measure of the third angle. *(Lesson 14–4)*

41. This statement is true: All rhombuses are parallelograms. Decide which of the following statements are also true. *(Lesson 14-7)*
    a. All parallelograms are rhombuses.
    b. If a quadrilateral is a parallelogram, then it is a rhombus.
    c. If a quadrilateral is not a parallelogram, then it is not a rhombus.
    d. If a quadrilateral is not a rhombus, then it is not a parallelogram.

Curriculum Connection Projects

- **Zoology** You decide to raise rabbits. Your first rabbit has a litter of x bunnies. Each of her offspring has identical size litters. Use a tree diagram to represent the three generations. Then write a polynomial representing the number of rabbits you have.

- **Design** Draw a floor plan of a one-story house. Use the scale 1 cm = 1 m. Let $x = 1$ cm, $y = 3$ cm, and $z = 5$ cm. Make the dimensions multiples of x, y, and z. Write a polynomial for the area of your house.

Read More About It

Duane, Diane. *High Wizardry.*
Sarnoff, Jane and Ruffins, Reynold. *The Chess Book.*
Weiss, Harvey. *Model Buildings and How to Make Them.*

Chapter 14 Test

Model each monomial or polynomial using area tiles or drawings.

1. $6x^2 + 3x$
2. $-5x^2 - 2x + 8$
3. $4x^2 + 7x - 2$
4. $x^2 - x$

Evaluate each polynomial if $r = -1$, $s = 2$, and $t = 4$.

5. $r^3t - s$
6. $s^3 - 3s + t$
7. $r^2st^2 - 3t$
8. Identify the polynomial represented by the model at the right.

Simplify each polynomial. Use area tiles or drawings if necessary.

9. $3x^2 + 5x + 4x^2 + 7x$
10. $5c^2 + 2c - 2c^2 + c$
11. $4x^2 + y - 6x^2 + y$
12. $2a^2 + 8a - 3a^2 + 4a - 1$

Find each sum or difference. Use area tiles or drawings if necessary.

13. $(8z^2 - 2z) - (4z^2 + 9z)$
14. $(5c^2 + 3c) + (-3c^2 + c)$
15. $(6n^2 - 5n + 1) - (3n - 4)$
16. $(-x^2 + 3x - 4) + (x^2 - 7x)$
17. $\begin{array}{r} 6r^2 - 5r + 4 \\ + \ r^2 - 4r - 8 \\ \hline \end{array}$
18. $\begin{array}{r} 9y^2 + 5y - 8 \\ -\ (6y^2 + 8y - 9) \\ \hline \end{array}$

Match each product with its corresponding model. Then state the product.

19. $(x + 3)(x + 4)$
20. $(x + 1)^2$
21. $(2x + 1)(x + 2)$
22. $(x + 1)(3x + 2)$

a.

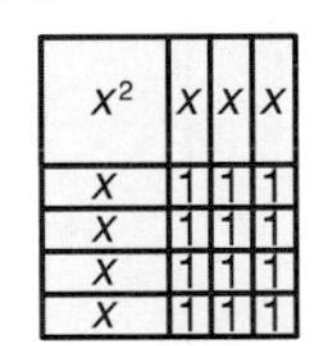

b.

c.

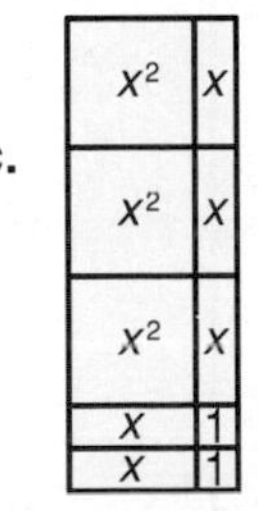

d.

Find each product. Use area tiles or drawings if necessary.

23. $4d(2d + 5)$
24. $x(5x + 3)$

25. Jacob, Amerette, Susan, and Lonny have after-school jobs. One works at the local Chicken Flicken, one is a stock person at the 24-hour grocery, one delivers fliers door-to-door, and one babysits. Amerette has nothing to do with food. Jacob and the babysitter are brothers. Jacob and the stock person do not know each other. What is each student's job?

Bonus Write a formula for finding the product of $ax + b$ and $cx + d$, where a, b, c, and d are whole numbers.

Chapter 14

Standard Format, Chapters 1-14

Academic Skills Test

Directions: Choose the best answer. Write A, B, C, or D.

1. If $2c - 5 = -3$, what is the value of c?

A -4 **B** -1
C 1 **D** 4

2. What is $0.\overline{39}$ written as a fraction?

A $\frac{1}{3}$ **B** $\frac{4}{10}$
C $\frac{13}{33}$ **D** $\frac{39}{100}$

3. If $0.3x = 5.28$, what is the value of x?

A 17.6 **B** 15.84
C 1.76 **D** 1.584

4. Each spinner is spun once. What is the probability of spinning a prime number and a vowel?

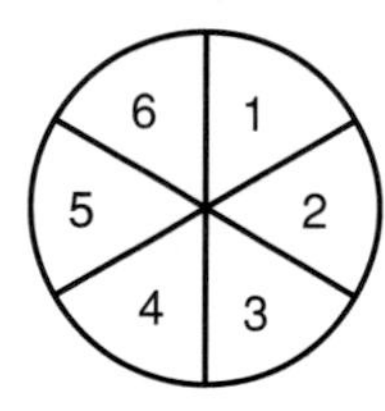

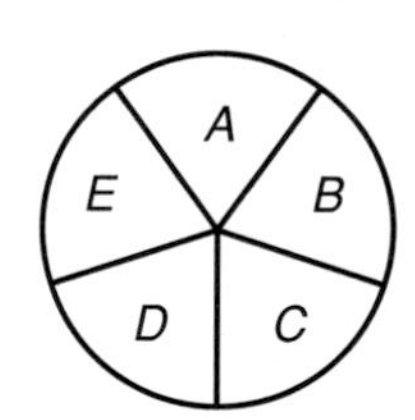

A $\frac{1}{5}$ **B** $\frac{3}{7}$
C $\frac{4}{15}$ **D** $\frac{9}{10}$

5. What is the value of x in the square shown below?

A 2 units
B $\sqrt{10}$ units
C 10 units
D none of these

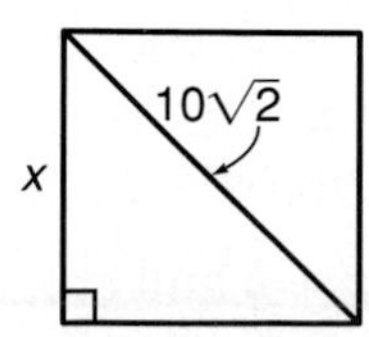

6. In a scale drawing of a room, 1 unit = 6 in. What are the scale dimensions of a 40 in. by 60 in. table?

A 4 by 6 **B** $6\frac{2}{3} \times 10$
C 8 by 12 **D** 24×36

7. 36.5% =

A 36.5 **B** 3.65
C 0.365 **D** 0.0365

8. What percent of 30 is 2.5?

A $8\frac{1}{3}$ % **B** 12%
C 75% **D** $83\frac{1}{3}$ %

9. Which is the graph of $y = 3x - 3$?

A

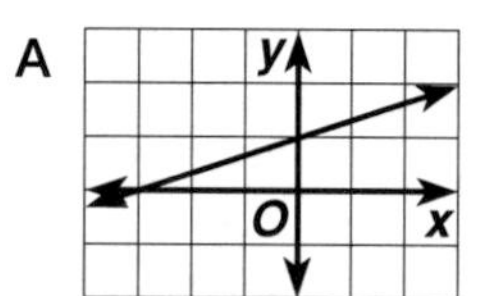

B

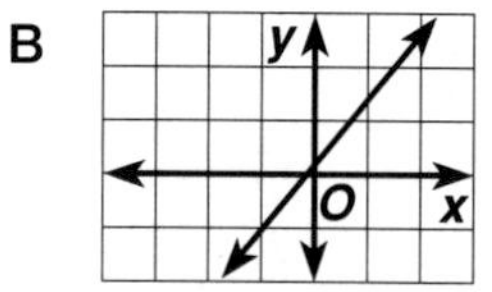

C

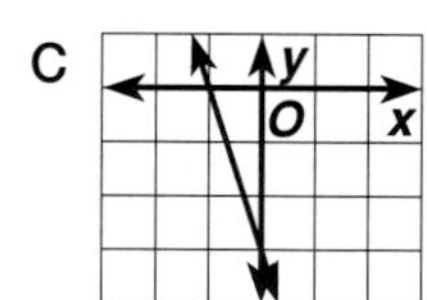

D

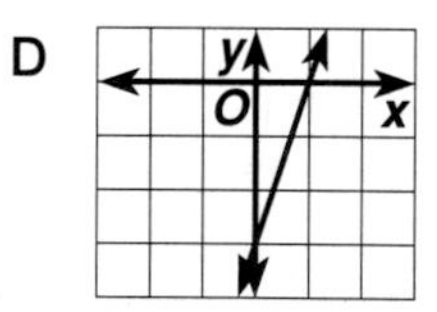

10. What are the coordinates of rectangle *MATH* translated by (2, -1) if $M(-3, 1)$, $A(2, 6)$, $T(6, 2)$, and $H(1, -3)$?

A (-3, -1), (2, -6), (6, -2), (1, 3)
B (-1, 0), (4, 5), (8, 1), (3, -4)
C (3, 1), (-2, 6), (-6, 2), (-1, -3)
D (5, 1), (4, 7), (8, 3), (3, 4)

11. Alison owns a tent in the shape of a pyramid. The base of the tent is 8' by 10'. The tent is 7' tall at the center. How can you find the volume of the air inside the tent?

A Find $8 \times 10 \times 7$ and divide by 3.

B Add $2\left(\frac{1}{2} \cdot 7 \cdot 8\right)$ and $2\left(\frac{1}{2} \cdot 7 \cdot 10\right)$.

C Find $7 \cdot 8 \cdot 10$.

D None of these

12. How many significant digits are there in 0.0250 cm?

A 5 B 4
C 3 D 2

13. Luis is redecorating his room. He has a choice of 4 colors of paint, 3 colors of carpet, and 2 colors of curtains. How many combinations of paint, carpet, and curtain colors can he use?

A 8 B 9
C 12 D 24

14. In how many ways can all four of the shapes be arranged in a row?

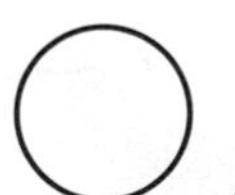  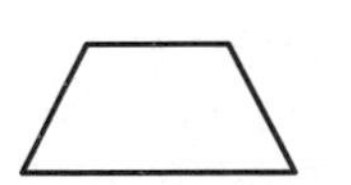 

A 4 B 12
C 24 D 36

15. How many combinations of 4 flowers can you choose from one dozen flowers?

A 11,880 B 495
C 48 D 24

16. $3x + 6x - 5x =$

A $9x - 5$ B $4x$
C $4x^3$ D $14x$

Test-Taking Tip

On standardized tests, guessing can improve your score if you make educated guesses. First, find out if there is a penalty for incorrect answers. If there is no penalty, a guess can only increase your score, or at worst, leave your score the same. If there is a penalty, try to eliminate enough choices to make the probability of a correct guess greater than the penalty for an incorrect answer. That is, if there is a quarter point penalty for an incorrect response and you have a choice of three responses, the probability of making a correct guess is 1 out of 3, which is better than the penalty.

17. In a random sample of 150 students, 60 ride the bus to school, 54 ride in car pools, and 36 walk. If there are to be 800 students next year, about how many will need bus transportation?

A 320 B 160
C 80 D 60

18. $(3a + 1) + (2a + 5) =$

A $4a + 7a$ B $5a + 6$
C $5a^2 + 5$ D $6a + 5$

19. $4(3x - 2) =$

A $12x^2 - 8$ B $12x - 2$
C $12x - 8$ D $12x + 8$

20. A triangle has an area of $6x^2 + x - 1$ square centimeters. Find the altitude and the base of the triangle if $x = 2$ centimeters.

A $4x + 2$ cm, $3x - 1$ cm B 10 cm, 5 cm
C A and B D none of these

EXTENDED PROJECTS HANDBOOK

To The Student

One of the goals of *Mathematics: Applications and Connections* is to give you the opportunity to work with the mathematics that you will likely encounter outside the classroom. This includes the mathematics demanded by many of the courses you will take in high school and by most jobs as well as the mathematics that will be required of a good citizen of the United States.

Equally important, the authors want you to approach the mathematics you will encounter in your life with curiosity, enjoyment, and confidence.

Hopefully, the **Extended Projects Handbook** reflects these goals.

Three of the most important "big" ideas you are working with throughout *Mathematics: Applications and Connections* are the following. The **Extended Projects** include these "big" ideas.

1. Proportional Reasoning

You probably have a great deal of experience with proportional reasoning. One example is a straight line in which the "rise" is proportional to the "run" and their ratio is the slope of the line. Other topics include ratio, rate, percent, similarity, scale drawings, and probability.

You have made connections among the various applications of proportional reasoning and in this way have seen that the various items listed above are all part of a very big idea — proportions.

2. Multiple Representations

Mathematics provides you with many ways to present information and relationships. These include sketches, perspective drawings, tables, charts, graphs, physical models, verbalizing, and writing. You can use a computer to make graphs, data bases, spreadsheets, and simulations.

You have represented information and relationships in many different ways to completely describe various kinds of situations using mathematics.

3. Patterns and Generalizations

Mathematics has been called the science of patterns. You have experience recognizing and describing simple number and geometric patterns. You will be asked to make, test, and then use generalizations about given information in order to help you solve problems.

You may have used an algebraic expression to generalize a number pattern or the idea of similarity to make a scale drawing.

Looking Ahead

Your **Extended Projects Handbook** contains the following projects.

Project 1
Why Do I Have To Learn This?

Project 2
Endangered Species

Project 3
High Energy

Project 4
A Taste Of America

Project 1 Slip Sliding Away

Everyone seems to like sliding down hills. You can see people trying it in cold weather with skis and sleds. Some people, like Tommy Moe and Picabo Street, take it to extremes. They won medals in the downhill skiing events in the 1994 Winter Olympics at Lillehammer, Norway.

In hot weather, people also take part in sliding activities. From sliding boards to water slides, people of all ages enjoy the thrill of this kind of motion.

Have you ever watched people on a water slide? What are some factors that affect sliding?

In this project, you will pose questions, state hypotheses (or theories), and analyze the results of experiments.

Getting Started

Think what will happen if you put some boxes on a ramp and then begin to raise one end of the ramp. Will all the boxes start to slip at the same time? Will a lightweight box slip before a heavier one? Will a rough box slip less than a smooth one?

An object's weight and surface texture seem to be important factors in determining if it slips on a raised ramp. Write a hypothesis or guess about how you think the steepness of a ramp at which a box just begins to slip down the ramp is related to the box's weight.

- What variable is the independent variable in your hypothesis?
- How will you measure it? How will you change it?
- What variable is the dependent variable in your hypothesis?
- How will you measure it?
- What variables must you control?

After thinking about the answers to these questions, write a procedure to test your hypothesis. Also, design a table for the data you will collect. Perform your experiment.

Now, write a hypothesis about how you think the steepness of the ramp at which a box just begins to slide is related to the box's surface texture.

- Answer the questions posed above for your new hypothesis.

Write a procedure to test your hypothesis and perform your experiment.

Analyze the results of your experiments. Share the results with your classmates.

- Did the results of your experiments support your hypotheses?
- Were there any surprises? Explain.
- Did you and your classmates have the same hypotheses?
- Did you and your classmates get the same results?
- What is the important factor in predicting how easily an object will begin sliding down a ramp?
- How does this factor relate to snow skiing and water slides?

Extension: Rolling Along

One way to reduce the surface texture of an object on a ramp is to put the object on wheels. Using a toy car, design an experiment to find the relationship between the height of the ramp and the distance a toy car will travel.

Project

2 Endangered Species

Wolves, whooping cranes, and grizzly bears are just three endangered species in North America. Many large zoos around the country work directly for conservation of animals by establishing captive breeding programs for endangered species. For example, the Minnesota Zoo has been successfully breeding snow leopards from the mountains of southern Asia.

Although this is an important first step, its value is lessened if some of the offspring cannot be successfully returned to the wild. The introduction and survival of captive-raised animals in their native habitat is much more difficult than you might expect.

In this project, you will research the controversies and issues surrounding the extinction of an animal.

Getting Started

Use resource materials to find out more about animals facing extinction. Gather background information about each animal and answer the following questions.

- What is the animal's natural habitat?
- What factors have caused endangerment of the animal?
- What efforts have been made to save the animal from extinction?
- What is the extended outlook for the survival of this animal?

Many people see little wildlife except when they visit a zoo. What problems do you think might arise when establishing a breeding population in a zoo? Do you think breeding in zoos will help or hurt the species in the long run? Consider the following difficulties:

- Under zoo conditions, captive animals cannot search for food and protect themselves from predators in their natural habitat. What do you think will happen to an animal who is returned to the wild after having been born and raised in a zoo?
- Under zoo conditions, captive animals lose their fear of man. Why is this disastrous for animals that may be hunted?
- How is an animal's behavior in a cage at the zoo different from his behavior in the wild?

Use your information to make a book about an endangered species, what is being done to protect this animal, and how you propose to help. You can include photos, interviews with a local zookeeper, background information about the history of the animal, and any attempts that have been made to breed the animal in captivity.

To make your own hardcover book, follow these steps.

- Handwrite or word process your story on an $8\frac{1}{2}$ by 11-inch sheet of paper. Staple the pages together using two staples along the side about $\frac{1}{4}$ inch from the left edge.
- To form the front and back cover, cut out two rectangles of cardboard that are $8\frac{3}{4}$ inches by $11\frac{1}{4}$ inches.
- Cut larger rectangles of paper to cover and decorate each piece of cardboard. You might use contact paper, old wallpaper, or construction paper. You could also use plain paper and then draw a design on the cover.
- Place the completed back cover on the table with the decorated side down. Place a thin line of glue on the left edge of the top side. Place your stapled story face-up onto the cover. Place a thin line of glue along the left edge of your stapled story. Then place the other cover, decorated side up, on top of the story. Be sure to align the two covers.
- You might want to use colored plastic tape to add a binding edge along the left side of your book.

Extension: Show Time

You can trade your books with other classmates and learn about different endangered species. You can also use the information from your book as a reference and share your information with younger students.

Create animal puppets and put on an informative puppet show for younger children. The more people know, understand, and enjoy animals, the more they will care about preserving them.

Project 3

High Energy

In physics, energy is the capacity for doing work. It may exist as *potential* energy or *kinetic* energy. Potential energy is energy of position. That is, because of its state, an object may have the ability to perform work. Examples of objects with potential energy are a lawn mower filled with gasoline or a baseball player waiting on-deck to bat.

Kinetic energy is energy in motion. The lawn mower cutting the grass displays kinetic energy. A ball player swinging a bat also shows kinetic energy.

In this project, you will pose questions, state hypotheses (or theories), and analyze the results of physics experiments.

Getting Started

Work in groups and have each member of the group bring in different kinds of balls, such as a tennis ball, a baseball, a Ping Pong ball, a styrofoam ball, a sponge ball, a croquet ball, a golf ball, and so on.

In the experiment, you will measure the energy of different balls under different conditions. In addition to the various balls, you will need a tape measure or yardstick taped to the wall and a piece of carpet. Each ball will be dropped and the height of the first bounce of the ball will be recorded. First you will drop the ball onto the bare floor, and then you will drop it onto the carpet.

Which ball will bounce the highest? Write a hypothesis, or guess, before you begin the experiment.

- Hold the ball at the top of the yardstick or tape measure. *This is potential energy.*
- Release the ball. *The falling ball is kinetic energy.*
- Observe the bounce. Record the height of the first bounce.
- Repeat these steps for each ball.

Will a ball bounce higher on carpet or the bare floor? Write a hypothesis.

- Repeat the steps above, but drop each ball onto the carpet square instead of the bare floor.
- Record the height of the first bounce.

Analyze the results of the experiment.

- When the ball hits the floor, what happens to the energy of the falling ball?
- Does the shape of the ball change as it is being bounced?
- How does the carpet affect the amount of "squish" in this ball?

Without conducting an experiment, discuss the following issues with your group. Write a plan how you would prove your hypotheses.

- Does a new tennis ball or an old one have a higher bounce?
- Does the drop height make a difference in the bounce height? Determine the ratio between the drop height and the bounce height. Does this ratio stay the same when a ball is dropped from different heights?
- Why do balls with a hard exterior such as golf balls and baseballs bounce? They are not "squishable" and would appear not to be able to spring back and bounce. Nevertheless, they can still bounce. How can you explain this?
- Does the weight of the ball affect the height of the bounce?

Extension: Wind It Up

Have you ever wondered what makes a wind-up toy move? What happens inside the toy to make it move?

Buy an inexpensive wind-up toy and take it apart. Inside you will find gears and a spiral spring. Energy is stored in the spring. Explain how it transfers stored energy to the toy to make it move.

Project

4 A Taste of America

The many kinds of foods available in major cities in this country are one reflection of the many different cultures that make up America's population. For example, in Boston, people can choose from many kinds of restaurants including Chinese, Thai, Indian, Ethiopian, Italian, Middle Eastern, French, Indonesian, Spanish, Hungarian, Brazilian, German, Caribbean, Japanese, French, Russian, Vietnamese, and Mexican.

In this project, you will use random samples to predict which kind of restaurant would be profitable to open in your community.

Discuss with your group the kinds of ethnic food you like to eat. Did any two people suggest the same type of food?

Since people's tastes vary, it may be difficult to decide on which type of restaurant to open with just the opinions of your group members. Therefore, you will need to conduct a survey for a random sample of people.

The number of people you poll will vary depending on the size of your community. You will want to question people in a neutral, public place, such as a library or grocery store. Ask them, "If a new restaurant were to open in this area in the near future, what kind of food would you like it to serve?" Have a list of types from which they may choose.

Record your data in a chart. Use your random samples to predict which kind of restaurant most people would like to see in the neighborhood.

Example:

$$P(\text{Italian}) = \frac{50}{100} \begin{array}{l} \leftarrow \text{number of people who chose Italian} \\ \leftarrow \text{total number of people surveyed} \end{array}$$

P(Italian) × number of people in community = expected number of people who prefer this kind of restaurant

After you have calculated the probability for each type of response, analyze your data and examine your results. Was there enough information from the responses to assist you in determining what kind of restaurant you would open? How many people did you poll? Do you think this was a fair sample?

Getting Started

Opening a new restaurant involves making many decisions. Make a list of the factors you should consider if you hope to have a successful business.

Examples:

- How large is the community where your restaurant will be located?
- What other kinds of food establishments are located in your neighborhood?
- How important is location?
- What kind of customers are you hoping to attract (families, business people, teenagers)?
- What would make people want to go to your restaurant?

Decide on a location and make a map of the surrounding area. Label the map with all of the restaurants in the vicinity of your planned site. If the area already has too many restaurants, you may want to modify your plans.

Once you have addressed these issues, you should try to locate some data about the size of your community, how often people go out to eat, and how much money they spend on a meal. Some possible sources are: a local Chamber of Commerce, a restaurant association, or local restaurant owners. Record your data in a chart.

Extension:

Create a menu from appetizers to desserts for your new restaurant. Be sure to include prices and descriptions of the kind of food you will be serving. You may need to research this information thoroughly before you begin writing. Make your menus as attractive as possible by using pictures from cooking magazines.

Discuss the possibilities of having a Multicultural Food Day and have each group prepare a food from your menus and share it with the class.

Trigonometric Ratios

Angle	sin	cos	tan	Angle	sin	cos	tan
0°	0.0000	1.0000	0.0000	45°	0.7071	0.7071	1.0000
1°	0.0175	0.9998	0.0175	46°	0.7193	0.6947	1.0355
2°	0.0349	0.9994	0.0349	47°	0.7314	0.6820	1.0724
3°	0.0523	0.9986	0.0524	48°	0.7431	0.6691	1.1106
4°	0.0698	0.9976	0.0699	49°	0.7547	0.6561	1.1504
5°	0.0872	0.9962	0.0875	50°	0.7660	0.6428	1.1918
6°	0.1045	0.9945	0.1051	51°	0.7771	0.6293	1.2349
7°	0.1219	0.9925	0.1228	52°	0.7880	0.6157	1.2799
8°	0.1392	0.9903	0.1405	53°	0.7986	0.6018	1.3270
9°	0.1564	0.9877	0.1584	54°	0.8090	0.5878	1.3764
10°	0.1736	0.9848	0.1763	55°	0.8192	0.5736	1.4281
11°	0.1908	0.9816	0.1944	56°	0.8290	0.5592	1.4826
12°	0.2079	0.9781	0.2126	57°	0.8387	0.5446	1.5399
13°	0.2250	0.9744	0.2309	58°	0.8480	0.5299	1.6003
14°	0.2419	0.9703	0.2493	59°	0.8572	0.5150	1.6643
15°	0.2588	0.9659	0.2679	60°	0.8660	0.5000	1.7321
16°	0.2756	0.9613	0.2867	61°	0.8746	0.4848	1.8040
17°	0.2924	0.9563	0.3057	62°	0.8829	0.4695	1.8807
18°	0.3090	0.9511	0.3249	63°	0.8910	0.4540	1.9626
19°	0.3256	0.9455	0.3443	64°	0.8988	0.4384	2.0503
20°	0.3420	0.9397	0.3640	65°	0.9063	0.4226	2.1445
21°	0.3584	0.9336	0.3839	66°	0.9135	0.4067	2.2460
22°	0.3746	0.9272	0.4040	67°	0.9205	0.3907	2.3559
23°	0.3907	0.9205	0.4245	68°	0.9272	0.3746	2.4751
24°	0.4067	0.9135	0.4452	69°	0.9336	0.3584	2.6051
25°	0.4226	0.9063	0.4663	70°	0.9397	0.3420	2.7475
26°	0.4384	0.8988	0.4877	71°	0.9455	0.3256	2.9042
27°	0.4540	0.8910	0.5095	72°	0.9511	0.3090	3.0777
28°	0.4695	0.8829	0.5317	73°	0.9563	0.2924	3.2709
29°	0.4848	0.8746	0.5543	74°	0.9613	0.2756	3.4874
30°	0.5000	0.8660	0.5774	75°	0.9659	0.2588	3.7321
31°	0.5150	0.8572	0.6009	76°	0.9703	0.2419	4.0108
32°	0.5299	0.8480	0.6249	77°	0.9744	0.2250	4.3315
33°	0.5446	0.8387	0.6494	78°	0.9781	0.2079	4.7046
34°	0.5592	0.8290	0.6745	79°	0.9816	0.1908	5.1446
35°	0.5736	0.8192	0.7002	80°	0.9848	0.1736	5.6713
36°	0.5878	0.8090	0.7265	81°	0.9877	0.1564	6.3138
37°	0.6018	0.7986	0.7536	82°	0.9903	0.1392	7.1154
38°	0.6157	0.7880	0.7813	83°	0.9925	0.1219	8.1443
39°	0.6293	0.7771	0.8098	84°	0.9945	0.1045	9.5144
40°	0.6428	0.7660	0.8391	85°	0.9962	0.0872	11.4301
41°	0.6561	0.7547	0.8693	86°	0.9976	0.0698	14.3007
42°	0.6691	0.7431	0.9004	87°	0.9986	0.0523	19.0811
43°	0.6820	0.7314	0.9325	88°	0.9994	0.0349	28.6363
44°	0.6947	0.7193	0.9657	89°	0.9998	0.0175	57.2900
45°	0.7071	0.7071	1.0000	90°	1.0000	0.0000	∞

Extra Practice

Extra Practice

Lesson 1-2 Use mental math to find each answer.

1. $2 + (24 \div 6)$
2. $24 \div (10 + 2)$
3. $4 \cdot (25 \cdot 9)$
4. $500 - 468$
5. $(7 \cdot 20) \cdot 5$
6. $462 + 195$
7. 5×25
8. $26 + 41 + 14$
9. $\$1.99 \times 3$
10. $5 \cdot (10 + 7)$
11. 6×35
12. $2 \cdot 84 \cdot 50$
13. $6 + 27 + 14$
14. $200 - 95$
15. $7 \cdot 19$
16. $1{,}762 + 124$
17. $5 \times 6 \times 2$
18. 12×25

Lesson 1-3 Estimate. Use an appropriate strategy.

1. $216 + 492$
2. $1{,}235 + 5{,}645$
3. $6{,}478 - 2{,}345$
4. $601 \div 6$
5. $298 + 109$
6. $8{,}710 - 610$
7. $364 \div 6$
8. $410 \div 7$
9. $0.245 + 0.256$
10. $17.985 - 9.001$
11. $11.75 \div 3$
12. $1{,}616 + 2{,}439$
13. $601 - 295$
14. $8.52 + 9.410$
15. $149 \div 5$
16. $39 + 41 + 40 + 38 + 39$
17. $1.12 + 0.9865 + 1.023 + 0.99 + 0.98$

Lesson 1-6 Complete each sentence.

1. 1 kg = ____ g
2. 632 mg = ____ g
3. 2.9 kL = ____ L
4. 400 mm = ____ cm
5. 30 g = ____ kg
6. 13.5 L = ____ kL
7. 0.3 km = ____ m
8. 38.6 kg = ____ g
9. 3.5 kL = ____ L
10. 4.8 cm = ____ mm
11. 9.5 mg = ____ g
12. 16 L = ____ mL
13. 12.6 g = ____ mg
14. 16.35 kL = ____ L
15. 415 m = ____ cm
16. 21 g = ____ mg
17. 63 L = ____ mL
18. 1.7 m = ____ cm
19. 1.02 kg = ____ g
20. 6.53 kL = ____ L
21. 45 cm = ____ mm

Lesson 1-7 Complete each sentence.

1. 7 ft = ____ in.
2. 5 T = ____ lb
3. 2 lb = ____ oz
4. 5 mi = ____ yd
5. $\frac{1}{4}$ lb = ____ oz
6. 31,680 ft = ____ mi
7. $\frac{1}{4}$ mi = ____ ft
8. 24 fl oz = ____ c
9. 8 pt = ____ c
10. 10 pt = ____ qt
11. 9 ft = ____ in.
12. 24 in. = ____ ft
13. 4 gal = ____ qt
14. 4 qt = ____ fl oz
15. 12 pt = ____ c
16. 5 yd = ____ ft
17. 15 qt = ____ gal
18. 4 pt = ____ c
19. 2 mi = ____ ft
20. 3 T = ____ lb
21. 6 lb = ____ oz

Lesson 1-9 Write each product using exponents.

1. $4 \cdot 4 \cdot 4 \cdot 4$
2. $3 \cdot 3$
3. $7 \cdot 7 \cdot 7 \cdot 7 \cdot 7 \cdot 7$

Evaluate each expression.

4. 4^3
5. 6^2
6. 2^6
7. $5^2 \times 6^2$
8. 3×2^4
9. $10^4 \times 3^2$
10. $5^3 \times 1^9$
11. $2^2 \times 2^4$
12. $2 \times 3^2 \times 4^2$
13. 7^3
14. $9^2 + 3^2$
15. 0.5^2

Lesson 2-1 Evaluate each expression.

1. $15 - 5 + 9 - 2$
2. $6 \times 6 + 3.6$
3. $12 + 20 \div 4 - 5$
4. $6 \times 3 \div 9 - 1$
5. $(4^2 + 2^3) \times 5$
6. $24 \div 8 - 2$
7. $3 \times (4 + 5) - 7$
8. $4.3 + 24 \div 6$
9. $(5^2 + 2) \div 3$
10. $27 \div 3^2 \times 2$
11. $4 \times 4^2 \times 2 - 8$
12. $12 \div 3 - 2^2 + 6$
13. $3^3 \times 2 - 5 \times 3$
14. $10 \times 2 + 7 \times 3$
15. $7 - 2 \times 8 \div 4$
16. $17 - 2^3 + 5$
17. $5 \times 6 \div 10 + 1$
18. $18 + 4 \div 2$
19. $(7 + 5 \times 4) \div 9$
20. $100 \div (28 + 9 \times 8)$
21. $6 \times 4 - 8 \times 3$

Lesson 2-2 Solve each equation.

1. $19 + 4 = y$
2. $5 \cdot 6 = n$
3. $q - 7 = 7$
4. $7 + a = 10$
5. $x - 3 = 12$
6. $2m = 8$
7. $4y = 24$
8. $36 = 6z$
9. $19 + j = 29$
10. $13 = 9 + c$
11. $p \div 4 = 4$
12. $6 = t \div 5$
13. $42 = 6n$
14. $\frac{m}{7} = 5$
15. $45 = 9d$
16. $24 = 14 + k$
17. $2a = 18$
18. $c \div 8 = 2$
19. $12 + f = 15$
20. $17 = 37 - g$
21. $25 = 5x$

Lesson 2-3 Solve each equation. Check your solution.

1. $g - 3 = 10$
2. $b + 7 = 12$
3. $a + 3 = 15$
4. $r - 3 = 4$
5. $t + 3 = 21$
6. $s + 10 = 23$
7. $9 + n = 13$
8. $13 + v = 31$
9. $s - 0.4 = 6$
10. $x - 1.3 = 12$
11. $18 = y + 3.4$
12. $7 + g = 91$
13. $63 + f = 71$
14. $0.32 = w - 0.1$
15. $c - 18 = 13$
16. $23 = n - 5$
17. $j - 3 = 7$
18. $18 = p + 3$
19. $12 + p = 16$
20. $25 = 50 - y$
21. $x + 2 = 4$

Lesson 2-4 Solve each equation. Check your solution.

1. $4x = 36$
2. $39 = 3y$
3. $4z = 16$
4. $t \div 5 = 6$
5. $100 = 20b$
6. $8 = w \div 8$
7. $10a = 40$
8. $s \div 9 = 8$
9. $420 = 5s$
10. $8k = 72$
11. $2m = 18$
12. $\frac{m}{8} = 5$
13. $0.12 = 3h$
14. $\frac{w}{7} = 8$
15. $18q = 36$
16. $9w = 54$
17. $4 = p \div 4$
18. $14 = 2p$
19. $12 = 3t$
20. $\frac{m}{4} = 12$
21. $6h = 12$

Lesson 2-6 Write each phrase or sentence as an algebraic expression.

1. 12 more than a number
2. 3 less than a number
3. a number divided by 4
4. a number increased by 7
5. a number decreased by 12
6. 8 times a number
7. 28 multiplied by m
8. 15 divided by a number
9. 54 divided by n
10. 18 increased by y
11. q decreased by 20
12. n times 41

Lesson 2-7 Solve each equation. Check your solution.

1. $2x + 4 = 14$
2. $5p - 10 = 0$
3. $5 + 6a = 41$
4. $\frac{x}{3} - 7 = 2$
5. $18 = 6(q - 4)$
6. $18 = 4m - 6$
7. $3(r - 1) = 9$
8. $2x + 3 = 5$
9. $0 = 4x - 28$
10. $3x - 1 = 5$
11. $3z + 5 = 14$
12. $3(x - 5) = 12$
13. $9a - 8 = 73$
14. $2x - 3 = 7$
15. $3t + 6 = 9$
16. $2y + 10 = 22$
17. $15 = 2y - 5$
18. $3c - 4 = 2$
19. $6 + 2p = 16$
20. $8 = 2 + 3x$
21. $4(b + 6) = 24$

Lesson 2-9 Find the perimeter and area of each figure.

1.
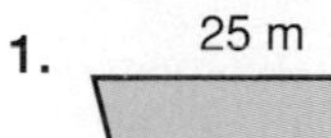
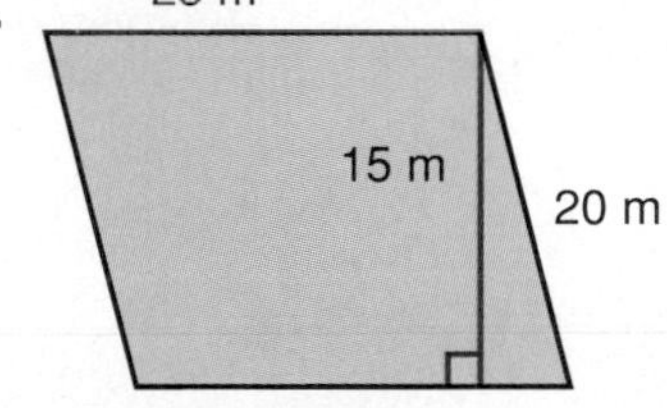

2.
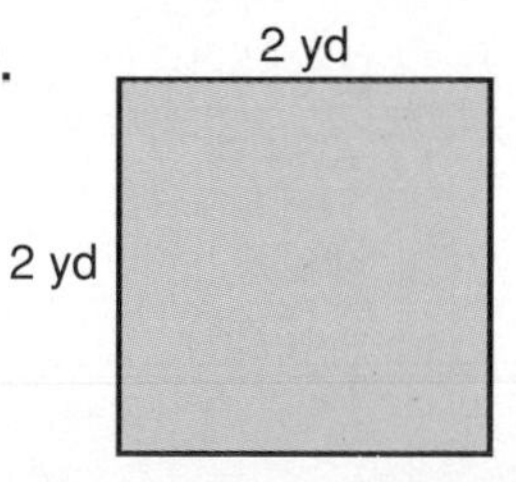

3.
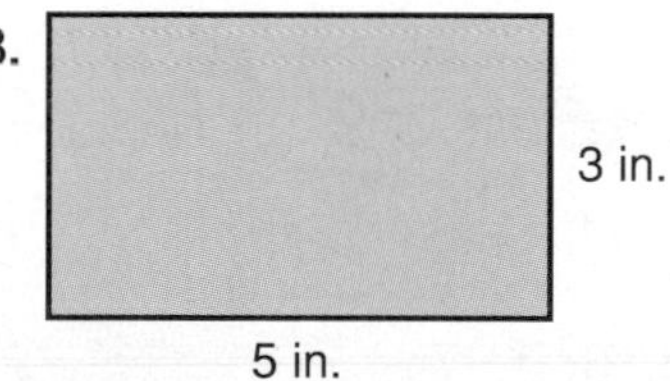

4.
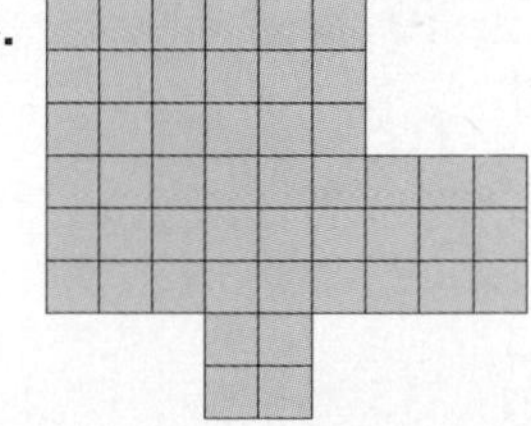

5.
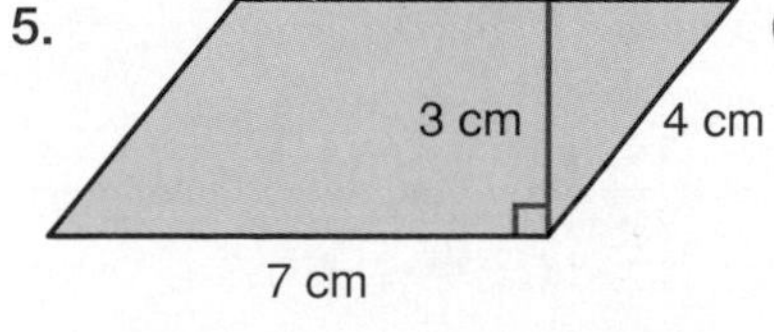

6.
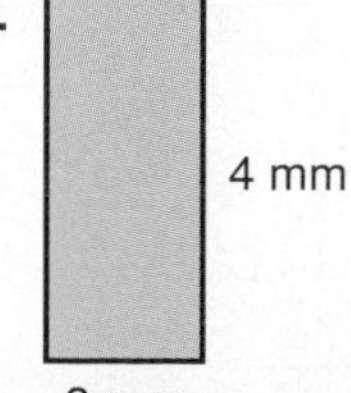

Lesson 2-10 Solve each inequality. Show the solution on a number line.

1. $y + 3 > 7$
2. $c - 9 < 5$
3. $x + 4 > 9$
4. $y - 3 < 15$
5. $t - 13 > 5$
6. $5p < 25$
7. $4x < 12$
8. $15 < 3m$
9. $\frac{d}{3} > 15$
10. $8 < r \div 7$
11. $2y + 5 > 15$
12. $16 < 5d + 6$
13. $3x + 2 > 11$
14. $\frac{a}{3} - 2 > 1$
15. $9g < 27$
16. $14 < 2x + 4$
17. $\frac{x}{2} - 6 > 0$
18. $k + 5 < 6$
19. $15 > c - 2$
20. $4p > 24$
21. $24 < 14 + k$

Lesson 3-1 Graph each set of numbers on a number line.

1. $\{-8, -9, -6, -10\}$
2. $\{-3, 2, 0, -1\}$
3. $\{5, 6, 8, 7, 9\}$

Find each absolute value.

4. $|-1|$
5. $|-92|$
6. $|3|$
7. $|160 + 32|$
8. $|80 - 100|$
9. $|0|$
10. $|7 - 3|$
11. $|3 - 7|$
12. $|-161|$
13. $|150|$
14. $|2 - 102|$
15. $|-116|$

Lesson 3-2 Replace each ● with $>$, $<$, or $=$.

1. -3 ● 0
2. -1 ● -2
3. -5 ● -4
4. 6 ● -7
5. 8 ● 10
6. -6 ● 6
7. -11 ● -20
8. -8 ● 2
9. -13 ● -12
10. 5 ● 2
11. 9 ● -8
12. 19 ● -19
13. $|-2|$ ● $|5|$
14. $|13|$ ● $|-19|$
15. $|-6|$ ● $|2|$
16. $|14|$ ● $|-14|$
17. $|0|$ ● $|-4|$
18. $|23|$ ● $|-20|$
19. $|-75|$ ● $|75|$
20. -71 ● 72
21. -15 ● -35

Lesson 3-3 Solve each equation.

1. $-7 + (-7) = h$
2. $k = -36 + 40$
3. $m = 18 + (-32)$
4. $47 + 12 = y$
5. $y = -69 + (-32)$
6. $-120 + (-2) = c$
7. $x = -56 + (-4)$
8. $14 + 16 = k$
9. $-18 + 11 = d$
10. $-42 + 29 = r$
11. $h = -13 + (-11)$
12. $x = 95 + (-5)$
13. $-120 + 2 = b$
14. $w = 25 + (-25)$
15. $a = -4 + 8$
16. $g = -9 + (-6)$
17. $42 + (-18) = f$
18. $-33 + (-12) = w$
19. $-96 + (-18) = g$
20. $-100 + 98 = a$
21. $5 + (-7) = y$

Lesson 3-4 Solve each equation. Check by solving another way.

1. $a = 7 + (-13) + 6 + (-7)$
2. $x = -6 + 12 + (-20)$
3. $4 + 9 + (-14) = k$
4. $c = -20 + 0 + (-9) + 25$
5. $b = 5 + 9 + 3 + (-17)$
6. $-36 + 40 + (-10) = y$
7. $(-2) + 2 + (-2) + 2 = m$
8. $6 + (-4) + 9 + (-2) = d$
9. $9 + (-7) + 2 = n$
10. $b = 100 + (-75) + (-20)$
11. $x = -12 + 24 + (-12) + 2$
12. $9 + (-18) + 6 + (-3) = c$
13. $(-10) + 4 + 6 = k$
14. $c = 4 + (-8) + 12$

Lesson 3-5 Solve each equation.

1. $3 - 7 = y$
2. $-5 - 4 = w$
3. $a = -6 - 2$
4. $12 - 9 = x$
5. $a = 0 - (-14)$
6. $a = 58 - (-10)$
7. $n = -41 - 15$
8. $c = -81 - 21$
9. $26 - (-14) = y$
10. $6 - (-4) = b$
11. $z = 63 - 78$
12. $-5 - (-9) = h$
13. $m = 72 - (-19)$
14. $-51 - 47 = x$
15. $-99 - 1 = p$
16. $r = 8 - 13$
17. $-2 - 23 = c$
18. $-20 - 0 = d$
19. $55 - 33 = k$
20. $84 - (-61) = a$
21. $z = -4 - (-4)$

Lesson 3-6 Solve each equation.

1. $5(-2) = d$
2. $-11(-5) = c$
3. $-5(-5) = z$
4. $x = -12(6)$
5. $b = 2(-2)$
6. $-3(2)(-4) = j$
7. $a = (-4)(-4)$
8. $4(21) = y$
9. $a = -50(0)$
10. $b = 3(-13)$
11. $a = 2(2)$
12. $d = -2(-2)$
13. $x = 5(-12)$
14. $2(2)(-2) = b$
15. $a = 6(-4)$
16. $x = -6(5)$
17. $-4(8) = a$
18. $3(-16) = y$
19. $c = -2(2)$
20. $6(3)(-2) = k$
21. $y = -3(12)$

Lesson 3-7 Solve each equation.

1. $a = 4 \div (-2)$
2. $16 \div (-8) = x$
3. $-14 \div (-2) = c$
4. $h = -18 \div 3$
5. $-25 \div 5 = k$
6. $n = -56 \div (-8)$
7. $x = 81 \div 9$
8. $-55 \div 11 = c$
9. $-42 \div (-7) = y$
10. $g = 18 \div (-3)$
11. $t = 0 \div (-1)$
12. $-32 \div 8 = m$
13. $81 \div (-9) = w$
14. $18 \div (-2) = a$
15. $x = -21 \div 3$
16. $d = 32 \div 8$
17. $8 \div (-8) = y$
18. $c = -14 \div (-7)$
19. $-81 \div 9 = y$
20. $q = -81 \div (-9)$
21. $-49 \div (-7) = y$

Lesson 3-9 Solve each equation. Check your solution.

1. $-4 + b = 12$
2. $z - 10 = -8$
3. $-7 = x + 12$
4. $a + 6 = -9$
5. $r \div 7 = -8$
6. $-2a = -8$
7. $r - (-8) = 14$
8. $0 = 6r$
9. $\frac{y}{12} = -6$
10. $m + (-2) = 6$
11. $3m = -15$
12. $c \div (-4) = 10$
13. $5 + q = 12$
14. $\frac{16}{x} = -4$
15. $-6f = -36$
16. $81 = -9w$
17. $t + 12 = 6$
18. $8 + p = 0$
19. $0.12 = -3h$
20. $12 - x = 8$
21. $14 + t = 10$

Lesson 3-10

Name the ordered pair for the coordinates of each point graphed on the coordinate plane below.

1. A
2. B
3. C
4. D
5. E
6. F
7. G
8. H
9. I

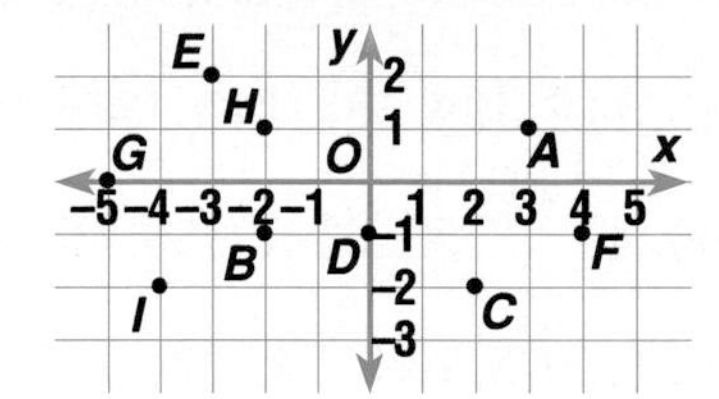

Graph each point on the same coordinate plane.

10. (3, -2)
11. (2, 4)
12. (-1, 6)
13. (0, 5)
14. (-2, -3)
15. (-4, 0)
16. (4, -4)
17. (0, 0)
18. (3, 1)
19. (-4, -1)

Lesson 4-2

Use the histogram below to answer each question.

1. How large is each interval?
2. Which interval has the most buildings?
3. Which interval has the least buildings?
4. Compared to the total, how would you describe the number of buildings over 70 feet tall?
5. How does the number of buildings between 61 and 80 feet tall compare to the number of buildings between 31 and 50 feet tall?

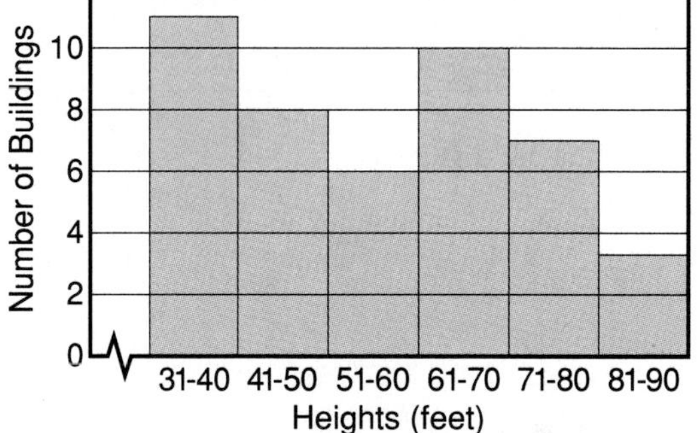

Lesson 4-3

Make a line plot for each set of data.

1. 8, 12, 10, 15, 11, 9, 12, 7, 14, 13, 8, 15, 17, 14, 11, 9, 8, 12, 15
2. 32, 41, 46, 38, 34, 51, 55, 49, 37, 42, 55, 46, 39, 58, 40, 35, 34, 52, 46
3. 161, 158, 163, 162, 165, 157, 159, 160, 163, 162, 158, 164, 161, 157, 166, 164
4. 78, 82, 83, 90, 58, 67, 95, 87, 97, 88, 75, 82, 78, 89, 86, 88, 79, 80, 91, 77
5. 49¢, 55¢, 77¢, 65¢, 51¢, 74¢, 68¢, 56¢, 49¢, 73¢, 62¢, 71¢, 54¢, 50¢, 70¢, 65¢
6. 2, 4, 12, 10, 2, 5, 7, 11, 7, 6, 3, 9, 12, 7, 5, 3, 7, 11, 2, 8, 7, 10, 8, 3, 7
7. 303, 298, 289, 309, 300, 294, 299, 301, 296, 308, 302, 289, 306, 308, 298, 299
8. 67, 73, 78, 61, 63, 77, 66, 75, 79, 66, 72, 69, 70, 74, 61, 63, 76, 64, 65, 78, 66

Lesson 4-4 Make a stem-and-leaf plot for each set of data.

1. 5.5, 6.2, 6.8, 5.9, 7.3, 8.6, 5.4, 6.3, 8.2, 7.5, 7.1, 5.7, 8.4, 5.9, 6.1, 8.8

2. 115, 153, 145, 119, 136, 154, 142, 137, 125, 121, 112, 156, 129, 133, 140, 123, 155

3. 55, 58, 45, 60, 47, 52, 53, 63, 47, 55, 49, 65, 56, 61

4. 71.3, 72.4, 74.8, 71.8, 73.5, 74.2, 71.9, 73.6, 73.9, 72.3, 73.7, 72.2

5. 415, 427, 412, 398, 407, 395, 422, 401, 393, 412, 427, 419, 424, 405, 391, 413

6. 11, 23, 27, 46, 36, 32, 17, 22, 49, 36, 19, 41, 26, 33, 15, 32, 35, 21

7. Make a back-to-back stem-and-leaf plot for the data in the table below.

Quiz 2 Scores	42	37	32	45	34	29	46	33	45	44	37	43	42	39	38	41	36
Quiz 3 Scores	48	49	40	50	44	39	45	49	37	47	49	41	46	48	50	42	45

Lesson 4-5 Find the mean, median, and mode for each set of data. Round to the nearest tenth.

1. 2, 7, 9, 12, 5, 14, 4, 8, 3, 10

2. 58, 52, 49, 60, 61, 56, 50, 61

3. 122, 134, 129, 140, 125, 134, 137

4. 25.5, 26.7, 20.9, 23.4, 26.8, 24.0, 25.7

5.

Stem	Leaf
3	6
4	1358
5	24667
6	045

5 | 2 means 52

6.

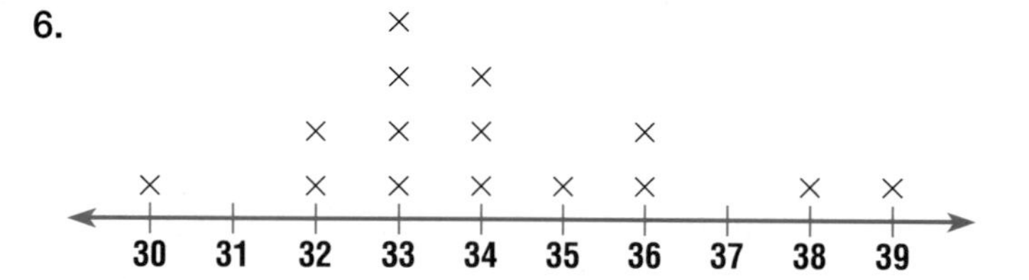

Lesson 4-6 Find the range of each set of data.

1. 15, 12, 21, 18, 25, 11, 17, 19, 20

2. 2, 10, 6, 13, 8, 6, 11, 4

3. 189, 149, 155, 190, 141, 152

4. 451, 501, 388, 428, 510, 480, 390

Find the median and upper and lower quartiles of each set of data.

5. 22, 18, 9, 26, 14, 15, 6, 19, 28

6. 245, 238, 251, 255, 248, 241, 250

7. 46, 45, 50, 40, 49, 42, 52

8. 128, 148, 130, 142, 164, 120, 152, 168

9. 8, 3, 2, 6, 4, 12, 10, 2, 6, 11

10. 88, 84, 92, 93, 90, 96, 87, 97

11. 378, 410, 370, 336, 361, 394, 345, 328, 388, 339

Lesson 4-7 Draw a box-and-whisker plot for each set of data.

1. 79, 70, 84, 66, 72, 64, 75, 82
2. 307, 313, 304, 306, 312, 301, 310
3. 7, 4, 10, 3, 2, 9, 6, 3, 7
4. 32, 39, 27, 40, 45, 30, 22, 36, 24, 48
5. 61, 67, 53, 56, 51, 66, 58, 63, 52
6. \$32, \$26, \$39, \$23, \$18, \$30, \$21, \$34

7.

Stem	Leaf
5	2459
6	1336
7	48
8	013457
9	1278

8 | 3 means 83

8. 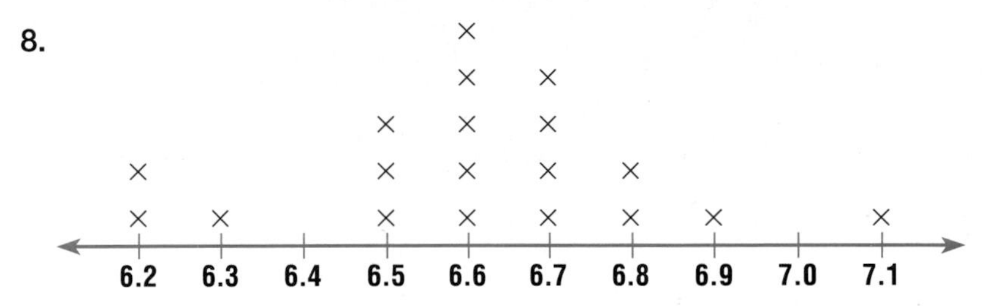

Lesson 4-8 Determine whether a scatter plot of the data below would show a positive, negative, or no relationship.

1. height and hair color
2. hours spent studying and test scores
3. income and month of birth
4. price of oranges and number available
5. size of roof and number of shingles
6. number of clouds and number of stars seen

7. 0 1 2 3 4 5 6 7 8 9 10

8. 0 2 4 6 8 10 12 14 16 18 20

9. 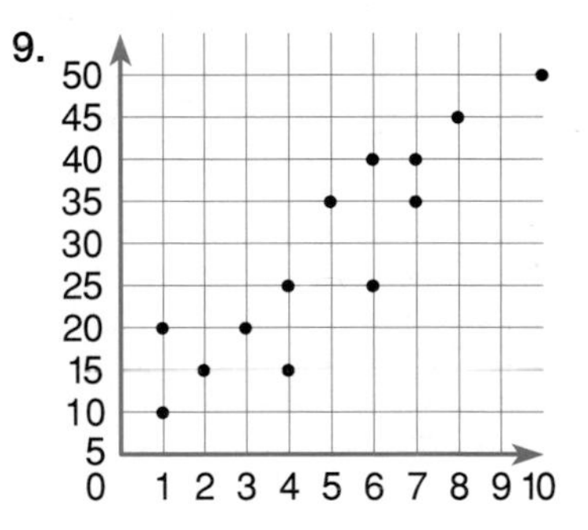

Lesson 5-1 Name the parallel segments, if any, in each figure.

1. A, B, C

2. 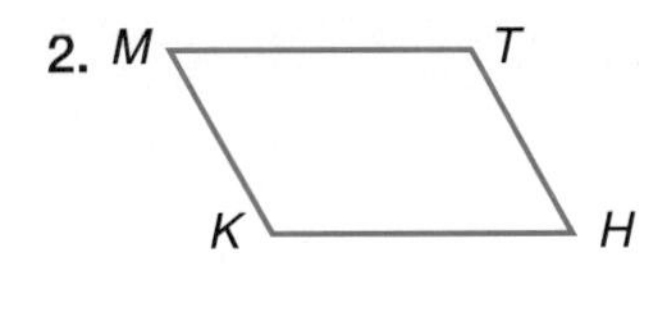

3. 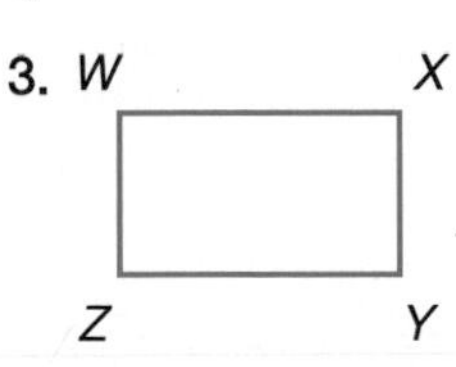

Use the figure at the right for Exercises 4–7.

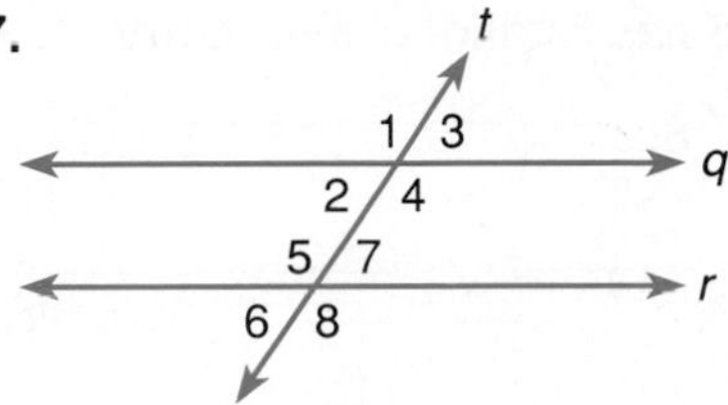

4. Find $m\angle 6$, if $m\angle 3 = 42°$.
5. Find $m\angle 4$, if $m\angle 7 = 71°$.
6. Find $m\angle 1$, if $m\angle 8 = 128°$.
7. Find $m\angle 7$, if $m\angle 2 = 83°$.

Lesson 5-3

Classify each triangle by its sides and by its angles.

1. 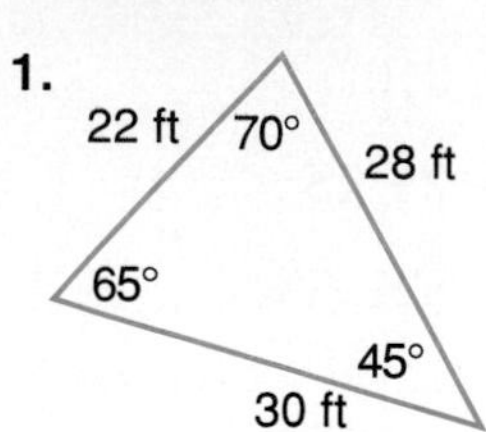

2. 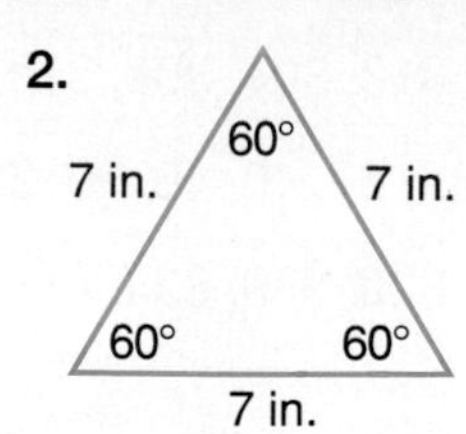

3. 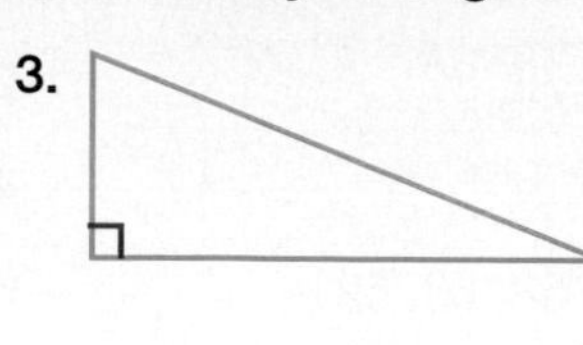

4. 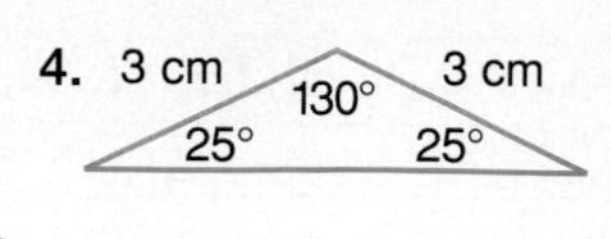

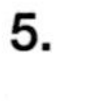

5. 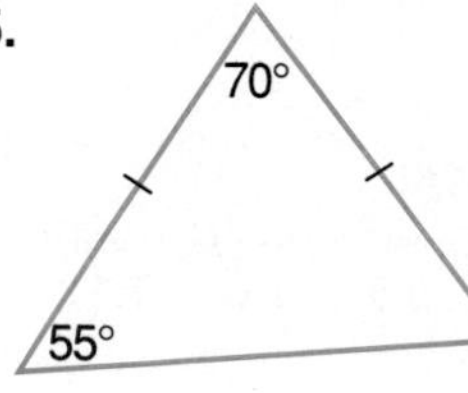

6. 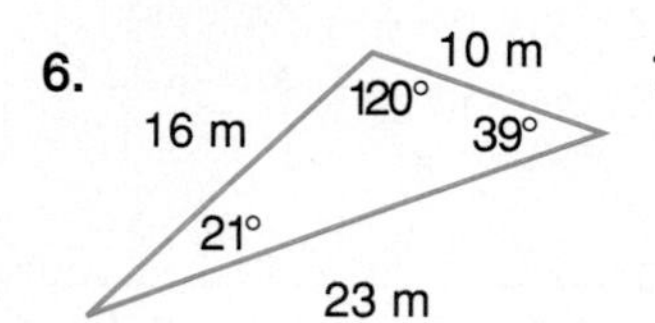

7. 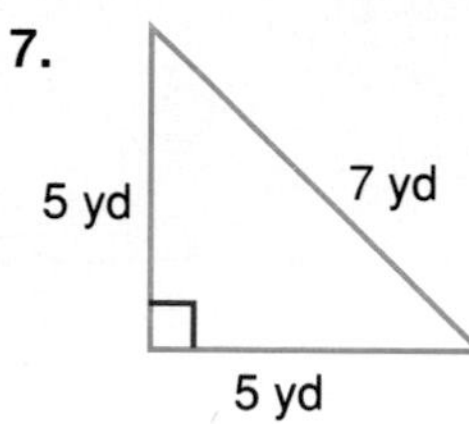

8. 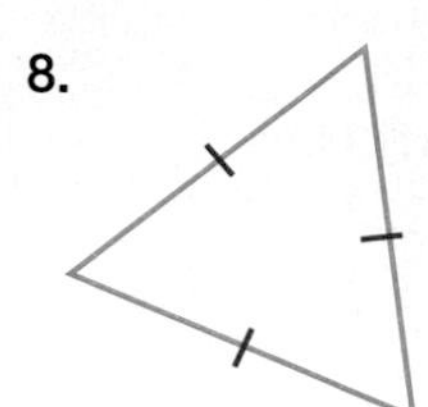

Lesson 5-4

Let Q = quadrilateral, P = parallelogram, R = rectangle, S = square, RH = rhombus, and T = trapezoid. Write all letters that describe each figure.

1. 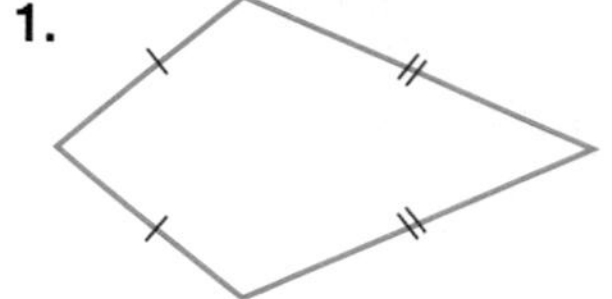

2. 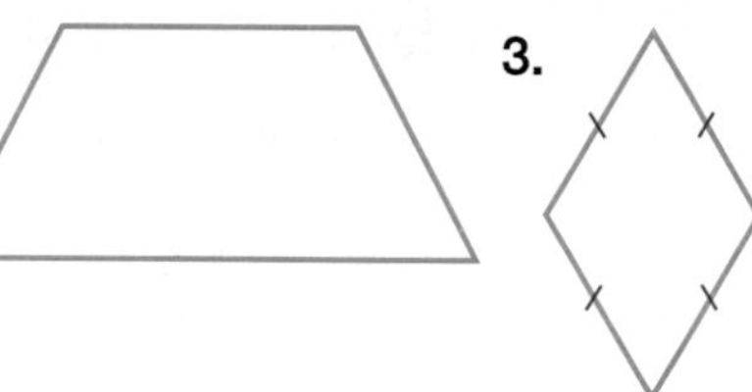

3.

4. 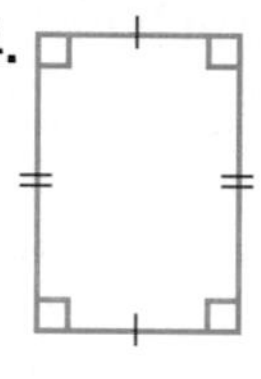

5. 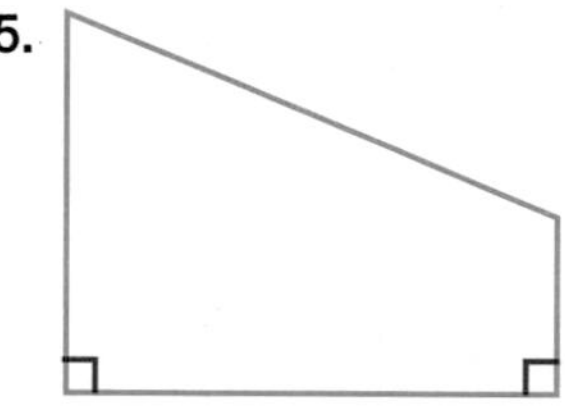

6. 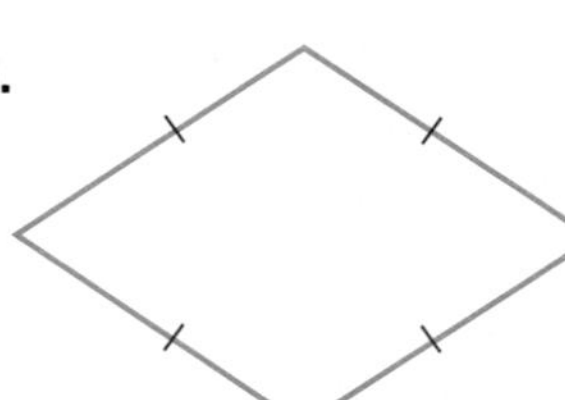

7. 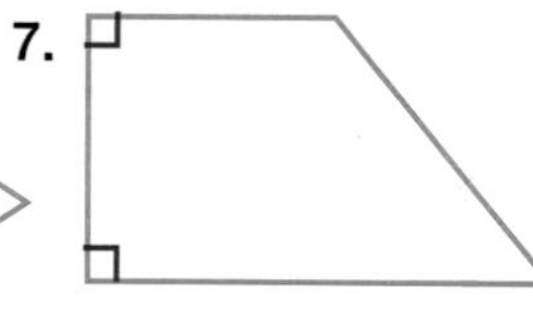

8. 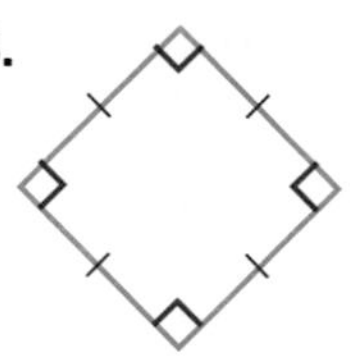

Lesson 5-5

Trace each figure. Determine if the figure has line symmetry. If so, draw the lines of reflection.

1. 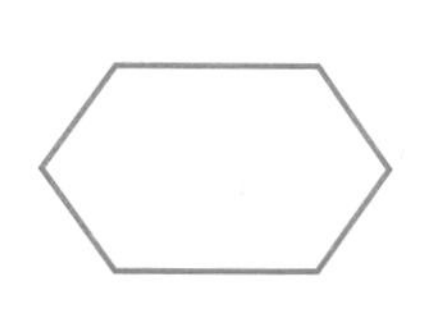

2. 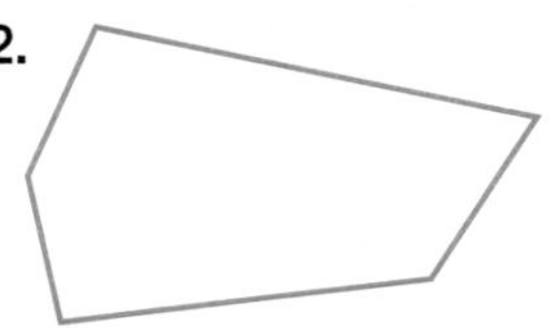

3. 

4. 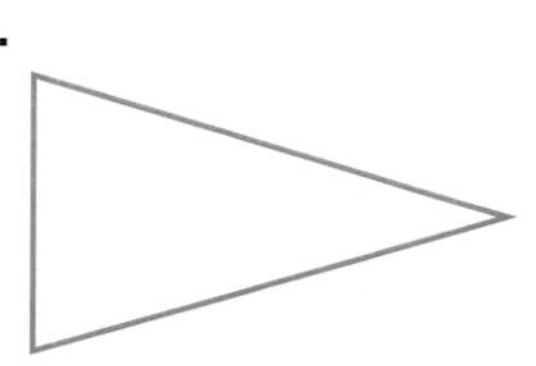

Determine if each figure has rotational symmetry.

5. 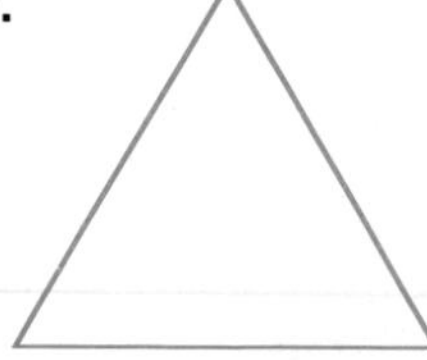

6. 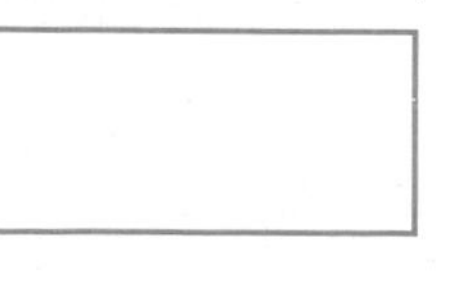

7. 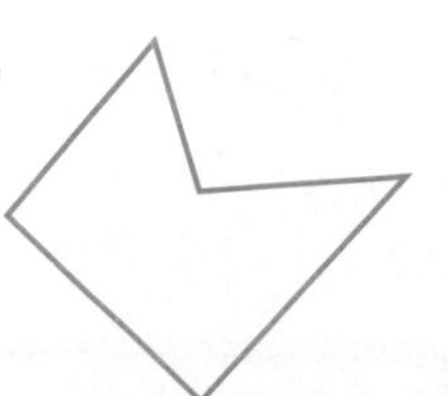

8. 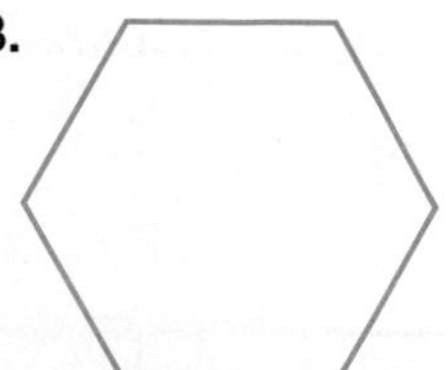

Lesson 5-6

Tell if each pair of figures is *congruent, similar,* or *neither.* Justify your answer.

1. 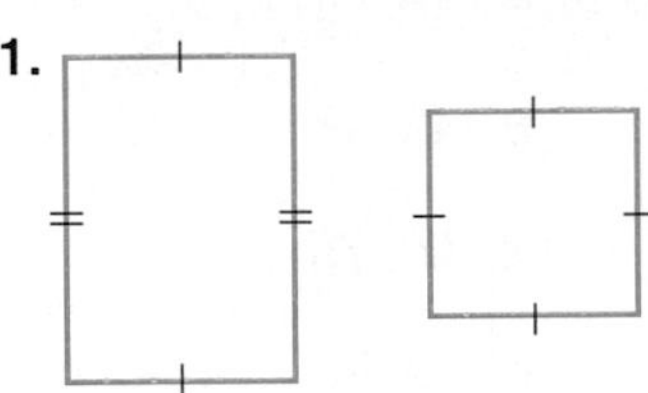

2. 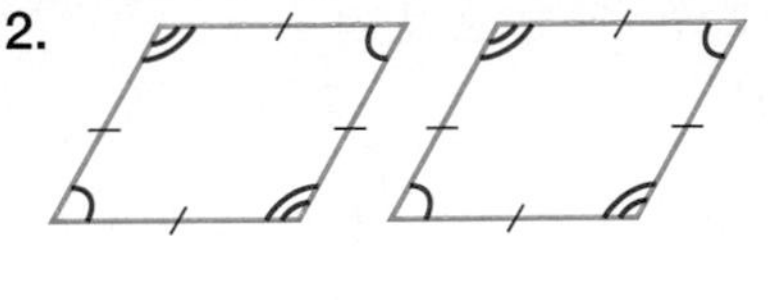

3. 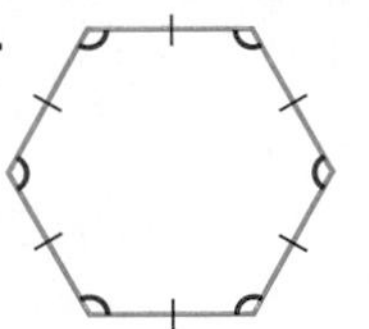

Find the value of *x* in each pair of figures.

4. $\triangle ABC \cong \triangle JKI$

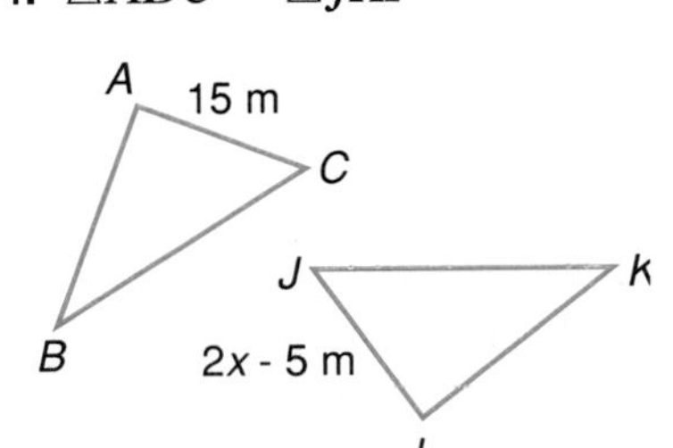

5. $\triangle LFR \sim \triangle GPC$

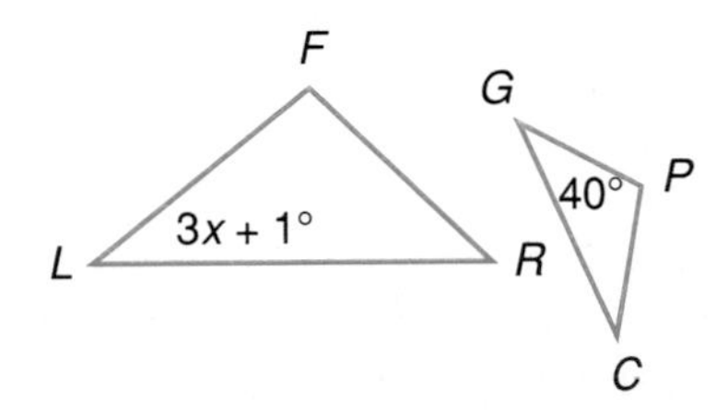

6. $\square XTDH \cong \square PEBL$

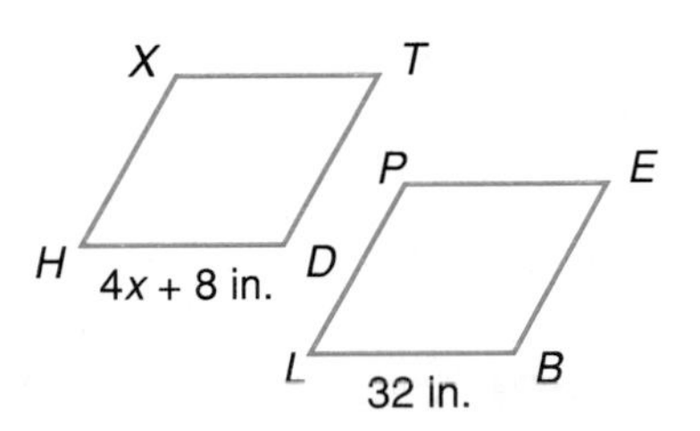

Lesson 6-1

Using divisibility rules, determine whether each number is divisible by 2, 3, 4, 5, 6, 8, 9, or 10.

1. 210
2. 614
3. 985
4. 756
5. 432
6. 96
7. 87
8. 113

Use divisibility rules to determine if the first number is divisible by the second number. Write *yes* or *no*.

9. 936; 6
10. 752; 6
11. 1,249; 8
12. 208; 4
13. 216; 9
14. 1,346; 2
15. 1,687; 3
16. 448; 8

Lesson 6-2

Determine whether each number is *prime, composite,* or *neither.*

1. 17
2. 1,258
3. 37
4. 483
5. 97
6. 0
7. 25
8. 61
9. -45
10. 419

Find the prime factorization of each number.

11. 20
12. 65
13. 52
14. 30
15. 28
16. 72
17. 155
18. 50
19. 96
20. 201
21. 1,250
22. 2,648

Lesson 6-4 Find the GCF for each set of numbers.

1. 8, 18
2. 6, 9
3. 4, 12
4. 18, 24
5. 8, 24
6. 17, 51
7. 65, 95
8. 42, 48
9. 64, 32
10. 72, 144
11. 54, 72
12. 60, 75
13. 16, 24
14. 12, 27
15. 25, 30
16. 48, 60
17. 16, 20, 36
18. 12, 18, 42
19. 30, 45, 15
20. 20, 30, 40
21. 81, 27, 108
22. 9, 18, 12

Lesson 6-5 Write each fraction in simplest form.

1. $\frac{12}{16}$
2. $\frac{28}{32}$
3. $\frac{75}{100}$
4. $\frac{8}{16}$
5. $\frac{6}{18}$
6. $\frac{27}{36}$
7. $\frac{16}{64}$
8. $\frac{8}{16}$
9. $\frac{50}{100}$
10. $\frac{24}{40}$
11. $\frac{32}{80}$
12. $\frac{8}{24}$
13. $\frac{20}{25}$
14. $\frac{4}{10}$
15. $\frac{3}{5}$
16. $\frac{14}{19}$
17. $\frac{9}{12}$
18. $\frac{6}{8}$
19. $\frac{15}{18}$
20. $\frac{9}{20}$
21. $\frac{8}{21}$
22. $\frac{10}{15}$
23. $\frac{9}{24}$
24. $\frac{6}{31}$
25. $\frac{18}{32}$

Lesson 6-6 Express each fraction as a decimal.

1. $\frac{2}{5}$
2. $\frac{3}{8}$
3. $-\frac{3}{4}$
4. $\frac{5}{16}$
5. $\frac{3}{4}$
6. $-\frac{7}{8}$
7. $\frac{17}{20}$
8. $\frac{14}{25}$
9. $\frac{7}{10}$
10. $\frac{7}{20}$

Express each decimal as a fraction or mixed number in simplest form.

11. 0.5
12. 0.8
13. 0.32
14. -0.75
15. 1.54
16. 0.38
17. -0.486
18. 20.08
19. -9.36
20. 10.18
21. 0.06
22. 1.75
23. -0.375
24. 0.79
25. 1.9

Lesson 6-7

Write the first ten decimal places of each decimal.

1. $0.\overline{09}$
2. $0.\overline{076923}$
3. $0.8\overline{4563}$
4. $0.987\overline{45}$
5. $0.\overline{254}$
6. $0.\overline{1470}$
7. $0.12\overline{7}$
8. $0.\overline{3}$

Express each decimal using bar notation.

9. 0.161616...
10. 0.12351235...
11. 0.6666...
12. 0.15151...
13. 0.125656...
14. 0.1254777...
15. 85.0124124...
16. 0.214111...

Express each repeating decimal as a fraction.

17. $0.\overline{3}$
18. $0.\overline{4}$
19. $0.\overline{27}$
20. $0.8\overline{3}$
21. $0.\overline{24}$
22. $0.58\overline{3}$
23. $0.7\overline{3}$
24. $0.\overline{8}$

Lesson 6-8

A date is chosen at random from the month of November. Find the probability of choosing each date.

1. The date is the thirteenth.
2. The date is Friday.
3. It is after the twenty-fifth.
4. It is before the seventh.
5. It is an odd-numbered date.
6. The date is divisible by 3.

November

S	M	T	W	T	F	S
		1	2	3	4	5
6	7	8	9	10	11	12
13	14	15	16	17	18	19
20	21	22	23	24	25	26
27	28	29	30			

Lesson 6-9

Find the LCM for each set of numbers.

1. 5, 6
2. 9, 27
3. 12, 15
4. 8, 12
5. 5, 15
6. 13, 39
7. 16, 24
8. 18, 20
9. 21, 14
10. 25, 30
11. 28, 42
12. 7, 13
13. 6, 30
14. 12, 42
15. 8, 10
16. 30, 10
17. 12, 18, 6
18. 15, 75, 25
19. 6, 10, 15
20. 3, 6, 9
21. 21, 14, 6
22. 12, 35, 10

Lesson 6-10 Replace each ● with a <, >, or = to make a true sentence.

1. -5.6 ● 4.2
2. 4.256 ● 4.25
3. 0.233 ● $0.\overline{23}$
4. $\frac{5}{7}$ ● $\frac{2}{5}$
5. $\frac{6}{7}$ ● $\frac{7}{9}$
6. $\frac{2}{3}$ ● $\frac{2}{5}$
7. $\frac{3}{8}$ ● 0.375
8. $-\frac{1}{2}$ ● 0.5
9. 12.56 ● $12\frac{3}{8}$

Order each set of numbers from least to greatest.

10. $0.24, 0.2, 0.245, 2.24, 0.25$
11. $0.\overline{3}, 0.3, 0.3\overline{4}, 0.\overline{34}, 0.33$
12. $\frac{2}{5}, \frac{2}{3}, \frac{2}{7}, \frac{2}{9}, \frac{2}{1}$
13. $\frac{1}{2}, \frac{5}{7}, \frac{2}{9}, \frac{8}{9}, \frac{6}{6}$

Lesson 6-11 Express each number in standard form.

1. 4.5×10^3
2. 2×10^4
3. 1.725896×10^6
4. 9.61×10^2
5. 1×10^7
6. 8.256×10^8
7. 5.26×10^4
8. 3.25×10^2
9. 6.79×10^5

Express each number in scientific notation.

10. 720
11. 7,560
12. 892
13. 1,400
14. 91,256
15. 51,000
16. 145,600
17. 90,100
18. 123,568,000,000

Lesson 7-1 Solve each equation. Write each solution in simplest form.

1. $\frac{17}{21} + \left(-\frac{13}{21}\right) = m$
2. $t = \frac{5}{11} + \frac{6}{11}$
3. $k = -\frac{8}{13} + \left(-\frac{11}{13}\right)$
4. $-\frac{7}{12} + \frac{5}{12} = a$
5. $\frac{13}{28} - \frac{9}{28} = g$
6. $b = -1\frac{2}{9} - \frac{7}{9}$
7. $r = \frac{15}{16} + \frac{13}{16}$
8. $2\frac{1}{3} - \frac{2}{3} = n$
9. $-\frac{4}{35} - \left(-\frac{17}{35}\right) = c$
10. $\frac{3}{8} + \left(-\frac{5}{8}\right) = w$
11. $s = \frac{8}{15} - \frac{2}{15}$
12. $d = -2\frac{4}{7} - \frac{3}{7}$
13. $-\frac{29}{9} - \left(-\frac{26}{9}\right) = y$
14. $2\frac{3}{5} + 7\frac{3}{5} = i$
15. $x = \frac{5}{18} - \frac{13}{18}$
16. $j = -2\frac{2}{7} + \left(-1\frac{6}{7}\right)$
17. $p = -\frac{3}{10} + \frac{7}{10}$
18. $\frac{4}{11} + \frac{9}{11} = e$

Lesson 7-2

Solve each equation. Write each solution in simplest form.

1. $r = \frac{7}{12} + \frac{7}{24}$
2. $-\frac{3}{4} + \frac{7}{8} = z$
3. $\frac{2}{5} + \left(-\frac{2}{7}\right) = q$
4. $d = -\frac{3}{5} - \left(-\frac{5}{6}\right)$
5. $\frac{5}{24} - \frac{3}{8} = j$
6. $g = -\frac{7}{12} + \frac{3}{4}$
7. $-\frac{3}{8} + \left(-\frac{4}{5}\right) = x$
8. $t = \frac{2}{15} + \left(-\frac{3}{10}\right)$
9. $r = -\frac{2}{9} - \left(-\frac{2}{3}\right)$
10. $a = -\frac{7}{15} - \frac{5}{12}$
11. $\frac{3}{8} + \frac{7}{12} = s$
12. $-2\frac{1}{4} + \left(-1\frac{1}{3}\right) = m$
13. $3\frac{2}{5} - 3\frac{1}{4} = v$
14. $b = \frac{3}{4} + \left(-\frac{4}{15}\right)$
15. $f = -1\frac{2}{3} + 4\frac{3}{4}$
16. $-\frac{1}{8} - 2\frac{1}{2} = n$
17. $p = 3\frac{2}{5} - 1\frac{1}{3}$
18. $y = 5\frac{1}{3} + \left(-8\frac{3}{7}\right)$

Lesson 7-3

Solve each equation. Write each solution in simplest form.

1. $\frac{2}{11} \cdot \frac{3}{4} = m$
2. $4\left(-\frac{7}{8}\right) = r$
3. $d = -\frac{4}{7} \cdot \frac{3}{5}$
4. $g = \frac{6}{7}\left(-\frac{7}{12}\right)$
5. $b = \frac{7}{8} \cdot \frac{1}{3}$
6. $\frac{3}{4} \cdot \frac{4}{5} = t$
7. $-1\frac{1}{2} \cdot \frac{2}{3} = k$
8. $x = \frac{5}{6} \cdot \frac{6}{7}$
9. $c = 8\left(-2\frac{1}{4}\right)$
10. $-3\frac{3}{4} \cdot \frac{8}{9} = q$
11. $\frac{10}{21} \cdot -\frac{7}{8} = n$
12. $w = -1\frac{4}{5}\left(-\frac{5}{6}\right)$
13. $a = 5\frac{1}{4} \cdot 6\frac{2}{3}$
14. $-8\frac{3}{4} \cdot 4\frac{2}{5} = p$
15. $y = 6 \cdot 8\frac{2}{3}$
16. $i = \left(\frac{3}{5}\right)^2$
17. $-4\frac{1}{5}\left(-3\frac{1}{3}\right) = h$
18. $-8 \cdot \left(\frac{3}{4}\right)^2 = v$

Lesson 7-4

Name the multiplicative inverse of each of the following.

1. 3
2. -5
3. $\frac{2}{3}$
4. $2\frac{1}{8}$
5. $\frac{a}{b}$
6. -8
7. $\frac{1}{15}$
8. 0.75
9. c
10. $-\frac{3}{5}$
11. $1\frac{1}{3}$
12. 0.5
13. $-2\frac{3}{7}$
14. $-\frac{1}{11}$
15. 12
16. $\frac{7}{9}$
17. $\frac{x}{y}$
18. $-1\frac{3}{7}$
19. $\frac{21}{5}$
20. $\frac{4}{5}$
21. 0.8
22. $-\frac{8}{15}$
23. $\frac{1}{m}$
24. $-6\frac{5}{11}$
25. $\frac{3}{11}$
26. $3\frac{1}{4}$
27. $-5\frac{3}{5}$
28. $\frac{24}{9}$
29. $-\frac{1}{b}$
30. $4\frac{5}{8}$

Lesson 7-6

State whether each sequence is *arithmetic, geometric,* or *neither.* Then write the next three terms of each sequence.

1. 1, 5, 9, 13, …
2. 2, 6, 18, 54, …
3. 1, 4, 9, 16, 25, …
4. 729, 243, 81, …
5. 2, -3, -8, -13, …
6. 5, -5, 5, -5, …
7. 810, -270, 90, -30, …
8. 11, 14, 17, 20, 23, …
9. 33, 27, 21, …
10. 21, 15, 9, 3, …
11. $\frac{1}{8}, -\frac{1}{4}, \frac{1}{2}, -1, \ldots$
12. $\frac{1}{81}, \frac{1}{27}, \frac{1}{9}, \frac{1}{3}, \ldots$
13. $\frac{3}{4}, 1\frac{1}{2}, 3, \ldots$
14. 2, 5, 9, 14, …
15. $-1\frac{1}{4}, -1\frac{3}{4}, -2\frac{1}{4}, -2\frac{3}{4}, \ldots$
16. 9.9, 13.7, 17.5, …
17. $\frac{1}{2}, 1\frac{1}{2}, 2\frac{1}{2}, 3\frac{1}{2}, \ldots$
18. 2, 12, 32, 62, …
19. 3, -6, 12, -24, …
20. 5, 7, 9, 11, 13, …
21. -0.06, 2.24, 4.54, …
22. 7, 14, 28, …
23. -5.4, -1.4, 2.6, …
24. -96, 48, -24, 12, …
25. 4, 12, 36, …
26. 20, 19, 18, 17, …
27. 768, 192, 48, …

Lesson 7-7

Find the area of each figure described below.

1. triangle: base, $2\frac{1}{2}$ in.; height, 7 in.
2. triangle: base, 12 cm; height, 3.2 cm
3. trapezoid: bases, 5 ft and 7 ft; height, 11 ft
4. trapezoid: bases, $4\frac{1}{4}$ yd and $3\frac{1}{2}$ yd; height, 5 yd

State the measures of the base(s) and the height of each triangle or trapezoid. Then find the area.

5.
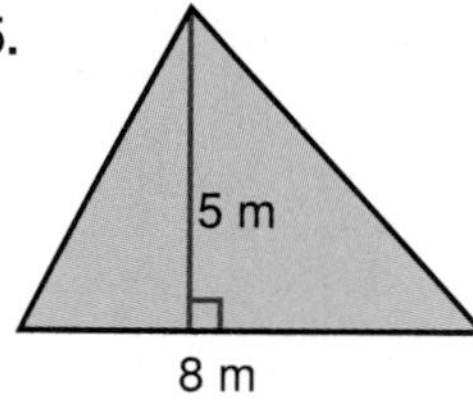
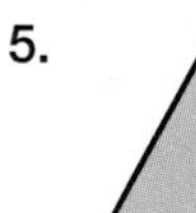

6.
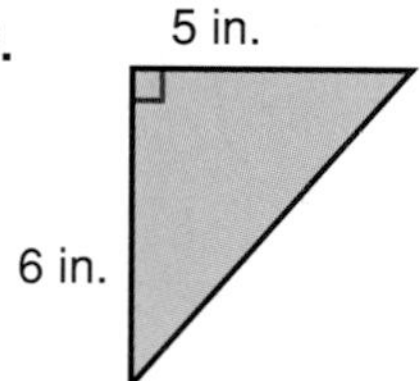

7.
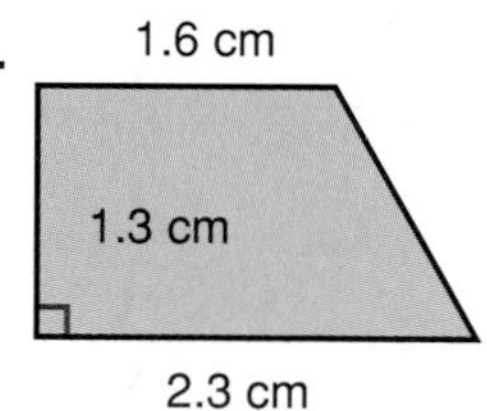

8.
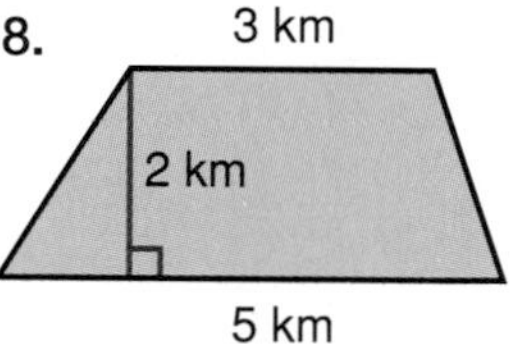

Lesson 7-8

Find the circumference of each circle described below. Round answers to the nearest tenth.

1. 14 mm, diameter
2. 18 cm, diameter
3. 24 in., radius
4. 42 m, diameter

5. 20 mm

6. 3.5 m

7.
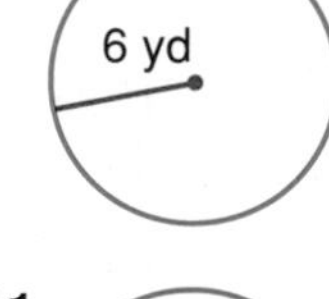

8.
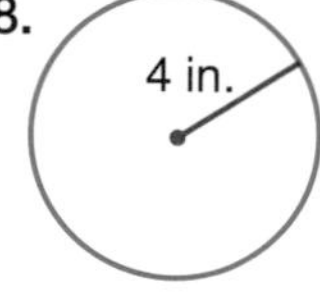

9.
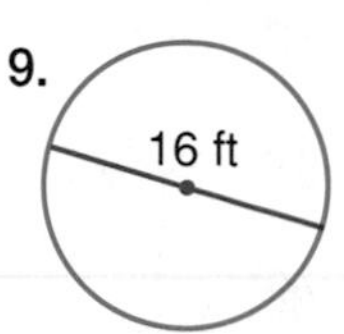

10.
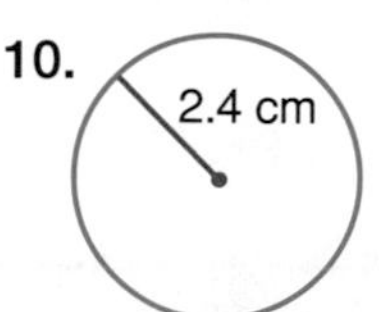

11.
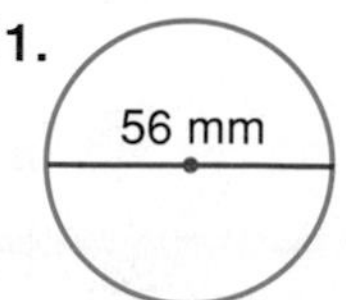

12. 35 in.

Lesson 7-9

Solve each equation. Write each solution in simplest form.

1. $x = \frac{2}{3} \div \frac{3}{4}$
2. $-\frac{4}{9} \div \frac{5}{6} = c$
3. $\frac{7}{12} \div \frac{3}{8} = q$
4. $m = \frac{5}{18} \div \frac{2}{9}$
5. $a = \frac{1}{3} \div 4$
6. $5\frac{1}{4} \div \left(-2\frac{1}{2}\right) = g$
7. $-6 \div \left(-\frac{4}{7}\right) = d$
8. $n = -6\frac{3}{8} \div \frac{1}{4}$
9. $p = \frac{6}{7} \div \frac{3}{5}$
10. $e = 3\frac{1}{3} \div (-4)$
11. $2\frac{5}{12} \div 7\frac{1}{3} = r$
12. $v = \frac{5}{6} \div 1\frac{1}{9}$
13. $\frac{3}{8} \div (-6) = b$
14. $i = \frac{5}{8} \div \frac{1}{6}$
15. $4\frac{1}{4} \div 6\frac{3}{4} = w$
16. $f = 4\frac{1}{6} \div 3\frac{1}{8}$
17. $t = 8 \div \left(-1\frac{4}{5}\right)$
18. $j = -5 \div \frac{2}{7}$
19. $\frac{3}{5} \div \frac{6}{7} = y$
20. $4\frac{8}{9} \div \left(-2\frac{2}{3}\right) = h$
21. $f = 8\frac{1}{6} \div 3$
22. $k = -\frac{3}{4} \div 9$
23. $s = 1\frac{11}{14} \div 2\frac{1}{2}$
24. $-2\frac{1}{4} \div \frac{4}{5} = z$

Lesson 7-10

Solve each equation. Check your solution.

1. $434 = -31y$
2. $6x = -4.2$
3. $\frac{3}{4}a = -12$
4. $-10 = \frac{b}{-7}$
5. $7.2 = \frac{3}{4}c$
6. $2r + 4 = 14$
7. $-2.4i = 7.2$
8. $7 = \frac{1}{2}d - 3$
9. $3.2n - 0.64 = -5.44$
10. $\frac{t}{3} - 7 = 2$
11. $\frac{3}{8} = \frac{1}{2}x$
12. $\frac{1}{2}h - 3 = -14$
13. $-0.46k - 1.18 = 1.58$
14. $4\frac{1}{2}s = -30$
15. $\frac{2}{3}f = \frac{8}{15}$
16. $\frac{2}{3}m + 10 = 22$
17. $\frac{2}{3}g + 4 = 4\frac{5}{6}$
18. $7 = \frac{1}{2}v + 3$
19. $\frac{g}{1.2} = -6$
20. $\frac{4}{7}z - 4\frac{5}{8} = 15\frac{3}{8}$
21. $-12 = \frac{1}{5}j$

Lesson 8-1

Find each square root.

1. $\sqrt{9}$
2. $\sqrt{0.16}$
3. $\sqrt{81}$
4. $\sqrt{0.04}$
5. $-\sqrt{625}$
6. $\sqrt{36}$
7. $-\sqrt{169}$
8. $\sqrt{144}$
9. $\sqrt{2.25}$
10. $\sqrt{961}$
11. $\sqrt{25}$
12. $\sqrt{225}$
13. $\sqrt{0.01}$
14. $-\sqrt{4}$
15. $-\sqrt{0.09}$
16. $\sqrt{529}$
17. $-\sqrt{484}$
18. $\sqrt{196}$
19. $\sqrt{0.49}$
20. $\sqrt{1.69}$
21. $\sqrt{729}$
22. $\sqrt{0.36}$
23. $\sqrt{289}$
24. $-\sqrt{16}$
25. $\sqrt{1,024}$
26. $\sqrt{\frac{289}{10,000}}$
27. $\sqrt{\frac{169}{121}}$
28. $-\sqrt{\frac{4}{9}}$
29. $-\sqrt{\frac{81}{64}}$
30. $\sqrt{\frac{25}{81}}$

Lesson 8-2 Estimate to the nearest whole number.

1. $\sqrt{229}$ 2. $\sqrt{63}$ 3. $\sqrt{290}$ 4. $\sqrt{27}$ 5. $\sqrt{1.30}$
6. $\sqrt{8.4}$ 7. $\sqrt{96}$ 8. $\sqrt{19}$ 9. $\sqrt{200}$ 10. $\sqrt{76}$
11. $\sqrt{17}$ 12. $\sqrt{34}$ 13. $\sqrt{137}$ 14. $\sqrt{540}$ 15. $\sqrt{165}$
16. $\sqrt{326}$ 17. $\sqrt{52}$ 18. $\sqrt{37}$ 19. $\sqrt{79}$ 20. $\sqrt{18.35}$
21. $\sqrt{71}$ 22. $\sqrt{117}$ 23. $\sqrt{410}$ 24. $\sqrt{25.70}$ 25. $\sqrt{333}$
26. $\sqrt{23}$ 27. $\sqrt{89}$ 28. $\sqrt{47}$ 29. $\sqrt{62}$ 30. $\sqrt{742}$

Lesson 8-3 Name the set or sets of numbers to which each real number belongs.

1. 6.5 2. $\sqrt{25}$ 3. $\sqrt{3}$ 4. -7.2 5. $-0.\overline{61}$

Find an approximation for each square root. Then graph the square root on the number line.

6. $-\sqrt{12}$ 7. $\sqrt{23}$ 8. $\sqrt{2}$ 9. $\sqrt{10}$ 10. $-\sqrt{30}$

Solve each equation. Round decimal answers to the nearest tenth.

11. $y^2 = 49$ 12. $x^2 = 225$ 13. $x^2 = 64$ 14. $y^2 = 79$
15. $x^2 = 16$ 16. $y^2 = 24$ 17. $y^2 = 625$ 18. $x^2 = 81$

Lesson 8-5 Find the missing measure for each right triangle. Round decimal answers to the nearest tenth.

1. *a*, 6 cm; *b*, 5 cm 2. *a*, 12 ft; *b*, 12 ft 3. *a*, 8 in.; *b*, 6 in. 4. *a*, 20 m; *c*, 25 m
5. *a*, 9 mm; *c*, 14 mm 6. *b*, 15 m; *c*, 20 m 7. *a*, 5 ft; *b*, 50 ft 8. *a*, 4.5 yd; *c*, 8.5 yd

Write an equation to solve for *x*. Then solve. Round decimal answers to the nearest tenth.

9.
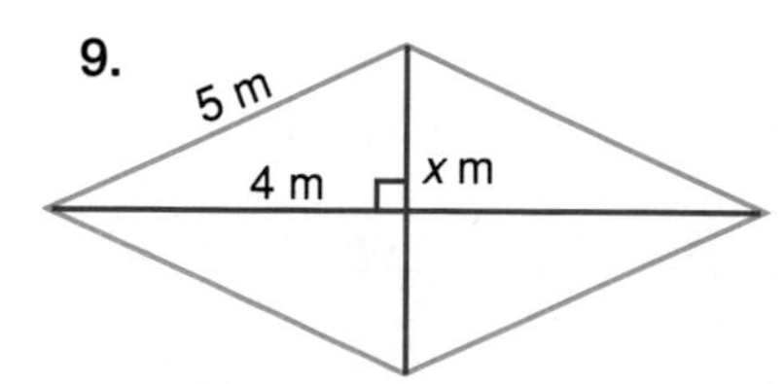

10.

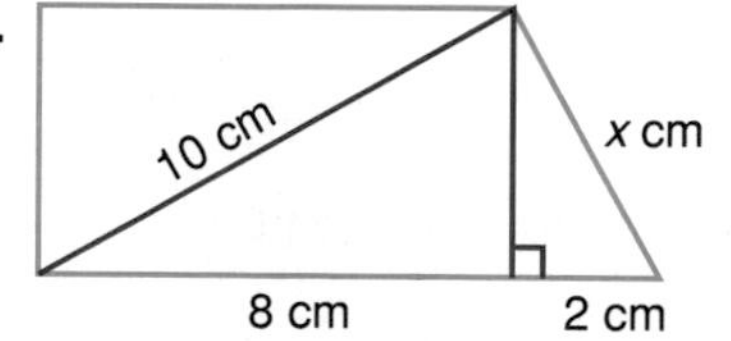

11.

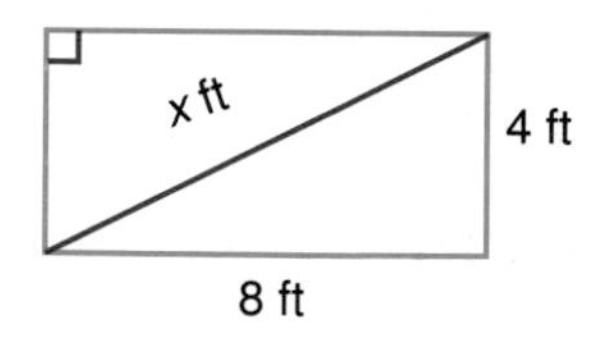

Determine whether each triangle with sides of a given length is a right triangle.

12. 15 m, 8 m, 17 m 13. 7 yd, 5 yd, 9 yd 14. 5 in., 12 in., 13 in.

Lesson 8-7 Find the distance between each pair of points whose coordinates are given. Round answers to the nearest tenth.

1. (1, 2); (-3, -3)

2. (-1, 4); (4, 1)

3. (0, 4); (7, 1)

Graph each pair of ordered pairs. Then find the distance between the points. Round answers to the nearest tenth.

4. (-4, 2); (4, 17)
5. (5, -1); (11, 7)
6. (-3, 5); (2, 7)
7. (7, -9); (4, 3)
8. (5, 4); (-3, 8)
9. (-8, -4); (-3, 8)
10. (2, 7); (10, -4)
11. (9, -2); (3, 6)

Lesson 8-8 Find the lengths of the missing sides. Round answers to the nearest tenth.

1.
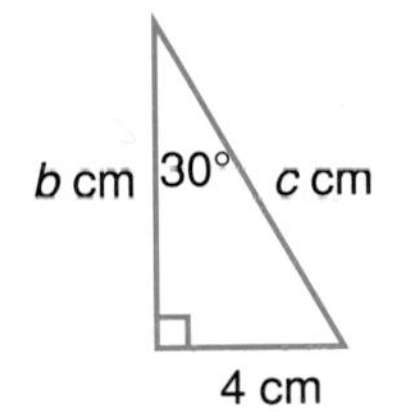

2.
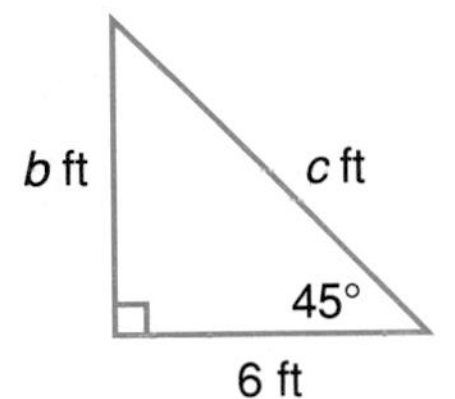

3.
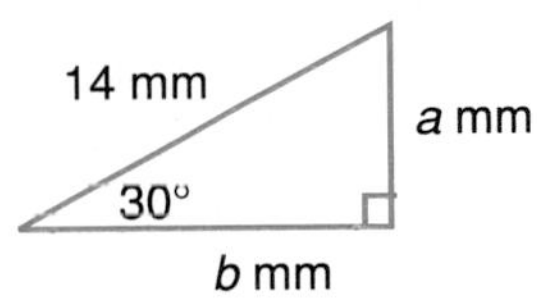

4.
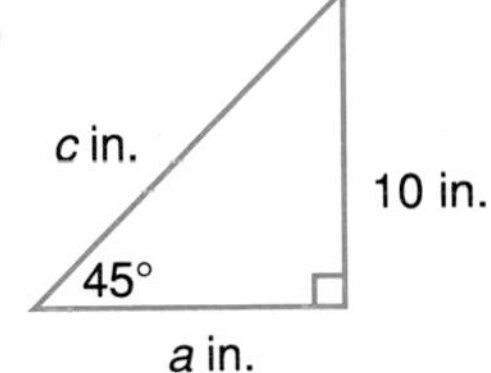

5.
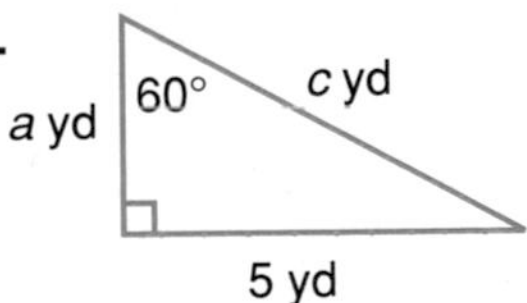

6.
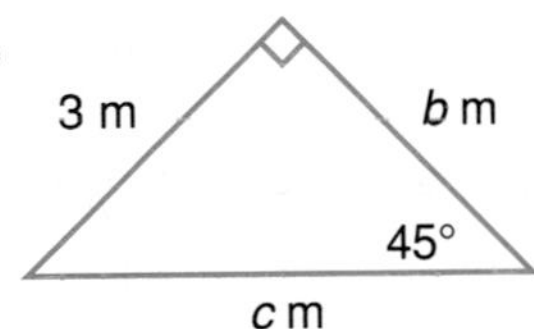

7.
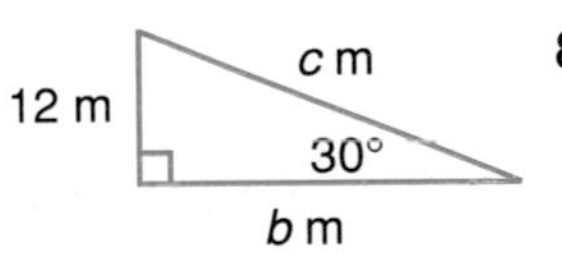

8.
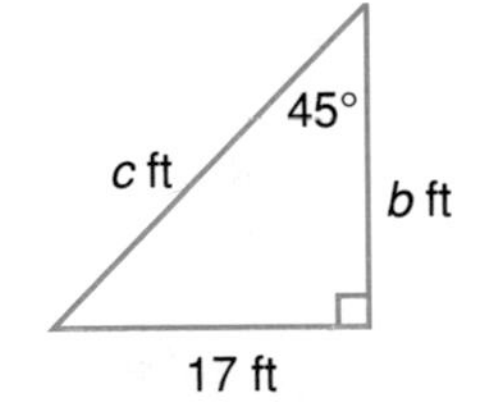

Lesson 9-1 Express each ratio or rate as a fraction in simplest form.

1. 27 to 9
2. 4 inches per foot
3. 16 out of 48
4. 10 : 50
5. 40 min. per hour
6. 35 is to 15
7. 16 wins, 16 losses
8. 7 out of 13
9. 5 out of 50

Express each as a unit rate.

10. $24 per dozen
11. 600 students to 30 teachers
12. 6 pounds gained in 12 weeks
13. $800 for 40 tickets
14. $6.50 for 5 pounds
15. 6 inches of rain in 3 weeks

Lesson 9-2

Tell whether each pair of ratios form a proportion.

1. $\frac{3}{5}, \frac{5}{10}$
2. $\frac{8}{4}, \frac{6}{3}$
3. $\frac{10}{15}, \frac{5}{3}$
4. $\frac{2}{8}, \frac{1}{4}$
5. $\frac{6}{18}, \frac{3}{9}$
6. $\frac{14}{21}, \frac{12}{18}$
7. $\frac{4}{20}, \frac{5}{25}$
8. $\frac{9}{27}, \frac{1}{3}$

Solve each proportion.

9. $\frac{2}{3} = \frac{a}{12}$
10. $\frac{7}{8} = \frac{c}{16}$
11. $\frac{3}{7} = \frac{21}{d}$
12. $\frac{2}{5} = \frac{18}{x}$
13. $\frac{3}{5} = \frac{n}{21}$
14. $\frac{5}{12} = \frac{b}{5}$
15. $\frac{4}{36} = \frac{2}{y}$
16. $\frac{3}{10} = \frac{z}{36}$
17. $\frac{2}{3} = \frac{t}{4}$
18. $\frac{9}{10} = \frac{r}{25}$
19. $\frac{16}{8} = \frac{y}{12}$
20. $\frac{7}{8} = \frac{a}{12}$

Lesson 9-3

Use a proportion to solve each problem.

1. On a radar screen the distance between two planes is $3\frac{1}{2}$ inches. If the scale is 1 inch on the screen to 2 miles in the air, what is the actual distance between the two planes.
2. A car travels 144 miles on 4 gallons of gasoline. At this rate, how many gallons are needed to drive 450 miles?
3. A park ranger stocks a pond with 4 sunfish for every three perch. Suppose 296 sunfish are put in the pond. How many perch should be stocked?
4. A furniture store bought 8 identical sofas for $4,000. How much did each sofa cost?

Lesson 9-5

Tell whether each pair of polygons are similar.

1.

2 cm
4 cm

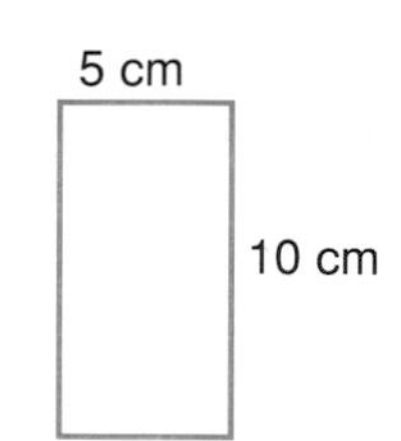

2.

5 m
4.6 m
2 m

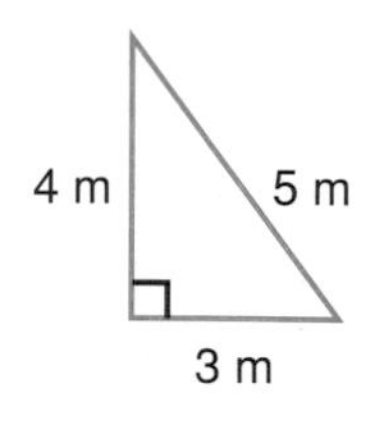

3.

$4\frac{1}{2}$ in.
6 in.
2 in.
$1\frac{1}{2}$ in.

4.

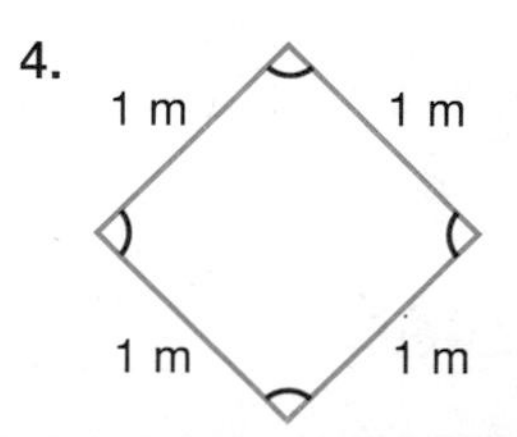

0.5 m
0.5 m
0.5 m
0.5 m

Lesson 9-6

Write a proportion to find the value of x. Assume the triangles are similar.

1.

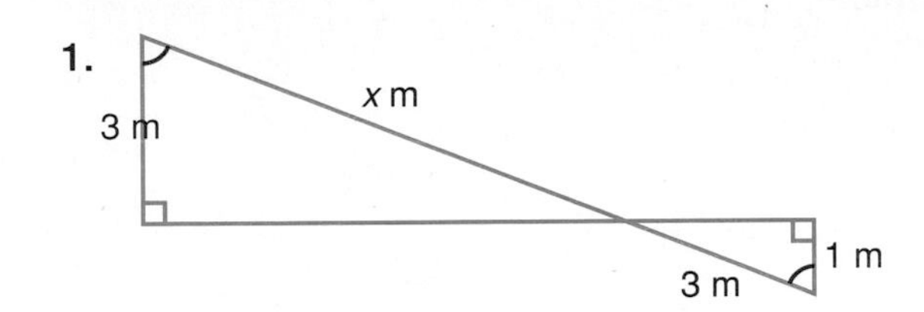

2.

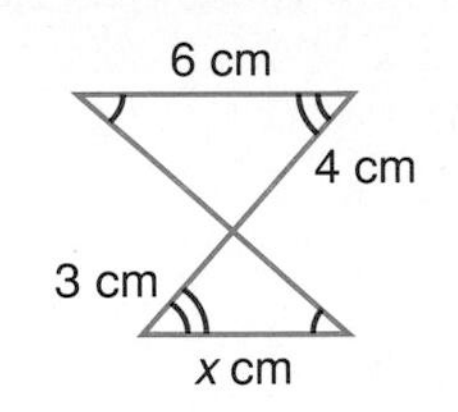

3.

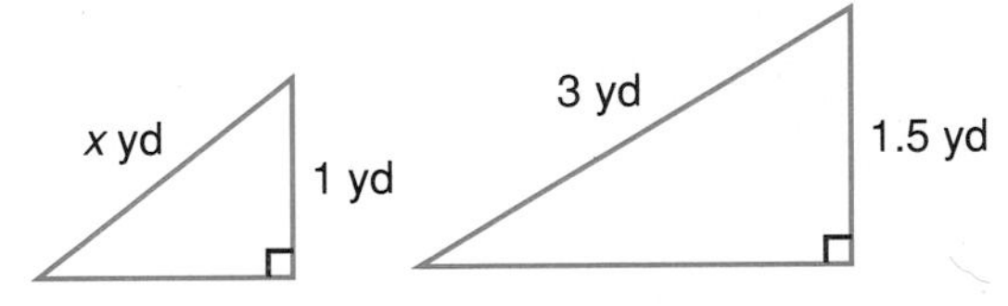

4.

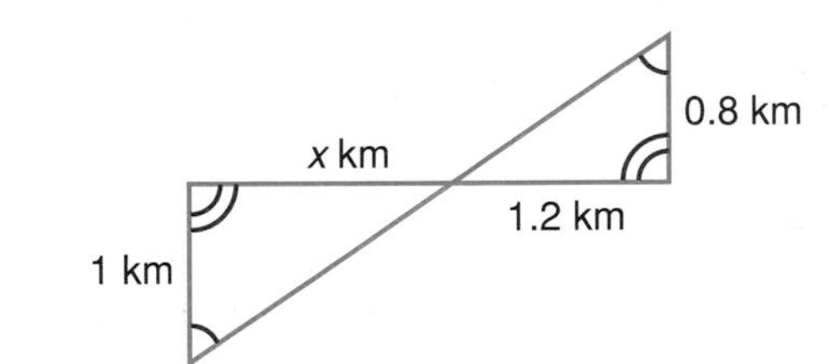

5.

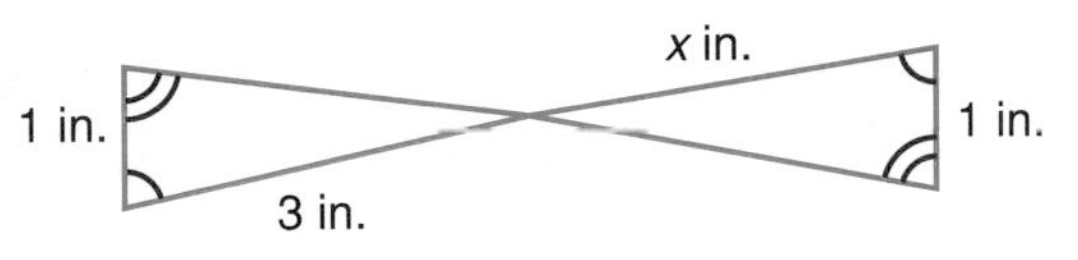

6.

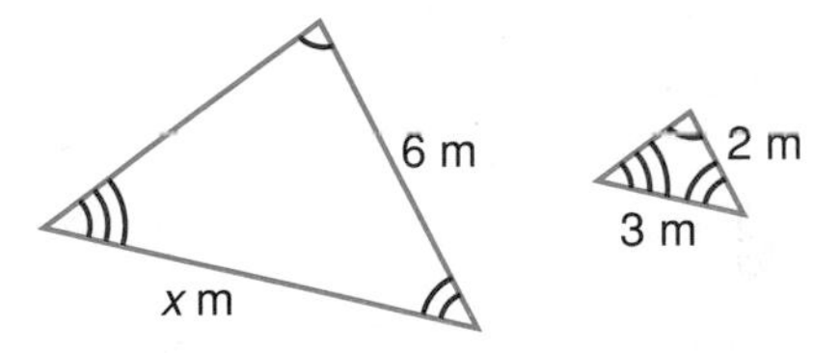

Lesson 9-7

The distance on a map is given. Find the actual distance, if the scale on the map is 1 cm : 50 miles.

1. 2 cm **2.** 0.5 cm **3.** 1 cm **4.** 5 cm

5. 1.5 cm **6.** 2.8 cm **7.** 3.2 cm **8.** 10 cm

9. 0.2 cm **10.** 4.5 cm **11.** 3 cm **12.** 6.4 cm

13. 7 cm **14.** 0.6 cm **15.** 8 cm **16.** 45 cm

Lesson 9-8

1. Graph segment PQ with $P(4, 4)$ and $Q(2, 0)$. Then graph its image for a dilation with a scale factor of 4.
2. Graph segment AB with $A(3, 6)$ and $B(0, -1)$. Then graph its image for a dilation with a scale factor of $\frac{1}{2}$.
3. Graph segment XY with $X(-2, -4)$ and $Y(1, 3)$. Then graph its image for a dilation with a scale factor of 3.

Triangle ABC has vertices $A(2, 2)$, $B(-1, 4)$, and $C(-3, -5)$. Find the coordinates of its image for a dilation with each given scale factor. Graph the original triangle and its dilation.

4. 1 **5.** 0.5 **6.** 2 **7.** 3 **8.** $\frac{1}{4}$

Lesson 9-9

Use the figures below. Write the ratios in simplest form. Find angle measures to the nearest degree.

1. Find $\tan X$.
2. Find $\tan Y$.
3. Find $\tan A$.
4. Find $\tan B$.
5. Find $m\angle X$.
6. Find $m\angle Y$.
7. Find $m\angle A$.
8. Find $m\angle B$.

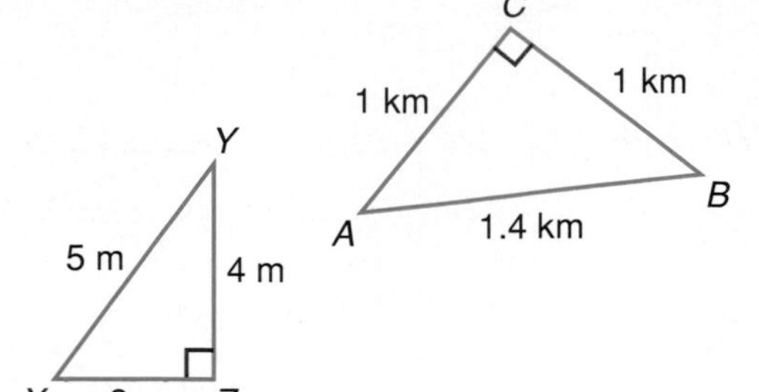

Lesson 9-10

Use the figures below. Write the ratios in simplest form. Find angle measures to the nearest degree.

1. Find $\cos A$.
2. Find $\sin A$.
3. Find $m\angle A$.
4. Find $\sin B$.
5. Find $\cos B$.
6. Find $m\angle B$.
7. Find $\cos Y$.
8. Find $\sin X$.
9. Find $m\angle X$.
10. Find $m\angle Y$.

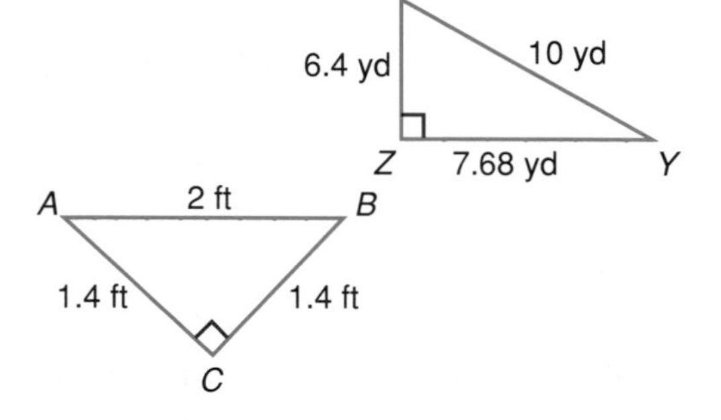

Lesson 10-1

Express each fraction as a percent.

1. $\frac{2}{100}$	2. $\frac{3}{25}$	3. $\frac{20}{25}$	4. $\frac{10}{16}$	5. $\frac{4}{6}$	6. $\frac{1}{4}$
7. $\frac{26}{100}$	8. $\frac{3}{10}$	9. $\frac{21}{50}$	10. $\frac{7}{8}$	11. $\frac{1}{3}$	12. $\frac{2}{3}$
13. $\frac{2}{5}$	14. $\frac{2}{50}$	15. $\frac{8}{10}$	16. $\frac{5}{12}$	17. $\frac{7}{10}$	18. $\frac{9}{20}$
19. $\frac{1}{2}$	20. $\frac{3}{20}$	21. $\frac{10}{25}$	22. $\frac{3}{8}$	23. $\frac{4}{20}$	24. $\frac{19}{25}$

Write a percent proportion to solve each problem. Round answers to the nearest tenth.

25. 39 is 5% of what number?
26. What is 19% of 200?
27. 28 is what percent of 7?
28. 24 is what percent of 72?
29. 9 is $33\frac{1}{3}\%$ of what number?
30. Find 55% of 134.

Lesson 10-2 Express each decimal as a percent.

1. 0.35 2. 14.23 3. 0.9 4. 0.13 5. 6.21
6. 0.23 7. 0.08 8. 0.036 9. 2.34 10. 0.39

Express each percent as a fraction in simplest form.

11. 40% 12. 24.5% 13. 42% 14. $33\frac{1}{3}\%$ 15. 81%
16. 8% 17. 55% 18. 4.5% 19. 16.5% 20. 2%

Express each percent as a decimal.

21. 2% 22. 25% 23. 29% 24. 6.2% 25. 16.8%
26. 14% 27. 23.7% 28. 42% 29. 25.4% 30. 98%

Lesson 10-3 Express each percent as a fraction or mixed number in simplest form.

1. 540% 2. $\frac{25}{50}\%$ 3. 0.02% 4. 620% 5. 0.7%
6. 111.5% 7. $\frac{7}{35}\%$ 8. 0.72% 9. 0.004% 10. 364%
11. 0.15% 12. 1,250% 13. $\frac{9}{10}\%$ 14. 730% 15. 100.01%

Express each percent as a decimal.

16. 0.07% 17. $5\frac{2}{3}\%$ 18. 310% 19. 6.05% 20. 7,652%
21. $\frac{12}{50}\%$ 22. 0.93% 23. 200% 24. 197.6% 25. 10.75%
26. 0.66% 27. 417% 28. 7.76% 29. 390% 30. $10\frac{7}{10}\%$

Lesson 10-5 Estimate.

1. 33% of 12 2. 24% of 84 3. 39% of 50 4. 1.5% of 135

Estimate the percent of each area shaded.

5. 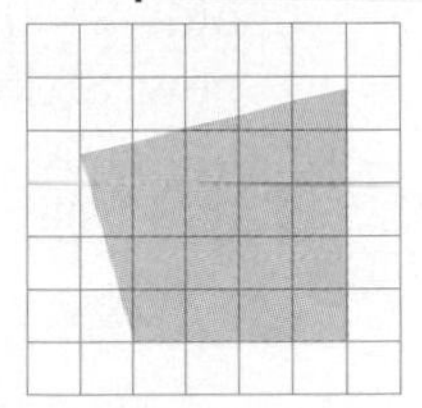

6. 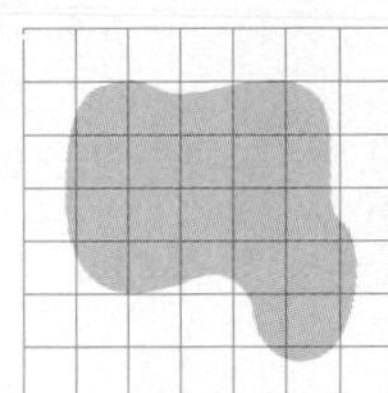

7. 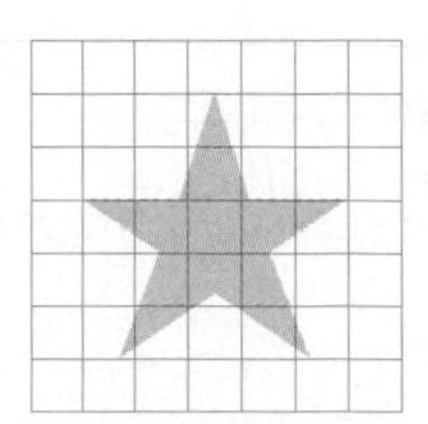

Estimate the percent.

8. 11 out of 99 9. 28 out of 89 10. 9 out of 20 11. 25 out of 270
12. 6 out of 25 13. 17 out of 65 14. 72 out of 280 15. 120 out of 181

Lesson 10-6 Solve.

1. Find 5% of $73.
2. What is 15% of 15?
3. Find 80% of $12.
4. What is 7.3% of 500?
5. Find 21% of $720.
6. What is 12% of $62.50?
7. Find 0.3% of 155.
8. What is 75% of $450
9. Find 7.2% of 10.
10. What is 10.1% of $60?
11. Find 23% of 47.
12. What is 89% of 654?
13. $20 is what percent of $64?
14. Sixty-nine is what percent of 200?
15. Seventy is what percent of 150?
16. 26 is 30% of what number?
17. 7 is 14% of what number?
18. $35.50 is what percent of $150?
19. $17 is what percent of $25?
20. 152 is 2% of what number?

Lesson 10-7 Make a circle graph for each set of data.

1.

Sporting Goods Sales	
Shoes	44%
Apparel	30%
Equipment	26%

2.

Energy Use in Home	
Heating/cooling	51%
Appliances	28%
Lights	21%

3.

Household Income	
Primary job	82%
Secondary job	9%
Investments	5%
Other	4%

4.

Students in North High School	
White	30%
Black	28%
Hispanic	24%
Asian	18%

Lesson 10-8 Estimate the percent of change.

1. old: $97
new: $149
2. old: $54
new: $64
3. old: $0.39
new: $1.20
4. old: $450
new: $325
5. old: $0.48
new: $1.02
6. old: $2.39
new: $2.59
7. old: $46.50
new: $37.99
8. old: 50¢
new: 95¢

Find the percent of change. Round to the nearest whole percent.

9. old: $35
new: $29
10. old: $550
new: $425
11. old: $72
new: $88
12. old: $25
new: $35
13. old: $28
new: $19
14. old: $46
new: $55
15. old: $78
new: $44
16. old: $120
new: $75

Lesson 10-9 Find the amount of discount and the sale price of each item.

1. $4,220 piano, 35% off
2. $14 scissors, 10% off
3. $29 book, 40% off
4. $38 sweater, 25% off
5. $45 pants, 50% off
6. $280 VCR, 25% off
7. $3,540 motorcycle, 30% off
8. $15.95 compact disc, 20% off

Find the percent of discount.

9. regular price: $250
   discount: $35
10. regular price: $15.50
    discount: $2.48
11. regular price: $27.50
    discount: $11

Lesson 10-10 Find the simple interest to the nearest cent.

1. $500 at 7% for 2 years
2. $2,500 at 6.5% for 36 months
3. $8,000 at 6% for 1 year
4. $1,890 at 9% for 42 months

Find the total amount in each account.

5. $300 at 10% after 3 years
6. $3,200 at 8% after 6 months
7. $20,000 at 14% after 20 years
8. $4,000 at 12.5% after 4 years

Find the annual rate of simple interest.

9. principal: $4,500; interest: $526.50; time: 3 years
10. principal: $7,400; interest: $878.75; time: 30 months

Lesson 11-1 Complete each function table.

1. $f(n) = -4n$

n	$-4n$	$f(n)$
−2		
−1		
0		
1		
2		

2. $f(n) = n + 6$

n	$n + 6$	$f(n)$
−6		
−4		
−2		
0		
2		

3. $f(n) = 3n + 2$

n	$3n + 2$	$f(n)$
−3.5		
−2.5		
−1.5		
0		
1.5		

4. $f(n) = 2n - 6$

n	$2n - 6$	$f(n)$
$-2\frac{1}{2}$		
−1		
$-\frac{1}{4}$		
0		
$\frac{1}{2}$		

5. $f(n) = -\frac{1}{2}n + 4$

n	$-\frac{1}{2}n + 4$	$f(n)$
−4		
−2		
0		
2.5		
6		

6. $f(n) = -5n + 1$

n	$-5n + 1$	$f(n)$
−4		
−2		
0		
1		
4		

Lesson 11-2

Write the ordered pairs for each function. Then graph the function.

1. $f(n) = 6n + 2$

n	$f(n)$	$(n, f(n))$
−3		
−1		
1		
$\frac{7}{3}$		

2. $f(n) = -2n + 3$

n	$f(n)$	$(n, f(n))$
−2		
−1		
0		
1		
2		

3. $f(n) = 4.5n$

n	$f(n)$	$(n, f(n))$
−4		
−2		
0		
1		
6		

Make a function table for each function. Then graph each function.

4. $f(n) = \frac{8}{n}$
5. $f(n) = \frac{2}{3}n + 1$
6. $f(n) = n^2 - 1$
7. $f(n) = 3.5n$
8. $f(n) = 4n - 1$
9. $f(n) = \frac{3}{5}n + \left(-\frac{1}{5}\right)$

Lesson 11-3

Copy and complete the table for each equation.

1. $y = 3x - 1$

x	y
−5	
−3	
−1	
0	
1	

2. $y = \frac{x}{4} + 2$

x	y
−8	
−4	
0	
4	

3. $y = -1.5x - 3$

x	y
−4	
−2	
2	
6	
10	

4. $y = 4x - 3$

x	y
−2	
$\frac{1}{2}$	
0	
$2\frac{1}{4}$	

Find four solutions for each equation. Write the solution set.

5. $y = -3x + 5$
6. $y = 2x - 1$
7. $y = \frac{2}{3}x + 4$
8. $y = -0.4x$
9. $y = 12x - 8$
10. $y = \frac{3}{4}x + 2$
11. $y = -2.4x - 3$
12. $y = 5x + 7$

Lesson 11-4

Graph each function.

1. $y = -5x$
2. $y = 10x - 2$
3. $y = -2.5x - 1.5$
4. $y = 7x + 3$
5. $y = \frac{x}{4} - 8$
6. $y = 3x + 1$
7. $y = 25 - 2x$
8. $y = \frac{x}{6}$
9. $y = -2x + 11$
10. $y = 7x - 3$
11. $y = \frac{x}{2} + 5$
12. $y = 4 - 6x$
13. $y = -3.5x - 1$
14. $y = 4x + 10$
15. $y = 8x$
16. $y = -5x + \frac{1}{2}$
17. $y = \frac{x}{3} + 9$
18. $y = -7x + 15$
19. $y = 10x - 2$
20. $y = 1.5x - 7.5$

Lesson 11-5 Solve each system of equations by graphing.

1. $y = x$, $y = -x + 4$
2. $y = -x + 8$, $y = x - 2$
3. $y = -3x$, $y = -4x + 2$
4. $y = x - 1$, $y = -x + 11$
5. $y = -x$, $y = 2x$
6. $y = -x + 3$, $y = x + 3$
7. $y = x - 3$, $y = 2x + 8$
8. $y = -x + 6$, $y = x + 2$
9. $y = -x + 1$, $y = x - 4$
10. $y = -3x + 6$, $y = x - 2$
11. $y = 3x - 4$, $y = -3x - 4$
12. $y = 2x + 4$, $y = 3x - 9$
13. $y = -x + 4$, $y = x - 10$
14. $y = -x + 6$, $y = 2x$
15. $y = x - 4$, $y = -2x + 5$
16. $y = 2x$, $y = -x + 3$

Lesson 11-7 Graph each quadratic function.

1. $y = x^2 - 1$
2. $y = 1.5x^2 + 3$
3. $f(n) = n^2 - n$
4. $y = 2x^2$
5. $y = x^2 + 3$
6. $y = -3x^2 + 4$
7. $y = -x^2 + 7$
8. $f(n) = 3n^2$
9. $f(n) = 3n^2 + 9n$
10. $y = -x^2$
11. $y = \frac{1}{2}x^2 + 1$
12. $y = 5x^2 - 4$
13. $y = -x^2 + 3x$
14. $f(n) = 2.5n^2$
15. $y = -2x^2$
16. $y = 8x^2 + 3$
17. $y = -x^2 + \frac{1}{2}x$
18. $y = -4x^2 + 4$
19. $f(n) = 4n^2 + 3$
20. $y = -4x^2 + 1$
21. $y = 2x^2 + 1$
22. $y = x^2 - 4x$
23. $y = 3x^2 + 5$
24. $f(n) = 0.5n^2$
25. $f(n) = 2n^2 - 5n$
26. $y = \frac{3}{2}x^2 - 2$
27. $y = 6x^2 + 2$
28. $f(n) = 5n^2 + 6n$

Lesson 11-8 Find the coordinates of the vertices of each figure after the translation described. Then graph the figure and its translation.

1. $\triangle ABC$ with vertices $A(-6, -2)$, $B(-1, 1)$, and $C(2, -2)$, translated by $(4, 3)$
2. $\triangle XYZ$ with vertices $X(-4, 3)$, $Y(0, 3)$, and $Z(-2, -1)$, translated by $(5, -2)$
3. rectangle $HIJK$ with vertices $H(1, 3)$, $I(4, 0)$, $J(2, -2)$, and $K(-1, 1)$, translated by $(-4, -6)$
4. rectangle $PQRS$ with vertices $P(-7, 6)$, $Q(-5, 6)$, $R(-5, 2)$, and $S(-7, 2)$, translated by $(9, -1)$
5. pentagon $DGLMR$ with vertices $D(1, 3)$, $G(2, 4)$, $L(4, 4)$, $M(5, 3)$, and $R(3, 1)$, translated by $(-5, -7)$

Lesson 11-9

Graph $\triangle CAT$ with vertices $C(2, 3)$, $A(8, 2)$, and $T(4, -3)$.

1. Find the coordinates of the vertices after a reflection over the y-axis.
2. Graph $\triangle C'A'T'$.

Graph trapezoid $TRAP$ with vertices $T(-2, 5)$, $R(1, 5)$, $A(4, 2)$, and $P(-5, 2)$.

3. Find the coordinates of the vertices after a reflection over the x-axis.
4. Graph trapezoid $T'R'A'P'$

Graph rectangle $ABCD$ with vertices $A(4, -1)$, $B(7, -4)$, $C(4, -7)$, and $D(1, -4)$.

5. Find the coordinates of the vertices after a reflection over the y-axis.
6. Graph rectangle $A'B'C'D'$.

Lesson 11-10

Triangle ABC has vertices $A(-2, -1)$, $B(0, 1)$, and $C(1, -1)$.

1. Graph $\triangle ABC$.
2. Find the coordinates of the vertices after a 90° counterclockwise rotation.
3. Graph $\triangle A'B'C'$.

Rectangle $WXYZ$ has vertices $W(1, 1)$, $X(1, 3)$, $Y(6, 3)$, and $Z(6, 1)$.

4. Graph rectangle $WXYZ$.
5. Find the coordinates of the vertices after a rotation of 180°.
6. Graph rectangle $W'X'Y'Z'$.

Lesson 12-1

Find the area of each circle whose radius or diameter is given. Round answers to the nearest tenth.

1. radius, 4 m
2. diameter, 6 in.
3. radius, 12 in.
4. diameter, 16 yd
5. diameter, 11 ft
6. radius, 5 in.
7. radius, 19 cm
8. diameter, 29 mm

9. 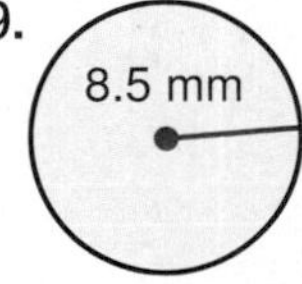

10. 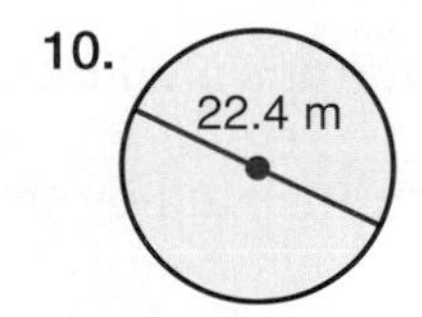

11. 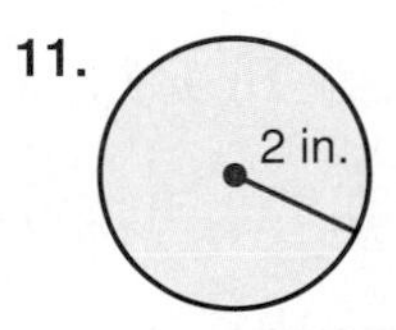

12. 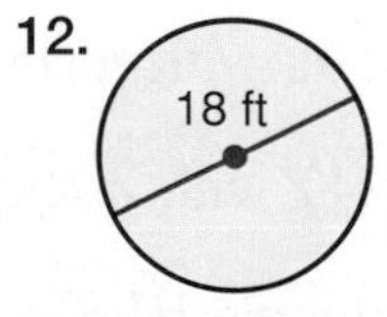

13. 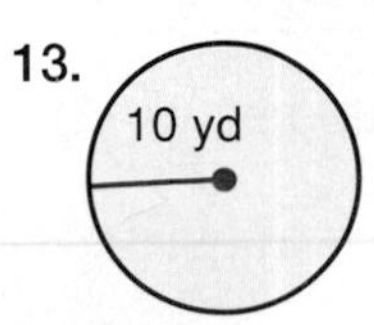

14. 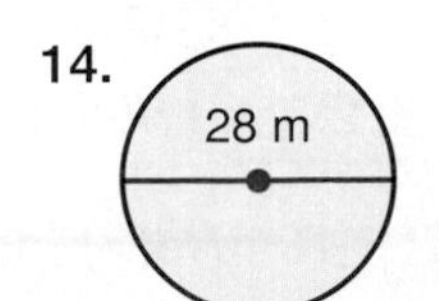

15. 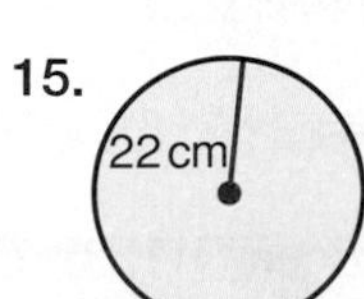

16. 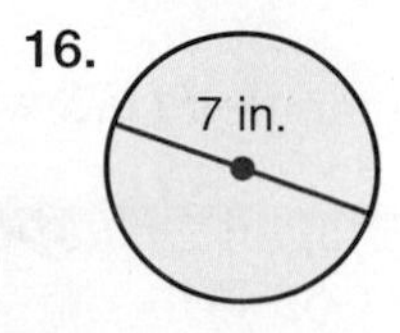

Lesson 12-4 Find the surface area of each prism.

1. length = 2 in.
   width = 1 in.
   height = 10 in.

2. length = 18 m
   width = 7 m
   height = 14 m

3. length = 2.5 cm
   width = 1 cm
   height = 4.5 cm

4. length = 6 m
   width = 4 m
   height = 10 m

5. length = 14 ft
   width = 7 ft
   height = 14 ft

6. length = 10 cm
   width = 10 cm
   height = 10 cm

7. length = 4.5 yd
   width = 3.6 yd
   height = 10.6 yd

8. length = 18 in.
   width = 12 in.
   height = 11 in.

9. length = 12.6 mm
   width = 6.8 mm
   height = 10.4 mm

Lesson 12-5 Find the surface area of each cylinder. Round answers to the nearest tenth.

1. 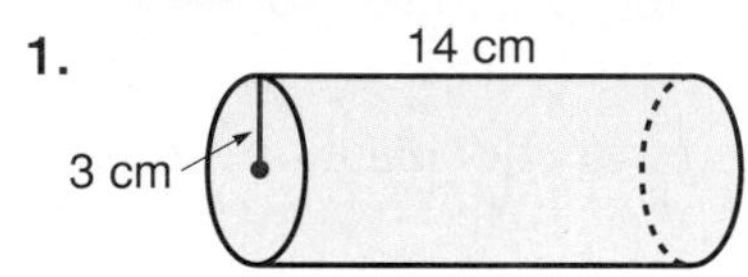

2. 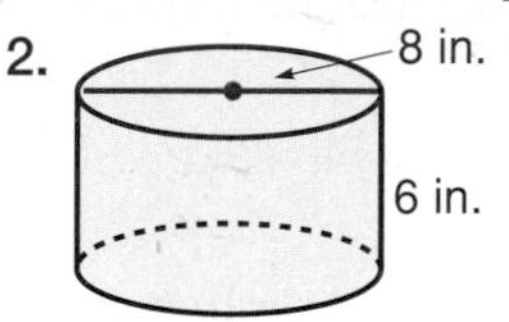

3. 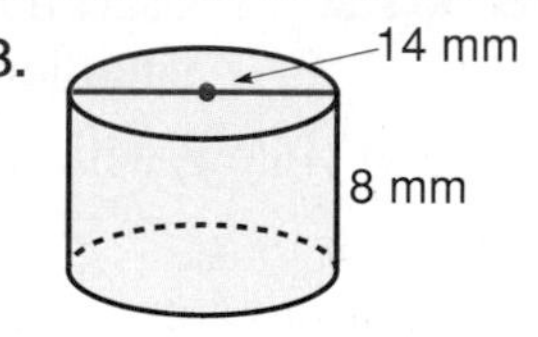

4. 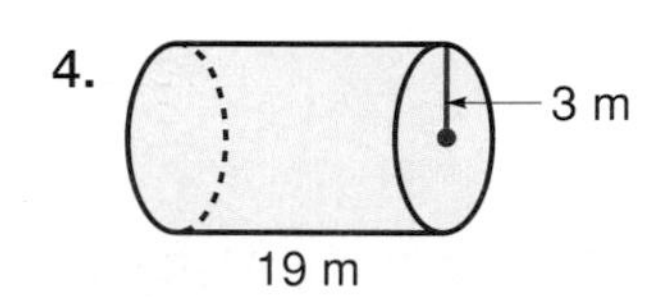

5. 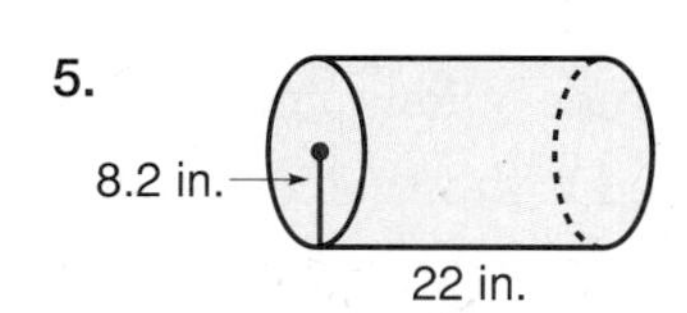

6. 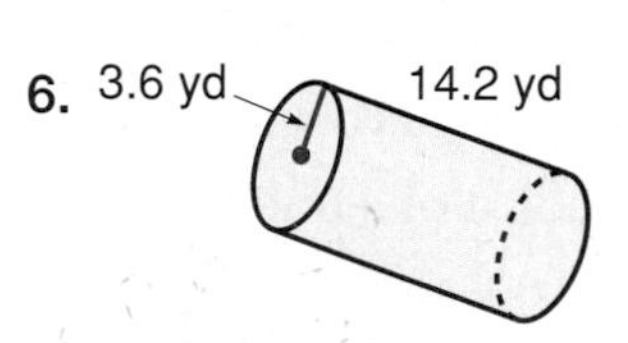

7. radius = 4.2 cm
   height = 12.4 cm

8. radius = 5 in.
   height = 10 in.

9. radius = 6.3 ft
   height = 4.6 ft

Lesson 12-6 Find the volume of each solid. Round answers to the nearest tenth.

1. 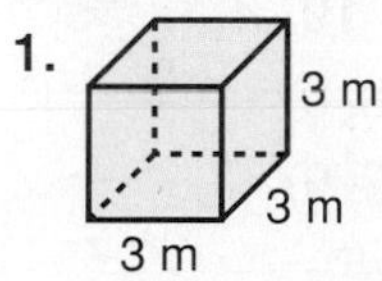

2. 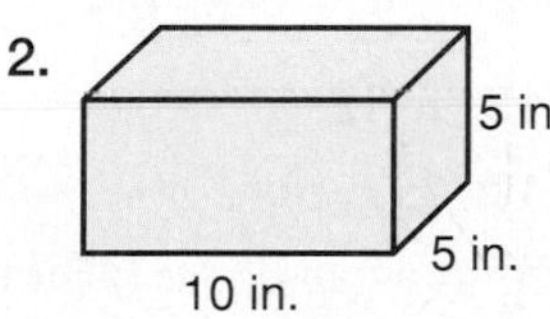

3. 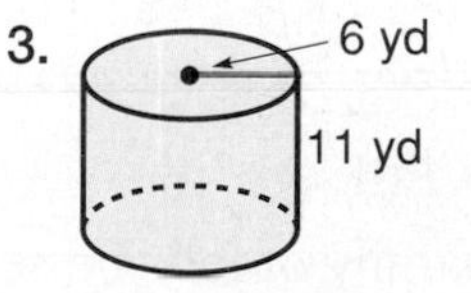

4. 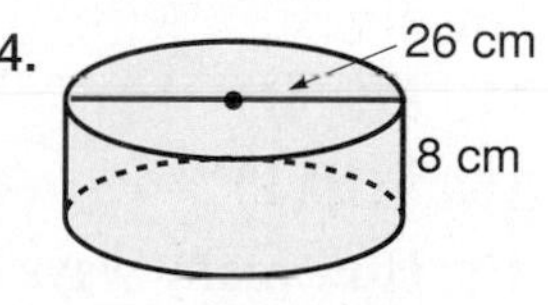

5. 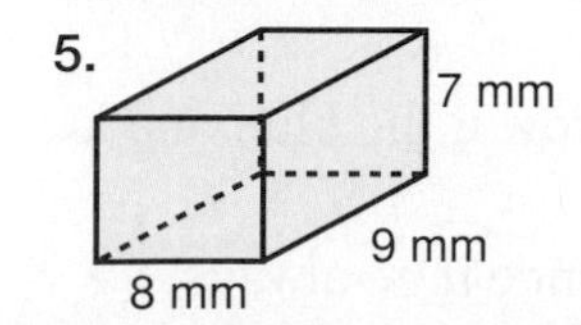

6. 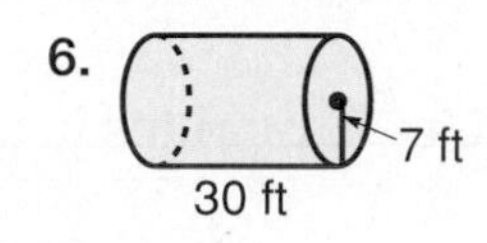

7. 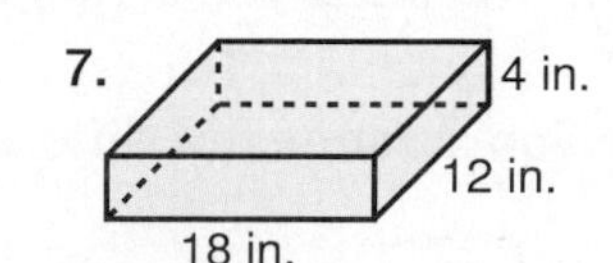

8. 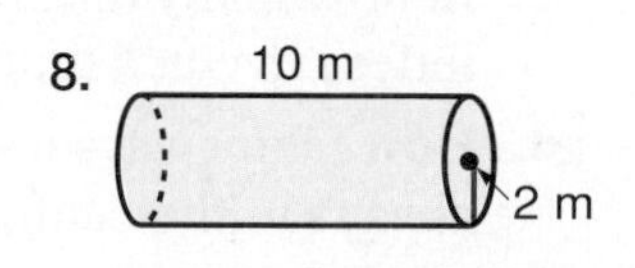

Lesson 12-7 Find the volume of each solid. Round answers to the nearest tenth.

1. 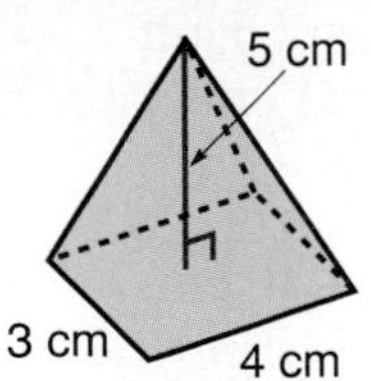

2. 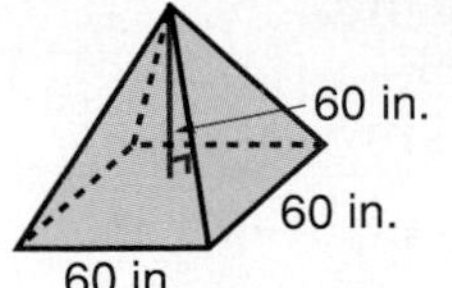

3. 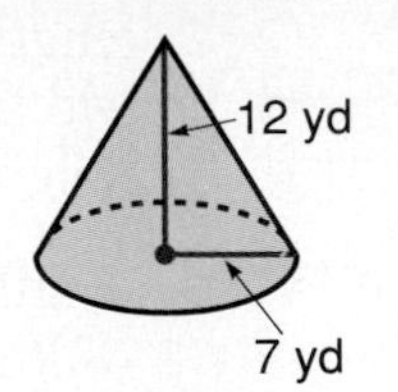

4. 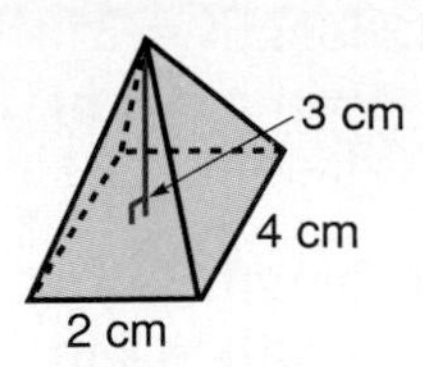

5. 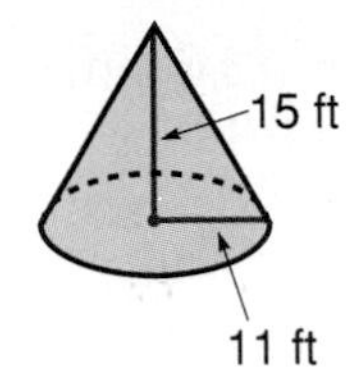

6. 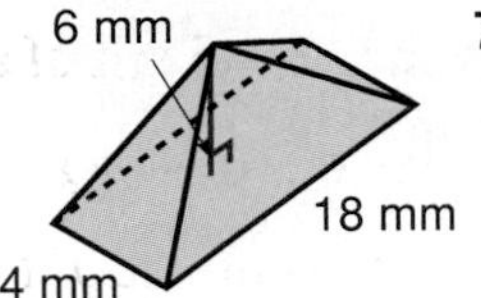

7. 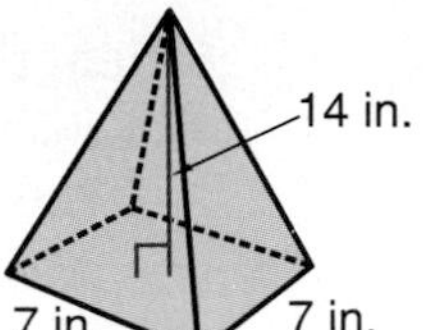

8. 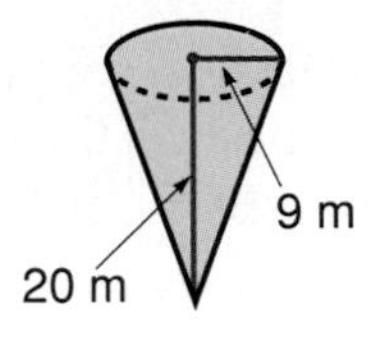

Lesson 13-1 Draw a tree diagram to find the number of possible outcomes for each situation.

1. A choice of yellow, white, chocolate, or marble cake with a choice of chocolate or vanilla frosting.
2. A particular car comes in white, black, or red with standard or automatic transmission and with a 4-cylinder or 6-cylinder engine.
3. A choice of roses or carnations in red, yellow, pink, or white.
4. A choice of a queen or king size bed with a firm or super firm mattress.
5. A pizza can be ordered with a regular or deep dish crust and with a choice of one topping, two toppings, or three toppings.
6. A choice of red, white, blue, or black women's shoes with a choice of high, medium, or low heels.

Lesson 13-2 Find each value.

1. 8!
2. 10!
3. 0!
4. 7!
5. 6!
6. 5!
7. 2!
8. 11!
9. 9!
10. 4!
11. $P(5, 4)$
12. $P(3, 3)$
13. $P(12, 5)$
14. $P(8, 6)$
15. $P(10, 2)$
16. $P(6, 4)$
17. $P(7, 6)$
18. $P(9, 9)$
19. How many ways can a family of four be seated in a car if the father is driving?
20. In how many different ways can you arrange the letters in the word *orange* if you take the letters five at a time?
21. How many ways can you arrange five different colored marbles in a row if the blue one is always in the center?
22. In how many different ways can Kevin listen to each of his ten CDs once if he always saves a certain CD for last?

Lesson 13-3 Find each value.

1. $C(8, 4)$
2. $C(30, 8)$
3. $C(10, 9)$
4. $C(7, 3)$
5. $C(12, 5)$
6. $C(17, 16)$
7. $C(24, 17)$
8. $C(9, 7)$
9. How many ways can you choose five compact discs from a collection of 17?
10. How many combinations of three flavors of ice cream can you choose from 25 different flavors of ice cream?
11. How many ways can you choose three books to read out of a selection of ten books?
12. How many ways can you choose seven apples out of a bag of two dozen apples?
13. How many ways can you choose two movies to rent out of ten possible movies?

Lesson 13-5 Two socks are drawn from a drawer which contains one red sock, three blue socks, two black socks, and two green socks. Once a sock is selected, it is not replaced. Find the probability of each outcome.

1. a black sock and then a green sock
2. a red sock and then a green sock
3. a blue sock two times in a row
4. a green sock two times in a row

There are three quarters, five dimes, and twelve pennies in a bag. Once a coin is drawn from the bag, it is not replaced. If two coins are drawn, find the probability of each outcome.

5. a quarter and then a penny
6. a nickel and then a dime
7. a dime and then a penny
8. a dime two times in a row

Lesson 13-8 Use the survey on favorite type of music to answer each question.

1. What is the size of the sample?
2. What is the mode?
3. What fraction prefers country music?
4. What fraction prefers rap music?

Favorite Type of Music	
Country	72
Heavy Metal	41
Rap	45
Light Rock	92

Use the survey on favorite fruit to answer each question.

5. What is the size of the sample?
6. If 7,950 people were to choose one of these three fruits, how many would you expect to choose an orange?

Favorite Fruit	
Apple	155
Orange	300
Banana	145

Lesson 14-1 Write a monomial or polynomial for each model.

1. 

2. 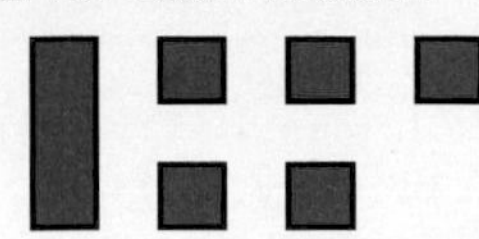

3. 

4. 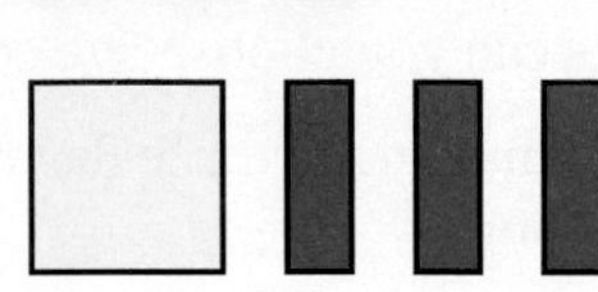

Model each monomial or polynomial using area tiles or drawings.

5. $-x^2 + 7$
6. $3x + 3$
7. $3x^2 - 2x + 1$
8. $-5x + 1$
9. $x + 2$
10. $-2x^2 + 3x$
11. $-x - 4$
12. $x^2 + 2x + 3$
13. $2x^2 - 2$
14. $2x - 5$
15. $-x^2 + 7x$
16. $2x^2 - 3x$

Lesson 14-2 Name the like terms in each list of terms.

1. $10x, 6y, y^2, 3x$
2. $-a, 2b^2, -3a, b^2$
3. $2m, n, -m, n$
4. $9, 2, 7m, -m^2$
5. $3y, 6, 5, y^2$
6. $4x, 6, 2x, x^2$

Write a polynomial in simplest form for each model.

7. 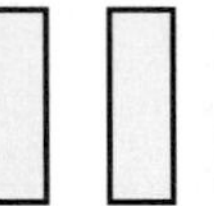  

8.  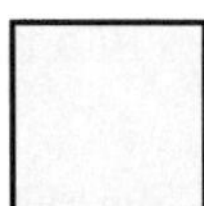

Simplify each polynomial. If necessary, use area tiles or drawings.

9. $-2y + 3 + x^2 + 5y$
10. $m + m^2 + n + 3m^2$
11. $a^2 + b^2 + 3 + 2b^2$
12. $1 + a + b + 6$
13. $x + x^2 + 5x - 3x^2$
14. $-2y + 3 + y - 2$

Lesson 14-3 Find each sum. Use area tiles or drawings, if necessary.

1. $\begin{array}{r} 2x^2 - 5x + 7 \\ \underline{x^2 - x + 11} \end{array}$

2. $\begin{array}{r} 2m^2 + m + 1 \\ \underline{-m^2 + 2m + 3} \end{array}$

3. $\begin{array}{r} 2a - b + 6c \\ \underline{3a - 7b + 2c} \end{array}$

4. $\begin{array}{r} 5a + 3a^2 - 2 \\ \underline{2a + 8a^2 + 4} \end{array}$

5. $\begin{array}{r} 3c + b + a \\ \underline{-c + b - a} \end{array}$

6. $\begin{array}{r} -z^2 + x^2 + 2y^2 \\ \underline{3z^2 + x^2 + y^2} \end{array}$

7. $(5x + 6y) + (2x + 8y)$
8. $(4a + 6b) + (2a + 3b)$
9. $(7r + 11m) + (4m + 2r)$
10. $(-z + z^2) + (-2z + z^2)$
11. $(3x - 7y) + (3y + 4x + 1)$
12. $(5m + 3n - 3) + (8m + 6)$
13. $(a + a^2) + (3a - 2a^2)$
14. $(3s - 5t) + (8t + 2s)$

Lesson 14-4 Find each difference. Use area tiles or drawings, if necessary.

1. $(5a - 6m) - (2a + 5m)$
2. $(2a - 7) - (8a - 11)$
3. $(3 + 2a + a^2) - (5 + 8a)$
4. $(9r^2 - 3) - (11r^2 + 12)$
5. $(7y + 9x) - (6x + 5y)$
6. $(5 - 2x) - (7 + 8x)$
7. $(9x + 3y) - (9y + x)$
8. $(3x^2 + 2x - 1) - (2x + 2)$
9. $(a^2 + 6a + 3) - (5a^2 + 5)$
10. $(5a + 2) - (3a^2 + a + 8)$
11. $(3x^2 - 7x) - (8x - 6)$
12. $(3m + 3n) - (m + 2n)$
13. $(3m - 2) - (2m + 1)$
14. $(x^2 - 2) - (x + 3)$
15. $(5x^2 - 4) - (3x^2 + 8x + 4)$
16. $(7z^2 + 1) - (3z^2 + 2z - 6)$

Lesson 14-5 Find each product. Use area tiles or drawings, if necessary.

1. $m(m + 2)$
2. $x(x - 1)$
3. $y(y - 2)$
4. $a(a - 3)$
5. $6(a + 3)$
6. $m(m - 7)$
7. $z(z + 3)$
8. $x(x + 10)$
9. $y(y - 5)$
10. $-2(x + 1)$
11. $m(m - 2)$
12. $3(y + 6)$
13. $3(m + 1)$
14. $z(z + 5)$
15. $b(b + 1)$
16. $-3(a + 2)$

Factor. Use area tiles or drawings, if necessary.

17. $3m + 3$
18. $-2x - 2$
19. $-4b - 4$
20. $2z + 2$
21. $z^2 - 5z$
22. $c^2 + 2c$
23. $z^2 + 3z$
24. $-6a - 6$
25. $a^2 + 3a$
26. $2m^2 + m$
27. $y^2 - 2y$
28. $n^2 + 4n$
29. $6b + 6$
30. $x^2 + 6x$
31. $5x + 5$
32. $7y - 7$

Lesson 14-6 Name the two binomials being multiplied and give their product.

1.

x^2	x^2	x	x	x
x	x	1	1	1

2.

x^2	x	x	x	x
x	1	1	1	1
x	1	1	1	1

Find each product. Use area tiles or drawings, if necessary.

3. $(r + 3)(r + 4)$
4. $(z + 5)(z + 2)$
5. $(3x + 7)(x + 1)$
6. $(x + 5)(2x + 3)$
7. $(c + 1)(c + 1)$
8. $(a + 3)(a + 7)$
9. $(b + 3)(b + 1)$
10. $(2y + 1)(y + 3)$
11. $(z + 8)(2z + 1)$
12. $(2m + 4)(m + 5)$
13. $(x + 3)(x + 2)$
14. $(c + 2)(c + 8)$
15. $(r + 4)(r + 4)$
16. $(2x + 4)(x + 4)$
17. $(6y + 1)(y + 3)$

Glossary

A

absolute value (87) The number of units a number is from zero on the number line.

acute angle (183) Any angle that measures between 0° and 90°.

addition property of equality (51) If you add the same number to each side of an equation, the two sides remain equal. If $a = b$, then $a + c = b + c$.

additive inverse (99) Two integers that are opposites of each other are called additive inverses. The sum of any number and its additive inverse is zero, $a + (-a) = 0$.

algebraic expression (45) A combination of variables, numbers, and at least one operation.

alternate exterior angles (176) In the figure, transversal t intersects lines ℓ and m. $\angle 1$ and $\angle 7$, and $\angle 2$ and $\angle 8$ are alternate exterior angles. If lines ℓ and m are parallel these angles are congruent.

alternate interior angles (176) In the figure, transversal t intersects lines ℓ and m. $\angle 3$ and $\angle 5$, and $\angle 4$ and $\angle 6$ are alternate interior angles. If lines ℓ and m are parallel these angles are congruent.

altitude (75, 278) A segment in a quadrilateral that is perpendicular to both bases, with endpoints on the base lines.

altitude (487) The segment that goes from the vertex of a pyramid to its base and is perpendicular to the base.

area (73) The number of square units needed to cover a surface.

arithmetic sequence (272) A sequence of numbers in which you can find the next term by adding the same number to the previous term.

associative property of addition (9) For any numbers a, b, and c, $(a + b) + c = a + (b + c)$.

associative property of multiplication (9) For any numbers a, b, and c, $(a \cdot b) \cdot c = a \cdot (b \cdot c)$.

B

back-to-back stem-and-leaf plot (141) Used to compare two sets of data. The leaves of one set of data are on one side of the stem and the leaves for the other set are on the other side of the stem.

bar notation (230) In repeating decimals, the line or bar placed over the digits that repeat. Another way to write 2.636363 is $2.\overline{63}$.

base (35) The number used as a factor. In 10^3, 10 is the base.

base (75) Any side of a parallelogram.

base (278) The parallel sides of a trapezoid.

base (381) In a percent proportion, the number to which the percentage is compared.

base (471) The bases of a prism are the two parallel, congruent sides.

binomial (560) A polynomial with two terms.

box-and-whisker plot (155) A diagram that summarizes data using the median, the upper and lower quartiles, and the extreme values. A box is drawn around the quartile value and the whiskers extend from each quartile to the extreme data points.

C

center (284) The middle point of a circle or sphere. The center is the same distance from all points on the circle or sphere.

circle (284) The set of all points in a plane that are the same distance from a given point called the center.

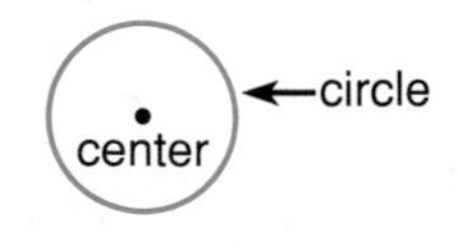

circle graph (403) A type of statistical graph used to compare parts of a whole.

circular cone (488) A shape in space that has a circular base and one vertex.

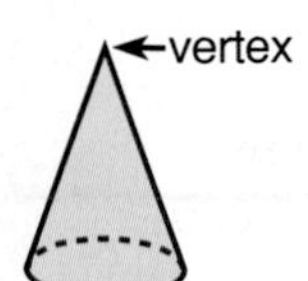

circular cylinder (478) A cylinder with two bases that are parallel, congruent circular regions.

circumference (284) The distance around a circle.

clustering (11) A method used to estimate decimal sums and differences by rounding a group of closely related numbers to the same whole number.

combination (512) Arrangements or listings where order is not important.

common difference (272) The difference between any two consecutive terms in an arithmetic sequence.

common ratio (273) The constant factor used to multiply consecutive terms in a geometric sequence.

commutative property of addition (8) For any numbers a and b, $a + b = b + a$.

commutative property of multiplication (8) For any numbers a and b, $a \cdot b = b \cdot a$.

compatible numbers (11) Two numbers that are easy to divide mentally. They are often members of fact families and can be used as an estimation strategy.

compensation (8) A math strategy in which you simplify a problem by making equivalent adjustments to each part.

$$\begin{array}{r} 397 + 3 \rightarrow 400 \\ + 103 - 3 \rightarrow \underline{100} \\ 500 \end{array}$$

composite number (215) Any whole number greater than one that has more than two factors.

congruent figures (197) Figures that are exactly the same size and shape.

converse (321) The converse of the Pythagorean Theorem can be used to test whether a triangle is a right triangle. If the sides of the triangle have lengths a, b, and c, such that $c^2 = a^2 + b^2$, then the triangle is a right triangle.

coordinate (86) A number associated with a point on a number line.

coordinate system (117) Two perpendicular number lines that intersect at their zero points form a coordinate system.

corresponding angles (177) Angles that hold the same position on two different parallel lines cut by a transversal.

cosine (368) If $\triangle ABC$ is a right triangle and A is an acute angle,

$$\cos A = \frac{\text{measure of the leg adjacent to } \angle A}{\text{measure of the hypotenuse}}.$$

cross product (344) If the cross products in a ratio are equal then the ratio forms a proportion. In the proportion $\frac{2}{3} = \frac{8}{12}$, the cross products are 2×12 and 3×8.

customary system (26) A system of weights and measures frequently used in the United States. The basic unit of weight is the pound, and the basic unit of capacity is the quart.

D

data analysis (130) To study data and drav conclusions from the numbers observed.

dependent event (522) Two or more events in which the outcome of one event does affect the outcome of the other event or events.

diameter (284) The distance across a circle through its center.

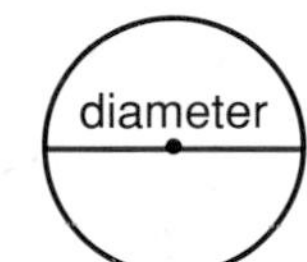

dilation (361) The process of reducing or enlarging an image in mathematics.

discount (409) The amount by which the regular price is reduced.

distributive property (8) The sum of two addends multiplied by a number is the sum of the product of each addend and the number. $a \cdot (b + c) = a \cdot b + a \cdot c$.

divisible (212) A number is divisible by another if the quotient is a whole number and the remainder is zero.

division property of equality (55) If each side of an equation is divided by the same nonzero number, then the two sides remain equal. If $a = b$, then $\frac{a}{c} = \frac{b}{c}$, $c \neq 0$.

domain (423) The set of input values in a function.

E

edge (471) The intersection of faces of three-dimensional figures.

equation (48) A mathematical sentence that contains an equals sign, =.

equilateral triangle (183) A triangle that has three congruent sides.

evaluate (35, 44) To find the value of an expression by replacing variables with numerals.

event (233) A specific outcome or type of outcome.

experimental probability (528) An estimated probability based on the relative frequency of positive outcomes occurring during an experiment.

exponent (35) The number of times the base is used as a factor. In 10^3, the exponent is 3.

F

face (471) Any surface that forms a side or a base of a prism.

factor (35) When two or more numbers are multiplied, each number is a factor of the product.

factorial (507) The expression $n!$ is the product of all counting numbers beginning with n and counting backwards to 1.

factoring (558) Finding the factors of a product.

factor tree (215) A diagram used to illustrate the factorization of a number.

frequency table (130) A table for organizing a set of data that shows the number of times each item or number appears.

front-end-estimation (11) A method used to estimate decimal sums and differences by adding or subtracting the front-end digits, then adjusting by estimating the sum or difference of the remaining digits, then adding the two values.

function (422) A relationship in which the output value depends upon the input according to a specified rule. For example, with a function $f(x) = 2x$, if the input is 5, the output is 10.

Fundamental Principle of Counting (504) If event M can occur in m ways and is followed by event N that can occur in n ways, then the event M followed by event N can occur in $m \times n$ ways.

G

geometric sequence (273) When consecutive terms of a sequence are formed by multiplying by a constant factor, the sequence is called a geometric sequence.

gram (23) The basic unit of mass in the metric system.

graph (86) A dot marking a point that represents a number on a number line or an ordered pair on a coordinate plane.

greatest common factor (GCF) (221) The greatest of the common factors of two or more numbers. The greatest common factor of 18 and 24 is 6.

H

height (73, 278) The length of the altitude of a quadrilateral.

histogram (133) A special kind of bar graph that displays the frequency of data that has been organized into equal intervals. The intervals cover all possible values of data, therefore there are no spaces between the bars of the graph.

hypotenuse (319) In a right triangle, the side opposite the right angle is called the hypotenuse.

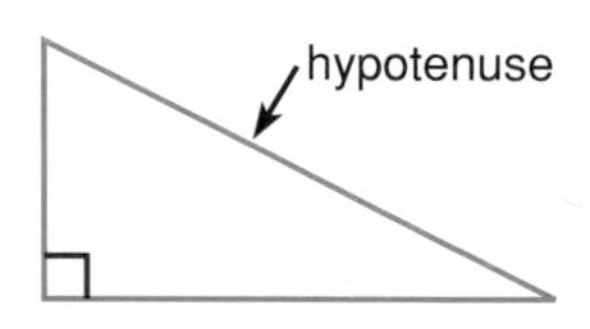

I

independent event (521) Two or more events in which the outcome of one event does not affect the outcome of the other event or events.

indirect measurement (356) A technique using proportions to find a measurement.

inequality (77) Any sentence that contains $>$, $<$, $\neq$, $\leq$, $\geq$.

integers (86) The whole numbers and their opposites. . . . , -3, -2, -1, 0, 1, 2, 3, . . .

interest (412) The amount charged or paid for the use of money.

interquartile range (151) The range of the middle half of data.

inverse operation (52) Pairs of operations that undo each other. Addition and subtraction are inverse operations. Multiplication and division are inverse operations.

irrational number (310) Numbers that cannot be expressed as $\frac{a}{b}$, where a and b are integers and $b \neq 0$.

isosceles triangle (183) A triangle that has two congruent sides.

L

leaf (141) The second greatest place value of data in a stem-and-leaf plot.

least common denominator (LCD) (241) The least common multiple of the denominators of two or more fractions.

least common multiple (LCM) (236) The least of the common multiples of two or more numbers, other than zero. The least common multiple of 2 and 3 is 6.

leg (319) Either of the two sides of a right triangle that form the right angle.

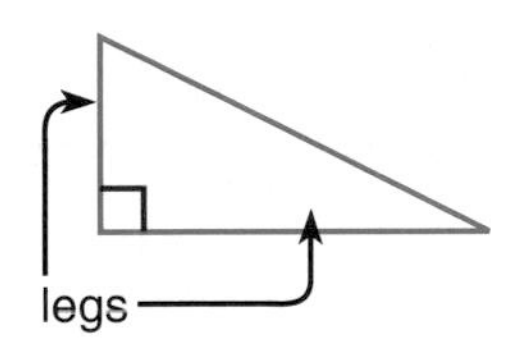

like terms (546) Expressions that contain the same variables, such as $3ab$ and $7ab$.

linear function (433) An equation in which the graphs of the solutions form a line.

line plot (136) A graph that uses an X above a number on a number line each time that number occurs in a set of data.

line symmetry (192) Figures that match exactly when folded in half have line symmetry.

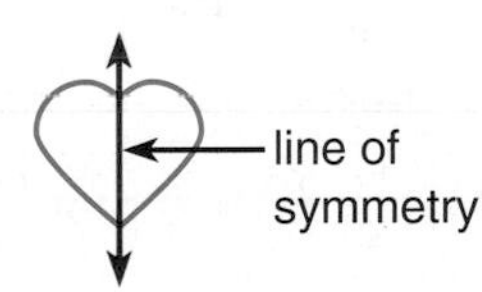

line of symmetry (451) A fold line on a figure that shows symmetry. Some figures can be folded in more than one way to show symmetry.

liter (23) The basic unit of capacity in the metric system. A liter is a little more than a quart.

lower quartile (152) The median of the lower half of data in an interquartile range.

M

mean (145) The sum of the numbers in a set of data divided by the number of pieces of data; the arithmetic average.

median (146) The number in the middle when the data are arranged in order. When there are two middle numbers, the median is their mean.

meter (23) The basic unit of length in the metric system.

metric system (23) A system of weights and measures based on tens. The meter is the basic unit of length, the kilogram is the basic unit of weight, and the liter is the basic unit of capacity.

mixed number (256) A number that shows the sum of a whole number and a fraction. $6\frac{2}{3}$ and $9\frac{1}{2}$ are mixed numbers.

mode (145) The number or item that appears most often in a set of data.

monomial (543) A number, a variable, or a product of a number and one or more variables.

multiple (236) The product of a number and any whole number.

multiplication property of equality (55) If each side of an equation is multiplied by the same number, then the two sides remain equal. If $a = b$, then $ac = bc$.

multiplicative inverse (265) A number times its multiplicative inverse is equal to 1. The multiplicative inverse of $\frac{2}{3}$ is $\frac{3}{2}$.

N

numerical expression (44) A mathematical expression that has a combination of numbers and at least one operation. $4 + 2$ is a numerical expression.

O

obtuse (183) Any angle that measures between 90° and 180°.

open sentence (48) An equation that contains a variable.

opposite (99) Two integers are opposites if they are represented on the number line by points that are the same distance from zero, but on opposite sides of zero. The sum of opposites is zero.

order of operations (44) The rules to follow when more than one operation is used. 1. Do all operations within grouping symbols first; start with the innermost grouping symbols. 2. Do all powers before other operations. 3. Do multiplication and division in order from left to right. 4. Do all addition and subtraction in order from left to right.

ordered pair (117) A pair of numbers where order is important. An ordered pair that is graphed on a coordinate plane is written in this form: (*x*-coordinate, *y*-coordinate).

origin (117) The point of intersection of the *x*-axis and *y*-axis in a coordinate system.

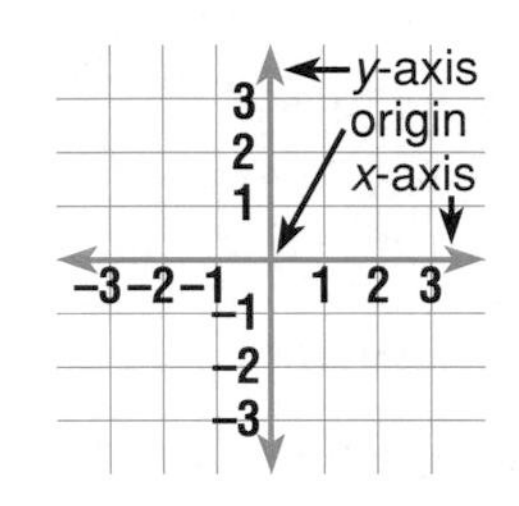

outcome (233, 504) One possible result of a probability event. 4 is an outcome when a die is rolled.

outliers (155) Data that are more than 1.5 times the interquartile range from the upper or lower quartiles.

P

parallel (176) Lines that are in the same plane but do not intersect.

parallelogram (73) A quadrilateral that has both pairs of opposite sides parallel.

Pascal's Triangle (515) A triangular arrangement of numbers in which each number is the sum of the two numbers to the right and to the left of it in the row above.

pentagon (353) A polygon having five sides.

percent (380) A ratio that compares a number to 100.

percent proportion (381)

$$\frac{\text{Percentage}}{\text{Base}} = \text{Rate or } \frac{P}{B} = \frac{r}{100}.$$

percentage (381) In a percent proportion, a number (P) that is compared to another number called the base (B).

perfect square (304) Squares of whole numbers.

perimeter (73) The distance around a geometric figure.

permutation (507) An arrangement, or listing, of objects in which order is important.

perpendicular lines (183) Two lines or line segments that intersect to form right angles.

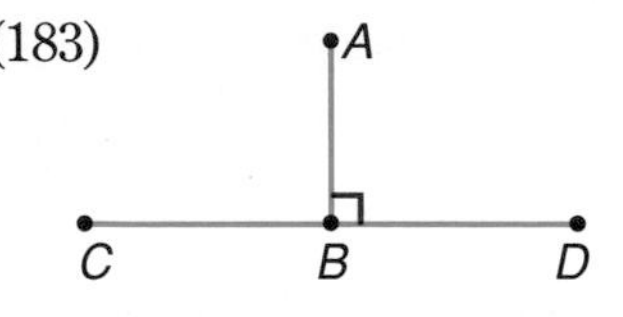

polygon (353) A simple closed figure in a plane formed by three or more line segments.

polynomial (543) The sum or difference of two or more monomials.

population (533) The entire group of items or individuals from which the samples under consideration are taken.

power (35) A number expressed using an exponent. The power 7^3 is read *seven to the third power,* or *seven cubed.*

precision (491) The precision of a measurement depends on the unit of measure. The smaller the unit the more precise the measurement is.

prime factorization (215) A composite number that is expressed as the product of prime numbers. The prime factorization of 12 is $2 \times 2 \times 3$.

prime number (215) A number that has exactly two factors, 1 and the number itself.

principal (412) The amount of an investment or a debt.

principal square root (304) A nonnegative square root.

prism (471) A three-dimensional figure that has two parallel and congruent bases in the shape of polygons.

probability (233) The ratio of the number of ways an event can occur to the number of possible outcomes; how likely it is that an event will occur.

proportion (344) A proportion is an equation that shows that two ratios are equivalent; $\frac{a}{b} = \frac{c}{d}, b \neq 0, d \neq 0$.

pyramid (487) A figure in space with three or more triangular faces and a base in the shape of a polygon.

Pythogorean Theorem (319) In a right triangle, the square of the hypotenuse is equal to the sum of the squares of the legs. $c^2 = a^2 + b^2$.

Pythagorean triple (323) A set of three integers that satisfy the Pythagorean Theorem.

Q

quadrant (117) One of the four regions into which two perpendicular number lines separate a plane.

quadratic function (442) A function in which the greatest power is 2.

quadrilateral (187) Any four-sided figure.

quartile (151) Values that divide data into four equal parts.

R

radical sign (304) The symbol used to represent a nonnegative square root is $\sqrt{}$.

radius (284) The distance from the center of a circle to any point on the circle.

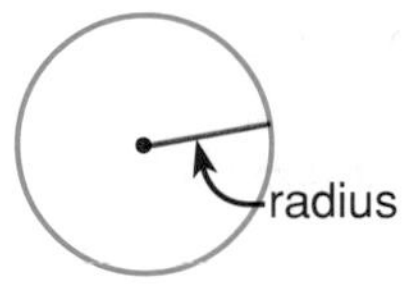

random (233) Outcomes occur at random if each outcome is equally likely to occur.

random (533) Making selections unsystematically where everything has an equal chance of being selected.

range (151) The difference between the greatest number and the least number in a set of data.

range (423) The set of output values in a function.

rate (338) A ratio of two quantities with different units.

rate (381) In a percent proportion, the ratio of a number to 100.

rate (412) The percent charged or paid for the use of money.

ratio (338) A comparison of two numbers by division. The ratio comparing 2 to 3 can be stated as 2 out of 3, 2 to 3, 2:3, or $\frac{2}{3}$.

rational number (224) Any number that can be expressed in the form $\frac{a}{b}$, where a and b are integers and $b \neq 0$.

real number (311) Irrational numbers together with rational numbers form the set of real numbers.

reciprocal (265) Another name for a multiplicative inverse.

rectangle (73) A parallelogram with all angles congruent.

rectangular prism (475) A prism with rectangles as bases.

reflection (192, 451) A mirror image of a figure across a line of symmetry.

repeating decimal (230) A decimal whose digits repeat in groups of one or more. Examples are 0.181818... and 0.83333... Using bar notation, these examples are written $0.\overline{18}$ and $0.8\overline{3}$.

replacement set (49) A set of numbers from which to choose the value of a variable for an equation.

rhombus (187) A parallelogram that has four congruent sides.

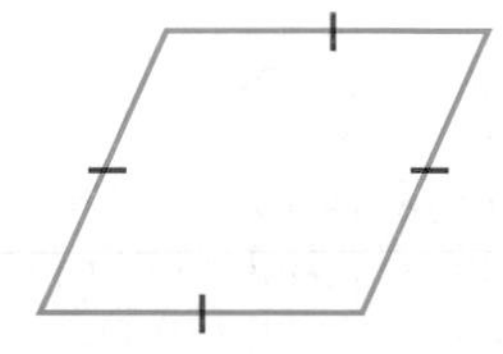

right angle (183) An angle that measures exactly 90°.

rotation (202, 454) Transformations of tessellations that involve a turn about the vertices of the base figure.

rotational symmetry (193) A figure has rotational symmetry if it can be turned less than 360° about its center and still looks like the original.

rounding (11) A method used to estimate by changing numbers to the nearest tens, hundreds, thousands, and so on. 387 in round numbers would be 390 to the nearest tens or 400 to the nearest hundreds.

S

sale price (409) The price after the discount has been subtracted.

sample (165, 533) A randomly selected group chosen for the purpose of collecting data.

scale factor (361) The ratio of a dilated image to the original image.

scalene triangle (183) A triangle with no congruent sides.

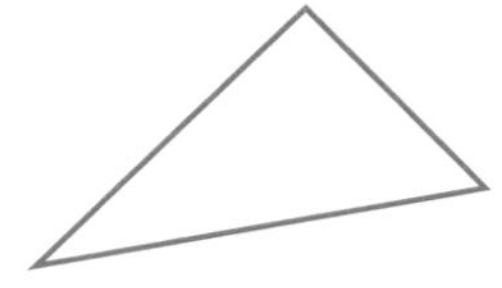

scatter plot (159) A graph that shows the general relationship between two sets of data.

scientific notation (245) A way of expressing numbers as the product of a number that is at least 1, but less than 10, and a power of 10. In scientific notation 5,500 is 5.5×10^3.

sequence (272) A list of numbers in a specific order.

significant digits (491) All of the digits of a measurement that are known to be accurate plus one estimated digit.

similar figures (198) Figures that have the same shape but may differ in size are similar.

similar polygons (353) Two polygons are similar if their corresponding angles are congruent and their corresponding sides are in proportion. They have the same shape but may not have the same size.

simplest form (224) The form of a fraction where the GCF of the numerator and denominator is 1.

simplest form (546) An expression that has no like terms in it. In simplest form $6x + 8x$ is $14x$.

simulation (525) The process of acting out a problem.

sine (368) If $\triangle ABC$ is a right triangle and A is an acute angle,

$$\sin A = \frac{\text{measure of the leg opposite } \angle A}{\text{measure of the hypotenuse}}.$$

solid (471) Three-dimensional figures.

solution (48) The value for a variable that makes an equation true. The solution for $10 + y = 25$ is 15.

square (73) A parallelogram with all sides congruent and all angles congruent.

square root (304) One of the two equal factors of a number. If $a^2 = b$, then a is the square root of b. The square root of 144 is 12 because $12^2 = 144$.

statistics (130) The branch of mathematics that deals with collecting, organizing, and analyzing data.

stem (141) The greatest place value of data in a stem-and-leaf plot.

stem-and-leaf plot (141) A system used to condense a set of data where the greatest place value of the data forms the stem and the next greatest place values forms the leaves.

subtraction property of equality (51) If you subtract the same number from each side of an equation, then the two sides remain equal. If $a = b$, then $a - c = b - c$.

substitute (44) To replace a variable in an algebraic expression with a number, creating a numerical expression.

surface area (475) The sum of the area of all the faces of a three-dimensional figure.

symmetric (451) Figures that can be folded into two identical parts.

system of equations (436) A common solution for two or more equations.

T

tangent (365) If $\triangle ABC$ is a right triangle and A is an acute angle,

$$\tan A = \frac{\text{measure of the leg opposite } \angle A}{\text{measure of the leg adjacent to } \angle A}.$$

term (546) A number, a variable, or a product of numbers and variables.

term of the sequence (272) A number in a sequence.

terminating decimal (227) A quotient in which the division ends with a remainder of zero. 0.25 and 0.125 are terminating decimals.

tessellation (202) A repetitive pattern of polygons that fit together with no holes or gaps.

theoretical probability (528) The long-term probability of an outcome based on mathematical principles.

time (412) When used to calculate interest, time is given in years.

transformation (202) Movements of geometric figures to modify tessellations.

translation (202) A method used to make changes in the polygons of tessellations by sliding a pattern to create the same change on opposite sides.

translation (447) To translate a point as described by an ordered pair, add the coordinates of the ordered pair to the coordinates of the point. (x, y) moved (a, b) becomes $(x = a, y = b)$.

transversal (176) A line that intersects two other lines to form eight angles.

trapezoid (187, 278) A quadrilateral with exactly two parallel sides.

tree diagram (504) A diagram used to show the total number of possible outcomes in a probability experiment.

triangular prism (476) A prism with triangles as bases.

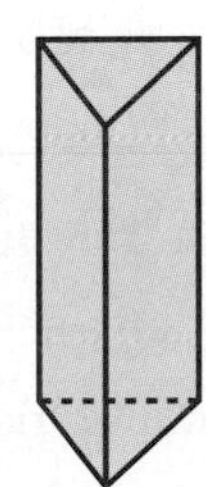

trigonometry (368) The study of triangle measurement.

U

unit rate (338) A rate that has a denominator of 1. This type of rate is frequently used when comparing statistics.

upper quartile (152) The median of the upper half of data in an interquartile range.

V

variable (45) A symbol, usually a letter, used to represent a number in mathematical expressions or sentences.

variation (151) The spread in the values in a set of data.

Venn diagram (181) A diagram using circles and rectangles to represent various types of mathematical sets and to show the relationship between them.

vertex (471) A vertex of a polygon is any point of intersection of the sides of a polygon.

vertex (487) The vertex of a pyramid is the point where all the faces except the base intersect.

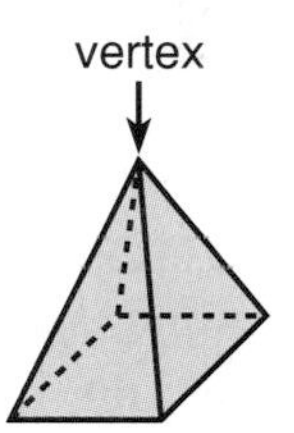

volume (482) The number of cubic units required to fill a space.

X

***x*-axis** (117) The horizontal line of the two number lines in a coordinate plane.

***x*-coordinate** (117) The first number of an ordered pair.

Y

***y*-axis** (117) The vertical line of the two perpendicular number lines in a coordinate plane.

***y*-coordinate** (117) The second number of an ordered pair.

Selected Answers

1 Tools for Problem Solving

Pages 6-7 Lesson 1-1
4. 12:30 A.M. **5.a.** beaker + sodium chloride = 84.8 grams; beaker = 63.3 grams **b.** subtract the two masses **c.** 21.5 grams **d.** yes
7. 3,000 ft **9.** $0.50 **11.** 2,000 ft **13.** 4 by 4 by 4

Pages 9-10 Lesson 1-2
4. 67 **5.** 145 **6.** 677 **7.** 320 **8.** 148
9. 93,000 **10.** 245 **11.** 4,300 **12.** 169
13. 153 **15.** 135 **17.** 5,300 **19.** 827
21. 100 **23.** 1,400 **25.** 79 **27.** 1,123
29. 300 **31.** $2.65 **32.** 3 hours
33. 28 cards

Pages 13-14 Lesson 1-3
4. 5,000 **5.** 700 **6.** 0.09 **7.** front-end estimation, 1,590 **8.** rounding, 1,000
9. rounding, 1,000 **10.** rounding, $9
11. rounding, 6 **12.** clustering, 10 **13.** 9,000, 8,900 **15.** 0.3, 0.25 **17.** 60
19. 20,000 **21.** 200 **23.** 60 **25.** 198 **27.** 42 pounds **28.** 96 pounds **29.** 315 **30.** 622
31. about 200,000 sheets of paper **33.a.** about 2,500 ft of fence

Pages 15-16 Lesson 1-4
3. No 45,000 ÷ 1,500 = 30, not 300 **4.** 200 crates
5. $10.00 **7.** They are equal. **9.** 50 sit-ups
11. 23¢ for each ounce over 1 ounce; more

Pages 18-19 Lesson 1-5
3. b **4.** c **5.** b **7.** 3981 or 1101 **9.** 10
11. a

Page 19 Mid-Chapter Review
1. $55.02 **3.** $7.97 **5.** 28 **7.** 1,000 **9.** 37

Pages 24-25 Lesson 1-6
4. centimeter **5.** milliliter **6.** kilogram
7. kilometer **8.** kiloliter **9.** millimeter
10. × by 1,000 = 10,000 mg **11.** × by 1,000 = 1,000,000 m **12.** ÷ by 1,000 = 0.00439 L
13. ÷ by 1,000 = 0.0015 L **14.** ÷ by 1,000 = 0.0593 kg **15.** ÷ by 1,000 = 0.00789 km
17. 0.525 **19.** 1,370 **21.** 9,240 **23.** 92.4
25. 0.0723 mm **27.** 947 mL **28.** 4 goldfish
29. 378, distributive property **30.** 4,000
31. 6

Pages 27-28 Lesson 1-7
4. × by 16 = 48 **5.** ÷ by 2,000 = 3 **6.** ÷ by 4 = 6 **7.** × by 36 = 90 **8.** ÷ by 5,280 ≈ 1.52
9. × by 8 = 80 **11.** 26,400 **13.** 56 **15.** 3.5
17. 42 **19.** $\frac{1}{4}$ gallon **21.** 144 in.
23. Subtract 5 from both numbers and subtract the new numbers; 251. **24.** 1,400 **25.** 279 cm
26. 32 coffee mugs **29.** 105 feet **31.** about 45 feet **33.** 9.72 s faster

Page 34 Lesson 1-8
4. 7 **5.** 14 **6.** 100 **7.** all but 0 **8.** 3
9. 5 years old **11.** $20.30

Pages 36-37 Lesson 1-9
5. 5^3 **6.** 10^4 **7.** 8^5 **8.** 125 **9.** 49
10. 144 **11.** 1,080,000 **13.** $6^3 \cdot 7^2$ **15.** 625
17. 32 **19.** 196 **21.** 1,000,000 **23.** 32
25. 342 **26.** $40 **27.** 5.734 kg **28.** 3 pints
29. 2.5 lb for $2.70 **33.** 2,048
35.a. 2,870,000,000 **b.** 10^7 and 10^8

Pages 38-40 Study Guide and Review
7. 27 **9.** 145 **11.** 517 **13.** $7.56
15. 84 **17.** 149 **19.** 3,400 **21.** 40
23. 20,000 **25.** 0.0034 **27.** 0.620
29. 0.00862 **31.** 4.5 **33.** 5 **35.** 64
37. $3^2 \cdot 8^3$ **39.** 16 **41.** 1,250,000
43. 2.25 min **45.** 80 yd

2 An Introduction to Algebra

Pages 46-47 Lesson 2-1
5. multiplication; 39 **6.** multiplication; 17
7. division; 1 **8.** addition in parentheses; 5
9. power in parentheses; 75 **10.** subtraction in parentheses; 6 **11.** 39 **12.** 8 **13.** 78 **14.** 42 **15.** 36 **16.** 144 **17.** 53 **19.** 10 **21.** 22
23. 2 **25.** 101 **27.** 0 **29.** 72 ÷ (6 + 3) = 8

31. $7 + 4^2 \div (2 + 6) = 9$ **33.** $4 + 8 - (7 - 5) = 10$ **34.** 3 **35.** 1,500 **36.** 1,725 meters **37.** 4,000 pounds **38.** 288 **39.** 68 grams of protein **41.** $(4 \cdot 7 - 3)(6 - 5 + 2 + 1)$

Pages 49-50 Lesson 2-2

4. 52 **5.** $0.80 **6.** 16 **7.** 99 **8.** 20 **9.** 50 **10.** $1.74 **11.** 15 **13.** $2.67 **15.** $6.06 **17.** 7 **19.** 31 **21.** 7 **23.** 7,800 **24.** 3,500 **25.** $5^2 \cdot 8^3$ **26.** 3 **27.** 10,800 ft^2 **29.** $99 **31.** less than 150 pounds

Pages 52-53 Lesson 2-3

5. 6 **6.** 30 **7.** 224 **8.** 123 **9.** 33 **10.** 25 **11.** 29 **12.** 0.3 **13.** 6 **15.** 90 **17.** 20.1 **19.** 254 **21.** 2.22 **23.** 9.8 **25.** 2.89 **26.** 444 **27.** 4,280 mg **28.** $2\frac{1}{2}$ gallons **29.** 52 **31.** 60°

Pages 55-56 Lesson 2-4

4. 3 **5.** 4 **6.** 8 **7.** 126 **8.** 92 **9.** 540 **10.** 92 **11.** 2.16 **12.** 40.5 **13.** 61 **15.** 350 **17.** 2.3 **19.** 200 **21.** $16 **23.** about 90 **24.** 28 **25.** 93 **27.** 350 items

Pages 58-59 Lesson 2-5

4. 1 **5.** 100 **6.** 198 **7.** $136.50 **9.** $52.80 **11.** $8,100

Page 59 Mid-Chapter Review

1. 40 **3.** 19 **5.** 9 m **7.** 19.1 **9.** 150

Pages 63-64 Lesson 2-6

4. $p + 17$ **5.** $\frac{x}{3}$ **6.** $6r$ **7.** $m - 4$ **8.** $2t + 3$ **9.** $6 - m = 25$ **10.** $8 - \frac{a}{4} = 19$ **11.** $5x + 2 = 37$ **13.** $18 - n$ **15.** $24 - 2x$ **17.** $s + \$200$ **19.** $\frac{24}{x} - 2 = 2$ **21.** $2p + 6$ **23.** $3c - 14 = 46$ **25.** 98 **26.** 0.0347 liters **27.** 56 ounces **28.** 32 **29.** 7

Pages 68-69 Lesson 2-7

4. subtract 3; 2 **5.** subtract 8; 3 **6.** add 5.3; 1 **7.** subtract 0.8; 4.5 **8.** add 3; 28 **9.** subtract 7; 2 **11.** 4 **13.** 2 **15.** 192 **17.** 0.5 **19.** 0.15 **21.** 0.5 **23.** $\frac{x}{6} - 7 = 12; 114$ **25.** about 12 **26.** 432 **27.** 112 **28.** $4x - 8$ **29.a.** $\frac{x}{625} = 24$ **b.** 15,000 cans **31.** $10 + 6x = 40; 5$

Pages 71-72 Lesson 2-8

4. $0.68 **5.** 5 **6.** 126 adult, 252 student **7.** 31.25 days **9.** 12 **11.** about 2 **13.** West-$29,900 more; midwest $27,300 less; northeast-$15,400 more; south-$36,600 less **15.a.** 39.8 grams **b.** 16-karat gold

Pages 75-76 Lesson 2-9

4. 26 m; 42.25 m^2 **5.** 22 yd; 30 yd^2 **6.** 36 ft; 66 ft^2 **7.** 18.4 in.; 18.6 in^2 **9.** 20 units; 24 $units^2$ **11.** 16 m; 12.4 m^2 **13.** 48 cm **15.** 13,000 lb **16.** 100 **17.** 12 **21.a.** the third one **b.** when the darkened squares are not squares of the same color on a checkerboard.

Pages 78-79 Lesson 2-10

5. $t > 5$ **6.** $a > 7$ **7.** $d < 9$ **8.** $y < 15$ **9.** $m > 5$ **10.** $p < 9$ **11.** $x < 17$ **13.** $b > 15$ **15.** $g < 9$ **17.** $a > 7$ **19.** $c > 3$ **21.** $a > 4$ **23.** $5x > 60; x > 12$ **25.** $4x + 15 > 13; x > 2$ **26.** 1,169 **27.** $\frac{x}{3} + 20 = 25$ **28.** 28 **29.** $P = 14$ cm; $A = 10$ cm^2 **33.b.** 2,4; 3; 0,1

Pages 80-82 Study Guide and Review

9. 30 **11.** 20 **13.** 56 **15.** 30 **17.** 113 **19.** 42 **21.** 5 **23.** 2.48 **25.** 6 **27.** $8 - y = 31$ **29.** $8 + 6u$ **31.** 55 **33.** 208 **35.** 7 **37.** 24 m, 28 m^2 **39.** $g > 6$ **41.** $a > 3$ **43.** $8, $12 **45.** 5 necklaces

3 Integers

Page 88 Lesson 3-1

5. A, -6; B, -1; C, 2; D, 5 **6.** 4 **7.** 3 **8.** 23 **9.** 129 **10.** 0

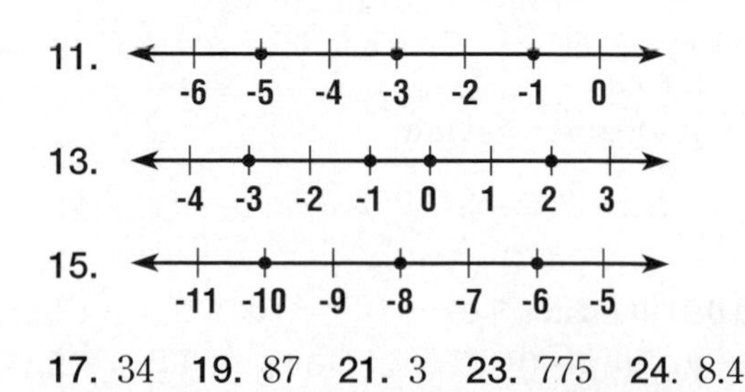

17. 34 **19.** 87 **21.** 3 **23.** 775 **24.** 8.4

25. $b > 6$ 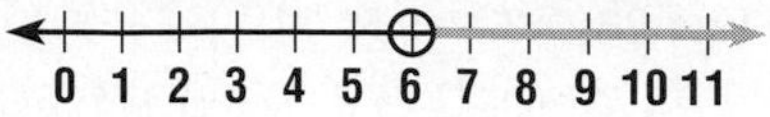

Page 90 Lesson 3-2

4. > 5. > 6. < 7. < 8. > 9. < 10. {-99, -7, -1, 0, 8, 34, 123} 11. {129, 78, 65, 34, 1, -6, -99, -665} 13. < 15. < 17. > 19. < 21. {-56, -33, -9, -7, 0, 34, 99} 23. 21 24. 24 25. 16

Pages 93-94 Lesson 3-3

5. – 6. – 7. + 9. 0 10. – 11. 20 12. -7 13. 41 14. -59 15. 0 16. 108 17. -33 19. 72 21. -130 23. -56 25. 54 27. 718 29. 50 31. -8 33. -5 35. 6 pounds 36. $4^2 \cdot 6^3$ 37. 7 38. $4x$ 39. < 41. They will lose $6.99 because they are selling the shoes for less than they bought them for. 43.a. about 1,200 b. No one could afford telephone service during the depression.

Pages 96-97 Lesson 3-4

4. 18 5. 14 6. -3 7. 22 8. 0 9. -3 10. 19 11. 15 12. 14 13. -4 15. 17 17. 20 19. 3 21. 12 23. 54 25. -36 27. -11 29. 2 31. 22 32. 36 oz 33. 32 34. $s < 9$ 35. 38 39.a. gained 3 yards b. lost 8 yards

Pages 100-101 Lesson 3-5

5. -10 6. 9 7. -30 8. 29 9. $4 + 7 = y$; 11 10. $n = -43 + (-99)$; -142 11. $p = -23 + 2$; -21 12. $53 + (-78) = z$; -25 13. $y = 14 + (-14)$; 0 14. $11 + 19 = p$; 30 15. $x = 17 + 26$; 43 16. $123 + 33 = n$; 156 17. $b = -345 + 67$; -412 19. 93 21. 1,313 23. -131 25. -8 27. 78 29. -735 31. 143 33. 52 35. 13 37. -3 39. -6 41. 125 42. 20 43. 64 cm^2 44. {128, 52, 15, 4, 0, -3, -22, -78} 45. 47 47.a. Asia: 9,247 m; S. America: 6,999 m; N. America: 6,280 m; Europe: 5,661 m; Africa: 6,050 m. b. -115 meters

Page 101 Mid-Chapter Review

1. 64 3. 4 5. {-7, -3, -2, 0, 5, 6, 8} 7. 0 9. 0 11. -114

Pages 104-105 Lesson 3-6

6. – 7. – 8. + 9. 0 10. -27 11. -150 12. -99 13. 21 14. 24 15. 96 16. -108 17. 120 18. -18 19. 64 20. 70 21. 35 23. -63 25. -81 27. 78 29. 504 31. 805 33. 441 35. -72 37. 54 39. -108 41. $55 42. 0.0394 km 43. 41

44. 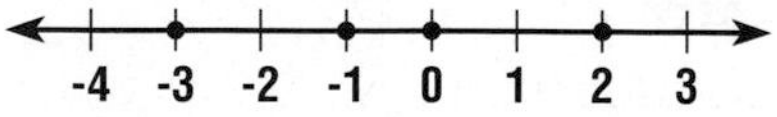

45. 83 47. Alaska; Hawaii

Pages 107-108 Lesson 3-7

4. – 5. – 6. + 7. – 8. -4 9. -73 10. 3 11. -49 12. -7 13. -2 14. 9 15. 28 16. 7 17. -31 19. -188 21. 59 23. -98 25. 33 27. -80 29. 5 31. -38 33. -31 35. 150 37. 186 38. 18 39. 32 inches 40. 11 41. 960 43. about $5,150

Pages 109-110 Lesson 3-8

3. not enough information 4. too much information; 143°F 5. too much information; about 2,500 feet 7. not enough information 9. 125 11. $560

Pages 115-116 Lesson 3-9

4. -45 5. -80 6. -448 7. -275 8. 6 9. -20 10. -48 11. -9 13. -318 15. 35 17. -600 19. 72 21. -72 23. 15 25. $-7x = 35$; -5 27. 8,000 28. 171 29. 60 30. -133 31. -8 33. averaged a 5 yard loss on each play 35.a. -$10 b. There was a loss.

Pages 118-119 Lesson 3-10

5. (-3, 5) 6. (3, 4) 7. (5, 0) 8. (3, -4) 9. (-3, -3) 10. (-5, 2) 17. (-2, 3) 19. (-3, -1) 21. (-3, -3) 23. (3, -2) 25. (3, 3) 34. 6 35. $4x < 20$; $x < 5$ 36. 217 37. -9 38. -20 41. (1, 5)

Pages 120-122 Study Guide and Review

8. 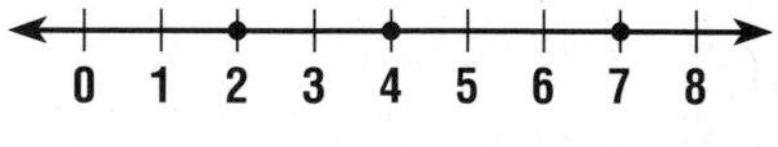

10. 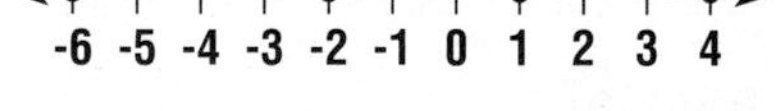

13. = 15. < 17. -282 19. 32 21. 81 23. 108 25. 10 27. -96 29. 80 31. -11 33. 32 35. 18 37. -122 43.a. Niko 6; Chuck -2; Rachel -4; Trenna 2; Amanda 0 45. 15 points 47. -33

4 Statistics and Data Analysis

Pages 131-132 Lesson 4-1

3.

Talking on Phone Time (hr)	Tally	Frequency
0–1	卌 |	6
2–3	卌 ||||	9
4–5	||||	4
6–7	|	1

5.

Video Game Prices	Tally	Frequency
0–10.99	|	1
11–20.99	|||	3
21–30.99	|||	3
31–40.99	卌 |	6
41–50.99	||||	4
51–60.99	|	1

5.a. 18 **b.** $40

7.

Subscriptions Sold	Tally	Frequency
10–19	||	2
20–29	||	2
30–39	|||	3
40–49	|||	3
50–59	||	2
60–69	|	1
70–79	卌 |	6
80–89	|	1

9.

Points Scored	Tally	Frequency
61–75	|||	3
76–90	卌 |	6
91–105	卌 ||	7
106–120	|||	3
121–135	|	1

Pages 134-135 Lesson 4-2

6. 10 scores **7.** 61-70 and 81-90 **8.** Intervals from 1-41 have been omitted.

9.

Science Test Scores	
Scores	**Frequency**
41–50	1
51–60	2
61–70	5
71–80	4
81–90	5
91–100	3

11.

Amount of Change		
¢	Tally	Frequency
61–70	||||	4
71–80	卌 |||	8
81–90	卌 |	6
91–100	||	2

12. 105 **13.** 14 units

Pages 137-138 Lesson 4-3

4.

5.a.

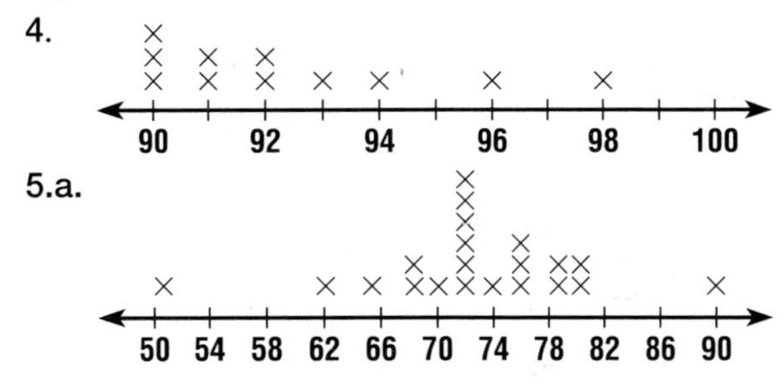

b. Most weigh 72 pounds.

7.

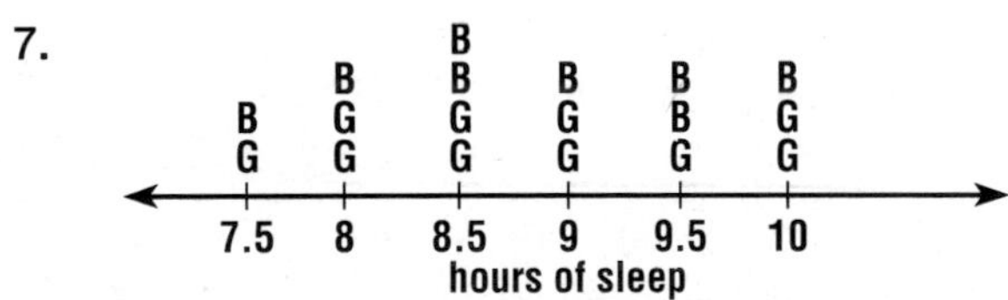

8. -241

9.

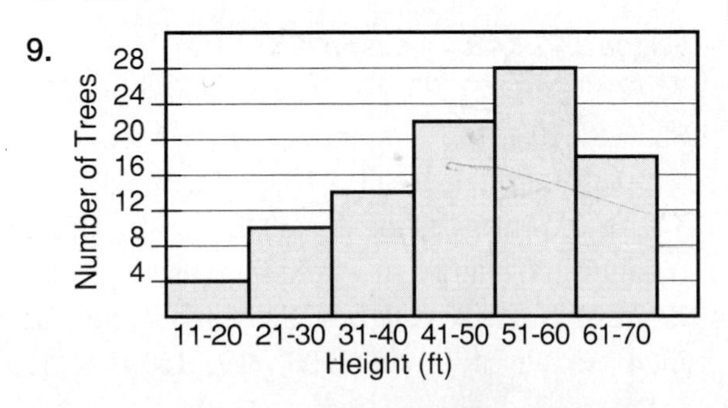

Pages 142-144 Lesson 4-4

4. 55 inches **5.** 72 inches

6.a.
5 | G G G B
6 | B B B B B B G G B G G G G
7 | G B B

b. You can see where most girls' and boys' heights fall.

c. You lose individual amounts

7.a. 0, 1, 2, 3, 4, 5, 6

b.
0 | 7 8 9
1 | 1 2 3 4 4 6 7 8 8 9
2 | 2 2 4 5
3 | 0 2 2 3 4 5
4 | 1 3 5 6
5 | 1 6 6 | 1 means 61.
6 | 1

c. 7, 61 **d.** teens

9. 0, 1, 2, 3, 4, 5
0 | 9
1 | 1 2 4
2 | 4
3 | 3
4 |
5 | 1 3 | 2 means 32.

11. 5, 6, 7, 8
5 | 49
6 | 3
7 | 15
8 | 6 5 | 4 means 5.4.

13.a. Bender Co. **b.** Sample answer: calculating expenses for health care benefits

14. 20 **15.** -16

16.

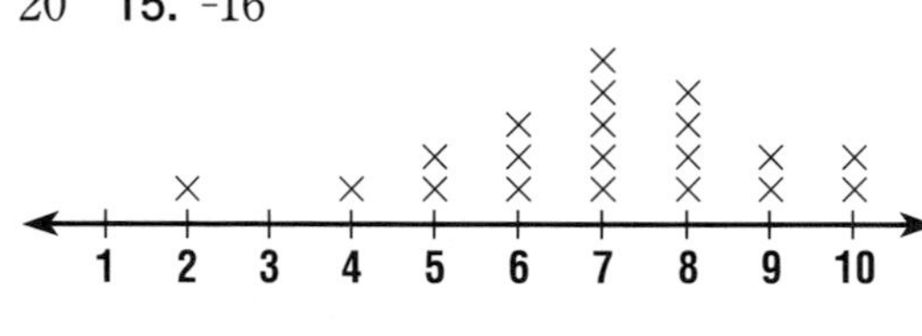

Pages 147-148 Lesson 4-5

5. 6, 7, 7 **6.** 27, 25, 25 **7.** $50; $50; $49, $50, $52 **8.** 10.4, 10.17, no mode **9.** 6.9; 6; 5, 6, 13 **11.** 157.8, 155, no mode **13.** 2.5, 2.5, 1.8 **15.** 68.2, 70, 75 **17.a.** 3.92, 3.6, 3.7 **b.** Only mean would change. It would increase.
c. Mean increases slightly; same mode; median increases slightly. **18.** 2.98 **19.** 152

20.
1 | 5 6 8
2 | 1 2 3 4 6 9
3 | 2 4 5 6 7
4 | 2 3 5 8
5 | 2 6 2 | 3 means 23.

Page 148 Mid-Chapter Review

1.

25-pt. History Test	
Score	**Frequency**
11–13	2
14–16	6
17–19	6
20–22	4
23–25	6

3. 14–16, 17–19, 23–25 **5.** Mode; because it is most representative of the central values.

Pages 153-154 Lesson 4-6

5. 7 **6.** 11 **7.** 63 **8.** 7, 8, 4 **9.** 17, 20.5, 14 **10.** 156, 177, 135.5 **11.** 41.5 **13.** 26 **15.** 7.5 **17.** 57 **19.** 33 **21.** 38, 29 **23.** 73 **24.** 960 **25.** 8.5, 8.5, 12 and 6

Pages 157-158 Lesson 4-7

5. same median, same lower extreme **6.** The top set of data is more widely dispersed than the lower one. **7.** Top plot; the box is longer.

8.

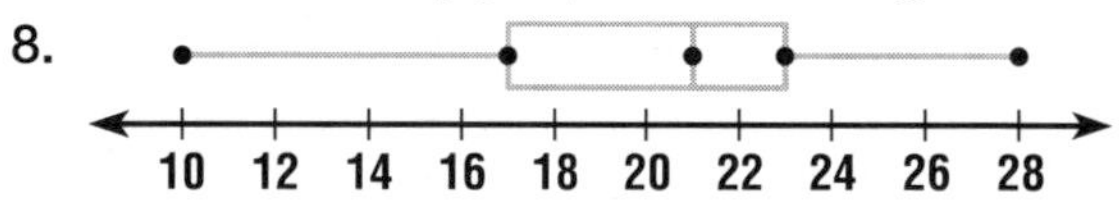

9. 21 **10.** 23 **11.** 17 **12.** 10 **13.** 6 **14.** no **15.** 8 and 32 **17.** 63 **19.** 67.5 **21.** 7.5 **23.** 48.75 and 78.75; no **25.** 10

27.

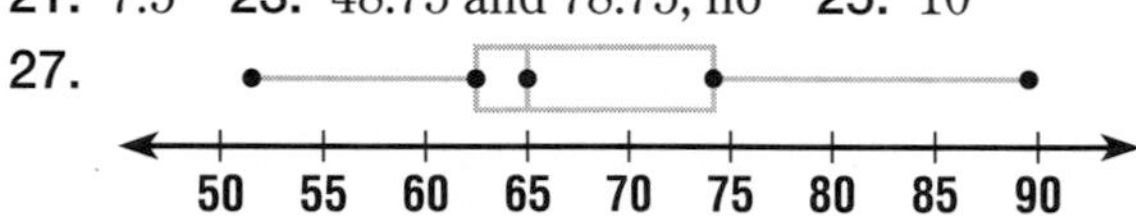

28. 3.5 gallons **29.** 35 **30.** {120, 18, 3, -24, -52, -186, -219} **31.** 184, 194, 172

35.

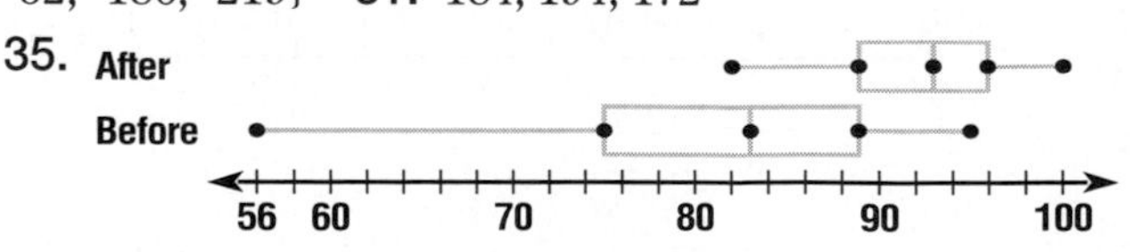

Pages 160-161 Lesson 4-8

3. negative **4.** positive **5.** negative **6.** positive **7.** no relationship **8.** positive **9.** no relationship **10.** negative **11.** positive

13. no relationship **15.** positive
17. negative **19.** 194 **20.** $x < 2$ **21.** -37
22.

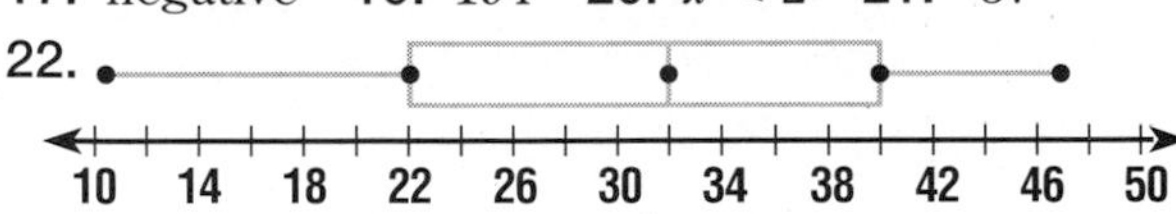

23.

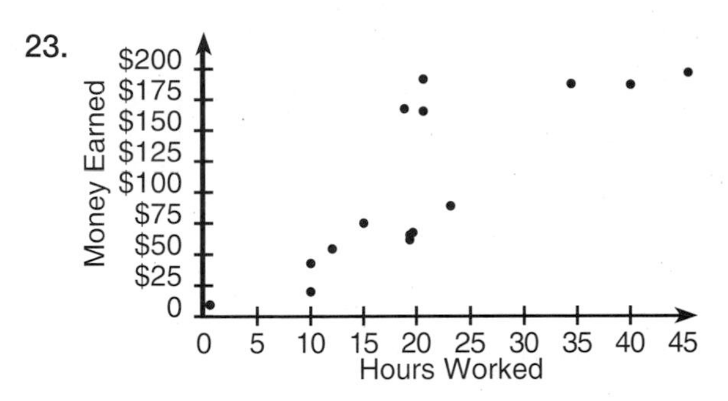

a. (hours worked, money earned) **b.** yes; positive

Pages 165-167 Lesson 4-9
4.a. increased by 2,000 **b.** relatively small change in numbers but the picture of the house changed; wider and taller **c.** yes
5. The mode, 88, because it's the highest.
7. mean **9.** no **11.** no **13.** 180 **14.** 32
15. No relationship; fishing is mostly chance.
19.a. No, samples should be more random.
b. Sample answer: What is the past history of the product or is it a new product?

Pages 168-170 Study Guide and Review
11.

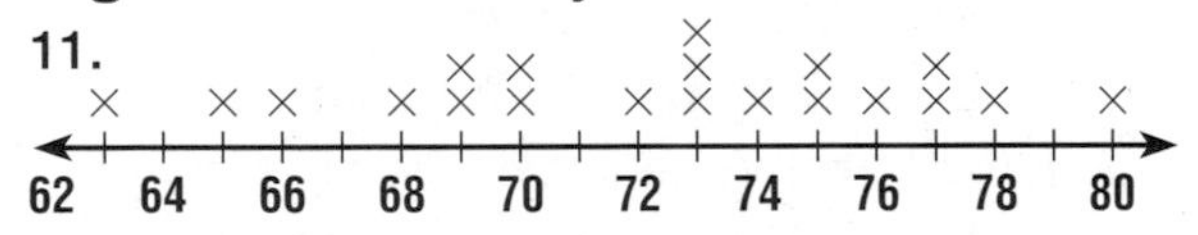

13. 7.7, 7.8, no mode **15.** 154.2, 157, no mode
17. 46, 150, 129 **19.** 8, 29, 23
21.

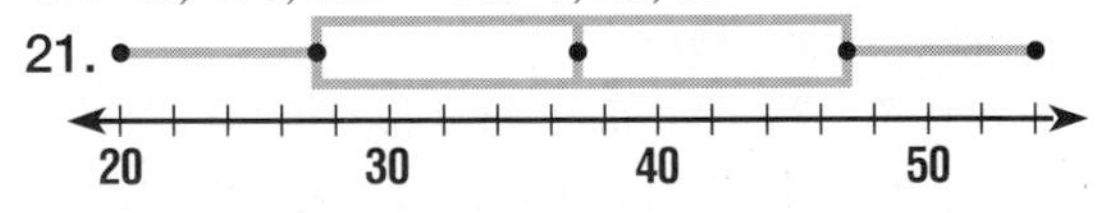

23. negative
25.a.

Cost of Air Fare	Tally	Frequency
$200–399	𝍸 ||||	9
$400–599	||	2
$600–799	𝍸	5
$800–999	||	2

b.

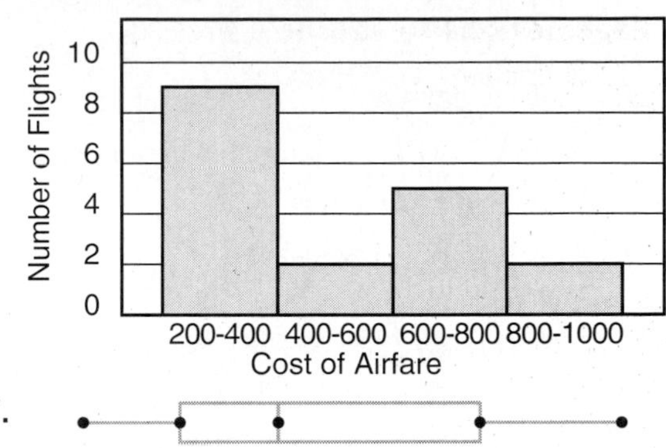

27.

6 7 8 9 10 11 12 13 14 15 16 17

5 Investigations in Geometry

Pages 178-179 Lesson 5-1
6. $\overline{PQ} \parallel \overline{SR}, \overline{PS} \parallel \overline{QR}$ **7.** $\overline{WX} \parallel \overline{ZY}, \overline{WZ} \parallel \overline{XY}$
8. $\overline{AB} \parallel \overline{ED}, \overline{BC} \parallel \overline{EF}, \overline{CD} \parallel \overline{FA}$ **9.a.** 60
b. 120 **c.** 120 **d.** 60 **e.** 60 **f.** 120 **g.** 120
11. none **13.** 35 **15.** 122 **17.** 4 **19.** 60
20. $2.51 **21.** 10 **22.** mean **23.a.** They are parallel. **b.** They are parallel. **c.** 90

Page 182 Lesson 5-2
3.a. 13 states **b.** 1 state
4.

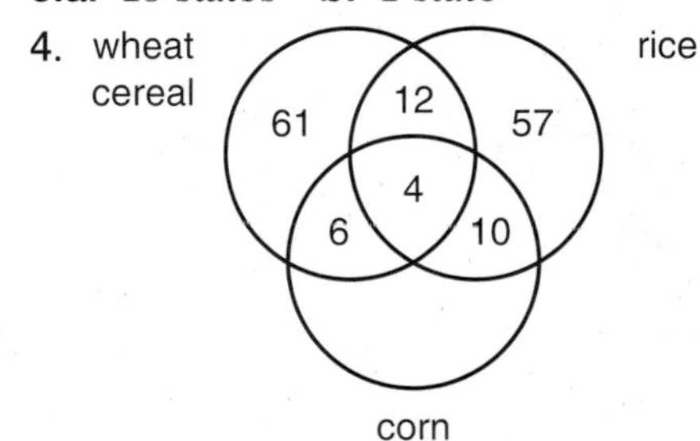

5.a.

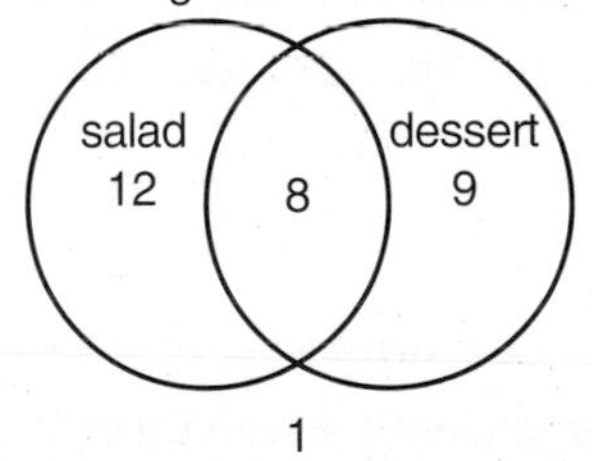

5.b. 12 people **c.** 1 person **7.** 35

Pages 184-186 Lesson 5-3
6. scalene, right **7.** scalene, acute
8. isosceles, acute **9.** equilateral, acute
10. scalene, obtuse **11.** isosceles, right
13. isosceles, right **15.** scalene, right
17. equilateral, acute **19.** true **21.** false

23. true **25.** 40 **26.** 92 **27.** -384

28. $\frac{3}{4}$ of the days

29. $\overline{HI} \parallel \overline{KJ}$ **31.b.** Triangles will not collapse as easily as other figures.

Pages 188-190 Lesson 5-4

5. Q, P, R, S, RH **6.** Q, T **7.** Q **8.** 60° **9.** Q, P **11.** Q, P, R, S, RH **13.** Q **15.** parallelogram, rectangle, square, rhombus **17.** square, rhombus **19.** false **21.** 70 **23.** 17 cm, 16.5 cm^2 **24.** IV **25.** 12

29.a.

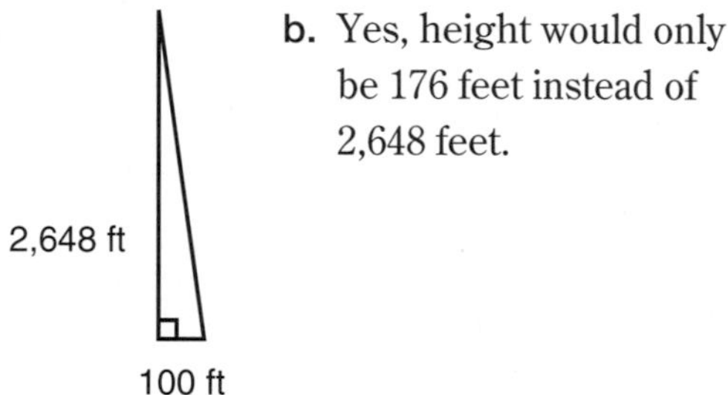

b. Yes, height would only be 176 feet instead of 2,648 feet.

Page 190 Mid-Chapter Review

1. $\overline{AB} \parallel \overline{CD}$ **3.** scalene, acute **5.** square

Pages 194-195 Lesson 5-5

4.

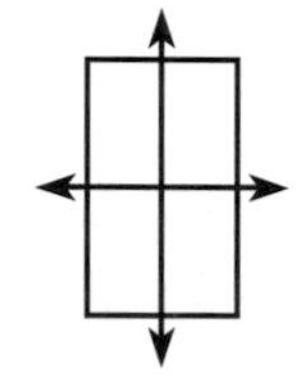

6.

7. all of them **9.** no line symmetry **11.** yes **13.** yes **15.** parallelogram, rectangle, square, rhombus **17.** -229 **18.** positive **19.** 30 **21.** A, B, C, D, E, H, I, K, M, O, T, U, V, W, X, Y

Pages 199-200 Lesson 5-6

6. similar **7.** congruent **8.** neither **9.** similar **11.** congruent **13.** similar **15.** 18 **16.** 8,000; 8,700 **17.** 27 **18.** yes **19.a.** \$28.50 **b.** Family Show

Pages 204-205 Lesson 5-7

7. 4: translation; 5: rotation; 6: translation **14.** 90 **16.** congruent

Pages 206-208 Study Guide and Review

11. $\overline{WX} \parallel \overline{ZY}, \overline{ZW} \parallel \overline{YX}$ **13.** scalene, right **15.** Q, P, RH

17.

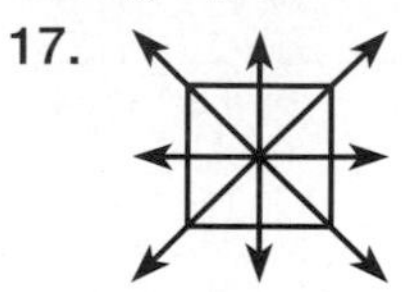

19. yes **21.** similar **25.** 4 pairs **27.** Cape May (C)

6 Patterns and Number Sense

Pages 213-214 Lesson 6-1

5. 2 **6.** 3, 9 **7.** 2, 3, 5, 6, 10 **8.** none **9.** 3, 5 **10.** yes **11.** no **12.** no **13.** yes **14.** Sample answer: $2 \cdot 63, 3 \cdot 42$ **15.** Sample answer: 6, 12 **17.** yes **19.** no **21.** yes **23.** no **25.** Sample answers: $2 \cdot 21{,}605$; $5 \cdot 8{,}642$ **27.** 486 **29.** 26, 378 **30.** \$1.89 **31.** -829 **32.** Probably no relationship

Page 217 Lesson 6-2

4. composite **5.** prime **6.** prime **7.** composite **8.** $3 \cdot 19$ **9.** $2^2 \cdot 3^2$ **10.** $2 \cdot 3^2 \cdot 5$ **11.** $2^2 \cdot 3^2 \cdot 5$ **13.** composite **15.** composite **17.** composite **19.** prime **21.** 3^4 **23.** 2^6 **25.** $2^4 \cdot 5^2$ **27.** $-1 \cdot 2^2 \cdot 3^3 \cdot 5^2$ **29.** $m < 6$ **31.** no symmetry **32.** yes **33.a.** 3, 5; 5, 7; 11, 13; 17, 19; 41, 43; 71, 73 **b.** 4, 6, 12, 18; Even numbers; yes; twin primes are both odd, therefore, the number in the middle is even. **35.** prime **37.** 41

Page 220 Lesson 6-3

3. 2 factors: 53; 3 factors: 49; 4 factors: 51, 5 factors: 81 **4.** 6; 123, 132, 231, 213, 312, 321 **5.** 81, 625, 2,401; n^4, where n is prime **7.** 12 teams **9.** 900 mL **11.** 20, 10, 5, 4, 2

Pages 222-223 Lesson 6-4

5. 4 **6.** 5 **7.** 14 **8.** 15 **9.** 3 **10.** 15 **11.** 9 **12.** 1 **13.** 2 **15.** 8 **17.** 25 **19.** 12 **21.** 5 **23.** 30 **25.** if the units digit is 0 or 5 **27.** 0.864 L **28.** 21 **29.** scalene, obtuse **30.** $2^3 \cdot 7$ **31.a.** 6 in. × 6 in. tile **b.** 20 tiles **33.** Sample answer: 7, 14, 21

Pages 225-226 Lesson 6-5

5. R **6.** R **7.** R **8.** W, I, R **9.** $-\frac{7}{8}$ **10.** simplest form **11.** $\frac{1}{2}$ **12.** $\frac{1}{4}$ **13.** simplest form **15.** R **17.** W, I, R **19.** $-\frac{1}{4}$ **21.** $\frac{4}{9}$ **23.** $\frac{1}{17}$ **25.** $-\frac{5}{8}$ **27.** $\frac{7}{33}$ **29.** 35 **30.** -12 **31.** 27 **32.** 36 **33.** yes; 2.5 or $2\frac{1}{2}$ **35.** $\frac{16}{25}$

Pages 228-229 Lesson 6-6

5. 0.4 **6.** -0.875 **7.** 3.25 **8.** -0.35 **9.** 3.56 **10.** $-\frac{2}{5}$ **11.** $\frac{3}{4}$ **12.** $\frac{17}{100}$ **13.** $3\frac{3}{25}$ **14.** $-5\frac{3}{8}$ **15.** -0.8 **17.** -0.28 **19.** -5.375 **21.** 7.25 **23.** -0.28125 **25.** $\frac{1}{20}$ **27.** $\frac{16}{25}$ **29.** $3\frac{17}{20}$ **31.** $-\frac{3}{40}$ **33.** terminating **35.** $\frac{y}{3} - 15$ **36.** $>$ **37.** 9, 10, 11, 12 **38.** I, R

Page 229 Mid-Chapter Review

1. yes **3.** yes **5.** $2^2 \cdot 3^2$ **7.** 2^7 **9.** 30°F **11.** 16 **13.** $\frac{1}{4}$ **15.** $\frac{1}{5}$ **17.** $-\frac{4}{5}$ **19.** -0.625

Pages 231-232 Lesson 6-7

5. 0.2642642642 **6.** 0.9222222222 **7.** 0.5082508250 **8.** 0.5082082082 **9.** 0.2161616161 **10.** 100 **11.** 10 **12.** 10 **13.** 100 **15.** $29.\overline{27}$ **17.** $-2.\overline{45}$ **19.** $-7.0\overline{74}$ **21.** 8.32 **23.** $-1\frac{7}{9}$ **25.** $\frac{6}{11}$ **27.** $2\frac{5}{9}$ **29.** $\frac{8}{33}$ **31.** $\frac{115}{333}$ **32.** $4^3 \cdot 5^2$ **33.** 15 **34.** 25 **36.** $8\frac{17}{25}$ **37.** $\frac{17}{57} \approx 0.298245614$

Pages 234-235 Lesson 6-8

6. $\frac{1}{2}$, 0.5 **7.** 0 **8.** $\frac{1}{3}$, $0.\overline{3}$ **9.** $\frac{1}{7}$, $0.\overline{142857}$ **10.** 1 **11.** 0 **12.** $\frac{3}{10}$ **13.** $\frac{7}{20}$ **14.** $\frac{4}{5}$ **15.** 1 **16.** $\frac{4}{5}$ **17.** $\frac{2}{11}$ **19.** 0 **21.** $\frac{3}{11}$ **23.** $\frac{24}{29}$ **25.** 1 **27.** $\frac{4}{49}$ **29.** 0 **30.** $P = 36$ in., $A = 60$ in^2 **31.** -13 **32.** neither **33.** $8\frac{8}{11}$ **35.a.** 150 cards **b.** 2,000 cards **c.** 400 cards **d.** 7,450 cards **37.** No; the number of ways something can occur cannot be negative.

Pages 237-238 Lesson 6-9

5. 0, 5, 10, 15, 20, 25 **6.** 0, 18, 36, 54, 72, 90 **7.** 0, 20, 40, 60, 80, 100 **8.** 0, n, $2n$, $3n$, $4n$, $5n$ **9.** 15 **10.** 105 **11.** 180 **12.** 840 **13.** 48 **14.** 60 **15.** 1,225 **16.** 72 **17.** 60 **19.** 420 **21.** 100 **23.** 24 **25.** 24 **27.** 8,449 **29.** no **30.** 559 **31.** 28 **32.** 608 **33.** no **34.** true **35.** $\frac{1}{2}$ **37.** when one number is a factor of the other **39.** $24n^2$

Pages 243-244 Lesson 6-10

3. 6 **4.** 40 **5.** 18 **6.** 100 **7.** $<$ **8.** $=$ **9.** $<$ **10.** $<$ **11.** $>$ **12.** $=$ **13.** $=$ **14.** -12, -5, -1, 2, 5 **15.** 0, $\frac{2}{5}$, $\frac{1}{2}$, $\frac{4}{5}$ **16.** 0.367, $\frac{3}{8}$, 0.376, $\frac{2}{5}$ **17.** $<$ **19.** $<$ **21.** $<$ **23.** $<$ **25.** $<$ **27.** 200 **29.** $-\frac{1}{3}$, $-\frac{1}{4}$, $\frac{1}{10}$, $\frac{1}{9}$ **31.** $0.\overline{18}$, 0.182, $0.18\overline{2}$, $0.182\overline{5}$ **32.** 14 quarts **33.** -11 **34.** 7, 7, 3 **35.** 120 **39.a.** 2.7 **b.** Asheville, NC

Page 247 Lesson 6-11

4. 34,500,000 **5.** 0.000089 **6.** 37,770 **7.** 1.23×10^7 **8.** 1.23×10^6 **9.** 1.23×10^{-4} **10.** 1.23×10^1 **11.** 5.6789×10^{-3} **12.** 8.29×10^2 **13.** 7.0×10^{-6} **14.** 1.0×10^{-6} **15.** -0.0000000999 **17.** 0.0000042 **19.** 0.096 **21.** 93,000,000 **23.** 8.542×10^7 **25.** 5.6×10^{-5} **27.** 4.0×10^{-8} **29.** 7.2×10^{-8} **30.** 29.2 **31.** $c > 8$ **32.** -221 **33.** 24 **35.a.** Jupiter **b.** 4.83×10^8 miles **c.** 3.901×10^8 miles

Pages 248-250 Study Guide and Review

7. no **9.** yes **11.** yes **13.** $2^4 \cdot 3$ **15.** $5^2 \cdot 7$ **17.** $3 \cdot 11$ **19.** 18 **21.** 7 **23.** 40 **25.** $\frac{3}{4}$ **27.** $-\frac{2}{3}$ **29.** $0.\overline{18}$ **31.** $6.\overline{6}$ **33.** $-\frac{7}{250}$ **34.** $-11\frac{3}{8}$ **37.** $-5\frac{28}{99}$ **39.** $6\frac{70}{111}$ **41.** $-\frac{1}{33}$ **43.** $\frac{3}{4}$ **45.** 60 **47.** 216 **49.** 870 **51.** $<$ **53.** $>$ **55.** 7.35×10^{-5} **57.** 6.8×10^{-4} **59.** 4th week

7 Rational Numbers

Pages 257-258 Lesson 7-1

4. $\frac{1}{2}$ **5.** $\frac{1}{5}$ **6.** $1\frac{1}{3}$ **7.** $1\frac{1}{8}$ **8.** $-\frac{1}{5}$ **9.** $\frac{4}{7}$ **11.** $1\frac{7}{9}$ **13.** $-\frac{1}{3}$ **15.** $-3\frac{1}{2}$ **17.** $-3\frac{1}{9}$

19. $\frac{2}{3}$ **21.** $\frac{1}{2}$ **23.** -3 **25.** 3 **27.** -3

29. n **31.** $3\frac{1}{2}$ tons **32.** 3

33. yes **34.** 5.238×10^7 **35.** 6 min **37.a.** $\frac{1}{3}$

b. $\frac{2}{3}$ **c.** 1

Pages 260-261 Lesson 7-2

3. 7 **4.** 5 **5.** 17 **6.** 11 **7.** $-\frac{1}{12}$ **8.** $1\frac{1}{6}$

9. $-\frac{3}{8}$ **10.** $1\frac{5}{18}$ **11.** $1\frac{2}{3}$ **12.** $6\frac{5}{6}$ **13.** $1\frac{3}{10}$

15. $-1\frac{1}{12}$ **17.** $-9\frac{3}{8}$ **19.** $8\frac{7}{8}$ **21.** $-10\frac{13}{18}$

23. $-10\frac{7}{10}$ **25.** $3\frac{71}{72}$ **27.** $3\frac{13}{36}$ **28.** 27

29. 132, 143, 124 **30.** $\frac{3}{4}$ **31.** $-2\frac{1}{3}$

Pages 263-264 Lesson 7-3

4. $\frac{3}{8}$ **5.** $\frac{8}{27}$ **6.** $\frac{8}{9}$ **7.** 3 **8.** $2\frac{2}{9}$ **9.** $-\frac{3}{4}$

10. $3\frac{3}{8}$ **11.** $-1\frac{1}{2}$ **12.** $4\frac{1}{5}$ **13.** $-\frac{3}{4}$

15. $-\frac{10}{27}$ **17.** $2\frac{13}{16}$ **19.** $-13\frac{1}{2}$ **21.** $27\frac{1}{2}$

23. $\frac{4}{9}$ **25.** $3\frac{1}{5}$ **27.** $-7\frac{1}{2}$ **29.** $\frac{15}{16}$

30. 20 cm, 25 cm^2 **31.** 38

32. $-\frac{1}{5}, -\frac{1}{10}, \frac{1}{8}, \frac{1}{3}, \frac{1}{2}$ **33.** $-4\frac{19}{24}$

Pages 266-267 Lesson 7-4

5. yes **6.** no **7.** no **8.** yes **9.** $\frac{1}{5}$

10. $-\frac{3}{2}$ **11.** 5 **12.** $\frac{5}{14}$ **13.** $-\frac{1}{12}$ **14.** $6\frac{2}{5}$

15. 22 **17.** $-\frac{5}{3}$ **19.** $\frac{9}{26}$ **21.** -1 **23.** $-\frac{1}{x}$

25. yes; $\frac{3}{10} \cdot \frac{10}{3} = 1$ **27.** $-4\frac{1}{2}$ **29.** $1\frac{5}{8}$

31. 0 **33.** $-2\frac{1}{12}$ **34.** 2,750 mm **35.** 30

36. 8 **37.** $-\frac{7}{36}$

Page 271 Lesson 7-5

3. 21 **4.** 14 **5.** 50 **7.** 36 exercises **9.** 6

Pages 274-275 Lesson 7-6

5. N; 25, 36, 49 **6.** G; $-\frac{1}{125}, \frac{1}{625}, -\frac{1}{3{,}125}$

7. G; 32, 64, 128 **8.** A; 97.4, 97, 96.6

9. A; -15, -21, -27 **10.** A; 36, 40, 44

11. A; 55, 44, 33 **12.** G; 81, -243, 724

13. A; 7.5, 9, 10.5 **14.** A and G; 89, 89, 89

15. A; 2, 4, 6 **16.** G; 32, -16, 8 **17.** 64, 55, 46

19. -3, -6, -9 **21.** -157, 78.5, -39.25

23. 11, 12.5, 14 **25.** 26, 37, 50

27. $\frac{1}{108}, \frac{1}{324}, \frac{1}{972}$ **29.** 55 **31.** 22, 20, 18, 16

33. 335 **34.** 46

35.

			×						
		×	×		×			×	×
×	×	×	×		×	×	×	×	×
16	17	18	19	20	21	22	23	24	25

36. -0.625 **37.** $2\frac{6}{7}$ **39.** \$6,246.40

Page 275 Mid-Chapter Review

1. $-\frac{1}{2}$ **3.** $6\frac{3}{11}$ **5.** $-10\frac{5}{24}$ **7.** -28

9. $12\left(3 + \frac{3}{4}\right) = 12 \cdot 3 + 12 \cdot \frac{3}{4}$

11. A; 32, 35, 38

Pages 280-281 Lesson 7-7

5. $2\frac{2}{3}$ ft, $3\frac{3}{4}$ ft, 5 ft^2 **6.** 5.8 cm, 2.2 cm, 3.6 cm, 14.4 cm^2 **7.** $4\frac{1}{2}$ in., $5\frac{1}{3}$ in., 6 in., $29\frac{1}{2}$ in^2

8. $8\frac{7}{16}$ ft^2 **9.** 11.7 cm^2 **10.** 35 in^2 **11.** 90 m^2

12. 0.08 km^2 **13.** $9\frac{3}{4}$ yd^2 **15.** $3\frac{1}{2}$ ft, $5\frac{1}{2}$ ft, $2\frac{2}{3}$ ft, 12 ft^2 **17.** 30 cm, 12 cm, 180 cm^2

19. 3 ft, 4 ft, 6 ft^2 **21.** 297 yd^2 **23.** $4\frac{31}{64}$ in^2

25. 112 mm^2 **27.** 255 ft^2 **29.** 68.88 km^2

31. 0.204 m^2 **32.** $2c + 3 = 15$ **33.** -500

34. 31.75, 33, 40 **35.** similar **36.** 64, 60, 56

37. 570 ft^2 **39.a.** $A \approx \frac{1}{2}(273)(219)$ **b.** about 29,894 mi^2

Pages 286-287 Lesson 7-8

4. 56.55 in. **5.** 8.17 m **6.** 15.7 cm

7. 43.98 in. **8.** 42.73 m **9.** 16.5 in or 16.49 in.

10. 21.99 km **11.** 3.14 ft **13.** 27.49 in. or $27\frac{1}{2}$ in. **15.** 42.6 m **17.** 50.27 ft **19.** 27.65 m

21. $28\frac{2}{7}$ yd or 28.27 yd **22.** 64

23.a.

Stem	Leaf
5	8
6	2 5 7
7	1 1 2 3 5 6 9
8	2 3 4 5 8
9	1 3 4 7

7 | 1 means 71.

b. 70–79

24. 15 yd, 20 yd, 10 yd, 175 yd^2 **25.** rectangle; 7.5 cm × 28.6 cm **27.** ≈ 235,933.6 miles

29.a. 56.55 **b.** about 4.2 inches

Pages 289-290 Lesson 7-9

6. $\frac{4}{5}$ **7.** $\frac{5}{12}$ **8.** $-3\frac{3}{4}$ **9.** $\frac{2}{27}$ **10.** $3\frac{5}{9}$

11. $\frac{3}{5}$ **12.** $5\frac{1}{2}$ **13.** $-\frac{17}{18}$ **14.** $\frac{9}{25}$ **15.** 4

17. -5 **19.** 6 **21.** 6 **23.** $\frac{2}{3}$ **25.** $\frac{3}{14}$

27. $2\frac{3}{4}$ or 2.75

28. $b > -8$

-10 -9 -8 -7 -6

29. 16 **30.** $15.\overline{36}$ **31.** 31.4 mm

33. 128 slices

Pages 292-293 Lesson 7-10

3. -5.5 **4.** $-\frac{52}{75}$ **5.** 3.24 **6.** $4\frac{1}{2}$ **7.** $2\frac{3}{10}$

8. $-2\frac{1}{2}$ **9.** 22.8 **10.** 84 **11.** $0.2\overline{27}$ **13.** -18

15. -15.2 **17.** 1.8875 **19.** -0.12 **21.** $-1\frac{1}{9}$

23. $-1\frac{4}{5}$ **25.** 1,200 **26.** positive **27.** 6

28. 90 **29.** $\frac{5}{8}$ **31.a** 35 mph **b.** 55 mph

c. 25 mph

Pages 294-296 Study Guide and Review

11. $-\frac{3}{4}$ **13.** $-\frac{4}{15}$ **15.** $-11\frac{5}{12}$ **17.** $-10\frac{1}{2}$

19. $2\frac{1}{3}$ **21.** $-\frac{3}{19}$ **23.** $2\frac{4}{5}$ **25.** G; $\frac{1}{3}, \frac{1}{9}, \frac{1}{27}$

27. A; 44, 40, 36 **29.** 51 cm^2 **31.** 15.08 m

33. 17.28 in. **35.** $\frac{3}{5}$ **37.** $\frac{5}{24}$ **39.** -17.34

41. 4.1 **43.** 18.85 cm

8 Real Numbers

Pages 301-302 Lesson 8-1

4. 7 **5.** 9 **6.** 11 **7.** -8 **9.** 20 **11.** -3

13. 25 **15.** -10 **17.** 0.4 **19.** $\frac{8}{10}$

21. 1.3 meters **22.** 7.5 yards **24.** 3 **25.** $-\frac{1}{3}$

27. $23.80

Page 305 Lesson 8-2

3. 7 **4.** 12 **5.** 5 **6.** 3 **7.** 5 **9.** 7 **11.** 7

13. 13 **15.** 20 **17.** 31 **19.** 6 **20.** $-1\frac{7}{12}$

21. 30 **23.** about 36.6 miles

Pages 308-309 Lesson 8-3

5. irrational **6.** rational **7.** integer, rational

8. rational **9.** 2.6 **10.** 2.8 **11.** 4.5

12. -1.4 **13.** 12, -12 **14.** 30, -30 **15.** 7.1,

-7.1 **17.** irrational **19.** rational **21.** rational

23. 2.4 **25.** 5.2 **27.** 8, -8 **29.** 19.0, -19.0

31. 1.2, -1.2 **33.** 12 **34.** -336 **35.** 59, 59, no

mode **36.** $2\frac{1}{8}$ **37.** 17 **39.** 36

Pages 311-312 Lesson 8-4

3. 135 miles **4.** -$2,005 **5.** 65 miles per hour

7. 720 ft^2 **9.a.** 4 miles **b.** about 20 seconds

Page 312 Mid-Chapter Review

1. 6 **3.** -5 **5.** 9 **7.** 5 **9.** natural, whole,

integer, rational **11.** rational **13.** 7, -7

15. 120 feet per second

Pages 317-318 Lesson 8-5

4. $c^2 = 144 + 81$, 15 **5.** $c^2 = 144 + 25$, 13

6. $1{,}681 = 81 + b^2$, 40 **7.** no **8.** yes

9. 7.9 ft **11.** 7.4 m **13.** 18.4 cm **15.** 9 in.

17. $x^2 = 100 + 100$, 14.1 **19.** $x^2 = 1 + 2$, 1.7

21. yes **23.** yes **25.** yes **26.** 1,250 grams

27. 32, 33, 34, 35 **28.** -0.0000347 **29.** 12, -12

31. 17 yards

Pages 321-322 Lesson 8-6

3. 20.1 ft, 15 ft **4.** 21.4 miles **5.** 14.7 ft

7. Sample answer: 16–30–34, 40–75–85

9. Sample answer: 18–80–82, 45–200–205

11. 8 **12.** no **13.** about 13.2 m

15. about 70 ft

Pages 326-327 Lesson 8-7

4. 5 units **5.** 10 units **6.** 5 units

7. 4.1 units **9.** 5.1 units **11.** 7.6 units

13. 10.3 units **15.** 4.2 units **16.** 42 **17.** -80

18. about 31.5 ft **19.b.** 34.6 units **21.** 9.4

miles

Pages 330-331 Lesson 8-8

4. 8.7 cm, 10 cm **5.** 8 ft, 11.3 ft **6.** 6 in.,

10.4 in. **7.** 15 cm **8.** 6 in. **9.** 34.6 ft, 40 ft

11. 9.5 in., 16.5 in. **13.** 3.75 in. **14.** 60, 64, 50

15. $\frac{2}{3}$ **16.** 10 units **17.** 12 feet

Pages 332-334 Study Guide and Review

11. -1.5 **13.** $\frac{7}{10}$ **15.** 2.3 **17.** 7 **19.** 19

21. 16 **23.** integer, rational **25.** rational

numbers **27.** 8.6 cm **29.** 4 in. **31.** 36 in.

33. 3.6 units **34.** 7.8 units **37.** 14 in., 24.2 in. **39.** 15 mm, 21.2 mm **41.** 1.25 m per second

9 Applications with Proportion

Pages 340-341 Lesson 9-1

4. $\frac{5}{7}$ **5.** $\frac{1}{8}$ **6.** $\frac{4}{5}$ **7.** $\frac{7}{11}$
8. 3 brown-eyed/2 blue-eyed **9.** $\frac{1}{3}$
10. 25 miles/1 hour; unit rate
11. 3 pounds/1 week; unit rate
12. 2 inches/15 days; not a unit rate
13. 34 passengers/3 minivans; not a unit rate
15. $\frac{7}{11}$ **17.** 1 win/1 loss **19.** $\frac{13}{21}$ **21.** $\frac{4}{1}$
23. $\frac{3}{2}$ **25.** $\frac{1}{4}$ **27.** $\frac{1}{3}$ **29.** \$2.50/disk
31. \$28/ticket **33.** \$0.08/egg **35.** 10
36. Mean is the arithmetic average. Median is the middle number when data are ordered from least to greatest. Mode is the most frequent data.
37. 1 out of 10 chances of winning a prize
38. $7\frac{2}{7}$, 7.35, $\frac{37}{5}$ **39.** 15 **40.** natural numbers, whole numbers, integers, rational numbers, real numbers **43.** Eisenhower Middle School
45. 0.320

Pages 345-346 Lesson 9-2

3. yes **4.** no **5.** yes **6.** no **7.** 6 **8.** 0.2 **9.** 300 **10.** 4.9 **11.** yes **13.** yes **15.** yes **17.** no **19.** 15 **21.** 85 **23.** 10.5 **25.** 3.5 **27.** -12 **28.** 40° **29.** 16 **30.** 40 shrimp/ 1 pound

Pages 348-349 Lesson 9-3

3. \$26.88 **4.** 14 pounds **5.** 3 cups
7. 262.5 min **9.** 18 cans **10.** 9.6 **11.** -8
12. 12 **13.** 21 pounds **15.** Alvarez 75,000; Cruz 90,000; Hoffman 55,000; Newton 30,000

Pages 351-352 Lesson 9-4

3. 49 seats **4.** 4 games **5.** -19
7. 28 handshakes **9.** 9 meters

Page 352 Mid-Chapter Review

1. $\frac{3}{4}$ **3.** \$0.80/dozen **5.** 10 **7.** 1,050 bushels

Pages 354-355 Lesson 9-5

4. no **5.** yes **6.** 12 **7.** 18 **8.** They would be half the original lengths. **9.** yes
11. $\angle EAD \cong \angle CAB, \angle AED \cong \angle ACB, \angle ADE \cong \angle ABC$ **13.** 12 **15.** $4^3 \cdot 8^2$
16. 51, 23, 0, -8, -16, -30, -51 **17.** $\frac{7}{9}$
18. 10 tablespoons **21.** 2.3 inches

Pages 357-358 Lesson 9-6

5. 35 feet **6.** 20 feet **7.** 12.5 km **9.** 45 miles
10. 45 **11.** 240 **12.** $\frac{AC}{DF}, \frac{AB}{DE}, \frac{BC}{EF}$
13. 169.25 feet **15.** 985.92 ft

Page 360 Lesson 9-7

4. 120 ft, 16 ft **5.** 180 miles **6.** 135 miles
7. 247.5 miles **8.** 37.5 miles **9.** 24 ft
11. 10 ft **13.** 6 **14.** 1,250 feet **15.** 384 cm

Pages 362-363 Lesson 9-8

4. (6, 8) **5.** (15, 20) **6.** 3 **7.** 25 **13.** 4
15. 712.5 mi **17.a.** rectangle **b.** 42 units, 108 units^2 **d.** 126 units, 972 units^2 **e.** $\frac{3}{1}$ **f.** $\frac{9}{1}$

Pages 366-367 Lesson 9-9

4. $\frac{3}{5}$ **5.** $\frac{5}{3}$ **6.** $\frac{9}{5}$ **7.** $\frac{5}{9}$ **8.** 31° **9.** 59°
10. 61° **11.** 29° **12.** 3.0 inches **13.** 5 ft
15. $\frac{12}{5}$, $\frac{5}{12}$, 67°, 23° **17.** 6.9 in. **19.** 3.6 km
22. $4\frac{13}{16}$ **25.** about 48 feet

Pages 370-371 Lesson 9-10

4. $\frac{3}{5}$ **5.** $\frac{4}{5}$ **6.** 53° **7.** $\frac{3}{5}$ **8.** $\frac{4}{5}$ **9.** 37°
10. $\frac{5}{13}$ **11.** $\frac{5}{13}$ **12.** 23° **13.** 67°
15. $\frac{4}{5}, \frac{3}{5}, \frac{3}{5}, \frac{4}{5}$, 53°, 37° **17.** 12.0 **19.** 45°
21. -180 **22.** 0.00092 **23.** $\frac{3}{4}$
25. about 39.2 feet **27.** 66 ft

Pages 372-374 Study Guide and Review

9. $\frac{1}{2}$ **11.** $\frac{36}{5}$ **13.** 15 **15.** 40.5 **17.** \$175
19. 25 pages **21.** similar **23.** 105 km
25. 56 km **27.** 147 km **31.** 0.4167
33. 0.8824 **34.** 0.4706 **37.** $\frac{7}{16}$

10 Applications with Percent

Pages 382-383 Lesson 10-1

4. 68% **5.** 35% **6.** 87.5% **7.** 90% **8.** 5% **9.** b **10.** a **11.** c **12.** 101.1 **13.** 11 **14.** 60% **15.** 9% **17.** 70% **19.** 46% **21.** $66\frac{2}{3}\%$ **23.** 80% **25.** 80.5 **27.** 12.5% **29.** 190 **31.** 112 **33.** 64% **34.** $67.20 **35.** 21; 23; 17 **36.** 80° **37.** $\frac{5}{9}$ **38.** 14.8 **39.** 23°

Pages 386-387 Lesson 10-2

4. 70% **5.** 60.5% **6.** 26% **7.** 2% **8.** $\frac{13}{20}$ **9.** $\frac{13}{200}$ **10.** $\frac{1}{8}$ **11.** $\frac{24}{25}$ **12.** 0.78 **13.** 0.09 **14.** 0.123 **15.** 0.084 **17.** 0.3 **19.** 18% **21.** 70.4% **23.** 55.3% **25.** $\frac{29}{50}$ **27.** $\frac{89}{200}$ **29.** $\frac{73}{400}$ **31.** 0.28 **33.** 0.8425 **35.** 0.384 **37.** 0.2, 20%, $\frac{1}{5}$ **38.** -21 **39.** $-2 \cdot 3^4$ **40.** -7.2 **41.** They will have the same shape. **42.** 70

Pages 389-390 Lesson 10-3

5. $\frac{1}{5{,}000}$ **6.** $\frac{1}{1{,}000}$ **7.** $1\frac{3}{4}$ **8.** $\frac{1}{500}$ **9.** 1.78 **10.** 2.012 **11.** 0.006 **12.** 0.0005 **13.** $>$ **14.** $<$ **15.** $\frac{3}{1{,}000}$ **17.** $7\frac{3}{5}$ **19.** $2\frac{43}{100}$ **21.** $\frac{1}{625}$ **23.** 2.12 **25.** 0.0003 **27.** 0.007 **29.** 0.00008 **31.** $\frac{3}{8}\%$, 67%, 0.8, 7 **32.** 6 **33.** 8.4×10^{-5} **34.** 14.5 feet **35.** $\frac{1}{27}$ **36.** $0.385 = \frac{77}{200}$ **37.** 0.0985 **39.** Yes, the purchase price of the camera is 100% + 5.5% tax = 105.5%.

Page 392 Lesson 10-4

3. 105 **4.** 20,100 **5.** 63 **6.** 18 **7.** 6.5% **9.** 50 calories **11.** 128,000 people **13.** $\frac{1}{4}$

Pages 394-396 Lesson 10-5

4. a **5.** b **6.** b **7.** a **8.** 30 **9.** 10 **10.** 54 **11.** 50% **12.** $66\frac{2}{3}\%$ **13.** 150% **14.** 28% **15.** 20% **17.** 9 **19.** 22 **21.** 5 **23.** 30% and 150 **25.** 64% **27.** 40% **29.** 20% **31.** 25% **33.** 58.5 ft^2 **34.** $\frac{1}{6}$ **35.** $0.32/ounce **36.** $\frac{1}{250}$ **39.** 17,600; 8,800; 7,200; 8,800; 880; 440

Page 396 Mid-Chapter Review

1. 24 **3.** 44% **5.** 62.5% **7.** 0.80, $\frac{4}{5}$ **9.** 0.375, $\frac{3}{8}$ **11.** 0.004, $\frac{1}{250}$ **13.** 0.00875, $\frac{7}{800}$ **15.** 6 cheerleaders **17.** b **19.** c

Pages 401-402 Lesson 10-6

4. 24.44 **5.** 44.8 **6.** 48.1% **7.** 0.06% **8.** $160 **9.** 82.1% **10.** $23\frac{1}{3}\%$ **11.** 9.9 **12.** 16,666.67 **13.** 9.072 **15.** 500 **17.** 2.96 **19.** 325 **21.** 120% **23.** $50 **25.** 24 **27.** $6,540 **29.** 40 **30.** $-a$ **31.** 11, -11 **32.** 18 **33.** $48,700 **37.** $242.64 **39.** 7%

Pages 404-405 Lesson 10-7

5. 100% **6.** 360° **11.** 28 **12.** 306 **13.** 37.5%

Pages 407-408 Lesson 10-8

4. 50% **5.** 10% **6.** $16\frac{2}{3}\%$ **7.** 33% **8.** 11% **9.** 13% **11.** 25% **13.** 17% **15.** 36% **17.** 11% **19.** 7% **20.** 4.5 pints **21.** 9.2 ft

Pages 410-411 Lesson 10-9

3. $9.80 **4.** $100 **5.** $3.44 **6.** $5.78 **7.** $809.10 **8.** $199.50 **9.** $28.46 **10.** $28 **11.** 38.5% **12.** 12.5% **13.** $7.49; $22.46 **15.** $2.18; $12.32 **17.** $39.83; $79.67 **19.** 25% **21.** 50% **23.** $3.41 **25.** 20% **27.** 20 **28.** $19\frac{1}{32}$ **29.** 7.2 **30.** 3.5% **31.a.** $4.56 **b.** $2.94 **c.** 60%

Pages 413-415 Lesson 10-10

4. $10.78 **5.** $90 **6.** $195.84 **7.** $2.57 **8.** $643.70 **9.** $142.95 **10.** $128.28 **11.** $256.95 **12.** 6% **13.** 12.75% **15.** $75 **17.** $674.56 **19.** $945 **21.** $210.13 **23.** $240.41 **25.** $10,650 **27.** 9% **29.** Number of people who watch 0–5 hours is greater than the number of people who watch 15–17 hours. **30.** $\frac{5}{54}$ **31.** 55 miles **32.** $26.70; $62.30 **33.** $1,064 **37.a.** to change the percent to a decimal **b.** 2330

c. A10: 4000; B10: $\frac{B2}{100}$; C10: C2; D10: A10 * B10 * C10; E10: A10 + D10

Pages 416-418 Study Guide and Review

11. 112 **13.** 82 **15.** 80 **17.** $\frac{2}{25}$ **19.** $\frac{7}{40}$ **21.** $\frac{1}{125}$ **23.** $\frac{3}{700}$ **25.** 20 **27.** 9 **29.** 28 **31.** 18.75% **33.** 150 **35.** 20% increase **37.** 25% increase **39.** \$9, \$51 **41.** \$21, \$63 **43.** \$74.52 **45.** \$22.50 **47.** \$29.99 **49.** 10% increase

11 Algebra: Functions and Graphs

Pages 423-424 Lesson 11-1

5.

n	-5n	f(n)
-4	-5(-4)	20
-2	-5(-2)	10
0	-5(0)	0
2.5	-5(2.5)	-12.5

6.

n	2n + (-6)	f(n)
-2	2(-2) + (-6)	-10
-1	2(-1) + (-6)	-8
0	2(0) + (-6)	-6
$\frac{1}{2}$	$2\left(\frac{1}{2}\right) + (-6)$	-5

7. 2 **8.** 14 **9.** 25

11.

n	3n	f(n)
-1	3(-1)	-3
0	3(0)	0
$\frac{2}{3}$	$3\left(\frac{2}{3}\right)$	2
1	3(1)	3

13.

n	-0.5n + 1	f(n)
-4	-0.5(-4) + 1	3
-2	-0.5(-2) + 1	2
0	-0.5(0) + 1	1
2.5	-0.5(2.5) + 1	-0.25
8	-0.5(8) + 1	-3

15. -8 **17.** 18.12

18. Sample answer:

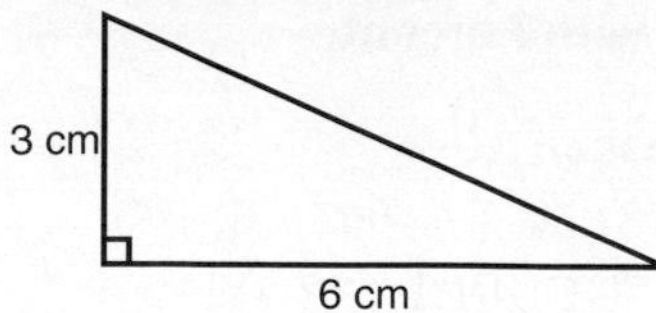

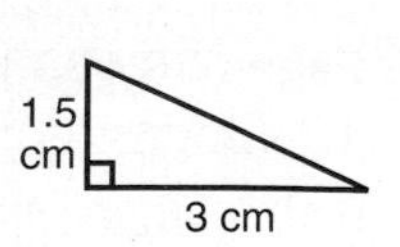

19. 6 cm **20.** \$304

Pages 426-427 Lesson 11-2

5.

n	f(n)	(n, f(n))
-1	3	(-1, 3)
1	5	(1, 5)
2	6	(2, 6)
4	8	(4, 8)
5	9	(5, 9)

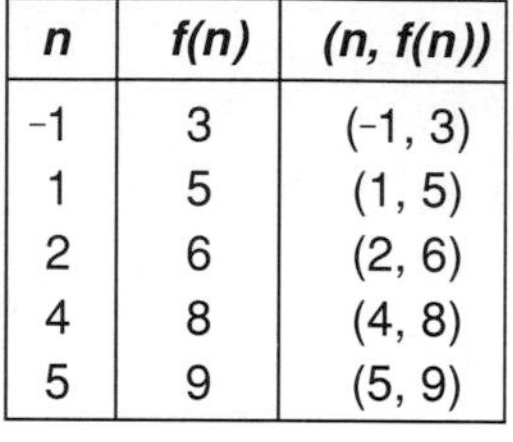

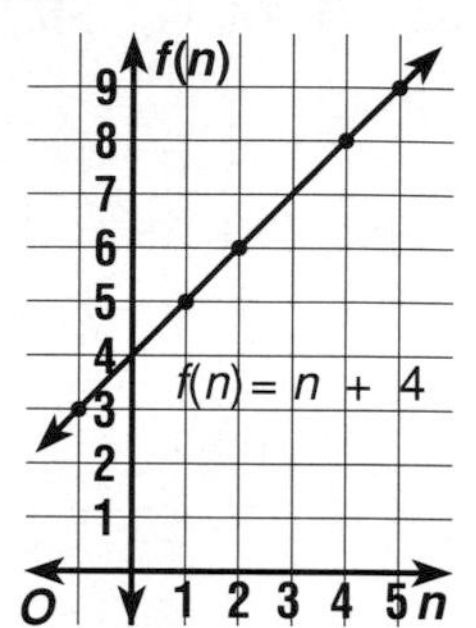

6.

n	f(n)	(n, f(n))
$\frac{1}{2}$	16	$(\frac{1}{2}, 16)$
2	4	(2, 4)
4	2	(4, 2)
8	1	(8, 1)
16	$\frac{1}{2}$	$(16, \frac{1}{2})$

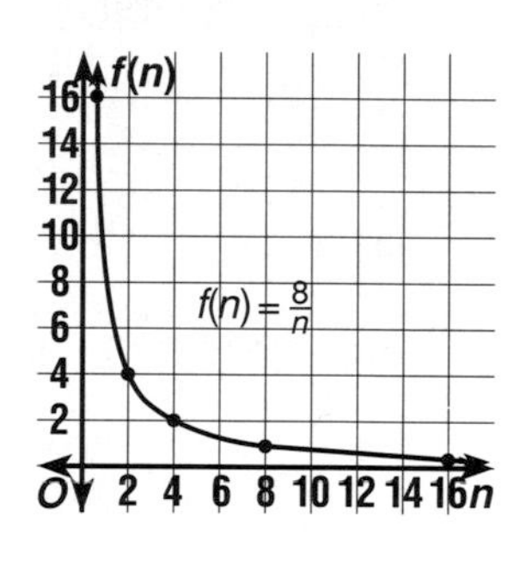

7.

n	f(n)	(n, f(n))
0	1	(0, 1)
1	4	(1, 4)
2	7	(2, 7)
3	10	(3, 10)

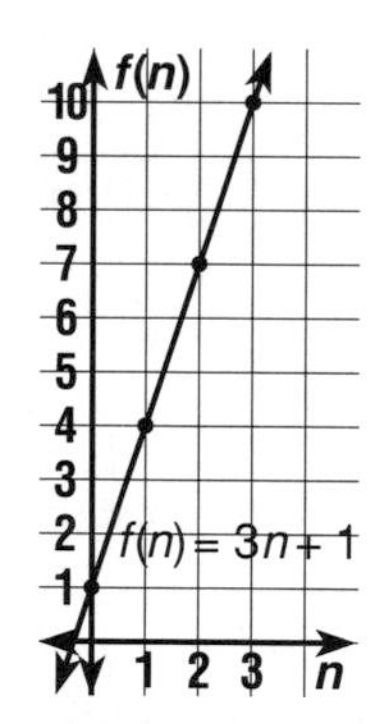

9.

n	f(n)	(n, f(n))
8	2	(8, 2)
2	8	(2, 8)
4	4	(4, 4)
$\frac{8}{3}$	6	$(\frac{8}{3}, 6)$

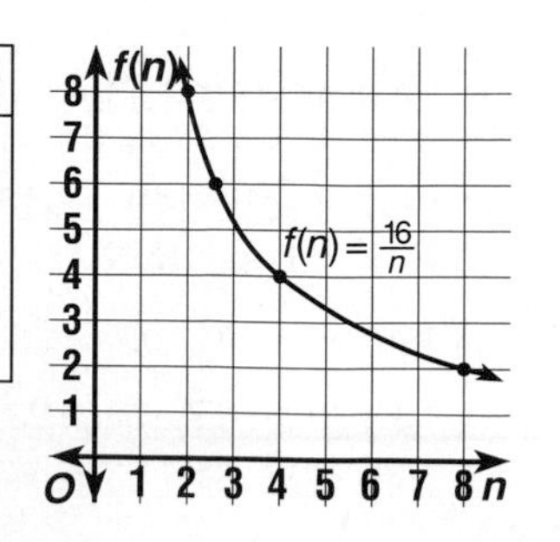

11.

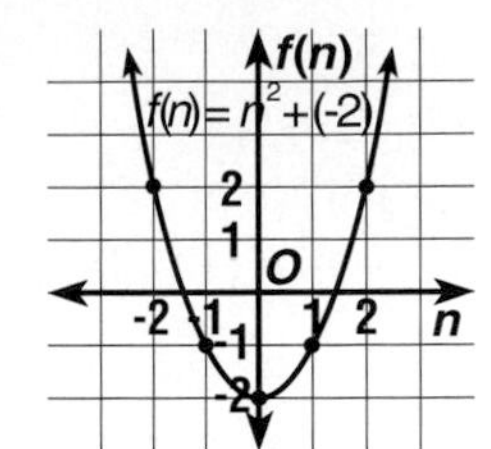

12. 16.96 mm
13. -7

15.a.

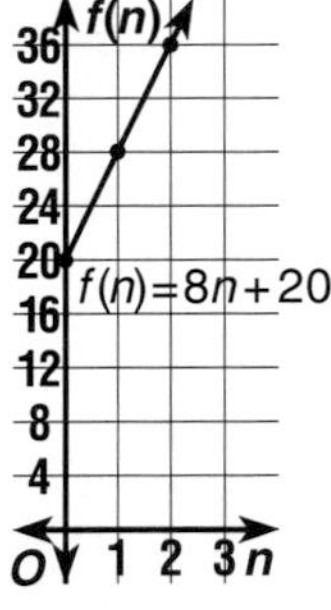

15.b. No, there is a safety device that only lets it rise so far; yes.

Pages 429-430 Lesson 11-3

4. -2, 0, 2, 3 **5.** 2, -1, 0, -3 **6.** -6, 8, $\frac{1}{3}$
7. Sample answers: (-4, -1), (-1, 2), (0, 3), (5, 8)
9. 3, 5, 6, $6\frac{2}{3}$ **11.** Sample answer: {(-2, -3), (-1, -1), (0, 1), (1, 3)} **13.** Sample answer: {(-4, 1), (-2, 2), (0, 3), (2, 4)}
15. Sample answer: {(-2, 7), (-1, 5), (0, 3), (2, -1)}
17. {(6, 2), (3, 5), (-45, 53)} **18.** 280 **19.** $2\frac{4}{9}$
20.

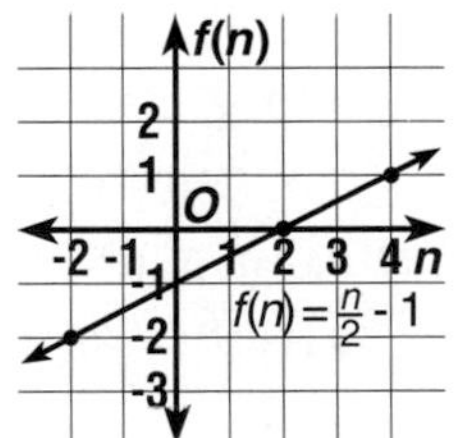

21.a. $y = x + 9$; (-10, -1), (0, 9), (1, 10)
b. $y = 2x + 3$; (0, 3), (1, 5), (-2, -1)
c. $x + y = 0$; (1, -1), (-2, 2), (0, 0)
23. 52, 63, 71, 87, 90

Pages 433-434 Lesson 11-4

4.

x	y	(x, y)
-4	-20	(-4, -20)
-1	-5	(-1, -5)
0	0	(0, 0)
2	10	(2, 10)

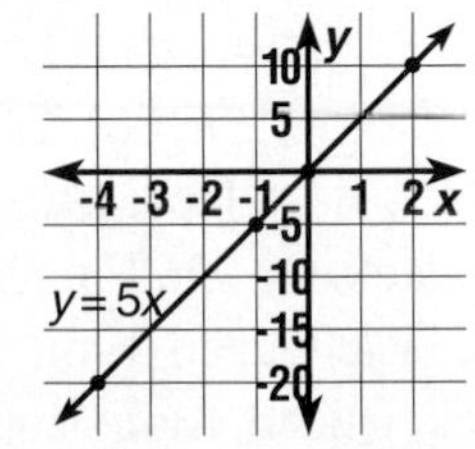

6.

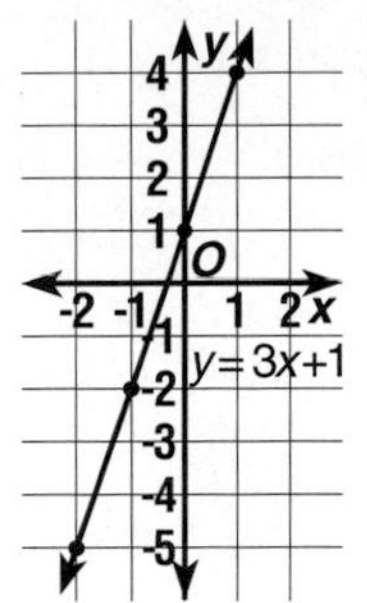

9.

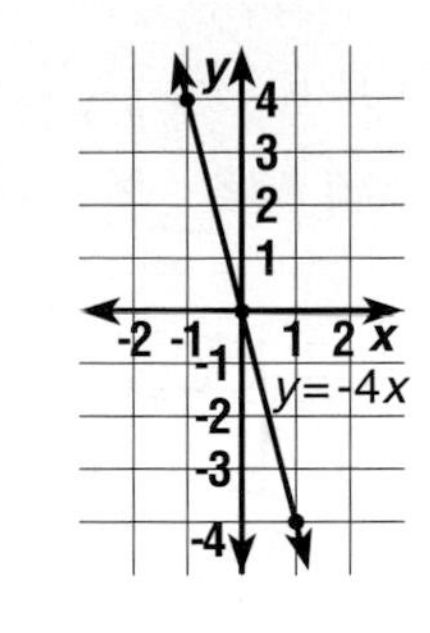

13.

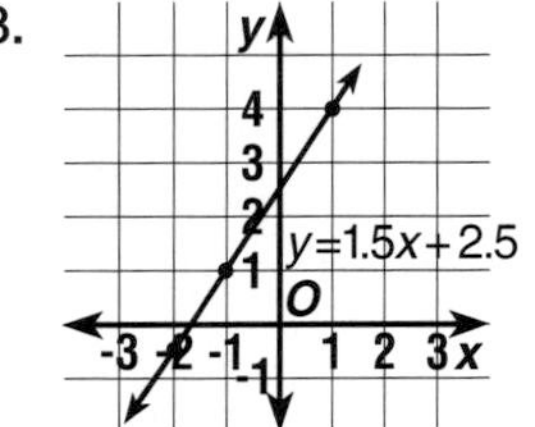

17. $A + 6$

18.

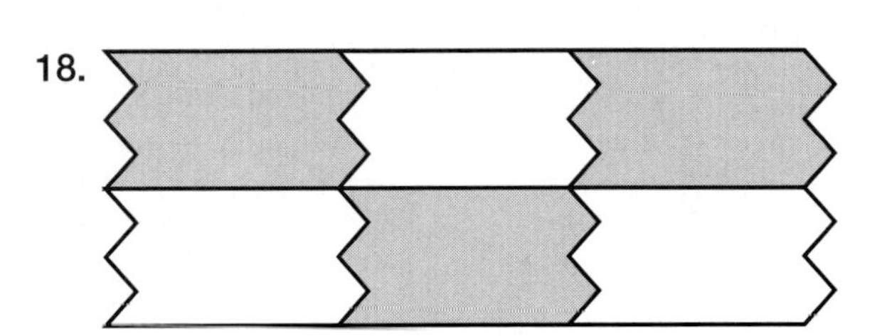

19. 4.375 in² **20.** Sample answer: {(0, 8), (3, 9), (-3, 7), (-6, 6)}
21.a. (0, 32), (100, 212) **c.** Find the coordinates of other points on the line.
22.

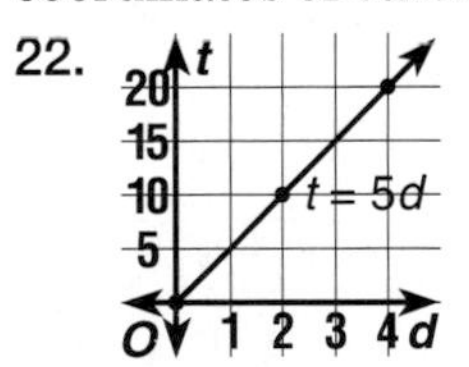

Pages 437-438 Lesson 11-5

6. (6, 2) **7.** (-4, -4) **8.** (1, -1) **9.** (0, 4)
10. (-3, 5) **11.** (-1, 2) **12.** (-2, 1)
13. no solution **15.** (2, 2) **17.** (6, 9)
19. (5, -6) **21.** (2, 1) **22.** positive **23.** >
24. about 8
25.

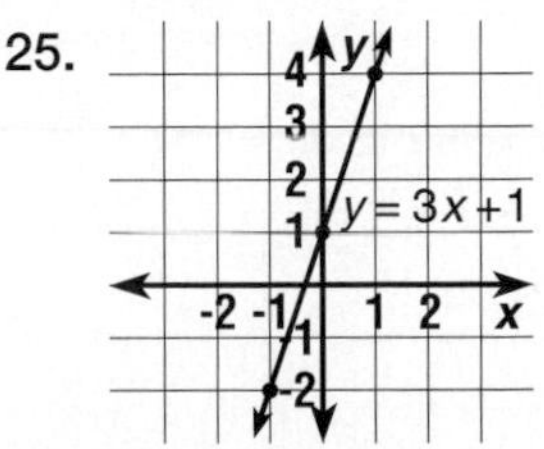

27.b. (2, 9) **27.c.** perpendicular lines

Page 438 Mid-Chapter Review

1. Sample answer:

n	*f(n)*	*(n, f(n))*
0	5	(0, 5)
1	3	(1, 3)
4	-3	(4, -3)

3. {(0, 4), (3, 5), (-3, 3), (-6, 2)} **5.** (3, 2)

Pages 440-441 Lesson 11-6

3. about $660 billion **4.** from March to April **5.** decrease in business loans and increase in government securities **7.** 15 **9.** 1955-1960 **11.** Foreign cars are taking over the majority of motor vehicle production.

Pages 444-445 Lesson 11-7

4.

x	*f(x)*	*(x, f(x))*
-2	4	(-2, 4)
-1	1	(-1, 1)
0	0	(0, 0)
1	1	(1, 1)
2	4	(2, 4)

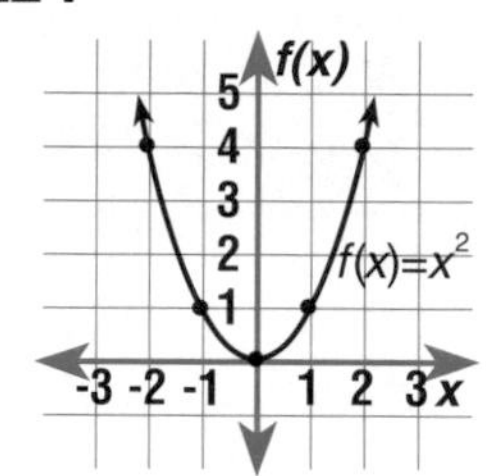

6.

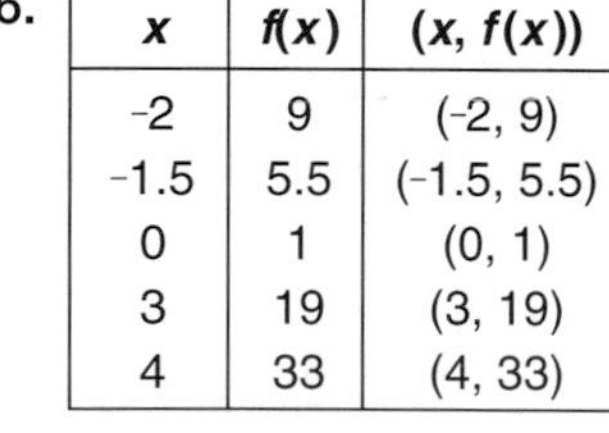

x	*f(x)*	*(x, f(x))*
-2	9	(-2, 9)
-1.5	5.5	(-1.5, 5.5)
0	1	(0, 1)
3	19	(3, 19)
4	33	(4, 33)

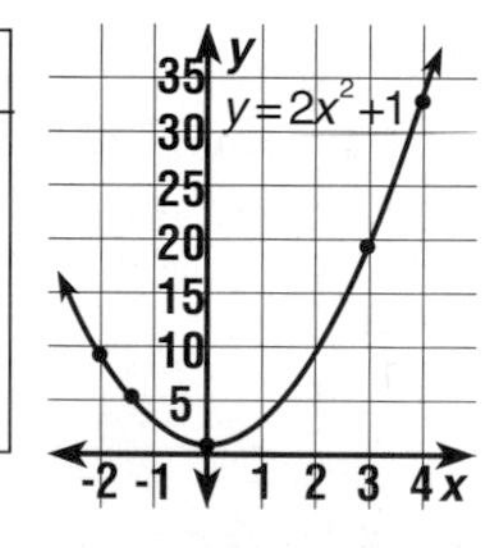

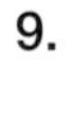

9.

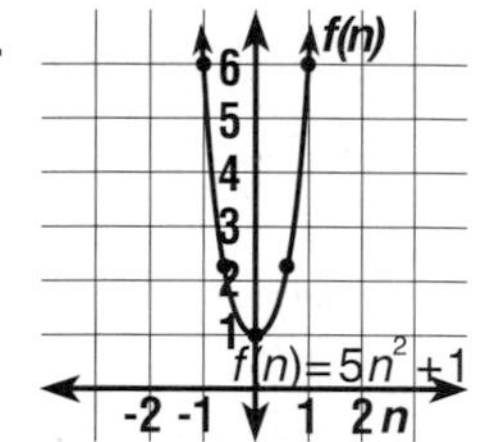

13.

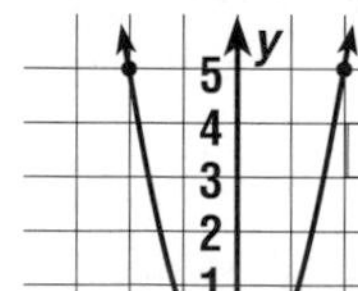

17. {(-2, 8), (0, -4), (1, -1)} **19.** 97 **20.** -31 **21.** $3\frac{7}{20}$ **22.** (3, 2)

25.

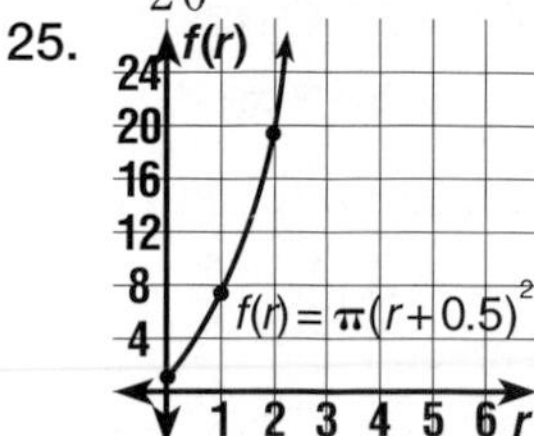

Pages 448-449 Lesson 11-8

4. (4, 5) **5.** (10, -5) **6.** *P*′(-6, 3), *Q*′(-9, -1), *R*′(-5, 6) **7.** *W*′(1, 4), *X*′(3, 6), *Y*′(1, 8), *Z*′(-1, 6) **9.** *P*′(-5, 5), *Q*′(1, 8), *R*′(2, 6), *S*′(-4, 3) **11.** (5, 2) **13.** yes

14.

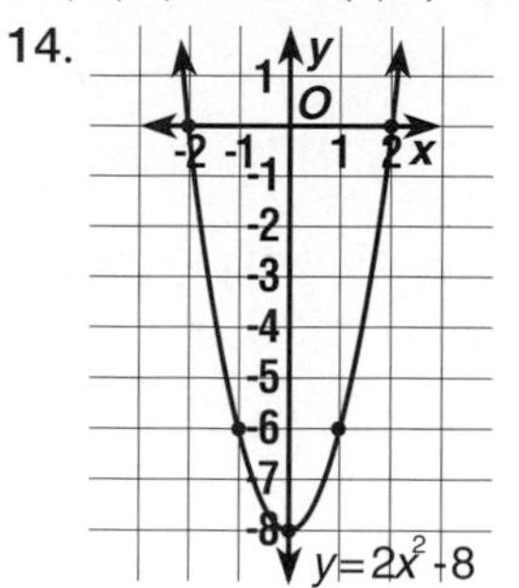

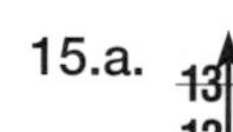

15.a.

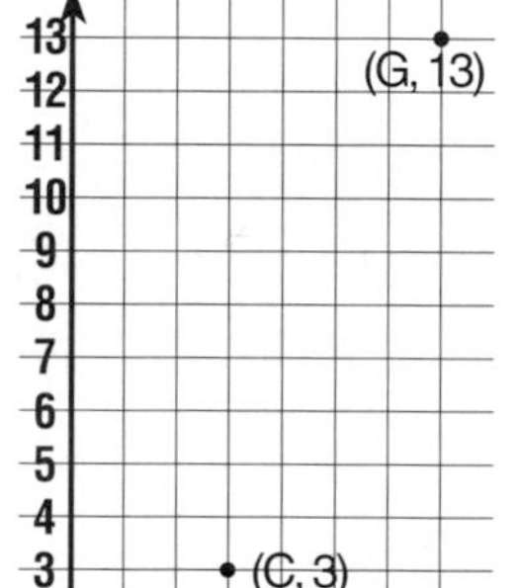

b. over 4, back 10

Pages 452-453 Lesson 11-9

4. *y*-axis **5.** *x*-axis **6.** *x*-axis **7.a.** *C*′(3, -3), *O*′(0, 0), *W*′(6, 1) **7.b.** *C*′(-3, 3), *O*′(0, 0), *W*′(-6, -1) **9.a.** *M*′(1, -2), *O*′(0, 0), *N*′(-5, 0), *Y*′(-4, -2) **11.** 28 **12.** *A*′(-1, 1), *B*′(1, 4), *C*′(4, 2)

Pages 456-457 Lesson 11-10

4. yes **5.** yes **6.** no **7.a.** *H*′(-4, -3), *A*′(-4, -5), *I*′(2, -5), *R*′(2, -3) **b.** *H*′(3, -4), *A*′(5, -4), *I*′(5, 2), *R*′(3, 2) **8.** Sample answers: square, regular hexagon **9.b.** *R*′(3, -1), *S*′(9, -6), *T*′(5, -8) **11.** (-4, 1), (-1, 4), (-5, -8) **13.** 105 **14.** 12 **15.a.** *T*′(1, 1), *R*′(3, 1), *A*′(3, 3), *B*′(1, 3) **b.** *T*′(-1, -1), *R*′(-3, -1), *A*′(-3, -3), *B*′(-1, -3) **17.** 2, 4, 10, Jack, Queen, King of hearts; 2, 4, 10, Jack, Queen, King of clubs; 2, 3, 4, 5, 6, 8, 9, 10, Jack, Queen, King, Ace of diamonds; 2, 4, 10, Jack, Queen, King of spades

Pages 458-460 Study Guide and Review

11.

n	$f(n)$	$(n, f(n))$
-1	-3	(-1, -3)
-0.5	-1	(-0.5, -1)
0	1	(0, 1)
0.5	3	(0.5, 3)
1	5	(1, 5)

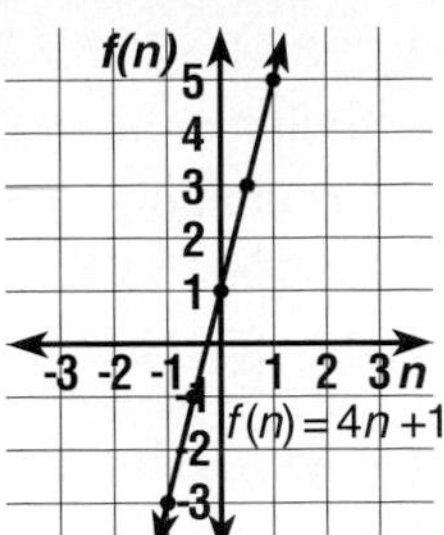

13.

n	$f(n)$	$(n, f(n))$
-1	3	(-1, 3)
-0.5	1.5	(-0.5, 1.5)
0	0	(0, 0)
0.5	-1.5	(0.5, -1.5)
1	-3	(1, -3)

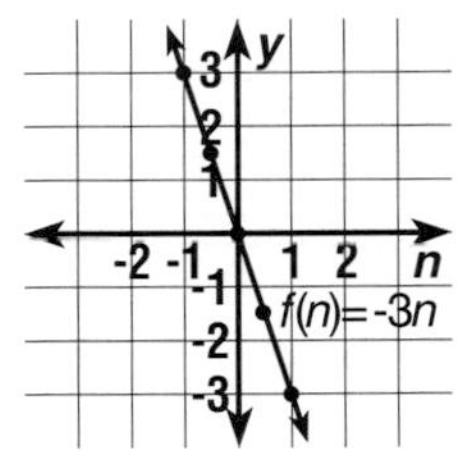

15. Sample answer: {(-2, 2), (-1, 3), (0, 4), (1, 5)}

17.

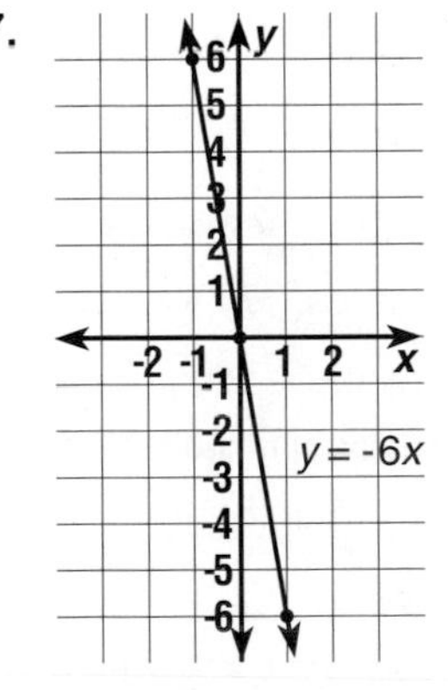

19.

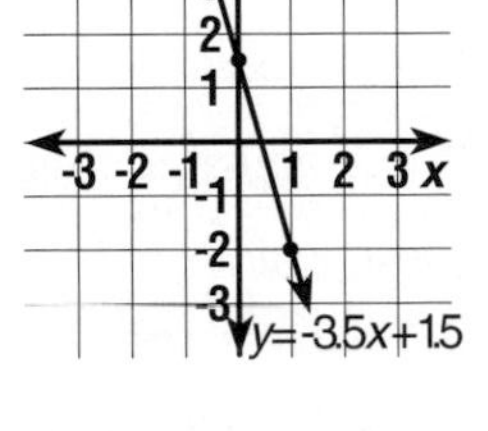

21. (1, 6)

23.

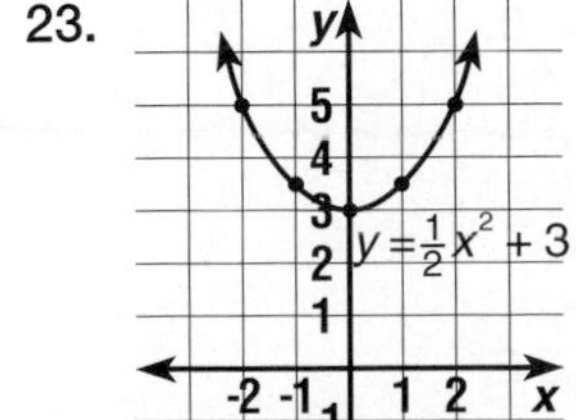

25.

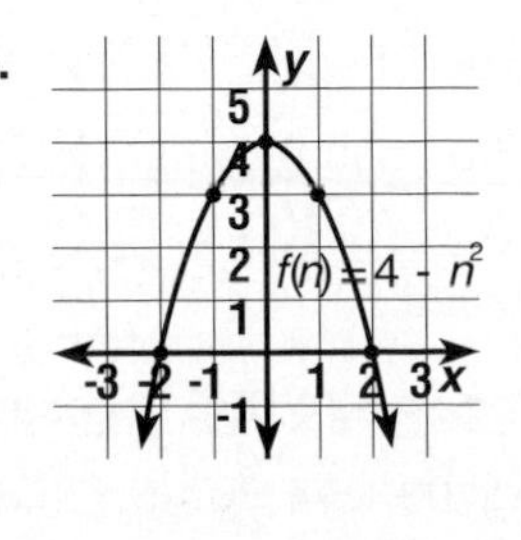

27. $R'(-1, 5), S'(1, 5), T'(1, 3), U'(-1, 3)$
29. $A'(2, -5), B'(6, -5), C'(6, -3), D'(2, -3)$
31. $L'(-4, 1), A'(-7, 4), T'(-4, 7), E'(-1, 4)$
33. It will never occur since the graphs of their profit equations never meet.

Area and Volume

Pages 466-467 Lesson 12-1

5. 153.9 ft^2 6. 32.2 km^2 7. 95.0 yd^2 8. $\frac{1}{100}$
9. 19.6 cm^2 11. 63.6 cm^2 13. 78.5 m^2
15. 7.7 m^2 17. 25.1 in^2 18. 12 19. 8.2 units
21.a. 28.3 in^2, 84.8 in^2, 141.4 in^2 b. $\frac{1}{3}$
23.a. 201.06 in^2, 226.19 in^2
b. L = \$0.0447126, M = \$0.0441645 c. L = \$0.04, M = \$0.04; same price per square inch

Page 469 Lesson 12-2

3. 12 cubes 5. \$75 7. c

Pages 472-473 Lesson 12-3

4. $2 \times 3 \times 5$ units 5.

7.

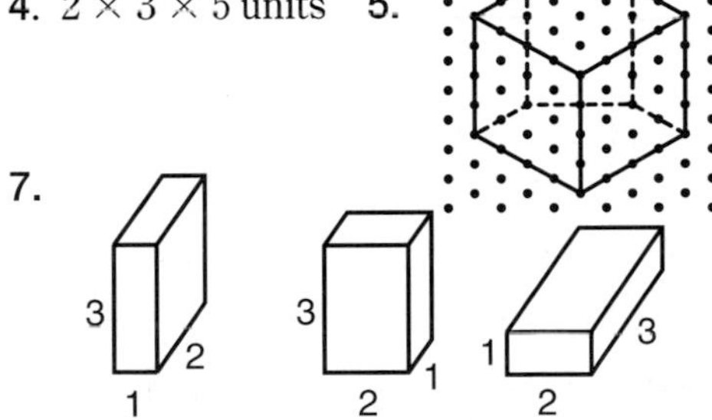

12. 7 13. 12 14. 120.8 m^2

Pages 476-477 Lesson 12-4

4. rectangular prism, 108 ft^2 5. triangular prism, 97.84 m^2 6. rectangular prism, 468 in^2
7. 528 yd^2 9. 527.4 cm^2 11. 64.5 m^2
13. 122 cm^2 15. 384 ft^2 16. 1,024 17. No, the last 2 digits are not divisible by 4.
19. Monica's cube—it has the greatest surface area 21. 8,100 in^2

Pages 479-481 Lesson 12-5

5. 251.3 m^2 6. 1,407.4 cm^2 7. 32.99 in^2
9. 280.8 cm^2 11. 150.8 in^2 13. 397.2 m^2
14. $b < 2$

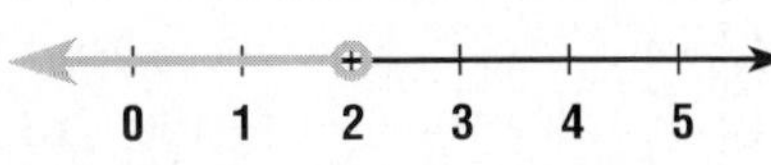

15. $17\frac{1}{2}$ 16. 0.352 17. 62 in^2
21. about 55 in^2

Page 481 Mid-Chapter Review

1. about 113.1 cm^2 **5.** 1,885 m^2 **7.** 1,188 in^2

Pages 484-485 Lesson 12-6

6. 120 m^3 **7.** 314.2 ft^3 **8.** 60 yd^3 **9.** 141.4 cm^3 **10.** 18 in^3 **11.** 1,125.7 cm^3 **13.** 1,032.2 cm^3 **15.** 1.728 cm^3 **17.** 5,513.5 mm^3 **19.** 300 ft^3 **22.** -9 **23.** 326.73 cm^2

Pages 488-489 Lesson 12-7

4. 40 in^3 **5.** 20.9 m^3 **6.** 56 yd^3 **7.** 339.3 m^3 **9.** 100.4 m^3 **11.** 80 yd^3 **13.** 45 m^3 **15.** 27, 33, 39, 45 **16.** $D'(6, 12)$, $E'(20, 16)$, $F'(18, 6)$ **17.** 502.65 $feet^3$ **19.** about 229 cm^3

Pages 492-493 Lesson 12-8

4.a. 3 **b.** closer to 5.5 ft than 5.51 ft **5.** 20.3, accurate to meter, 3 tenths is estimate; 4,200 accurate to thousands, 200 is estimate; 0.00251, accurate to 0.0025, 1 hundred thousandth is estimate; 0.0580, accurate to 0.058, 0 ten thousandths is an estimate. **6.** 34.3 oz; tenths of an ounce is a more precise unit than an ounce or a pound **7.** Accurate to $212 billion, 9 hundred million is estimate; it is closer to $212.9 billion than to $212.8 or $213.0 billion. **15.** 305 feet **18.** 0.000094 **19.** 153.94 m^3 **21.** nearest tenth of a second **23.** accurate to next to last digit of input

Pages 494-496 Study Guide and Review

11. 50.3 m^2 **17.** 54 cm^2 **19.** 4,523.89 yd^2 **21.** 378 m^3 **23.** 6 mm^3 **25.** 3 **27.** 4 **29.** accurate to 5.0 miles, 2 hundredths is estimate **31.** 55 cubes

13 Discrete Math and Probability

Pages 505-506 Lesson 13-1

5.b. 24 **c.** 6 **d.** 4 **e.** 1 **f.** $\frac{1}{4}$ **6.** 32 **7.** 12 **9.** 6 **11.** 16 **13.** 24 **15.** 27 **16.** 8 **17.** 22 **18.** 3 **19.** 9

Pages 508-509 Lesson 13-2

3. 24 **4.** 11,880 **5.** 6 **6.** 5,040 **7.** 120 **8.** 24 **9.** 720 **10.** 6 **11.** 120 **13.** 1,680 **15.** 720 **17.** 120 **19.** 120 **21.** 24 **22.** -32 **23.** 14 **24.** 60% **25.** 243 **27.** 60

Pages 513-514 Lesson 13-3

5. 10 **6.** 1 **7.** 3 **8.** 792 **9.** permutation **10.** combination **11.** combination **12.** permutation **13.** 4 **15.** 6 **17.** 2,042,975 **19.** 2,598,960 **21.** 56 **23.** 0.078 liters **24.** 80° **25.** 225 miles **26.** 336 **29.** 28

Pages 517-518 Lesson 13-4

4. 10 **5.** 7 **6.** 56 **7.** 1 **8.** 56 **9.** 35 **10.** 8 **11.** 28 **13.** 4 **15.** 210 **17.** 8 **19.** 6 **21.** combination of 4 things taken two at a time **22.** 157.7, 158 **23.** $9\frac{7}{18}$ **24.** Sample answer: {(1, 13), (2, 16), (3, 19), (4, 22)} **25.** 120 **29.** 252 **31.** 35

Page 518 Mid-Chapter Review

1. 12 outcomes **3.** 720 **5.** 20

Pages 523-524 Lesson 13-5

4. $\frac{1}{30}$ **5.** $\frac{1}{30}$ **6.** $\frac{1}{10}$ **7.** 0 **8.** $\frac{1}{5}$ **9.** Independent; one outcome does not affect the other. **10.** $\frac{5}{28}$ **11.** $\frac{1}{28}$ **12.** $\frac{5}{56}$ **13.** $\frac{3}{14}$ **14.** $\frac{5}{28}$ **15.** Dependent; one outcome affects the other. **17.** $\frac{1}{6}$ **19.** $\frac{5}{42}$ **21.** $\frac{1}{30}$ **23.** $\frac{1}{15}$ **25.** 0 **27.** $\frac{9}{20}$ **29.** $\frac{1}{10}$ **30.** 49 **31.** {-157, -28, -3, 2, 18, 226} **32.** no **33.** 15 **35.** 67.5%

Pages 526-527 Lesson 13-6

4. Sample answer: 3 **5.** Sample answer: $\frac{4}{10}$ **6.** 1 half dollar, 1 quarter, 4 dimes **7.** 72 pages **9.** Sample answer: $\frac{1}{45}$ **11.** Math—Mrs. Gossell; Music—Ms. Alvarez; Social Studies—Mrs. Yamaguchi **13.** -3 **15.** 30 students

Pages 529-530 Lesson 13-7

3.a. $\frac{3}{8}$ **b.** $\frac{1}{2}$ **c.** $\frac{3}{40}$ **4.a.** $\frac{3}{8}$ **b.** $\frac{3}{8}$ **c.** $\frac{1}{8}$ **5.a.** $\frac{217}{400}$ or 0.5425 **b.** yes **c.** conduct more experiments **7.** -39 **8.** 3 **9.** $\frac{1}{4}$ **13.a.** Sample answer: $P(A) = 0.33$, $P(B) = 0.22$, $P(C) = 0.45$ **b.** Replace 0.67 with 0.5 in lines 30 and 50.

Pages 534-535 Lesson 13-8

3. 400 **4.** $\frac{9}{20}$ **5.** 15,880 people **6.** false;

327:273 < 2:1 **7.** true; $\frac{327}{600} = 54.5\%$ **8.** False; this cannot be determined from this survey. **9.** 72 **10.** cola **11.** $\frac{1}{6}$ **12.** 2,500 colas **13.** yes; randomly taken **15.** 225 red, 90 green, 18 yellow, 117 blue **17.** 92 **19.** 8.36 **20.** (4, 1) **21.** $\frac{1}{2}$

Pages 536-538 Study Guide and Review

11. 6 outcomes **13.** 720 ways **15.** 15 pairs **17.** 35 combinations **19.** 84 **21.** $\frac{1}{6}$ **23.c.** $\frac{1}{4}, \frac{1}{2}$ **25.** $\frac{9}{50}$ **27.** 362,880 orders

29.

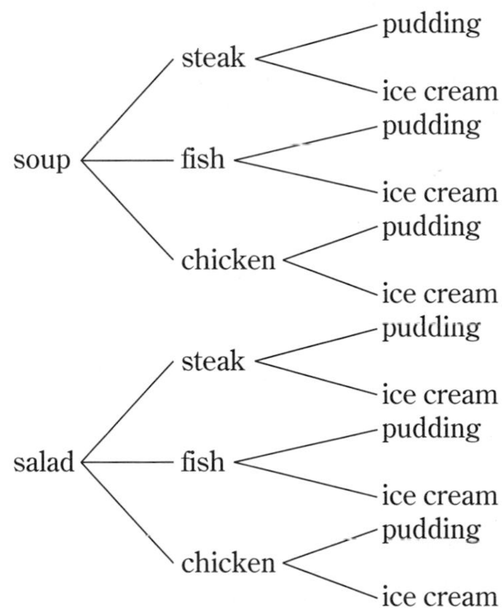

12 outcomes

14 Algebra: Investigations Polynomials

Pages 544-545 Lesson 14-1

5. $-x^2 + 3x - 4$ **6.** $-4x + 2$ **10.** 6 **11.** $-4x$ **13.** $-2x^2 + 9$ **21.** 11 **23.** 0 **25.** 51.52 mm **26.** 50 **27.** 700 ginger ales

Pages 548-549 Lesson 14-2

5. $3x^2, -2x^2$; $4x, 10x$ **6.** $4y, -2y, -3y$; 8, 9 **7.** none **8.** $-a^2, 4a^2$ **9.** $x^2 + x - 1$ **10.** $a^2 + 3a$ **11.** $6y, -11y$ **13.** none **15.** $6x$ **17.** $5x + 5$ **19.** $2x^2 + 4x - 4$ **21.** $-7a$ **23.** $9a + 14b$; 80 **25.** $a + 9b$; 61 **27.** Subtract the lower quartile from the upper quartile. **28.** \$1.20/foot **30.** 19 **31.a.** $4q + 8d + 5n$; \$2.05 **33.** \$176.26

Pages 551-552 Lesson 14-3

3. $-x + 6$ **4.** $3a^2 - 10a$ **5.** $-7x^2 + x + 1$ **6.** $-4y^2 + y$; -3 **7.** $6x^2 + 2x + 2$ **9.** $5y^2 + 3y + 5$ **11.** $12x + 12y$ **13.** $9s + 2t$ **15.** $10a^2 + 2a - 3$ **17.** $5x^2 - 3x + 4$ **19.** $7c - d + 1$; 52 **21.** $16c - d + 1$; 124 **22.** 32 **23.** 1.3, -1.3 **24.** 50.27 m³ **25.** $4x^2 + 4x$ **27.a.** $8x + 24$ **b.** $6x + 6$ **c.** $8x + 34$

Pages 554-555 Lesson 14-4

4. $-x$ **5.** $5x^2$ **6.** -4 **7.** $-10x^2$; $-3x$ **8.** $3x + 2$ **9.** $-x^2 + x$ **10.** $4x + 1$ **11.** $2x^2 - 3x + 5$ **12.** $x^2 - 4x$ **13.** $-4x^2 - x + 2$ **15.** $2a^2 - 5a - 9$ **17.** $-5a + 6$ **19.** $4m - 5n$ **21.** $2p^2 - p + 1$ **23.** $8a^2 + 2ab - 3b^2$ **24.** 1,050 **25.** $2^4 \cdot 3$ **26.** 336 **27.** $-4m^2 + 10m + 9$

Page 555 Mid-Chapter Review

3. $8x^2 - 7$ **5.** $7x^2 - 4x$ **7.** $2n + 7$

Pages 558-559 Lesson 14-5

5. $2x + 10$ **6.** $x^2 + 3x$ **7.** $5y + 45$ **8.** $y^2 + 2y$ **9.** $a^2 + 2a$ **10.** $4(z + 1)$ **11.** $y(y + 5)$ **13.** $4b + 12$ **15.** $d^2 + 15d$ **17.** $4x^2 + 2x$ **19.** $3x(2x + 1)$ **21.** $8(a^2 + 1)$ **23.** $4(1 + 5x)$ **25.** $2(y + 4)$ **27.** 83 **28.** 28% increase

29.

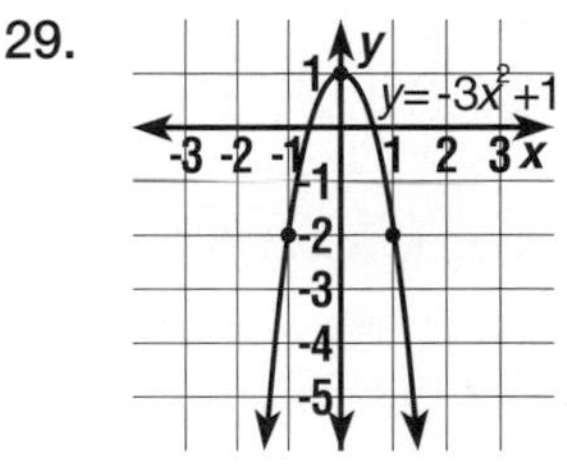

30. $3x - 4$ **31.b.** $2x(x + 3)$; $2x^2 + 6x$ **c.** 260 ft²

Pages 561-562 Lesson 14-6

4. $x^2 + 3x + 2$ **5.** $4x^2 + 10x + 6$ **6.** $m^2 + 7m + 12$ **7.** $6a^2 + 11a + 3$ **8.** $2x^2 + 5x + 3$ **9.** $12z^2 + 19z + 5$ **11.** d **13.** a **15.** $x^2 + 5x + 4$ **17.** $2x^2 + 5x + 2$ **19.** $x^2 + 2x + 1$ **21.** $\frac{3}{4}$ **22.** 8% **23.** $2x^2 + 3x$

Page 565 Lesson 14-7

3. 9 minutes **4.** 4 mini-buses **5.** b; 3 cubed is 27 so 123 cubed will have seven in the ones place. **7.** $x^2 + 6x + 8$

Pages 566-568 Study Guide and Review

13. $15m^2 + 8m$ **15.** $2x^2 + 6x$ **17.** $14m^2 + 2m$ **19.** $14d + 8$ **21.** $4b^2 - 3b - 4$ **23.** $3g + 2$ **25.** $3s^2 + 5$ **27.** $2p^2 - 5p - 4$ **29.** $4z^2 + 12z$ **31.** $3t^2 + 18t$ **33.** $2x^2 + 8x + 6$ **35.** $x^2 + 7x + 10$ **37.** $(2x + 1)$ meters by x meters **39.** $12x + 32$ yards **41.** c

Index

Index

Photo Credits

Cover: (tl) Dick Frank/The Stock Market, (tc) Myron J. Dorf/The Stock Market, (tr) Matt Meadows, (bl) C. & J. Walker/Liaison International, (br) J. Sekowski

iii, Robert Mullenix; **viii,** (t) Ken Frick, (b) First Image; **ix,** (t) Robert Mullenix, (b) National Audubon Society/Photo Researchers, Inc.; **x,** (m) KS Studio, (t) Tim Courlas, (b) FPG International, Inc., (bkgd) Mark Gibson; **xi,** Pictures Unlimited; **xii,** (t) file photo, (b) Angabe A. Schmidecker/FPG International, Inc.; **xiii,** (t) F. Grehan/Photo Researchers, Inc., (b) MAK-I; **xiv,** (t) Phil Degginger, (b) file photo, (bkgd) file photo; **xv,** (t) Comstock, Inc./Jack K. Clark, (b) Pictures Unlimited; **xvi,** Ivor Sharp/The Image Bank; **xvii, xviii, xix, xx** Matt Meadows; **2,** (t) Studiohio, (l) John Swart/Allsport, (bm) Yoram Kahana/Shooting Star, (br) International Tennis Hall of Fame and Tennis Museum at The Newport Casino, Newport, Rhode Island; **3,** (m) Duomo/Paul J. Sutton, (bl) courtesy of Texas Instruments, (bm) Ken Frick, (br) Lori Adamski Peek/Tony Stone Worldwide; **4,** MAK-I; **6,** Ken Frick; **8,** ©FPG International, Inc.; **11,** Doug Martin; **14,** Ken Frick; **15,** KS Studio; **16,** NASA; **17,** Doug Martin; **20-21,** Robert Mullenix; **21,** (tl) Tomas del Amo/Profiles West, (tr) Robert Mullenix, (bm) Dimaggio/Kalish/The Stock Market; **25,** MAK-I; **28,** Ken Frick; **32,** Pictures Unlimited; **37,** Jet Propulsion Laboratory; **40,** Doug Martin; **42,** courtesy of the Singer Company; **43,** (bl) Archive Photos/Welgos, (br) D.Goldberg/Sygma; **44,** Niki Mareschal/The Image Bank; **47,** Ken Frick; **48,** ©FPG International, Inc.; **49,** Ken Frick; **53,** UPI/The Bettmann Archive; **54,** Doug Martin; **57,61,** Ken Frick; **62,** Wesley Bocxe/Photo Researchers, Inc.; **63,65,** Ken Frick; **66,** file photo; **69,** MAK-I; **70,** Doug Martin; **71,** First Image; **72,** Keith Turph; **82,** Robert Mullenix; **84,** (t) Archive Photos/Hirz, (l) Joe Munroe/Life Magazine/©1959 Time Inc., (br) Robert Mullenix; **85,** (t) Robert Mullenix, (bl) Barbie doll photo, ©1959 Mattel,Inc. used with permission, (br) Phil Degginger; **86,** Jeff Gnass/The Stock Market; **88,** Comstock, Inc.; **89,** Doug Martin; **91,** MAK-I; **94,** (t) ©Allen B. Seiden/FPG International, (m) Robert Mullenix, (b) MAK-I; **95,** ©H. Gris/FPG International, Inc.; **97,** Ken Frick; **98,** Frank Wong/The Image Bank; **100,** Ken Frick; **102,** John Kelly/The Image Bank; **105,** Phil Lauro/ProFiles West; **108,** Peter Beck/The Stock Market; **109,** Paolo Koch/Photo Researchers, Inc.; **110,** Ken Frick; **116,** Bud Fowle; **117,** Comstock, Inc.; **119,** Ken Frick; **122,** ©FPG International, Inc.; **126,** (bl) Alan Kearney/Viesti Associates,Inc., (br) Historical Pictures Service; **127,** (t) KS Studios, (b) Culver Pictures, Inc.; **128,** Doug Martin; **130,** ©Michael Tamborrino/FPG International, Inc.; **132,** (t) ©FPG International, (b) Doug Martin; **135,** Pete Saloutos/The Stock Market; **136-137,** Doug Martin; **140,** Aluis Upitis/The Image Bank; **140,** (t) Cobalt, (b) Jan Halaska/Photo Researchers, Inc.; **141,** MAK-I; **143,** ©Travelpix/FPG International, Inc.; **144,** (t) Ken Frick, (b) ©FPG International, Inc.; **145,** Doug Martin; **146,** ©FPG International, Inc.; **149,150,** Ken Frick; **151,** MAK-I; **153,** (l) John R. Ramey/Stockphotos/The Image Bank, (r) Pictures Unlimited; **154,** (l) Kunio Owki/The Stock Market, (r) Pictures Unlimited; **157,** MAK-I; **158,** (t) Elaine Shay, (b) Luis Villota/The Stock Market; **161,** Doug Martin; **162-163,** Robert Mullenix; **164,** Pictures Unlimited; **165,** MAK-I; **166,** National Audubon Society/Photo Researchers, Inc.; **167,** (t) ©Dennis Halliman/FPG International, (b) Tim Courlas; **170,** Richard Steedman/The Stock Market; **172,** (t) Nik Wheeler/Westlight, (b) Steve Lissau, (bm) The Far Side cartoon by Gary Larson is reprinted by permission of Chronicle Features, San Francisco, CA.; **173,** (l) Robert Mullenix, (r) Jack Hoehn Jr./Profiles West; **176,** (t) Duomo/William R. Sallaz, (b) Duomo/Steven E. Sutton; **179,** ©E.A. McGee/FPG International, Inc.; **181,** (t) J.R. Schnelzer, (m) Steve Niedorf/The Image Bank, (b) Comstock, Inc.; **183,** MAK-I; **186,** Harald Sund/The Image Bank; **187,** Ed Bock/The Stock Market; **188,** Pictures Unlimited; **191,** Doug Martin; **192,** Pictures Unlimited; **193,** Skip Comer; **195,** Doug Martin; **196,** Joseph Dichello; **197,** Ken Frick; **208,** Doug Martin; **210,** (t) Warner Bros/Shooting Star, (b) NASA; **211,** (t) Movie Still Archives, (bl) The Far Side Cartoon by Gary Larson is reprinted by permission of Chronicle Features, San Francisco, CA., (b) NASA/Westlight; **212,214,** Doug Martin; **215,** Pictures Unlimited; **216,** Bud Fowle; **218,** Duomo/Dan Helms; **220,** (t) Doug Martin, (b) Pictures Unlimited; **221,** Bill Clark/Stock South; **223,** (t) Ken Frick, (b) Doug Martin; **224,** (t) ©FPG International, (b) Michael Melford/The Image Bank; **227-231,** Doug Martin; **233,** (t) Ken Frick, (b) MAK-I; **234,** Doug Martin; **236,** Comstock, Inc./Russ Kinne; **238,** (l) MAK-I, (r) Gabe Palmer/The Stock Market; **240,** Doug Martin; **241,** Ken Frick; **244,** Pictures Unlimited; **245,** KS Studio; **247,** NASA; **250,** Elaine Shay; **250,** Studiohio; **254,** (t) National Baseball Library, Cooperstown, NY, (bl) Archive Photos/Lambert, (bm) Historical Pictures Service (br) National Baseball Library, Cooperstown, NY; **255,** (t) H. Armstrong Roberts, (b) National Baseball Library, Cooperstown, NY; **256,** Tom McHugh/Photo Researchers, Inc.; **258,** MAK-I; **261,** Ken Frick; **264,** Naoki Okamoto/The Stock Market; **265,** Leonard Lee Rue/Photo Researchers, Inc.; **267,** MAK-I; **268,** Robert Mullenix; **269,** (t) KS Studios, (b) Robert Mullenix; **270,** ©Neal & Mary Jane Mishler/FPG International, Inc.; **271,** Ken Frick; **272,** The Bettmann Archive; **275,** Ken Frick; **276,** Superstock, Inc.; **285,** file photo; **287,** (t) Ken Frick, (b) The Bettmann Archive; **288,** Frank P.

Rossotto/The Stock Market; **290,** MAK-I; **291,** Doug Martin; **293,** Jack Sullivan/Photo Researchers, Inc.; **296,** (t) David Brownell, (b) Will & Deni McIntyre/Photo Researchers, Inc.; **298,** (t) ©Angabe A. Schmidecker/FPG International, Inc., (l) Dr. E.R. Degginger, (bm) Animals Animals/Ted Levin, (br) Robert Mullenix; **299,** (ml) DR. E.R. Degginger/Animals Animals/Earthscenes, (mr) E.Delaney/Profiles West, (b) Bob Winsett/Profiles West; **300,** Luis Villota/The Stock Market; **301,** Derek Berwin/The Image Bank; **302,** (t) Lisl Dennis/The Image Bank, (b) Ken Frick; **303,** Pictures Unlimited; **304,** MAK-I; **305,** ©Joe Baker/FPG International, Inc.; **309,** Comstock, Inc.; **310,** Ken Cooper/The Image Bank; **311,** Comstock, Inc.; **312,** Keith Kent/Science Photo Library/Photo Researchers, Inc.; **315,** ©FPG International, Inc.; **318,** Andreqa Pistolesi/The Image Bank; **319,** History of Mathematics, Vol.1 by David Eugene Smith, © 1951 by Eva May Luse Smith, Dover Publications, Inc., New York, New York; **322,** Matchncer; **323,** Historical Pictures Service; **328,** Duomo/Daniel Forster; **331,** MAK-I; **334,** Duomo/David Madison; **336,** (l) Viesti Associates,Inc./Joe Viesti, (r) Comstock, Inc; **337,** (t) Reprinted with special permission of North American Syndicate, (m) Tomas del Amo/Profiles West, (b) Brian Vikander; **338,** Duomo/Bryan Yablonsky; **339,** file photo; **341,** H. Armstrong Roberts; **343,** (t) Pictures Unlimited, (l) F. Grehan/Photo Researchers, Inc., (m) SYGMA , (r) Archivi Alinari/Art Resource, New York; **344,** Gregory Heisler/The Image Bank; **345,** Doug Martin; **347,** Pictures Unlimited; **348,** MAK-I; **349,** Comstock, Inc.; **350-351,** MAK-I; **353,** ©Peter Gridley/FPG International, Inc.; **355-356,** MAK-I; **358,** Brent Petersen/The Stock Market; **359,** Ken Frick; **360,** Alvin E. Staffan; **361,** Kay Chernush/The Image Bank; **363,** Photo by: Albert Chong, courtesy Bernice Steinbaum Gallery, New York; **364,** MAK-I; **365,** Ken Frick; **368,** Gian Berto Vanni/Art Resource, New York; **371,** ©Chris Michaels/FPG International, Inc.; **374,** Pictures Unlimited; **378,** (m) GARFIELD reprinted by permission of United Features Syndicate, Inc., (b) Larry Lefever/Grant Heilman Photography Inc.; **379,** (t) J.Chenet/Woodfin Camp, (b) Mark Gibson; **380-381, 383,** Pictures Unlimited; **384** Ken Frick; **387,** file photo; **388,** The Bettmann Archive; **390,** First Image; **391,** Blair Seitz/Photo Researchers, Inc.; **392,** MAK-I; **393,** BLT Production; **393,** Doug Martin; **395, 397,** MAK-I; **398,** Robert Mullenix; **399,** (t) KS Studios, (b) Robert Mullenix; **401,** Doug Martin; **402,** MAK-I; **403,** ©Travelpix/FPG International, Inc.; **405,** MAK-I; **407,** Pictures Unlimited; **408,** Pictures Unlimited; **409,** Comstock, Inc.; **411,** Ken Frick; **412-413,** Pictures Unlimited; **415,** Doug Martin; **418,** ©Chris Michaels/FPG International, Inc.; **420,** (t) Charles Seaborn/Woodfin Camp, (m) Australia Picture Library/Westlight, (bl) The Bettmann Archive, (br) Robert Mullenix; **421,** Sygma; **422,** MAK-I; **425,** Tim Courlas; **427,** Randy Scheiber; **428,** Doug Martin; **430,** Comstock, Inc./Jack K. Clark; **432,** Pictures Unlimited; **434,** Comstock, Inc.; **435,** Pictures Unlimited; **438,** Doug Martin; **441,** (t) MAK-I, (b) Tim Courlas; **442,** ©S.M. Estvanik/FPG International, Inc.; **445,** Jeff Adamo/The Stock Market; **446,** JPH Images/The Image Bank; **449,** Ken Frick; **451,** The Stock Market; **454,** Comstock, Inc.; **457,460,** MAK-I; **462,** (bl) reprinted by permission of Tribune Media Services, (b) Dr.E.R.Degginger; **463,** KS Studios; **464,** Werner Bokelberg/The Image Bank; **466,** Pictures Unlimited; **468,** Robert Mullenix; **469,** Pictures Unlimited; **470,** National Museum of American Art/Art Resource, New York; **473,** (t) Robert Mullenix, (b) The Bettmann Archive; **476,** Ellen Schuster/The Image Bank; **478,** Ken Frick; **480,482,** Robert Mullenix; **486,** Comstock, Inc.; **487,** Steve Krongard/The Image Bank; **491,** Duomo/David Madison; **493,** Jean Miele/The Stock Market; **496,** Pictures Unlimited; **500,** (t) Robert Mullenix, (b) GARFIELD reprinted by permission of United Features Syndicate, Inc.; **501,** Robert Mullenix; **502,** MAK-I; **503,** Pictures Unlimited; **506,** Elaine Shay; **507,** Pictures Unlimited; **508,** MAK-I; **509,** (t) The Bettmann Archive, (b) ©FPG International, Inc.; **510,** Rafael Macia/Photo Researchers Inc.; **511,** K.L.Giese/Profiles West; **512,** ©R. Pleasant/FPG International, Inc.; **513,** Pictures Unlimited; **514,** MAK-I; **515,** Robert Mullenix; **516,** Bud Fowle; **518,** MAK-I; **521,** (l) ©Gveracy Cubitt/FPG International, Inc., (r) ©T. Quing/FPG International, Inc.; **523,** Pictures Unlimited; **524,** ©Gerard Fritz/FPG International, Inc.; **525,** Gref Davis/The Stock Market; **526,** Duomo/Steven E. Sutton; **527,** ©William D. Adams/FPG International, Inc.; **528,** Pictures Unlimited; **532,** (l) First Image, (r) file photo; **533,** David Frazier/Tony Stone Worldwide; **534,** Doug Martin; **538,** Ivor Sharp/The Image Bank; **540,** (l) Robert Mullenix, (bl) Mike Dobel/Masterfile, (br) Reprinted with special permission of King Features Syndicate; **541,** (t) Porterfield/Chickering/Photo Researchers Inc., (b) Allen Russell/Profiles West; **543,** Tim Davis/Science Source/Photo Researchers, Inc.; **545-546,** Robert Mullenix; **549,** Gary Cralle/The Image Bank; **550,** ©Michael Krasowitz/FPG International, Inc.; **552,** Naideau/The Stock Market; **557,** Robert Mullenix; **559,** Pictures Unlimited; **561,** ©Lee Kuhn/FPG International, Inc.; **562,564,** Pictures Unlimited; **565,** ©FPG International, Inc.; **568,** Studiohio; **572,** KS Studio; **573,** Ken Frick; **574,** Allsport USA/Vandystadt/Zoom; **575,** Mark Burnett; **576,** (l) Roy Morsch/The Stock Market, (r) Animals Animals/Patricia Caulfield; **577,** Allen Russell/Profiles West; **578,** Globus Brothers/The Stock Market; **579,** (t) Jon Feingersh/The Stock Market, (m) Ed Bock/The Stock Market, (b) Tony Duffy/Allsport; **580,** Philip Kretchmar/The Image Bank; **581,** (l) Jan Cobb/The Image Bank, (r) Claudia Parks/The Stock Market; **658, 660, 663, 664, 665,** Elaine Shay; **666,** David W. Hamilton/Image Bank; **667,** William R. Sallaz/Duomo; **668,** Elaine Shay.

TECHNOLOGY ACTIVITIES & DATA BANK

Technology Activities

Data Bank

TECHNOLOGY ACTIVITY 1:

Evaluating Expressions with a Graphing Calculator

Use with Lesson 2-1, pages 44-47

Graphing calculators observe the order of operations when an expression is evaluated. So there is no need to perform each operation in the expression separately. You can enter the expression just as it is written to evaluate it. The calculators also have parentheses that you key into the calculator in the same way as parentheses are written in an expression. You can also use parentheses to indicate multiplication. For example, 3(2) or (3)(2) can be entered for 3×2.

The expression appears as you enter it in a graphing calculator. On TI calculators, the multiplication and division signs do not appear on the screen as they do on the keys. Instead, the calculator displays symbols used in computer language. That is, * means multiplication, and / means division.

Example

Evaluate $3(x - 6) \div 2 + (x^2 - 15)$ if $x = 8$.

The [x^2] *key is located above the* [$\sqrt{\ }$] *key on the Casio fx-7700. Press* [SHIFT] [$\sqrt{\ }$] *to access the x^2 function.*

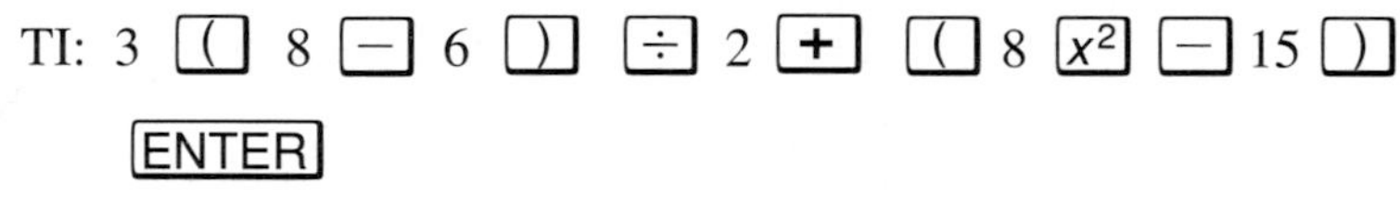

Casio: 3 [(] 8 [−] 6 [)] [÷] 2 [+] [(] 8 [x^2] [−] 15 [)] [EXE]

$3(x - 6) \div 2 + (x^2 - 15) = 52$ if $x = 8$.

If you get an error message or discover that you entered the expression incorrectly, you can use the REPLAY feature to correct your error and reevaluate without reentering your expression. Follow the steps below to use the REPLAY feature.

TI: On the TI-81, press [▲]. On the TI-82, press [2nd] [ENTRY]. Use the arrow keys to move to the location of the correction. Then type over, use [INS], or use [DEL] to make the correction. Then press [ENTER] to evaluate. You don't have to move the cursor to the end.

Casio: Press [⇨] or [⇦]. The answer disappears and the cursor goes to the beginning or end of the expression. Make changes and then press [EXE] to evaluate.

Exercises

Use a graphing calculator to evaluate each expression if $x = 4$, $y = 7$, and $z = 9$.

1. $17 - z$
2. $x^2 + 9$
3. xy^2
4. $\frac{2(z - x)}{(y - 2)^2}$
5. $\frac{3xz}{2} - 12$
6. $x(y + z) - 7$

TECHNOLOGY ACTIVITY 2:

Finding a Mean with a Spreadsheet

Use with Lesson 4-5, pages 145-148

Mrs. Roberts uses tests, quizzes, homework, and class participation to determine the final grades in her U.S. History course. In order to determine the test portion of the grade, she will find the mean of each student's four test scores. A portion of the spreadsheet she set up to find the grades is shown below.

	A	B	C	D	E	F
1	Student	Test 1	Test 2	Test 3	Test 4	Mean
2	Kenneth	78	76	81	83	(B2+C2+D2+E2)/4
3	Rena	84	82	85	88	(B3+C3+D3+E3)/4
4	Kelly	72	83	85	83	(B4+C4+D4+E4)/4
5	Anthony	88	92	90	91	(B5+C5+D5+E5)/4
6	Umeko	90	88	87	92	(B6+C6+D6+E6)/4

The formulas in the cells in column F find the mean of the scores that are entered in the cells in columns B, C, D, and E. The formula first finds the sum of the scores, then divides the sum by 4 to find the average. The printout below shows the results when the calculations are complete.

	A	B	C	D	E	F
1	Student	Test 1	Test 2	Test 3	Test 4	Mean
2	Kenneth	78	76	81	83	79.5
3	Rena	84	82	85	88	84.75
4	Kelly	72	83	85	83	80.75
5	Anthony	88	92	90	91	90.25
6	Umeko	90	88	87	92	89.25

Exercises

Use the spreadsheet above to answer each question.

1. Suppose Theo's test grades are 92, 84, 89, and 95. What is his test average?
2. How would you alter the spreadsheet to find the mean of seven quiz scores?

TECHNOLOGY ACTIVITY 3:

Distance Between Two Points on a Graphing Calculator

A graphing calculator can be used to write and run programs just like larger computers. The program below will find the distance between two points in the coordinate plane. In order to use the program, you must first enter the program into the calculator's memory. You may want to refer to the User's Guide for your graphing calculator. To access the TI-81 program memory, use the following keystrokes.

Enter: [PRGM]  1

Enter the program exactly as it is shown. Consult the User's Guide for the locations of commands in the menus.

Example

Find the distance between the points (-3, 5) and (9, 1) on a graphing calculator. Round your answer to the nearest tenth.

Run the program.

Enter the coordinates of the points as the program asks for them.

Enter: [(-)] 3 [ENTER]
5 [ENTER] 9
[ENTER] 1
[ENTER]

The calculator automatically rounds the distance to the nearest tenth of a unit. The distance between (-3, 5) and (9, 1) is approximately 12.6 units.

```
Prgm 1: DISTANCE
:Fix 1
:Disp "ENTER THE X-
 COORDINATE OF POINT 1"
:Input A
:Disp "ENTER THE Y-
 COORDINATE OF POINT 1"
:Input B
:Disp "ENTER THE X-
 COORDINATE OF POINT 2"
:INPUT C
:DISP "ENTER THE Y-
 COORDINATE OF POINT 2"
:Input D
: √ ((A-C)^2+(B-D)^2) → E
:Disp "THE DISTANCE IS"
:Disp E
```

The program is written for use on a TI-81 graphing calculator. If you have a different type of programmable calculator, consult your User's Guide to adapt the program for use on your calculator.

Exercises

Use the program to find the distance between each pair of points to the nearest tenth.

1. (9, 1), (-2, 1)
2. (3, 3), (-7, -1)
3. (-12, 1), (15, -5)
4. (-3, -2), (-19, -9)
5. (-8, 3), (-2, -4)
6. (2.4, 6.1), (0.2, 0.3)
7. (0, 3), (8, -4)
8. (-47, 21), (125, 72)

TECHNOLOGY ACTIVITY 4:

Proportions with a Spreadsheet

Use with Lesson 9-3, pages 347-349

The Elegant Eatery is catering the annual fall festival picnic. Their chili dip recipe is shown on the card at the right.

Chili Dip
1 c. cottage cheese $\frac{1}{4}$ *c. chili sauce*
$\frac{1}{4}$ *t. onion powder* $\frac{1}{4}$ *c. skim milk*
3 T. grated Parmesan cheese
Mix all ingredients in blender until smooth. Chill. Serves 20.

There are 120 people expected at the picnic. In order to determine how much of each ingredient to use, Rosa set up the spreadsheet below.

CHILI DIP RECIPE

	A	B	C
1	People To Serve =	B1	
2	Batches needed =	B1/20	
3	INGREDIENT	NUMBER	
4	cottage cheese	B2	cups
5	chili sauce	B2/4	cups
6	onion powder	B2/4	t.
7	skim milk	B2/4	cups
8	parmesan cheese	B2 * 3	T.

In order to use the spreadsheet, Rosa entered the number of people expected at the picnic, 120, in cell B1. The spreadsheet told her to use 6 cups of cottage cheese, $1\frac{1}{2}$ cups of chili sauce, $1\frac{1}{2}$ teaspoons of onion powder, $1\frac{1}{2}$ cups of skim milk, and 18 tablespoons of parmesan cheese.

The spreadsheet uses proportions to determine the amount of each ingredient that should be used in order to make enough chili dip for a group. Solving the proportion $\frac{\textit{number of people expected}}{\textit{number of servings per batch}} = \frac{\textit{number of batches needed}}{\textit{1 batch}}$ will give the number of batches needed. Since B1 is the number of people and the number of servings per batch is 20, we can rewrite the proportion as $\frac{\text{B1}}{20} = \frac{x}{1}$. Thus, the number of batches is $\frac{\text{B1}}{20}$. The formula in cell B2 uses B1 ÷ 20 to find the number of batches that need to be made.

Exercises

1. Use the spreadsheet to find the amount of each ingredient need to make enough chili dip for 180 people.

2. How could you alter the spreadsheet if one batch of chili dip served 12 people?

3. Kevin wants to add $\frac{1}{8}$ teaspoon of Tabasco sauce to each batch of chili dip. Write the formula to enter in a cell to find the number of teaspoons of Tabasco sauce to add to a batch for a group of people.

TECHNOLOGY ACTIVITY 5:

Discounts with a Spreadsheet

Use with Lesson 10-9, pages 409-411

The sale price of an item can be found by multiplying the percent paid by the original price. A spreadsheet like the one below can be used to generate a table of sale prices for various original prices.

Suppose you are the manager of the casual clothes department of a local department store. The store has frequent sales when many items are the same percentage off. You have the spreadsheet below to generate signs for each sale.

	A	B	C
1	Discount Rate =	B1	
2	Item	Original Price	Sale Price
3	Cotton Sweaters	29.99	(100-B1)/100 * B3
4	Denim Jackets	36.29	(100-B1)/100 * B4
5	Team Sweatshirts	24.89	(100-B1)/100 * B5
6	Sport Socks 3-Pack	6.59	(100-B1)/100 * B6
7	T-Shirts	7.99	(100-B1)/100 * B7

The discount rate is entered into cell B1. Then the formulas in the cells in column C determine the sale prices. The formulas find the percent paid by subtracting the percentage off from 100, then dividing by 100. Then the percentage paid is multiplied by the price of each item.

Exercises

Use the spreadsheet to answer each question.

1. At the Midnight Madness Sale, all items were 25% off. What was the sale price of a denim jacket?

2. What is the sale price of a cotton sweater if the discount rate is 33%?

3. What is the discount on a T-shirt if the discount rate is 40%?

4. Suppose you wanted to add a new row to the spreadsheet for a $99.59 suede jacket. List each of the cell entries (A8, B8, and C8) that you would enter.

TECHNOLOGY ACTIVITY 6:

Linear Equations on a Graphing Calculator

Use with Lesson 11-4, pages 432-434

A graphing calculator is a powerful tool for studying functions. Any of the graphing calculators will graph linear functions, but the procedure for graphing is slightly different for each one. On any of the calculators, you must set an appropriate range before you can graph a function. A viewing window of [-10, 10] by [-10, 10] with a scale factor of 1 on both axes denotes the values $-10 \leq x \leq 10$ and $-10 \leq y \leq 10$. The scale factor of 1 indicates that the tick marks on both axes are one unit apart. This is called the **standard viewing window.**

Example

Graph $y = 2x - 4$ in the standard viewing window.

Be sure that your calculator is in the correct mode for graphing functions.

Casio fx-7000: [MODE] [+]

Casio fx-7700: [MODE] [+] [MODE] [MODE] [+]

TI: Press the [MODE] key. If "Function" and "Rect" are not highlighted, use the arrow and [ENTER] keys to highlight them.

Press [2nd] [QUIT] to return to the home screen.

Now graph the function.

Casio fx-7000: [Graph] 2 [ALPHA] [X] [−] 4 [EXE]

Casio fx-7700: [Graph] 2 [X,θ,T] [−] 4 [EXE]

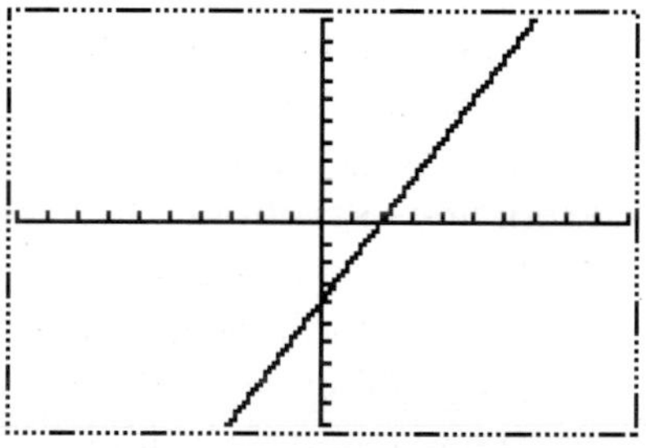

TI: [Y=] 2 [X|T] [−] 4 [GRAPH]

On the TI-82, x is entered using the [X,T,θ] key.

Exercises

Graph each function on a graphing calculator. Then sketch the graph on a piece of paper.

1. $y = 5 - x$
2. $y = 4x - 2$
3. $y = 3$
4. $y = \frac{1}{2}x - 1$
5. $y = -2x + 3$
6. $y = -0.5x + 2$

You will need to clear the graphics screen before you can graph a second function. To clear the screen on a Casio, press [SHIFT] [Cls] [EXE]. Changing the range before entering a new function to be graphed on a Casio calculator will also clear the graphics screen. To clear the graphics screen on a TI, press [Y=] and use the arrow and [CLEAR] keys to clear the Y= list.

TECHNOLOGY ACTIVITY 7:

Probability on a Graphing Calculators

Use with Lesson 13-7, pages 528-530

The graphing calculator program below will generate random numbers. You can use the program to simulate real events like rolling a number cube or tossing a coin. To use the program, you must first enter the program into the calculator's memory. You may want to refer to the User's Guide for your graphing calculator. To access the TI-81 program memory, use the following keystrokes.

Enter: PRGM ▶ 1

Example

Use the graphing calculator program to simulate rolling a number cube fifty times. Make a table to show the results.

```
Prgm 1: RAND.NUM
:ClrHome
:Disp "LEAST INTEGER"
:Input S
:Disp "GREATEST
 INTEGER"
:Input L
:Disp "NUMBER OF
 VALUES TO GENERATE"
:Input A
: 0 → B
:Lbl 1
:B + 1 → B
:Int ((L − S + 1)Rand + S)
 → R
:Disp R
:Pause
:If A ≠ B
:Goto 1
```

Run the program.

The program will generate integers between two numbers that you enter. Since the numbers on a number cube are 1 through 6, enter 1 as the least integer and 6 as the greatest integer.

Enter: 1 ENTER 6 ENTER

You want the calculator to generate fifty numbers.

Enter: 50 ENTER

The calculator will display the number generated and wait for you to press ENTER before it continues.

The program is written for use on a TI-81 graphing calculator. If you have a different type of programmable calculator, consult your User's Gude to adapt the program for use on your calculator.

Exercises

Use the table of results from your graphing calculator experiment to answer each question.

1. Do you think that each number on a number cube has an equal chance of occurring when you throw the cube?

2. How could you use the graphing calculator program to simulate spinning a game spinner that had seven equal-sized regions 15 times?

POSTAGE RATES AND PLANETS

First Class Postage Rates (1994)

Weight not exceeding (ounces)	First Class Cost
1	$0.29
2	0.52
3	0.75
4	0.98
5	1.21
6	1.44

Minimum Size:
All pieces must be at least 0.007 inch thick. Pieces that are $\frac{1}{4}$ inch or less thick must be: (1) rectangular in shape, (2) at least $3\frac{1}{2}$ inches high, and (3) at least 5 inches long.

Planets of our Solar System

Planets	Diameter	Average Distance from Sun	Number of Moons	1 Rotation*	Orbit*
Mercury	3,100 miles (4,987.0 km)	36 million miles (75.9 million km)	0	59 days	88 days
Venus	7,500 miles (12,067.5 km)	67 million miles (107.8 million km)	0	243 days	225 days
Earth	7,926 miles (12,752.0 km)	93 million miles (149.6 million km)	1	24 hours	365 days
Mars	4,218 miles (6,786.8 km)	14.2 million miles (228.5 million km)	2	24.4 hours	687 days
Jupiter	89,400 miles (143,844.6 km)	483 million miles (777.1 million km)	16	10 hours	11.86 years
Saturn	75,000 miles (120,675 km)	886 million miles (14,235.6 km)	20	10.4 hours	29.46 years
Uranus	32,300 miles (51,970.7 km)	1.8 billion miles (2.9 billion km)	15	17 hours	84 years
Neptune	30,000 miles (48,270 km)	2.8 billion miles (4.5 billion km)	3	18-22 hours	165 years
Pluto	1900 miles (3057.1 km)	3.7 billion miles (5.95 billion km)	1	6.4 days	248 years

**Hours, days, and years are Earth Time.*

HOME BUYING AND MILEAGE

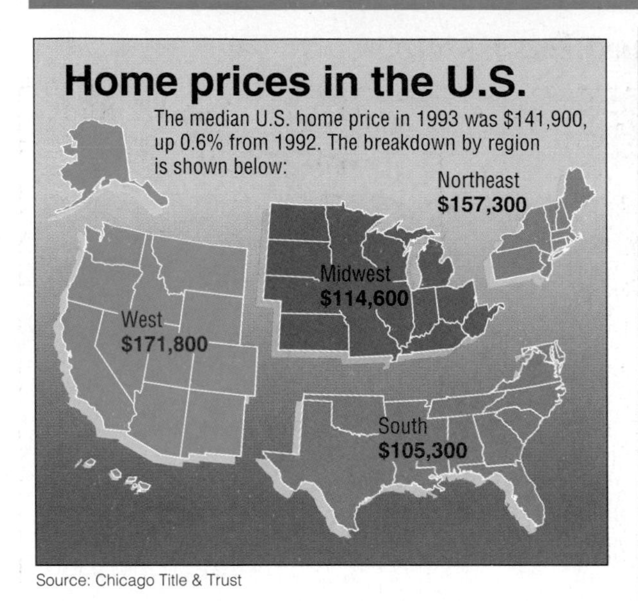

Source: Chicago Title & Trust

Incomes of Home Buyers

Household income of home buyers in 1993

$61,000 or more: 41%

Less than $30,000: 10%

$31,000-$41,000: 15%

$41,000-$50,000: 18%

$51,000-$60,000: 16%

Source: Chicago Title & Trust

Mileage Chart

From: / To:	Atlanta	Birmingham	Charlotte	Columbia	Jackson	Jacksonville	Memphis	Miami	Nashville	Orlando
Atlanta, GA		150	240	214	399	313	382	663	246	426
Birmingham, AL	150		391	362	245	463	255	754	194	526
Charlotte, NC	240	391		94	632	393	630	740	421	534
Columbia, SC	214	362	94		602	296	616	643	437	437
Jackson, MS	399	245	632	602		609	213	912	414	700
Jacksonville, FL	313	463	393	296	609		694	353	563	140
Memphis, TN	382	255	630	616	213	694		997	209	776
Miami, FL	663	754	740	643	912	353	997		910	229
Nashville, TN	246	194	421	437	414	563	209	910		688
Orlando, FL	426	546	534	437	700	140	776	229	688	

TEMPERATURES, PLAYING FIELDS, AND HELVETICA FONT

Record High and Low Temperatures

State	Low	High
AL	-27	112
AK	-80	100
AZ	-40	127
AR	-29	120
CA	-45	134
CO	-61	118
CT	-32	105
DE	-17	110
FL	-2	109
GA	-17	112
HI	12	100
ID	-60	118
IL	-35	117
IN	-35	116
IA	-47	118
KS	-40	121
KY	-34	114

State	Low	High
LA	-16	114
ME	-48	105
MD	-40	109
MA	-35	107
MI	-51	112
MN	-59	114
MS	-19	115
MO	-40	118
MT	-70	117
NE	-47	118
NV	-50	122
NH	-46	106
NJ	-34	110
NM	-50	116
NY	-52	108
NC	-34	110
ND	-60	121

State	Low	High
OH	-39	113
OK	-27	120
OR	-54	119
PA	-42	111
RI	-23	104
SC	-19	111
SD	-58	120
TN	-32	113
TX	-23	120
UT	-69	117
VT	-50	105
VA	-30	110
WA	-48	118
WV	-37	112
WI	-54	114
WY	-63	114

Playing Field Dimensions

Sport	Dimensions
Baseball	90 x 90 feet (diamond)
Basketball	26 x 15 meters
Football	360 x 160 feet
Olympic Swimming	50 x 21 meters
Soccer	100 x 73 meters
Tennis	78 x 36 feet (doubles)
Volleyball	18 x 9 meters

Helvetica Font

A B C D E F G H I J K L M N
O P Q R S T U V W X Y Z

D A T A B A N K

OLYMPIC RACES AND BIRTHDAYS

Olympic 1,500 Meter Race Winners

Women's 1,500-Meter Race		
Year	**Winner, Country**	**Time**
1972	Lyudmila Bragina, USSR	4 m 1.4 s
1976	Tatyana Kazankina, USSR	4 m 5.5 s
1980	Tatyana Kazankina, USSR	3 m 56.6 s
1984	Gabriella Dorio, Italy	4 m 3.3 s
1988	Paula Ivan, Romania	3 m 54.0 s
1992	Hassiba Boulmerka, Algeria	3 m 55.3 s

Men's 1,500-Meter Race		
Year	**Winner, Country**	**Time**
1972	Pekka Vasala, Finland	3 m 36.3 s
1976	John Walker, New Zealand	3 m 39.2 s
1980	Sebastian Coe, Great Britain	3 m 38.4 s
1984	Sebastian Coe, Great Britain	3 m 32.5 s
1988	Peter Rono, Kenya	3 m 36.0 s
1992	Fermin Cacho Ruiz, Spain	3 m 40.1 s

Likelihood of Sharing a Birthday

This graph shows the probability of two people in a group sharing the same birthday. For example, if there are 20 people in a group, there is a 40% probability that two of them have the same birthday.

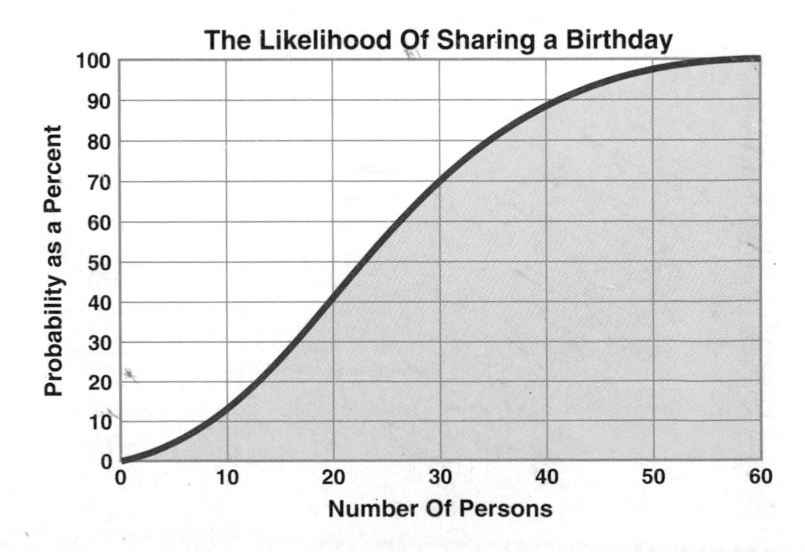